The early Mesozoic era was a critical period in the evolution of life on land, when most of today's major groups of terrestrial vertebrates (mammals, turtles, lizards, frogs, salamanders, etc.) arose and dinosaurs and pterosaurs rose to prominence. In recent years this period in the earth's history has received a great deal of attention from paleontologists, and, in particular, it is now believed that the small vertebrates that lived in the shadows of the first dinosaurs can tell us a great deal about the evolution of what might be termed modern terrestrial ecosystems.

This book is the result of a workshop held in Front Royal, Virginia, and features contributions by experts with international reputations in their fields. There are chapters on the taxonomy and phylogeny of the principal vertebrate groups (amphibians, lepidosaurs, crocodylomorphs, and mammals), followed by a section dealing with the most significant early Mesozoic tetrapod assemblages worldwide. The final section looks at how faunal turnover at that time is measured and examines the possibility of mass extinctions.

In the shadow of
the dinosaurs

In the shadow of the dinosaurs

Early Mesozoic tetrapods

Edited by

Nicholas C. Fraser
Virginia Museum of Natural History

Hans-Dieter Sues
Royal Ontario Museum

CAMBRIDGE UNIVERSITY PRESS
Cambridge, New York, Melbourne, Madrid, Cape Town,
Singapore, São Paulo, Delhi, Tokyo, Mexico City

Cambridge University Press
The Edinburgh Building, Cambridge CB2 8RU, UK

Published in the United States of America by Cambridge University Press, New York

www.cambridge.org
Information on this title: www.cambridge.org/9780521458993

First published 1994
First paperback edition 1997

A catalogue record for this publication is available from the British Library

Library of Congress Cataloguing in Publication Data
In the shadow of the dinosaurs: early Mesozoic tetrapods / edited by
Nicholas C. Fraser, Hans-Dieter Sues.
    p.   cm.
Includes index.
ISBN 0–521–45242–2
1. Vertebrates, Fossil – Congresses.   2. Paleontology – Mesozoic –
Congresses.   I. Fraser, Nicholas C.   II. Sues, Hans-Dieter.
QE841. 15   1994
566–dc20   93-14566
                                                      CIP

ISBN 978-0-521-45242-7 Hardback
ISBN 978-0-521-45899-3 Paperback

Contents

Contents   vi

Acknowledgments

We should like to thank all participants in the 1991 symposium at Front Royal for making it such a success. The chapters of this volume are largely based on presentations at the meeting. Each contribution was reviewed by at least two referees, and the volume has been greatly improved by the constructive criticisms and suggestions of the reviewers. For their careful reviews of individual chapters, we are indebted to D. Baird, M. J. Benton, J. R. Bolt, R. L. Carroll, J. A. Clack, J. M. Clark, P. Dodson, R. F. Dubiel, C. J. Duffin, S. E. Evans, J. A. Gauthier, A. Hallam, J. A. Hopson, L. L. Jacobs, J. A. Lillegraven, Z. Luo, J. G. Maisey, A. R. Milner, D. B Norman, M. A. Norell, K. Padian, J. M. Parrish, O. Rieppel, D. B. Weishampel, and R. Wild. We extend special thanks to K. Padian for his gargantuan effort in critically reading the entire text; his comments have greatly improved the final version. E. B. Sues provided helpful comments on our introductory sections.

P. E. Olsen expertly led the participants of the symposium to Triassic vertebrate sites in Virginia and North Carolina, and we thank the Culpeper Stone Company and the Solite Corporation for granting us access to their working quarries.

At Cambridge University Press we are particularly grateful to R. Smith for his encouragement and support and to A. Sigal and K. Lamoza for all their input.

E. Compton-Gooding assisted with redrafting some of the figures for publication, and C. Skrabec with retyping some copy for the press.

The staff of the 4H Center made our stay at Front Royal particularly enjoyable and provided a most convivial atmosphere for the symposium. The Virginia Museum of Natural History provided financial support for the event, and we are especially indebted to M. Hager, R. Knighton, and K. Sickles for their encouragement and organizational skills. H.-D. S. gratefully acknowledges support by NSF grant EAR-9016677 during the symposium and preparation of this volume.

Contributors

Michael J. Benton
Department of Geology
University of Bristol
Bristol BS8 1RJ, U.K.

Robert L. Carroll
Redpath Museum and
Department of Biology
McGill University
Montréal, Québec H3A 2K6, Canada

James M. Clark
Department of Vertebrate Paleontology
American Museum of Natural History
New York, NY 10024

Susan E. Evans
Department of Anatomy and
Developmental Biology
University College London
London WC1E 6BT, U.K.

David E. Fastovsky
Department of Geology
University of Rhode Island
Kingston, RI 02881

Nicholas C. Fraser
Virginia Museum of Natural History
Martinsville, VA 24112

Gerhard Hahn
Institut für Geologie und Paläontologie
Fachbereich Geowissenschaften
Philipps-Universität
D-35032 Marburg, Germany

René Hernández
Instituto de Geologia
Universidad Nacional Autonoma de Mexico
Delegacion Coyoacan, D.F., Mexico

James A. Hopson
Department of Organismal Biology
and Anatomy
The University of Chicago
Chicago, IL 60637

Adrian P. Hunt
New Mexico Museum of Natural History
and Science
Albuquerque, NM 87104

Farish A. Jenkins, Jr.
Museum of Comparative Zoology
Harvard University
Cambridge, MA 02138

Annika K. Johansson
Lamont-Doherty Earth Observatory
Columbia University
Palisades, NY 10964

Andrew L. A. Johnson
Department of Earth Sciences
University of Derby
Derby DE22 1GB, U.K.

Fran Tannenbaum Kaye
Museum of Paleontology
University of California
Berkeley, CA 94720

Kenneth A. Kermack
Department of Biology
University College London
London WC1E 6BT, U.K.

Peter A. Kroehler
Department of Paleobiology
National Museum of Natural History
Washington, DC 20560

Spencer G. Lucas
New Mexico Museum of Natural History and Science
Albuquerque, NM 87104

Zhexi Luo
Department of Biology
College of Charleston
Charleston, SC 29424

Sara J. Metcalf
Department of Geology
University of Bristol
Bristol BS8 1RJ, U.K.

Andrew R. Milner
Department of Biology
Birkbeck College
London WC1E 7HX, U.K.

Marisol Montellano
Instituto de Geologia
Universidad Nacional Autonoma de Mexico
Delegación Coyoacán, D.F., Mexico

Andrew J. Newell
Department of Geology
University of Bristol
Bristol BS8 1RJ, U.K.

Paul E. Olsen
Lamont-Doherty Earth Observatory
Columbia University
Palisades, NY 10964

Kevin Padian
Department of Integrative Biology and
Museum of Paleontology
University of California
Berkeley, CA 94720

Olivier Rieppel
Department of Geology
Field Museum of Natural History
Chicago, IL 60605

Alastair H. Ruffell
School of Geosciences
The Queen's University of Belfast
Belfast BT7 1NN, Northern Ireland

Neil H. Shubin
Department of Biology
University of Pennsylvania
Philadelphia, PA 19104

Denise Sigogneau-Russell
URA 12 CNRS, Institut de Paléontologie
Muséum National d'Histoire Naturelle
F-75005 Paris, France

Michael J. Simms
Department of Geology
University of Bristol
Bristol BS8 1RJ, U.K.

Patrick S. Spencer
Department of Geology
University of Bristol
Bristol BS8 1 RJ, U.K.

Hans-Dieter Sues
Department of Vertebrate Palaeontology
Royal Ontario Museum
Toronto, Ontario M5S 2C6, Canada
and
Department of Zoology
University of Toronto
Toronto, Ontario M5S 1A1, Canada

Rachel J. Walker
Department of Geology
University of Bristol
Bristol BS8 1RJ, U.K.

Geoffrey Warrington
British Geological Survey
Keyworth, Nottingham NG12 5GG, U.K.

Rupert Wild
Paläontologische Abteilung
Staatliches Museum für Naturkunde
D-70191 Stuttgart, Germany

Xiao-chun Wu
Royal Tyrrell Museum of Palaeontology
Drumheller, Alberta T0J 0Y0, Canada

Introduction

NICHOLAS C. FRASER AND HANS-DIETER SUES

During the past two decades, the Cretaceous–Tertiary (K/T) boundary has attracted much attention in the field of paleontology, reflecting enormous interest in the extinction of the dinosaurs (other than birds) and many other organisms, as well as speculation regarding the causation of that event. Consequently, the K/T boundary has been examined in great detail in western North America and other regions of the world, and we now have a much more detailed picture of the tempo and mode of the biotic changes at the end of the Cretaceous. Furthermore, evidence for an extraterrestrial impact at the K/T boundary is now very strong, and its possible implications for the extinctions continue to be subjects of vigorous debate. Usually groups of organisms generate interest because of their evolutionary success. Thus it is perhaps unfortunate that so much attention has been paid to the demise of the nonavian dinosaurs. Yet they were immensely successful by every biological measure, and increasing attention is now being paid to their origin and initial radiation in the Triassic.

Although research on Late Triassic and Early Jurassic biotas and biotic changes has, by comparison with research on the K/T boundary, been neglected, it promises to be as rewarding and as important as the study of the latter. Not only was the Triassic the time of the first appearance of dinosaurs and pterosaurs, as revealed by the fossil record, but it also witnessed the first appearance of most major groups of present-day continental vertebrates, including lissamphibians, turtles, lepidosaurs, and mammaliaform synapsids. Furthermore, we have evidence for at least one mass-extinction event, an unambiguous record of a giant impact structure generated by a bolide in Québec, Canada, and extensive sections of fossiliferous continental strata worldwide. The tremendous significance of that time interval for the evolution of vertebrates is now increasingly becoming appreciated.

A reassessment of the stratigraphic ages of many early Mesozoic vertebrate-bearing continental strata by Olsen and Galton (1977) brought about dramatic changes in our ideas concerning faunal diversity and biotic change at the Triassic–Jurassic boundary. Initially, the previous interpretations of the end of the Triassic as a period of mass extinction (Colbert, 1958) were challenged by Olsen and Galton's new age assignments, and a more gradual faunal replacement was postulated. But clearly the data used to support those theories were rather poor, and numerous efforts to document the Triassic–Jurassic biotic changes more thoroughly have been made in recent years. This additional information has resulted in a general shift back toward the hypothesis of an abrupt extinction event at the end of the Triassic. The publication of a volume edited by Kevin Padian, *The Beginning of the Age of Dinosaurs* (1986), completely updated our knowledge of continental vertebrate assemblages across the Triassic–Jurassic boundary, and since then there has been increasing interest in that time interval. Additional new localities for early Mesozoic tetrapods have been discovered and explored, and these have provided valuable additional data. In particular, the standard techniques for collecting the skeletal remains of small vertebrates, developed by Hibbard (1949), Henkel (Kühne, 1971), and others (e.g., McKenna, 1962; Ward, 1984), have been consistently employed in recent years, at both previously known sites and new sites. Much of the information gleaned from such efforts had remained scattered throughout the primary literature or had never been published. Thus, there was a growing need to compile all the available information on early Mesozoic small terrestrial vertebrates, and in response to that need we organized a workshop with an international group of active investigators at Front Royal, Virginia, in May 1991. This volume represents a collection of contributions presented at that meeting and focuses on the small tetrapods.

There is an obvious question that must be addressed at the outset: What is a small tetrapod? The term "microvertebrate" is often used by American authors to designate skeletal elements smaller than 5 mm in greatest dimension (Ward, 1984). Clearly the entire animal would have been much larger than that. In this volume, the contributors mainly discuss taxa of continental tetrapods with total body lengths ranging up to about 50 cm. Furthermore, many moderate-size animals with overall lengths ranging from 50 cm to 2 m may also be represented by their smallest elements (such as teeth) and/or fragments of bone in fossil assemblages dominated by smaller forms. For this reason, some of the contributions discuss tetrapods that normally would not be regarded as small (e.g., many of the early Mesozoic amphibians reviewed by Milner). There were three criteria for inclusion of tetrapod taxa in this volume: (1) They occur or might be expected to occur in assemblages dominated by small vertebrates. (2) Their bones and/or teeth appear in residues resulting from standard screening techniques aimed at recovering small tetrapod remains. (3) Their exclusion would have resulted in an incomplete picture of the phylogeny and paleobiology of the small forms and their paleobiogeographic and stratigraphic context.

For background information concerning the paleoecology of early Mesozoic terrestrial biotas and the ecological impact of the Triassic–Jurassic extinctions, we refer the interested reader to a recent review by Wing and Sues (1992).

This volume is not intended to be an exhaustive review of the biodiversity and evolution of continental vertebrates during the early Mesozoic, but we hope that it will provide an adequate update regarding the latest information on small tetrapods, which often have been overlooked in previous discussions of the Triassic–Jurassic transition, and will place them in the context of the major biotic changes during the early Mesozoic.

References

Colbert, E. H. 1958. Tetrapod extinctions at the end of the Triassic period. *Proceedings of the National Academy of Sciences* 44: 973–977.

Hibbard, C. W. 1949. Techniques of collecting microvertebrate fossils. *Contributions from the Museum of Paleontology, University of Michigan* 8: 7–19.

Kühne, W. G. 1971. Collecting vertebrate fossils by the Henkel process. *Curator* 14: 175–179.

McKenna, M. C. 1962. Collecting small fossils by washing and screening. *Curator* 3: 221–235.

Olsen, P. E., and P. M. Galton. 1977. Triassic–Jurassic extinctions: are they real? *Science* 197: 983–986.

Padian, K. (ed.). 1986. *The Beginning of the Age of Dinosaurs: Faunal Change across the Triassic–Jurassic Boundary*. Cambridge University Press.

Ward, D. J. 1984. Collecting isolated microvertebrate fossils. *Zoological Journal of the Linnean Society* 82: 245–259.

Wing, S. L., and H.-D. Sues (rapporteurs). 1992. Mesozoic and early Cenozoic terrestrial ecosystems. Pp. 327–416 in A. K. Behrensmeyer, J. Damuth, W. A. DiMichele, R. Potts, H.-D. Sues, and S. L. Wing (eds.), *Terrestrial Ecosystems through Time: Evolutionary Paleoecology of Terrestrial Plants and Animals*. University of Chicago Press.

PART I

Phylogeny

Part I of this volume focuses on several major groups of tetrapods that have become much better known in recent years; there have been considerable advances in our understanding of their relationships as a result of both more rigorous phylogenetic analyses and the discovery and examination of new fossil material. The individual chapters do not necessarily discuss the merits of the various competing phylogenetic schemes currently available. This does not mean that viewpoints other than those favored by the contributors are ignored, but one of the focal points of Part I is the discussion of morphological characters of potential value for identification of dissociated and fragmentary skeletal remains of early Mesozoic small tetrapods. Many assemblages of such forms contain specimens that have diagnostic features but do not readily conform to anything previously described. It can be invaluable in such instances to be aware of what conceivably might be present in a given assemblage. For instance, the earliest undisputed lizards are Late Jurassic in age (Gauthier, Estes, and de Queiroz, 1988; Rieppel, Chapter 2); yet, given the well-corroborated sister-group relationship between Squamata and Sphenodontia, the fossil record of true lizards ought to extend well into the Late Triassic.

A persistent problem in the identification of basal taxa of well-known groups is their recognition in practice. Although various forms, such as *Palaeagama*, *Paliguana*, and *Kuehneosaurus*, have traditionally been assigned to the Squamata (e.g., Carroll, 1977), the debate continues as to their phylogenetic positions, because these taxa previously had been characterized largely by plesiomorphies (Gauthier et al., 1988). It is to be expected that the early members of the Squamata will lack many of the derived characters used to diagnose later constituent taxa of this group, and early representatives are likely to exhibit a very generalized structural pattern. As Rieppel argues in Chapter 2,

there may be few, if any, clear-cut characters that would allow unambiguous identification of fragmentary and dissociated skeletal remains such as occur in many assemblages of small tetrapods. Such considerations are emphasized in many of the contributions presented here.

In view of their potential significance, the phylogeny of the Sphenodontia is examined in some detail in this volume by Wu in Chapter 3, and for the sake of completeness, the earliest aquatic representatives of this group from the Lower Jurassic of Germany are reviewed by Carroll and Wild (Chapter 4). Since the erection of the order Rhynchocephalia by Günther (1867), its constitution has been debated. Although originally erected solely for the extant genus *Sphenodon*, a considerable variety of fossil diapsids subsequently came to be added to this order. Later the group again became restricted to comprise just the sphenodontids, pleurosaurs, and rhynchosaurs – taxa that supposedly shared the trait of a rostral "beak" formed by the premaxillae. More recent research has shown that, in fact, the rhynchosaurs belong to the archosauromorph clade, and thus they have been removed from the Rhynchocephalia (Carroll, 1977; Gauthier et al., 1988). As used by Rieppel and several other authors in this volume, the taxon Rhynchocephalia comprises the Early Jurassic *Gephyrosaurus* and the Sphenodontia, following the definition of Gauthier et al. (1988). Because of the erroneous association of the Rhynchosauria with the Rhynchocephalia in the past, and the similarity in names, many recent authors have chosen to abandon the term Rhynchocephalia altogether to avoid any confusion, and some contributors to this volume prefer a more inclusive use of the name Sphenodontia Williston, 1925 (which is used here because of its priority over Sphenodontida Estes, 1983) to refer to both *Gephyrosaurus* and the more derived forms.

References

Carroll, R. L. 1977. The origin of lizards. Pp. 359–396 in S. M. Andrews, R. S. Miles, and A. D. Walker (eds.), *Problems in Vertebrate Evolution*. Linnean Society Symposium Series 4. London: Academic Press.

Gauthier, J. A., R. Estes, and K. de Queiroz. 1988. A phylogenetic analysis of Lepidosauromorpha.

Pp. 15–98 in R. Estes and G. Pregill (eds.), *Phylogenetic Relationships of the Lizard Families: Essays Commemorating Charles L. Camp*. Stanford University Press.

Günther, A. 1867. Contribution to the anatomy of *Hatteria* (*Rhynchocephalus*, Owen). *Philosophical Transactions of the Royal Society of London* 157: 595–629.

1

Late Triassic and Jurassic amphibians: fossil record and phylogeny

ANDREW R. MILNER

Introduction

Most of the late Paleozoic groups at the amphibian grade of organization have not been found in post-Permian rocks and apparently became extinct, either during the Late Permian or at the end-Permian extinction event. No nectrideans, aïstopods, microsaurs, anthracosaurs, or seymouriamorphs appear to have survived into the Triassic, and only the temnospondyl-lissamphibian clade is represented in post-Permian assemblages. I have recently (Milner, 1990) published a provisional hypothesis for the relationships among the temnospondyl families and the Lissamphibia and have fitted the resulting cladogram onto a stratigraphic table to argue for particular patterns of cladogenesis and extinction, notably a large end-Permian extinction, followed by diversification of some surviving lineages into the vacant niche space. I have concluded that as few as five lineages of the temnospondyl-lissamphibian clade may have survived the end-Permian event. The minimum number of lineages that need to have survived were the Brachyopidae, Plagiosauridae, and Micropholidae, which show little Triassic diversification, together with the stems of the larger clades Stereospondyli and Lissamphibia, both of which must have diversified substantially in the early Mesozoic. The Micropholidae are known only from the Early Triassic, but the other groups all have Late Triassic and Jurassic representatives in the fossil record and provide the basis for this review. It appears that, by the Late Triassic, the brachyopid, stereospondyl, and plagiosaurid temnospondyls were restricted to large adult body size (more than 500 mm), whereas the Lissamphibia, hardly ever found as fossils from the Triassic, were restricted to small body size (500 mm or less). Not only are the two groups different in size, but also they occur in different circumstances. The late temnospondyls occur largely in floodplain channel and lacustrine deposits, whereas

most early lissamphibians occur in microvertebrate assemblages in low-energy coastal lagoon deposits. No amphibians have yet been described from the "Rhaeto"-Liassic fissure or cave assemblages. At present, then, the amphibians are of limited occurrence and hence of limited stratigraphic value, but their potential could be immense. It seems likely that the lissamphibians were undergoing their basic diversification, and hence considerable changes with time, through the Triassic and Jurassic, and a rich lissamphibian record could be of great stratigraphic use.

The groups conventionally placed in the grade Temnospondyli are considered in the first part of this review, and the Lissamphibia are considered in the second part.

Temnospondyli

Introduction

Temnospondyl amphibians, from the Late Triassic and subsequent times, grew to more than 500 mm in length when adult, and the group spans the range from microvertebrates to medium-sized vertebrates. I have included a review of all temnospondyl families, for three reasons: First, however large the adults may have been, all temnospondyls may be presumed to have started life as small larvae or juveniles [e.g., capitosaurids (Warren and Hutchinson, 1988) and brachyopids (Watson, 1956)]. These larvae had well-ossified dermal elements and well-developed teeth, even if some parts of the skeleton were poorly ossified, and thus might be expected to contribute to microvertebrate assemblages. Second, the larger temnospondyls are frequent in and characteristic of several aquatic facies in the Triassic–Jurassic and thus contribute fragments to microvertebrate assemblages (e.g., Buffetaut and Wouters, 1986; Shishkin, 1987; Nessov,

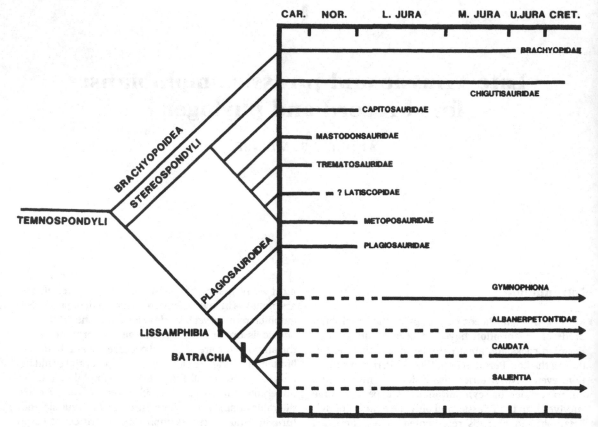

Figure 1.1. Cladogram showing ranges and interrelationships of the post-Ladinian temnospondyl families and the major subgroups of lissamphibians. No pre-Carnian taxa are included.

1988). Third, in recent years temnospondyls have been invoked both as stratigraphically significant taxa and as components of extinction events (Olsen and Sues, 1986; Benton, 1989), and a thorough documentation of their distribution in space and time will be necessary to qualify such discussions.

This review follows my recent provisional hypothesis of temnospondyl family-level relationships (Milner, 1990), and thus the characters used to support it are not comprehensively relisted here. The relationships of the various families are discussed later under the higher taxonomic headings preceding the family reviews. Of the eight temnospondyl families found in post-Middle Triassic rocks, all but the Latiscopidae (= Almasauridae) have representatives from the Middle Triassic or earlier. The Latiscopidae appear to be the sister-group of the Metoposauridae (Milner, 1990), which are first known from the Ladinian and can therefore also be assumed to have had a Ladinian or earlier origin. Thus, there was no family-level diversification of temnospondyls taking place by the Late Triassic. That had happened in the Early and Middle Triassic, and so the record of the Late Triassic and subsequent times is composed of a series of terminal lineages (Figure 1.1). In the following part of this review, I look at these eight families of post-Ladinian temnospondyls. For each family, I briefly review the characteristics, the post-Ladinian records, and the post-Ladinian chronological and geographic ranges that can logically be inferred from the known fossil record.

Brachyopoidea

The clade Brachyopoidea is here taken to comprise the family Brachyopidae and its Early Triassic sister-family, the Tupilakosauridae, but not the Chigutisauridae. The earliest brachyopoid is the Late Permian brachyopid *Bothriceps* from Australia (Watson, 1956). The most immediate relatives of the brachyopoids appear to be the Saurerpetontidae from the Pennsylvanian and Lower Permian of North America. Coldiron (1978), Milner (1990), and Foreman (1990) have argued for this relationship on the basis of the similarity of the Early Permian saurerpetontid *Acroplous* to the brachyopids. Whether the saurerpetontids are the brachyopoid stem-group (Coldiron, 1978; Milner, 1990) or sister-group (Foreman, 1990), the entire

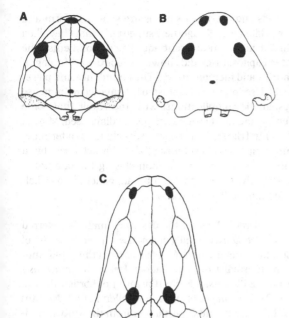

Figure 1.2. Skulls of Triassic–Jurassic temnospondyls in dorsal aspect: (A) *Sinobrachyops* (Brachyopidae), (B) *Pelorocephalus* (Chigutisauridae), (C) *Cyclotosaurus* (Capitosauridae).

clade is recognizable as far back as the Middle Pennsylvanian (*Saurerpeton* from Linton, Ohio) and clearly is not part of the Stereospondyli. This clade appears to be an early offshoot of the Temnospondyli prior to the dichotomy of the stereospondyl stem and the eryopid-dissorophoid stem. If so, then by the Late Triassic the surviving brachyopids were the most distant relatives of all other amphibians (Figure 1.1).

Brachyopidae

Characteristics. The brachyopids were a family of specialized aquatic temnospondyls growing to about 1.5 m in length, though most must have been under 1 m in length. They had short, deep skulls, large, dorsally situated orbits, and prominent dermosensory canals, but they show no trace of tympanic notches and undoubtedly were suction-gulpers, probably feeding in open water or at the surface. They first appeared in the Late Permian, and, throughout, their record indicates that they were never abundant or diverse. Earlier brachyopids coexisted with capitosaurids and rhytidosteids, but hardly ever with plagiosaurids. Later brachyopids have not been found with other temnospondyls.

Record. A mere decade ago the Brachyopidae were believed to be predominantly a Late Permian and

Early Triassic family, with *Hadrokkosaurus* from the Moenkopi Formation of Arizona as a large terminal genus from the Middle Triassic. Over the past ten years, three Jurassic brachyopids have been described or reported, all from central or eastern Asia. Dong (1985) described the skull of *Sinobrachyops placenticephalus* (Figure 1.2A) from the Xiashaximiao Formation (Middle Jurassic) of Sichuan, China. Nessov recorded (1988) and described (1990) temnospondyl fragments as *Ferganobatrachus riabinini* from the Balabansay Formation (Callovian) of Kirghizia. He identified them as capitosauroid, but the type clavicle appears to belong to a brachyopid (Shishkin, 1991). Finally, Shishkin (1991) has described, as *Gobiops desertus*, some brachyopid material from the Upper Jurassic (stage unspecified) of Mongolia. In addition to those records, Kitching and Raath (1984) have reported brachyopid material from the upper Elliot Formation of South Africa, although pending its description there remains a possibility that this material is chigutisaurid.

Post-Ladinian range. Carnian to Callovian (based on three or four records). One can legitimately generalize that brachyopids survived in the Gondwanan region until at least the end of the Middle Triassic and may have remained in that region up to the end of the Triassic, if the South African record can be confirmed. However, in the East Asian region of Laurasia, they persisted throughout most, if not all, of the Jurassic. The presence or absence of the family Brachyopidae is of no stratigraphic value within the Triassic and Jurassic. The later constituent genera were too few, and their relationships and distribution too poorly understood, for them to be of stratigraphic value.

Stereospondyli

The clade Stereospondyli at one time comprised almost all Triassic temnospondyls, and it still includes most of them. The name Stereospondyli is now a misnomer, as only some members of the Mastodonsauridae and Metoposauridae show the stereospondylous condition of a massive intercentrum and small pleurocentra or absence of pleurocentra. The remaining stereospondyls are neorhachitomous. Two families, Brachyopidae and Plagiosauridae, now appear not to be related to the stereospondyls (Milner, 1990), but to share only some presumably neoteny-related convergences with them [for arguments of these cases, see Coldiron (1978) for the Brachyopidae and Panchen (1959) for the Plagiosauridae]. The basal members of the stereospondyl crown-group probably resembled the Rhinesuchidae.

Chigutisauridae

Characteristics. The Chigutisauridae were a family of specialized aquatic temnospondyls growing to about 2.5 m in length. They had short, deep skulls, anterodorsally situated orbits, and prominent dermosensory canals; they retained characteristic tympanic notches (Figure 1.2B) and undoubtedly were suction-gulpers, probably living in open water. They first appeared in the Early Triassic. Of the Late Triassic stereospondyl families, the Chigutisauridae probably were most distantly related to the others. They appear to have been neotenous members of a lydekkerinid-rhytidosteid-chigutisaurid clade that separated off from other stereospondyls in the basal Triassic or even the Late Permian (Milner, 1990).

Record. The Chigutisauridae were originally based on a restricted suite of specimens from the Middle Triassic of Argentina, but that has been expanded by Warren to include several temnospondyls from across the Southern Hemisphere. Late chigutisaurids include *Pelorocephalus ischigualastensis* Bonaparte, 1975 from the Ischigualasto Formation (Carnian) of Argentina, undescribed material from the upper Maleri Formation (Norian) of India (Kutty and Sengupta, 1990), *Siderops kehli* Warren and Hutchinson, 1983 from the Evergreen Formation (Lower Jurassic) of Queensland, Australia, and undescribed material from the Strzelecki Formation (Lower Cretaceous) of Victoria, Australia. Jupp and Warren (1986) published a preliminary report on the first fragment of the latter, which Milner (1989) interpreted as a plagiosaurid. Further material, collected subsequently, has shown it to be a large chigutisaurid (Warren et al., in press, and pers. commun.). As noted in the section on Brachyopidae, some "brachyopid" material from the Triassic–Jurassic of South Africa may also prove to belong to this family.

Post-Ladinian range. Carnian to Early Cretaceous (based on four records). All records (including those from the Early and Middle Triassic) are from the Gondwanan region of Pangaea, and chigutisaurids may have been restricted to that region. They may have been restricted to Australia from the Early Jurassic onward. The later members of the family are so widely spread chronologically as to be of little stratigraphic value.

Capitosauridae

Characteristics. The Capitosauridae comprise one of the best-known and most abundant families of Triassic temnospondyls. They were large, superficially crocodilelike temnospondyls growing to 3 m in length. The snout was elongate and triangular to broadly parallel-sided, giving an alligatorlike shape, with the eyes set well back. Capitosaurids probably were semiaquatic piscivores; they had prominent dermosensory canals and poorly ossified carpals and tarsals. Capito-

saurids and mastodonsaurids are similar in form and are seldom found together, suggesting that they filled similar niches in different aquatic environments. The later capitosaurids did, however, coexist with plagiosaurids and metoposaurids. The abundance and piecemeal descriptions of capitosaurids have led to immense systematic complications, with the result that some genera and most species are poorly diagnosed. Most of the Late Triassic material clearly belongs to a terminal subgroup – the cyclotosaurs – characterized by a tabular-squamosal connection closing the otic notch posteriorly, together with a broad, generally parallel-sided muzzle (Figure 1.2C).

Record. Most of the described material referred to *Cyclotosaurus* comes from the Upper Triassic of Germany and includes *C. robustus* from the Schilfsandstein (Carnian) of Feuerbacher Heide, *C. ebrachensis* from the Blasensandstein (Carnian) of Ebrach (Kuhn, 1932), *C. carinidens* from the Knollenmergel (Norian) of Halberstadt, and *C. posthumus* from the Stubensandstein (Norian) of Pfaffenhofen (Fraas, 1913) and Sindelfingen (undescribed material in the collections of the Staatliches Museum für Naturkunde Stuttgart). The only well-documented cyclotosaur from outside Germany is a specimen from the Huai Hin Lat Formation (Norian) of Thailand, referred to *C. posthumus* by Ingavat and Janvier (1981), although Dutuit (1976) has reported cyclotosaur material from the Carnian of Morocco. Other Late Triassic capitosaurid fragments of doubtful affinities are *Calamops paludosus* from the Stockton Formation (Carnian) of Pennsylvania (Olsen, 1980), the original *Capitosaurus arenaceus* from the lower Keuper (Carnian) of Benk, Germany (Broili, 1915), "*Capitosaurus*" *polaris* from the Upper Triassic of Mt. Congress, Spitsbergen (Wiman, 1914), unnamed probable capitosaurid material from the lower Elliot Formation (upper Carnian) of South Africa (Hopson, 1984), and *Promastodonsaurus bellmanni* from the Ischigualasto Formation (Carnian) in Argentina (Bonaparte, 1963). The latter specimen has often been reported as a mastodonsaurid, but its visible characteristics are those of a capitosaurid, as argued by Paton (1974, p. 283). Nessov (1988, 1990) identified temnospondyl material from the Callovian of Kirghizia as a "capitosauroid," which he named *Ferganobatrachus riabinini*. Milner (1989) followed that identification, but it now appears that this material is either brachyopid (q.v.) or indeterminate.

Post-Ladinian range. Carnian to Norian (based on 12 records). Capitosaurids are found, with certainty, through the Carnian and Norian, but are not unambiguously documented from later beds. Their distribution is still patchy, but the European, Thai, and Argentinian records suggest a global range during the Late Triassic. Careful comparative study of the European

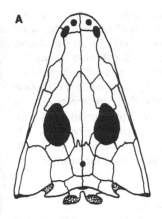

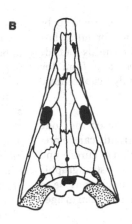

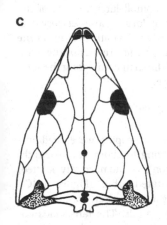

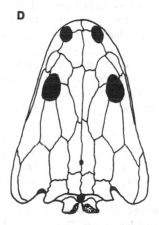

Figure 1.3. Skulls of Triassic–Jurassic temnospondyls in dorsal aspect: (A) *Mastodonsaurus* (Mastodonsauridae), (B) *Tertrema* (Trematosauridae), (C) *Almasaurus* (Latiscopidae), (D) *Metoposaurus* (Metoposauridae).

species of *Cyclotosaurus* may permit them to be of stratigraphic use. Ingavat and Janvier (1981) dated the Huai Hin Lat Formation as Norian, in part on the resemblance of the included cyclotosaur to *C. posthumus*, which is found in the Norian, but not the Carnian.

The benthosuchid descendants

The classification of the "capitosauroid" stereospondyls is in a state of flux, but it increasingly appears that the Benthosuchidae are a critical group, probably a grade of early relatives of various families. Some benthosuchids, such as the Heylerosaurinae (*Eocyclotosaurus*, *Meyerosuchus*, and *Odenwaldia*), may be early relatives of the Mastodonsauridae, whereas others, such as the Thoosuchinae and *Vyborosaurus*, may be relatives of the Trematosauridae and other families.

Mastodonsauridae

Characteristics. The mastodonsaurids were large, superficially crocodilelike temnospondyls growing to

6 m in length. The snout was elongate and generally triangular in outline, giving a crocodilelike shape, with the relatively large orbits set well back (Figure 1.3A). Mastodonsaurids probably were semiaquatic piscivores; they had prominent dermosensory canals and poorly ossified carpals and tarsals. Mastodonsaurids and capitosaurids appear superficially similar, and they are seldom found together, suggesting that they filled similar niches in different aquatic environments. Mastodonsaurids did, however, coexist with plagiosaurids.

Record. The Mastodonsauridae are represented principally by material referable to the genus *Mastodonsaurus* (Figure 1.3A) from numerous localities in the Middle Triassic of Europe east to the Urals. Ochev and Shishkin (1989) argued for *Mastodonsaurus* sensu stricto (excluding the Anisian "*Heptasaurus*") to be a diagnostic genus for the Ladinian. However, at least one relict occurs in the Carnian, namely, *Mastodonsaurus keuperinus* from the Schilfsandstein of Feuerbacher Heide, near Stuttgart, Germany (Fraas, 1889), and so some of the Russian material may yet prove to be Carnian in age. There is no evidence for other late mastodonsaurids, nor for mastodonsaurids outside that region of Eurasia. *Promastodonsaurus* from the Carnian of Argentina is almost certainly a capitosaurid (q.v.).

Post-Ladinian range. Carnian (based on one record). Relict Mastodonsauridae currently can be predicted to be present only in the Carnian of Europe and the west Asian part of Russia. The only stratigraphic value of such relict members of the family may be in indicating a horizon to be no later than Carnian in age.

The trematosaurid-latiscopid-metoposaurid clade
The families in this group share features such as a characteristic pattern of elongation of the skull table, small, laterally situated orbits in adults, ectopterygoid entering into the margin of the interpterygoid vacuity, a parasphenoid with a deep, narrow cultriform process (broadened in metoposaurids), a weakly developed oblique ridge on the pterygoid, large posterior Meckelian fenestra, and an exoccipital-pterygoid suture. Many of these characters (but not the last) also occur in some benthosuchids (Getmanov, 1989; Novikov, 1990), and this clade must have emerged from the benthosuchid grade of organization early in the Triassic.

Trematosauridae

Characteristics. The trematosaurids were medium-sized temnospondyls growing to 2 m in length. The Early Triassic genera were very diverse in terms of skull shape, varying from slender-snouted, gharial-like forms to forms with short, triangular snouts. The skull table (Figure 1.3B) was always elongate, and the eyes small and laterally situated in adults. Trematosaurids probably were aquatic piscivores; they had prominent dermosensory canals and poorly ossified postcranial skeletons. The Early Triassic trematosaurids coexisted with capitosaurids, rhytidosteids, and lydekkerinids, whereas the only Carnian form, as noted later, was found with metoposaurid material.

Record. The Trematosauridae are known almost entirely from the Early Triassic, and they had a global distribution. One Carnian form has recently been identified, however. Hellrung (1987) has shown that *Hyperokynodon keuperinus* Plieninger, 1852 from the Schilfsandstein (Carnian) of Heilbronn, Germany, is actually a snout impression of a large *Tertrema*-like trematosaurid (Figure 1.3B depicts *Tertrema*). A second undescribed fragment of this form, also from Heilbronn, is present in the collections of the Staatliches Museum für Naturkunde, Stuttgart, and there can be no doubt of its identity. No other Late Triassic trematosaurid material has been reported.

Post-Ladinian range. Carnian (based on one record). Trematosaurids can, with certainty, be stated to be or predicted to be present only in the Carnian of Europe. However, their global distribution during the Early Triassic leaves open the possibility that relict forms might be found in any part of the world. One could argue conservatively that the presence of a trematosaurid would support a pre-Norian age for an assemblage, but though the Carnian representation is based on a single locality, it would be unwise to use late trematosaurids as a basis for stratigraphic argument.

The latiscopid-metoposaurid clade
The families Latiscopidae and Metoposauridae are the latest to make their first appearances in the fossil record, the former in the Carnian and the latter in the late Ladinian. They appear to have been immediately related to one another, sharing not only the pattern of skull-table elongation and small lateral orbits of the trematosaurs but also the large, close-set terminal nares and an extremely reduced palatine ramus of the pterygoid, with the posteromedial ramus of the ectopterygoid forming part of the strut connecting the skull margin with the braincase.

Latiscopidae (= Almasauridae?)

Characteristics. The latiscopids were medium-sized temnospondyls growing to about 1m in length. The muzzle was slightly elongate and narrowly pointed, with small, laterally situated orbits (Figure 1.3C). Latiscopids probably were aquatic piscivores; they had prominent dermosensory canals. The latiscopids coexisted with metoposaurids. The poorly preserved *Latiscopus* has been interpreted as a late trematosaurid (Warren and Black, 1985), but it appears to resemble *Almasaurus*, which lacks the parasphenoid underplating of the occiput, a diagnostic trematosaurid feature.

Record. The Latiscopidae are represented by the poor holotypic skull of *Latiscopus disjunctus* from the Dockum Formation of Howard County, Texas (Wilson, 1948) and a second undescribed skull from the Dockum of Post, Garza County, Texas (Chatterjee, 1986, p. 145). The former appears to be Carnian, and the latter locality is either Carnian (Olsen and Sues, 1986) or Norian in age (Chatterjee, 1986). The Almasauridae are represented by numerous specimens of *Almasaurus habbazi* (Figure 1.3C) from the Carnian t.5 beds of Alma, Morocco (Dutuit, 1972). These two taxa closely resemble each other and almost certainly represent a single family. The senior name Latiscopidae is used here.

Post-Ladinian range. Carnian to Norian (based on three records). The presence of latiscopids in the

Carnian/Norian of North America and Africa demonstrates a minimum range across equatorial Pangaea. Most material is of Carnian age, but the Texas locality is not firmly placed stratigraphically, and so this family is not yet of stratigraphic value beyond indicating a Late Triassic age.

Metoposauridae

Characteristics. The metoposaurids were large, superficially crocodilelike temnospondyls growing to 2 m in length. The snout was short, and the posterior portion of the skull was elongate and broadly parallel-sided, giving an alligatorlike shape, but with the eyes set well forward (Figure 1.3D). Metoposaurids probably were semiaquatic piscivores; they had prominent dermosensory canals and poorly ossified carpals and tarsals. In different localities, metoposaurids coexisted with capitosaurids, plagiosaurids, trematosaurids, and latiscopids.

Record. The Metoposauridae are almost entirely known from the Upper Triassic, although there are two Ladinian specimens from Germany. The well-preserved specimens show consistent variation in a number of characters, and phylogenetic analysis of their relationships is possible. Hunt (1989) has discussed these characters and produced a hypothesis for the relationships of some metoposaurs. I disagree with the polarity of one of Hunt's major characters, namely, the relationship of the lacrimal to the orbital margin. Although the primitive temnospondyl condition is for the lacrimal to enter the orbital margin, this is modified in the Stereospondyli, and the primitive stereospondyl condition is to have the lacrimal excluded from the orbital margin. The range of lacrimal configurations in metoposaurs represents progressive reversal to the primitive temnospondyl condition, probably associated with paedomorphic snout abbreviation.

The diagnosable metoposaurid species can be discussed as a series of morphologically definable grades and clades.

Grade 1 Metoposaurus. Basal grade defined by primitive features: (1) lacrimal separated from orbital margin by broad prefrontal-jugal suture, (2) small, widely spaced nares, and (3) a continuous lateral-line loop behind the orbit.

In southern Germany, *Metoposaurus* has been described from the Schilfsandstein (Carnian) of Feuerbacher Heide, Hanweiler, Heilbronn, and Huhnediehaide (Fraas, 1889, 1913), the Blasensandstein (Carnian) of Ebrach (Kuhn, 1932), and the Lehrbergstufe (Norian) of Stuttgart-Sonnenberg (Fraas, 1913). Records of this grade outside Germany include *M. bakeri* from Carnian strata of the Dockum Formation of Texas (Figure 1.3D)

and the Wolfville Formation of Nova Scotia (Baird, 1986).

Clade 1A New "dwarf" metoposaur. Hunt (1989) has noted the existence of a small metoposaur in the Upper Triassic of Arizona, New Mexico, and Texas. It retains the primitive character-states of Grade 1, but is characterized by unusually elongate intercentra, absence of pleurocentra, and a skull length of 25 cm or less. The intercentrum shape indicates that this is not merely a juvenile of one of the well-known larger metoposaurid genera. This form is rare in the Carnian beds, where larger metoposaurs predominate, but it is the common metoposaur in unequivocally Norian horizons in the American Southwest. The only definitely metoposaurid material from the Norian of Germany is *M. stuttgartiensis* from the Lehrbergstufe of Stuttgart-Sonnenberg, and the holotype and only specimen of this form (a clavicle and interclavicle) also belonged to a relatively small animal.

Grade 2 Metoposaurus. As Grade 1, except that (1) the lacrimal is longer and is separated from the orbital margin by narrow prefrontal-jugal contact. Characters 2 and 3 retain the primitive state. The only metoposaurids belonging to this grade are three species of *Metoposaurus* from Carnian t.5 beds of Morocco (Dutuit, 1976), namely, *M. azerouali, M. lyazidi,* and *M. ouazzoui.*

Grade 3 Buettneria. As Grade 1, except that (1) there is an elongated lacrimal entering the orbital margin. Characters 2 and 3 retain the primitive state. This grade includes much of the material from the Dockum Formation of Texas (*Buettneria perfecta* and *B. howardensis*) and probably that from the Chinle Formation of Arizona and New Mexico.

Clade 4 Anaschisma. In this group, which appears to be a terminal clade, the character-states are (1) an elongated lacrimal entering the orbital margin, (2) large, closely spaced nares, and (3) a lateral-line canal loop broken behind the orbit. This group includes *Anaschisma browni* and *Koskinonodon princeps* (= *Borborophaqus wyomingensis*) from the Popo Agie Formation of Wyoming and "*M.*" *maleriensis* from the lower Maleri Formation of India (Chowdhury, 1965; Kutty and Sengupta, 1990). As Hunt (1989) has pointed out, the supposedly notchless condition of *Anaschisma* is based on plaster reconstructions.

Many metoposaurid taxa and specimens are indeterminate or probably so. These include "*M.*" *santaecrucis* from the Upper Triassic Raibl Beds of South Tyrol, Italy (Koken, 1913), *Dictyocephalus elegans* from the Upper Triassic of North Carolina, and indeterminate specimens from the Stockton, Lockatong, and Oxford for-

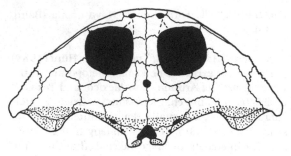

Figure 1.4. Skull of Triassic–Jurassic temnospondyl in dorsal aspect: *Gerrothorax* (Plagiosauridae).

mations of Pennsylvania and New Jersey (Baird, 1986). There are also some poor metoposaurid scraps from the Upper Triassic Isalo Formation of Madagascar (Dutuit, 1978).

Most of the described large metoposaurids are Carnian, but some large material is Norian, and Hunt and Lucas (1989, p. 90) and Hunt (1989, p. 296) mention some possible scraps from the Norian Revuelto Creek locality in New Mexico. The small metoposaurid with elongated centra characterizes the Norian formations of Arizona, New Mexico, and Texas (Davidow-Henry, 1989; Hunt, 1989). Some material of this new form was reported by Gregory (1980) as *Anaschisma* sp. from the Redonda Formation of Apache Canyon, near Tucumcari, New Mexico. Murry (1987), followed by Milner (1989), suggested that this might be as late as Liassic in age, but this has subsequently been disputed by Hunt and Lucas (1989, 1990), who argue for an early Norian age for this site.

Post-Ladinian range. Carnian to Norian (based on more than 30 records). Metoposaurids have been described from western Europe, North America, North Africa, and India, with attributed scraps from Madagascar. They may well have been global in distribution. The metoposaurids are potentially of considerable stratigraphic value within the Upper Triassic, as they are abundant in the prevailing facies, and the populations are variable. If this variability can be shown to have a strictly chronological basis rather than a local ecological basis, it may be possible to characterize assemblages as Carnian or Norian on the basis of the constituent metoposaurids.

Plagiosauroidea

The Plagiosauroidea comprise the Triassic Plagiosauridae and the Late Permian *Peltobatrachus* from Tanzania. *Peltobatrachus* shares several derived characters with the Plagiosauridae and appears to be their most immediate Permian relative (Panchen, 1959; Warren, 1985). The relationships of the Plagiosauroidea to other temnospondyls are less certain, but

Peltobatrachus appears to have branched from the lissamphibian stem after the eryopid plesion but before the various dissorophoid families (Milner, 1990). If its closest relatives are forms of the eryopid-dissorophoid grade, then the Plagiosauroidea are not stereospondyls and share no common ancestor with stereospondyls later than the Late Carboniferous. Their origin within the eryopid-dissorophoid grade implies a closer relationship to the lissamphibians, which generally are held also to be immediately related to Paleozoic dissorophoids (Milner, 1988, 1990).

Plagiosauridae

Characteristics. The plagiosaurids were specialized aquatic temnospondyls growing occasionally to about 2.5 m in length, although most must have been under 1m. They had short, very wide, shallow skulls, very large, dorsally situated orbits, and prominent dermosensory canals. They undoubtedly were suction-gulpers, probably living on the bottom. They first appeared in the Early Triassic. Plagiosaurids coexisted with mastodonsaurids, capitosaurids, and metoposaurids, but hardly ever with brachyopids.

Record. Until recently the Plagiosauridae were known largely from material from the Triassic of Europe and European Russia. Late Triassic forms from Europe comprise *Gerrothorax franconicus* from the Blasensandstein (Carnian) of Ebrach (Kuhn, 1932), *Gerrothorax* spp. (Figure 1.4) from several localities in the Norian (including "Rhaetian") of Sweden and Germany, *Plagiosaurus depressus* from the Knollenmergel (Norian) of Halberstadt, Germany, *Plagiosaurus* sp. from the Keuper of St.-Nicolas-de-Port (Buffetaut and Wouters, 1986), unnamed material from the Keuper of Luneville, France, and an unnamed specimen from the Carnian of Bjørnøya, Spitsbergen (Lowy, 1949; Panchen, 1959). Most belong to the subfamily Plagiosaurinae, characterized by pustulate dermal ornament, but the specimen from Spitsbergen appears to belong to the second subfamily Plagiosterninae. A recent record extends the plagiosaurids farther geographically. Suteethorn, Janvier, and Morales (1988) have reported a plagiosaurid scute from the Huai Hin Lat Formation (Norian) of Thailand. A temnospondyl jaw fragment from the Lower Cretaceous of Victoria, Australia, was interpreted as a plagiosaurid mandible by Milner (1989, 1990), but subsequently collected material has shown this relict temnospondyl to be a chigutisaurid (q.v.).

Post-Ladinian range. Carnian to Norian (based on nine records). It is certain that plagiosaurids are distributed throughout the Carnian and Norian. The Thai record and an Early Triassic Australian record suggest that this group must have been much more widespread than the abundant European material might have led one to presume. The plagiosaurids may ultimately be of stratigraphic value, the genera *Gerrothorax* and *Plagiosaurus* being thus far restricted to the Upper Triassic, whereas other genera are restricted to Middle Triassic horizons. Again, far more well-dated records are needed before we can be confident of the stratigraphic value of these genera.

Miscellaneous fragments

Numerous Upper Triassic localities have produced miscellaneous indeterminate fragments of temnospondyls. These are too numerous to record and would add nothing of significance to our understanding of the distribution of the group in space and time. Post-Triassic records are still relatively rare, however, and there are three records of indeterminate temnospondyls that are worth documenting.

The first post-Triassic temnospondyl to be described was *Austropelor wadleyi* Longman, 1941, based on a partial mandibular ramus from the Lower Jurassic Marburg Sandstone of Queensland, Australia. It may have belonged to a brachyopid or a chigutisaurid (Warren and Hutchinson, 1983), both of which are known from post-Triassic beds.

A series of neorhachitomous vertebrae from the Fengjiahe (Lufeng) Formation of Yunnan, China, were described by Sun (1962). The productive horizon is now recognized as Lower Jurassic (Sun and Cui, 1986), but the material remains indeterminate within the Temnospondyli. Vertebral centra and skull fragments were described by Ishing (1987) from the Middle Jurassic Wucaiwan Formation and the Upper Jurassic Shishugao Formation of Xinjiang, China. The material was named *Superstogyrinus ultimus*, but it does not appear to be diagnostic.

These records, together with *Siderops*, *Sinobrachyops*, *Ferganobatrachus*, *Gobiops*, and the Strzelecki chigutisaurid, demonstrate the survival of relict temnospondyls well into the Jurassic of East Asia, central Asia, and Australia, and even the Early Cretaceous of Australia.

Latest appearances and extinctions

Until about a decade ago, a less extensive temnospondyl record was widely interpreted as indicating that the group became extinct at the end of the Triassic, along with many amniote groups (Colbert, 1958). That seemed so clear that one Jurassic temnospondyl fragment (*Austropelor wadleyi*) was assumed to have been reworked simply because it was Jurassic in age (Colbert, 1967). The discoveries of recent years have shown that to be an oversimplified view. So far as is currently known, the temnospondyl families listed earlier make

their last certain appearances in the fossil record as follows (see Figure 1.1):
Carnian: Mastodonsauridae, Trematosauridae
Carnian/Norian(?): Latiscopidae
Norian: Capitosauridae, Metoposauridae, Plagiosauridae
Callovian: Brachyopidae
Early Cretaceous: Chigutisauridae
Several observations can be made from these data.

First, of the eight latest-appearance records given here, three have changed during the past decade: Trematosauridae (Hellrung, 1987), Brachyopidae (Dong, 1985; Shishkin, 1991), and Chigutisauridae (Jupp and Warren, 1986; A. A. Warren, pers. commun). A 37 percent turnover of data points in one decade suggests that these latest appearances in the record are no more than that and will be liable to considerable further modifications in future decades. This very ephemeral status should caution against their use as presumed extinction dates (e.g., Olsen and Sues, 1986; Benton, 1989).

Second, the use of secondary, rather than primary, sources has resulted in erroneous assessments of last appearances in some previous discussions. Two examples of this may be given:

1. The last record of the Mastodonsauridae was quoted as Norian by Olsen and Sues (1986, p. 343, fig. 25.2). The material was not identified, but the record appears to have been taken either from the range chart given by Carroll (1977, fig. 2) or from that given by Anderson and Cruickshank (1978, chart 3.2). The upward extension of the mastodonsaurid range in Carroll's figure presumably was a drafting error, as the primary data in the associated appendix (Carroll and Winer, 1977, p. 5) gave the latest mastodonsaurid as Carnian, as recognized here. The "Norian mastodonsaurid" record in Anderson and Cruickshank's chart is *Promastodonsaurus bellmanni* from Ischigualasto, here considered to be a Carnian-age capitosaurid, as discussed earlier.

2. Olsen and Sues (1986, p. 343, figs. 25.2, 25.4) argued for a Carnian extinction of the Metoposauridae, but at the time that was written, one of the minor records of metoposaurids in Germany fell indisputably within the Norian Stubensandstein, namely *M.* "*stuttgartiensis*" (Fraas, 1913), probably a nomen dubium, but certainly a metoposaurid.

Third, it now appears that two lineages of neotenous suction-gulping temnospondyls crossed the Triassic–Jurassic boundary in at least some regions of the world and were not subject to the competition or extinction events that apparently affected the more crocodilelike temnospondyls. The slender-snouted trematosaurs appear to have been replaced by the phytosaurs during the Middle Triassic, whereas the broader-muzzled capitosaurids and metoposaurids appear to have been replaced by the crocodiles by the Jurassic. The relict suction-gulpers, such as the brachyopids in East Asia and the chigutisaurids in Australia, may ultimately have been replaced by teleostean fish such as the larger osteoglossids or silurids.

Fourth, for most of these families, the latest records are rare, chronologically widely spaced, single-locality occurrences, which could be argued to suggest survival of odd relict populations after a peak in diversity, rather than a sudden extinction of a diverse and successful group. Most post-Ladinian families are known from only one to four records, and this poverty of records must qualify any statement made about their geographic range or date of extinction. The only possible exceptions are the last appearances in the Norian of the Capitosauridae, Metoposauridae, and Plagiosauridae. Their apparent sudden demise, in particular in Europe, might support a regional Norian extinction event, but that still must be qualified by the relative scarcity of Early Jurassic freshwater assemblages.

Lissamphibia

General relationships

The clade Lissamphibia comprises the Salientia (frogs and toads), Caudata (salamanders and newts), Gymnophiona (the caecilians) and Albanerpetontidae (an extinct group of enigmatic relationships). The problems of the origin(s) and relationships of the Lissamphibia have been much discussed in recent years; see Milner (1988) for a recent review. There is an increasing consensus (1) that the Lissamphibia are a monophyletic group relative to the Amniota and (2) that within the Lissamphibia, Salientia and Caudata are immediately related (as Batrachia) (Milner, 1988). However, there are two distinct theories in contention concerning the relationship of lissamphibians to Paleozoic amphibians. One is that the entire Lissamphibia arose from the dissorophoid grade of temnospondyl amphibians, with the Early Permian genus *Doleserpeton* as the probable closest known relative (Milner, 1988, and earlier work cited therein). The alternative theory is identical for the origin and relationships of the Batrachia (Salientia and Caudata), but derives the Gymnophiona from the tuditanomorph microsaurs, with the Early Permian genus *Rhynchonkos* as the closest known relative of the Gymnophiona (Carroll and Currie, 1975). As far as the fossil record is concerned, the significant difference between these two theories will ultimately be tested by the Permo-Triassic relatives of the Gymnophiona, when they are discovered. Skeletal evidence favors a monophyletic origin from temnospondyls (Milner, 1988), and that interpretation is followed here, but it must be noted that there are character contradictions, and the evidence is by no means overwhelming.

The early Mesozoic fossil record of lissamphibians is extremely poor, comprising just three Triassic specimens and a metaphorical handful of Jurassic forms. The Triassic material and its significance are dealt with first, followed by the Jurassic representatives of each of the major lissamphibian subgroups.

Triassic Lissamphibia

Despite the rich record of continental tetrapods throughout the Triassic, only three Triassic lissamphibian specimens have been described or reported, and two specimens have been too poor to be of much value. The good specimen is that of the stem-salientian *Triadobatrachus massinoti* from the Lower Triassic of Madagascar (Figure 1.5A), which has recently been redescribed by Rage and Roček (1989). It is undoubtedly an early member of the salientian clade, possessing a partly fused frontoparietal, toothless dentary, abbreviated ribs, an elongated ilium, and elongated tarsals. Its presence means that the fundamental dichotomy between the Salientia and the Caudata (and hence also that between the Batrachia and the Gymnophiona) had already occurred and that at least stem-members of the Caudata and the Gymnophiona must also have been present by the Early Triassic. The presence of a bipartite atlas centrum in *Triadobatrachus* implies that the vertebrae may still have been endochondral, rather than composed of membrane bone as in the crown-group Salientia. If so, then isolated vertebrae of early stem-Salientia may not be readily recognized by comparison with those of their later relatives.

The other probable lissamphibian that has been described is *Triassurus sixtelae* Ivakhnenko, 1978 from the Upper Triassic of Kirghizia in central Asia. This is a poorly preserved, tiny specimen with a length of about 12 mm from snout to pelvis. It may fall within the batrachian clade, having a broad skull, apparently with a large squamosal-pterygoid bar supporting the quadrate, about 20 presacral vertebrae, short ribs, and long humerus and femur. Estes (1981) suggested that it might be a larval temnospondyl, but the slender limbs do not match those of any known Late Triassic temnospondyl. Ivakhnenko argued it to be a urodele, but there are no specifically urodelan features visible, and it simply documents the presence of lissamphibians in the Upper Triassic of central Asia.

Sues and Olsen (1990) have recently reported a possible lissamphibian fragment from the Carnian of the Richmond Basin of Virginia. It is a piece of jaw with tooth-bases resembling those of lissamphibians, but unfortunately it is not critically diagnostic.

The Triassic record of lissamphibians is thus extremely disappointing. It demonstrates that they were present in the Early Triassic and, by phylogenetic inference, had differentiated into salientians, gymnophionans, and caudates. It also shows that they were present in places as far apart as Madagascar, central Asia, and possibly North America and hence probably

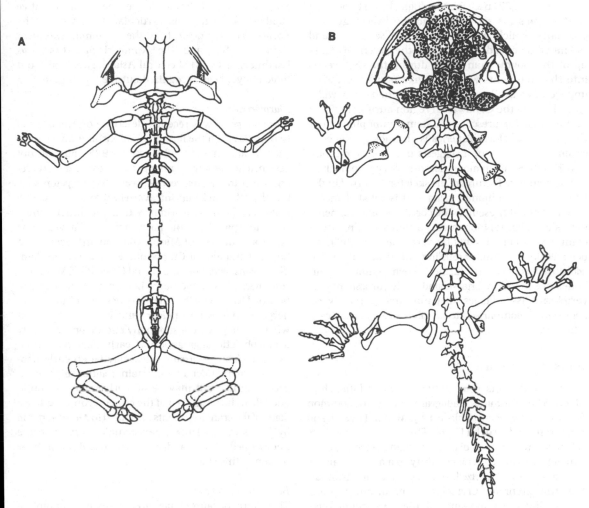

Figure 1.5. Early lissamphibians: (A) *Triadobatrachus* (Salientia), (B) *Karaurus* (Caudata). (A after Rage and Roček, 1989; B drawn from photograph in Ivakhnenko, 1978.)

were widespread in Pangaea during the Triassic. The Triassic record provides no other direct evidence concerning the pattern and rate of diversification of the Lissamphibia.

Jurassic Gymnophiona

There is no consensus on the interrelationships of modern gymnophionans and hence no agreed cladogram to which to relate fossil taxa. Conflicting theories of relationships have been published by Lescure, Renous, and Gasc (1986) and Duellman and Trueb (1986). This lack of agreement on the interrelationships of crown-group gymnophionans is not yet an acute problem for early Mesozoic paleontological studies, as the only relevant fossils appear to represent a stem-gymnophionan.

No pre-Cretaceous gymnophionan material has yet been described, but potentially valuable specimens have recently been reported from the Lower Jurassic Kayenta Formation of Arizona (Jenkins and Walsh, 1990). These comprise several articulated or associated skeletons. The taxon that they represent possesses numerous gymnophionan synapomorphies, including a tentacular fossa, an os basale (fused exoccipitals, otic capsules, and parasphenoid), and a mandible comprising a pseudangular and pseudodentary. It retains, however, a circumorbital series of bones, a quadratojugal, an interglenoid tubercle on the atlas, intercentra, and reduced limbs. All these characters place it outside the crown-group gymnophionans, whatever their internal relationships, and speculatively suggest that the gymnophionan crown-group diversification may not yet have commenced in the Early Jurassic. Case and Wake (1977) used immunological data to propose relationships for six geographically diverse gymnophionans, which were congruent in sequence, and estimated times of divergence with the pattern of break-up of the Gondwanan region during the Cretaceous into the continents where the taxa now occur. That implied that the basal crown-group had been widespread across the Gondwanan region during the Early Cretaceous. The presence of a stem-gymnophionan in the Jurassic of North America is chronologically consistent with that, but it suggests that the stem-group may have been even more widespread geographically.

The primitive features of the skeleton reported in the Kayenta gymnophionan mean that isolated elements may not closely resemble the corresponding elements in modern gymnophionans. In particular, the presence of intercentra implies that the vertebrae may still have been endochondral, paralleling the situation in *Triadobatrachus*. When further stem-gymnophionan bones are found in Triassic and Early Jurassic microvertebrate assemblages, they may not be readily recognized by comparison to the skeletons of their living relatives.

Jurassic Caudata

There is little consensus on the internal relationships of the extant Caudata. Cladograms with some common features have been published by Milner (1983) and Duellman and Trueb (1986). This is not yet of acute relevance to the study of Jurassic salamanders, as none can yet be assigned with certainty to a modern family. Milner (1983) theorized, on biogeographic grounds, that the Sirenidae, Cryptobranchoidea, and a group comprising the remaining salamanders might have differentiated by the end of the Jurassic; however, none of the Jurassic fossils can be firmly assigned to these groups. Two of the described Jurassic taxa to be discussed later, the Karauridae and *Marmorerpeton*, appear to be earlier offshoots of the caudate stem, as they lack the diagnostic crown-group caudate character of intravertebral nerve foramina, even in the atlas. Indisputable Caudata first appear in the Bathonian, and at least three major morphological types of caudates are known from Middle and Late Jurassic assemblages.

Karauridae
The Karauridae appear to be the most primitive offshoot within the certain Caudata (i.e., excluding Albanerpetontidae, q.v.) and hence the sister-group of all other well-defined caudates (Estes, 1981; Milner, 1988). Karaurids retain a squamosal with dermal ornament, whereas all other caudates have a smooth dorsal surface to the squamosal, over which the superficial internal mandibular adductor muscle passes to insert on the exoccipitals or the cervical vertebrae.

They are thus believed to be the most primitive caudates known from articulated material. The Karauridae are known from the Bathonian (*Kokartus* Nessov, 1988) and the Kimmeridgian (*Karaurus* Ivakhnenko, 1978) of central Asia (Figure 1.5B), but have not yet been recognized outside that region.

Marmorerpeton
Marmorerpeton was recently described on the basis of two species from the Bathonian of Kirtlington, Oxfordshire (Evans, Milner, and Mussett, 1988), but was not assigned to a family. *Marmorerpeton* bears a general resemblance to the Cretaceous–Eocene Scapherpetontidae, but the atlas lacks an intravertebral foramen, which suggests a form more primitive than any other known caudate, apart from the karaurids. Similar vertebrae occur at other Forest Marble (Bathonian) localities in England and also at Guimarota, Portugal (Oxfordian) (S. E. Evans, pers. commun.), and Como Bluff, Wyoming (Kimmeridgian–Tithonian). Even before the description of *Marmorerpeton*, the relationships of the Scapherpetontidae were unclear, and it was not certain whether they were a monophyletic taxon or merely a polyphyletic aggregation of early neotenous salamanders. Thus *Marmorerpeton* may be an early neotenous salamander of uncertain relationships, even pre-karaurid, or it may be an early scapherpetontid and fall in the basal area of the crown-group caudates. Some of the cranial elements attributed to *Marmorerpeton* by Evans et al. (1988) may actually pertain to the undescribed "salamander A" discussed in a later section of this chapter.

Batrachosauroididae
The Batrachosauroididae are an extinct family of neotenous salamanders, similar to the Proteidae and sometimes argued to be related to them (Estes, 1981; Duellman and Trueb, 1986). If that were so, they would be crown-group salamanders, but it is not clear to what extent the similarities to the Proteidae are convergent, larval features shared by two neotenous lineages. Until recently, the published record of the Batrachosauroididae extended from the Coniacian to the Pliocene. A recently discovered microvertebrate assemblage from the Tithonian of Purbeck, Dorset, appears to include a batrachosauroidid (Ensom, 1988; Ensom, Evans, and Milner, 1991).

Miscellaneous
Other caudate material, distinct from the preceding, but not yet determinate, is also present in the Bathonian of Britain ("salamander B" of Evans and Milner, Chapter 18) and in the Kimmeridgian–Tithonian Morrison Formation of Como Bluff, Wyoming (*Comonecturoides marshi* Hecht and Estes, 1960, based on a femur and some damaged vertebrae) (Estes, 1981; Evans and Milner, 1993).

Jurassic Salientia

Cladistic analyses of the interrelationships of frog families include those by Lynch (1973), Sokol (1977), and Duellman and Trueb (1986). All of these follow the older view that the most primitive living frog families/superfamilies are the Leiopelmatidae, Discoglossidae, Pipoidea, and Pelobatidae. The monophyly of the Pipoidea is widely accepted, and all of the authors cited earlier and Clarke (1988) treat the Leiopelmatidae and Discoglossidae as monophyletic groups as well. Lynch (1973) treats the Leiopelmatidae, Discoglossidae, and Pipoidea as ranked successive monophyletic sister-groups of all other anurans, whereas Sokol (1977) and Duellman and Trueb (1986) treat the Leiopelmatidae and Discoglossidae as a clade, Discoglossoidei, which was the sister-taxon to all other anurans. Cannatella and Ford (unpublished data) argue that the Leiopelmatidae represent a paraphyletic basal grade of the Anura, whereas the Discoglossidae represent the next grade up and are also paraphyletic. While such a diversity of interpretations of the relationships and phylogeny of modern families prevails, the paleontologist can draw only limited phylogenetic conclusions from the early fossil material.

Evans et al. (1990) reviewed the record of Jurassic frogs from eight localities, one of which, Aveyron, had produced only one doubtful fragment (Seiffert, 1969b; Estes and Reig, 1973), while potentially valuable material from another site, Guimarota, remains undescribed. Evans et al. (1990) also followed earlier workers in recording the frog material from the Sierra del Montsech, Lerida Province, Spain, as a possible Late Jurassic or basal Cretaceous assemblage. It is, however, increasingly accepted that the locality is Berriasian–Valanginian in age within the Early Cretaceous, based on ostracods (Brenner, Goldmacher, and Schroeder, 1974) and palynological evidence (Barale et al., 1984). The discoglossid and palaeobatrachid material from Montsech, therefore, does not contribute to our understanding of Jurassic frogs. Our systematic knowledge of Jurassic frogs is based on material from the remaining five localities.

Vieraella

The only described Early Jurassic frog is the single tiny specimen of *Vieraella herbstii* from the Roca Blanca Formation of Argentina (Reig, 1961; Casamiquela, 1965; Estes and Reig, 1973). It has variously been made the type of the Vieraellidae (Reig, 1961), referred to the extinct Notobatrachidae (Casamiquela, 1965), and placed in the Ascaphidae (now Leiopelmatidae) (Estes and Reig, 1973). As noted by Estes and Reig, it is most similar to modern leiopelmatids, but this resemblance is phenetic rather than cladistic. It certainly falls near the base of the crown-group frogs and could be a stem-frog, a primitive leiopelmatid or even possibly a primitive discoglossid. It demonstrates only that crown-group frogs, or forms close to the crown-group, were present by the Liassic.

Leiopelmatidae (= Ascaphidae)

The Leiopelmatidae are today represented by the two relict genera *Leiopelma* from New Zealand and *Ascaphus* from northwestern North America. A Late Jurassic frog usually assigned to this family is *Notobatrachus degiustoi*, from the La Matilde Formation of Patagonia, Argentina. Principal descriptions are by Reig (1957), Casamiquela (1961), and Estes and Reig (1973). As with *Vieraella*, most of the resemblances to the modern leiopelmatids are primitive, but Estes and Reig have argued that it has derived resemblances to *Leiopelma* (shape of squamosal and clavicle) and to *Ascaphus* (shape of urostyle), and they also have shown that it has primitive features not found in modern leiopelmatids (well-developed middle ear) or even in modern frogs (carpal structure). *Notobatrachus* is thus a confusing mosaic of characters, demonstrating that convergence is frequent and/or that we do not fully understand character polarity in early frog evolution.

Discoglossidae

The Discoglossidae comprise five extant genera (*Discoglossus, Alytes, Baleaphryne, Barbourula,* and *Bombina*) and today are restricted to Eurasia and North Africa. However, for most of its history, this family appears to have been a Laurasian group, and there are many records of discoglossids from the Cretaceous and Tertiary of Eurasia and North America. The only Jurassic discoglossid described from numerous elements is *Eodiscoglossus oxoniensis*, from the Bathonian of Kirtlington (Evans et al., 1990) and, from undescribed material, several other English Bathonian localities (Evans and Milner, Chapter 18). Other Jurassic discoglossid material comprises an undescribed ilium from the Kimmeridgian–Tithonian Morrison Formation of Como Bluff, Wyoming (Evans and Milner, 1993), and undescribed material from the Tithonian Purbeck Formation of Dorset, England (Ensom et al., 1991). The holotype of *Comobatrachus aenigmatis*, a broken humerus also from Como Bluff, may be discoglossid (Estes and Sanchíz, 1982a), but it is not diagnostic.

Pipoidea

The Pipoidea include the extant families Rhinophrynidae and Pipidae and the extinct family Palaeobatrachidae. The pipids (e.g., *Xenopus, Hymenochirus,* and *Pipa*) and the palaeobatrachids are similar, largely aquatic frogs, and the two families appear to be sister-taxa. The pipoid families Palaeobatrachidae and Pipidae are both represented by Early Cretaceous fossils, the former by *Neusibatrachus* from Montsech and the latter by *Cordicephalus* and *Shomronella* from Israel. The superfamily clearly had diversified into the

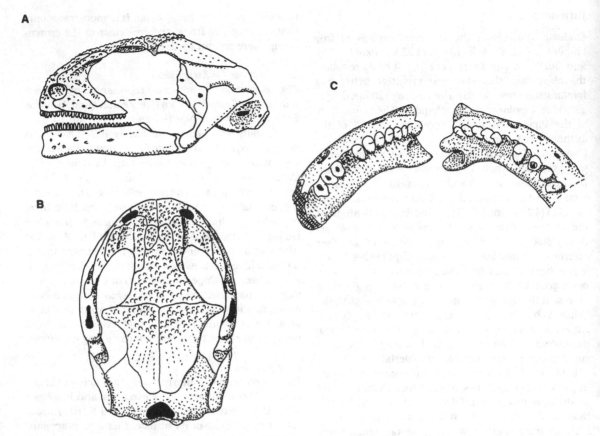

Figure 1.6. *Albanerpeton* (Albanerpetontidae): (A) skull in lateral view, (B) skull in dorsal view, (C) mandibular symphysis with characteristic articulation. (A and B after Estes and Hoffstetter, 1976; C after Fox and Naylor, 1982.)

Laurasian palaeobatrachids and the Gondwanan pipids by the earliest Cretaceous, and one can infer the presence of pipoids globally during the Late Jurassic. The only suggested pipoid fossil from the Jurassic is *Eobatrachus agilis*, represented by a broken humerus from the Kimmeridgian–Tithonian Morrison Formation at Como Bluff, Wyoming, and most recently reviewed by Estes and Reig (1973). However, work by Evans and Milner (1993) on frog material from Como Bluff suggests that this humerus falls within the range of variation of discoglossid humeri and is indeterminate at present. Thus no diagnostic pipoid fossils have yet been found in Jurassic strata.

Pelobatidae

No Jurassic pelobatids have been reported thus far, but an ilium from the Morrison Formation at Como Bluff (Evans and Milner, 1993) is that of a pelobatid frog. This specimen takes the history of this family back from the Middle Cretaceous to the Late Jurassic.

Summary

In all published cladograms of frog relationships based on modern material, the Leiopelmatidae, Discoglos-

sidae, and Pelobatidae are produced by the basal four or five dichotomies, even though there is no consensus as to the precise relationships. These are the three groups that have been recognized in the Jurassic record and may genuinely reflect the "state of the art" in frog evolution.

Albanerpetontidae and other unplaced forms

Unrecognized twenty years ago, the albanerpetontids are now known to have been major components of microvertebrate assemblages in the Northern Hemisphere from the Middle Jurassic onward. Initially identified as prosirenid salamanders by Estes and Hoffstetter (1976), they were later separated completely from the type specimens of the Prosirenidae by Fox and Naylor (1982), who erected the Albanerpetontidae for them. Fox and Naylor went further and suggested that the Albanerpetontidae were not caudates, but represented a fourth group of lissamphibians, which they called the Allocaudata. That interpretation was disputed by Estes and Sanchíz (1982b). Albanerpetontids have several unique characters, such as the symphyseal hinge-joint (Figure 1.6C), and fragments of them can be readily recognized in microvertebrate assemblages, but these unique features have no bearing on the extrinsic relationships of the family. Milner (1988) noted that of the few albanerpetontid characters that had acute bearing on their relationships, some supported a position within the Caudata, whereas others suggested a position outside the Batrachia and hence outside the Caudata. This group needs further study, both to understand its structure more completely and to attempt to determine its phylogenetic position.

Albanerpetontids are first known from the Bajocian (Middle Jurassic), represented by a single diagnostic atlas centrum from Aveyron, France (Seiffert, 1969a). They are present at several Bathonian localities in Britain (Evans and Milner, Chapter 18) and in later Jurassic assemblages in the Callovian of Kirghizia (Nessov, 1988), the Oxfordian of Guimarota, Portugal (S. E. Evans, pers. commun.), and the Tithonian of Britain (Ensom, 1988; Ensom et al., 1991). Albanerpetontids have been described from the Cretaceous of North America, Europe, and central Asia and presumably were found across Laurasia for most of their Mesozoic history. Middle Jurassic material appears to show the full suite of albanerpetontid characteristics, suggesting at least some earlier history. If albanerpetontids fall outside the salientian–caudate dichotomy, then they certainly will have a distinct history back at least to the Early Triassic. Even if they prove to fall within the Caudata, they probably have a history extending back to the Triassic.

Evans and Milner (Chapter 18) are working on a further enigmatic lissamphibian ("salamander A") from the Bathonian of Oxfordshire. This is the commonest large lissamphibian in the Kirtlington assemblage, but is not certainly a member of the Caudata. Study of this form is still at a preliminary stage.

Conclusions

The temnospondyls of the Late Triassic and subsequent time comprise eight families, each representing a terminal lineage. Most are very rare, and only the Capitosauridae, Metoposauridae, and Plagiosauridae are represented by more than four Late Triassic records. These three families are the only ones available in sufficient quantity to be of potential value for stratigraphic or extinction studies.

It may be inferred that the Lissamphibia have been adaptively radiating from the Early Triassic onward. At present, Triassic and Jurassic lissamphibians are of no stratigraphic value, as they are known from an extremely poor fossil record, represented by three Triassic and ten Jurassic occurrences. Only one group, the Albanerpetontidae, is known from at least five sites and might be of stratigraphic potential in the foreseeable future.

Acknowledgements

I should particularly like to thank Nick Fraser and Hans-Dieter Sues for the opportunity to present this review at the workshop. I also thank Susan Evans for invaluable discussions about Jurassic lissamphibians, and Sandy Sequeira for assistance with the illustrations. My travel was supported by a Birkbeck College Research Conference Grant.

References

Anderson, J. M., and A. R. I. Cruickshank. 1978. The biostratigraphy of the Permian and the Triassic. 5. A review of the classification and distribution of Permo-Triassic tetrapods. *Palaeontologia Africana* 21: 15–44.

Baird, D. 1986. Some Upper Triassic reptiles, footprints, and an amphibian from New Jersey. *The Mosasaur* 3: 125–153.

Barale, G., C. Blanc-Louvel, E. Buffetaut, B. Courtinat, B. Peybernes, L. Via Boada, and S. Wenz. 1984. Les gisements de calcaires lithographiques du Crétacé inférieur du Montsech (Province de Lérida, Espagne). Considerations paléoécologiques. *Géobios, Mémoire Spécial* 8: 275–283.

Benton, M. J. 1989. Mass extinctions among tetrapods and the quality of the fossil record. *Philosophical Transactions of the Royal Society of London* B325: 369–386.

Bonaparte, J. F. 1963. *Promastodonsaurus bellmanni* n.g. et n.sp., capitosáurido del Triásico Medio de Argentina (Stereospondyli – Capitosauroidea). *Ameghiniana* 3: 67–78.

1975. Sobre la presencia del laberintodonte *Pelorocephalus* en la formacion de Ischigualasto y su significado estratigraphico (Brachyopoidea – Chigutisauridae). *Actas Primer Congreso Argentino de Paleontologia y Bioestratigrafia* 1: 537–544.

Brenner, P., W. Goldmacher, and R. Schroeder. 1974. Ostrakoden und Alter der Plattenkalke von Rubies (Sierra de Montsech, Prov. Lerida, NE-Spanien). *Neues Jahrbuch für Geologie und Paläontologie, Monatshefte* 1974: 513–524.

Broili, F. 1915. Ueber *Capitosaurus arenaceus* Münster. *Centralblatt für Mineralogie, Geologie und Paläontologie* 1915: 569–575.

Buffetaut, E., and G. Wouters. 1986. Amphibian and reptile remains from the Upper Triassic of Saint-Nicolas-de-Port (eastern France) and their biostratigraphic significance. *Modern Geology* 10: 133–145.

Carroll, R. L. 1977. Patterns of amphibian evolution: an extended example of the incompleteness of the fossil record. Pp. 405–437 in A. Hallam (ed.), *Patterns of Evolution.* Amsterdam: Elsevier.

Carroll, R. L., and P. J. Currie. 1975. Microsaurs as possible apodan ancestors. *Zoological Journal of the Linnean Society* 57: 229–247.

Carroll, R. L., and L. Winer. 1977. Privately produced taxonomic appendix to Carroll (1977). Available from authors.

Casamiquela, R. M. 1961. Nuevos materiales de
    Notobatrachus degiustoi Reig. Revista del Museo de La
    Plata (Nueva Serie), Paleontologia 4: 35–69.
    1965. Nuevo material de Vieraella herbstii Reig. Revista
    del Museo de La Plata (Nueva Serie), Paleontologia 4:
    265–317.
Case, S. M., and M. H. Wake. 1977. Immunological
    comparisons of caecilian albumins (Amphibia:
    Gymnophiona). Herpetologica 33: 94–98.
Chatterjee, S. 1986. The Late Triassic Dockum vertebrates:
    their stratigraphic and palaeobiogeographic
    significance. Pp. 139–150 in K. Padian (ed.), The
    Beginning of the Age of Dinosaurs. Cambridge
    University Press.
Chowdhury, T. R. 1965. A new metoposaurid amphibian
    from the Upper Triassic Maleri formation of Central
    India. Philosophical Transactions of the Royal Society
    of London B250: 1–52.
Clarke, B. T. 1988. Evolutionary relationships of the
    discoglossoid frogs – osteological evidence. Ph.D.
    thesis, City of London Polytechnic.
Colbert, E. H. 1958. Tetrapod extinctions at the end
    of the Triassic period. Proceedings of the National
    Academy of Sciences USA 44: 973–977.
    1967. A new interpretation of Austropelor, a supposed
    Jurassic labyrinthodont amphibian from
    Queensland. Memoirs of the Queensland Museum 15:
    35–41.
Coldiron, R. W. 1978. Acroplous vorax Hotton (Amphibia,
    Saurerpetontidae) restudied in light of new
    material. American Museum Novitates 2662: 1–27.
Davidow-Henry, B. 1989. Small metoposaurid amphibians
    from the Triassic of western North America and
    their significance. Pp. 278–292 in S. G. Lucas and
    A. P. Hunt (eds.), Dawn of the Age of Dinosaurs in the
    American Southwest. Albuquerque: New Mexico
    Museum of Natural History.
Dong Z. 1985. [A Middle Jurassic labyrinthodont
    (Sinobrachyops placenticephalus gen. et sp. nov.) from
    Dashanpu, Zigong, Sichuan Province.] Vertebrata
    Palasiatica 23: 301–306. [in Chinese, with English
    summary]
Duellman, W. E., and L. Trueb. 1986. Biology of Amphibians.
    New York: McGraw-Hill.
Dutuit, J.-M. 1972. Un nouveau genre de stégocéphale du
    Trias supérieur marocain: Almasaurus habbazi.
    Bulletin du Muséum National d'Histoire Naturelle, Paris
    (3) 72: 73–81.
    1976. Introduction à l'étude paléontologique du Trias
    continental marocain. Description des premiers
    Stégocéphales recueillis dans le couloir d'Argana
    (Atlas occidental). Mémoires du Muséum National
    d'Histoire Naturelle, Sér. C 36: 1–253.
    1978. Description de quelques fragments osseux
    provenant de la région de Folakara (Trias supérieur
    malgache). Bulletin du Muséum National d'Histoire
    Naturelle, Paris (3) 69: 79–89.
Ensom, P. C. 1988. Excavations at Sunnydown Farm,
    Langton Matravers, Dorset: Amphibians discovered
    in the Purbeck Limestone Formation. Proceedings of
    the Dorset Natural History and Archaeological Society
    109: 148–150.

Ensom, P.C., S. E. Evans, and A. R. Milner. 1991.
    Amphibians and reptiles from the Purbeck
    Limestone Formation (Upper Jurassic) of Dorset. 5th
    Symposium on Mesozoic Terrestrial Ecosystems and
    Biota, Extended Abstracts. Contributions of the
    Palaeontological Museum, University of Oslo 364:
    19–20.
Estes, R. 1981. Gymnophiona, Caudata. Pp. 1–115 in
    P. Wellnhofer (ed.), Handbuch der Paläoherpetologie,
    Vol. 2. Stuttgart: Gustav Fischer Verlag.
Estes, R., and R. Hoffstetter. 1976. Les urodèles du
    Miocène de La Grive-St. Alban (Isère, France).
    Bulletin du Muséum National d'Histoire Naturelle,
    Paris (3) 57: 297–343.
Estes, R., and O. A. Reig. 1973. The early fossil record of
    frogs: a review of the evidence. Pp. 1–63 in J. L. Vial
    (ed.), Evolutionary Biology of the Anurans:
    Contemporary Research on Major Problems.
    Columbia: University of Missouri Press.
Estes, R., and B. Sanchíz. 1982a. New discoglossid and
    palaeobatrachid frogs from the Late Cretaceous of
    Wyoming and Montana, and a review of other frogs
    from the Lance and Hell Creek Formations. Journal
    of Vertebrate Paleontology 2: 9–20.
    1982b. Early Cretaceous lower vertebrates from Galve
    (Teruel), Spain. Journal of Vertebrate Paleontology 2:
    21–39.
Evans, S. E., and A. R. Milner. 1993. Frogs and
    salamanders from the Upper Jurassic Morrison
    Formation (Quarry Nine, Como Bluff) of North
    America. Journal of Vertebrate Paleontology 13:
    24–30.
Evans, S. E., A. R. Milner, and F. Mussett. 1988. The
    earliest known salamanders (Amphibia: Caudata): a
    record from the Middle Jurassic of England. Géobios
    21: 539–552.
    1990. A discoglossid frog (Amphibia: Anura) from the
    Middle Jurassic of England. Palaeontology 33:
    299–311.
Foreman, B. C. 1990. A revision of the cranial morphology
    of the Lower Permian temnospondyl amphibian
    Acroplous vorax Hotton. Journal of Vertebrate
    Paleontology 10: 390–397.
Fox, R. C., and B. G. Naylor. 1982. A reconsideration of
    the relationships of the fossil amphibian
    Albanerpeton. Canadian Journal of Earth Sciences 19:
    118–128.
Fraas, E. 1889. Die Labyrinthodonten der schwäbischen
    Trias. Palaeontographica 36: 1–158.
    1913. Neue Labyrinthodonten aus der schwäbischen
    Trias. Palaeontographica 60: 275–294.
Getmanov, S. N. 1989. [Triassic amphibians of the East
    European platform (family Benthosuchidae
    Efremov).] Trudy Paleontologicheskogo Instituta
    Akademiya Nauk SSSR 236: 1–102. [in Russian]
Gregory, J. T. 1980. The otic notch of metoposaurid
    labyrinthodonts. Pp. 125–136 in L. L. Jacobs (ed.),
    Aspects of Vertebrate History. Flagstaff: Museum of
    Northern Arizona Press.
Hecht, M. K., and R. Estes. 1960. Fossil amphibians from
    Quarry Nine. Postilla 46: 1–19.
Hellrung, H. 1987. Revision von Hyperokynodon keuperinus

Plieninger (Amphibia: Temnospondyli) aus dem Schilfsandstein von Heilbronn (Baden-Württemberg). *Stuttgarter Beiträge zur Naturkunde, Ser. B* 136: 1–28.

Hopson, J. A. 1984. Late Triassic traversodont cynodonts from Nova Scotia and southern Africa. *Palaeontologia Africana* 25: 181–201.

Hunt, A. P. 1989. Comments on the taxonomy of North American metoposaurs and a preliminary phylogenetic analysis of the family Metoposauridae. Pp. 293–300 in S. G. Lucas and A. P. Hunt (eds.), *Dawn of the Age of Dinosaurs in the American Southwest*. Albuquerque: New Mexico Museum of Natural History.

Hunt, A. P., and S. G. Lucas. 1989. Late Triassic vertebrate localities in New Mexico. Pp. 72–101 in S. G. Lucas and A. P. Hunt (eds.), *Dawn of the Age of Dinosaurs in the American Southwest*. Albuquerque: New Mexico Museum of Natural History.

1990. The status of the "Jurassic" metoposaurs in the American Southwest. *Stegocephalian Newsletter* 1:16–17.

Ingavat, R., and P. Janvier. 1981. *Cyclotosaurus* cf. *posthumus* Fraas (Capitosauridae: Stereospondyli) from the Huai Hin Lat Formation (Upper Triassic), northeastern Thailand with a note on capitosaurid biogeography. *Géobios* 14: 711–725.

Ishing [I.V.P.P. staff]. 1987. [Vertebrate palaeontology and stratigraphy of Xinjiang.] Pp. 1–61 in [Geological Department of Academica Sinica on "The evolution of the Junggar Basin and the formation of its petroleum and gas fields".] [in Chinese]

Ivakhnenko, M. F. 1978. [Urodelans from the Triassic and Jurassic of Soviet Central Asia.] *Paleontologicheskiy Zhurnal* 1978(3): 84–89. [in Russian]

Jenkins, F. A., Jr., and D. M. Walsh. 1990. During the Jurassic, caecilians had limbs. *Journal of Vertebrate Paleontology* 10 (Suppl. to 3):29A. [abstract]

Jupp, R., and A. A. Warren 1986. The mandibles of the Triassic temnospondyl amphibians. *Alcheringa* 10: 99–124.

Kitching, J. W., and M. A. Raath. 1984. Fossils from the Elliot and Clarens formations (Karroo sequence) of the northeastern Cape, Orange Free State and Lesotho, and a suggested biozonation based on tetrapods. *Palaeontologia Africana* 25: 111–125.

Koken, E. 1913. Beiträge zur Kenntnis der Schichten von Heiligkreuz. *Abhandlungen der kaiserlich-königlichen geologischen Reichsanstalt, Wien* 16(4): 1–43.

Kuhn, O. 1932. Labyrinthodonten und Parasuchier aus dem mittleren Keuper von Ebrach in Oberfranken. *Neues Jahrbuch für Mineralogie, Geologie und Paläontologie, Abt. B*, 69: 94–144.

Kutty, T. S., and D. P. Sengupta. 1990. The Late Triassic formations of the Pranhita-Godavari Valley and their vertebrate succession – a reappraisal. *Indian Journal of Earth Science* 16: 189–206.

Lescure, J., S. Renous, and J. P. Gasc. 1986. Propositions d'une nouvelle classification des amphibiens Gymnophiones. *Mémoires de la Société Zoologique de France* 43: 145–177.

Longman, H. 1941. A Queensland fossil amphibian (*Austropelor*). *Memoirs of the Queensland Museum* 12: 29–32.

Lowy, J. 1949. A labyrinthodont from the Trias of Bear Island, Spitsbergen. *Nature* 163: 1002.

Lynch, J. D. 1973. The transition from archaic to advanced frogs. Pp. 133–182 in J. L. Vial (ed.), *Evolutionary Biology of the Anurans: Contemporary Research on Major Problems*. Columbia: University of Missouri Press.

Milner, A. R. 1983. The biogeography of salamanders in the Mesozoic and early Cainozoic: a cladistic-vicariance model. Pp. 431–468 in R. W. Sims, J. H. Price, and P. E. S. Whalley (eds.), *Evolution, Time and Space: The Emergence of the Biosphere*. London: Academic Press.

1988. The relationships and origin of living amphibians. Pp. 59–102 in M. J. Benton (ed.), *The Phylogeny and Classification of the Tetrapods, Vol. 1*. Oxford: Clarendon Press.

1989. Late extinctions of amphibians. *Nature* 338: 117.

1990. The radiations of temnospondyl amphibians. Pp. 321–349 in P. D. Taylor and G. P. Larwood (eds.), *Major Evolutionary Radiations*. Oxford: Clarendon Press.

Murry, P. A. 1987. Notes on the stratigraphy and paleontology of the Upper Triassic Dockum Group. *Journal of the Arizona-Nevada Academy of Science* 22: 73–84.

Nessov, L. A. 1988. Late Mesozoic amphibians and lizards of Soviet Middle Asia. *Acta Zoologica Cracoviensis* 31: 475–486.

1990. [The latest labyrinthodonts (Amphibia, Labyrinthodontia) and other relict groups of vertebrates from northern Fergana.] *Paleontologicheskiy Zhurnal* 1990(3): 82–90. [in Russian]

Novikov, I. V. 1990. [New Early Triassic labyrinthodonts from the Middle Timan.] *Paleontologicheskiy Zhurnal* 1990(1): 87–100. [in Russian]

Ochev, V. G., and M. A. Shishkin. 1989. On the principles of global correlation of the continental Triassic on the tetrapods. *Acta Palaeontologica Polonica* 34: 149–173.

Olsen, P. E. 1980. A comparison of the vertebrate assemblages from the Newark and Hartford Basins (early Mesozoic, Newark Supergroup) of eastern North America. Pp. 35–53 in L. L. Jacobs (ed.), *Aspects of Vertebrate History*. Flagstaff: Museum of Northern Arizona Press.

Olsen, P. E., and H.-D. Sues. 1986. Correlation of continental Late Triassic and Early Jurassic sediments, and patterns of the Triassic–Jurassic tetrapod transition. Pp. 321–351 in K. Padian (ed.), *The Beginning of the Age of Dinosaurs*. Cambridge University Press.

Panchen, A. L. 1959. A new armoured amphibian from the Upper Permian of East Africa. *Philosophical Transactions of the Royal Society of London* B242: 207–281.

Paton, R. L. 1974. Capitosauroid labyrinthodonts from the Trias of England. *Palaeontology* 17: 253–290.

Rage, J.-C., and Z. Roček 1989. Redescription of

Triadobatrachus massinoti (Piveteau, 1936), an anuran amphibian from the early Triassic. *Palaeontographica* A206: 1–16.

Reig, O. A. 1957. Los anuros del Matildense. In P. N. Stipanicic and O. A. Reig (eds.), El complejo porfírico de la Patagonia extraandina y su fauna de anuros. *Acta Geologica Lilloana* 2: 231–297.

1961. Noticia sobre un nuevo anuro fósil del Jurásico de Santa Cruz (Patagonia). *Ameghiniana* 2: 73–78.

Seiffert, J. 1969a. Urodelen-Atlas aus dem obersten Bajocien von SE-Aveyron (Südfrankreich). *Paläontologische Zeitschrift* 43: 32–36.

1969b. Sternalelement (Omosternum) eines mitteljurassischen Anuren von SE-Aveyron/ Südfrankreich. *Zeitschrift für zoologische Systematik und Evolutionsforschung* 2: 145–153.

Shishkin, M. A. 1987. [Evolution of early amphibians (Plagiosauroidea).] *Trudy Paleontologicheskogo Instituta Akademiya Nauk SSSR* 225: 1–143. [in Russian]

1991. [A labyrinthodont from the late Jurassic of Mongolia.] *Paleontologicheskiy Zhurnal* 1991(1): 81–95. [in Russian]

Sokol, O. M. 1977. A subordinal classification of frogs (Amphibia: Anura). *Journal of Zoology (London)* 182: 505–508.

Sues, H.-D., and P. E. Olsen. 1990. Triassic vertebrates of Gondwanan aspect from the Richmond Basin of Virginia. *Science* 249: 1020–1023.

Sun, A. 1962. Discovery of neorhachitomous vertebrae from Lufeng, Yunnan. *Vertebrata Palasiatica* 6: 109–110.

Sun, A., and Cui K. 1986. A brief introduction to the Lower Lufeng saurischian fauna (Lower Jurassic; Lufeng, Yunnan, People's Republic of China).

Pp. 275–278 in K. Padian (ed.), *The Beginning of the Age of Dinosaurs.* Cambridge University Press.

Suteethorn, V., P. Janvier, and M. Morales. 1988. Evidence for a plagiosauroid amphibian in the Upper Triassic Huai Hin Lat Formation of Thailand. *Journal of Southeast Asian Earth Science* 2: 185–187.

Warren, A. A. 1985. Triassic Australian plagiosauroid. *Journal of Paleontology* 59: 236–241.

Warren, A. A., and T. Black. 1985. A rhytidosteid (Amphibia, Labyrinthodontia) from the Early Triassic Arcadia Formation of Queensland, Australia, and the relationships of Triassic temnospondyls. *Journal of Vertebrate Paleontology* 5: 303–327.

Warren, A. A., and M. N. Hutchinson. 1983. The last labyrinthodont? A new brachyopoid (Amphibia, Temnospondyli) from the early Jurassic Evergreen Formation of Queensland, Australia. *Philosophical Transactions of the Royal Society of London* B303: 1–62.

1988. A new capitosaurid amphibian from the early Triassic of Queensland, and the ontogeny of the capitosaur skull. *Palaeontology* 31: 857–876.

Warren, A. A., L. Kool, M. Cleeland, T. H. Rich, and P. V. Rich. (in press). Early Cretaceous labyrinthodont. *Alcheringa.*

Watson, D. M. S. 1956. The brachyopid labyrinthodonts. *Bulletin of the British Museum (Natural History), Geology* 2: 315–392.

Wilson, J. A. 1948. A small amphibian from the Triassic of Howard County, Texas. *Journal of Paleontology* 22: 359–361.

Wiman, C. 1914. Ueber die Stegocephalen aus der Trias Spitzbergens. *Bulletin of the Geological Institution of the University of Upsala* 13: 1–34.

2

The Lepidosauromorpha: an overview with special emphasis on the Squamata

OLIVIER RIEPPEL

Introduction

The classification of reptiles has undergone a renaissance following the application of cladistic techniques (Gauthier, 1984; Benton, 1985; Evans, 1988). The Synapsida (including Mammalia) have been recognized as the sister-taxon of all other Amniota, and some poorly understood taxa such as millerettids, procolophonids, and mesosaurs have been relegated to the "Parareptilia" (of questionable monophyletic status) (Gauthier, Kluge, and Rowe, 1988a). This has provided the basis for a complete and logically stringent subordination of reptilian subgroups, the currently accepted classification being as follows: (Anapsida (Romeriida (Diapsida (Neodiapsida (Sauria (Archosauromorpha (Lepidosauromorpha)))))))) (Gauthier, Kluge, and Rowe, 1988a,b). A major unresolved problem concerns the classification of Mesozoic marine reptiles such as ichthyosaurs and euryapsids, which have been recognized as diapsids (in fact, subordinated to the Neodiapsida) (Sues, 1987; Rieppel, 1989a; Massare and Callaway, 1990; Storrs, 1991) but have not been satisfactorily placed with respect to the archosauromorph versus the lepidosauromorph clade.

One major corollary of the revision of reptilian classification is a shift in emphasis away from patterns of temporal fenestration and toward a host of additional characters, including both cranial and postcranial features. This ought to open an avenue for identification and classification of incomplete, disarticulated, and poorly preserved fossil material and their insertion into the currently accepted hierarchy of homologies. Such fossils, in turn, may influence the currently accepted hierarchy of groups within groups.

This chapter is not meant to be a critical reassessment of the phylogenetic interrelationships of lepidosauromorphs, but rather relates to the current lepidosauromorph classification in the light of small fossil vertebrates that potentially could be found in Mesozoic deposits. It is designed to be of use as a guide toward the recognition of the potential affinities of fossils. The discussion will center on the possibilities, and difficulties, of relating fossil lepidosauromorphs to extant groups and subgroups on the basis of synapomorphy rather than overall similarity (Patterson and Rosen, 1977). Commensurate with that goal, the main emphasis will be on those characters of the lepidosauromorph hierarchy of homologies that are potentially useful in the identification of disarticulated fossil material, omitting those that require complete and articulated specimens, as well as characters with a high degree of irregularity of distribution (incongruence) across extant taxa. In some cases the distribution of incongruent characters will be outlined, if they are likely to be observed, even in incomplete fossils, though they may not provide unambiguous clues to fossil affinities. Furthermore, the discussion will be restricted to Mesozoic fossils. The characters discussed are those compiled by Gauthier, Estes, and de Queiroz (1988) and Evans (1988). I shall refer to these characters without quoting those two major sources of information in each instance. A discussion of the potential meaning of fossil squamates for the analysis of squamate interrelationships follows the systematic account.

The Lepidosauromorpha

The Lepidosauromorpha include the Younginiformes, the enigmatic genera *Palaeagama*, *Paliguana*, and *Saurosternon*, the Kuehneosauridae, and the Lepidosauria (Gauthier, Estes, and de Queiroz, 1988). This clade is the sister-group of the Archosauromorpha. More recently, Laurin (1991) has urged that the Younginiformes be placed outside the Sauria (i.e., outside the archosauromorph–lepidosauromorph dichotomy) (Gauthier, Kluge, and Rowe, 1988a).

The Younginiformes include the well-known genus *Youngina, Acerosodontosaurus,* and the aquatic Tangasauridae (Currie, 1982) from the Upper Permian and Lower Triassic of Africa and Madagascar. The humerus in lepidosauromorphs shows a complete ectepicondylar foramen. The interclavicle is T-shaped or cruciform, but this is also true for at least some prolacertiforms (Rieppel, 1989b) and thalattosaurs (*Clarazia*) (Rieppel, 1987a), and thus is not a character unique to lepidosauromorphs. The sternum has been treated as a lepidosauromorph feature, but this character is notoriously poorly fossilized. There is, indeed, a continuing debate whether the Younginiformes should be included within the Lepidosauromorpha or should be classified outside that group (Zanon, 1990; Laurin, 1991).

Palaeagama, Paliguana, and *Saurosternon* are isolated genera from the Upper Permian and Lower Triassic of southern Africa, representing Carroll's (1975) (paraphyletic) "Paliguanidae." Their isolated positions and incompletely known structure has caused considerable debate as to their classification. Because these taxa are based on plesiomorphy, "it is not possible to refer other specimens to the taxon in question with any assurance" (Gauthier, Estes, and de Queiroz, 1988, p. 21).

The Lepidosauriformes

Gauthier, Estes, and de Queiroz (1988, p. 28) introduced the Lepidosauriformes as a monophyletic taxon including *Paliguana, Saurosternon,* Kuehneosauridae, and Lepidosauria. An important character that was used to diagnose this group is the structure of the quadrate, which bears a prominent lateral conch supporting a tympanum. The lateral conch on the quadrate was formerly thought to be diagnostic of the Squamata alone, because it is lacking in *Sphenodon*, but recent investigations have supported the assumption that this represents a secondary loss within sphenodontids (Evans, 1980; Wu, Chapter 3). Treating the Younginiformes as the sister-group of the Archosauromorpha and Lepidosauromorpha renders this quadrate structure diagnostic of the Sauria (J. A. Gauthier, pers. commun.). In correlation with the development of an impedance-matching middle ear and a large tympanic membrane supported by the quadrate, the quadrate is bowed (convex anteriorly, concave posteriorly) in lateral view, the stapes is slender, and the retroarticular process is prominent (serving as the ventral site of attachment of the tympanic membrane). The cleithrum is lost in the pectoral girdle in Lepidosauriformes, and the interclavicle tends to be of a more gracile shape.

The Kuehneosauridae and the Lepidosauria form a monophyletic taxon of lower rank on the basis of a number of features (Evans, 1988; Gauthier, Estes, and de Queiroz, 1988), including the loss of the tabulars and postparietals, the loss of the teeth on the transverse flange of the pterygoid, the reduced lateral exposure of the angular on the lower jaw, and the development of a thyroid fenestra in the pelvis. A fenestrated pelvis has evolved independently in archosaurs and some thalattosaurs (Evans, 1988).

Characters of the Lepidosauria

Sphenodon has long been recognized as being closely related to the lizards, and in fact it was first described as a member of the "Agamidae" by J. E. Gray. It was left to Richard Owen to recognize its differences from lizards and its similarities to Mesozoic fossils such as *Homoeosaurus*. Modern analysis shows *Gephyrosaurus* and the sphenodontians to form a monophyletic group, the Rhynchocephalia (sensu Gauthier, Estes, and de Queiroz, 1988), which is the sister-group of the Squamata, the two together constituting the Lepidosauria.

The Lepidosauria are diagnosed by a suite of shared derived characteristics (Benton, 1985; Evans, 1988; Gauthier, Estes, and de Queiroz, 1988), some of which are of potential use in the identification of disarticulated fossil material. The teeth are attached superficially to the tooth-bearing elements [see discussions of this character by Gauthier, Estes, and de Queiroz (1988, pp. 37ff.) and later in this chapter; exceptions are the deeply socketed teeth of mosasaurs and the shallowly socketed teeth of snakes]. The mandibular condyle is formed only by the articular. Caudal autotomy septa are present. [They are lost in pleurosaurs (Carroll, 1985), perhaps as a consequence of the tail being used as a propulsive organ in subaqueous locomotion; however, caudal autotomy seems to persist in the aquatic mesosaurs from the Paleozoic. Caudal autotomy is also absent from varanoid lizards and some other lizards (Bellairs and Bryant, 1985).] The ectepicondylar foramen is complete in the embryo (a complete ectepicondylar foramen in the adult is a lepidosauromorph feature) (Gauthier, Estes, and de Queiroz, 1988). The pelvic elements fuse late during ontogeny. The ilium bears a pubic flange, and the pubis has an outturned anteromedial projection. The flattened and pointed proximal epiphysis of the fibula meets the femoral condyle in a distal recess. A dorsal process of the fourth distal tarsal meets a recess in the astragalo-calcaneum (fused in adults). The fifth metatarsal is hooked, with its proximal head deflected to articulate with the fourth distal tarsal. The fifth distal tarsal fails to differentiate; its homology with an epiphyseal center on the proximal head of the fifth metatarsal must be rejected on developmental and morphological grounds (Rieppel, 1992b). The fifth metatarsal is also hooked in archosauromorphs, but it is flattened in the plane of the pes (not inflected about its long axis as in lepidosaurs) and lacks lepidosaur specializations such as the offset outer process and the lateral and medial plantar tubercles. Developmental studies in a number of lizards (Rieppel, 1992b) show that these lepidosaurian specializations are correlated with the potential to form secondary ossification centers in epiphyses and apophyses (Haines, 1969). In fact, the lizard (and, by implication, the sphenodontid) fifth metatarsal bears two proximal epiphyses, one opposing the fourth distal tarsal, and the other capping the outer process, contributing to its distinctiveness. The lateral and medial plantar tubercles ossify as separate apophyses. Specialized intervertebral articulations have been listed as a lepidosaurian synapomorphy, but their structure is diverse, and their distribution irregular among squamates.

Some characters of lepidosaurians are problematical and deserve further discussion. The developmental studies cited earlier showed a distinct delay of ossification of the fifth metatarsal, as compared with the other metatarsals. This character may thus be masked in aquatic lepidosaurians showing skeletal reductions. Rieppel (1989b) described lateral and plantar tubercles and a distinct outer process in a fully ossified specimen of *Macrocnemus*, a prolacertiform currently classified with the Archosauromorpha. The same specimen also shows a probable epiphyseal ossification on the proximal end of the femur (Rieppel, 1989b, fig. 6b), while another specimen shows a sesamoid ossification in the knee joint. Other prolacertiforms do not show these features, but it remains to be seen to what extent this might be due to aquatic habits. Again, *Prolacerta* (Gow, 1975) and *Macrocnemus* (Rieppel, 1989b) show a bifurcated second sacral rib, a character also known in *Clevosaurus* (Fraser, 1988) and some lizards (Hoffstetter and Gasc, 1969). A thyroid fenestra is developed in the pelvis in kuehneosaurs, lepidosaurs, advanced prolacertiforms, and some thalattosaurs (Rieppel, 1987a; Evans, 1988). Obviously, character incongruences indicate the need for further scrutiny in the analysis of the Lepidosauromorpha, and caution must guide their use in the assessment of the affinities of fragmentary fossils.

Rhynchocephalia

The Rhynchocephalia, as defined by Gauthier, Estes, and de Queiroz (1988), include *Gephyrosaurus* and its sister-group, the Sphenodontia. Erected by Günther (1867), the Rhynchocephalia comprised *Sphenodon*, which was removed from the "Agamidae." The inclusion of Triassic rhynchosaurs within the Rhynchocephalia [e.g., Rhynchocephalia s. str. of Zittel (1887–90, p. 589), Rhynchocephalia vera of Boulenger (1891, p. 172)] was later shown to be unjustified (Carroll, 1977). The contents and relationships of the Rhynchocephalia have most recently been reviewed by Fraser (1982, 1986, 1988), Fraser and Benton (1989), Fraser and Walkden (1984), Carroll (1985), Carroll

and Wild (Chapter 4), Whiteside (1986), Evans (1988), and Gauthier, Estes, and de Queiroz (1988). The group includes a number of genera and species that remain very poorly known and hence are of doubtful relationships beyond their inclusion within the Sphenodontia (Gauthier, Estes, and de Queiroz, 1988). Discussion will here be restricted to the better-known genera.

A row of enlarged teeth on the palatine, running parallel to the maxillary tooth row, is a general sphenodontid synapomorphy, shared by *Gephyrosaurus*; the same is true of the posterior process of the dentary underlying the glenoid fossa. Gauthier, Estes, and de Queiroz (1988) include *Gephyrosaurus* and sphenodontians within a more inclusive taxon of higher rank, the Rhynchocephalia. Recognition of these relationships resulted in a reinterpretation of acrodonty, the shift from pleurodont to acrodont tooth attachment having occurred within rhynchocephalians, and even within sphenodontians, because *Diphydontosaurus* retains pleurodonty in the anterior part of the maxilla and dentary. Pleurodont tooth attachment is not recapitulated during the ontogeny of *Sphenodon*.

The nonreplaced or only partially replaced series of alternating-size acrodont teeth on the maxilla and dentary, characterizing sphenodontians (but not *Gephyrosaurus*), represents the second- and third-generation teeth, sensu Harrison (1901, 1901–2). The feature of nonreplaced acrodont teeth on the posterior parts of maxilla and dentary is also shared by the Chamaeleonidae (sensu Frost and Etheridge, 1989). Embryonic and early postembryonic stages of chamaeleonines (*Bradypodion pumilus* and *Chamaeleo hoehnelii*) show these alternating-size teeth to represent two successive tooth generations, with no replacement even during the early stages of development [contra Edmund (1969, p. 161) and Whiteside (1986, p. 407)], and becoming ankylosed to the dorsomedial edge of the maxilla and dentary, thus supporting the interpretation of that character by Gauthier, Estes, and de Queiroz (1988) and Estes, de Queiroz, and Gauthier (1988, pp. 161ff.).

Recent advances in our understanding of rhynchocephalian and sphenodontian relations have also altered the interpretation of other characters, most notably of the lower temporal arcade. The jugal does not contact the quadratojugal in *Gephyrosaurus* (Evans, 1980), nor in some (juvenile?) specimens of *Diphydontosaurus* (Whiteside, 1986), *Planocephalosaurus* (Fraser, 1982), and *Clevosaurus* (Fraser, 1988) nor in *Palaeopleurosaurus* and pleurosaurs (Carroll, 1985; Carroll and Wild, Chapter 4). Contra Robinson (1973), Fraser (1988) described an overlap of the posterior ramus of the jugal in some specimens of *Clevosaurus*, where the connection seems to have been a loose overlap, rather than sutured, just as in a hatchling *Sphenodon* (Rieppel, 1992a). During the ontogeny of *Sphenodon*, the posterior ramus of the jugal ossifies in an anteroposterior direction (Howes and Swinnerton, 1901), and it eventually meets the quadratojugal in a position (behind the lower temporal fenestra) that is topographically nonequivalent to that observed in other diapsids, such as *Youngina* and *Petrolacosaurus* (Whiteside, 1986). This suggests that the complete lower temporal arcade, where present in sphenodontids, is a secondary development (Whiteside, 1986); this conclusion is supported by the topology of the cladogram with kuehneosaurs (characterized by an incomplete lower temporal bar) as the sister-group of the Lepidosauria, and *Clevosaurus* being the sister-group of later sphenodontians (Gauthier, Estes, and de Queiroz, 1988, pp. 49–50). Fraser (1988, p. 168) speculated that in those *Clevosaurus* specimens with an incomplete jugal–quadratojugal contact, a ligament may have bridged the gap. This would be the homologue of the quadrato-maxillary ligament of squamates, which is absent from *Sphenodon* and also from crocodilians (which have a complete lower temporal arcade).

Fraser (1988, pp. 173ff.) made the interesting suggestion that sphenodontians may be subdivided on the basis of dermal skull roof morphology. The parietal table is broad and flat in *Polysphenodon* (Carroll, 1985), *Diphydontosaurus*, *Homoeosaurus* (considered paedomorphic by Gauthier, Estes, and de Queiroz, 1988, p. 26), *Planocephalosaurus*, and *Gephyrosaurus*, but narrow in *Clevosaurus*, and it forms a parietal crest in *Sphenodon*, *Kallimodon*, *Sapheosaurus*, and pleurosaurs. A narrow and/or crestlike parietal table is correlated with the increase in laterally descending flanges of the parietal from which parts of the jaw adductor musculature take their origin. Lateral parietal downgrowths, providing a surface for jaw adductor muscle origins, are also observed in some squamates such as non-scleroglossan lizards (with exceptions). However, all squamate parietals showing this condition fuse during ontogeny and hence differ from the sphenodontid condition (among squamates, the parietals remain paired in most Gekkota and Xantusiidae, with a ventral attachment of the jaw adductor musculature) (Estes et al., 1988).

The ontogeny of *Sphenodon* (Howes and Swinnerton, 1901; Rieppel, 1992a) shows the parietal table to develop in a manner comparable to that in lizards. The parietals originate as paired elements, ossifying first along their margins and mapping a broad and flat skull table. The pineal foramen, where it persists, is a remnant of the original fontanelle between the diverging posterior parts of the incomplete frontal ossifications and anterior parts of the parietal ossifications. Its position is between the parietals in all rhynchocephalians (sensu Gauthier, Estes, and de Queiroz, 1988) and most squamates (exceptions are discussed by Estes et al., 1988, p. 148).

The reduction of the parietal table width and the transformation of the parietals into a parietal crest occur only later during development. This shows the broad and flat parietal table to represent the more generalized condition of form, a conclusion corroborated by outgroup comparison. An interesting aspect of parietal structure in sphendontians is that narrowing of the parietal skull table is not correlated with a fusion of the median skull roof elements: The frontals and parietals are paired in *Clevosaurus*, *Polysphenodon*, *Sphenodon*, *Homoeosaurus*, and *Kallimodon*, but are fused in *Planocephalosaurus* and *Diphydontosaurus* (Fraser, 1988, p. 173; contra Whiteside, 1986), as well as in *Gephyrosaurus*. Treating narrow parietals as a less general characteristic implies placement of *Palaeopleurosaurus* and pleurosaurs (with a narrow parietal skull table) within the sphenodontians (Evans, 1988; Gauthier, Estes, and de Queiroz, 1988, p. 97), rather than as their sister-group (Carroll, 1985; Carroll and Wild, Chapter 4).

The supratemporal bone has long been a point of contention in discussions of rhynchocephalian classification. The bone is absent from *Gephyrosaurus* and all sphenodontians except for *Clevosaurus* (Fraser, 1988). In view of this conflicting character distribution, Whiteside (1986, p. 415) referred to Baur's (1896) hypothesis that the squamosal of *Sphenodon* (and, by implication, of *Gephyrosaurus*, *Diphydontosaurus*, and *Planocephalosaurus*) may incorporate the supratemporal in its ascending process, but he found no supporting evidence for that hypothesis in Howes and Swinnerton's (1901) study. A hatchling *Sphenodon* at hand (Rieppel, 1992a) lends some support to Baur's (1896) hypothesis in showing a two-tongued dorsal process of the squamosal. The two processes are of subequal lengths, one lying deeper and somewhat anterior to the other, and both lying superficial to the supratemporal process of the parietal. The two processes remain fully separated from one another along their entire length before merging into the body of the squamosal at the level of its anterior process. The deeper process of the two might represent the supratemporal, becoming incorporated into the dorsal process of the squamosal later during ontogeny (the alternative interpretation is that it represents a neomorph).

Squamata

Most of the major squamate clades made their appearances in the Late Jurassic or Cretaceous, indicating that major cladogenetic events must have occurred even earlier. Their representation in earlier Mesozoic microvertebrate samples can therefore be expected, but whereas it may be relatively easy to diagnose a squamate as such, more specific identifications are rendered difficult by the generalized morphology of the early representatives. There are few clear-cut characters that would allow unambiguous diagnosis of incomplete and fragmentary fossils at the familial level. As will become apparent in the following discussion, identification of fossil squamates at the familial (and sometimes even suprafamilial) level often is based on overall similarity as judged by authors with wide experience.

Squamate phylogeny and classification were most recently reviewed by Estes et al. (1988) and Rieppel (1988a). The monophyly of the group is highly corroborated (Estes et al., 1988, list 84 characters), but basic interrelationships within the group still remain poorly supported. In particular, there is general agreement that on the basis of current knowledge, "lizards" become paraphyletic if amphisbaenians and snakes are excluded; however, the precise classifications of the latter two groups, as well as those for some enigmatic forms such as the burrowing Dibamidae, within "lizard" subgroups remain equivocal and/or controversial, and the case for a monophyletic Lacertilia is questionable (Rieppel, 1988b).

Squamate synapomorphies of potential use for analysis of fossil material are mostly those of "lizards," with reductions and/or secondary modifications in amphisbaenians and snakes. With the exception of the Chamaeleonidae (sensu Frost and Etheridge, 1989, including "agamids," and hence equivalent to the Acrodonta of Estes et al., 1988), the teeth on the maxillary and dentary show a pleurodont or "subpleurodont" (with shallow and sloping sockets) tooth attachment. The skull roof is characterized by a more or less transverse orientation of the frontoparietal suture (in "lizards," in superficial dorsal view), which is broader than the nasofrontal suture; in some squamates ("lizards"), the parietal bears anterolateral lappets fitting into articular facets on the frontal bone(s). The squamosal bone has lost its ventral ramus (except in the Chamaeleoninae sensu Frost and Etheridge, 1989). The braincase shows distally expanded paroccipital processes, which participate in the suspension of the quadrates, a subdivision of the metotic fissure, and a complete enclosure, even posterior to the basipterygoid processes, of the Vidian canal within the parasphenoid-basisphenoid complex. The epipterygoid is slender and lacks an expanded ventral base. In the lower jaw, the angular does not reach the mandibular condyle, and the coronoid eminence is formed by the coronoid bone (superficially overlapped, to a greater or lesser degree, by a process of the dentary in some groups). Sacral and caudal "ribs" fuse with their respective centra during ontogeny – a character that implies some problems of homology. Recent investigations of ossification patterns in the skeleton of squamates indicate the necessity to review the question of serial homology of the dorsal and sacral ribs and the caudal ribs or transverse processes, respectively; in adults, however, the distinction becomes irrelevant for identification purposes. Gastralia are absent. The ventral symphysis between the pubes is narrow, and the thyroid fenestra is enlarged. Squamate limbs, if not reduced, are long and slender, and the humerus has lost the entepicondylar foramen. Many aspects of articulations between limb elements show specializations (Estes et al., 1988), which, however, will be more difficult to identify in incomplete and disarticulated fossils.

The Squamata split into two major clades: the relatively generalized Iguania (sensu Frost and Etheridge, 1989) and the less generalized Scleroglossa (claimed to include amphisbaenians and snakes by Estes et al., 1988; see Rieppel, 1988b). In view of their generalized morphology, osteological synapomorphies of the Iguania are few. Frost and Etheridge (1989), in the most recent comprehensive treatment of the group, list the ontogenetic fusion of the frontal bones, their constriction between the orbits (reversed in some groups), and a broad frontal shelf underlying the nasals as diagnostic features. Estes (1983) would include the fusion of the parietals, which show a dorsal origin of the jaw adductor musculature (later considered a symplesiomorphy), and the position of the parietal foramen on the frontoparietal suture as additional features. (The parietals are transformed to a casquelike structure in chamaeleonines; a dorsal origin of the jaw adductors from the parietal also occurs in teiids, burrowing scincomorphs, and varanoids.) It should be remembered, however, that the frontal and parietal bones begin ossification along their lateral margins, so that a wide fontanelle persists between the diverging posterior parts of the frontals and the anterior parts of the parietals until fairly late stages of postembryonic development. The position of a pineal foramen on the frontoparietal suture might thus represent a more generalized condition of form, rather than a synapomorphic feature.

The earliest possible iguanian, *Pristiguana* Estes and Price, 1973, is from the Upper Cretaceous of Brazil. The controversy surrounding its placement within the Squamata, summarized by Estes (1983), only highlights the problem of assessing phylogenetic relationships among these early and generalized "lizards." *Pristiguana* was classified within the Iguania by Estes (1983) mainly on the basis of tooth-crown structure and a relatively less open Meckelian sulcus in the lower jaw, which are characters of overall similarity, rather than synapomorphy.

The "Agamidae" (nonchamaeleonine Chamaeleonidae sensu Frost and Etheridge, 1989) have been characterized by their dentition, the anteriormost teeth being pleurodont, and the more posterior teeth acrodont, with tooth wear becoming more and more apparent in an anteroposterior gradient (Augé, 1990). The Euposauridae from the Upper Jurassic of France have been referred to "agamids" (Cocude-Michel, 1963) based on the acrodont tooth implantation and the lack of caudal autotomy. This conclusion, again, remains controversial (Estes, 1983), particularly in view of the paired frontal and parietal bones (Hoffstetter, 1964), which might suggest sphenodontian affinities. The figures of *Euposaurus cerinensis* given by Cocude-Michel (1963) indicate a broad and flat parietal skull table, with a ventral origin of the jaw adductor musculature, a relatively narrow constriction of the frontal bones between the orbits, and a large pineal foramen on the frontoparietal suture. Other fossil "agamids" from the Upper Cretaceous of Mongolia are *Priscagama*, *Pleurodontagama*, and *Mimeosaurus*, again with few diagnostic features beyond the acrodont dentition, but including the meeting of the maxillae on the ventral midline of the palate, between the premaxillae and the vomers, and the braincase characteristics as outlined by Borsuk-Bialynicka and Moody (1984). Borsuk-Bialynicka and Moody (1984) erected the subfamily Priscagaminae for these fossils, but as pointed out by Frost and Etheridge (1989, p. 32), this subfamily is based on plesiomorphy.

With the exception of some teiids, scleroglossan lizards (Estes et al., 1988) can be identified, if appropriately preserved, by the hockey-stick-shaped squamosal (which lost its dorsal process, except in some teiids) (Robinson, 1967), the pronounced ventral downgrowths of the frontal bones, and the obliquely twisted posterior border of the retroarticular process (not in the Lacertoidea nor in the Pygopodidae); Gauthier, Estes, and de Queiroz (1988, p. 200) would add additional characteristics. The Scleroglossa include the Gekkota and the Autarchoglossa. Extant Gekkota include the Gekkonidae and Pygopodidae. [Kluge (1987) suggested that pygopodids are nested within the Gekkonidae, as the sister-group of the Diplodactylinae. Eublepharid gekkos may be treated as a family of their own (Grismer, 1988).] Their monophyly is, again, corroborated by an extensive suite of characters, some of which are paedomorphic (Estes, 1983; Estes et al., 1988). In fact, paedomorphosis seems to have been a dominant factor in gekkotan or at least gekkonid diversification (Rieppel, 1988a) and has been involved in the differentiation of a number of diagnostic features of the group. Problems arise with the inclusion of the "Ardeosauridae" from the Upper Jurassic of Germany (*Ardeosaurus*) and Manchuria (*Yabeinosaurus*) within the Gekkota (Camp, 1923; Hoffstetter, 1964), which because of their generalized morphology would result in a reduced list of synapomorphies for the group as a whole. The upper temporal and postorbital arches are complete in ardeosaurids, but not in gekkotans, probably again due to paedomorphosis. The parietals are paired in gekkotans, with the exception of eublepharid gekkos, but they are fused in *Ardeosaurus*. Indeed, scincomorph affinities of the "Ardeosauridae" have been suggested on the basis of a new specimen of *Ardeosaurus brevipes* described by Mateer (1982), which shows distinct impressions of large scales or dermal osteoscutes on the skull roof, whereas gekkotans typically are characterized by small cephalic scales and the absence of osteodermal incrusting, leaving the parietal skull table with a smooth surface. Indeed, Mateer (1982) referred this genus to the Gekkota for conservative reasons only, highlighting its scincomorph affinities.

Eichstaettisaurus is another genus from the Solnhofen lithograpic limestones (Upper Jurassic) referred to the Ardeosauridae by Estes (1983), and it is closely similar to *Ardeosaurus*, yet approaching the gekkonid structure more closely than the latter genus (Cocude-Michel, 1961; Hoffstetter, 1964) in showing a broad and rounded snout, a gekkonid dentition composed of many small and simple teeth, and a suture persisting between at least the anterior parts of the parietals. The upper temporal arch is complete, and an upper temporal fenestra persists, but the completeness of the postorbital arch is questionable. A revision of *Ardeosaurus*, *Yabeinosaurus*, and *Eichstaettisaurus* might well call into question the monophyly of the "Ardeosauridae," as was already anticipated by Hoffstetter (1964).

The "Bavarisauridae" are yet another group of Late Jurassic lizards from Germany, with gekkotan affinities, as shown by the broad, flat, and rounded snout and the numerous small and conical teeth. The vertebrae are amphicoelous, suggesting that synovial procoely is synapomorphic within the Gekkota (Kluge, 1987). Estes (1983), however, emphasized the very long tail in the specimen of *Bavarisaurus* from the stomach contents of *Compsognathus* (Ostrom, 1978), which again recalls the plesiomorphic (scincomorph) condition. In summary, Mesozoic Gekkota are difficult to place because of their generalized morphology, but seem to show some similarities that the Gekkota share with the Scincomorpha.

Palaeolacerta is a small and poorly known fossil from the Upper Jurassic of Germany, considered an adult by Hoffstetter (1964). The nasals and frontals are paired and the parietals unpaired, and the presence of a pineal foramen on the frontoparietal suture is uncertain. Excluding lacertoid relationships, and on the basis of amphicoelous vertebrae, Hoffstetter (1964) classified *Palaeolacerta* with the Gekkota, realizing, however, that this allocation is based on weak evidence. Amphicoely is primitive at the level of the Lepidosauria, but derived at the level of the Scleroglossa, and hence is diagnostic of the Gekkota (with procoely being derived for eublepharid gekkotans).

The Scincomorpha, together with the Anguimorpha, constitute the monophyletic Autarchoglossa, which, as a group, is supported only by few and problematical characters. The Scincomorpha show a separation of the nasals from the prefrontals by pronounced anterolateral processes of the frontals, pointed lateral downgrowths of the parietal at the level of the epipterygoid, a narrow lateral exposure of the coronoid bone, and an extremely elongated anteroventral (symphyseal) process of the pubis. However, none of these characters is present in all scincomorphs and only in scincomorphs.

The Scincomorpha comprise two monophyletic subgroups: the Lacertoidea and Scincoidea. The Lacertoidea, including the Xantusiidae, Lacertidae, Teiidae, and Gymnophthalmidae, share the extension of the posterior jaw adductor far anteriorly into Meckel's canal in the lower jaw. In the Lacertiformes (all Lacertoidea without Xantusiidae), this character is correlated with an inflation of the widely open adductor fossa in the lower jaw. This character (or, in fact, its absence) was one of the reasons for referring *Pristiguana* to the iguanids rather than teiids. Among the Lacertoidea, only the Teiidae have an extensive fossil record in the Mesozoic, most prominently represented by the large and generalized Polyglyphanodontinae from the Upper Cretaceous of North America and Mongolia. The group closely resembles the Teiinae, except for the occasional fusion of the supratemporal with the squamosal in very large individuals, and the very large postorbital (Estes, 1983). The skull in polyglyphanodontines shares with that in teiines the deep lower jaw, with a prominent crest delineating the facet for the insertion of the superficial jaw adductor musculature on the lateral surface of the surangular and angular. The Teiinae proper include some fossil genera from the Upper Cretaceous of North America (*Leptochamops, Meniscognathus, Peneteius,* and *Chamops*) (Estes, 1983).

The Scincoidea are characterized by the ossification of palpebral bones, by the presence of (compound) osteoderms, and by characters of the lower jaw such as a dentary process overlapping the lateral surface of the coronoid process and a medially inflected retroarticular process, which is broadened posteriorly and bears a tubercle or a small flange on its medial margin. Again, none of these characteristics occurs only in Scincoidea and in all Scincoidea. The two families included within the Scincoidea are the Cordylidae and Scincidae. The Scincoidea (as in xantusiids, gymnophthalmids, and some anguids) show a tendency to roof over the upper temporal fossa, this being accomplished primarily by the postorbital in the Cordylidae and by the postfrontal in the Scincidae. The monophyly of the Scincidae remains poorly supported (see Rieppel, 1988a, for a discussion), which is of no great concern in this context in view of their poor Mesozoic fossil record. *Slavoia darevskii* has been reported as a Late Cretaceous scincomorph lizard from Mongolia (Sulimski, 1984). A relationship to scincids is suggested by the scroll-like palatines forming a rudimentary secondary palate, but the sturdy skull and, in particular, the deep lower jaw recall the teiine condition.

Included with the Cordylidae within the superfamily Cordyloidea have been the Paramacellodidae from the Upper Jurassic of England, Portugal, and North America

(Estes, 1983). They are believed to be related to cordylids on the basis of mandibular and scutellation characteristics (presence of noncompound dorsal and compound ventral osteoscutes). In his discussion of the affinities of *Paramacellodus* from the Upper Jurassic Purbeck beds of England, Estes (1983, p. 114) emphasized that "I have found all of the cordyloid resemblances cited by Hoffstetter to occur in some skinks as well." *Contogenys* is a genus first known from the Upper Cretaceous of Montana that has been referred to the Scincidae by Estes (1983), although it is "not clearly separable from cordyloids on presently known material."

The Mesozoic fossil record for the Anguimorpha indicates that, in fact, this squamate clade was diversified to the highest degree by the end of the Cretaceous. Fossil anguimorphans can be identified on the basis of mandibular characteristics, including the subdivision of Meckel's canal near the posterior end of the dentary tooth row by a well-developed intramandibular septum, the medioventral opening of Meckel's canal anterior to the anterior alveolar foramen, and a reduction of the overlap of postdentary bones by the dentary. The Anguimorpha include the monophyletic Varanoidea and the plesiomorphic (paraphyletic?) Anguioidea. The replacement teeth in anguioids (i.e., nonvaranoid anguimorphs) are positioned posteromedially to the functional tooth and develop outside the (small) resorption pits until late during the replacement cycle (Rieppel, 1978). The Varanoidea are diagnosed by a suite of additional shared derived characteristics to be discussed later.

The Dorsetisauridae have been identified as (poorly known) anguimorphs from the Upper Jurassic of Portugal, England, and the western United States. *Dorsetisaurus* lacks the intramandibular septum, and it resembles a variety of anguioid genera, but also some scincomorphs, in skull (including braincase) characteristics. A "convincing case for anguimorph relationship has not been established" for these early squamates (Estes, 1983, p. 133).

Exostinus, a genus extending back to the Late Cretaceous, has been identified as a xenosaurid lizard. McDowell and Bogert (1954) believed *Exostinus* to be more closely related to the Chinese genus *Shinisaurus* than to the North American genus *Xenosaurus*, an assumption refuted by Estes (1983, and literature cited therein) and Gauthier (1982). Gauthier (1982) erected a new genus, *Restes*, to include *Exostinus rugosus* that is distinguished by the pattern of osteodermal incrusting on the skull roof. More recently, Borsuk-Bialynicka (1985) proposed a new family, the Carusiidae, to include the two genera *Carusia* and *Shinisauroides* from the Upper Cretaceous of Mongolia. Although the two lizards share some xenosaurid characters, such as the constricted frontals and the osteodermal crust, they also share nongekkotan scleroglossan (i.e., scincomorphan) characters such as the fingerlike ventral parietal flange, the posterior extent of the dentary, and the widely open Meckelian groove. Late Cretaceous fossils might indicate a scincomorph-anguimorph relationship (Borsuk-Bialynicka, 1983), which in fact corroborates the monophyletic Autarchoglossa (Borsuk-Bialynicka, 1985). The concept of the "preanguimorphan grade" to include the Late Cretaceous genera *Paravaranus* and *Bainguis* (Borsuk-Bialynicka, 1984) highlights the problem of subordinal affiliation of autarchoglossan lizards at that geological time.

The Anguidae, a well-diagnosed group with an extensive Tertiary fossil record, are characterized by a suite of (additional) mandibular characters such as the anterior extent of the surangular, the reduction of the angular process of the dentary, which forms the free posteroventral margin of the intramandibular septum, and the entry of the dentary into the dorsal and anterior border of the anterior inferior alveolar foramen (Estes, 1983, p. 134). The only anguid subfamily extending back into the Late Cretaceous is the Glyptosaurinae, with *Odaxosaurus piger* from the Upper Cretaceous of Wyoming. Interestingly enough, this species shares only tooth and osteoscute shape similarities with other members of the subfamily, as well as the closure of the fenestra between premaxilla and maxilla (Gauthier, 1982; Estes, 1983) (closure of the premaxillary fenestra is polarized as a synapomorphic feature by outgroup comparison). Other synapomorphies characterizing the Glyptosaurinae evolved only at a later date – they are present in *Proxestops jepseni* from the upper Paleocene of Wyoming (Estes, 1983, p. 148). A few isolated Late Cretaceous fossils have tentatively been referred to the gerrhonotine *Gerrhonotus*, but this allocation is based on overall similarity of tooth and dentary shape, rather than on synapomorphy (Estes, 1983, p. 165).

The Varanoidea are well characterized by the development of plicidentine, resulting in superficial striation of the broad tooth base, and a lack of resorption pits for the replacement teeth, which are in a posteromedial position (McDowell and Bogert, 1954; Rieppel, 1978). The lower jaw shows at least a tendency to develop an intramandibular hinge joint (see Gauthier, 1982, for structural details and problems of homology). The premaxilla bears a long, slender nasal process. The anterior portion of the maxilla usually is depressed to a greater or lesser degree (in correlation with the posterior extension of the external naris). The nasals usually are slender and fused. The jaw adductor musculature takes its origin from the dorsal surface of the parietal. Three families have been included within the Varanoidea: "Necrosauridae," Helodermatidae, and Varanidae.

The "Necrosauridae" are of rather generalized structure, with an unretracted external naris, a maxilla with a fairly steeply ascending anterior dorsal margin, and frontal bones that fail to meet below the olfactory

tracts; the dentition is that of a predaceous animal, the teeth being laterally compressed and recurved, and bearing basal striations on the inflated base. Borsuk-Bialynicka (1984) introduced the "necrosaurian grade" to include platynotans of generalized morphology from the Upper Cretaceous of Mongolia (*Proplatynotia*, *Parviderma*, and *Gobiderma*), and *Colpodontosaurus* is a poorly known necrosaurid from the Upper Cretaceous of Wyoming, which, however, lacks the predaceous tooth specializations. These are present in *Parasaniwa* from the same age and locality (Estes, 1983). *Paraderma* is a third genus represented from the Upper Cretaceous of Wyoming that seems to be difficult to refer unequivocally to either the Helodermatidae or the "Necrosauridae" but which has been recognized as a helodermatid by Pregill, Gauthier, and Greene (1986). The latter authors conclude (1986, p. 179) that whereas "the monophyly of Recent varanoids and varanids are highly corroborated hypotheses, considerable work remains to be done in the character analysis of 'Necrosauridae'."

The Varanidae have an extensive Tertiary fossil record and extend back into the Late Cretaceous with *Palaeosaniwa* from Alberta, Canada. The laterally compressed teeth with widely expanded bases and complex dentine infolding, as well as the marked precondylar constriction of the vertebral centra, indicate varanid affinities, but other characters remain more generalized than would be typical for the group. The chevrons do not appear to articulate with distinct peduncles on the posterior part of the centrum, the neural spines are not expanded, and skull bones indicate the presence of a regular pattern of dermal osteoscutes resembling that of anguids. *Cherminotus* is the Late Cretaceous sister-taxon to the extant *Lanthanotus* (Borsuk-Bialynicka, 1984), genera now recognized as a separate subfamily, the Lanthanotinae, within the Varanidae. Pregill et al. (1986, p. 179) consider it "safe to say that by the Late Cretaceous Helodermatidae was represented by *Paraderma*, Lanthanotinae by *Cherminotus*, and Varaninae by *Saniwides* and *Telmasaurus*."

The Varanoidea also include the marine mosasaurs (Russell, 1967), as well as the rather imperfectly known aigialosaurs, small marine lizards with an elongated body and reduced limbs. The dolichosaurs are fossils that, from their discovery, have always been linked to the problem of snake origins (reviewed by Rieppel, 1988b). Haas (1980a) tentatively assigned the lower Cenomanian genus *Ophiomorphus* to dolichosaurs. The tendencies toward some elongation of the trunk and reductions of limbs are already documented in the genus *Colubrifer* from the Lower Triassic of South Africa (Carroll, 1982), thought to be related to the Squamata by its describer, but now acknowledged to be of uncertain affinities (Estes, 1983; Evans, 1988).

The earliest snake fossils, three vertebrae identified as *Lapparentophis defrennei* Hoffstetter, come from the

Lower Cretaceous of North Africa (Rage, 1984). *Coniophis* is an anilioid genus from the Upper Cretaceous; *Coniophis cosgriffi*, however, is a controversial species based on one vertebra from the Upper Cretaceous of New Mexico, that was identified as anilioid by Armstrong-Ziegler (1978), but probably represents a booid snake (Estes, Berberian, and Meszoely, 1969; Rage, 1984, p. 13). Another booid genus extending back into the Late Cretaceous is *Madtsoia*, a representative of the subfamily Madtsoiinae (Hoffstetter, 1961). Snakes as a group are well characterized by a complete closure of the lateral braincase wall (correlated with an intracranial course of the opthalmicus profundus branch of the trigeminal nerve, which emerges from the optic foramen in the posteroventral part of the orbit), the emergence of the maxillary and mandibular branches of the trigeminal nerve from a subdivided foramen between parietal and prootic (except in scolecophidians, where the foramen remains undivided), medial ventral pillars of the frontal bones separating the olfactory tracts from one another and forming the support for the prokinetic joint (except in scolecophidians), the meeting of the exoccipitals dorsal to the foramen magnum, and the cover of the footplate of the stapes by a crista circumfenestralis. Vertebral structure is critical for the identification of fossil snakes, because vertebrae are among the most easily fossilized parts of ophidians. However, "there is no fundamental difference between the vertebral morphology of snakes and that of the other modern Lepidosauria" (Hoffstetter and Gasc, 1969, p. 283). There are several characteristics in detailed vertebral structure, in particular relating to the zygosphene–zygantrum articulation, that help to distinguish snake vertebrae from other squamate vertebrae (Hoffstetter and Gasc, 1969). Very poorly known fossils from the Cenomanian (lower part of the Upper Cretaceous), sometimes referred to ophidians, but excluded from that group by Rage (1984), are *Mesophis* Bolkay, 1925 and *Pachyophis* Nopcsa, 1923.

Again, however, major uncertainties prevail in the interpretation of Mesozoic ophidians. *Pachyrhachis* is a snakelike reptile from the lower Cenomanian of Israel (Haas, 1979, 1980b) that shows a curious mixture of varanoid and ophidian characteristics. The sutural contact between premaxilla and maxillae seems to have been lost (as indicated by the rounded anterior tips of the maxillae). The relatively elongated postorbital portion of the skull indicates an ophidian-type braincase, and the supratemporals (squamosals of Haas, 1979) are excluded from the braincase wall as in advanced (macrostomatan) snakes. The shape and suspension of the quadrate are snakelike. The frontoparietal suture appears not to be straight as in mesokinetic lizards; however, a joint may have existed between the frontals and the parietal. A pineal foramen may have been present. The mandible bears a

strong and steeply ascending coronoid process. The fossil was used by Haas (1979) in his argument that simoliopheids are, in fact, varanoid lizards. Rage (1984) did not follow Haas (1979, 1980b) in the inclusion of *Pachyrhachis* in the Simoliopheidae and continued to treat *Simoliophis* from the lower part of the Upper Cretaceous of France as the only member of its family, known only from vertebrae. He acknowledged, however (Rage, 1984, p. 74), the modified structure of the midtrunk vertebrae, but denied their similarity to lacertilian vertebrae and suggested a derivation from the snakelike *Lapparentophis*.

Perhaps the most interesting fossil snake is *Dinilysia* from the Upper Cretaceous of Patagonia, known from an almost complete skull (Estes, Frazzetta, and Williams, 1970) and a number of vertebrae (Hecht, 1982). The skull combines some distinctly lizardlike (plesiomorphic) features, such as the rudiment of a jugal bone, the morphology of the palatine bone (which retains a choanal groove), and the absence of medial frontal pillars (also absent from scolecophidian snakes), with typical snakelike characteristics, such as the lack of a sutural contact of maxilla and premaxilla, the complete closure of the braincase, the dorsal contact of the exoccipitals, and the mode of quadrate suspension. New material of *Dinilysia*, not yet described, will mandate reconsideration of some of the characters, such as the position of the optic foramen and the supposed absence of a crista circumfenestralis around the vestibular fenestra (Rage, 1984). Although all authors agree that *Dinilysia* is a snake, its position within the Ophidia remains controversial (Rage, 1984; Rieppel, 1988b).

Discussion and summary

Looking back on this brief review of Mesozoic Lepidosauromorpha, and Squamata in particular, the pattern of phylogenetic diversification that becomes apparent is one of successive individualization of subordinated inclusive taxa through time (Figure 2.1). In this context, the "individualization" of monophyletic taxa does not imply ontological judgments, but simply refers to the identification of taxa on the basis of homology (synapomorphy), tested by similarity, congruence, and conjunction (Patterson, 1982). The Lepidosauromorpha first appeared in the Late Permian. The unnamed taxon, subordinated to the Lepidosauriformes and including the Kuehneosauridae and the Lepidosauria, is first recorded from the Upper Triassic [the earliest sphenodontian described is *Palacrodon* Broom from the Lower Triassic of South Africa, but Evans (1988, p. 232) notes its uncertain, perhaps procolophonid, affinities]. The Squamata, subordinated to the Lepidosauria, are first (unequivocally) documented in the Late Jurassic (Evans, 1988, p. 231). All major subclades of the Squamata appear to have

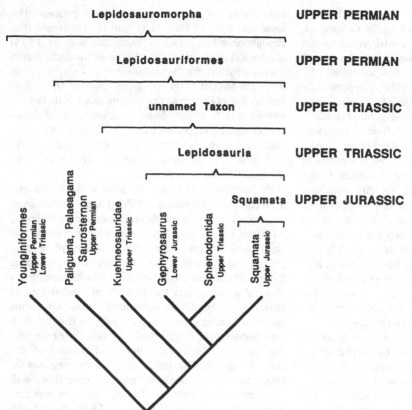

Figure 2.1. Cladistic interrelationships within the Lepidosauromorpha, indicating the geological time of individualization of inclusive taxa.

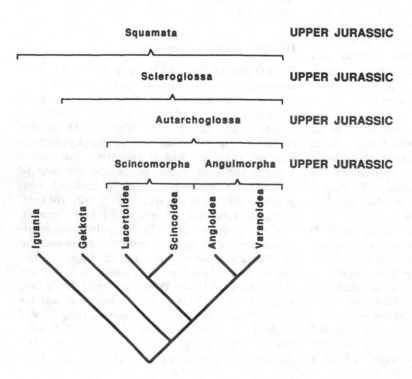

Figure 2.2. Cladistic interrelationships within the Squamata, indicating the geological time of individualization of inclusive taxa.

individualized in the Late Jurassic (Figure 2.2), with fossils referred to ardeosaurids, bavarisaurids, paramacelloideans, and dorsetisaurids representing the Scleroglossa; the earliest possible iguanian is *Pristiguana* from the Upper Cretaceous of Brazil. The fossil record and taxonomic resolution improve for the Late Cretaceous, particularly among the Varanoidea. Indeed, the Anguimorpha, with a good fossil record, seem to show an early individualization of subordinated groups, as compared with other squamates. There is an interesting delay in the individualization of subordinated taxa: Their sister-group relation (to the Sphenodontia) indicates that the Squamata must date back to the Late Triassic, but the first unequivocal representatives are Late Jurassic in age. And resolution at the subordinal level within the Squamata improves only for the Late Cretaceous. The identification of many nonvaranoid squamate families first becomes possible for the Tertiary.

The pattern of successive individualization of monophyletic groups within groups may be evaluated in the context of some long-standing assumptions about the general pattern of diversification among major clades of early origin. In his discussion of the fossil record, Rensch (1954) attempted to demonstrate that major clades show an early phase of "explosive differentiation" once their general body plan has become established, whereas some equilibrium of origination and extinction rates seems to be approached at later stages of (sub)clade diversification; similar views had been held by a variety of earlier authors quoted by Rensch (1954); see also Valentine (1990) for a more recent review. Gould, Gilinsky, and German (1987) subjected this phenomenon to a statistical analysis, and they introduced the term "bottom-heavy clades" to characterize the early diversification within a clade following major morphological innovation. Their analysis indicated that bottom-heaviness is characteristic of clades with early origins, whereas (sub)clades of later origin may be symmetrical or top-heavy. Analyses of patterns of taxic diversity over time, such as the one presented by Gould et al. (1987), have been subject to various criticisms and are, in some respects, not comparable to the approach pursued in this review.

Gould et al. (1987) claim to have analyzed patterns of clade diversity over time, but they acknowledge that in fact they designate as "clades" nothing but conventionally accepted taxonomic units (families or genera), irrespective of their demonstrated monophyletic status. This, of course, raises the question whether or not bottom-heaviness of early originating clades is just an artifact of paraphyly: Early clades may appear bottom-heavy because later descendants have been incorporated into other taxa (Patterson and Smith, 1987; Valentine, 1990). Another problem faced in the analysis of taxic diversity over time is the "pull of the Recent," due to the fact that the volumes of sedimentary rock exposed for study, as well as the numbers and quality of preservation of the fossils contained therein, decrease with increasing age. A greater species diversity may thus seem to be indicated for younger deposits, although there may have been no real increase in the number of species over time; for a more complete discussion, see Raup (1988). The pull of the Recent seems particularly appreciable for Late Cretaceous and Tertiary squamates. On the whole, however, this review does not focus on the diversity of species or more inclusive taxa over time, but on the differentiation of homologies as expressed in the taxic hierarchy of extant squamates and in relation to the geological time scale. It might be argued that the failure to identify many modern squamate families before the Tertiary is due to the fact that fossils of Mesozoic age belong to different families, poorly understood because of the fragmentary fossil record of the Mesozoic. Although that is true, it can still be argued that if a modern clade, such as the Scleroglossa, is diagnosed by certain characters, and if sister-group relations indicate that this clade dates back to the Late Jurassic, then the corresponding synapomorphy, or combination of synapomorphies, should be differentiated at that time on the appropriate bones of a fossil even if it belongs to an unknown family, subordinated to that more inclusive clade.

One problem of this approach to fossil squamates is that synapomorphy (homology) is a relational concept, which means that the phylogenetic information content of a character emerges not from the character in itself but from the relations of a character with respect to all other characters known (i.e., from the test of congruence). Analyzing and classifying fossil squamates on the basis of synapomorphies characterizing the hierarchy of homologies observed among extant forms was stated to be a methodological prerequisite in the analysis of fossils (Patterson and Rosen, 1977). But that was later claimed to be a misguided approach, because the analysis of an all-inclusive data set might result in the revision of currently accepted diagnoses of squamate taxa (Gauthier, Kluge, and Rowe, 1988b). In other words, the addition of fossil taxa to an all-inclusive data set might result in a revision of currently accepted notions of squamate phylogeny and classification and thus might lead to the recognition of clades more properly identifiable in the fossil record.

One of the major problems in the analysis of extant squamates is rampant incongruence, indicating a high degree of convergence and/or reversal, which at the same time renders the identification of disarticulated and incomplete fossil material difficult. Adding fossil taxa to an all-inclusive data set can be expected to increase the amount of incongruence, as indeed would also result from the addition of extant taxa. But incongruence is the flip side of congruence, and parsimony analysis counts only congruence as the signal, dismissing incongruence as noise.

The evaluation of alternative hypotheses of interrelationships on the basis of parsimony (i.e., by the test of congruence) highlights the fact that homology (i.e., similarity to be explained by common ancestry) cannot be established a priori (e.g., by character weighting), nor does it reside in any character per se, but instead resides in the relation of any character to all others known. This, in turn, implies that characters need not, indeed cannot, be weighted on an a priori basis, but weigh themselves by congruence (Patterson, 1982). The characters with the greatest information content will also be the most congruently distributed ones. Extant taxa provide a great many more characters than fossils ever can, including soft anatomy and molecular data. In the light of these theoretical premises, fossil data may be added to an all-inclusive data matrix in two different ways. One is with the deletion of all soft anatomy and molecular data, so as to provide a more balanced comparison of extant and fossil forms. The consequence will be that the homology of osteological features can no longer be tested by their relation to *all* other characters known. The test of congruence is skewed toward osteological data, which means that the osteological characters are a priori weighted more heavily than soft anatomy or molecular data.

If, however, the fossils are entered into an all-inclusive data matrix with soft anatomy and molecular data, all the latter will have to be coded as unknown for the fossil. The same happens if incomplete fossils are coded along with complete osteological data gathered from extant organisms. Computer analysis will assign character states to the unknown characters in search for a most parsimonious solution, and it will do these assignments in the light of the most parsimonious matching of all those characters positively known. Not only will such a procedure incorporate character assignments that actually are unknowable, but also the test of congruence will be skewed toward the characters positively known, which implicitly will thereby be assigned greater weight.

A major impact from the inclusion of fragmentary fossil squamates into an all-inclusive data set for reanalysis of squamate interrelationships can be expected if a fossil displays a character combination crucial for the overturning of some weakly supported node in the cladogram. [Overall support for the most parsimonious hierarchy inherent in the data matrix compiled by Estes et al. (1988) is rather weak, the consistency index ranging from 0.41 to 0.42, depending on character interpretation (Kluge, 1989), and increasing to 0.56 after deletion of amphisbaenids, snakes, and *Dibamus*.] For example, only three synapomorphies are listed as diagnostic for the Autarchoglossa by Estes et al. (1988, p. 207), one of which is the osteodermal incrusting on the skull roof, a character qualified as of "doubtful significance" (Estes et al., 1988, p. 207). A reanalysis of the Ardeosauridae might be expected to bear on this point.

On the other hand, many of the fragmentary fossils will have to be analyzed in the light of what is known about extant taxa. The discussion presented earlier shows that character combinations, diagnosing less inclusive entities among extant squamates (such as taxa ranked at the family level), are largely absent in the Mesozoic (except for varanoid lizards). Mesozoic squamates are, by and large, of a rather generalized morphology, often combining characters that among extant squamates are *dissociated* in the diagnosis of extant clades. Whatever their controversial status in terms of monophyly, ardeosaurids, bavarisaurids, dorsetisaurids, preanguimorphan- and necrosaurid-grade taxa, as well as *Pachyrhachis* and *Dinilysia*, all provide character combinations that preclude their unequivocal assignment to monophyletic extant taxa even of higher rank.

Acknowledgments

I thank Drs. N. C. Fraser, H.-D. Sues, and R. L. Carroll, who all read an earlier draft of this chapter, offering helpful advice and criticism. I particularly acknowledge the efforts of Dr. J. A. Gauthier to provide helpful criticisms in spite of the fact that he and I appear to be living in different universes.

References

Armstrong-Ziegler, G. 1978. An aniliid snake and associated vertebrates from the Campanian of New Mexico. *Journal of Paleontology* 52: 480–483.

Augé, M. 1990. La faune des lézards et d'amphisbaenes de l'Eocène inférieur de Condé-en-Brie (France). *Bulletin du Muséum National d'Histoire Naturelle* (4)C, 12: 11–141.

Baur, G. 1896. The paroccipital of the Squamata and the affinities of the Mosasauridae once more. A rejoinder to Professor E. D. Cope. *American Naturalist* 30: 143–147.

Bellairs, A. d'A., and S. V. Bryant. 1985. Autotomy and regeneration in reptiles. Pp. 301–410 in C. Gans et al. (eds.), *Biology of the Reptilia, Vol. 15.* New York: Academic Press.

Benton, M. J. 1985. Classification and phylogeny of diapsid reptiles. *Zoological Journal of the Linnean Society* 84: 97–164.

Bolkay, S. J. 1925. *Mesophis nopcsai* n.g.n.sp., ein neues, schlangenähnliches Reptil aus der unteren Kreide (Neocom) von Bilek-Selišta (Ost-Hercegovina). *Glasnik Zemaljski Muzej u Bosni i Hercegovini* 37: 125–135.

Borsuk-Bialynicka, M. 1983. The early phylogeny of Anguimorpha as implicated by craniological data. *Acta Palaeontologica Polonica* 28: 31–42.

——— 1984. Anguimorphans and related lizards from the Late Cretaceous of the Gobi Desert, Mongolia. *Palaeontologia Polonica* 46: 5–105.

1985. Carolinidae, a new family of xenosaurid-like lizards from the Upper Cretaceous of Mongolia. *Acta Palaeontologica Polonica* 30: 151–176.

Borsuk-Bialynicka, M., and S. Moody. 1984. Priscagaminae, a new subfamily of the Agamidae (Sauria) from the Late Cretaceous of the Gobi desert. *Acta Palaeontologica Polonica* 29: 51–81.

Boulenger, G. A. 1891. On British remains of *Homoeosaurus*, with remarks on the classification of the Rhynchocephalia. *Proceedings of the Zoological Society of London* 1891: 167–172.

Camp, C. L. 1923. Classification of the lizards. *Bulletin of the American Museum of Natural History* 48: 289–481.

Carroll, R. L. 1975. Permo-Triassic "lizards" from the Karroo. *Palaeontologia Africana* 18: 71–87.

1977. The origin of lizards. Pp. 359–396 in S. M. Andrews, R. Miles, and A. D. Walker (eds.), *Problems in Vertebrate Evolution*. London: Academic Press.

1982. A short limbed lizard from the *Lystrosaurus* Zone (Lower Triassic) of South Africa. *Journal of Paleontology* 56: 183–190.

1985. A pleurosaur from the Lower Jurassic and the taxonomic position of the Sphenodontida. *Palaeontographica* A189: 1–28.

Cocude-Michel, M. 1961. Les sauriens des calcaires lithographiques de Bavière, d'âge Portlandien. *Bulletin de la Société Géologique de France* 7(2)6: 707–710.

1963. Les rhynchocéphales et les sauriens des calcaires lithographiques (Jurassique supérieur) d'Europe occidentale. *Nouveaux Archives du Muséum d'Histoire Naturelle, Lyon* 7: 1–187.

Currie, P. J. 1982. The osteology and relationships of *Tangasaurus mennelli* Haughton (Reptilia; Eosuchia). *Annals of the South African Museum* 86: 247–265.

Edmund, A. G. 1969. Dentition. Pp. 117–200 in C. Gans and T. S. Parsons (eds.), *Biology of the Reptilia, Vol. 1*. London: Academic Press.

Estes, R. 1983. *Sauria terrestria, Amphisbaenia*. Pp. 1–249 in P. Wellnhofer (ed.), *Encyclopedia of Paleoherpetology, Vol. 10A*. Stuttgart: Gustav Fischer Verlag.

Estes, R., P. Berberian, and C. A. M. Meszoely. 1969. Lower vertebrates from the Late Cretaceous Hell Creek Formation. McCone County, Montana. *Breviora* 337: 1–33.

Estes, R., K. de Queiroz, and J. A. Gauthier. 1988. Phylogenetic relationships within Squamata. Pp. 119–281 in R. Estes and G. Pregill (eds.), *Phylogenetic Relationships of the Lizard Families*. Stanford University Press.

Estes, R., T. H. Frazzetta, and E. E. Williams. 1970. Studies on the fossil snake *Dinilysia patagonica* Woodward. Part I. Cranial morphology. *Bulletin of the Museum of Comparative Zoology, Harvard University* 140: 25–74.

Estes, R., and L. Price. 1973. Iguanid lizard from the Upper Cretaceous of Brazil. *Science* 180: 748–751.

Evans, S. E. 1980. The skull of a new eosuchian reptile from the Lower Jurassic of South Wales. *Zoological Journal of the Linnean Society* 70: 203–264.

1981. The postcranial skeleton of the Lower Jurassic eosuchian *Gephyrosaurus bridensis*. *Zoological Journal of the Linnean Society* 73: 81–116.

1988. The early history and relationships of the Diapsida. Pp. 221–260 in M. J. Benton (ed.), *The Phylogeny and Classification of the Tetrapods. Vol. 1: Amphibians, Reptiles, Birds*. Oxford: Clarendon Press.

Fraser, N. C. 1982. A new rhynchocephalian from the British Upper Trias. *Palaeontology* 25: 709–725.

1986. New Triassic sphenodontids from South-West England and a review of their classification. *Palaeontology* 29: 125–186.

1988. The osteology and relationships of *Clevosaurus* (Reptilia: Sphenodontida). *Philosophical Transactions of the Royal Society of London* B321: 125–178.

Fraser, N. C., and M. J. Benton. 1989. The Triassic reptiles *Brachyrhinodon* and *Polysphenodon* and the relationships of the sphenodontids. *Zoological Journal of the Linnean Society* 96: 413–445.

Fraser, N. C., and G. M. Walkden. 1984. The postcranial skeleton of the Upper Triassic sphenodontid *Planocephalosaurus robinsonae*. *Palaeontology* 27: 575–595.

Frost, D. L., and R. Etheridge. 1989. A phylogenetic analysis and taxonomy of iguanian lizards (Reptilia: Squamata). *The University of Kansas Museum of Natural History, Miscellaneous Publication* 81: 1–65.

Gauthier, J. A. 1982. Fossil xenosaurid and anguid lizards from the early Eocene Wasatch Formation, southeast Wyoming, and a revision of the Anguioidea. *Contributions to Geology, University of Wyoming* 21: 7–54.

1984. A cladistic analysis of the higher systematic categories of the Diapsida. Ph.D. dissertation, University of California, Berkeley.

Gauthier, J. A., R. Estes, and K. de Queiroz. 1988. A phylogenetic analysis of Lepidosauromorpha. Pp. 15–81 in R. Estes and G. Pregill (eds.), *Phylogenetic Relationships of the Lizard Families*. Stanford University Press.

Gauthier, J. A., A. G. Kluge, and T. Rowe. 1988a. The early evolution of the Amniota. Pp. 103–155 in M. J. Benton (ed.), *The Phylogeny and Classification of the Tetrapods. Vol. 1: Amphibians, Reptiles, Birds*. Oxford: Clarendon Press.

1988b. Amniote phylogeny and the importance of fossils. *Cladistics* 4: 105–209.

Gould, S. J., N. L. Gilinsky, and R. Z. German. 1987. Asymmetry of lineages and the direction of evolutionary time. *Science* 236: 1437–1441.

Gow, C. E. 1975. The morphology and relationships of *Youngina capensis* Broom and *Prolacerta broomi* Parrington. *Palaeontologia Africana* 18: 89–131.

Grismer, L. L. 1988. Phylogeny, taxonomy, classification and biogeography of eublepharid geckos. Pp. 369–469 in R. Estes and G. Pregill (eds.)., *Phylogenetic Relationships of the Lizard Families*. Stanford University Press.

Günther, A. 1867. Contribution to the anatomy of *Hatteria* (*Rhynchocephalus*, Owen). *Philosophical Transactions of the Royal Society* 157: 595–629.

Haas, G. 1979. On a new snakelike reptile from the Lower Cenomanian of Ein Jabrud, near Jerusalem. *Bulletin du Muséum National d'Histoire Naturelle* (4)C, 1: 51–64.

1980a. Remarks on a new ophiomorph reptile from the Lower Cenomanian of Ein Jabrud, Israel. Pp. 177–192 in L. L. Jacobs (ed.), *Aspects of Vertebrate History.* Flagstaff: Museum of Northern Arizona Press.

1980b. *Pachyrhachis problematicus* Haas, snakelike reptile from the Lower Cenomanian: ventral view of the skull. *Bulletin du Muséum National d'Histoire Naturelle* (4)C, 2: 87–104.

Haines, R. 1969. Epiphyses and sesamoids. Pp. 81–116 in C. Gans and T. S. Parsons (eds.), *Biology of the Reptilia, Vol. 1.* London: Academic Press.

Harrison, H. S. 1901. Development and succession of the teeth in *Hatteria punctata. Quarterly Journal of Microscopical Sciences (N.S.)* 44: 161–213.

1901–2. *Hatteria punctata,* its dentition and incubation period. *Anatomischer Anzeiger* 20: 145–158.

Hecht, M. K. 1982. The vertebral morphology of the Cretaceous snake *Dinilysia patagonica* Woodward. *Neues Jahrbuch für Geologie und Paläontologie, Monatshefte* 1982: 523–532.

Hoffstetter, R. 1961. Nouveaux restes d'un serpent boïdé (*Madtsoia madagascariensis* nov. sp.) dans le Crétacé supérieur de Madagascar. *Bulletin du Muséum National d'Histoire Naturelle* (2) 33: 152–160.

1964. Les Sauria du Jurassique supérieur ét spécialement les Gekkota de Bavière et de Mandchourie. *Senckenbergiana Biologica* 45: 281–324.

Hoffstetter, R., and J. Gasc. 1969. Vertebrae and ribs of modern reptiles. Pp. 201–310 in C. Gans and T. S. Parsons (eds.), *Biology of the Reptilia, Vol. 1.* London: Academic Press.

Howes, G. B., and H. H. Swinnerton. 1901. On the development of the skeleton of the tuatara, *Sphenodon puntatus,* with remarks on the egg, on the hatchling, and on the hatched young. *Transactions of the Zoological Society of London* 16: 1–86.

Kluge, A. G. 1987. Cladistic relationships in the Gekkonoidea. *Miscellaneous Publications of the Museum of Zoology, University of Michigan,* 173: 1–54.

1989. Progress in squamate classification. *Herpetologica* 45: 368–379.

Laurin, M. 1991. The osteology of a Lower Permian eosuchian from Texas and a review of diapsid phylogeny. *Zoological Journal of the Linnean Society* 101: 59–95.

McDowell, S. B., and C. M. Bogert. 1954. The systematic position of *Lanthanotus* and the affinities of anguinomorphan lizards. *Bulletin of the American Museum of Natural History* 105: 1–142.

Massare, J. A., and J. M. Callaway. 1990. The affinities and ecology of Triassic ichthyosaurs. *Bulletin of the Geological Society of America* 102: 409–416.

Mateer, N. J. 1982. Osteology of the Jurassic lizard *Ardeosaurus brevipes* (Meyer). *Palaeontology* 25: 461–469.

Nopcsa, F. 1923. *Eidolosaurus* und *Pachyophis,* zwei neue Neocom-Reptilien. *Palaeontographica* 55: 97–154.

Ostrom, J. H. 1978. The osteology of *Compsognathus longipes* Wagner. *Zitteliana* 4: 73–118.

Patterson, C. 1982. Morphological characters and homology. Pp. 21–74 in K. A. Joysey and A. E. Friday (eds.), *Problems of Phylogenetic Reconstruction.* London: Academic Press.

Patterson, C., and D. E. Rosen. 1977. Review of ichthyodectiform and other Mesozoic teleost fishes, and the history and practice of classifying fossils. *Bulletin of the American Museum of Natural History* 158: 81–172.

Patterson, C., and A. B. Smith. 1987. Is the periodicity of extinctions a taxonomic artefact? *Nature* 330: 248–251.

Pregill, G. K., J. A. Gauthier, and H. W. Greene. 1986. The evolution of helodermatid squamates, with description of a new taxon and an overview of Varanoidea. *Transactions of the San Diego Society of Natural History* 21: 167–202.

Rage, J.-C. 1984. *Serpentes.* In P. Wellnhofer (ed.), *Handbuch der Paläoherpetologie, Vol. 11.* Stuttgart: Gustav Fischer Verlag.

Raup, D. M. 1988. Testing the fossil record for evolutionary progress. Pp. 293–317 in M. H. Nitecki (ed.), *Evolutionary Progress.* University of Chicago Press.

Rensch, B. 1954. *Neuere Probleme der Abstammungslehre,* 2nd ed. Stuttgart: Ferdinand Enke.

Rieppel, O. 1978. Tooth replacement in anguinomorph lizards. *Zoomorphology* 91: 77–90.

1987a. *Clarazia* and *Hescheleria*: a reinvestigation of two problematical reptiles from the Middle Triassic of Monte San Giorgio, Switzerland. *Palaeontographica* A195: 101–129.

1987b. The development of the trigeminal jaw adductor musculature and associated skull elements in the lizard *Podarcis sicula. Journal of Zoology (London)* 212: 131–150.

1988a. The classification of the Squamata. Pp. 261–293 in M. J. Benton (ed.), *The Phylogeny and Classification of the Tetrapods. Vol. 1: Amphibians, Reptiles, Birds.* Oxford: Clarendon Press.

1988b. A review of the origin of snakes. Pp. 37–130 in M. K. Hecht, B. Wallace, and G. T. Prance (eds.), *Evolutionary Biology, Vol. 22.* New York: Plenum Press.

1989a. A new pachypleurosaur (Reptilia: Sauropterygia) from the Middle Triassic of Monte San Giorgio, Switzerland. *Philosophical Transactions of the Royal Society of London* B323: 1–73.

1989b. The hind limb of *Macrocnemus bassanii* (Reptilia: Diapsida): development and functional anatomy. *Journal of Vertebrate Paleontology* 9: 373–387.

1992a. The skull in a hatchling *Sphenodon punctatus. Journal of Herpetology* 26: 80–84.

1992b. Studies on skeleton formation in reptiles. III. Patterns of ossification in the skeleton of *Lacerta vivipara* Jacquin (Reptilia, Squamata). *Fieldiana Zoology* 68: 1–25.

Robinson, P. L. 1967. The evolution of the Lacertilia. *Colloques Internationaux du Centre National de Recherche Scientifique* 163: 395–407.

1973. A problematic reptile from the British Upper Trias. *Journal of the Geological Society of London* 129: 457–479.

Russell, D. A. 1967. Systematics and morphology of American mosasaurs. *Bulletin of the Peabody Museum of Natural History, Yale University* 23: 1–240.

Simpson, G. G. 1944. *Tempo and Mode in Evolution.* New York: Columbia University Press.

Storrs, G. W. 1991. Anatomy and relationships of *Corosaurus alcovensis* (Diapsida: Sauropterygia) and the Triassic Alcova Limestone of Wyoming. *Bulletin of the Peabody Museum of Natural History, Yale University* 44: 1–151.

Sues, H.-D. 1987. Postcranial skeleton of *Pistosaurus* and the interrelationships of the Sauropterygia (Diapsida). *Zoological Journal of the Linnean Society* 90: 109–131.

Sulimski, A. 1984. A new Cretaceous scincomorph from Mongolia. *Palaeontologia Polonica* 46: 143–155.

Valentine, J. W. 1990. The macroevolution of clade shape. Pp. 128–150 in R. M. Ross and W. D. Allmon (eds.), *Causes of Evolution: A Paleontological Perspective.* University of Chicago Press.

Whiteside, D. I. 1986. The head skeleton of the Rhaetian sphenodontid *Diphydontosaurus avonis* gen. et sp. nov. and the modernizing of a living fossil. *Philosophical Transactions of the Royal Society of London* B312: 379–430.

Zanon, R. T. 1990. The sternum of *Araeoscelis* and its implications for basal diapsid phylogeny. *Journal of Vertebrate Paleontology* 9: 51A.

Zittel, K. A. von 1887–90. *Handbuch der Palaeontologie. 1. Abt. Palaeozoologie. Bd. 3: Vertebrata.* Leipzig: R. Oldenbourg.

3

Late Triassic–Early Jurassic sphenodontians from China and the phylogeny of the Sphenodontia

XIAO-CHUN WU

Introduction

In contrast with all other extant reptile groups, sphenodontians are represented by only a single living genus and two species: *Sphenodon punctatus* and *S. guentheri* (Daugherty et al., 1990). These two species survive on a few isolated islands off the New Zealand coast and have no fossil record. However, the sphenodontians as a group were common and widespread during a period from about 230 million years ago to 100 million years ago.

Since the 1970s, a number of new genera of sphenodontians (including several as yet unnamed) have been described from the early Mesozoic terrestrial deposits of England (Evans, 1980, 1981; Fraser, 1982, 1986; Whiteside, 1986), North America (Olsen, 1980; Murry, 1987), and southern Africa (Gow and Raath, 1977; Rich, Molnar, and Rich, 1983), and several earlier-known fossils have been restudied (Robinson, 1973; Fabre, 1981; Throckmorton, Hopson and Parks, 1981; Carroll, 1985; Fraser, 1988; Fraser and Benton, 1989). Undoubtedly, these new finds and revisions will be very important to our knowledge of the morphology and history of the early sphenodontians, but unfortunately, none of the fossils so far known have well-preserved braincase material, and most of them are based on disarticulated (if abundant) bones or very fragmentary specimens. Hence the cranial structure (especially in the temporal or middle ear region) of the fossil sphenodontians is not well known, and the origin and phylogenetic relationships of the constituent genera have been subject to dispute.

The sole specimen of *Dianosaurus petilus* Young, 1982 was the first Late Triassic–Early Jurassic sphenodontian material from Asia (Figure 3.1E, F). It was recovered from the top of the Dull Purplish Beds of the Lower Lufeng Formation of the Lufeng basin, Yunnan Province, southwest China (Figure 3.2). As with the majority of other small tetrapod fossils of the formation, the specimen is preseved in a nodule. It was not adequately prepared prior to the initial description. Three additional skulls of two previously unrecognized sphenodontians (Figure 3.1A–D) were collected by the author from the middle of the Dark Red Beds of the Lower Lufeng Formation of the same basin in 1984 (Figure 3.3). Together with the specimen of *Dianosaurus petilus*, they are referred to three different species of the genus *Clevosaurus*, originally described from the Late Triassic to Early Jurassic fissure deposits of southwest Britain (Swinton, 1939; Robinson, 1973; Fraser, 1988).

The purpose of this chapter is to describe the head anatomy of the three species and to establish group relationships within the Sphenodontia by making use of the procedure of cladistic analysis (Hennig, 1966; Eldredge and Cracraft, 1980; Wiley, 1981).

Most workers have included *Sphenodon* and its fossil relatives in a single family, the Sphenodontidae, although some authors have referred them to a suborder, Sphenodontia (Williston, 1925; Hoffstetter, 1955; Evans, 1984; Benton, 1985) or Sphenodontoidea (Kuhn, 1969; Fraser, 1982; Whiteside, 1986). The name of the order Rhynchocephalia was erected by Günther in 1867 to include the genus *Sphenodon*, and the rhynchosaurs were added later (Chatterjee, 1974). Now that it is judged necessary to separate the two groups (Carroll, 1976, 1985, 1988a; Benton, 1985; Evans, 1988; Gauthier, Estes, and de Queiroz, 1988), the name Rhynchocephalia might logically be retained for *Sphenodon* and its allies. I also place *Gephyrosaurus* Evans, 1980 and the pleurosaurs in this group, but I prefer to retain Sphenodontia as the ordinal name in order to avoid confusion between Rhynchocephalia and Rhynchosauria.

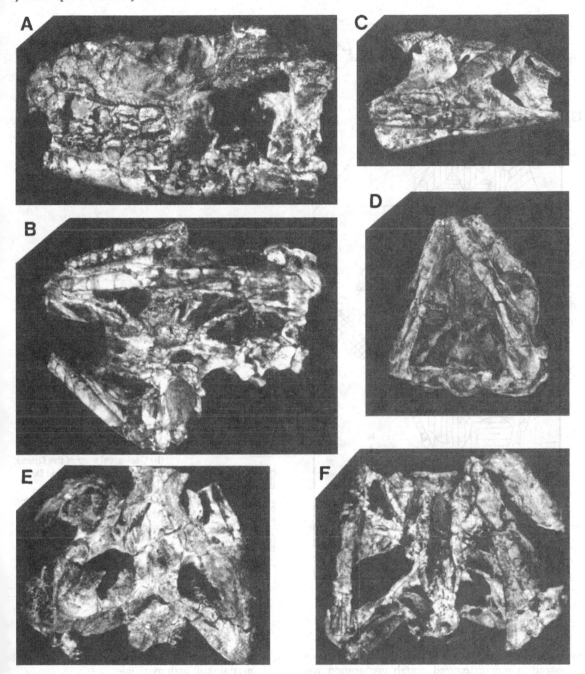

Figure 3.1. Lateral (A) and ventral (B) views of the skull of *Clevosaurus wangi* sp. nov. (V.8271, × 3). Lateral (C) and ventral (D) views of the skull of *Clevosaurus mcgilli* sp. nov. (V.8272, × 3). Dorsal (E) and ventral (F) views of the skull of *Clevosaurus petilus* (V.4007, × 2.5).

Materials and methods

As early as 1941, the sedimentary rocks of the Lufeng Basin were demonstrated to contain a richly varied vertebrate fauna (Bien, 1941). Young (1948) described the vertebrate assemblage, later naming it the Lufeng Saurischian Fauna (Young, 1951). Since then, many new taxa have been added to the fauna, and recently two separate assemblages have been recognized (Sun et al., 1985). The three sphenodontian taxa studied here are present in the Lower Lufeng Formation. The geology of the Lufeng Basin has been reviewed by Luo and Wu (Chapter 14).

All of the specimens described here are preserved in

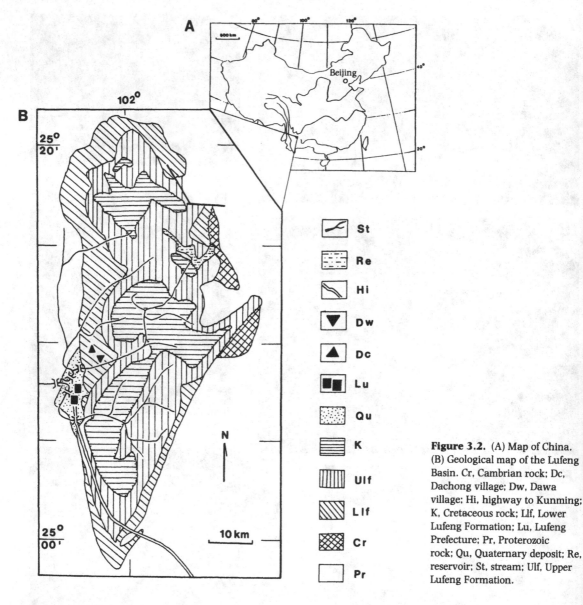

Figure 3.2. (A) Map of China. (B) Geological map of the Lufeng Basin. Cr, Cambrian rock; Dc, Dachong village; Dw, Dawa village; Hi, highway to Kunming; K, Cretaceous rock; Llf, Lower Lufeng Formation; Lu, Lufeng Prefecture; Pr, Proterozoic rock; Qu, Quaternary deposit; Re, reservoir; St, stream; Ulf, Upper Lufeng Formation.

very hard, ferruginous nodules. Before mechanical preparation was attempted, rough preparation was made using 40 percent acetic acid. Fine sable-hair brushes were dipped in the acid to brush the working areas, which were subsequently washed after each brushing. The acid treatment continued until bone was exposed.

All the specimens are housed in the Institute of Vertebrate Paleontology and Paleoanthropology (IVPP), Academia Sinica, Beijing, China:

1. IVPP V.4007, *Clevosaurus petilus*, holotype; partial skull with mandible.
2. IVPP V.8271, *Clevosaurus wangi*, holotype; almost complete skull with mandible.
3. IVPP V.8272, *Clevosaurus mcgilli*, holotype; partial skull with mandible.
4. IVPP V.8273, *Clevosaurus mcgilli*, paratype; partial skull with mandible.

Systematic paleontology

Lepidosauromorpha Gauthier, 1984

Lepidosauria Haeckel, 1866

Sphenodontia Williston, 1925

Clevosaurus Swinton, 1939

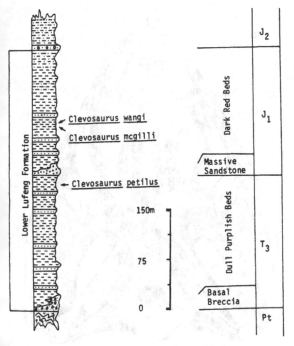

Figure 3.3. Sphenodontian-bearing beds in the Lower Lufeng Formation.

Type species. *Clevosaurus hudsoni* Swinton, 1939

Included species. *C. minor* Fraser, 1988; *C. wangi* sp. nov.; *C. mcgilli* sp. nov.; *C. petilus* (*Dianosaurus petilus* Young, 1982).

Revised diagnosis. Sphenodontians uniquely sharing the following features: The premaxilla has a posterodorsal process that separates the maxilla from the external naris (unknown in *C. petilus*); the suborbital fenestra is enclosed by only the palatine and ectopterygoid, and the supratemporal is present. In addition, the following characters are also shared with certain other genera: absence of the premaxillary process of the maxilla (unknown in *C. petilus*), a relatively short snout (unknown in *C. petilus*), the parietal table narrower than the interorbital width, a complete lower temporal bar (unknown in *C. petilus*), the relatively broad orbital portion of the maxilla and two rows of pterygoid teeth (all the characters are unknown in *C. minor*; Fraser,1988).

Comments. This chapter incorporates some of the results of a study presented as part of my Ph.D. dissertation, in which *C. wangi*, *C. mcgilli*, and *C. petilus* were previously treated as belonging to separate genera: *Asiacephalosaurus wangi*, *Rarojugalosaurus mcgilli*, and *Dianosaurus petilus*. Additional material recently collected from Nova Scotia, Canada, and southwest Britain reveals that *Clevosaurus* possesses considerable variation in structure (N. C. Fraser and H.-D. Sues, pers. commun.). On the basis of the current specimens, the skeletal structures are not sufficiently different to warrant erection of three new genera for the Chinese sphenodontians. Consequently, *Asiacephalosaurus*, *Rarojugalosaurus*, and *Dianosaurus* are herein considered junior synonyms of *Clevosaurus*.

Clevosaurus wangi sp. nov.

Holotype. V.8271, an almost complete but cracked skull with mandible.

Locality. Near Dawa village, Lufeng County, Yunnan Province, southwest China.

Stratigraphic position. Near the middle of the upper part (Dark Red Beds) of the Lower Lufeng Formation (stratum 6 in Figure 14.3; see Luo and Wu, Chapter 14).

Age. Early Jurassic.

Etymology. In honor of Mr. Zheng-ju Wang, for his contributions to local vertebrate paleontology and paleoanthropology during the past thirty years.

Diagnosis. Differing from other species of *Clevosaurus* as follows: relatively small supratemporal fenestra; parietal width between the supratemporal passages broader than the interorbital width; paroccipital process fitting into a fossa on the undersurface of the squamosal; squamosal resting in a deep depression on the cephalic head of the quadrate; supratemporal underlying the posteromedial margin of the squamosal; dorsal process of the jugal sloping backward to overlap the squamosal at a point beyond the midpoint of the supratemporal bar; pterygoid with elongate central region between three rami.

The epipterygoid is broad dorsally, but narrow ventrally, a condition that may or may not be unique to *C. wangi*, because it is unknown in the other species.

Description. The skull of the single known specimen is laterally compressed but the left side posterior to the orbit is almost complete. The left mandibular ramus is articulated with the skull. The tip of the rostrum, the right side of the skull, and the right mandibular ramus are damaged. Palatal elements are present, but incomplete. Despite being crushed, the braincase is generally well preserved.

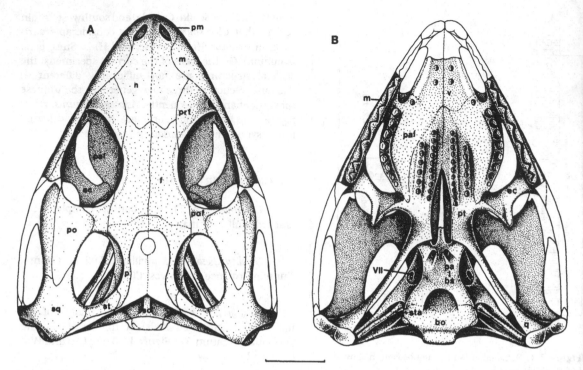

Figure 3.4. *Clevosaurus wangi* sp. nov. Reconstruction of the skull. Dorsal (A) and ventral (B) views. Scale equals 5 mm.

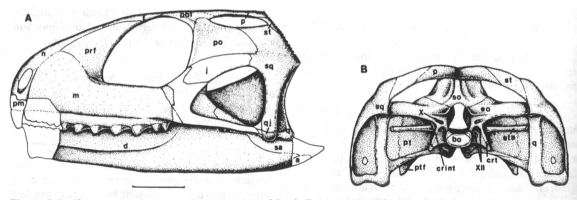

Figure 3.5. *Clevosaurus wangi* sp. nov. Reconstruction of the skull. Lateral (A) and occipital (B) views. Scale equals 5 mm.

General features of the skull. The skull is approximately 3 cm long (Figures 3.4 and 3.5). It is very similar in configuration to that of *C. hudsoni* (Fraser, 1988). In dorsal view, the maxilla does not enter the margin of the external naris. The suborbital fenestra is banana-shaped and enclosed by only the palatine and ectopterygoid. The nasals, frontals, and probably the parietals are paired. A supratemporal is retained. The lacrimal is absent. The lower temporal bar is bowed laterally.

Although the inferior temporal fenestra is comparable to the orbit in size, the preservation indicates that the oval supratemporal fenestra is much smaller.

In *C. hudsoni* the three openings are almost of the same size.

In lateral view, the lower jaw shows a pronounced coronoid process posterior to the dentition, and it probably had a prominent retroarticular process posteriorly. The maxilla is deep dorsoventrally, with an obtuse posterior termination. The beaklike structure of the snout is restored on the basis of *C. hudsoni*.

In ventral view, the lateral row of palatine teeth parallels the maxillary tooth row anteriorly, but gradually curves toward the midline posteriorly, as in most early sphenodontians. There are two rows of pterygoid teeth. The pterygoid does not enter the suborbital fenestra as in *Sphenodon* and most fossil sphenodontians. There is a semicircular depression on the ventral surface of the parabasisphenoid that probably reaches the basioccipital posteriorly.

In contrast to that of *Sphenodon* and *C. hudsoni*, the middle ear region is very similar in structure to that of modern lizards: The middle ear cavity is well developed, an occipital recess is present on the basal tubercle, the metotic fissure is divided into two openings (fenestra rotunda and vagus foramen), and the vena jugularis recess is evident on the anterolateral wall of the braincase, although it is of limited extent. However, an unossified gap is present anteroventral to the ventral process of the opisthotic, as described in *Sphenodon* and *C. hudsoni* (Fraser, 1988). Viewed posteriorly, the medial wall of the otic capsule is fully ossified. This is in contrast to the membranous condition in *Sphenodon*. The quadrate is much deeper than that of *Sphenodon*, reaching about 59 percent of the height of the skull. The posttemporal fenestra is small, and the braincase is shallow.

Dermal bones of the skull roof. The premaxillae are badly damaged, but appear to be similar to those of *C. hudsoni* (Figures 3.6 and 3.7). The left premaxilla bears a posterodorsal process that separates the maxilla from the external naris, as described in *C. hudsoni*. The posterior process is overlapped by the maxilla.

The left maxilla is almost complete. Its anterodorsal process extensively overlaps the premaxilla anteriorly and the nasal and prefrontal dorsally. Ventromedial to the orbital margin, the maxilla forms a shelflike palatal process that meets the prefrontal, palatine, and ectopterygoid. Its palatal process disappears along the posterior half of the orbital margin, where the maxilla sheathes the anterior process of the jugal. As in *Sphenodon*, the lateral aspect of the maxilla is concave above the tooth row. Although incomplete, its posterior process indicates that the maxilla bows laterally posterior to the tooth row around the adductor chamber.

The nasals have been eroded on both sides, except for the region near the suture with the frontals. Nevertheless, their configuration can be reconstructed from impressions. They are slightly different from those of *C. hudsoni* in having a combined W-shaped suture with the frontals. The nasal overlaps the frontal posteriorly. Damage to the nasal obscures its relationship with the premaxilla. The oval external naris appears to be moderate in size.

The prefrontal is well preserved on the left side, although its surface is worn in certain areas. As in most sphenodontians, the bone is triangular in outline. It contributes to the anterodorsal margin of the orbit and has extensive contacts with the nasal and frontal. Its ventral process contacts the palatine at the anteroventral corner of the orbit. Because there is no independent lacrimal bone, the lacrimal foramen is laterally

exposed and bounded ventrolaterally by the maxilla.

The frontals meet along the midline without the complex interdigitation found in *C. hudsoni* (Fraser, 1988, fig. 8). The left frontal is complete except for the area articulating with the parietal. It is broader posteriorly than anteriorly. Laterally, the bone enters the orbital margin for a relatively short distance. The suture between the frontals and parietals appears to have been somewhat curved.

The postfrontal is present on the left side, but its ventral process and the area near the suture with the parietal are damaged. As in most sphenodontians, the postfrontal has an extensive contact with the lateral margin of the frontal and probably a short posterior process that overlaps the anterolateral edge of the parietal.

The large, triangular postorbital is nearly complete on the left side, but it has been fractured by dorsoventral compression and is damaged anteriorly. It overlaps the postfrontal dorsally and the anterior process of the squamosal. It contributes significantly to the supratemporal bar.

The jugal is represented by the anterior and dorsal processes on the left side. The anterior process is short and largely hidden in lateral view by the maxilla. A facet on its medial surface is contiguous with the maxilla and probably received the anterolateral process of the ectopterygoid. The dorsal process extends backward for more than half of the length of the supratemporal bar to overlap the anterior process of the squamosal. As a result, the postorbital is excluded from the margin of the inferior temporal fenestra, as is the case in all other fossil sphenodontians. It is difficult to determine whether or not the lower temporal bar was complete.

The squamosal is well preserved on the left side. It is a large, more or less quadrilateral bone. Its anterior process forms the posterior part of the supratemporal bar. Its descending process runs down along the lateral edge of the quadratojugal, contributing to the lateral rim of the tympanic crest to a point just above the condyle. Anteroventrally, the process is broken, which makes it impossible to obtain any direct evidence of its relationship with the jugal. There is no indication that it tapers distally, suggesting that the squamosal may have made contact with the jugal at the posteroventral corner of the inferior temporal fenestra, as it does in some individuals of *Planocephalosaurus* (Fraser, 1982) and *Sphenodon*, and contributed to the formation of a complete lower temporal bar. The descending process of the squamosal in *C. hudsoni* tapers ventrally, but it is thought that the process made weak contact with the jugal (Fraser, 1988). The posteromedial edge of the squamosal extensively overlaps the posterolateral extension of the supratemporal. This is in sharp contrast to *C. hudsoni*, where the supratemporal overlaps the posteromedial margin of the squamosal. Near

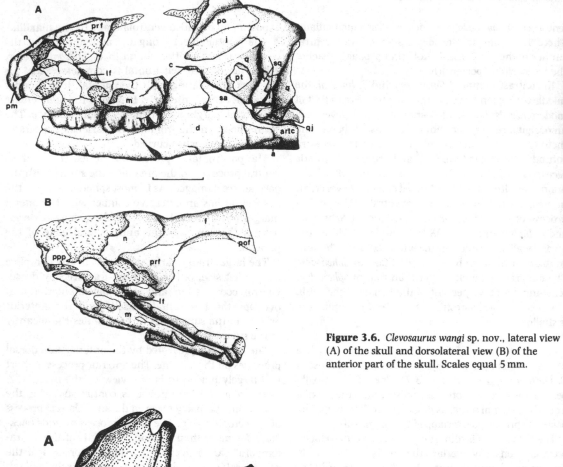

Figure 3.6. *Clevosaurus wangi* sp. nov., lateral view (A) of the skull and dorsolateral view (B) of the anterior part of the skull. Scales equal 5 mm.

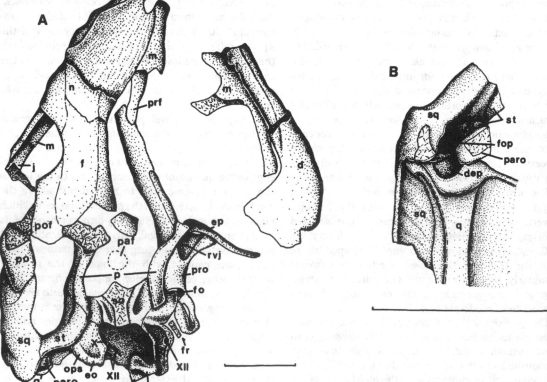

Figure 3.7. *Clevosaurus wangi* sp. nov. (A) Dorsal view of the skull. (B) Relationships of the quadrate, paroccipital process, squamosal, and supratemporal. Scales equal 5 mm.

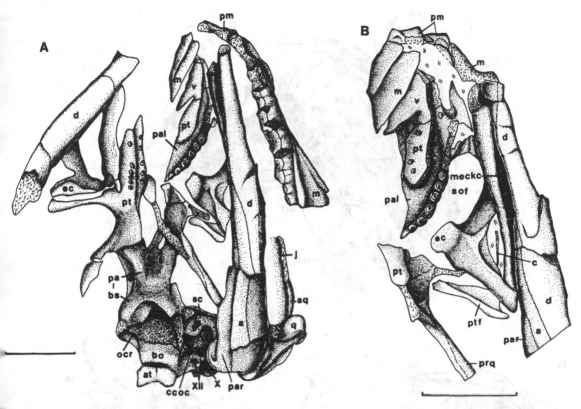

Figure 3.8. *Clevosaurus wangi* sp. nov. (A) Ventral view of the skull. (B) Ventral view of the anterior part of the skull. Scales equal 5 mm.

the posterodorsal tip of the squamosal, a deep fossa can be seen on the undersurface of the bone for the reception of the thickened lateral termination of the paroccipital process. The squamosal fits into a basin-like depression on the cephalic head of the quadrate, forming a ball-and-socket articulation (Figure 3.7B).

The supratemporal is preserved on the left side. It is trap-shaped and is positioned between the squamosal and the parietal. Medially, it extensively overlaps the posterolateral process of the parietal along the posteromedial margin of the supratemporal fenestra. In contrast with *C. hudsoni*, the bone does not reach the cephalic head of the quadrate.

The parietals are not well preserved on either side, but they are reconstructed as paired elements following the paired nature of the frontals. As the size of the supratemporal fenestra indicates, the bones are relatively large in comparison with that of *C. hudsoni*, and their width between the supratemporal passages probably was greater than the interorbital width. The exact configuration of the parietal foramen is uncertain, but its anterior margin would not have reached the level between the anterior edges of the two supratemporal fenestrae.

Palatoquadrate. The quadrate complex is almost complete on the left side (Figures 3.6A, 3.7B, 3.8A, and 3.9A). The posterolateral surface of the complex forms a conchlike basin, bordered laterally by the tympanic crest (which is strengthened by the overlap of the descending process of the squamosal) and medially by a posterior column that runs from the cephalic head of the quadrate down to the articulating surface for the lower jaw. As in other sphenodontians, the double condyle of the quadrate shows a large medial part and a smaller lateral portion. The broad pterygoid ramus is curved medially and expands anteromedially to overlap the quadrate ramus of the pterygoid, forming a deep, concave lateral wall in the middle ear cavity. The extensive contacts with both the squamosal and the pterygoid provide rigidity between the quadrate complex and the rest of the skull. The quadratojugal is fused to the quadrate, with no trace of a suture. Because of poor preservation of its anteroventral surface, the relationship of the quadratojugal to the jugal remains uncertain.

The epipterygoid is well preserved on the right side, although it was somewhat twisted dorsally. The bone is very distinctive in having a shape that shows a reverse configuration to that of the epipterygoid of

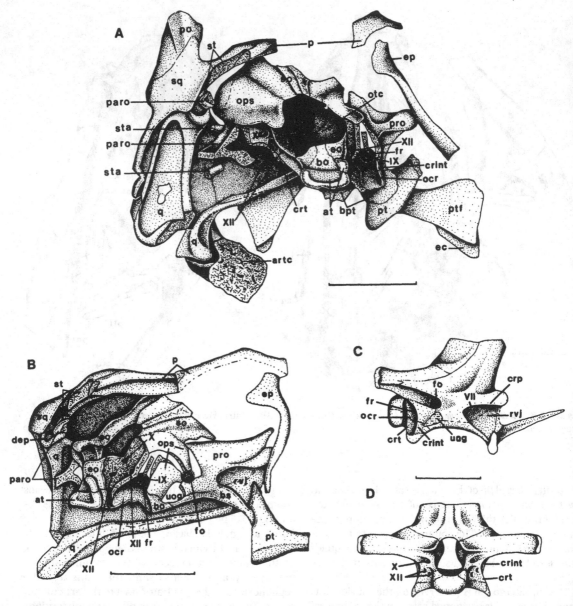

Figure 3.9. *Clevosaurus wangi* sp. nov. (A) Occipital view of the skull. (B) Lateral (slightly posterior) view of the posterior part of the skull. Lateral (C) and occipital (D) views of the reconstruction of the braincase. Scales equal 5 mm.

Sphenodon (i.e., an expanded dorsal end connected with the parietal, and a narrow ventral tip that meets the pterygoid).

    Dermal bones of the palate. The vomers can be recognized as paired triangular bones, although they are damaged both anteriorly and posteriorly (Figure 3.8). The right vomer is displaced over the left. Erosion has made it difficult to discern details.

    The palatines are present on both sides, and the left

is almost complete. The maxillary process is a very robust shelf that slots into the maxilla. The palatine forms the posterior border of the internal choana and slopes away from its anterior edge to form a posterior shelf to the nasal vacuity. An opening on this shelf is interpreted as a foramen for the superior alveolar nerve and artery. The palatine has extensive medial contact with the pterygoid and tapers posteriorly along the medial margin of the suborbital fenestra to wedge into the pterygoid. The enlarged lateral tooth row consists of approximately eight teeth that decrease in size posteriorly. In addition, a single palatine tooth lies anteromedial to the lateral row.

The pterygoids are relatively complete, and each can be divided into three rami. The large, flat anterior ramus forms the lateral border of a moderate-sized interpterygoid vacuity, and with its counterpart passes anteriorly between the palatines. The transverse flange is deep and thickened distally. It extensively overlaps the ectopterygoid anteroventrally. The quadrate ramus is a large, fan-shaped lamina that is almost vertically oriented. The central area between the three rami is elongated as in *Sphenodon* and *Palaeopleurosaurus*. The pterygoid articulates firmly with the basipterygoid process.

The ectopterygoids are well preserved on both sides, but are not entirely visible because the mandibles are still occluded. The element is robust and resembles the twisted H shape described by Whiteside (1986) in *Diphydontosaurus*. It follows the same pattern as *C. hudsoni*.

Braincase. The basioccipital is almost complete, except for the lateral surface (Figures 3.7A, 3.8A, 3.9, and 3.10A). It forms the posteroventral part of the floor of the cranial cavity and the midpart of the occipital condyle that is still articulated with the atlas. Anteroventrally, its suture with the parabasisphenoid is indistinct because of superficial erosion, but the basioccipital appears to be joined anteriorly by a tongue-and-groove suture to the parabasisphenoid. Anterolaterally, the basioccipital is expanded and turns dorsally to suture with the ventral process of the opisthotic and with the crista interfenestralis and the crista tuberalis (Oelrich, 1956) of the exoccipital, where it is drawn ventrally into an incomplete basal tubercle. The basal tubercle is deeply excavated dorsally into an occipital recess. The occipital recess faces posterolaterally and forms part of the floor of the middle ear cavity. The crista tuberalis of the basioccipital is partially preserved on the left side. When complete, the crista tuberalis would have been broadly expanded laterally and, together with the exoccipital portion, would have formed the posterior wall of the occipital recess, as in most lizards (Figure 3.9C,D). The basioccipital continuation of the crista interfenestralis contributes to the anterior wall of the occipital recess. As in *Sphenodon*, an unossified gap posteroventral to the fenestra ovalis remains between the basioccipital, the parabasisphenoid, the opisthotic, and the prootic. A similar gap was described by Fraser in *C. hudsoni* (1988), but in that species it apparently does not involve the prootic.

The parabasisphenoid complex is well preserved. Posteriorly, it contributes to the anterior floor of the cranial cavity, and anteriorly it becomes a long, stout

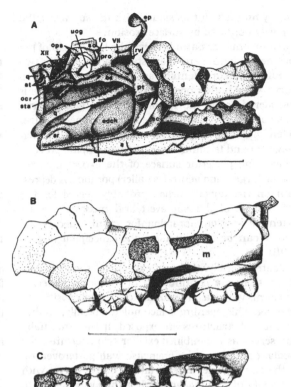

Figure 3.10. *Clevosaurus wangi* sp. nov. (A) Lateral (slightly ventral) view of the skull, showing the medial aspect of the left lower jaw. Lateral (B) and occlusal (C) views of the left maxilla. Scales equal 5 mm.

median cultriform process missing its anterior end. The ventral surface of the complex is constricted just posterior to the basipterygoid processes and broadens posteriorly. It extends laterally into a pair of processes that reach the basal tubera and overlap the basioccipital posteriorly. The paired basipterygoid processes are short, but broad and laterally directed. The geometry of their articulating surface with the pterygoids suggests little movement at the basipterygoid joint. A pair of foramina for the internal carotid arteries can be seen between the basipterygoid processes. Posterolaterally, the sutures of the complex with the prootic and the basioccipital are very clear, but anterolaterally the articulation between the complex and the prootic is indistinct. Based on the condition in *Sphenodon* and lizards, it may reach the inferior process of the prootic.

The exoccipitals are incompletely fused to the opisthotics. The right exoccipital is the better-preserved, except for the loss of the paroccipital process and the posterior surface. However, these areas are preserved on the left side. The exoccipital forms the lateral portion of the condyle and the posteroventral third of the otic capsule. Its articulation with the basioccipital is clearly marked, but its suture with the supraoccipital is partly obscured by surface erosion.

An incomplete basin between the left margin of the foramen magnum and the crista tuberalis exhibits two openings that are identified as the foramina for cranial nerve XII. Another opening just dorsal to the basin is considered to be the foramen for the vagus nerve (cranial nerve X). Dorsal to the vagus foramen, a ridge, extending from the lateral margin of the foramen magnum and terminating on the paroccipital process, divides the posterior surface of the exoccipital into dorsal (larger) and ventral (smaller) portions. A depression on the dorsal portion probably served for the attachment of the supravertebral muscles. Despite extensive erosion, foramina for cranial nerves XII and X can be seen in the right exoccipital (Figure 3.9B).

Ventrolaterally, the right exoccipital shows a deep cavity that is a continuation of the occipital recess and forms with it a large atrium to a dorsally positioned opening. This opening is incomplete dorsally, so that its internal structures are exposed. It runs internally and serves as a combined exit for two apertures; the medial one pierces the braincase wall posteroventral to the otic capsule, and the lateral one connects with the cavity of the otic capsule. It is thus very similar to the fenestra rotunda in extant lizards. The medial aperture is considered to have served as the exit for cranial nerve IX, and the lateral one presumably was associated with the perilymphatic system of the inner ear. The space between the opening and the two internal apertures is thus interpreted as the perilymphatic sac. As mentioned earlier the vagus nerve exits from the braincase through an independent foramen. This indicates that the metotic fissure is divided into two openings – the vagus foramen and the fenestra rotunda – as is the case in most extant lizards.

The central cavity of the right otic capsule is exposed because of the loss of the paroccipital process. The crista tuberalis and crista interfenestralis of the exoccipital are also incomplete. Anteriorly, the bone is medially compressed and disarticulated from the prootic, so that the fenestra ovalis has lost its original configuration. Its appearance is further distorted by the damage to the paroccipital process. Ventral to the fenestra rotunda and the fenestra ovalis the ventral process of the opisthotic borders the unossified gap anteroventrally. From the ventral process the crista interfenestralis arises between the fenestra rotunda and the fenestra ovalis, as in the majority of extant lizards.

The left paroccipital process is damaged. Anteriorly, it must have been overlapped by the prootic and formed, together with the latter, a ventral paroccipital recess for the fenestra ovalis. Its broken lateral ter-mination is isolated and preserved on the cephalic head of the quadrate.

The supraoccipital is nearly complete apart from the anterior and posterior margins. It is hexagonal in configuration and dorsally convex. The element "roofed" the posterior part of the cranial cavity and formed the dorsal rim of the foramen magnum. It contributed to the anterodorsal third of the otic capsule. Its suture with the exoccipital is indistinct. Anteriorly, it is disarticulated from the prootic. Dorsally, it probably joined the parietals through a cartilaginous junction. The supraoccipital bears a moderate median ridge that presumably served for the attachment of the neck ligaments. On each side of the ridge is a large depression that received the insertions of epaxial muscles.

The prootic is well preserved on the right side, although it has lost its posterodorsally directed process, from which a crest forming the anterodorsal wall of the paroccipital recess is derived. The portion forming the anterolateral surface of the cranial wall is slightly crushed dorsoventrally and is covered dorsally by matrix, thus reducing its exposure in lateral view. The bone makes up the anteroventral third of the otic capsule. Its articulation with the parietal may be similar to that in *Sphenodon* and extant lizards. The prootic probably made contact with the dorsal end of the epipterygoid by a membranous junction. Posteroventrally, the bone forms the anterior rim of the unossified gap, and farther dorsally it shows a more or less semicircular notch that probably formed the anteroventral margin of the fenestra ovalis. The crista alaris is not apparent anterodorsally. As in *Sphenodon*, the trigeminal notch is shallow, but the inferior process is very pronounced at the probable position of the suture with the parabasisphenoid. This contrasts with the situation in *C. hudsoni*, where the trigeminal notch is very deep as a result of a pronounced pila antotica (Fraser, 1988, fig. 17b). From the inferior process a pronouced ridge, the crista prootica, extends posteriorly and becomes indistinct more than halfway to the fenestra ovalis. Below the crista prootica, the prootic is deeply excavated into a fossa, the recess for the vena jugularis, which expands ventrally onto the parabasisphenoid. The recess is distinct from that in extant lizards in that it does not expand posterodorsally to join the paroccipital recess. Near the posterodorsal margin of the recess for the vena jugularis there is a foramen through which cranial nerve VII goes forward into the orbital region.

Mandible. The left dentary is almost complete, except for the symphysial region and the posterior tip, and it generally resembles that in other sphenodontians (Figures 3.6A, 3.8, and 3.10A). It bears a deep coronoid process and a distinctive posterior process that tapers abruptly below the jaw articulation. At the midpoint between the last tooth and the jaw articulation, the small mandibular foramen for the recurrent nerve is located at the dentary–surangular suture. Ventral to the tooth row the dentary is thickened. On its medial surface, a Meckelian groove runs through the length of the bone. Anterior to the coronoid, the canal is bordered ventrally by the angular.

The left angular is incompletely preserved. As in *C. hudsoni* and *Palaeopleurosaurus*, the angular has more lateral exposure than in *Sphenodon*. It curves posteriorly from its lateral contact with the dentary and surangular onto the ventral surface of the mandible, and then extends anterodorsally between the prearticular above, and the dentary below, onto the medial surface of the mandible. Anteriorly, the angular tapers along the ventral margin of the Meckelian canal.

Much of the left coronoid is covered medially by the transverse flange of the pterygoid. The coronoid forms the medial part of the coronoid process. It has a small lateral exposure. Ventrally, the bone becomes very thin. Presumably, as in *Sphenodon*, the coronoid makes contacts with the anterior end of the prearticular and the dorsal margin of the angular to form the anterior border of the adductor chamber, although this region is not well exposed.

The articular, prearticular, and surangular form a co-ossified complex, although the suture between the latter two is partly preserved. The complex, missing only the retroarticular process, is well preserved on the right side.

The surangular forms most of the lateral surface posterior to the mandibular foramen and dorsal to the posterior process of the dentary. In medial aspect, the surangular extends from the dorsal region of the coronoid back toward the articular to form the dorsal margin of the adductor chamber that is walled laterally by the dentary.

The position of the prearticular is indicated by a straplike structure on the medial surface. Its posterior part expands dorsomedially to support the robust articular portion of the complex. It curves posteriorly from the contact with the angular onto the medial aspect, and then runs forward to contribute to the ventral margin of the adductor chamber.

The articular is short but massive. It forms most of the articular facet (with a lateral contribution from the surangular) and contributes to the posterior margin of the adductor chamber. The dorsal condylar surface is divided by a ridge into a deep posteromedial fossa and a shallow anterolateral fossa, corresponding to the double quadrate condyle. The massive broken surface posteriorly suggests that the retroarticular process probably was well developed, as in other early sphenodontians.

Stapes. The left stapes is represented by two fragments (Figures 3.9A and 3.10A). The proximal fragment is displaced with the distortion of the braincase, and its footplate is covered by the paroccipital process. Toward the footplate the rodlike stump becomes thicker. The distal portion is positioned ventral to the paroccipital process and has an approximate diameter of 0.5 mm.

Dentition. No premaxillary teeth are preserved (Figures 3.6A, 3.8A, and 3.10). The left maxilla bears seven acrodont teeth somewhat damaged at the crowns. The first is small and subconical. It probably represents the remnant of the hatchling dentition (Robinson, 1976). The four subsequent teeth progressively increase in size and are flanged posteriorly. They belong to the first series of the additional teeth (Robinson, 1976). The last two of the seven teeth are very small and conical. The first displays a very rudimentary flange posteriorly, and the last is not fully ankylosed to the bone. They probably represent the second series of the additional teeth. The maxillary tooth row shows considerable tooth wear lingually.

The left dentary displays the most posterior six teeth in lingual view. The last two are smaller than the rest. The first is damaged anteriorly. All six teeth are flanged on both anterior and posterior sides. The anterior flanges are more pronounced than the posterior, as described by Fraser (1988) in *C. hudsoni*. The labial aspect of the teeth is covered by the occlusion of the upper jaw. The dentition of the dentary has a pattern similar to that of the maxilla: The first four teeth belong to the first series of the additional teeth, and the two smaller posterior teeth probably represent the second series.

Postcranial skeleton. The only preserved elements of the postcranial skeleton (Figures 3.8A, 3.9A,B, and 3.10A) are the intercentrum of the atlas and the ventral part of its right neural arch, which are still articulated with the occipital condyle.

The atlantal intercentrum is robust and very similar in appearance to that of *Sphenodon*. It is bowed slightly upward on both sides, so that its flat base is transversely convex ventrally. Its anterior and posterior margins are curved, the latter more so than the former. The posterior surface is round, sloping, and concave to receive the intercentrum of the axis. The anterior surface of the intercentrum, for articulation of the basioccipital condyle, appears to be similar in appearance to the posterior surface, as in *Sphenodon*, except for the degree and direction of the sloping.

The fragment of the neural arch lies on the dorsolateral surface of the intercentrum. Its concave medial and bulging lateral faces show that the neural arches on both sides may have formed a ring-shaped structure around the foramen magnum.

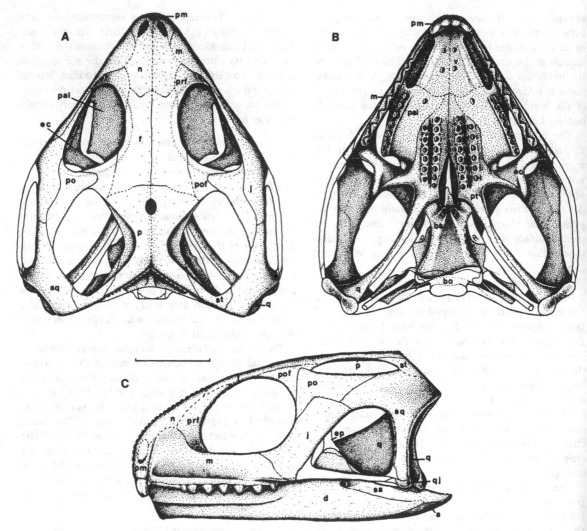

Figure 3.11. *Clevosaurus mcgilli* sp. nov. Reconstruction of the skull. Dorsal (A), ventral (B), and lateral (C) views. Scale equals 5 mm.

Clevosaurus mcgilli sp. nov.

Holotype. V.8272, a skull with mandible, but damaged in both the snout and occipital region.

Paratype. V.8273, a skull lacking the postorbital region, with the anterior parts of the mandible.

Locality. Near Dawa village, Lufeng County, Yunnan Province, southwest China.

Stratigraphic position. About the middle of the upper part (Dark Red Beds) of the Lower Lufeng Formation (stratum 6 in Figure 14.3).

Age. Early Jurassic.

Etymology. A testimonial to McGill University, Montreal, where I completed my Ph.D. training under the guidance of Professor Robert L. Carroll.

Diagnosis. Differing from the other species of *Clevosaurus* as follows: a large row of palatine teeth that is subparallel to the marginal teeth of the maxilla; palatine becoming broad posteriorly, resulting in an L-shaped suborbital fenestra; jugal with a long anterior process that meets the ventral process of the prefrontal at the anteroventral corner of the orbit and a dorsal process that is expanded laterally at its extremity; postorbital strongly curved ventrally, showing a T-shaped lateral outline; supratemporal asymmetrically V-shaped; very short snout that is less than a quarter of the skull length.

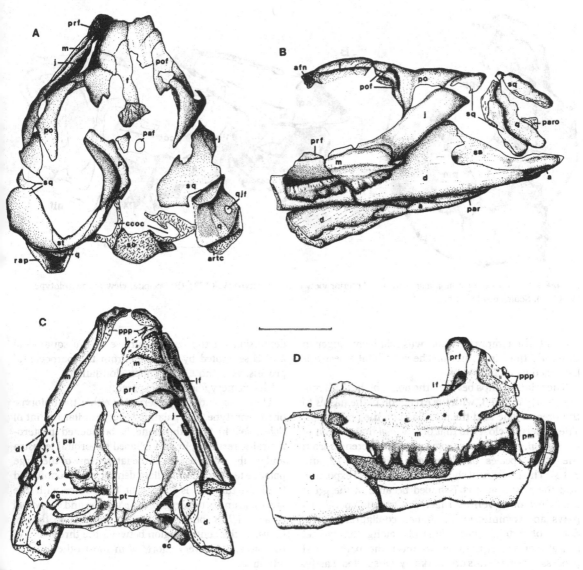

Figure 3.12. *Clevosaurus mcgilli* sp. nov. Dorsal (A) and lateral (B) views of the holotype (V.8272). Dorsal (C) and lateral (D) views of the paratype (V.8273). Scale equals 5 mm.

Description. The holotype skull has been dorsoventrally compressed posterior to the orbit, so that the braincase has been crushed. The basioccipital and antorbital regions have been lost. The maxilla and jugal are not preserved on the right side. The skull table is heavily eroded along the midline. The mandible is almost complete except for the symphyseal region. In the paratype, the skull roof has been damaged, but the maxillae, the anterior tip of the snout, part of the right prefrontal, and the elongated anterior process of the right jugal are preserved. The mandible is complete anterior to the coronoid process. Of the palatal elements, the palatine, pterygoid, and ectopterygoid are present in both specimens.

The reconstruction of the skull is based on both specimens. As in *C. wangi* a complete lower temporal bar is indicated by the expanded extremity of the descending process of the squamosal and its great ventral extension along the lateral side of the quadratojugal.

The skull (Figure 3.11) is about 2 cm long in the holotype and about 2.3 cm long in the paratype. It has a relatively short and broad appearance in comparison with the other species of *Clevosaurus*. Its snout is about 23 percent of the skull length. In this aspect, *C. mcgilli* resembles *Brachyrhinodon* (Fraser and Benton, 1989). The supratemporal fenestra is rhombic in outline, and it probably was longer than the orbit. As

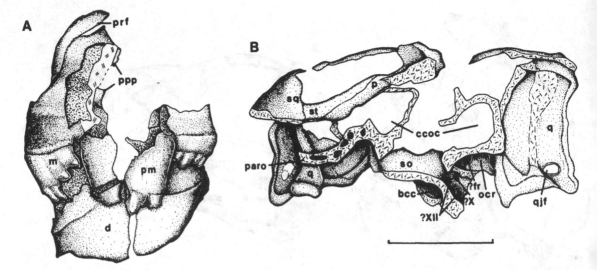

Figure 3.13. *Clevosaurus mcgilli* sp. nov. (A) Anterior view of the paratype (V.8273). (B) Occipital view of the holotype (V.8272). Scale equals 5 mm.

a result, the temporal region was relatively longer in *C. mcgilli* than in *C. wangi*. The suborbital fenestra is L-shaped as in *Sphenodon*.

The relationships between the bones of the skull roof are clearly marked, with the exception of the nasals to the premaxillae and the maxillae and the prefrontals. The pattern is closer to that of *C. wangi* than to that of *C. hudsoni*, but there are some notable differences from the other species of *Clevosaurus* (Figures 3.12 and 3.13). The prefrontal shows a limited lateral exposure, and the postorbital is T-shaped because of the great curvature of its ventral edge. The supratemporal displays an asymmetrical V-shaped configuration and does not extend posterolaterally along the medial margin of the squamosal to meet the paroccipital process. The parietals are markedly constricted. Lastly, the jugal possesses an elongated anterior process that runs forward along the ventral edge of the orbit to contact the ventral process of the prefrontal and a relatively short dorsal process with an expanded extremity that fits in the concave ventral margin of the postorbital. As the size of the supratemporal fenestra indicates, the parietal table is relatively narrower than in *C. wangi*, but broader than in *C. hudsoni*. In addition, the supratemporal bar is not as broad as in *C. wangi* because of the expansion of the supratemporal fenestra. Although the parietals were damaged along the midline, the position and size of the parietal foramen can be determined by an infilling that is a different color from that of the surrounding matrix. Unlike *C. hudsoni*, the foramen does not reach the line between the anterior margins of the supratemporal fenestrae.

The quadrate (Figures 3.12A,B, 3.13B, and 3.14B) is quite distinct from that of *C. wangi* in that the cephalic head is convex and fits into a matching

depression on the ventral surface of the squamosal and is separated by the latter from the paroccipital process, as in most other sphenodontians.

The epipterygoid is not exposed.

The vomers are not preserved in either the holotype or the paratype. The palatine is very similar to that of *Sphenodon* in that the bone is broadened posterolaterally, resulting in an L-shaped suborbital fenestra and a subparallel course of the lateral row of the large palatine teeth to the maxillary dentition. A pronounced toothlike tubercle is formed at the suture with the pterygoid and ectopterygoid in *Sphenodon* and *C. hudsoni* (Fraser, 1988, pl. 3, fig. 15). In contrast to that of *C. wangi*, the central region between the three rami of the pterygoid is very short, as in most other sphenodontians.

The braincase is very fragmentary (Figures, 3.12A, 3.13B, and 3.14B). Only the parabasisphenoid is well enough preserved to warrant description. It is relatively long in comparison with that of the other species and does not have any depressions or fossae like that of *C. wangi* on the ventral surface. The left paroccipital process shows on the broken surface a series of pneumatic spaces. As a result of poor preservation, the identifications of the fenestra ovalis, the fenestra rotunda, and the foramina for cranial nerves shown in the figures are tentative.

The mandible (Figures 3.12B–D, 3.13A, and 3.14) shows no significant differences from that in most other sphenodontians (except for *C. wangi*, in which the posterior process of the dentary tapers abruptly beyond the mandibular foramen). The symphysis of the dentary (not preserved in *C. wangi*) is very broad, as in *C. hudsoni*. The retroarticular process (damaged in *C. wangi*) is well developed. On its dorsal surface a

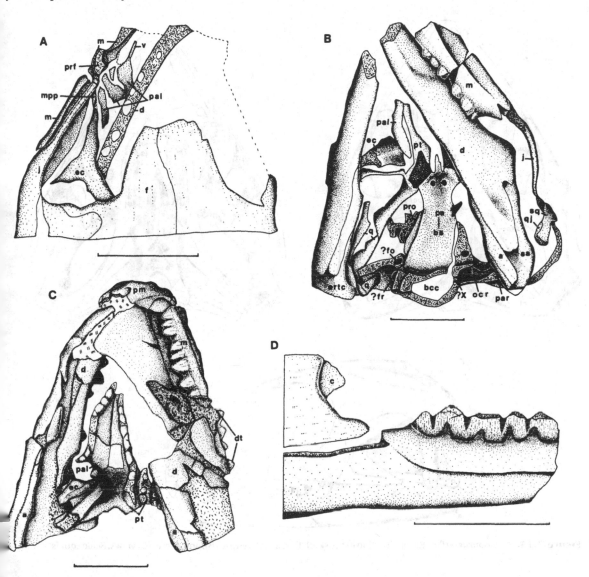

Figure 3.14. *Clevosaurus mcgilli* sp. nov. (A) Dorsal view of the antorbital region of the holotype (V.8272), showing the relationships between the jugal and prefrontal and between the palatine and ectopterygoid. (B) Ventral view of the holotype (V.8272). (C) Ventral view of the paratype (V.8273). (D) Lateral view of the middle part of the left dentary in the holotype (V.8272). Scales equal 5 mm.

entral depression is rimmed by a raised margin. The lateral margin of the rim contributes to the tympanic crest, which was for the attachment of the tympanum, while the medial margin served as the insertion for the M. pterygoideus typicus. The posterior extremity of the process for the attachment of the M. depressor mandibulae is missing.

The premaxilla of the paratype bears two teeth on each side, but their crowns on the right side have been damaged (Figures 3.12D, 3.13A, and 14C). The teeth are conical and of similar sizes.

As in *Sphenodon* and *C. hudsoni*, the teeth overhang the tip of the mandible to form a beaklike structure.

The maxillary dentition (Figures 3.12B,D, 3.13A, and 3.14), seen in the paratype, is slightly different from that of *C. wangi* in that there is an additional series of five equal-sized teeth. The right dentary of the holotype bears the posteriormost five teeth. As in *C. hudsoni* and *Palaeopleurosaurus*, the lateral aspects of the teeth display vertical score lines that continue across the surface of the dentary.

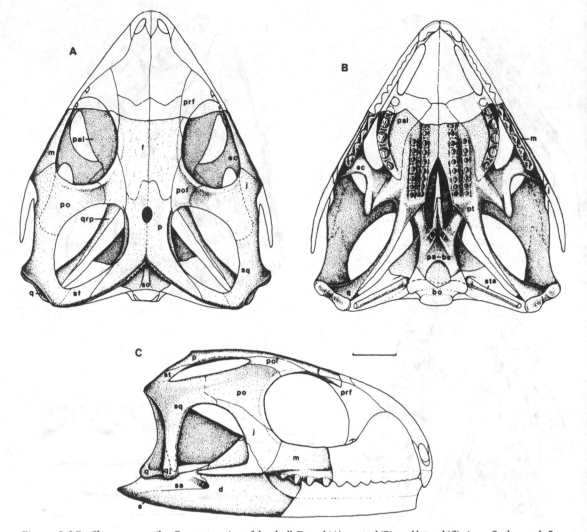

Figure 3.15. *Clevosaurus petilus*. Reconstruction of the skull. Dorsal (A), ventral (B), and lateral (C) views. Scale equals 5 mm.

Clevosaurus petilus (*Dianosaurus petilus* Young, 1982)

Holotype. V.4007, skull with mandible, damaged anterior to the orbits.

Locality. Close to Dachong village, Lufeng County, Yunnan Province, southwest China.

Stratigraphic position. Near the top of the lower part (Dull Purplish Beds) of the Lower Lufeng Formation (stratum 4 in Figure 14.3).

Age. Probably latest Triassic.

Diagnosis. Differing from the other species of *Clevosaurus* as follows: oval supratemporal fenestra is diagonally oriented; parietal narrow anteriorly; dorsal margin of the maxilla slopes upward toward its contact with the jugal; jugal with short anterior process; basipterygoid process very slender; prearticular extends anteriorly beyond the ventral edge of the coronoid.

Discussion. The material of *Clevosaurus petilus* was described as *Dianosaurus petilus* by Young (1982) in a posthumously published paper. Young believed that "*D. petilus*" did not possess an inferior temporal fenestra, and consequently he placed it in the Euryapsida. Having excluded an aquatic life style, Young referred it to the Protorosauria. Because "*D. petilus*" has no dental features comparable to those of the Trilophosauridae, which were supposed to be the only Late Triassic family of the Protorosauria, and differs in many other aspects from the known protorosaurs, Young doubted his systematic assessment of the species in the same paper and pointed out that

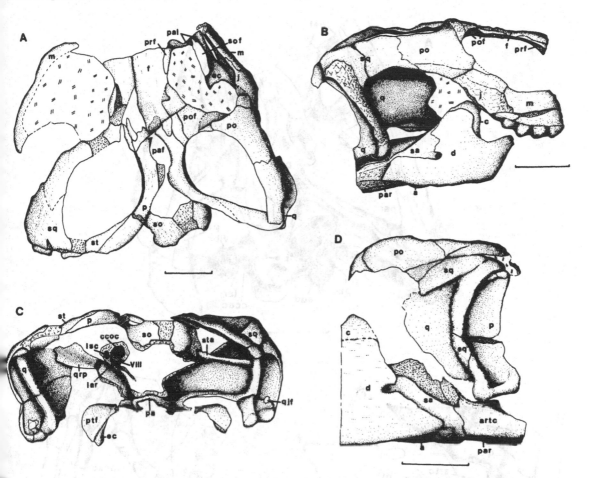

Figure 3.16. *Clevosaurus petilus.* Dorsal (A), lateral (B), occipital (C), and left lateral (D) views of the skull. Scales equal 5 mm.

"*D. petilus*" might represent a new taxon of reptiles thus far unknown from the Lufeng area.

The current study of the material reveals that Young erred in his description of a number of significant systematic characters and clearly indicates sphenodontian affinities for "*D. petilus*" and reassignment as a species of *Clevosaurus*. Therefore, *Dianosaurus petilus* is renamed *Clevosaurus petilus*.

Description

The left side of the skull is not well preserved. The right supratemporal fenestra is distorted. The lower temporal bars are broken. The braincase was damaged, especially in the posterior region.

The skull (Figure 3.15) is approximately 3.5 cm long. The supratemporal fenestra is relatively much larger than that of *C. wangi*. Its major axis is not parallel to but rather oblique to the midline, although comparable in both size and configuration to those of *C. mcgilli* and *C. hudsoni*. In this aspect, *C. petilus* can be compared with *Brachyrhinodon* (Fraser and Benton, 1989). It is difficult to determine whether or not the lower temporal bar is complete. The antorbital region of the skull is reconstructed on the basis of those of the other two species of *Clevosaurus* from the Lufeng Basin.

The parietal (Figure 3.16A), unlike those of the other species of *Clevosaurus*, is not broad anteriorly, but the parietal width between the supratemporal passages is narrower than the interorbital region, as in *C. hudsoni*. Because the bone does not form a moderate-sized medial shelf of the supratemporal fenestra, the parietal table is relatively broader than those of *C. hudsoni* and *C. mcgilli*, but relatively narrower than that of *C. wangi*. The suture of the parietals to the frontals is more or less W-shaped. The small size of the posterolateral process of the parietal suggests that there was a supratemporal bone. The jugal (Figure 3.16) is distinct from that in the other clevosaurs in that its restricted anterior process forms only a small portion of the posteroventral margin of the orbit. The maxilla is similar to that of *Sphenodon* in having a dorsal edge that runs upward along the posteroventral margin of the orbit.

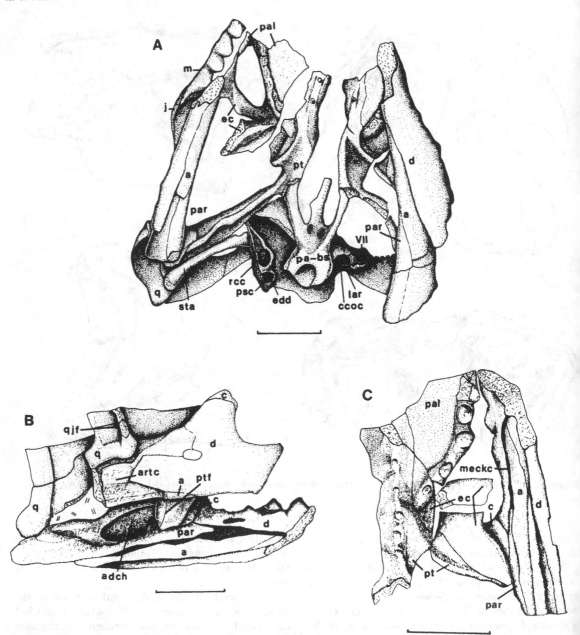

Figure 3.17. *Clevosaurus petilus.* Ventral (A) and lateral (slightly ventral) (B) views of the skull. (C) Ventral view of the suborbital region (with part of the left lower jaw). Scales equal 5 mm.

Poor preservation of the quadrate obscures its relationships to the adjacent elements. On the left side the suture between the quadrate and the quadratojugal is distinguishable below the quadratojugal foramen.

The epipterygoid is not preserved.

The palate (Figure 3.17A,C) is similar to that of *C. mcgilli* and *C. hudsoni.*

Of the braincase elements (Figures 3.16A,C and 3.17A), only the parabasisphenoid, supraoccipital, and prootic are represented. Although incomplete, the parabasisphenoid clearly differs from those of the other species of *Clevosaurus* and other genera. It is characterized by a pair of extremely long and slender basipterygoid processes. On the ventral surface of the bone, a semicircular depression is present, as in *C. wangi.* The supraoccipital of *C. petilus* is similar to those of *C. wangi* and *C. mcgilli,* but some details of the internal structures of the inner ear, which are not available in the other two Lufeng species, are exposed in *C. petilus.* These structures are recognized as the foramina for the openings of the posterior semicircular canal and the endolymphatic duct and the recessus crus communis with which both vertical semicircular canals connect. The preserved part of the left prootic is the portion forming the anteroventral third of the otic capsule. On its transversely broken surface, the opening of the lateral semicircular canal, the central cavum capsularis, the lagenar recess, and the foramen for one of the branches of cranial nerve VIII are well exposed. In addition, the foramen for cranial nerve VII can be identified in lateral aspect.

The mandible (Figures 3.16B,D and 3.17) is distinct from those of other species of *Clevosaurus* in having a prearticular that extends forward beyond the ventral edge of the coronoid. As a result, the coronoid is prevented from contacting the angular. Among sphenodontians, this configuration has been illustrated only in *Gephyrosaurus* by Evans (1980, figs. 39 and 40), although she did not mention it in her description. As in *C. mcgilli* and *C. hudsoni*, the dentary has a gently tapered posterior process.

The right maxilla bears four acrodont teeth posteriorly (Figures 3.16B and 3.17). They are similar in size and shape. Their posterior flanges are not as conspicuous as those in the other species. This may be attributable to poor preservation. The anterior flanges of the three teeth of the left dentary are no more pronounced than the posterior flanges. This is in contrast to the condition in other species of *Clevosaurus*.

Discussion

Comparisons within *Clevosaurus*

Because *C. minor* (Fraser, 1988) is based on fragmentary material, it will not be discussed in the following comparison between the species of *Clevosaurus*.

Similarities among *C. mcgilli*, *C. petilus*, *C. wangi*, and *C. hudsoni* are quite obvious, especially the following: the maxilla is excluded from the external naris by the posterodorsal process of the premaxilla (unknown in *C. petilus*) (Figure 3.18F–I), the suborbital fenestra is solely enclosed by the palatine and ectopterygoid (Figure 3.19G–J), the pterygoid shows two rows of tiny teeth, and a supratemporal is retained. On the other hand, many differences can be recognized among the four species:

1. In *C. wangi* the supratemporal is separated by the paroccipital process from the cephalic head of the quadrate and was laterally overlapped by the squamosal. In *C. hudsoni* it is strap-shaped and extends posterolaterally along the posteromedial margin of the squamosal and contacts the cephalic head of the quadrate, as in lizards.

It displays a peculiar V-shaped configuration and no contact with the paroccipital process in *C. mcgilli*. Judging from the weak posterolateral process of the parietal, the supratemporal most probably is present in *C. petilus*, but its relationships with the bones lateral to it are not clear.

2. The parietal region is very broad in *C. wangi* and *C. mcgilli*, as in most early sphenodontians, but less so in *C. mcgilli*. In contrast, the width of the parietal region is less than the interorbital distance in both *C. hudsoni* and *C. petilus*. In *C. hudsoni* the parietal table even shows a tendency to form a median ridge as in *Palaeopleurosaurus*, *Kallimodon*, *Sapheosaurus*, and *Sphenodon*.

3. The descending process of the squamosal in *C. wangi* and *C. mcgilli* has a broad ventral termination, as in *Brachyrhinodon* and *Sphenodon*, unlike the tapered condition in *C. hudsoni*. The configuration of the ventral tip of the descending process is unknown in *C. petilus*.

4. A semicircular depression is formed on the ventral surface of the basioccipital in both *C. wangi* and *C. petilus*. This structure clearly is absent from *C. hudsoni*, *C. mcgilli*, and other sphenodontians.

5. The supratemporal fenestrae of *C. wangi* are relatively small, about 80 percent of the length of the orbits. They are almost as large as the orbits in *C. hudsoni*, *C. mcgilli*, and *C. petilus*. Thus the temporal region in *C. wangi* is relatively shorter than those in the other three species, and this is regarded as a primitive condition. The anteromedial–posterolateral orientation of the supratemporal fenestra in *C. petilus* is unique. The different orientations of the fenestrae certainly indicate different patterns of adductor jaw musculature in *C. petilus* and the other three species.

6. The skull of *C. hudsoni* has a relatively narrow temporal region, but those of the three Chinese species are somewhat broader, with *C. mcgilli* having the broadest temporal region.

7. The frontals form an interdigitating articulation along the midline in *C. hudsoni* whereas the frontal suture is straight in the other species of the genus. In addition, the frontals are relatively elongate in *C. hudsoni*, as in most other early sphenodontians.

8. The antorbital region is relatively short (less than one-third of the skull length) in both *C. wangi* and *C. hudsoni*. This region is even shorter in *C. mcgilli*, where it is less than one-fourth of the skull length. This region is not preserved in *C. petilus*.

9. The central portion of the pterygoid joining the three rami is elongated in *C. wangi*. A similar condition occurs in *Palaeopleurosaurus* and *Sphenodon*. By contrast, the central region of the pterygoid is very short in the other three species of *Clevosaurus*, and this is more typical of sphenodontians.

10. A ball-and-socket articulation between the paroccipital process and the squamosal and between the squamosal and the cephalic head of the quadrate has not been described in any sphenodontian except

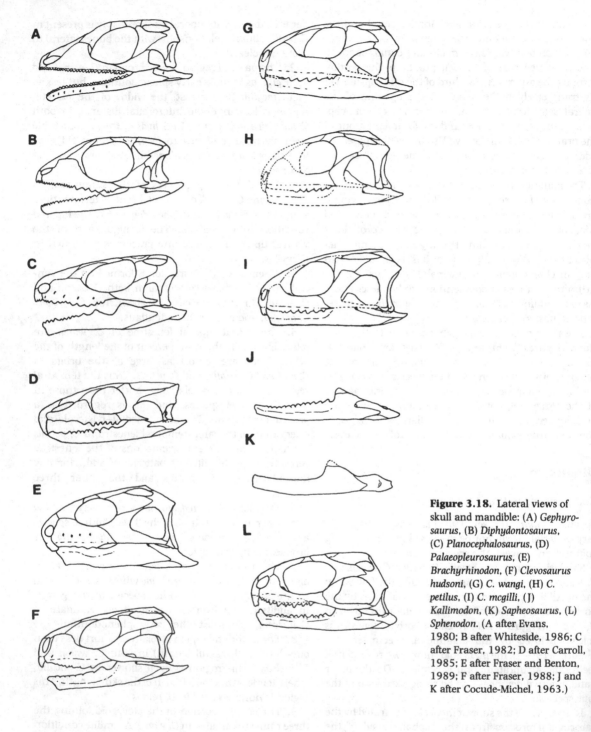

Figure 3.18. Lateral views of skull and mandible: (A) *Gephyrosaurus*, (B) *Diphydontosaurus*, (C) *Planocephalosaurus*, (D) *Palaeopleurosaurus*, (E) *Brachyrhinodon*, (F) *Clevosaurus hudsoni*, (G) *C. wangi*, (H) *C. petilus*, (I) *C. mcgilli*, (J) *Kallimodon*, (K) *Sapheosaurus*, (L) *Sphenodon*. (A after Evans, 1980; B after Whiteside, 1986; C after Fraser, 1982; D after Carroll, 1985; E after Fraser and Benton, 1989; F after Fraser, 1988; J and K after Cocude-Michel, 1963.)

for *C. wangi*. The other three species show a condition common to sphenodontians.

11. The posterior process of the dentary tapers gradually in *C. hudsoni*, *C. mcgilli*, and *C. petilus*, as is typically the case in other sphenodontians, but it terminates abruptly below the jaw articulation in *C. wangi*.

12. The braincase in *C. wangi* is comparable to that in lizards. If the restoration of the braincase in *C. hudsoni* is correct (Fraser, 1988), then it is similar to that of *Sphenodon* in the middle ear region, but differs in that the prootic forms a pronounced pila antotica, as in archosauromorphs. No further comparison of the braincase with those of other species can be made because of poor preservation. Nevertheless, a fenestra rotunda within the occipital recess may have been present

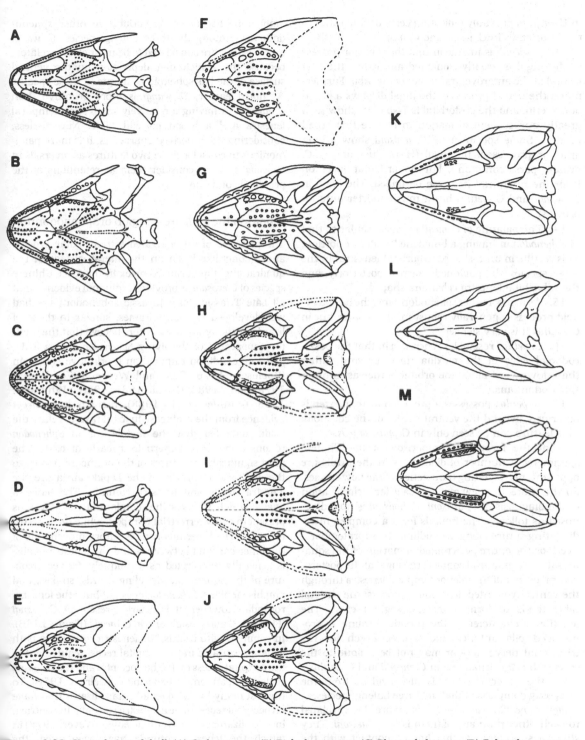

Figure 3.19. Ventral view of skull: (A) *Gephyrosaurus*, (B) *Diphydontosaurus*, (C) *Planocephalosaurus*, (D) *Palaeopleurosaurus*, (E) *Brachyrhinodon*, (F) *Polysphenodon*, (G) *Clevosaurus hudsoni*, (H) *C. wangi*, (I) *C. petilus*, (J) *C. mcgilli*, (K) *Kallimodon*, (L) *Sapheosaurus*, (M) *Sphenodon*. (A after Evans, 1980; B after Whiteside, 1986; C after Fraser, 1982; D after Carroll, 1985; E and F after Fraser and Benton, 1989; G after Fraser, 1988; K and L after Cocude-Michel, 1963.)

in *C. mcgilli*, probably indicating certain similarities to the braincases in *C. wangi* and *C. mcgilli*.

13. *C. mcgilli* is unique in that the anterior process of the jugal is greatly elongated and meets the prefrontal at the anteroventral corner of the orbit. Furthermore, the dorsal process of the jugal displays a broad dorsal end, and the postorbital is T-shaped, showing a greatly curved ventral margin to receive this broad process. Some specimens of *C. hudsoni* show a long anterior process of the jugal (H.-D. Sues and N. C. Fraser, pers. commun.), but no articular facet on it for the prefrontal has been described. The jugal in *C. wangi* and *C. petilus* has a much shorter anterior process.

14. The palatine of *C. mcgilli* is very similar to that of *Sphenodon* in having a broadened posterior portion. This results in an L-shaped suborbital fenestra. In the other species, the palatine is narrow posteriorly, and the suborbital fenestra is banana-shaped.

15. As in most other sphenodontians, the basipterygoid process is generally robust in *Clevosaurus*, but in *C. petilus* it is very slender.

16. *C. petilus* resembles *Sphenodon* in that the posterodorsal edge of the maxilla turns upward rather than downward along the orbital border as in other sphenodontians.

17. *C. petilus* possesses a prearticular that extends anteriorly beyond the ventral border of the coronoid. This feature can be seen only in *Gephyrosaurus*.

From the foregoing comparison of the Chinese specimens, a number of differences in the skulls are apparent, necessitating the erection of the new species. Furthermore, the new material calls for a clarification of the diagnosis of *C. hudsoni*. *C. hudsoni* is now diagnosed as follows: The frontals form a complex interdigitating suture along the midline; the supratemporal overlaps the entire posteromedial margin of the squamosal; the squamosal contacts the jugal at the posteroventral corner of the inferior temporal fenestra through the ventrally pointed descending process; the parietal table tends to form a median sagittal crest. The additional character of the prootic forming a pronounced pila antotica and a deep notch for the trigeminal nerve may or may not be unique to the species, but it is unknown in *C. mcgilli* and *C. petilus*.

On the other hand, it is also evident from the foregoing comparison that the three Lufeng species, *C. wangi*, *C. mcgilli*, and *C. petilus*, are more closely related to each other than any of them is to *C. hudsoni*. They share a supratempoal that has no contact with the cephalic head of the quadrate, a broad ventral extremity of the descending process of the squamosal, and a relatively short and broad frontal. In these aspects, *C. hudsoni* shows the primitive condition. A relatively broad postorbital region of the skull probably is an additional derived character shared by the three Lufeng species. However, the derived state of this character is

ambiguous because it is variable in other sphenodontians. Among the three Lufeng species, *C. wangi* and *C. petilus* are considered to be most closely related, because the semicircular depression on the ventral surface of the basisphenoid is unique to these species. As noted earlier, *C. wangi* is more primitive than *C. hudsoni* in having a relatively small supratemporal fenestra and a broad parietal table. Nevertheless, considering other derived characters, it is more parsimonious to consider these two features as reversals in *C. wangi* than to consider them as retentions of the primitive condition.

A note on the middle ear region of Sphenodontia

The presence of an elongated retroarticular process and a conchlike basin on the lateral surface of the quadratojugal-quadrate complex in the three Chinese species of *Clevosaurus* provides additional evidence that all Late Triassic–Early Jurassic sphenodontians had well-developed tympanic frames, similar to those of lizards. Although many authors believe that the auditory apparatus of the living *Sphenodon* shows a degenerate condition rather than a primitive condition (Robinson, 1976; Gans and Wever, 1976; Wever, 1978; Gans, 1983; Carroll, 1985, 1988a; Whiteside, 1986; Gauthier et al., 1988), there is little direct evidence from the braincase in fossil taxa to refute the traditional idea that the middle ear in *Sphenodon* retains a primitive pattern from early diapsids. The most significant structures of the braincase related to the auditory apparatus in the Lepidosauria are the fenestra ovalis and fenestra rotunda. The fenestra ovalis, which receives the footplate of the stapes, is present in every terrestrial tetrapod, while the fenestra rotunda, which permits compensatory movements of the inner ear fluid between it and the fenestra ovalis, is generally interpreted as a uniquely derived structure of the Squamata, including lizards, snakes, and amphisbaenians. *Sphenodon* does not have the fenestra rotunda. However, it is clearly present in *C. wangi* (Figure 3.9) and possibly also in *C. mcgilli* (Figure 3.13B). In common with lizards, the fenestra rotunda in both genera is situated in the occipital recess (equivalent to the occipital recess on the basioccipital and the recessus scalae tympani on the exoccipital) (Oelrich, 1956). To the best of my knowledge, all extant lizards that have an occipital recess have a fenestra rotunda. In contrast, in those lizards (such as chameleons) (Wever, 1978) in which the fenestra rotunda has degenerated, the occipital recess is absent, as in *Sphenodon*. Therefore, the middle ear apparatus in *C. wangi* and *C. mcgilli* is comparable to that in most lizards in both configuration and function. This not only questions the traditional assumption of the primitive pattern of the middle ear of *Sphenodon*, but it also raises the possibility that a lizardlike, impedance-matching middle ear may have been established in the common ancestor of the Sphenodontia and Squamata.

Although the basic pattern of the middle ear apparatus may be very similar in some early members of the Sphenodontia and lizards, only the lizards possess a highly mobile streptostylic quadrate. In the most primitive taxa of the Sphenodontia, *Gephyrosaurus* and *Diphydontosaurus* (Evans, 1988; Fraser and Benton, 1989), the lower temporal bar is incomplete, as in lizards, but the quadrate is fixed to the rest of the skull. Based on the present study, the lizardlike impedance-matching middle ear could have been considered a synapomorphy of the Lepidosauria. If so, was the fixed quadrate of the Sphenodontia or the streptostylic quadrate of the Squamata retained from their common ancestor? It is difficult to answer this question on the basis of the available fossil record. Because the quadrate has special relationships with both the middle ear and the feeding apparatus in the Lepidosauria, functional analysis of the jaw musculature of both *Sphenodon* and squamates might shed some light on this question.

Paleogeographic considerations

China, like Southeast Asia, is composed of a number of tectonic blocks or cratons. The North and South China Blocks were portions of Gondwana during the Paleozoic (Lin, Fuller, and Zhang, 1985). The North China Block accreted to Laurasia in the Late Permian, when the South China Block was still isolated. The date of accretion of the South China Block to the North China Block remains uncertain, although many reconstructions of the paleogeography of the block have been suggested (Lin et al., 1985; Metcalfe, 1987; Sengör et al., 1987). According to Lin et al. (1985), the South China Block may have joined with the North China Block during a period from the Middle Triassic to the Early Jurassic. This suggestion is, to some degree, in concordance with the fossil record of the Sphenodontia.

The current study shows that the oldest known Chinese sphenodontian is *C. petilus* from the latest Triassic of the Lufeng basin. Paleogeographically, the Lufeng basin was situated close to the southwest margin of the South China Block. All early sphenodontians were small, lizardlike reptiles. It is unlikely that such animals would have had the ability to cross large stretches of water. Therefore, the occurrence of sphenodontians on the South China Block in the latest Triassic presumably indicates that the collision of that block with the North China Block must have taken place prior to the Jurassic. The new sphenodontians also potentially provide additional evidence for the correlation of the Late Triassic–Early Jurassic Lufeng vertebrate faunas with those from the Newark Super-

group and Kayenta Formation in North America and the fissure fillings of Britain.

Interrelationships within the Sphenodontia

Recently, several authors have applied cladistic methods to the analysis of sphenodontian interrelationships (Whiteside, 1986; Evans, 1988; Fraser, 1988; Gauthier et al., 1988; Fraser and Benton, 1989). There are several points of controversy regarding the phylogenetic relationships of *Homoeosaurus*, *Palaeopleurosaurus*, and *Clevosaurus*. With new information from the Chinese species, we can reevaluate the phylogenetic patterns of sphenodontians.

It is well established that the closest relatives of the Sphenodontia are the Squamata. As in the analyses by Whiteside (1986) and Evans (1988), Squamata serve here as the out-group for determining the polarity of characters that vary within the Sphenodontia.

In my analysis of sphenodontian relationships, only skull characters are considered, because well-preserved postcranial skeletons are available for only a few genera. Thirty-six cranial characters have been used for this analysis (Tables 3.1 and 3.2). These characters have been selected primarily on the basis of the data set used by Fraser and Benton (1989). Their data set has twenty-four cranial characters. Two have been omitted in my analysis, and four have been modified. Character 6: "Frontals and parietals: separate (0), fused (1)" must be divided into two separate characters because the frontals are fused but the parietals are separate in *Diphydontosaurus*. The reverse condition occurs in *Palaeopleurosaurus*. Character 15: "Flanges or ridges on palatal tooth row: absent (0), posterolateral ridges or flanges present on some palatine teeth (1)." This character can be determined in *Sphenodon* and in a few fossil taxa, but it cannot be satisfactorily appraised in most of the other genera. Therefore, it has not been considered here. Character 17: "Antorbital region: elongate (0), shortened (1)" has been modified and is considered here as two separate characters: (a) the ratio antorbital region/skull length from the tip of the snout to the line between the posterior edges of the squamosals (Table 3.3): more than one-fourth (0), less than one-fourth (1); (b) the ratio antorbital region/skull length: more than one-third (0), less than one-third (1). Character 18: "Temporal region: short (0), elongate (1)." This character has been modified to be the ratio of the length of the supratemporal fenestra to the length of the orbit: less than three-fourths (0), more than three-fourths (1). This is a particularly useful measure because the supratemporal fenestra and the orbit are preserved in most known genera, and they can be restored with more confidence than other portions of the skull. Character 19: Parietal table: broader than interorbital width (0), narrower (1). The parietal table is the flat area between the two supratemporal fenestrae.

Table 3.1. *Cranial characters used for cladistic analysis*

1 Premaxillary process of maxilla: elongate (0), very weak or reduced (1)
2 Posterior portion of maxilla: gradually tapering off or very narrow (0), dorsoventrally broad (1)
3 (5) Lacrimal: present (0), absent (1)
4 (6 in part) Frontals: separate (0), fused (1)
5 (6 in part) Parietals: separate (0), fused (1)
6 (19 modified) Parietal width between supratemporal fenestrae broader than interorbital width (0), narrower (1)
7 (20) Parietal crest: absent (0), present (1)
8 Posterior edge of parietal(s): greatly emarginated (0), slightly emarginated (1), convex (2)
9 Dorsal process of jugal: broad and short (0), narrow and elongate (1)
10 Parietal foramen: posterior to line between anterior margins of supratemporal fenestrae (0), reaches or crosses line (1) (modified from Gauthier et al., 1988)
11 (17 modified) Antorbital region/skull length ratio: more than one-fourth (0), less than one-fourth (1)
12 Antorbital region/skull length ratio: more than one-third (0), less than one-third (1)
13 (18 modified) Supratemporal fenestra/orbit length ratio: less than three-fourths (0), more than three-fourths (1)
14 Lower temporal fenestra/skull length (from rostrum to posterior edge of parietal) ratio: less than one-fourth (0), more than one-fourth (1)
15 (2) Posterior process of dentary: short (0), elongate (1)
16 Coronoid process of dentary: absent or weak (0), pronounced (1)
17 (3) Lower temporal bar: aligned exactly with maxillary tooth row (0), bows away laterally beyond limit of adductor chamber (1)
18 Lower temporal bar: incomplete (0), secondarily complete (1)
19 (25) Retroarticular process: pronounced (0), reduced (1)
20 (26) Quadratojugal-quadrate conch: pronounced (0), reduced (1)
21 (7) Dentition: pleurodont (0), a degree of acrodont (1)
22 (8) Premaxillary teeth: more than seven acrodont teeth (0), seven or fewer (1)
23 (13) Premaxillary teeth: more than four (0), four or fewer (1)
24 (14) Premaxillary teeth: more than three (0), three or fewer (1)
25 (9) Premaxillae: individual teeth remain discrete in adults (0), premaxillae develop into chisel-like structure in mature individuals (1)
26 (10) Posterior maxillary teeth: simple conical structures (0), presence of a posteromedial ridge or flanges (1)
27 (11) Lateral and medial wear facets on marginal teeth: absent or poorly developed (0), well-established on both maxillary and mandibular teeth (1)
28 (12) Ridges or flanges on dentary teeth: absent (0), anterolabial ridges or flanges on at least one dentary tooth (1)
29 (28) Extensive posterolingual flanges on some maxillary teeth (at least as long again as main tooth cone): absent (0), present (1)
30 (1) Lateral palatine tooth row: small (0), enlarged (1)
31 (16) Palatine tooth row: more than one tooth row (0), a single large lateral tooth row (1)
32 Pterygoid teeth: more than two tooth rows (0), two tooth rows or absent (1) (modified from Evans, 1988)
33 Palatine: tapered posteriorly (0), becomes relatively wide posteriorly (1)
34 Central region of pterygoid between three rami: short (0), elongate (1)
35 Pterygoid: enters into suborbital fenestra (0), precluded from suborbital fenestra (1)
36 (24) Jaw movement: precision-shear bite (0), propalinal (1)

Note: Primitive state is indicated by 0, derived states by 1 and 2. The character numbers in parentheses at the beginning are from Fraser and Benton (1989).

It should not be confused with the parietal width between the two supratemporal passages that incorporates the median shelves (derived from the lateral sides of the parietal table) of the fenestrae. In *Gephyrosaurus* and *Brachyrhinodon*, the parietal table is not as broad as the interorbital region, but it was often coded conversely in previous analyses by neglecting the presence of the narrow median shelf of the supratemporal fenestrae. According to the preservation of the parietal region in known genera, it is better to use the ratio of parietal width between supratemporal fenestrae/interorbital breadth as a cranial character.

Table 3.2. *Character-state data on 11 sphenodontian taxa and the Squamata (the designated out-group)*

	1	2	3		
Taxon	1234567890	1234567890	1234567890	123456	%
Squamata	00000000V0	000V000000	0000000000	000000	94
Gephyrosaurus	0001100000	0000101000	0000000001	000000	100
Diphydontosaurus	0011000000	0100111000	1100000001	000000	100
Planocephalosaurus	0011100000	0000111100	1110010101	000000	100
Polysphenodon	?110000010	110?1111??	1111111??1	110?10	81
Homoeosaurus	0110000V01	010011111?	1111111111	111010	94
Brachyrhinodon	1110000010	111111110?	1111111??1	1100?0	89
Clevosaurus	11100V0010	V111111100	1111111111	11VV10	89
Kallimodon	?1100111?1	001011110?	1111111111	1?1010	89
Palaeopleurosaurus	0010111200	0010110000	1111111111	110100	100
Sphenodon	0110011111	0111111111	1111111101	111111	100
Sapheosaurus	01?0011101	011011110?	NNNN1NNNNN	NN10??	58

Note: Characters are listed in Table 3.1. Abbreviations: 0, primitive state; 1 and 2, derived states; ?, unknown; N, not applicable; V, variable.

Table 3.3. *Main ratios of skull measurements in 14 sphenodontians*

Taxon	Length ATO/skull[a]	Length STF/orbit
1 *Gephyrosaurus*	0.342	0.63
2 *Diphydontosaurus*	0.304	0.57
3 *Planocephalosaurus*	0.344	0.52
4 *Palaeopleurosaurus*	0.343	1.35
5 *Brachyrhinodon*	0.225	0.93
6 *Polysphenodon*	—[c]	0.58
7 *Clevosaurus hudsoni*	0.303	1.00
8 *C. wangi*	0.306	0.80
9 *C. petilus*	—	1.05
10 *C. mcgilli*	0.233	1.19
11 *Homoeosaurus*[b]	0.345	0.42
12 *Kallimodon*	0.334	0.98
13 *Sapheosaurus*	0.270	1.09
14 *Sphenodon*	0.283	0.83

Note: ATO, antorbital region; STF, supratemporal fenestra.
[a] Skull length is measured from the tip of the snout to the line between the posterior margins of the squamosals.
[b] Based on *Homoeosaurus maximiliani* (Fabre, 1981).
[c] Not reported.
Sources: 1, Evans (1980); 2, Whiteside (1986); 3 and 7, Fraser (1982, 1988); 4, Carroll (1985); 5 and 6, Fraser and Benton (1989); 12 and 13, Cocude-Michel (1963).

Character 27: Breadth of marginal teeth: approximately equal to the length (0), greatly expanded medio-laterally (1). Among all known sphenodontians, only *Eilenodon* and *Toxolophosaurus* show a derived condition of this character. These two genera are not considered here, because their skulls are unknown, and hence this character is omitted in the present analysis.

Among the remaining characters, numbers 10 and 16 are derived from those used by Gauthier et al. (1988), and character 32 is modified from that used by Evans (1988). The characters that are newly added to this analysis are as follows:

Character 1: Premaxillary process of maxilla elongate (0), very weak or reduced (1). In common with squamates and other lepidosauromorphs, most sphenodontians show a maxilla in which the premaxillary process is elongate and extends forward along the ventral edge of the external naris. The reduction of the premaxillary process of the maxilla is certainly a derived pattern in the group.

Character 2: Posterior portion of maxilla gradually tapering or very narrow below the orbit (0), dorsoventrally broad (1). The posterior portion of the maxilla is elongated and is pointed posteriorly along the ventrolateral margin of the orbit in lizards and early lepidosauromorphs such as *Youngina* (Gow, 1975; Carroll, 1988b). This condition is retained in most early sphenodontians (Figure 3.18). The width and shortness of the posterior portion of the maxilla at the orbital region in other members of the Sphenodontia are considered here derived.

Character 8: Posterior edge of parietal(s) greatly

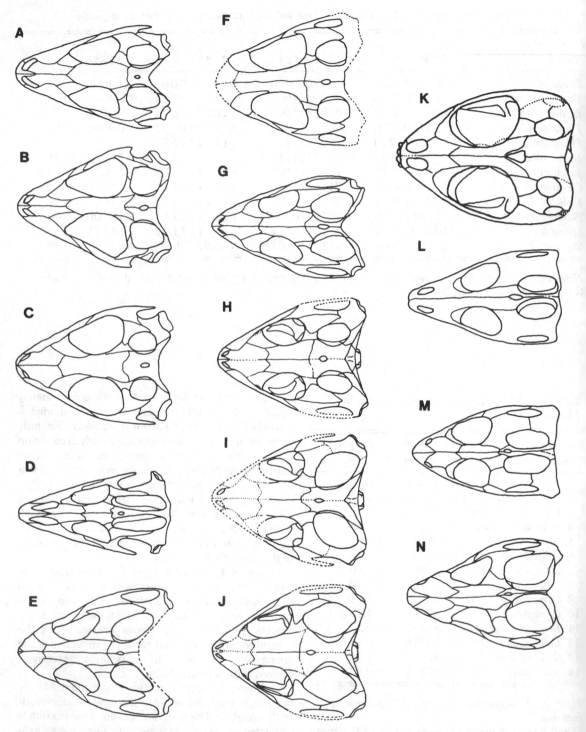

Figure 3.20. Dorsal views of the skull: (A) *Gephyrosaurus*, (B) *Diphydontosaurus*, (C) *Planocephalosaurus*, (D) *Palaeopleurosaurus*, (E) *Brachyrhinodon*, (F) *Polysphenodon*, (G) *Clevosaurus hudsoni*, (H) *C. wangi*, (I) *C. petilus*, (J) *C. mcgilli*, (K) *Homoeosaurus maximiliani*, (L) *Kallimodon*, (M) *Sapheosaurus*, (N) *Sphenodon*. (A after Evans, 1980; B after Whiteside, 1986; C after Fraser, 1982; D after Carroll, 1985; E and F after Fraser and Benton, 1989; G after Fraser, 1988; K after Fabre, 1981; L and M after Cocude-Michel, 1963.)

emarginated (0), slightly emarginated (1), convex (2). The primitive condition of the character is present in lizards, *Youngina*, and most early sphenodontians (Figure 3.20). The derived state can be seen in some Late Jurassic sphenodontians. This character is variable in *Homoeosaurus*, the derived state being present in the holotype of *H. maximiliani* (Cocude-Michel, 1963), while the primitive state is shown by a specimen of *H.* aff. *H. solnhofensis* (Fabre, 1981).

Character 9: Dorsal process of jugal broad and short (0), narrow and elongate (1). In *Youngina*, the dorsal process of the jugal is very low and forms only the ventral half of the anterior border of the inferior temporal fenestra, as is also the case in the primitive diapsid *Petrolacosaurus* (Reisz, 1981). This character shows considerable variation in lizards. The primitive state can, for example, be observed in chameleons; in iguanid lizards, the dorsal process is narrow and extends posterodorsally along the anterior and anterodorsal borders of the inferior temporal fenestra. In most early sphenodontians, the dorsal process of the jugal is broad and sheetlike and forms the anterior border of the inferior temporal fenestra (Figure 3.18). A narrow, elongate dorsal process of the jugal in certain members of the Sphenodontia and the Squamata is separately derived.

Character 14: Ratio of the length of the lower temporal fenestra to skull length (from tip of rostrum to posterior edge of parietal) less than one-fourth (0), more than one-fourth (1). In common with *Youngina* and other early lepidosauromorphs, most early sphenodontians exhibit the primitive state for this character. In lizards this character is also variable. A relatively small infratemporal fenestra is present in chameleons.

Character 18: Lower temporal bar incomplete (0), complete (1). A complete lower temporal bar is certainly the primitive character state. However, in common with all squamates and some primitive diapsid groups, certain early sphenodontians show that the lower temporal bar was incomplete posteriorly. Whiteside (1986) argued that in the common ancestor of the Squamata and the Sphenodontia the lower temporal bar may have been incomplete, and a complete lower temporal bar in later sphenodontians could have been secondarily derived. The different sutural patterns of the lower temporal bar demonstrate that the completeness of the bar is not homologous between the late sphenodontians and other diapsids. Evidence from the ontogeny of *Sphenodon* also supports this argument. Consequently, the complete lower temporal bar is considered derived within the Sphenodontia.

Character 33: Palatine tapered posteriorly (0), relatively broad (1). A posteriorly tapered palatine occurs in early lepidosauromorphs and squamates. Similarly, most early sphenodontians have a palatine that gradually tapers posteriorly. However, the great width of the posterior part of the palatine in *Petrolacosaurus* (Reisz, 1981) indicates that the derived state of this character may be secondarily derived in some sphenodontians, probably associated with the development of a propalinal jaw movement. Although some squamates (e.g., *Sceloporus*) possess a palatine that is not tapered posteriorly, this condition cannot be homologous with that seen in advanced sphenodontians.

Character 34: Central region of the pterygoid joining the three main rami short (0), elongate (1). In most sphenodontians the central region of the pterygoid is very short. This is a common feature in lizards and other lepidosauromorphs. Some lizards, such as *Elgaria* (Rieppel, 1980), also display an elongate central region of the pterygoid.

Character 35: Pterygoid forming the posteromedial margin of the suborbital fenestra (0), excluded from the suborbital fenestra (1). In common with archosauromorphs, the palatine and the ectopterygoid in lepidosauromorphs are separated by the pterygoid along the margin of the suborbital fenestra. The exclusion of the pterygoid from the margin of the suborbital fenestra within certain members of the Sphenodontia (Figure 3.19) should therefore be considered to be derived.

About twenty sphenodontian genera have so far been described. However, only those eleven with more than 58 percent of the total character set have been considered in my analysis (Table 3.2). With respect to *Homoeosaurus*, the data are based primarily on *H. maximiliani* (Cocude-Michel, 1963; Fabre, 1981).

The analysis was carried out using the branch-and-bound option of PAUP, version 3.0 n. The multistate character 9 was entered as unordered. This program generated only one most parsimonious tree, which shows a broad sequence of sister-group relationships: from the Squamata, through *Gephyrosaurus*, *Diphydontosaurus*, *Planocephalosaurus*, *Palaeopleurosaurus*, the subgroup clevosaurs (Gauthier et al., 1988), *Homoeosaurus*, *Kallimodon*, and *Sapheosaurus*, to *Sphenodon* (Figure 3.21). This cladogram resembles other recent cladistic analyses in that *Gephyrosaurus*, *Diphydontosaurus*, and *Planocephalosaurus* form successively closer sister-groups to the more advanced sphenodontian taxa (Whiteside, 1986; Evans, 1988; Fraser and Benton, 1989). The primary differences between this and other analyses concern the relationships of *Homoeosaurus* and *Palaeopleurosaurus* to other sphenodontians.

Unlike previous cladistic analyses (Fraser, 1986; Evans, 1988; Fraser and Benton, 1989), the present data indicate that *Homoeosaurus* is more derived than *Palaeopleurosaurus*. Two unequivocal characters – the parietal foramen reaching or crossing the line between the anterior margins of the supratemporal fenestrae (character 10), and the relatively widened posterior part of the palatine (character 33) – link *Homoeosaurus* with *Kallimodon*, *Sapheosaurus*, and *Sphenodon* as a monophyletic grouping (Figure 3.21). Two additional

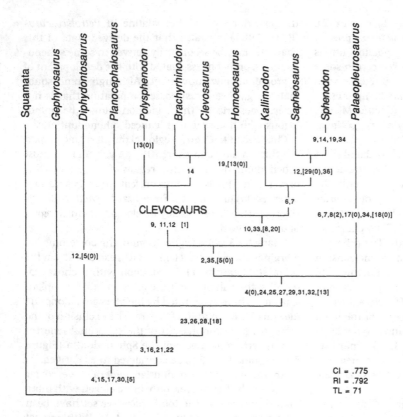

Figure 3.21. A most parsimonious tree yielded by the branch-and-bound algorithm of PAUP (version 3.0n) based on 36 characters in Table 3.1, depicting phylogenetic inter-relationships of the Sphenodontia. Ambiguous characters are in square brackets. CI, consistency index; RI, retention index; TL, tree length.

features – the posterior edge of the parietal(s) slightly emarginated (character 8), and reduction of the quad-ratojugal-quadrate conch (character 20) – have also been highlighted by PAUP as synapomorphies of the subgroup, but they are ambiguous because of variation in *Homoeosaurus* (character 8) or uncertainty (character 20) in certain members of the grouping. Furthermore, these four genera are united with the clevosaurs in sharing two unequivocal characters, a dorsoventrally broad posterior portion of the maxilla (character 2), and exclusion of the pterygoid from the sub-orbital fenestra (character 35), and one equivocal character, the separate parietals (the reversal of cha-racter 5).

As shown in Figure 3.21, the relationships among *Homoeosaurus*, *Kallimodon*, *Sapheosaurus*, and *Sphenodon* have also been established in this analysis. The last three genera form a monophyletic group. They share two unequivocal characters: narrow parietal width (character 6) and crested parietal table (character 7). Character 12 (ratio antorbital region/skull length less than one-third) unequivocally links *Sapheosaurus* and *Sphenodon* as a less inclusive subgroup) (Fig. 3.21) within the Sphenodontia. Equivocal characters of this clade include the absence of the extensive postero-lingual flanges on some maxillary teeth (the reversal of character 29) and development of propalinal jaw movement (character 36).

Characters 4, 24, 25, 27, 29, 31, and 32 show that *Clevosaurus* is more derived than *Gephyrosaurus*, *Diphy-dontosaurus*, and *Planocephalosaurus*. According to Fraser and Benton (1989), *Clevosaurus* is even more derived than *Brachyrhinodon* and *Polysphenodon* in having a parietal table that is narrower than the interorbital width. However, the relationships of *Clevosaurus* with the other derived sphenodontians could not be clarified in their analysis. In the present analysis, *Clevosaurus* has been associated with *Brachyrhinodon* and *Poly-sphenodon* as a monophyletic grouping that can be diagnosed by two unequivocal characters: narrow and elongate dorsal process of the jugal (character 9) and the ratio antorbital region/skull length less than one-third (character 12). The very short or reduced pre-maxillary process of the maxilla (character 1) and the ratio antorbital region/skull length less than one-fourth (character 11) are two additional synapo-morphies postulated for this group. However, they are ambiguous because of the uncertainty regarding char-acter 1 in *Polysphenodon* and the variation of character 11 in *Clevosaurus*. On the basis of character 14, *Clevosaurus* and *Brachyrhinodon* are considered more closely related to one another than either is to *Poly-sphenodon*. For convenience, this grouping is here informally referred to as the clevosaurs. This usage contrasts with that of Gauthier et al. (1988), in which only *Clevosaurus* and *Planocephalosaurus* are included.

However, most recent cladistic analyses (including this one) have questioned such a close relationship between these two genera and have demonstrated that *Planocephalosaurus* is more primitive than *Clevosaurus* (Whiteside, 1986; Evans, 1988; Fraser and Benton, 1989).

Acknowledgments

I thank Dr. R. L. Carroll (Department of Biology, McGill University, Montreal), under whose guidance this project was undertaken. I am indebted to Drs. R. L. Carroll and R. Holmes (Redpath Museum, McGill University) for their critical reading and editing of early drafts of this manuscript and for improving it considerably. The field work was assisted by Mrs. Z. J. Wang, H. Wang (The Culture Museum of Lufeng County, Yunnan, China), and F. Yi (Dawa Elementary School, Lufeng, China).

I also thank Drs. R.L. Carroll, R. Holmes, N.C. Fraser (Virginia Museum of Natural History), and H.-D. Sues (Royal Ontario Museum) for their comments and invaluable suggestions. I have particularly benefited from sharing unpublished information with Drs. H.-D. Sues and N. C. Fraser. They reviewed the final draft of this manuscript. My thanks are also due to Mr. M. R. Lamarche (Department of Biology, McGill University) for photographic work. This research was supported by grants from the Natural Sciences and Engineering Research Council of Canada to Dr. R. L. Carroll.

References

Baird, I. L. 1970. The anatomy of the reptilian ear. Pp. 193–275 in C. Gans and T. S. Parsons (eds.), *Biology of the Reptilia Vol. 2.* New York: Academic Press.

Benton, M. J. 1983. The Triassic reptile *Hyperodapedon* from Elgin: functional morphology and relationships. *Philosophical Transactions of the Royal Society of London,* B302: 605–720.

—— 1984. The relationships and early evolution of the Diapsida. Pp. 573–596 in M.W. J. Ferguson (ed.), *The Structure, Development and Evolution of Reptiles.* London: Academic Press.

—— 1985. Classification and phylogeny of the diapsid reptiles. *Zoological Journal of the Linnean Society* 84: 97–164.

Bien, M. N. 1941. "Red Beds" of Yunnan. *Bulletin of the Geological Society of China* 21: 159–168.

Bramble, D. M. 1978. Origin of the mammalian feeding complex; models and mechanisms. *Paleobiology* 4: 271–301.

Brinkman, D. 1981. The origin of the crocodiloid tarsi and the interrelationships of thecodontian reptiles. *Breviora* 464: 1–23.

Carroll, R. L. 1976. The oldest known rhynchosaur. *Annals of the South African Museum* 72: 37–57.

—— 1977. The origin of lizards. Pp. 359–396 in S. M. Andrews, R. S. Miles, and A. D. Walker (eds.), *Problems in Vertebrate Evolution.* London: Academic Press.

—— 1985. A pleurosaur from the Lower Jurassic and the taxonomic position of the Sphenodontida. *Palaeontographica* A189: 1–28.

—— 1988a. *Vertebrate Paleontology and Evolution.* San Francisco: Freeman.

—— 1988b. Late Palaeozoic and early Mesozoic lepidosauromorphs and their relation to lizard ancestry. Pp. 99–118 in R. Estes and G. Pregill (eds.), *Phylogenetic Relationships of the Lizard Families.* Stanford University Press.

Chatterjee, S. 1974. A rhynchosaur from the Upper Triassic Maleri Formation of India. *Philosophical Transactions of the Royal Society of London* B267: 209–260.

Cocude-Michel, M. 1963. Les rhynchocéphales et les sauriens des calcaires lithographiques (Jurassique Supérieur) d'Europe occidentale. *Nouvelles Archives du Muséum d'Histoire Naturelle, Lyon* 7: 1–187.

Colbert, E. H. 1966. A gliding reptile from the Triassic of New Jersey. *American Museum Novitates* 2230: 1–23.

—— 1970. The Triassic gliding reptile *Icarosaurus. Bulletin of the American Museum of Natural History* 143: 89–142.

Cong, L. Y., L. H. Hou, and X. C. Wu. 1984. [Age variation in the skull of *Alligator sinensis* Fauvel in topographic anatomy.] *Acta Herpetologica Sinica* 3(2): 1–14. [in Chinese, with English abstract]

Daugherty, C. H., A. Cree, J. M. Hay, and M. B. Thompson. 1990. Neglected taxonomy and continuing extinctions of tuatara (*Sphenodon*). *Nature* 347: 177–179.

Dawbin, W. H. 1982. The tuatara *Sphenodon punctatus* (Reptilia; Rhynchocephalia), a review. Pp. 149–182 in D. G. Newman (ed.), *New Zealand Herpetology.* Wellington: Victoria University.

De Beer, G. R. 1985. *The Development of the Vertebrate Skull.* University of Chicago Press.

Eldredge, N., and J. Cracraft. 1980. *Phylogenetic Patterns and the Evolutionary Process.* New York: Columbia University Press.

Estes, R. 1983. *Sauria terrestria, Amphisbaenia.* In *Handbuch der Paläoherpetologie, Part 10A.* Stuttgart: Gustav Fischer Verlag.

Evans, S. E. 1980. The skull of a new eosuchian reptile from the Lower Jurassic of South Wales. *Zoological Journal of the Linnean Society* 70: 203–264.

—— 1981. The postcranial skeleton of the Lower Jurassic eosuchian *Gephyrosaurus bridensis. Zoological Journal of the Linnean Society* 73: 81–116.

—— 1984. The classification of the Lepidosauria. *Zoological Journal of the Linnean Society* 82: 87–100.

—— 1986. The braincase of *Prolacerta broomi* (Reptilia: Triassic). *Neues Jahrbuch für Geologie und Paläontologie, Abhandlungen* 173: 181–200.

—— 1987. The braincase of *Youngina capensis* (Reptilia: Diapsida; Permian). *Neues Jahrbuch für Geologie und Paläontologie, Monatshefte* 1987: 193–203.

—— 1988. The early history and relationships of the Diapsida. Pp. 221–260 in M. J. Benton (ed.), *The Phylogeny and Classification of the Tetrapods. Vol. 1: Amphibians, Reptiles, Birds.* Oxford: Clarendon Press.

Fabre, J. 1981. *Les rhynchocéphales et les ptérosauriens à crête pariétale du Kiméridgien supérieur-Berriasien d'Europe occidentale. Le gisement de Canjuers (Var-France) et ses abords.* Paris: Editions de la Fondation Singer-Polignac.

Fraser, N. C. 1982. A new rhynchocephalian from the British Upper Trias. *Palaeontology* 25: 709–725.

1985. Vertebrate faunas from Mesozoic fissure deposits of South West Britain. *Modern Geology* 9: 273–300.

1986. New Triassic sphenodontids from south-west England and a review of their classification. *Palaeontology* 29: 165–186.

1988. The osteology and relationships of *Clevosaurus* (Reptilia: Sphenodontida). *Philosophical Transactions of the Royal Society of London* B321: 125–178.

Fraser, N. C., and M. J. Benton. 1989. The Triassic reptiles *Brachyrhinodon* and *Polysphenodon* and the relationships of the sphenodontids. *Zoological Journal of the Linnean Society* 96: 413–445.

Fraser, N. C., and G. M. Walkden. 1984. The postcranial skeleton of *Planocephalosaurus robinsonae*. *Palaeontology* 27: 575–595.

Gaffney, E. S. 1980. Phylogenetic relationships of the major groups of amniotes. Pp. 593–610 in A. L. Panchen (ed.), *The Terrestrial Environment and the Origin of Land Vertebrates*. London: Academic Press.

Gans, C. 1983. Is *Sphenodon punctatus* a maladapted relict? Pp. 613–620 in A. J. G. Rhodin and K. Miyata (eds.), *Advances in Herpetology and Evolutionary Biology*. Cambridge, MA: Museum of Comparative Zoology.

Gans, C., and E. G. Wever. 1976. Ear and hearing in *Sphenodon punctatus*. *Proceedings of the National Academy of Sciences USA* 78: 4244–4246.

Gauthier, J. A. 1984. A cladistic analysis of the higher systematic categories of the Diapsida. Ph.D. thesis, Department of Paleontology, University of California, Berkeley. (University Microfilms #85-12825, Ann Arbor, Michigan.)

Gauthier, J. A., R. Estes, and K. de Queiroz. 1988. A phylogenetic analysis of Lepidosauromorpha. Pp. 15–98 in R. Estes and G. Pregill (eds.), *Phylogenetic Relationships of the Lizard Families*. Stanford University Press.

Goin, C. J., O. B. Goin, and G. R. Zug. 1978. *Introduction to Herpetology*. San Francisco: Freeman.

Gorniak, G. C., H. I. Rosenberg, and C. Gans. 1982. Mastication in the tuatara *Sphenodon punctatus* (Reptilia: Rhynchocephalia): structure and activity of the motor system. *Journal of Morphology* 171: 321–353.

Gow, C. E. 1975. The morphology and relationships of *Youngina capensis* Broom and *Prolacerta broomi* Parrington. *Palaeontologia Africana* 18: 89–131.

Gow, C. E., and M. A. Raath. 1977. Fossil vertebrate studies in Rhodesia: Sphenodontid remains from the Upper Triassic of Rhodesia. *Palaeontologia Africana* 20: 121–122.

Günther, A. 1867. Contribution to the anatomy of *Hatteria* (*Rhynchocephalus*, Owen). *Philosophical Transactions of the Royal Society of London* 157: 595–629.

Haas, G. 1973. Muscles of the jaws and associated structures in the Rhynchocephalia and Squamata. Pp. 285–483 in C. Gans (ed.), *Biology of the Reptilia, Vol. 4*. London: Academic Press.

Hennig, W. 1966. *Phylogenetic Systematics*. Urbana: University of Illinois Press.

Hoffstetter, R. 1955. Rhynchocephalia. Pp. 556–576 in J. Piveteau (ed.), *Traité de Paléontologie, Vol. 5*. Paris: Masson et Cie.

1964. Les Sauria du Jurassique supérieur et spécialement les Gekkota de Baviere et de Mandchourie. *Senckenbergiana Biologica* 45: 281–324.

Huene, F. von. 1910. Ueber einen echten Rhynchocephalen aus der Trias von Elgin, *Brachyrhinodon taylori. Neues Jahrbuch für Mineralogie, Geologie und Paläontologie* 1910: 29–62.

1946. Die grossen Stämme der Tetrapoden in den geologischen Zeiten. *Biologisches Zentralblatt* 65: 266–275.

Kuhn, O. 1969. Proganosauria-Protorosauria. In O. Kuhn (ed.), *Handbuch der Paläoherpetologie, Part 9*. Stuttgart: Gustav Fischer Verlag.

Lin, J., M. Fuller, and W. Zhang. 1985. Preliminary Phanerozoic polar wander paths for the North and South China Blocks. *Nature* 312: 444–449.

Metcalfe, I. 1987. Origin and assembly of South-East Asian continental terranes. Pp. 101–118 in M. G. Audley-Charles and A. Hallam (eds.), *Gondwana and Tethys*. London: Geological Society Special Publication 37.

Murry, P. A. 1987. New reptiles from the Upper Triassic Chinle Formation of Arizona. *Journal of Paleontology* 61: 773–786.

Newman, D. G. 1982. Tuatara, *Sphenodon punctatus*, and burrows, Stephens Island. Pp. 213–224 in D. G. Newman (ed.), *New Zealand Herpetology*. Wellington: Victoria University.

Oelrich, T. M. 1956. The anatomy of the head of *Ctenosaura pectinata* (Iguanidae). *Miscellaneous Publications, Museum of Zoology, University of Michigan* 94: 1–122.

Olsen, P. E. 1980. A comparison of vertebrate assemblages from the Newark and Hartford Basins (early Mesozoic, Newark Supergroup) of eastern North America. Pp. 35–53 in L. L. Jacobs (ed.), *Aspects of Vertebrate History*. Flagstaff: Museum of Northern Arizona Press.

Owen, R. 1845. Description of certain fossil crania discovered by A. G. Bain. Esq., in the southeast extremity of Africa, referable to different species of an extinct genus of Reptilia (*Dicynodon*), and indicative of a new tribe or suborder of Sauria. *Transactions of the Geological Society of London* (2) 7: 59–84.

Patterson, C. 1982. Morphological characters and homology. Pp. 21–74 in K.A. Joysey and E. Friday (eds.), *Problems of Phylogenetic Reconstruction*. London: Academic Press.

Reisz, R. 1981. A diapsid reptile from the Pennsylvanian of Kansas. *Special Publication of the Museum of Natural History, University of Kansas*, 7: 1–74.

Rich, T. H. V., R. E. Molnar, and P.V. Rich. 1983. Fossil vertebrates from the Late Jurassic or Early Cretaceous Kirkwood Formation, Algoa Basin, southern Africa. *Transactions of the Geological Society of South Africa* 86: 281–291.

Rieppel, O. 1980. The perilymphatic system of the skull of *Typhlops* and *Acrochordus*, with comments on the origin of snakes. *Journal of Herpetology* 14: 105–108.

1985. The recessus scalae tympani and its bearing on the classification of reptiles. *Journal of Herpetology* 19: 373–384.

Rieppel, O., and R.W. Gronowski. 1981. The loss of the

lower temporal arcade in diapsid reptiles. *Zoological Journal of the Linnean Society* 72: 203–217.

Robinson. P. L. 1962. Gliding lizards from the Upper Keuper of Great Britain. *Proceedings of the Geological Society of London* 106: 137–146.

1973. A problematic reptile from the British Upper Trias. *Journal of the Geological Society of London* 129: 457–479.

1976. How *Sphenodon* and *Uromastyx* grow their teeth and use them. Pp. 43–64 in A. d'A. Bellairs and C. B. Cox (eds.), *Morphology and Biology of Reptiles*. London: Academic Press.

Romer, A. S. 1966. *Vertebrate Paleontology*, 3rd ed. University of Chicago Press.

Sengör, A. M. C., D. Altiner, A. Cin, T. Ustaomer, and K. J. Hsü. 1987. Origin and assembly of the Tethyside orogenic collage at the expense of Gondwana Land. Pp. 119–181 in M. G. Audley-Charles and A. Hallam (eds.), *Gondwana and Tethys*. London: Geological Society Special Publication 37.

Simmons, D. J. 1965. The non-therapsid reptiles of the Lufeng Basin, Yunnan, China. *Fieldiana, Geology* 15: 1–93.

Sun, A., G. Cui, Y. Li, and X. Wu. 1985. [A verified list of Lufeng Saurischian Fauna.] *Vertebrata Palasiatica* 23 (1): 1–12. [in Chinese]

Swinton, W. E. 1939. A new Triassic rhynchocephalian from Gloucestershire. *Annals and Magazine of Natural History* (11) 4: 591–594.

Throckmorton, G. S., J. A. Hopson, and P. Parks. 1981. A redescription of *Toxolophosaurus cloudi* Olson, a Lower Cretaceous herbivorous sphenodontid reptile. *Journal of Paleontology* 55: 586–597.

Thulborn, R.A. 1980. The ankle joint of archosaurs. *Alcheringa* 4: 241–261.

Toerien, M. J. 1962. The sound-conducting systems of lizards without tympanic membranes. *Evolution* 17: 540–547.

Tumarkin, A. 1955. On the evolution of the auditory conducting apparatus: a new theory based on functional considerations. *Evolution* 9: 221–243.

Walls, G. Y. 1982. Provisional results from a study of the feeding ecology of the tuatara (*Sphenodon punctatus*) on Stephens Island. Pp. 271–276 in D. G. Newman (ed.), *New Zealand Herpetology*. Wellington: Victoria University.

Wang, H. Z., and B. P. Liu. 1980. [History of the Earth.] Beijing: Geology Press. [in Chinese]

Wever, E. G. 1978. *The Reptile Ear*. Princeton University Press.

Whiteside, D.I. 1986. The head skeleton of the Rhaetian sphenodontid *Diphydontosaurus avonis* gen. et sp. nov. and the modernizing of a living fossil. *Philosophical Transactions of the Royal Society of London* B312: 379–430.

Wiley, E. O. 1981. *Phylogenetics: The Theory and Practice of Phylogenetic Systematics*. New York: Wiley.

Williston, S. W. 1925. *The Osteology of Reptiles*. Cambridge, MA: Harvard University Press.

Wu, X. 1986. [A new species of *Dibothrosuchus* from Lufeng Basin.] *Vertebrata Palasiatica* 24: 43–62. [in Chinese, with English summary]

Young, C. C. 1948. A review of Lepidosauria from China. *American Journal of Science* 246: 711–719.

1951. The Lufeng saurischian fauna in China. *Palaeontologia Sinica*, N. S., C 13: 1–96.

1982. [A new fossil reptile from the Lufeng Basin, Yunnan, China]. Pp. 36–37 in [*Selected Works of C. C. Young.*] Beijing: Science Press. [in Chinese]

4

Marine members of the Sphenodontia

ROBERT L. CARROLL AND RUPERT WILD

Introduction

The extant genus *Sphenodon*, the tuatara, now restricted to small islands off the coast of New Zealand, is frequently cited as an example of a "living fossil." The presence of complete lateral and ventral temporal bars, the absence of an impedance-matching middle ear, and the generally primitive nature of the postcranial skeleton led to the idea that *Sphenodon* retained a level of evolution equivalent to that of the Permian and Triassic ancestors of squamates, crocodiles, and dinosaurs (Romer, 1956).

Recent work by Whiteside (1986), Fraser (1988), and Wu (1991) has demonstrated that several important cranial features of *Sphenodon* evolved within the Sphenodontidae and that their Late Triassic and Early Jurassic precursors more closely resembled primitive squamates. Several early genera show structures of the middle ear that indicate an impedance-matching function similar to that of modern lizards. Many early members of the Sphenodontia have an incomplete lower temporal bar, and this condition may have been primitive for the group. The common possession of an impedance-matching middle ear apparatus, an incomplete lower temporal bar, and separate bony epiphyses in early sphenodontids and lizards suggests a sister-group relationship between the Squamata and Sphenodontia.

No certainly identified members of the Sphenodontia are known from earlier than the Late Triassic (Carnian or early Norian) (Fraser and Benton, 1989). Putative members of the Squamata (Carroll, 1977) and more primitive lepidosauromorphs are known from the Late Permian and Early Triassic, but the details of the relationships of the immediate ancestors of squamates and sphenodontians remain unknown.

Benton (1985), Gauthier, Estes, and de Queiroz (1988), and Carroll and Currie (1991) have provided extensive evidence that the lepidosaurs belong to a larger assemblage, the Lepidosauromorpha, including eosuchians (younginiforms) and probably the marine nothosaurs and plesiosaurs, which compose the sister-group of a second major diapsid assemblage, the Archosauromorpha. The common ancestry of diapsids is to be found among Pennsylvanian and Permian genera such as *Petrolacosaurus* (Reisz, 1981) and *Apsisaurus* (Laurin, 1991).

Members of the Sphenodontia are certainly not remnants of the ultimate ancestors of diapsids, and several significant cranial changes have occurred between the Triassic sphenodontians and their living descendants. On the other hand, most features of the skeleton have remained very conservative from the Late Triassic to the present.

Authors such as Gould (1982) and Eldredge and Stanley (1984) have argued that stasis is an extremely important factor in evolution that cannot be explained within the framework of Darwinian selection theory. Rather, they argue that factors within the genome and all-pervasive developmental processes constrain evolution to very narrow limits that are broken only by unusual circumstances. This seems to be the case with terrestrial sphenodontians, which have persisted for more than 200 million years. During that time, mammals appeared and radiated, angiosperms evolved and came to dominate the vegetation of the globe, and dinosaurs evolved to a position of dominance and then became extinct. Through all of that, sphenodontians spread throughout the world and gave rise to a score or more species, but most have retained nearly all the skeletal characters of their Triassic ancestors.

This is not the entire picture of sphenodontian evolution, however. In the early history of this group, there was one lineage that broke the mold, escaped the genetic and developmental confines of the ancestral stock, entered a new environment, and became sufficiently modified in body form and details of the skeleton that their taxonomic position has long been subject to dispute: the pleurosaurs. They were marine reptiles, known primarily from the Late Jurassic, with very elongate bodies and diminutive limbs. Pleurosaurs were recognized in the early nineteenth century. Münster (1839) compared them to chameleons. Meyer (1831) provided the first description, emphasizing their distinction from all extant reptiles. Fitzinger (1843) placed them in a distinct order, Pleurosauria, but Wagner (1860) again compared them to lizards. With the recognition of the Rhynchocephalia (Günther, 1867) (based originally on *Sphenodon*, but quickly extended to include the Triassic rhynchosaurs), pleurosaurs were included in that assemblage. Both Lydekker (1880) and Zittel (1887–90) allied *Pleurosaurus* with the sphenodontians, but Boulenger (1893) and Watson (1914) cited the reduction of the lower temporal bar to suggest affinities with lizards. Broili (1926) suggested recognition of a distinct order, but placed pleurosaurs closer to rhynchocephalians than to squamates. Huene (1952) considered them to represent a subfamily close to the sphenodontian *Homoeosaurus*. Hoffstetter (1955) resurrected Fitzinger's name Pleurosauria and argued for ordinal status, but with descent from the Rhynchocephalia.

Romer (1956) placed them in a distinct family within the Order Rhynchocephalia, along with the Sapheosauridae (clearly a member of the Sphenodontia) and the Claraziidae, whose affinities are now thought to lie with the thalattosaurs (themselves of uncertain relationships) (Rieppel, 1987). At that time, Rhynchocephalia was considered to embrace both sphenodontians and rhynchosaurs, which are now considered to belong to two widely divergent diapsid lineages (Benton, 1985). Romer's emphasis on the common presence of an acrodont dentition implies affinities with sphenodontians to the exclusion of rhynchosaurs. Both Cocude-Michel (1963) and Fabre (1981) allied the Pleurosauridae with the Sphenodontidae. On the other hand, Fraser and Benton (1989) questioned the affinities of these groups, at least indirectly, in excluding *Pleurosaurus* from their consideration of relationships among sphenodontians.

It is clear from this controversy that Late Jurassic pleurosaurs are sufficiently distinct from terrestrial sphenodontians to demonstrate a major evolutionary change between the two groups, if in fact they are related.

Strong evidence for a relationship between primitive terrestrial sphenodontians and the Late Jurassic pleurosaurs was provided by the discovery of an elongate sphenodontian from the Lower Jurassic (Liassic) of Holzmaden, Germany (Carroll, 1985). The taxon, *Palaeopleurosaurus posidoniae*, seems to demonstrate an intermediate morphology between Late Triassic and Early Jurassic terrestrial sphenodontians and the Late Jurassic pleurosaurs. Fraser and Benton (1989) included *Palaeopleurosaurus* in their analysis of relationships within the Sphenodontia, but did not believe that there was sufficient information regarding the Late Jurassic pleurosaurs to include them in their study.

This chapter will concentrate on two further souces of evidence that these groups are closely related. One is the description of a new specimen of *Palaeopleurosaurus*. The second is a redescription of the skull of the Late Jurassic *Pleurosaurus*.

Palaeopleurosaurus

The cranial anatomy revealed by previously described material for *Palaeopleurosaurus posidoniae* is clearly comparable to that of terrestrial sphenodontians, demonstrating all the synapomorphies cited by Evans (1988), Fraser and Benton (1989), and Carroll and Currie (1991). Fraser and Benton had no difficulty in allying it with previously recognized sphenodontian genera, placing it in most of their cladograms close to the present-day *Sphenodon*.

The skull differs from those of primitive terrestrial sphenodontians only in its elongation of the antorbital and temporal regions and the nasal opening, a further reduction of the lower temporal bar, loss of the pterygoid teeth, and narrowing of the skull table. In the latter two features it resembles both the Late Jurassic sphenodontian *Kallimodon* and the extant genus *Sphenodon*.

The postcranial skeleton, in contrast, appears superficially very different from that of any of the terrestrial genera: There are thirty-seven presacral vertebrae. The neural spines are square, rather than triangular, and occupy nearly the entire length of the segment. The limbs are much shorter relative to the total length of the column, and as measured relative to the length of individual vertebrae. Their ossification is reduced, but they retain most anatomical features of their terrestrial ancestors. The girdles are relatively smaller, and the scapula and coracoid are ossified as separate elements. The pelvis, in contrast, retains and even amplifies the distinctively sphenodontian long posterior ischiadic process.

The material described earlier demonstrates unequivocally the evolution of a distinct lineage of elongate early sphenodontians. Unfortunately, both of the previously described skeletons were prepared by methods that destroyed much of the surface detail. Fortunately, an additional specimen has recently been discovered in southern Germany (Figures 4.1 and 4.2). Only limited preparation had been undertaken before the specimen was loaned to the Staatliches Museum für Naturkunde, Stuttgart. Further work by Pamela Gaskill has exposed superbly preserved details of the trunk, girdles, and limbs. This information further documents the anatomy of this particular level of

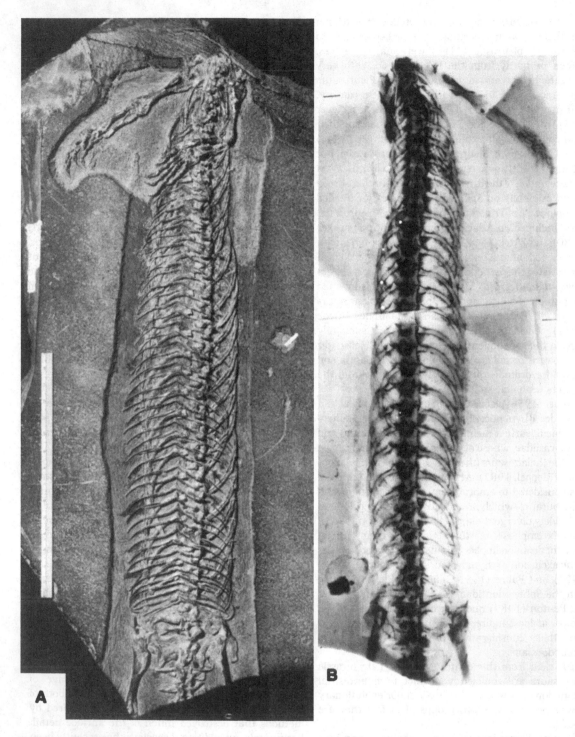

Figure 4.1. Newly discovered specimen of *Palaeopleurosaurus* from Kerkhofen, Bavaria, southern Germany: (A) skeleton as prepared; (B) x-ray taken from opposite surface.

aquatic evolution and demonstrates a number of additional features in which this genus is intermediate between Late Triassic terrestrial members of the Sphenodontia and the Late Jurassic pleurosaurs.

The skeleton was collected by Jürgen Schüssel during construction of the Rhein-Main-Donau canal near the village of Kerkhofen near Freyung in the Oberpfalz, Bavaria. It was found about 1.2 km south-southeast of Kerkhofen, approximately 300 m east of the bridge on the road between Kerkhofen and Erasbach (topographic map 1:50,000 Sheet L 3934 Beilngries, coordinates R = 4456450, H = 5447400). The stratigraphic level of the new find of *Palaeopleurosaurus* lies in the lower part of the zone of *Dactylioceras tenuicostatum* (Young and Bird), probably in the subzone of *Dactylioceras semicelatum* (Simpson) (Figure 4.3). It corresponds to the stratigraphic occurrence of the holotype and paratype specimens (Riegraf, Werner, and Lörcher, 1984, p. 31; Carroll, 1985).

The new specimen is considerably larger than the previously described material. The greatest comparable dimension is that between the glenoid and the acetabulum, which is 335 mm in the new specimen, and 263 mm in the holotype. The humerus is 38.5 mm long, compared with 26 mm, and the femur is 52 mm long, compared with 35 mm. The skull and cervical vertebrae are missing, as is the anterior margin of the shoulder girdle. Thirty presacral vertebrae are present. Other specimens of *Palaeopleurosaurus* have a total of 37 presacrals, so approximately seven cervicals have been lost.

Much more detail of the trunk vertebrae is visible than in the previously described specimens. The most conspicuous feature is the retention of a clearly defined suture between the neural arch and centrum (Figure 4.4). This is a common feature in secondarily aquatic reptiles, including nothosaurs, plesiosaurs, and ichthyosaurs. In contrast, even very small specimens of early terrestrial sphenodontians generally show complete fusion of the arch and centrum. In *Palaeopleurosaurus*, the suture begins anteriorly from a triangular area of unfinished bone at the margin of the centrum. It passes posteriorly beneath the transverse process, extends sharply dorsally behind the process, and then descends at a small angle to the end of the centrum, beneath a long thin extension of the pedicel that runs to the posterior extremity of the centrum. In *Pleurosaurus*, the suture runs through the transverse process, separating a large dorsal portion from a small oval ventral area. It is difficult to attribute a functional reason for the latter pattern, because it would have been impossible for any movement to have occurred along this suture in the living animal without disruption of the head of the rib, which would have been contiguous with the entire articulating surface of the transverse process. During development in squamates, the suture between the arch and centrum also runs

through the transverse process (O. Rieppel, pers. commun.).

The presence of a persistent suture between the arch and centrum provides additional evidence that *Palaeopleurosaurus* was aquatic and was similar to the Late Jurassic pleurosaurs. As in terrestrial sphenodontians and *Pleurosaurus*, large intercentra are retained throughout the trunk region. They were not recognized in the previously described material of *Palaeopleurosaurus*.

The sacral vertebrae are somewhat obscured by elements of the pelvic girdle, but reveal their articulating surfaces for the sacral ribs. The more anterior resembles the surface of the transverse process of the trunk vertebrae in angling down toward the tip of the intercentrum, but is somewhat larger. The second is longer from back to front and is more or less triangular, with the base nearly horizontal. The suture between the arch and centrum runs through the middle of the articulating surface. Two caudal vertebrae are preserved. They differ from the trunk vertebrae in having relatively narrower neural spines. The first two caudal ribs are suturally attached, rather than fused as in terrestrial sphenodontians. The short transverse processes are horizontal and bilobed, with distinct but confluent surfaces for attachment of two rib heads. The surfaces are divided horizontally by the suture between the arch and centrum. Intercentra, resembling those in the trunk region, accompany the first two caudals. Behind the second is the cross-piece of the first haemal arch. This is also the position of the first haemal arch in primitive terrestrial sphenodontians (Fraser, 1988, fig. 26).

The sacral ribs resemble those of other pleurosaurs in being suturally attached, rather than fused to the vertebrae. (This is also a feature of other aquatic reptiles, such as choristoderes.) The first has a bilobate articulating surface and a short, paddle-shaped blade. The second has a more massive proximal articulating surface and is divided distally into a large anterior plate for attachment to the ilium and a smaller process angling posteriorly. It should be noted that the first caudal rib was incorrectly identified as the second sacral in the original description of *Palaeopleurosaurus* (Carroll, 1985, fig. 8). The caudal ribs have distinctly bilobate articulating surfaces. The distal ends are bluntly pointed. In primitive terrestrial sphenodontians the caudal ribs are fused to the vertebrae without trace of suture (Fraser, 1988). They remain articulated in the holotype of *P. posidoniae* and were said to be fused, but are disarticulated in this specimen.

The gastralia are superbly displayed in this specimen. There are 50 units, spanning 25 vertebrae from presacral 12 to 36. Forty-seven were recognized in the slightly disarticulated material of the type specimen. Each has a large medial chevron and a more slender pair of lateral bones. One extra lateral element is visible on the left side, in association with vertebra 34. The

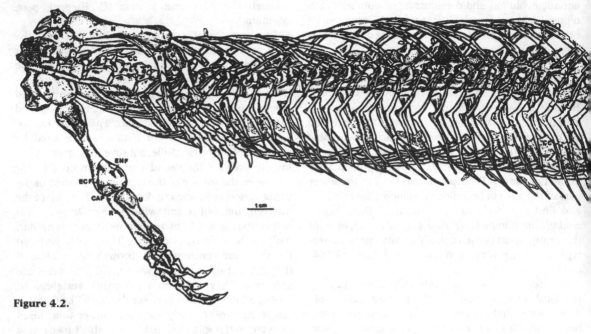

Figure 4.2.

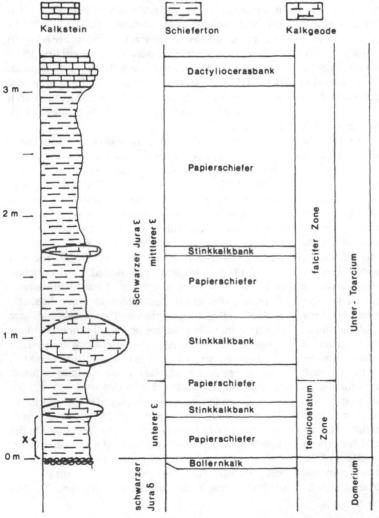

Figure 4.3. Section of the lower Toarcian Posidonienschiefer exposed during excavation of the Rhein-Main-Donau canal south of Kerkhofen, Oberpfalz. Based on a section drawn by Jürgen Schüssel (Nürnberg, Germany), who discovered the specimen; × marks the stratigraphic level of the new specimen of *Palaeopleurosaurus*.

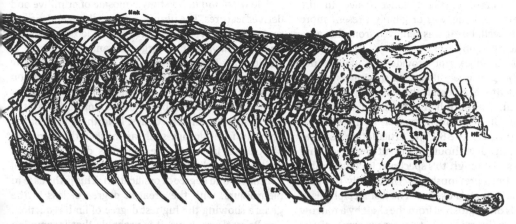

Figure 4.2. *Palaeopleurosaurus*, drawing of skeleton. Abbreviations: A and B, calcified cartilage thought to link ends of ribs 35 and 36 to gastralia; CAP, capitellum; CC, calcified cartilage; COF, coracoid foramen; COR, coracoid; CR, caudal rib; ECF, ectepicondylar foramen; ENF, entepicondylar foramen; EX, extra lateral gastralia; F, femur; H, humerus; HE, haemal arch; i, intermedium; IC, interclavicle; ic, intercentrum; IL, ilium; IS, ischium; IT, internal trochanter; l, lateral centrale; link, calcified cartilage linking rib and gastralia; m, medial centrale; P, pubis; PP, posterior process of ischium; R, radius; SC, scapula; SR_{i-ii}, first and second sacral ribs; U, ulna; ul, ulnare; i–v, metacarpals.

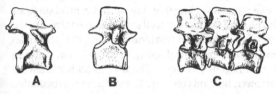

Figure 4.4. Vertebrae of members of the Sphenodontia. (A) Isolated trunk vertebra of the Late Triassic terrestrial sphenodontian *Planocephalosaurus*, redrawn from Fraser and Walkden (1984). (B) Reconstruction of a single trunk vertebra of the new specimen of *Palaeopleurosaurus*. (C) Posterior trunk vertebrae of *Pleurosaurus* (Bayerische Staatssammlung für Paläontologie und historische Geologie, München, 1978-I-7). Not to scale.

slight hooking at the end of the normal unit suggests that this extra piece may have lain beside it. The total width of the gastralia decreases slightly to the rear, but is nearly as great adjacent to the margin of the pelvis as in midtrunk.

From the twentieth through the thirty-fourth segment, small pieces of calcified cartilage are visible overlying the ribs on the left side. They are plausibly interpreted as linking the ends of the ribs with every alternate unit of the gastralia. One might expect that the ends of alternate gastralia would differ in order to accommodate these linking elements, but this is not evident. Lateral to the ends of the gastralia covering presacral vertebrae 35 and 36 are four larger pieces of calcified cartilage. The more medial are nearly straight and parallel the distal end of the ribs associated with presacral vertebra 33. The more lateral are in the shape of an arch. It is possible that these pieces linked the more posterior gastralia with the short ribs 35 and

36. All the exposed ribs end bluntly (the last pair of trunk ribs are obscured by the pubes).

The scapula and coracoid are separately ossified and somewhat disarticulated relative to one another. The scapula forms the anterior margin of the glenoid, which is nearly vertical and appears at a right angle to the articulating surface of the coracoid. The coracoids are roughly circular, each with a broadly oval area for attachment to the scapula and a thickened ridge for articulation with the humerus. Anterior to this ridge is a small coracoid foramen. This pattern resembles that of the Late Jurassic sphenodontian *Kallimodon* (Cocude-Michel, 1963), as well as that of *Pleurosaurus*.

The stem of the interclavicle extends posteriorly the length of one centrum behind the coracoids. The sternum is represented by an area of calcified cartilage extending from the posterior margin of the right coracoid for the length of $2\frac{1}{2}$ centra. The anterior surface is thickened adjacent to the coracoid, but the surface is convex, rather than concave as it is in the primitive lepidosauromorph *Thadeosaurus*, in which the two coracoids apparently rotated relative to the sternum as in modern lizards (Jenkins and Goslow, 1983). A short rod of calcified cartilage probably represents the tissue connecting the sternum and the ends of the thoracic ribs as in terrestrial sphenodontians.

Among the terrestrial sphenodontians from Solnhofen, the sternum is present, but poorly preserved. The sternum of this specimen of *Palaeopleurosaurus* appears to be in the process of reduction from the fully elaborated state seen in early terrestrial lepidosauromorphs and typical lizards and its absence in later pleurosaurs. Other areas of calcified cartilage appear as irregular patches between the girdles. A similar pattern is seen in Late Jurassic pleurosaurs. Figure 4.1

shows the calcified cartilage concentrated in the midtrunk region. Some was originally present more posteriorly as well, but it was removed from this area to expose the overlying ribs.

The humerus does not provide any details not evident in the previously described specimens of this genus. Most features resemble those of early terrestrial sphenodontians, although the shaft is considerably thicker than that in much smaller forms such as *Clevosaurus* (Fraser, 1988). The articulating surfaces are fully developed, without trace of sutures separating the epiphyses. Although this specimen is considerably larger than those previously described, only the very base of the olecranon is ossified. The distal epiphysis of the left ulna is distinguished from the shaft by a narrow groove. The right carpus is telescoped; some of the elements are missing, and the remainder is jumbled. There is some telescoping of the ulna and radius over the left carpus, but the exposed elements retain more or less their normal positions. They are better ossified and more nearly complete then their counterparts in the previously described specimens. The intermedium is by far the largest element and retains a deep notch for the perforating artery. The ulnare is a smaller, flat, squarish bone, partially covered by the end of the ulna. The area where a pisiform would lie is completely obscured by the base of the ulna in the left wrist. There is no bone in this position on the right, but the postmortem loss of other bones makes it impossible to determine if the pisiform might originally have been present. This bone is not preserved in other pleurosaurs. The absence of this bone would limit the capacity for supination, flexion, and extension of the manus, relative to the pattern in terrestrial sphenodontians (Holmes, 1977). The third and fourth distal carpals are very small, spherical elements. The fourth is much smaller than in terrestrial members of the Sphenodontia, but it is also small in other pleurosaurs. Two larger bones are identified as the medial and lateral centrale. On neither side is the radiale visible. On the left, the distal end of the radius is contiguous with the first metacarpal. On the right, there is a gap between these two bones. The pattern of the metacarpals is almost identical with that of *Sphenodon*. The similarities of these bones are particularly striking in regard to the relative widths and nature of overlap of the proximal articulating surfaces and their relative lengths. The metacarpals of the early sphenodontian *Clevosaurus* are relatively much longer (Fraser, 1988). The remainder of the manus is also very similar to that of *Sphenodon* in its proportions and configuration of the elements. The distal phalanges are sharp and recurved, with well-defined lateral grooves. The penultimate phalanges are longer than those immediately preceding, suggesting the presence of long ligaments to flex the claws. The proximal articulating surface extends far ventrally, indicating a wide range of flexion.

The wrist and hand show a mosaic of primitive and derived features. The ulnare, intermedium, and manus resemble those of primitive terrestrial lepidosauromorphs and *Sphenodon* in their large size and high degree of ossification. The possible loss of the pisiform and the small size of the distal tarsals resemble the condition in later pleurosaurs. Most secondarily aquatic reptiles do not use the forelimbs in locomotion, but, like the marine iguana and crocodiles, hold them against the body. In fossil groups such as most nothosaurs and placodonts, in which the forelimbs do not seem to contribute greatly to locomotion, the carpals are reduced in number and degree of ossification. This is the case in pleurosaurs, with the greatest degree of loss in the genera showing the highest degree of limb reduction. For *Palaeopleurosaurus*, it is probable that locomotion on land was still necessary, at least for reproduction, as in the aigialosaurid ancestors of the mosasaurs (Carroll and deBraga, 1992). In these animals, selection may have led to the retention of a fully developed manus, even though the function of the wrist was changing.

The elements of the pelvic girdle are exposed in a single plane, revealing their lateral and ventral surfaces. The blade of the ilium is well exposed, in contrast to the previously described material. It broadly resembles that of early terrestrial sphenodontians. The anterior margin is gently convex. The distal end is somewhat narrower, but ends bluntly. The surface for articulation with the pubis extends anteriorly as a narrow process. In contrast to the early terrestrial sphenodontians, the area for attachment of M. pubotibialis is not clearly distinguished from the anterior margin of the pubis, but forms a continuous arc with the articulating surface for the ilium. The ischium shows the conspicuous posterior process that is characteristic of both terrestrial sphenodontians and pleurosaurs. The pelvis show no trace of large plates of bone such as are present in the holotype.

The right femur is nearly complete, but the distal end is badly crushed. The left is truncated just distal to the midpoint of the shaft. The shaft is nearly as narrow, relative to the length, as that of primitive terrestrial sphenodontians such as *Clevosaurus* (Fraser, 1988). The internal trochanter is much deeper, and its proximal end is farther separated from the head. The articulating surface of the head appears somewhat smaller than those in terrestrial species, but the bone is somewhat damaged and may give an erroneous impression.

Pleurosaurus

Further comparisons will now be made with *Pleurosaurus goldfussi* itself. This taxon has been described many times, by Meyer (1861), Lortet (1892), Broili (1926), Huene (1952), Cocude-Michel (1963), and, most recently, Fabre (1981), but none of those authors

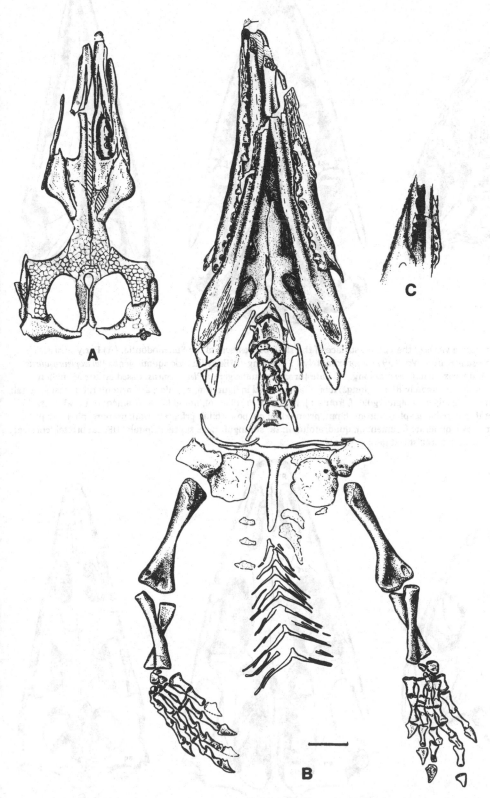

Figure 4.5. *Pleurosaurus goldfussi*, Bayerische Staatssammlung für Paläontologie und historische Geologie, München, 1925-I-18. (A) Dorsal view of skull roof. (B) Ventral view of anterior portion of skeleton. (C) Oblique view beneath left jaw showing palatine tooth row.

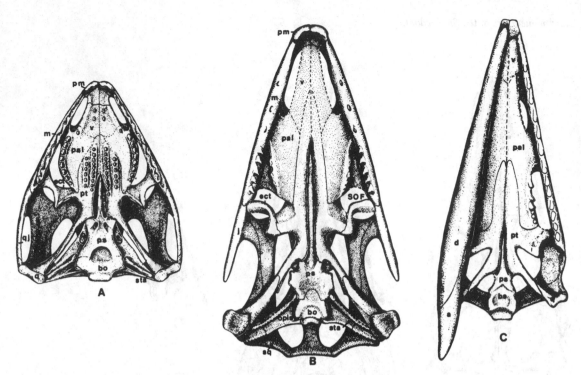

Figure 4.6. Comparative views of the ventral surface of the skull of members of the Sphenodontia. (A) Early Jurassic terrestrial genus *Clevosaurus*, from Wu (1991), approximately 15 mm long. (B) Early Jurassic aquatic genus *Palaeopleurosaurus*, from Carroll (1985), approximately 60 mm long. (C) Late Jurassic aquatic genus *Pleurosaurus*, based on the München specimen, 1925-I-18, approximately 86 mm long. Abbreviations used in figures: a, angular; art, articular; bo, basioccipital; d, dentary; ect, ectopterygoid; ept, epipterygoid; f, frontal; j, jugal; m, maxilla; mf, mandibular foramen; n, nasal; opis, opisthotic; p, parietal; pal, palatine; pf, postfrontal; pm, premaxilla; po, postorbital; prf, prefrontal; ps, parasphenoid; pt, pterygoid; q, quadrate; qf, quadrate foramen; qj, quadratojugal; sa, surangular; so, supraoccipital; SOF, suborbital fenestra; sq, squamosal; st, supratemporal; sta, stapes; v, vomer.

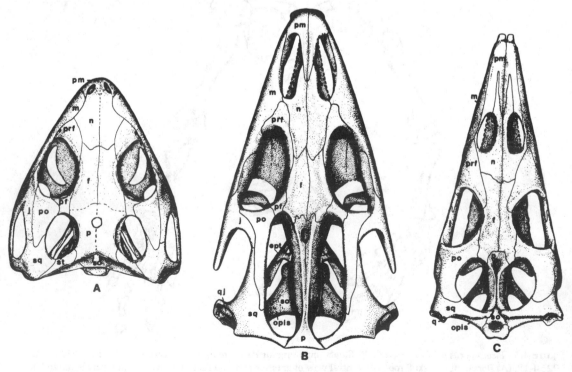

Figure 4.7. Comparative views of the dorsal surface of the skull for members of the Sphenodontia. (A) *Clevosaurus*, from Wu (1991). (B) *Palaeopleurosaurus*, from Carroll (1985). (C) *Pleurosaurus*, based on the München specimen 1925-I-18. Abbreviations as in Figure 4.6.

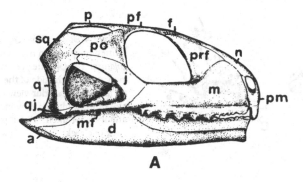

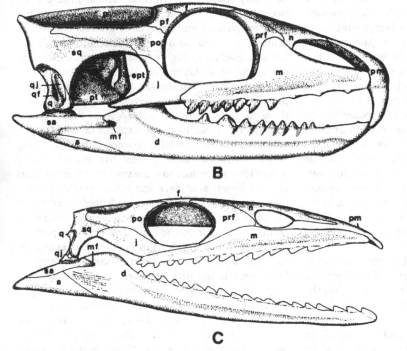

Figure 4.8. Comparative views of the lateral surface of the skull and lower jaw for members of the Sphenodontia. (A) *Clevosaurus*, from Wu (1991). (B) *Palaeopleurosaurus*, from Carroll (1985). (C) *Pleurosaurus*, based on the München specimen 1925-I-18. Abbreviations as in Figure 4.6.

prepared a skull completely, so that many aspects of its structure remained unknown. Some features remain unknown, but recent work has revealed many additional characters in common with terrestrial sphenodontians (Figures 4.5–4.11). This description is based primarily on specimen 1925–I–18 in the Bayerische Staatssammlung für Paläontologie und historische Geologie, München. The skull roof resembles that of *Palaeopleurosaurus*, except for the greater relative length of the antorbital region, the more posterior placement of the external nares, and the unquestioned fusion of the postorbital and postfrontal (this may have occurred in *Palaeopleurosaurus*, but it is uncertain in that genus). The temporal region is proportionately

shorter, closer to the pattern of primitive sphenodontians. The occipital condyle extends slightly behind the level of the supraoccipital. As exposed, the palate is largely obscured by the lower jaws, but by turning the specimen on edge, one can see the lateral row of palatine teeth that characterize all sphenodontians. The marginal teeth are anteroposteriorly elongated, as in many Late Jurassic sphenodontians. There is a single pair of enlarged premaxillary teeth. In lateral view, it can be noted that the posterior process of the jugal is completely lost. The lower jaw is strikingly similar to that of terrestrial sphenodontians in the great posterior extent of the dentary and the presence of a mandibular foramen. There is a long retroarticular

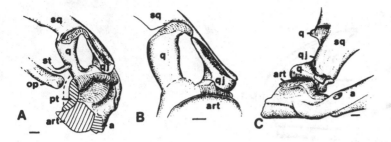

Figure 4.9. Quadrate region of the München specimen of *Pleurosaurus goldfussi*. (A) Oblique posterodorsal view of the right side of the occipital surface. (B) Posterior view of quadrate and quadratojugal, showing huge quadrate foramen. (C) Lateral view of back of cheek.

process, as in many aquatic reptiles. As pointed out by Wu (1991), the retroarticular process in primitive sphenodontians is longer than in *Sphenodon*, presumably in relation to the support of a tympanum, as in lizards. The quadrate and a slip of the quadratojugal can be seen behind and beneath the squamosal. The strikingly *Sphenodon*-like appearance of this area can be seen in occipital view, in which the large size of the quadrate foramen is evident (Figure 4.9).

The teeth are not particularly well exposed in this specimen, but they are superbly preserved in another pleurosaurid specimen (Figure 4.10). They are elongate at the base and have a wrinkled appearance, much as in the terrestrial sphenodontians of the Late Jurassic. In a further specimen (Figure 4.11) of a very large, and presumably old, individual, there are deep grooves in the dentary left by the occlusion of the maxillary teeth. Similar grooves are present in several terrestrial sphenodontians.

The postcranial skeleton in Late Jurassic pleurosaurs is divergent relative to early terrestrial sphenodontians, in relation to an aquatic way of life. It does retain the long posterior ischiadic process of terrestrial sphenodontians.

Other, conservative features shared with primitive lepidosauromorphs, but not unique to members of the Sphenodontia, are the retention of conspicuous ventral gastralia, intercentra throughout the column, close attachment but sutural division of the astragalus and calcaneum, and a hooked fifth metatarsal. A distinct sternum is not present, but a thin, irregular layer of calcified cartilage extends for most of the length of the trunk. This may have served to add weight in a uniform manner to the trunk region so as to decrease buoyancy in the water.

Discussion

Description of a new specimen of *Palaeopleurosaurus posidoniae* and redescription of the skull of *Pleurosaurus* should put to rest any lingering suspicion that pleurosaurs might have evolved from any group other than primitive terrestrial sphenodontians.

While terrestrial sphenodontians may have remained extremely conservative in their skeletal structure between the Late Triassic and today, the pleurosaurs show progressive and apparently gradual changes in skull, trunk, and limb proportions throughout the Jurassic, leading to highly modified forms by the end of that period. If there were genetic or developmental factors that restricted the amount or nature of evolutionary change among terrestrial sphenodontians, those factors clearly did not apply to their aquatic descendants. It seems more plausible to attribute the conservatism of terrestrial sphenodontians to stabilizing selection within a relatively constant environment and way of life, rather than to some undefinable constraints of their genetics or development. A similar pattern may be noted among lizards. Quadrupedal, terrestrial members of this assemblage have retained similar postcranial features from the Late Jurassic to the present, but many divergent lineages have greatly elongated their trunks and reduced or lost their limbs, and the Cretaceous mosasaurs achieved an extremely high degree of specialization for life in the water. As long as the environment and way of life remained relatively constant, there was little skeletal change. When the environment and/or way of life changed, selection acted on the available genetic differences, and morphological changes occurred. The extent of such changes depended on the longevity of the group and the amount of change necessary to accommodate to the new way of life.

The relationships of the pleurosaurs also raise a problem in classification. Before 1960, most biologists were comfortable with the concept of one taxon evolving from another (the dinosaurs evolved from thecodonts, the amphibians evolved from the rhipidistians, birds evolved from dinosaurs, etc.). Since the widespread use of phylogenetic systematics, it has been argued that no properly defined group could give rise to any other. Rather, the ancestral and descendant groups should be included in a single monophyletic assemblage. Any putative ancestral group is designated a paraphyletic group and is considered to be no more than an arbitrary and incorrect concep-

Figure 4.10. Detail of the left orbital region of a specimen of *Pleurosaurus* from Daiting, Mörnsheimer Schichten (Malm Zeta 3). Bayerische Staatssammlung für Paläontologie und historische Geologie, München, 1978-I-7. Teeth are elongated and show the wrinkled enamel common to Late Jurassic terrestrial sphenodontians. The anterior teeth are solidly ankylosed to the bone; the posterior teeth appear to be set in sockets. Note small size of manus.

tion of systematists. Benton, in the 1993 updating of *Fossil Record*, designated the Sphenodontidae as a paraphyletic group, because it appears to include the ancestors of pleurosaurs. Pleurosaurs and sphenodontids were generally accepted as distinct families, and there is general acceptance that the most primitive pleurosaur, *Palaeopleurosaurus*, fits within the cladogram of terrestrial sphenodontians somewhere between the terrestrial genera of Late Triassic and Early Jurassic age and the Late Jurassic and subsequent genera.

Evans (1988) placed *Palaeopleurosaurus* within an assemblage including *Clevosaurus, Homoeosaurus, Pio-*

cormus, Sapheosaurus, Kallimodon, and Leptosaurus. Fraser and Benton (1989) established a position within what they referred to as the crown-group of the Sphenodontia, including *Clevosaurus, Kallimodon, Sapheosaurus, Piocormus, Toxolophosaurus, Eilenodon,* and *Sphenodon*, as the sister-group of either *Kallimodon* or *Sphenodon*. Wu (1991) placed it among the more primitive sphenodontians, between *Planocephalosaurus* and the clevosaurs.

Evans recognized the assemblage including pleurosaurs on the basis of prominently flanged and striated teeth on the maxilla. Fraser and Benton united *Palaeopleurosaurus* with *Kallimodon* and *Sphenodon* on the basis of the narrow, crested parietal. According to Wu, *Palaeopleurosaurus* is distinguished from other Triassic and Early Jurassic members of the Sphenodontia in showing tooth wear and having a single pair of chisellike premaxillary teeth, posterolingual flanges on the marginal teeth, a single row of palatine teeth, and no pterygoid teeth, as well as elongation of the supratemporal fenestra. It is more primitive than the Late Jurassic terrestrial genera and *Sphenodon* in lacking a complete lower temporal bar and in having the pterygoid contributing to the margin of the suborbital fenestra. Wu considers the narrow, crested parietal convergent with *Kallimodon* and *Sphenodon*. Some incongruities must be accepted no matter what the relationships of the pleurosaurs, but it is clear that they lie somewhere between the Late Triassic genera, which have numerous unmodified teeth in the premaxilla, more than one row of teeth on the palatine, and teeth on the pterygoid, and the Late Jurassic and later genera, which have a solid lower temporal bar.

By itself, *Palaeopleurosaurus* can be considered a member of the Sphenodontidae, but it also forms the sister-group of the genus *Pleurosaurus*, which has been

Figure 4.11. Skull of *Pleurosaurus* from Daiting, Staatliches Museum für Naturkunde, Stuttgart, no. 56604. Note vertical grooves in lower jaw caused by occlusion of maxillary teeth.

placed in a separate family for the past 150 years. This problem can be solved at one level by simply including all pleurosaurs in the same family as *Brachyrhinodon*, *Homoeosaurus*, and *Sphenodon*. Among living lizards, similarly divergent morphologies are including within individual families. On the other hand, the Mosasauridae are distinguished at the family level from terrestrial varanoids primarily because of their high degree of aquatic adaptation.

Alternatively, one might seek to divide up the Sphenodontia not only between terrestrial sphenodontians and aquatic pleurosaurs but also between the terrestrial genera that appeared before the divergence of pleurosaurs and those that evolved subsequent to that event. A designation of plesion might then be applied to each of the older terrestrial genera. This, too, is a possible solution.

We consider it more informative to recognize that animals with a particular anatomical pattern, such as the terrestrial sphenodontians, gave rise to the animals termed "pleurosaurs" than to use a systematic methodology that can imply that two very different kinds of animals belong to a single taxonomic group, or that both families recognized subsequent to the speciation event had evolved divergently from a third, distinct biological category.

Acknowledgments

We thank Mr. Jürgen Schüssel of Nürnberg for the loan of the new specimen of *Palaeopleurosaurus posidoniae* to the Staatliches Museum für Naturkunde, Stuttgart, and for permission to continue its preparation and study in Canada. Without his discovery, a very informative specimen would have been lost to science. We express our appreciation to Pamela Gaskill for her extremely careful preparation of this specimen and her skillful illustration of this and other pleurosaurs. We thank Dr. Peter Wellnhofer, Bayerische Staatssammlung für Paläontologie und historische Geologie, München, for the loan of specimens of *Pleurosaurus goldfussi* and for permission to prepare further the specimen originally described by Broili. This research was supported by grants from the Natural Sciences and Engineering Research Council of Canada.

References

Benton, M. J. 1985. Classification and phylogeny of diapsid reptiles. *Zoological Journal of the Linnean Society* 84: 97–164.

Boulenger, G. A. 1893. On some newly-described Jurassic and Cretaceous lizards and rhynchocephalians. *Annals and Magazine of Natural History* (6) 11: 204–210.

Broili, F. 1926. Ein neuer Fund von *Pleurosaurus* aus dem Malm Frankens. *Abhandlungen der Bayerischen Akademie der Wissenschaften*, *mathematisch-naturwissenschaftliche Abteilung* 30: 1–48.

Carroll, R. L. 1977. The origin of lizards. *Linnean Society Symposium Series* 4: 359–396.

———. 1985. A pleurosaur from the Lower Jurassic and the taxonomic position of the Sphenodontida. *Palaeontographica* A189: 1–28.

Carroll, R. L., and P. J. Currie. 1991. The early radiation of diapsid reptiles. Pp. 354–424 in L. Trueb and H.-P. Schultze (eds.), *Origins of the Higher Groups of Tetrapods: Controversy and Consensus*. Ithaca: Cornell University Press.

Carroll, R. L., and M. de Braga. 1992. Aigialosaurs: Mid-Cretaceous varanoid lizards. *Journal of Vertebrate Paleontology* 12: 66–86.

Cocude-Michel, M. 1963. Les rhynchocéphales et les sauriens des calcaires lithographique d'Europe occidentale. *Nouvelles Archives du Muséum d'Historie Naturelle, Lyon* 7: 1–187.

Eldredge, N, and S. M. Stanley (eds.) 1984. *Living Fossils*. Berlin: Springer-Verlag.

Evans, S. 1988. The early history and relationships of the Diapsida. Pp. 221–260 in M. J. Benton (ed.), *The Phylogeny and Classification of the Tetrapods. Vol. 1: Amphibians, Reptiles, Birds*. Oxford: Clarendon Press.

Fabre, J. 1981. *Les Rhynchocéphales et les Ptérosauriens à Crête Pariétale du Kiméridgien Supérieur-Berriasien d'Europe Occidentale*. Paris: Editions de la Fondation Singer-Polignac.

Fitzinger, L. J. 1843. *Systema Reptilium*. Vienna.

Fraser, N. C. 1988. The osteology and relationships of *Clevosaurus* (Reptilia: Sphenodontida). *Philosophical Transactions of the Royal Society of London* B321: 125–178.

Fraser, N. C. and M. J. Benton. 1989. The Triassic reptiles *Brachyrhinodon* and *Polysphenodon* and the relationships of the sphenodontids. *Zoological Journal of the Linnean Society* 96: 413–445.

Fraser, N. C., and G. M. Walkden. 1984. The postcranial skeleton of the Upper Triassic sphenodontid *Planocephalosaurus robinsonae*. *Palaeontology* 27: 575–595.

Gauthier, J., R. Estes, and K. de Queiroz. 1988. A phylogenetic analysis of Lepidosauromorpha. Pp. 15–98 in R. Estes and G. Pregill (eds.), *Phylogenetic Relationships of the Lizard Families*. Stanford University Press.

Gould, S. J. 1982. The meaning of punctuated equilibrium and its role in validating a hierarchical approach to macroevolution. Pp. 83–104 in R. Milkman (ed.), *Perspectives on Evolution*. Sunderland, Mass.: Sinauer Associates.

Günther, A. 1867. Contribution to the anatomy of *Hatteria* (*Rhynchocephalus*, Owen). *Philosophical Transactions of the Royal Society of London* 157: 595–629.

Hoffstetter, R. 1955. Rhynchocephalia; Pp. 556–576 in J. Piveteau (ed.), *Traité de Paléontologie. Vol. 5, Amphibiens, Reptiles, Oiseaux*. Paris: Masson et Cie.

Homes, R. 1977. The osteology and musculature of the pectoral limb of small captorhinids. *Journal of Morphology* 152: 101–140.

Huene, F. von 1952. Revision der Gattung *Pleurosaurus* auf Grund neuer und alter Funde. *Palaeontographica* A101: 167–200.

Jenkins, F. A., Jr., and G. E. Goslow. 1983. The functional anatomy of the savannah monitor lizard (*Varanus exanthematicus*). *Journal of Morphology* 175: 195–216.

Laurin, M. 1991. The osteology of a Lower Permian eosuchian from Texas and a review of diapsid phylogeny. *Zoological Journal of the Linnean Society* 101: 59–95.

Lortet, L. 1892. Les reptiles fossiles du bassin du Rhône. *Archives du Muséum d'Histoire Naturelle, Lyon* 5: 29–73, 80–90.

Lydekker, R. 1880. *Catalogue of the Fossil Reptilia and Amphibia in the British Museum (Natural History).* Part 1. Trustees of the British Museum, London.

Meyer, H. von. 1831. [*Pleurosaurus goldfussi.*] Nova Acta Academiae Caesareae Leopoldino–Carolinae germanicae naturae curiosorum 15: 194–195.

1861. Zu *Pleurosaurus goldfussi* aus dem lithographischen Schiefer von Daiting. *Palaeontographica* 10: 37–45.

Münster, G. 1839. Ueber einige neue Versteinerungen in den lithographischen Schiefern von Baiern. *Neues Jahrbuch für Mineralogie, Geologie, und Palaeontologie* 1839: 676–682.

Reisz, R. R. 1981. A diapsid reptile from the Pennsylvanian of Kansas. *Occasional Papers, Museum of Natural History, University of Kansas* 7: 1–74.

Riegraf, W., G. Werner, and F. Lörcher. 1984. *Der Posidonienschiefer. Biostratigraphie, Fauna und Fazies des südwestdeutschen Untertoarciums (Lias)*. Stuttgart: Enke.

Rieppel, O. 1987. *Clarazia* and *Hescheleria*: a re-investigation of two problematical reptiles from the Middle Triassic of Monte San Giorgio (Switzerland). *Palaeontographica* A195: 101–129.

Romer, A. S. 1956. *Osteology of the Reptiles.* University of Chicago Press.

Wagner, A. 1860. Vergleichung der urweltlichen Fauna des lithographischen Schiefers von Cirin mit der gleichnamigen Ablagerung im Fränkischen Jura. *Gelehrter Anzeiger der königlichen bayerischen Akademie der Wissenschaften, München* 48: 390–412.

Watson, D. M. S. 1914. *Pleurosaurus* and the homologies of the bones of the temporal region of the lizard's skull. *Annals and Magazine of Natural History* (8) 14: 84–95.

Whiteside, D. I. 1986. The head skeleton of the Rhaetian sphenodontid *Diphydontosaurus avonis* gen. et sp. nov. and the modernizing of a living fossil. *Philosophical Transactions of the Royal Society of London* B312: 379–430.

Wu X.-c. 1991. The comparative anatomy and systematics of Mesozoic sphenodontians. Ph.D. dissertation, McGill University, Montreal.

Zittel, K. A. von. 1887–90. *Handbuch der Palaeontologie.* Bd. 3: Vertebrata. München: Oldenburg.

5

Patterns of evolution in Mesozoic Crocodyliformes

JAMES M. CLARK

Introduction

Following their first appearance in the Late Triassic, crocodyliforms (including three taxa traditionally placed within the Crocodylia: Protosuchia, Mesosuchia, and Eusuchia) occur with remarkable regularity in fluvial and nearshore marine sedimentary deposits worldwide. This unusually good fossil record has not received the attention it deserves, however, in part because of the difficulty of studying the disparate fossil collections and in part because of the incompleteness of many of the fossils. A more important factor, perhaps, is the widespread but mistaken notion that the evolutionary history of crocodyliforms was un-eventful and that crocodyliforms have changed little over the past 200 million years. Given the inadequacy of our knowledge of crocodyliform relationships and the necessity of a well-corroborated phylogeny for studying evolutionary phenomena, generalizations about crocodyliform evolution in the current literature must be considered tentative at best.

The basic taxa considered in this chapter often have been placed within the order Crocodylia. Recently I have applied the name Crocodyliformes to this group (in Benton and Clark, 1988). In the past, taxa typically were included or excluded from the Crocodylia based upon how crocodilelike they were, but such subjective distinctions served only to express an intuitive measure of similarity. I chose, instead, to use the taxonomic hierarchy to reflect phylogenetic relationships (de Queiroz and Gauthier, 1990), and I therefore recognize the Crocodylia as including only the closest relatives of extant Crocodylia (the crocodylian crown-group). A name is therefore needed to identify the group traditionally termed the Crocodylia, and I have proposed the term Crocodyliformes for this group.

The recent flurry of cladistic analyses breaking apart the "Thecodontia" and identifying the monophyletic

groups within Archosauria has generated a new taxonomy that has not yet stabilized (Gauthier, 1984, 1986; Benton, 1985; Benton and Clark, 1988; Sereno and Arcucci, 1990; Sereno, 1991). Crocodylians and birds compose the Archosauria, and I follow Gauthier (1986) in considering this group to be limited to the descendants of the last common ancestor of the two groups (crown-group).

The group of archosaurs that are more closely related to extant crocodylians than to extant birds has been given several names (in part because of different conceptions of group membership). Gauthier (1986) emended the taxon Pseudosuchia to comprise this group, Benton (Benton and Clark, 1988) applied to it the term Crocodylotarsi, and Sereno and Arcucci (1990) termed it the Crurotarsi. Among the taxa previously placed in the "Thecodontia," those that are now considered to belong to the crocodylian closed-descent community are Phytosauridae, Stagonolepi-didae, "Rauisuchidae," Poposauridae, *Gracilisuchus*, and, following Sereno and Arcucci (1990), Ornitho-suchidae.

Following the discovery of *Sphenosuchus acutus* in the early part of this century, it and similar forms were recognized as being closely related to crocodylians (e.g., Broom, 1927). Notable among these are several recently discovered Late Triassic and Early Jurassic taxa (Clark, 1986; Mattar, 1989; Parrish, 1991). The detailed description of *Sphenosuchus acutus* by Walker (1990) provides a firm basis for understanding the anatomy of these taxa, and detailed descriptions of other taxa are currently in preparation. The close relationship between these "sphenosuchians" and crocodyliforms prompted Walker (1968) to erect the Crocodylomorpha (emending a taxon named by Hay, 1930, that was later ignored). Clark (Benton and Clark, 1988) argued that "sphenosuchians" are para-phyletic with respect to other crocodylomorphs. Parrish

(1991) criticized the characters upon which that hypothesis was based, and he presented further evidence in support of it. Recent studies, however, have questioned this conclusion and instead have argued for sphenosuchian monophyly (P. C. Sereno, pers. commun.; X.-c. Wu, pers. commun.).

Phylogenetic relationships among crocodyliforms were analyzed by Benton and Clark (1988), and this chapter serves to present the complete data set used in that analysis updated in light of subsequent information and reinterpretations. The character matrix therefore has been modified from that used earlier, to exclude characters of questionable validity and to include characters discussed by Clark and Norell (1992).

Prior to my previous work there was no comprehensive cladistic analysis of the Crocodyliformes. Consequently, many of the higher taxa within this group had never been adequately diagnosed. My work presented a cladistic hypothesis in which many of these traditional taxa are not monophyletic. More important, I pointed out the ramifications of what I call here the "longirostrine problem," namely, that many features possibly related to the possession of a long rostrum ally thalattosuchians with other longirostrine taxa. Here I concentrate mainly on this latter phenomenon, which has important implications for the phylogeny of all crocodyliforms.

Materials and methods

Specimens of all of the sphenosuchian taxa and all of the crocodyliform taxa except *Sokotosuchus* (Buffetaut, 1981) and *Leidyosuchus* (Erickson, 1976) were examined firsthand. Unfortunately, I was not able to study one important group of longirostrine eusuchians – thoracosaurs – at the time of this study. The observations of specimens involved in generating this data set are summarized by Clark (1986), Benton and Clark (1988), Clark and Norell (1992), and Clark (unpublished data).

The characters presented in Appendix 5.1 were analyzed using cladistic analysis as implemented by the PAUP computer program (Swofford, 1990). Characters were given equal weight, and reversal was treated as being as likely as convergence. Those multistate characters that formed a morphocline were treated as ordered; those that did not were treated as unordered, as indicated in Appendix 5.1. Because of the large number of taxa involved, the general heuristic search option of PAUP was utilized (rather than the exhaustive search option).

The cladograms were rooted using the closest outgroups to crocodyliforms, including (following Benton and Clark, 1988) the best-known sphenosuchians, the poposaurid *Postosuchus kirkpatricki* Chatterjee, 1985, and *Gracilisuchus stipanicicorum* Romer, 1972. Following Clark (Benton and Clark, 1988), the relation-

ships of these groups were constrained as the first, second, and third out-groups of Crocodyliformes, respectively. Because the relationships among sphenosuchians are poorly resolved and controversial, their relationships to each other (i.e., sphenosuchian monophyly or paraphyly) were not constrained.

To investigate how two alternative phylogenetic hypotheses differ from the most parsimonious cladogram, the PAUP program was executed with the position of thalattosuchians constrained either as the sister-group to the remaining Crocodyliformes or as the sister-group to the remaining Mesoeucrocodylia. To address the hypothesis that characters related to the longirostrine condition are correlated, the PAUP analysis was repeated with all but one of the characters that evolved convergently in *Gavialis* on the one hand and thalattosuchians, dyrosaurs, and pholidosaurs on the other (characters 3, 14, 18, 20, 57, 69, 77, 79, and 81) omitted from the character matrix.

Results

The results of these analyses will be discussed using the terms of the traditional taxonomy of crocodylians (Table 5.1). "Longirostrine" crocodyliforms (i.e., those with a long, narrow rostrum) include thalattosuchians, pholidosaurids, dyrosaurids ("tethysuchians"), and *Gavialis*. The longirostrine condition is approached by some other taxa (e.g., *Tomistoma*), but several characters typical of taxa with an extremely long rostrum (e.g., basioccipital tubera) are not present.

Forty equally most parsimonious cladograms were found by PAUP (Figures 5.1 and 5.2). These cladograms are all quite similar, and differences involve the positions of *Eopneumatosuchus, Gobiosuchus, Orthosuchus, Baurusuchus, Bernissartia,* and *Goniopholis* and the relationships among eusuchians. The most notable points about these cladograms are as follows: (1) The "Protosuchia" are not monophyletic, but *Protosuchus, Hemiprotosuchus,* and an unnamed taxon from the Kayenta Formation of Arizona form a clade. (2) The Mesosuchia are paraphyletic, with some being more closely related to the Eusuchia (together forming the clade Mesoeucrocodylia of Whetstone and Whybrow, 1983). (3) The Thalattosuchia are a well-supported monophyletic group (although all of the apomorphies of this group were not included in this analysis; Benton and Clark, 1988). (4) The Sebecosuchia are polyphyletic. (5) The Notosuchia are polyphyletic. (6) The Eusuchia are monophyletic. (7) The Atoposauridae are monophyletic. (8) *Bernissartia* is not always the sister-group of Eusuchia, but atoposaurs are less closely related to Eusuchia. (9) All longirostrine forms except *Gavialis* form a clade nested well within the Mesoeucrocodylia.

The results of this analysis are essentially those reported by Benton and Clark (1988), with a few

Table 5.1. *Classifications of crocodylians*

Traditional classification[a]	Classification derived from preferred phylogeny of Clark[b]	Classification derived from the strict consensus of the most parsimonious-cladogram of this study
Order Crocodylia	Crocodyliformes	Crocodyliformes
Suborder Sphenosuchia	Protosuchidae	*Eopneumatosuchus*
Pseudhesperosuchus	*Protosuchus*	*Gobiosuchus*
Saltoposuchus	*Hemiprotosuchus*	*Orthosuchus*
Sphenosuchus	Kayenta *Edentosuchus*-like form	Protosuchidae
Dibothrosuchus	Unnamed taxon	Kayenta *Edentosuchus*-like form
Suborder Protosuchia	*Orthosuchus*	Protosuchinae
Protosuchus	Unnamed taxon	*Protosuchus*
Hemiprotosuchus	*Gobiosuchus*	*Hemiprotosuchus*
Orthosuchus	Mesoeucrocodylia	Mesoeucrocodylia
Eopneumatosuchus	Thalattosuchia	Fruita form
Gobiosuchus	*Pelagosaurus*	Unnamed taxon
Suborder Mesosuchia	Unnamed taxon	*Notosuchus*
Infraorder Notosuchia	Teleosauridae	*Baurusuchus*
Notosuchus	Metriorhynchidae	Unnamed taxon
Araripesuchus	Metasuchia	*Libycosuchus*
Libycosuchus	Fruita form	Unnamed taxon
Infraorder Sebecosuchia	Unnamed taxon	*Sebecus*
Baurusuchus	*Notosuchus*	Unnamed taxon
Sebecus	Unnamed taxon	*Araripesuchus*
Infraorder Thalattosuchia	*Baurusuchus*	Neosuchia (revised content)
Teleosauridae	*Libycosuchus*	Atoposauridae
Metriorhynchidae	Unnamed taxon	*Alligatorium*
Pelagosaurus	*Araripesuchus*	*Theriosuchus*
Infraorder Metamesosuchia	Unnamed taxon	Unnamed taxon
Atoposauridae	*Sebecus*	*Goniopholis*
Alligatorium	Neosuchia	*Bernissartia*
Theriosuchus	Atoposauridae	Unnamed taxon
Goniopholididae	*Alligatorium*	*Eutretauranosuchus*
Goniopholis	*Theriosuchus*	Unnamed taxon
Eutretauranosuchus	Unnamed taxon	Dyrosauridae
Pholidosauridae	Pholidosauridae	*Sokotosuchus*
Pholidosaurus	Goniopholididae	*Dyrosaurus*
Bernissartidae	Dyrosauridae	Unnamed taxon
Bernissartia	Unnamed taxon	*Pholidosaurus*
Infraorder Tethysuchia	*Bernissartia*	Thalattosuchia
Sokotosuchus	Eusuchia	*Pelagosaurus*
Dyrosaurus	*Hylaeochampsa*	Unnamed taxon
Suborder Eusuchia	Crocodylia	Metriorhynchidae
Hylaeochampsa	*Leidyosuchus*	Teleosauridae
Crocodylidae	Crocodylidae	Eusuchia
Leidyosuchus	Alligatoridae	*Leidyosuchus*
Alligatoridae	*Gavialis*	Crocodylidae
Gavialidae		Alligatoridae
		Gavialidae

[a]Based on Buffetaut (1982).
[b]Benton and Clark (1988).

differences due to the exclusion of some questionable characters from the data set presented here. The main differences are that *Orthosuchus* and *Gobiosuchus* are more distantly related to Mesoeucrocodylia than are protosuchids, and that the relationships of *Bernissartia* are not clearly with Eusuchia. The validity of one of the major features of this cladogram is suspect, however. The close relationships of thalattosuchians to other longirostrine taxa nested well within the Mesoeucrocodylia rest upon characters that may be func-

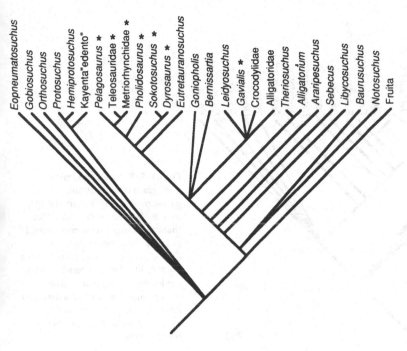

Figure 5.1. Strict consensus cladogram of the 40 most parsimonious cladograms resulting from the cladistic analysis. Longirostrine taxa are indicated by an asterisk.

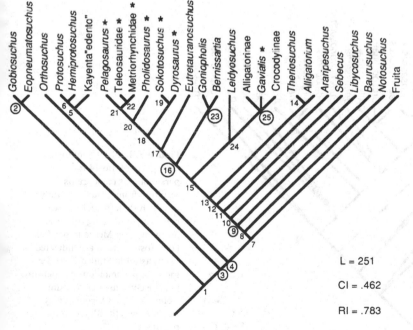

L = 251

CI = .462

RI = .783

Figure 5.2. One of the 40 equally most parsimonious cladograms, chosen arbitrarily. Longirostrine taxa are indicated by an asterisk; unresolved nodes in strict consensus cladogram circled. Abbreviations: CI, consistency index; L, length; RI, ensemble retention index.

tionally related to the long-snouted condition. This was investigated using two alternative hypotheses of thalattosuchian relationships for which there is character support, and which had been suggested by previous authors (e.g., Antunes, 1967; Buffetaut, 1982).

When thalattosuchians are constrained to be the sister-group of other crocodyliforms (Figure 5.3), the most parsimonious cladograms are nine steps longer

than the most parsimonious cladograms without such constraint and have the following features: (1) The same three protosuchians that are monophyletic in the first analysis are monophyletic. (2) The relationships of *Baurusuchus, Libycosuchus, Sebecus*, and *Notosuchus* are less well resolved. (3) *Pholidosaurus* and dyrosaurs form a monophyletic group, and *Gavialis* forms a clade with *Pholidosaurus* within this group. (4) This

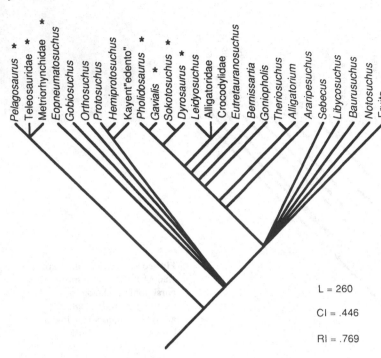

Figure 5.3. Strict consensus cladogram of the 136 most parsimonious cladograms resulting from a cladistic analysis in which thalattosuchians were constrained as being the sister-group of other Crocodyliformes. Longirostrine taxa are indicated by an asterisk. Statistics are for the most parsimonious cladograms, not the consensus tree. Abbreviations: CI, consistency index; L, length; RI, ensemble retention index.

L = 260

CI = .446

RI = .769

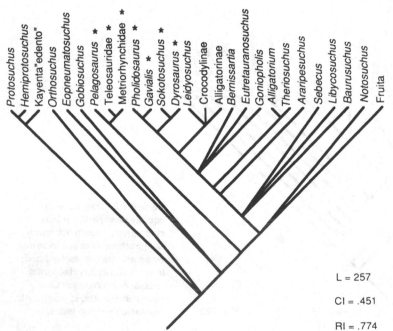

Figure 5.4. Strict consensus cladogram of the 24 most parsimonious cladograms resulting from a cladistic analysis in which thalattosuchians were constrained as being the sister-group of other Mesoeucrocodylia. Longirostrine taxa are indicated by an asterisk. Statistics are for the most parsimonious cladograms, not the consensus cladogram. Abbreviations: CI, consistency index; L, length; RI, ensemble retention index.

L = 257

CI = .451

RI = .774

pholidosaur-dyrosaur-*Gavialis* clade is the sister-group to other eusuchians.

When thalattosuchians are constrained to be the sister-group of other Mesoeucrocodylia (Figure 5.4), the cladograms are six steps longer than the most parsimonious cladograms, and the results are similar to those of the first analysis except that (1) *Eopneu-*

matosuchus and *Gobiosuchus* are the sister-taxa to Mesoeucrocodylia, (2) the positions of *Libycosuchus*, *Baurusuchus*, and *Sebecus* are unresolved, and (3) the relationships among neosuchians are as in the second analysis. Because of the changes to the character matrix, these results differ somewhat from those of Benton and Clark (1988), in which the Protosuchia were paraphyletic, and *Gavialis* formed a clade with pholidosaurs and dyrosaurs.

To further investigate the hypothesis that correlated characters are responsible for the placement of thalattosuchians with most other longirostrine taxa in the most parsimonious cladograms, the characters that evolved convergently in another longirostrine taxon, *Gavialis*, were combined into a single character. The consensus cladogram of the resulting most parsimonious cladograms is essentially the same as the consensus of most parsimonious cladograms.

Discussion

The results of this analysis are most notable for the strong support they evince for an unorthodox phylogeny in which all long-snouted crocodyliforms except *Gavialis* form a clade nested well within the Mesoeucrocodylia. Although there are reasons to suspect this phylogenetic conclusion, other, more traditional phylogenies are significantly less parsimonious. Furthermore, this new study suggests that when the relationships of thalattosuchians are constrained to follow two more orthodox hypotheses, the relationships of *Gavialis* lie closer to the pholidosaurid and dyrosaurid "mesosuchians" than to other extant crocodylians, a hypothesis that is also quite unorthodox.

That the characters found in longirostrine taxa may be correlated is suggested by two features of the most parsimonious cladogram and by comparison of the cladogram with the fossil record. First, the placement of thalattosuchians with other longirostrines requires a great deal of homoplasy, especially character reversal, at the base of thalattosuchians (Appendix 5.3, node 21). In essence, the most parsimonious cladogram requires the thalattosuchians to have lost many mesoeucrocodylian synapomorphies, especially in the braincase. Second, many of the synapomorphies of the longirostrine clade evolved independently in the longirostrine taxon *Gavialis*.

When the fossil record of crocodyliforms is taken into account, the position of the thalattosuchians nested well within the group is discordant, considering that they appeared in the Early Jurassic, long before the appearance of most mesoeucrocodylians. This hypothesis therefore requires nearly all of the other clades of Mesoeucrocodylia to have been present throughout the Jurassic, in spite of the absence of fossils indicating such ranges.

If the characters that unite thalattosuchians with most other longirostrines are correlated, then this may be evident from their distributions. The most parsimonious hypotheses indicate that one long-snouted taxon, *Gavialis*, evolved independently of other longirostrine taxa, suggesting that those features that evolved convergently in *Gavialis* on the one hand and other longirostrine taxa on the other are related to the longirostrine condition. The fact that the deletion of all but one of these characters does not affect the relationships of the longirostrine taxa therefore indicates that if character correlations are responsible for the placement of thalattosuchians with most other long-snouted forms, then more characters are involved in this character complex than those that evolved together independently in *Gavialis*.

What, then, are we to conclude about the relationships of crocodyliforms? In spite of their counterintuitive set of relationships, the most parsimonious cladograms must be accorded the respect due to hypotheses that best explain the data. Future work must address the unorthodox relationships of longirostrine forms common to these hypotheses and the pattern of character distributions underlying them. I strongly suspect that the addition of more taxa to this analysis would alter the relationships suggested in the most parsimonious cladograms of this analysis, but until such work is done, it will be difficult to justify the acceptance of other, less parsimonious hypotheses.

Acknowledgments

The research for this chapter derived from my dissertation work at the University of Chicago under the tutelage of James Hopson. This research was supported by the Department of Anatomy, University of Chicago, and a dissertation improvement grant from the National Science Foundation, and this chapter was written at the Smithsonian Institution and the American Museum of Natural History. Many people at many institutions provided help that was essential for the completion of this work; at the great risk of slighting them, I refer readers to my dissertation and to my monograph submitted to the *Bulletin of the American Museum of Natural History* for a complete listing. Finally, I sincerely thank the organizers of this workshop for creating a congenial atmosphere conducive to scientific exchange and for their patience in dealing with my lethargy.

References

Antunes, M. T. 1967. Um mesosuquiano do Liásico de Tomar (Portugal). Considerações sobre a origem dos crocodilos. *Mémoires des Services Géologiques du Portugal, N. S.* 13: 7–66.

Benton, M. J. 1985. Classification and phylogeny of the diapsid reptiles. *Zoological Journal of the Linnean Society* 84: 97–164.

Benton, M. J., and J. M. Clark. 1988. Archosaur phylogeny and the relationships of the Crocodylia. Pp. 295–338 in M. J. Benton (ed.), *The Phylogeny and Classification of the Tetrapods, Vol. 1.* Oxford: Clarendon Press.

Broom, R. 1927. On *Sphenosuchus*, and the origin of the crocodiles. *Proceedings of the Zoological Society of London* 1927: 359–370.

Buffetaut, E. 1981. *Sokotosuchus ianwilsoni* and the evolution of the dyrosaurid crocodilians. *Nigerian Field Society Monograph* 1: 31–41.

1982. Radiation évolutive, paléoécologie et biogéographie des crocodiliens mésosuchiens. *Mémoires de la Société Géologique de France*, N.S. 60 (142): 1–88.

Chatterjee, S. 1985. *Postosuchus*, a new thecodontian reptile from the Triassic of Texas and the origin of tyrannosaurs. *Philosophical Transactions of the Royal Society of London* B309: 395–460.

Clark, J. M. 1985. A new crocodylomorph from the Late Jurassic Morrison Formation of western Colorado, with a discussion of relationships within the "Mesosuchia." M.A. thesis, University of California, Berkeley.

1986. Phylogenetic relationships of the crocodylomorph archosaurs. Ph. D. dissertation, University of Chicago.

Clark, J. M., and M. A. Norell. 1992. The Early Cretaceous crocodylomorph *Hylaeochampsa vectiana* from the Wealden of the Isle of Wight. *American Museum Novitates* 3032: 1–19.

de Queiroz, K., and J. A. Gauthier. 1990. Phylogeny as a central principle in taxonomy; phylogenetic definitions of taxon names. *Systematic Zoology* 39: 307–322.

Erickson, B. R. 1976. Osteology of the early eusuchian crocodile *Leidyosuchus formidabilis*, sp. nov. *Monographs of the Science Museum of Minnesota* (*Paleontology*) 2: 1–61.

Gauthier, J. A. 1984. A cladistic analysis of the higher order relationships of Diapsida. Ph.D. dissertation, University of California, Berkeley.

1986. Saurischian monophyly and the origins of birds. *Memoirs, California Academy of Sciences* 8: 1–55.

Gauthier, J. A., A. G. Kluge, and T. Rowe. 1988. Amniote phylogeny and the importance of fossils. *Cladistics* 4: 105–209.

Hay, O. P. 1930. *Second Bibliography and Catalogue of the Fossil Vertebrata of North America*, 2 vols. Washington, D. C.: Carnegie Institution of Washington.

Mattar, L. C. B. 1989. Descrição osteológica do crânio e segunda vértebra cervical de *Barbarenasuchus brasiliensis* Mattar, 1987 (Reptilia, Thecodontia) do Mesotriássico do Rio Grande do Sul, Brasil. *Anais da Academia brasileira de Ciências* 61: 319–333.

Parrish, J. M. 1991. A new specimen of an early crocodylomorph (cf. *Sphenosuchus* sp.) from the Upper Triassic Chinle Formation of Petrified Forest National Park, Arizona. *Journal of Vertebrate Paleontology* 11: 198–212.

Romer, A. S. 1972. The Chañares (Argentina) Triassic reptile fauna. XIII. An early ornithosuchid pseudosuchian, *Gracilisuchus stipanicicorum*, gen. et sp. nov. *Breviora* 389: 1–24.

Sereno, P. C. 1991. Basal archosaurs: phylogenetic relationships and functional implications. *Society of Vertebrate Paleontology Memoir* 2: 1–53.

Sereno, P. C., and A. Arcucci. 1990. The monophyly of crurotarsal archosaurs and the origin of bird and crocodile ankle joints. *Neues Jahrbuch für Geologie und Paläontologie, Abhandlungen* 180: 21–52.

Swofford, D. L. 1990. PAUP: Phylogenetic Analysis Using Parsimony, version 3.0. Computer program distributed by the Illinois Natural History Survey, Champaign.

Walker, A. D. 1968. *Protosuchus, Proterochampsa*, and the origin of phytosaurs and crocodiles. *Geological Magazine* 105: 1–14.

1990. A revision of *Sphenosuchus acutus* Haughton, a crocodylomorph reptile from the Elliot Formation (late Triassic or early Jurassic) of South Africa. *Philosophical Transactions of the Royal Society of London* B330: 1–120.

Whetstone, K., and P. Whybrow. 1983. A "cursorial" crocodilian from the Triassic of Lesotho (Basutoland), southern Africa. *Occasional Papers of the University of Kansas Museum of Natural History* 106: 1–37.

Young, C. C. 1973. [A new fossil crocodile from Wuerho.] *Memoirs of the Institute of Vertebrate Paleontology and Paleoanthropology, Academia Sinica* 11: 37–45. [in Chinese]

Appendix 5.1: Characters

Bone surface

1. External surfaces of cranial and mandibular bones smooth (0) or heavily ornamented, with deep grooves and pits (1)

Rostrum

2. Rostrum narrow anterior to orbits, broadening abruptly at orbits (0) or broad throughout (1)
3. Rostrum higher than wide (0) or nearly tubular (1) or wider than high (2) (unordered)
4. Premaxilla forms at least ventral half of internarial bar (0) or forms little, if any, of internarial bar (1)
5. Premaxilla narrow anterior to nares (0) or broad, similar in breadth to the part lateral to nares (1)
6. Dorsal part of premaxilla vertical, nares laterally oriented (0), or dorsal part of premaxilla nearly horizontal, nares dorsolaterally or dorsally oriented (1)
7. Palatal parts of premaxillae do not meet posterior to incisive foramen (0) or meet posteriorly along contact with maxillae (1)
8. Premaxilla loosely overlies maxilla on face (0), or premaxilla and maxilla sutured together along butt joint (1)
9. Premaxilla and maxilla with broad contact on face, rostrum does not narrow at contact (0), or broad, laterally open notch between maxilla and premaxilla (1), or rostrum constricted at contact with premaxilla and maxilla, forming narrow slit (2), or rostrum constricted at contact with premaxilla and maxilla, forming broad, laterally directed concavity (3) (unordered)
10. Posterior ends of maxillae do not meet on palate anterior to palatines (0), or ends do meet (1)
11. Nasals contact lacrimal (0) or do not (1)
12. Lacrimal contacts nasal along medial edge only (0) or on medial and anterior edges (1)
13. Nasal takes part in narial border (0) or does not (1)

14. Nasal contacts premaxilla (0) or does not (1)
15. Descending process of prefrontal does not contact palate (0), or contacts palate (1), or contacts palate in robust suture (2) (ordered)

Temporal region

16. Postorbital anterior to jugal on postorbital bar (0), postorbital medial to jugal (1), or postorbital lateral to jugal (2) (unordered)
17. Anterior part of jugal as broad as posterior part (0) or about twice as broad as posterior part (1)
18. Jugal transversely flattened beneath lateral temporal fenestra (0) or rod-shaped beneath fenestra (1)
19. Quadratojugal narrows dorsally, contacting only a small part of postorbital (0), or quadratojugal extends dorsally as a broad sheet contacting most of postorbital portion of postorbital bar (1)

Supratemporal roofing bones

20. Frontals narrow between orbits (similar in breadth to nasals) (0) or are broad, about twice nasal breadth (1)
21. Frontals paired (0) or fused (1)
22. Dorsal surface of frontal and parietal flat (0) or with narrow midline ridge (1)
23. Frontal extends well into supratemporal fossa (0) or extends only slightly or not at all (1)
24. Supratemporal roof with complex dorsal surface (0), or dorsally flat "skull table" developed, with squamosal and postorbital with flat shelves extending laterally beyond quadrate contacts (1)
25. Postorbital bar weak, lateral surface sculpted (if skull sculpted) (0), or postorbital bar robust, unsculpted (1)
26. Postorbital bar transversely flattened, unsupported by ectopterygoid (0), or postorbital bar columnar, supported by ectopterygoid (1)
27. Vascular opening on lateral edge of dorsal part of postorbital bar absent (0) or present (1)
28. Postorbital without anterolateral process (0) or with anterolateral process (1)
29. Dorsal part of postorbital with anterior and lateral edges only (0) or with anterolaterally facing edge (1)
30. Dorsal end of postorbital bar broadens dorsally, continuous with dorsal part of postorbital (0), or dorsal part of postorbital bar constricted, distinct from dorsal part of postorbital (1)
31. Bar between orbit and supratemporal fossa broad and solid, with broadly sculpted dorsal surface (0), or bar narrow, with sculpturing on anterior part only (1)
32. Parietal without broad occipital portion (0) or with broad occipital portion (1)
33. Parietal with broad, sculpted region separating fossae (0) or with sagittal crest between supratemporal fossae (1)
34. Postparietal (dermosupraoccipital) a distinct element (0) or not distinct (fused with parietal?) (1)
35. Posterodorsal corner of squamosal squared off, lacking extra "lobe" (0) or with unsculpted "lobe" (1)
36. Posterior edge of squamosal nearly flat (0), or posterolateral edge of squamosal extends posteriorly as a long process (1)

Palate

37. Palatines do not meet on palate below narial passage (0), or form palatal shelves that do not meet (1), or meet ventral to narial passage, forming part of secondary palate (2) (ordered)
38. Pterygoid restricted to palate and suspensorium, joints with quadrate and basisphenoid overlapping (0), or pterygoid extends dorsally to contact laterosphenoid and form ventrolateral edge of trigeminal foramen, strongly sutured to quadrate and laterosphenoid (1)
39. Choana opens ventrally from palate (0) or opens posteriorly into midline depression (1)
40. Palatal surface of pterygoid smooth (0) or sculpted (1)
41. Pterygoids separate posterior to choanae (0) or are fused (1)
42. Choana moderate in size, less than one-fourth of skull breadth (0), or choana extremely large, nearly half of skull breadth (1)
43. Pterygoids do not enclose choanae (0) or enclose choanae (1)
44. Choanae situated near anterior edge of pterygoids (or anteriorly) (0) or in middle of pterygoids (1)

Suspensorium

45. Quadrate without fenestrae (0), or with single fenestra (1), or with three or more fenestrae on dorsal and posteromedial surfaces (2) (unordered)
46. Posterior edge of quadrate broad medial to tympanum, gently concave (0), or posterior edge narrow dorsal to otoccipital contact, strongly concave (1)
47. Dorsal, primary head of quadrate articulates with squamosal, otoccipital, and prootic (0) or with prootic and laterosphenoid (1)
48. Ventrolateral contact of otoccipital with quadrate very narrow (0) or broad (1)
49. Quadrate, squamosal, and otoccipital do not meet to enclose cranioquadrate passage (0), enclose passage near lateral edge of skull (1), or meet broadly lateral to passage (2) (ordered)
50. Pterygoid ramus of quadrate with flat ventral edge (0) or with deep groove along ventral edge (1)
51. Ventromedial part of quadrate does not contact otoccipital (0) or contacts otoccipital to enclose carotid artery and form passage for cranial nerves IX-XI (1)

Braincase

52. Eustachian tubes not enclosed between basioccipital and basisphenoid (0) or entirely enclosed (1)
53. Basisphenoid rostrum (cultriform process) slender (0) or dorsoventrally expanded (1)
54. Basipterygoid process prominent, forming movable joint with pterygoid (0), or basipterygoid process small or absent, with basipterygoid joint closed suturally (1)
55. Basisphenoid similar in length to basioccipital, with flat or concave ventral surface (0), or basisphenoid shorter than basioccipital (1)
56. Basisphenoid exposed on ventral surface of braincase (0) or virtually excluded from ventral surface by pterygoid and basioccipital (1)

57. Basioccipital without well-developed bilateral tuberosities (0) or with large, pendulous tubera (1)
58. Otoccipital without laterally concave descending flange ventral to subcapsular process (0) or with flange (1)
59. Cranial nerves IX–XI pass through common large foramen vagi in otoccipital (0), or cranial nerve IX passes medial to nerves X and XI in separate passage (1)
60. Otoccipital without large ventrolateral part ventral to paroccipital process (0) or with large ventrolateral part (1)
61. Crista interfenestralis between fenestrae pseudorotunda and ovalis nearly vertical (0) or horizontal (1)
62. Supraoccipital forms dorsal edge of foramen magnum (0), or otoccipitals broadly meet dorsal to the foramen magnum, separating supraoccipital from foramen (1)
63. Mastoid antrum does not extend into supraoccipital (0) or extends through transverse canal in supraoccipital to connect middle ear regions (1)
64. Posterior surface of supraoccipital nearly flat (0) or with bilateral posterior prominences (1)

Palpebral

65. One small palpebral present in orbit (0), or two large palpebrals present (1), or one large palpebral present (2) (unordered)

Cranial openings

66. External nares divided (0) or confluent (1)
67. Antorbital fenestra as large as orbit (0), or about half the diameter of the orbit (1), or much smaller than orbit (2), or absent (3) (ordered)
68. Supratemporal fenestrae much longer than orbits (0) or equal in length to or shorter than orbits (1)
69. Choanae confluent (0) or divided by septum (1)

Mandible

70. Dentary extends posteriorly beneath mandibular fenestra (0) or does not extend beneath fenestra (1)
71. Retroarticular process very short and robust (0), or absent (1), or short, robust, and ventrally situated (2), or posterodorsally curving and elongate (3), or posteroventrally projecting and paddle-shaped (4), or posteriorly projecting from ventral part of mandible and attenuating (unordered)
72. Prearticular present (0) or absent (1)
73. Articular without medial process articulating with otoccipital and basisphenoid (0) or with process (1)
74. Dorsal edge of surangular flat (0) or arched dorsally (1)
75. Mandibular fenestra present (0) or absent (1)
76. Insertion area for M. pterygoideus posterior does not extend onto lateral surface of angular (0) or extends onto lateral surface of angular (1)
77. Splenial not involved in symphysis (0), or involved slightly in symphysis (1), or involved extensively in symphysis (2) (ordered)

Dentition

78. Posterior two premaxillary teeth similar in size to anterior teeth (0) or much longer (1)
79. Maxillary teeth homodont, with lateral edge of maxilla straight (0), or teeth enlarged in the middle of tooth row, with edge of maxilla extending outward at these loci (1), or teeth enlarged and edge of maxilla curved in two waves ("festooned") (2) (unordered)
80. Anterior dentary teeth opposite premaxilla–maxilla contact no more than twice the length of other dentary teeth (0) or more than twice the length (1)
81. Dentary teeth posterior to tooth opposite premaxilla–maxilla contact homodont (0) or enlarged opposite smaller teeth in maxillary tooth row (1)

Pectoral girdle

82. Anterior and posterior scapular edges symmetrical in lateral view (0), or anterior edge more strongly concave than posterior edge (1)
83. Coracoid no more than half the length of scapula (0) or about equal in length to scapula (1)

Pelvic girdle and hindlimb

84. Anterior process of ilium similar in length to posterior process (0) or one-quarter or less the length of posterior process (1)
85. Pubis rodlike, without expanded distal end (0) or with expanded distal end (1)
86. Pubis forms anterior half of ventral edge of acetabulum (0), or pubis at least partially excluded from the acetabulum by an anterior process of the ischium (1)
87. Distal end for femur with large lateral facet for fibula (0) or with very small facet (1)
88. Fifth pedal digit with (0) or without (1) phalanges.

Vertebral column

89. Atlas intercentrum broader than long (0) or as long as broad (1)
90. Neural spines on posterior cervical vertebrae as broad as those on anterior cervical vertebrae (0) or anteroposteriorly narrow, rodlike (1)
91. Cervical vertebrae without well-developed hypapophyses (0) or with well-developed hypapophyses (1)
92. Cervical vertebrae amphicoelous or amphiplatyan (0) or procoelous (1)
93. Trunk vertebrae amphicoelous or amphiplatyan (0) or procoelous (1)
94. All caudal vertebrae amphicoelous or amphiplatyan (0), or first caudal vertebra biconvex, with other caudal vertebrate procoelous (1), or all caudal vertebrae procoelous (2) (unordered)

Osteoderms

95. Dorsal osteoderms rounded, ovate (0), or rectangular, broader than long (1), or square (2) (unordered)
96. Dorsal osteoderms with straight anterior edge (0) or with anterior process laterally on anterior edge (1)
97. Dorsal osteoderms in two parallel, longitudinal rows (0) or in more than two longitudinal rows (1)

98. Some or all osteoderms imbricated (0), or osteoderms sutured to one another (1)
99. Tail with dorsal osteoderms only (0) or completely surrounded by osteoderms (1)
100. Osteoderms absent from ventral part of trunk (0) or present (1)
101. Osteoderms with longitudinal keels on dorsal surfaces (0) or without keels (1)

Appendix 5.2: Character distributions

V, variable within the taxon; A, character cannot be scored in taxon; ?, unknown for the taxon.

Out-groups

	10	20	30	35	40	50
Gracilisuchus	00000 0?10? 0000?	00000 00?00	00000 0000?	0??0?	0?000 00A00	0???0

	60	70	80	90	100
	000A0 ???00 000?1	00000 ?0000	0000? 0???0	000?0 10000	0

	10	20	30	35	40	50
Postosuchus	000?0 00010 0000?	00011 00?00	00001 0111?	00000	00000 00A00	00000

	60	70	80	90	100
	000A0 ?0?0? 001?0	10000 ?0000	000?? 00100	000?? ?????	?

	10	20	30	35	40	50
Pseudhesperosuchus	000?0 0???? ?0001	?0??0 00100	0?00? 011??	0??0?	0?000 00A00	0??00

	60	70	80	90	100
	000A0 ???0? ?01??	1?0?? ??00?	000?? ?0??0	000?? ?????	?

	10	20	30	35	40	50
Saltoposuchus	000?? 0???0 ?0000	000?0 00100	00000 01110	00000	00000 00A00	0??0?

	60	70	80	90	100
	000A0 000?? ?01??	?000? ?0?00	0000? 010?0	000?0 10000	0

	10	20	30	35	40	50
Sphenosuchus	000?0 00010 1?000	000?0 00000	00000 01110	00000	00000 0?A00	00000

	60	70	80	90	100
	000A0 0000? 001??	1?000 ??000	0000? ?1??0	??0?? ???0?	?

	10	20	30	35	40	50
Dibothrosuchus	000?0 00010 1?001	?00?1 11100	0?000 01110	00?00	00002 00A00	00000

	60	70	80	90	100
	000A0 ?000? 000?0	10000 ?0000	0000? ????0	000?0 1????	0

	10	20	30	35	40	50
Kayenta "Spheno"[a]	000?0 00010 ?000?	?0??? 00100	??0?? 0101?	00??0	?0000 00?0?	?????

	60	70	80	90	100
	?0?A? 0000? 001??	???00 ?0000	0?00? ????0	??0?1 1?0??	?

Crocodyliformes

	10	20	30	35	40	50
Protosuchus	10000 00110 ?0000	11010 00110	00001 00010	00101	00002 01001	11111

	60	70	80	90	100
	00101 01101 010?1	20110 01101	01000 11100	000?1 10011	0

	10	20	30	35	40	50

Hemiprotosuchus
```
             10        20         30   35   40         50
?00?? 0?1?? ????? 10010 ?0??0 0?001 0011? ???01 ??002 0?00? 11?11
             60        70         80        90        100
00101 ??1?1 ?10?? 2?11? ???01 ????? ????0 ????1 ?00?1 ?
```

Kayenta "Edento"[b]
```
             10        20         30   35   40         50
10000 0?110 ?0000 10010 ?0??0 0???? 0?0?? ?1111 00002 01001 11111
             60        70         80        90        100
00001 011?1 010?0 20110 0110? 0??0? ???00 0?0?1 10010 0
```

Eopneumatosuchus
```
             10        20         30   35   40         50
1???? ????? ????? ????1 0001? ????? ?0010 0?1?? ????? ?1?0? 11?11
             60        70         80        90        100
0000? 0100? ??1?? ????? ????? ????? ????? ????? ????? ?
```

Orthosuchus
```
             10        20         30   35   40         50
100?0  00110  1?000 10010  00110 00001  00010 00?00 00000  01100 11111
             60        70         80        90        100
00??1  ?1?01  010?0  ?000?  00100 01000 11100  000?1  10010 0
```

Gobiosuchus
```
             10        20         30   35   40         50
110?0 01100  1?001 10011 00?10 0?0?1 00??0 ?0?00 00000  01100  11?11
             60        70         80        90        100
00001  ????? 01???  ??0?? ??10? ?10?? ????0  ????? ??11? ?
```

Pelagosaurus
```
             10        20         30   35   40         50
111?0 11101 00110 20101 00000 00000 00010 02110 10000 00110 11110
             60        70         80        90        100
01001 ?1000 1210? 30000 12000 01101 11?00 00001 10001 1
```

Teleosauridae
```
             10        20         30   35   40         50
0V1?0 11101 00110 10100 10000 00000 00110 02110 1000? 00110 11110
             60        70         80        90        100
01011 ?1?00 1210? 3?000 12000 00101 111?0 000?1 10001 0
```

Metriorhynchidae
```
             10        20         30   35   40         50
011?0 11101 0011? 20101 10000 00000 00110 02110 ?000? 00110 11110
             60        70         80        90        100
01011 ?1?00 1210? 30001 12000 00101 ?11?0 000?A AAAA0 ?
```

Fruita[c]
```
             10        20         30   35   40         50
110?0 00110 01000 10010 10010 00001 10010 021?1 ?0002 0112? 1????
             60        70         80        90        100
?0??0 ??1?1 030?? ????? 01111 01011 ?1?00 01111  1?0?? 1
```

Notosuchus
```
             10        20         30   35   40         50
?10?0 01101 01001 11010 11?11 0?011 00000 02??0 ?0002 01120 11?10
             60        70         80        90        100
00?10 ?110? ?10?? 5?010 ?110? 0??1? ????0 0?0?? ????? ?
```

	10	20	30	35	40	50

Baurusuchus

```
              10        20        30   35   40        50
Baurusuchus   00000 01111 ??00? 11010 1??11 0?011 0???? ?2?10 11001 11120 11?10
              60        70        80        90        100
              00??0 ?11?? 03?01 5?010 11101 0???? ????0 ????? ????? ?

              10        20        30   35   40        50
Sebecus       1000? 01101 0000? 11000 10?11 1?0?1 00010 02?10 1100? 11120 11110
              60        70        80        90        100
              00??0 ?1??1 03000 51000 11000 0???? ????0 ?00?? ????? ?

              10        20        30   35   40        50
Libycosuchus  ?10?0 01101 ??00? 11010 10?11 0?011 000?0 02??? ??00? 11120 11?10
              60        70        80        90        100
              10??0 ???0? 0?0?? 5?000 1?000 0???? ????0 0?0?? ????? ?

              10        20        30   35   40        50
Araripesuchus 11010 11121 01002 11010 10111 10011 0?0?? ?2?10 10001 1112? 11?10
              60        70        80        90        100
              00??0 ?11?1 01000 5?000 11010 1???? ????0 ?00?? ????? ?

              10        20        30   35   40        50
Theriosuchus  11010 1?131 01001 10000 11011 11001 10011 02110 10001 ?11?0 11110
              60        70        80        90        100
              00??? ??1?1 0200? 4?001 01010 11011 11??0 011?1 10010 0

              10        20        30   35   40        50
Alligatorium  11010 1?13? 0000? 10000 10111 1?00? 100?1 0???0 ??00? ?11?? 1??10
              60        70        80        90        100
              00??? ????1 0?0?? 4?001 01?10 11011 11100 ????1 ?0010 0

              10        20        30   35   40        50
Eutretauranosuchus 110?? 11101 00102 11000 10011 1?00? 00010 01110 ?000? 11120 11?10
              60        70        80        90        100
              10??0 ?1?0? 12010 4?000 01010 111?? ?1??0 ??0?1 ????? ?

              10        20        30   35   40        50
Goniopholis   110?1 1?131 ?010? 11000 10111 1?001 00010 02?10 1000? 11120 11?10
              60        70        80        90        100
              10?10 ?1?01 1301? 4?001 ?1010 11?1? ?1??0 ?00?1 100?1 1

              10        20        30   35   40        50
Pholidosaurus 101?? 11101 ??11? 11101 10011 1?001 00010 0211? 10000 11120 11?10
              60        70        80        90        100
              1??10 ?100? 1300? 300?? ?2?0? ??11? 1???0 ??0?? 1?0?? ?

              10        20        30   35   40        50
Sokotosuchus  111?1 11101 ??10? ???00 1001? ??101 001?0 12?1? ????? 1112? 11?1?
              60        70        80        90        100
              11??0 ???1? 1?1?? ????? ??01? ????? ????? ????? ????? ?
```

	10	20	30	35	40	50
Dyrosaurus	011??	1?101	?010?	11?00	10001	1?101	00110	12?10	10100	11120	11?10

	60	70	80	90	100
	11?10	?1010	1311?	3?00?	?2?00	0????	????0	?00??	????1	?

	10	20	30	35	40	50
Bernissartia	110?1	11131	??00?	11000	?0?11	1?001	000?0	02???	??000	1112?	11?10

	60	70	80	90	100
	100?0	?1???	1?0??	4?001	0?010	11?1?	11??0	?0021	01011	0

	10	20	30	35	40	50
Leidyosuchus	110??	11131	00102	11000	10011	1?001	00010	02110	10111	11120	11110

	60	70	80	90	100
	10010	?110?	13000	31000	11010	11111	11111	1111?	010?0	0

	10	20	30	35	40	50
Gavialis	101A1	11101	00112	11101	10111	11001	00010	02110	10110	11120	11110

	60	70	80	90	100
	11010	11101	13000	31000	12000	00111	11101	11112	01110	0

	10	20	30	35	40	50
Crocodylidae	11001	11131	00V2	11000	1VV11	11001	00010	02110	V0111	11120	11110

	60	70	80	90	100
	10010	11101	V30V0	31000	1V010	10111	11101	11112	01110	0

	10	20	30	35	40	50
Alligatoridae	1101V	1V1?1	V0V02	11000	10111	11001	000V0	02110	10111	11120	11110

	60	70	80	90	100
	10010	11101	V30V0	31000	11010	10111	11111	11112	0111V	0

[a]An unnamed "sphenosuchian" from the Lower Jurassic Kayenta Formation of Arizona, described by Clark (1986).
[b]An unnamed protosuchid from the Lower Jurassic Kayenta Formation of Arizona, similar to *Edentosuchus* Young, 1973, described by Clark (1986).
[c]An unnamed primitive "mesosuchian" from the Upper Jurassic Morrison Formation of Fruita, Colorado, described by Clark (1985).

Appendix 5.3: Diagnoses

Following are the apomorphies for the cladogram arbitrarily chosen from among the equally most parsimonious cladograms (Figure 5.2). Node numbers are those indicated on Figure 5.2. ACCTRANS optimization was used, and those characters that are ambiguous are indicated with an asterisk. Unless otherwise indicated, derived states are state number 1.

Node

1: 1, 8, 24, 32(0), 45(2), 47, 62, 67, 78
2: 2*, 7*, 9(0)*, 15*, 20, 23(0)*, 100*
3: 63*, 68(0)
4: 11(0)*, 37*, 39*, 40*, 74, 77, 80
5: 48(0), 50, 73
6: 37(0)*, 39(0)*, 58, 70*, 100
7: 2, 12*, 21, 37(2), 41*, 49(2), 55(0)*, 59*, 60(0), 61*, 67(3)*, 71(5)*, 72*, 84, 85
8: 7, 9(0)*, 10, 15, 17, 25, 29, 40(0)*, 76*, 94(2)*
9: 12(0)*, 42*, 46
10: 74(0), 78(0), 80(0)
11: 19(0)*, 26

12:   4, 6, 9(2)*, 42(0)*, 45, 67(2)*, 79, 81
13:   9(3)*, 27*, 29(0), 71(4), 76(0)*, 92*, 93*
14:   17(0), 31, 35, 75
15:   5, 13, 56, 66, 67(3)*, 83
16:   27(0)*, 45(0), 63(0)*, 69, 72(0)*, 92(0)*, 93(0)*, 100
17:   9(0), 23(0), 65(0)*, 94(0)*, 99(0)*
18:   3, 15(0)*, 18*, 57, 68*, 71(3), 76*, 77(2), 79(0)*, 81(0)
19:   28, 33, 36, 43*, 64
20:   5(0)*, 14, 20, 69(0)
21:   16(2), 17(0), 24(0), 25(0), 26(0), 30(0), 46(0), 47(0), 49, 56(0), 60, 67(2), 84(0)
22:   1(0), 33, 82(0)
23:   75
24:   15(2), 43, 44, 71(3), 76*, 90, 91, 95(0), 96(0), 97
25:   82(0), 98

Hemiprotosuchus: 33
Protosuchus: 17
Fruita: 23(0), 31, 79, 92, 93, 101
Notosuchus: 22, 34(0), 67*
Baurusuchus: 1(0), 2(0), 9*, 45, 70
Libycosuchus: 56
Sebecus: 2(0)
Araripesuchus: 12*, 15(2), 19*, 67
Theriosuchus: 12*, 22, 23(0)
Eutretauranosuchus: 15(2), 37, 67(2)
Goniopholis: 101
Sokotosuchus: 79*
Dyrosaurus: 1(0), 24(0)
Pholidosaurus: 2(0), 68(0)*
Pelagosaurus: 21(0), 59(0), 101
Metriorhynchidae: 75, 100(0)
Teleosauridae: 20(0)
Bernissartia: 13(0), 93*, 96(0), 97
Leidyosuchus: 23(0), 89
Gavialis: 2(0), 3, 9(0), 14, 18, 20, 45(0), 57, 77(2), 79(0), 81(0)
Crocodylidae: 4(0)
Alligatoridae: 89

6

Sister-group relationships of mammals and transformations of diagnostic mammalian characters

ZHEXI LUO

Introduction

The origin of Mammalia is characterized by many important transformations in dentition and skull structure. The interpretation of the phylogenetic transformations of these features, however, depends on our understanding of the phylogenetic relationships of advanced nonmammalian cynodonts and early mammals. The postulation of the sister-taxon to mammals is perhaps one of the most crucial issues concerning the evolutionary origin of mammals. Historically, four cynodont families have been proposed as being closely related to the common ancestry of all mammals: Thrinaxodontidae (Hopson, 1969; Hopson and Crompton, 1969; Barghusen and Hopson, 1970); Probainognathidae (Romer, 1970; Crompton and Jenkins, 1979), Tritheledontidae (Hopson and Barghusen, 1986; Shubin et al., 1991; Crompton and Luo, 1993), and Tritylodontidae (Kemp, 1982, 1983; Rowe, 1986, 1988; Wible, 1991).

Two competing hypotheses concerning the sister-taxon of Mammalia (Mammaliaformes of Rowe, 1988) have emerged from the recent debate on the phylogenetic relationships of nonmammalian cynodonts and early mammals: the tritheledontid-mammal hypothesis versus the tritylodontid-mammal hypothesis. The tritheledontid-mammal hypothesis was first proposed by Hopson and Barghusen (1986) and then gained support from studies by Shubin et al. (1991) and Crompton and Luo (1993). The tritylodontid-mammal hypothesis was first implied by Kemp (1983) and then greatly expanded by the more extensive studies of Rowe (1986, 1988) on a much larger data base. The early versions of the tritylodontid-mammal hypothesis did not explicitly state the phylogenetic position of tritheledontids (e.g., Kemp, 1983, fig. 11; Rowe, 1988, figs. 3 and 4). More recently, various authors have argued that tritylodontids and mammals are more

closely related to each other than either group is to tritheledontids (Wible, 1991, fig. 4; Wible and Hopson, 1993).

Both the tritheledontid-mammal hypothesis and the tritylodontid-mammal hypothesis are supported by large numbers of apomorphies in dentition, cranium, and postcranial skeleton. Yet both are also contradicted by a substantial amount of anatomical evidence. At the very best, only one of them can be correct. In order to make a convincing choice between the two competing hypotheses, we need additional morphological information from new transitional taxa, as well as the taxa documented by incomplete or poor fossil materials.

Recent studies of new material of *Sinoconodon* from the Lower Jurassic of Yunnan (Crompton and Sun, 1985; Crompton and Luo, 1993) have shown that this genus possesses the intermediate transformational steps for several diagnostic characters of other Liassic mammals. *Adelobasileus* Lucas and Hunt, 1990, based on an incomplete skull from the Carnian of Texas, has many diagnostic mammalian basicranial features. The character evidence for *Adelobasileus* and *Sinoconodon* suggests that these two taxa and all other mammals form a monophyletic group (Figure 6.1). Within the monophyletic Mammalia, however, *Sinoconodon* and *Adelobasileus* represent the intermediate clades between nonmammalian cynodonts (such as tritylodontids and tritheledontids) and other Liassic mammals (such as morganucodontids, triconodontids, and kuehneotheriids). Recent studies on the dentition and quadrate suspension of tritheledontids have revealed more apomorphic features in this group (Allin and Hopson, 1991; Shubin et al., 1991; Luo and Crompton, in press). The description of *Haldanodon* (Lillegraven and Krusat, 1991) has provided additional data on docodontids.

The first goal of this chapter is to evaluate the alternative hypotheses concerning the relationships

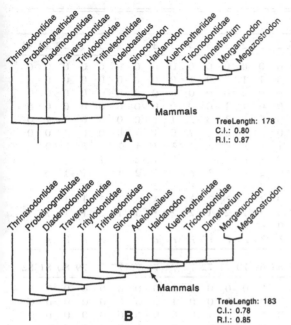

Figure 6.1. Hypotheses on the tritheledontid-mammal sister-group relationships. (A) Preferred phylogenetic tree of this study. (B) Strict consensus tree of PAUP (Swofford, 1990); transformational steps (TreeLength), consistency index (C.I.), and retention index (R.I.) for both trees. Based on parsimony analyses of eighty-two characters among fourteen relevant taxa (Table 6.1, Appendix 6.1). Multiple character states are treated as unordered data; no character is weighted. Based on character matrix (Table 6.1, Appendix 6.1), PAUP (3.0) has identified eight fully resolved phylogenetic trees that are equally parsimonious, and each of the eight trees has TreeLength = 187, C.I. = 0.77, R.I. = 0.83. Strict-consensus tree B is based on all eight equally parsimonious trees. Although multituberculates, monotremes, and later therians are not included in the trees, it is assumed that they belong to mammals.

of advanced nonmammalian cynodonts and early mammals in light of the new information on tritheledontids, *Sinoconodon, Adelobasileus,* and *Haldanodon.* The second goal is to examine the phylogenetic transformations of several functionally important mammalian characters, such as the postcanine roots, the temporomandibular joint, the quadrate, and the bony housing for the cochlea.

Data and methods

The emphasis in this analysis is on the interrelationships of Tritylodontidae, Tritheledontidae, *Sinoconodon, Adelobasileus, Haldanodon, Megazostrodon, Morganucodon,* Triconodonta, and Kuehneotheriidae; thus these are technically treated as in-group taxa (Wiley, 1979). Polarity of character states among these taxa was

determined using out-groups, including *Probainognathus,* Traversodontidae, Diademodontidae, and *Thrinaxodon.* A character state present in at least two consecutive out-groups is designated an ancestral character state (Maddison, Donaghue, and Maddison, 1984).

The relationships of multituberculates to monotremes and therian and other mammals have constituted an important issue in the systematics of mammals. The phylogenetic positions of multituberculates and monotremes relative to other mammals have direct bearings on the definition and diagnosis of the Mammalia (Rowe, 1987, 1988). However, because of limitations of space, monotremes, multituberculates, *Vincelestes,* and other therians are not included in this analysis. The phylogenetic relationships of monotremes, multituberculates, and *Vincelestes* were addressed in several recent papers (Miao, 1988; Rowe, 1988; Luo, 1989; Wible, 1991; Rougier, Wible, and Hopson, 1992).

Information on the distribution of morphological characters is the substance of phylogenetic hypotheses. Much of the current debate on the relationships of the mammalian sister-taxon can be attributed to different interpretations of the anatomy and distribution of morphological characters in relevant taxa (e.g., Kemp, 1983; Sues, 1985; Rowe, 1988; Wible, 1991). Table 6.1 and Appendix 6.1 present my interpretation of the anatomy and distributions of 82 cranial and dental characters for 14 taxa. Many characters in Table 6.1 and Appendix 6.1 have been evaluated by Kemp (1983), Hopson and Barghusen (1986), Rowe (1988), Wible (1991), Lillegraven and Krusat (1991), Crompton and Luo (1993), and Lucas and Luo (1993). I shall provide some new information from my own observations of the original specimens of many relevant taxa, including *Sinoconodon* (IVPP 4729, 8683, 8689, 8691, etc.), *Morganucodon* (IVPP 4727, 8682, 8684, etc.), *Adelobasileus* (NMMNH P-12971) (Lucas and Hunt, 1990; Lucas and Luo, 1993), *Kayentatherium* (MCZ 8812) (Sues, 1986), *Bienotheroides* (IVPP 7909, 7913, etc.), *Yunnanodon* (IVPP 5071, 8696, etc.; *Yunnania* of Cui, 1976; *Yunnanodon* of Sun and Cui, 1986, 1987). The information on *Megazostrodon* is based on Crompton (1974), Gow (1986b) and Rowe (1986), supplemented by my own observations of some fragmentary materials in the MCZ collection. Information on *Dinnetherium* and the Cloverly Formation triconodont is based on Crompton and Jenkins (1979), Jenkins (1984), Crompton and Sun (1985), and Crompton and Luo (1993), supplemented by my own observation of the MCZ collection made by F. A. Jenkins, Jr. The information on *Haldanodon* is based on Lillegraven and Krusat (1991).

Because the published information on tritheledontids, *Sinoconodon, Adelobasileus,* and *Haldanodon* comes primarily from dentitions and skulls, this study focuses on the cranial and dental characters. Ideally, an interpretation of phylogenetic relationships should be *rs for 14 taxa of advanced nonmammalian cynodonts and early mammals considered in this study*

8	9	10	11	12	13	14	15	16	17	18	19	20	21	22	23	24	25	26	27	28	29	30	31	32	33	34	35	36	37	38	39	40	41
0	0	0	0	0	0	0	?	0	0	0	0	0	0	0	0	0	0	?	0	0	0	0	0	0	0	0	0	0	0	0	0	0	0
0	0	0	0	0	0	0	?	0	0	0	0	0	0	0	0	0	0	?	1	0	1	1	1	0	1	0	0	0	0	0	0	0	0
0	0	0	0	1	1	3	?	1	4	?	0	0	0	0	0	0	0	?	0	0	0	0	0	0	0	0	0	0	0	0	0	0	0
0	0	0	0	3	1	3	?	2	4	?	0	0	0	0	0	0	0	?	0	0	0	1	1	0	0	0	0	0	0	0	0	0	0
0	4	0	0	3	1	3	?	2	4	?	0	1	0	0	1	0	0	?	0	0	0	1	3	0	2	1	1	0	0	0	0	1	1	0
0	1	0	0	0	0	0	?	1	1	1	1	0	0	0	1	1	1	2	0	2	1	2	1	2	0	1	0	0	0	0	1	0	0
?	?	?	?	?	?	?	?	?	?	?	?	?	?	?	?	?	?	?	1	?	?	?	?	?	?	1	1	1	?	?	?	0	1
0	2	0	0	3	0	0	0	0	0	0	1	0	2	0	1	3	0	3	0	2	?	?	?	?	?	?	1	2	1	1	1	1	1
1	3	1	1	1	1	1	0	1	2	2	2	1	1	1	1	3	1	4	0	2	?	?	?	?	?	?	1	2	1	?	1	1	0	1
1	3	1	3	3	0	1	0	1	2	2	2	0	1	2	1	2	1	4	0	3	?	?	?	?	?	?	1	3	1	2	1	1	0	1
1	3	1	1	1	0	1	1	1	2	2	2	0	1	2	1	2	1	4	0	3	?	?	?	?	?	2	1	3	1	2	1	1	1	1
1	3	1	1	2	0	1	0	1	2	1	2	0	1	2	1	2	1	4	0	3	1	2	1	2	2	1	3	1	2	1	1	1	1
1	3	1	1	1	0	2	0	1	2	1	2	0	1	2	1	2	1	4	0	?	?	?	?	?	?	?	1	2	1	?	1	1	1	1
1	3	1	2	1	2	2	0	1	3	2	2	0	0	2	1	2	1	?	?	?	?	?	?	?	?	?	?	?	?	?	?	?	?

49	50	51	52	53	54	55	56	57	58	59	60	61	62	63	64	65	66	67	68	69	70	71	72	73	74	75	76	77	78	79	80	81	82
0	0	0	0	0	0	0	0	0	0	0	0	0	0	0	0	0	0	0	0	0	0	0	0	0	0	0	0	0	0	0	0	0	0
0	0	0	0	0	1	0	0	0	1	0	0	0	1	0	1	1	1	0	0	0	0	0	0	0	1	0	0	0	0	0	0	0	0
0	0	0	0	0	1	0	0	0	0	1	0	0	0	0	0	0	0	0	0	0	0	0	1	0	0	0	0	0	0	0	0	0	0
0	0	1	1	2	1	0	0	0	1	0	0	0	0	1	1	1	0	0	0	0	0	1	1	0	0	0	0	0	0	0	0	0	0
0	0	2	1	2	2	1	2	1	1	2	1	0	1	0	1	1	0	0	0	1	0	1	1	0	0	0	0	0	0	0	0	0	1
0	0	2	0	0	3	1	1	0	0	1	0	1	0	1	1	1	0	1	1	1	1	0	0	1	1	0	0	0	0	0	?	0	1
0	1	1	2	1	?	1	?	?	?	2	1	1	?	1	1	1	1	1	?	1	?	1	0	0	2	3	1	1	1	1	1	?	?
?	1	2	1	2	3	1	2	1	1	2	1	0	2	1	1	1	1	1	1	1	1	1	0	0	3	2	1	1	1	1	1	0	1
1	0	2	2	?	3	1	?	?	1	2	1	1	2	?	1	0	1	1	2	1	2	2	0	0	3	2	0	1	1	?	1	?	1
1	1	2	2	1	3	1	2	1	1	2	1	0	2	2	1	1	1	2	2	1	1	1	0	0	3	2	1	1	1	1	1	1	1
1	1	2	2	1	?	1	?	?	1	2	?	?	2	?	1	1	1	?	?	1	?	1	0	0	3	2	1	1	1	1	1	?	?
1	1	2	2	1	3	1	2	1	1	2	1	0	2	2	1	1	1	1	1	1	1	1	0	0	3	2	1	1	1	1	1	1	1
1	0	?	2	1	?	1	?	?	1	2	1	0	2	?	1	1	1	1	?	1	1	1	?	0	3	2	1	1	1	1	1	?	?
?	?	?	?	?	?	?	?	?	?	?	?	?	?	?	?	?	?	?	?	?	?	?	?	?	?	?	?	?	?	1	?	?	?

es are coded 0, 1, 2, 3, 4; "?" represents the character states that are not preserved, or are not applicable, or have
iptions and definitions of these codes are listed in Appendix 6.1. Multiple character states are treated as unordered
cter is weighted.

Table 6.2. *Association of mammalian apomorphies by anatomical systems*[a]

Apomorphy association (anatomical system)	Apomorphies shared by tritheledontids and mammals	Apomorphies shared by tritylodontids and mammals
Orbitotemporal region	(1) slender zygomatic arch	(1) large ascending process of palatine in orbit (2) separate orbital openings for greater and lesser palatine nerves (3) palatine expands to participate in subtemporal border of orbit (4) orbitosphenoid contributing to the orbital wall
Temporomandibular joint and lower jaw	(1) posterior enlargement of lateral ridge of dentary (2) anteroventrally facing concavity in squamosal receiving the dentary (3) mobile symphysis (4) dorsomedial movement of lower jaw during occlusion	none
Palate	(1) greater width of posterior palate (2) pterygopalatine ridges reach basisphenoid	none
Petrosal	(1) greater separation of round window and jugular foramen (2) incipient separation of jugular and hypoglossal foramina	(1) presence of the hyoid ("stapedial") fossa (?) (2) bifurcation of the paroccipital processes (?)
Quadrate and its suspension	(1) orientation of dorsal plate is most similar to that of *Morganucodon* (2) rounded dorsal margin of the dorsal plate	(1) presence of "stapedial process" (2) quadrate contacts anterior paroccipital process of petrosal (?)
Dentition	(1) wear facets on lingual side of the upper teeth and labial side of lower teeth (?)[b]	(1) alveolar implantation of postcanine roots (?) (2) completely divided roots (?)

[a] Apomorphies shared by both tritheledontids and tritylodontids are excluded.
[b] ?, Equivocal apomorphies that are absent from one or more early mammalian taxa, or present in one or more out-groups.

based on as many characters as possible and as many sources of information as possible. Among the previous phylogenetic studies of advanced nonmammalian cynodonts and early mammals, Rowe (1986, 1988) has used the largest suite of characters from the dentition, the skull, and the postcranial skeleton. Most other studies of cynodont–mammal evolutionary transition have been limited to the information from dentition and skull (e.g., Hopson and Barghusen, 1986; Wible, 1991; Crompton and Luo, 1993). The practical emphasis on skull and dentition, although far from ideal, is necessitated by the availability of the best fossil evidence. It is justifiable only if it continues to generate insight into the important evolutionary

changes in these systems. In any case, the hypothesis established on cranial and dental evidence can always be tested further by data from postcranial skeletons.

In the treatment of discrete character states in Table 6.1, a question mark indicates that the character is not present or is not preserved, or else it indicates that the anatomical interpretation is uncertain. The character states of multiple-state characters are treated as unordered, in order to avoid the bias of a priori interpretation of character evolution. The character matrix (Table 6.1) serves as the data base for two phylogenetic programs: MacClade (Maddison and Maddison, 1986) and PAUP (Swofford, 1990). No character weighting was used with these phylogenetic algorithms.

The main results of these analyses are presented in Figure 6.1.

In this analysis, the main criterion for evaluating conflicting relationship hypotheses is the minimal number of transformational steps (Farris, 1982; Swofford, 1990). This criterion was first introduced in an analysis of phylogenetic relationships of non-mammalian cynodonts and mammals by Rowe (1988). Since then, similar parsimony analyses based on character matrices have been used in several studies (Wible, 1991; Lucas and Luo, 1993; Wible and Hopson, 1993). Using the minimal number of transformational steps as the criterion for identifying a preferred hypothesis is an objective approach that can best avoid the biased weighting of characters.

Parsimony analysis, however, is not infallible, for two reasons: First, selection of a preferred hypothesis by a minimum of transformational steps may be quite arbitrary if there is only a small difference in transformational steps between the competing hypotheses. Second, the criterion of minimal number of transformational steps is meaningful only if the number of characters used in the analysis truly reflects the morphological complexity. By giving equal weights to all characters, parsimony analysis precludes any consideration of the biological merits of characters (Patterson, 1982). Evaluation of conflicting hypotheses can be reduced to comparison of numbers of characters, which may not be equal in their informational consistency (Hall, 1991). By oversplitting apomorphies in its favor, one hypothesis can dominate over its rival without gaining any biological insight. One way to guard against this fallacy is to show how the apomorphies in support of a given hypothesis are biologically associated. If apomorphies can be grouped by their anatomical association, that should show whether or not the number of characters truly reflects the complexity of the anatomy. That will expose idiosyncratic oversplitting of apomorphies. In this study, I categorize apomorphies supporting each alternative hypothesis by the anatomical areas to which they belong (Table 6.2). By identifying the anatomical correlations among the apomorphies in support of each hypothesis, I wish to provide an evaluation of the degree to which the number of characters reflects the complexity of the anatomy (Table 6.2).

Grouping apomorphies by their anatomical association can help to demonstrate the functional correlation between individual apomorphies. Although the functional significance of a character was not considered in the analysis of phylogenetic relationships, an a posteriori evaluation of the functional implications of each hypothesis was discussed. By grouping the apomorphies into such character sets as the temporomandibular joint, the occlusal characteristics of the jaws, and the promontorium, the functional significance of these apomorphies can be better appreciated.

Tritylodontid-mammal hypothesis

The hypothesis of a sister-group relationship between tritylodontids and mammals is strongly supported by the similarities in the orbital wall between tritylodontids and early mammals (Figure 6.2). In tritylodontids, the palatine has an orbital process that contacts the frontal dorsally and forms an extensive area of the orbital wall (Kühne, 1956; Sun, 1984; Sues, 1986). By contrast, although an orbital process of the palatine is reported in tritheledontids (Crompton, 1958, fig. 3B), it is not large enough to contribute to the closure of the orbital vacuity. In all tritylodontids (Young, 1947; Kühne, 1956; Sun, 1984; Gow, 1986a; Rowe, 1986; Sues, 1986). The palatine expands along the subtemporal margin of the orbit and displaces the transverse flange of the pterygoid posteriorly, a condition present in early mammals but absent from other nonmammalian cynodonts. The orbitosphenoid forms an extensive part of the orbital wall anterior to the ascending process of the alisphenoid (epipterygoid) in tritylodontids (Hopson, 1964; Sun, 1984; Sues, 1986), Sinoconodon (Crompton and Luo, 1993), Adelobasileus (Lucas and Luo, 1993), and Morganucodon (Kermack, Mussett, and Rigney, 1981). In Haldanodon, the orbital wall anterior to the orbital fissure is completely ossified (Lillegraven and Krusat, 1991). The ossified area that forms the anterior border of the orbital fissure was tentatively termed a "possible ossified interorbital fascia" (Lillegraven and Krusat, 1991, fig. 15). I speculate that this "ossified interorbital fascia" may be a part of the orbitosphenoid, based on the consistent positional correlation of the bone to the orbital fissure and optical foramen in most mammals. By contrast, in Diademodon (Brink, 1955, 1963) and Exaeretodon (Bonaparte, 1962, 1966; personal observation of MCZ 4505, MCZ 4781), the orbital element of the orbitosphenoid is small and does not contribute to the enclosure of the orbital vacuity. An ossified orbitosphenoid is altogether absent from other nonmammalian cynodonts (Fourie, 1974). In correlation with the development of the orbital processes of the palatine and orbitosphenoid, the large orbital vacuity present in other nonmammalian cynodonts (Figue 6.2A) is reduced in Tritylodon (Rowe, 1986). The reduced orbital vacuity, termed the sphenoorbital fissure, becomes an even smaller "sphenopalatine foramen" (= greater palatine foramen) in such derived tritylodontids as Bienotheroides (Sun, 1984) and Kayentatherium (Sues, 1986).

Tritylodontids and mammals are similar in that the orbital opening for the lesser palatine nerve is separate from that of the sphenopalatine ("greater palatine") foramen for the greater palatine nerve. Separate orbital openings of the sphenopalatine foramen and the lesser palatine foramen are present in Sinoconodon (Crompton and Luo, 1993), Morganucodon (Kermack et al., 1981; Z. Luo, personal observation on IVPP 4729),

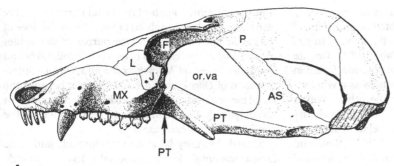

A

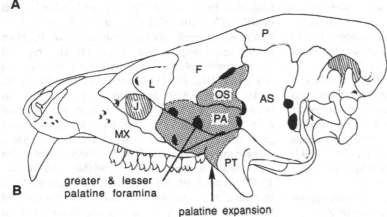

greater & lesser
palatine foramina

palatine expansion

B

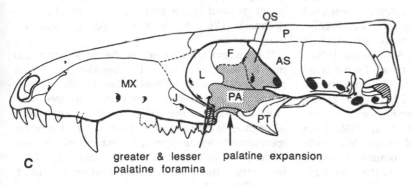

greater & lesser   palatine expansion
palatine foramina

C

Figure 6.2. Skulls of two cynodonts and *Sinoconodon* in lateral view. (A) *Probainognathus* (modified from Romer, 1970): The orbital vacuity is present; the palatine participates in the subtemporal margin but is not visible in the lateral view; the palatine does not displace the pterygoid along the subtemporal border of the orbit. (B) *Kayentatherium* (modified from Sues, 1986): The orbital wall is formed; separate orbital foramina for the greater and lesser palatine nerves are present; a large palatine displaces the pterygoid posteriorly along the subtemporal border of the orbit. (C) *Sinoconodon*. Abbreviations: AS, ascending process of the alisphenoid (epipterygoid); F, frontal; J, jugal; L, lacrimal; MX, maxilla; or.va, orbital vacuity, OS, orbitosphenoid; P, parietal; PA, palatine; PT, pterygoid. Stippled areas represent the complete medial wall of the orbit. Hatched areas represent cut surfaces of bones.

Ornithorhynchus (Zeller, 1989), multituberculates (Miao, 1988), *Didelphis* (Z. Luo, personal observation), *Leptictis* (Novacek, 1986), and *Canis* (Miller, Christensen, and Evans, 1964). The diagnostic mammalian condition of the orbital opening with a sphenopalatine foramen is the foramen being enclosed within the orbital process of the palatine. The foramen may be variable in relation to the surrounding bones in eutherian mammals. In *Leptictis*, the sphenopalatine foramen appears to be positioned on the suture of the orbitosphenoid and the ascending process of the palatine (Novacek, 1986, fig. 10). In multituberculates, the position of this foramen is also variable among different genera (Miao, 1988). By contrast, in tritylodontids, the foramen is completely encircled by the palatine: *Bienotheroides* (Sun, 1984),

Tritylodon (Rowe, 1986), *Kayentatherium* (Z. Luo, personal observation of MCZ 8812). Despite this minor difference from multituberculates and some eutherians, the sphenopalatine ("greater palatine") foramen in tritylodontids is very similar to that in *Sinoconodon*, *Morganucodon*, and other mammals.

An independent orbital opening for the lesser palatine foramen has not been observed in other nonmammalian cynodonts. By contrast, the orbital opening of the lesser palatine foramen is encircled by the palatine in tritylodontids, *Sinoconodon* (IVPP 8483), *Morganucodon* (IVPP 4729), *Leptictis* ("postpalatine foramen" of Novacek, 1986), and *Didelphis*. This foramen may be variable in some eutherians. In *Canis* (Miller et al., 1964), *Erinaceus* (Z. Luo, personal observation), and

Procyon (Story, 1951), the foramen is represented by a posteriorly open notch in the subtemporal border of the palatine. In rodents, this foramen may vary in size and number (Wahlert, 1974; Miao, 1988). Despite these variations, a separate lesser palatine foramen seems to be diagnostic of the Mammalia as a whole.

Undoubtedly, the similarities in the orbital region in tritylodontids and early mammals support the tritylodontid-mammal sister-group relationship proposed by Rowe (1988) and Wible (1991; Wible and Hopson, 1993). However, support from other areas of the skull and dentition for the tritylodontid-mammal hypothesis is, at best, very weak and inconclusive. Several similarities that give crucial support to the affinities of tritylodontids and mammals are based on direct comparison of tritylodontids to *Morganucodon* (Kemp, 1983; Rowe, 1988; Wible, 1991; Wible and Hopson, 1993). Some of these similarities between tritylodontids and *Morganucodon* may no longer be regarded as synapomorphies in light of the new information on the primitive condition in early mammals represented by *Sinoconodon* (Crompton and Luo, 1993; Luo and Crompton, in press) and *Adelobasileus* (Lucas and Hunt, 1990; Lucas and Luo, 1993).

According to Kühne (1956, fig. 13), *Oligokyphus* has two exits for branches of the facial nerve. Kühne's interpretation implies that the palatine branch of the facial nerve and the geniculate ganglion were enclosed by the floor of the cavum epiptericum in *Oligokyphus*. This was regarded as the diagnostic condition of tritylodontids by Rowe (character 49 of Rowe, 1988). However, there is only a single foramen for the facial nerve in the petrosal in *Tritylodon* (Z. Luo, personal observation), *Yunnanodon* (IVPP 7291, IVPP 8696), *Kayentatherium* (Sues, 1986), *Bocatherium* (Clark and Hopson, 1985), and *Bienotheroides* (Sun, 1984). Thus most tritylodontids retain the primitive cynodont condition, in which only one facial nerve foramen is present. In this case, the palatine branch and the geniculate ganglion were ventral to the floor of the cavum epiptericum. Even if Kühne (1956) had interpreted the anatomy of *Oligokyphus* correctly, it would not represent the diagnostic condition of tritylodontids. In this analysis I interpret tritylodontids as having the primitive condition (Table 6.1, Appendix 6.1).

The condition of *Oligokyphus*, as interpreted by Kühne (1956), is similar to that of *Morganucodon* in that the geniculate ganglion and the palatine branch of the facial nerve are enclosed by the floor of the cavum epiptericum (Crompton and Sun, 1985; Luo, 1989). The condition of *Morganucodon*, however, is different from that of *Adelobasileus* (Lucas and Hunt, 1990), nor is it characteristic of the later therian mammals (Luo, 1989). There is only a single facial foramen in the lateral trough of the petrosal anterior to the fenestra ovalis in *Adelobasileus*. *Adelobasileus* is similar to nearly all nonmammalian cynodonts in

that the palatine branch of the facial nerve (probably also the geniculate ganglion) lay ventral to the floor of the cavum epiptericum. Thus the purported similarities in the facial nerve features between tritylodontids and mammals based on *Morganucodon* and Kühne's interpretation of *Oligokyphus* are not tenable, because the facial nerve canals of *Morganucodon* may not represent the diagnostic condition of mammals in view of the condition in *Adelobasileus*.

Shared similarity between tritylodontids and some Liassic mammals in the stapedial ("hyoid") muscle fossa may be equivocal. The "pit for levator hyoidei muscle" of *Morganucodon* of Kermack et al. (1981; Luo, 1989) was termed "stapedial muscle fossa" by Miao (1988) and Crompton and Luo (1993). The stapedial muscle fossa not only is absent from *Adelobasileus* (Lucas and Luo, 1993) and monotremes (Edgeworth, 1935; Zeller, 1989) but also shows variation in triconodontids. The "stapedial" (hyoid) muscle fossa is clearly enclosed by the posterior paroccipital process in *Megazostrodon* (Gow, 1986b; Rowe, 1986), *Morganucodon oehleri* (IVPP 8682, 8684), *Sinoconodon* (IVPP 8683), and multituberculates (Miao, 1988; Luo, 1989). By contrast, the stapedial (hyoid) muscle fossa identified in *Trioracodon* ("pit for hyoid muscle?" of Kermack, 1963) is represented by a shallow depression that opens posteriorly. This "pit for hyoid muscle" is not enclosed by the posterior paroccipital process (Kermack, 1963, fig. 3; Z. Luo, personal observation).

It is unclear whether or not the hyoid (stapedial) muscle fossa is enclosed by the posterior paroccipital process in *Morganucodon watsoni*. Because the posterior paroccipital process is incomplete in many specimens of *M. watsoni* from the Welsh Liassic fissure deposits (Crompton and Luo, 1993), it appears that in these specimens the levator hyoid ("stapedial") muscle fossa opens posteriorly (Kermack et al., 1981, fig. 72E). By contrast, the completely preserved petrosal of *M. oehleri* (IVPP 8482) shows that the fossa is well enclosed by a ridge joining the posterior paroccipital process (Figure 6.3). Although the hyoid muscle fossa of *Trioracodon* resembles that of *M. watsoni*, I tentatively assign "?" for this character in triconodontids in the character analysis, based on the difference in *Trioracodon* and *Morganucodon oehleri*. The hyoid ("stapedial") muscle fossa in triconodontids deserves further study. Lillegraven and Krusat (1991) described a "ventrolateral petrosal pit" in *Haldanodon*. They noted that the pit was somewhat similar to the pit for the levator hyoid ("stapedial") muscle in *M. watsoni* (Kermack et al., 1981). However, Lillegraven and Krusat expressed some reservation about the homology of the pit in *Haldanodon* to the pit of the levator hyoid muscle in *Morganucodon*. The ventrolateral pit in *Haldanodon* is posteromedial to the ridge that separates the oval window and the round window and medial to the

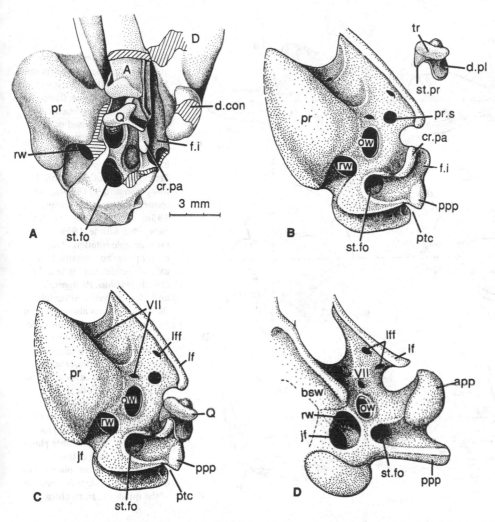

Figure 6.3. *Morganucodon oehleri* and a tritylodontid, ventral views of the left side of the basicranium. (A) *Morganucodon oehleri* (IVPP 8684, ventral lateral view): basicranium with quadrate, articular complex, and posterior part of the dentary. The quadrate is slightly shifted from the crista parotica and the fossa incudis. The posterior paroccipital process is broken. (B) Restoration of the isolated petrosal and quadrate: The quadrate, the paroccipital processes, the crest of the crista parotica and fossa incudis are based on IVPP 8682 and IVPP 8684. (C) Restoration of the articulation of the quadrate and the petrosal. (D) Generalized tritylodontid (based on stereophotographs of *Oligokyphus* and *Tritylodon*). Abbreviations: A, articular; app, anterior paroccipital process; bsw, basisphenoid wing; cr.pa, crest of crista parotica; D, dentary; d.con, dentary condyle; d.pl, dorsal plate (lamina) of the quadrate; f.i, fossa incudis; jf, jugular foramen; lf, lateral flange; lff, vascular foramen on lateral flange; ow, oval window; ppp, posterior paroccipital process; pr, promontorium; pr.s., tympanic opening for the prootic sinus; ptc, posttemporal canal; Q, quadrate; rw, round window; st.fo, "stapedial" fossa ("pit for levator hyoidei muscle" of Kermack et al., 1981); st.pr, stapedial process of quadrate; tr, trochlea of quadrate; VII, facial nerve foramina. Hatched areas represent breakage.

posterior paroccipital process (Lillegraven and Krusat, 1991, fig. 8). This is quite different from the stapedial ("levator hyoid") muscle fossa ("pit") in *Sinoconodon*, *Morganucodon oehleri*, and tritylodontids. For these reasons I also assign "?" to the stapedial muscle fossa in *Haldanodon*.

The internal acoustic meatus in tritylodontids is also difficult to interpret. Kemp (1983) and Rowe (1988) suggested that *Oligokyphus* resembles *Morganucodon* in

possessing a walled internal acoustic meatus. To the contrary, Sues (1985) pointed out that in *Morganucodon* the internal acoustic meatus is shallow, and the facial nerve foramen (VII) is separated from the vestibulo-cochlear foramina, therefore differing from the much deeper and pitlike meatus in tritylodontids. While I agree with Sues (1985) on the distinction between *Morganucodon* and tritylodontids, I note that the issue is further complicated by *Sinoconodon* (IVPP 8691), in

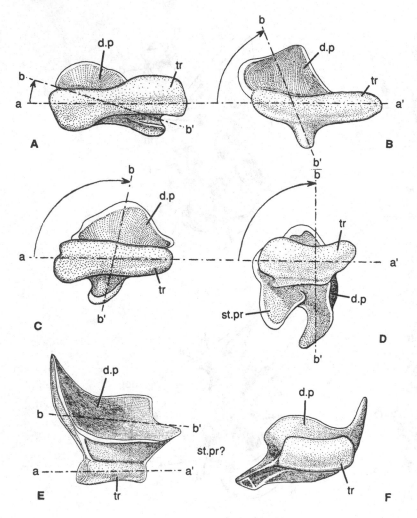

Figure 6.4. Structure of the quadrate in some cynodonts and *Morganucodon* (left side, ventral view). (A) *Thrinaxodon* (based on an unpublished specimen). (B) *Probainognathus* (based on several MCZ specimens). (C) *Pachygenelus* (based on a camera lucida drawing provided by Professor A. W. Crompton). (D) *Morganucodon*. (E) *Oligokyphus* (posterior view, after Kühne, 1956). (F) *Oligokyphus* (ventral view, after Kühne, 1956). Considerable rotation of the dorsal plate (bb') relative to the axis of trochlea (aa') is found in *Probainognathus*, *Pachygenelus*, and *Morganucodon*. Although the dorsal plate is also rotated in *Oligokyphus*, its orientation is so different from those of any other nonmammalian cynodont and *Morganucodon* that rotational relationships can be seen only in the posterior view (instead of ventral view as in other cynodonts and *Morganucodon*). Abbreviations: aa', axis of the trochlea; bb', approximate plane of the dorsal plate of the quadrate; d.p, dorsal plate of the quadrate; st.pr, stapedial process of the quadrate; tr, trochlea.

which the internal acoustic meatus appears to be pitlike, as in *Oligokyphus*.

The structure of the quadrate and the quadrate–skull contact does not provide unilateral and unambiguous support for the tritylodontid-mammal sister-taxon hypothesis. Kemp (1983) suggested that the quadrate of *Oligokyphus* is very similar to that of *Morganucodon*. However, according to recent work by Crompton and Luo, many similarities in the quadrate between *Oligokyphus* and *Morganucodon* are also present in *Probainognathus*, *Massetognathus*, and *Pachygenelus* (Luo and Crompton, in press).

Several features contribute to the general similarities in the quadrate between *Oligokyphus* and *Morganucodon*: the concave contacting facet of the dorsal plate, the rotation of the dorsal plate relative to the axis of the trochlea, the constricted neck that separates the dorsal plate from the trochlea, and the presence of a stapedial process. The concave contacting facet is present in *Probainognathus*, *Massetognathus* and *Pachygenelus*, and probably also in *Exaeretodon* (Figure 6.4). The

rotation of the dorsal plate relative to the trochlea is also characteristic of *Pachygenelus*, *Probainognathus*, and *Massetognathus*. In fact, *Pachygenelus* resembles *Morganucodon* more closely than it resembles tritylo-dontids in the orientation of the dorsal plate and the shape of its dorsal margin. The dorsal plate of the quadrate ("dorsal lamina" of Kermack et al., 1981) is rotated relative to the axis of the trochlea by more than 90° in *Pachygenelus* and *Morganucodon*. By contrast, in *Oligokyphus* (Figure 6.5) the dorsal plate is oriented almost horizontally and nearly parallel to the axis of the trochlea. Tritylodontids are so different from *Pachygenelus* and *Morganucodon* (Figure 6.5) in the orientation of the dorsal plate to the trochlea that the nearly parallel orientation of the dorsal plate to the trochlea can be seen only in posterior view, not in ventral view. The dorsal margin of the quadrate is rounded and flaring in both *Pachygenelus* and *Morganucodon* (Figure 6.4); the dorsal margin in *Yunnanodon* (IVPP 5071), *Oligokyphus* (Kühne, 1956; Crompton, 1964), and *Bienotherium* (Hopson, 1966)

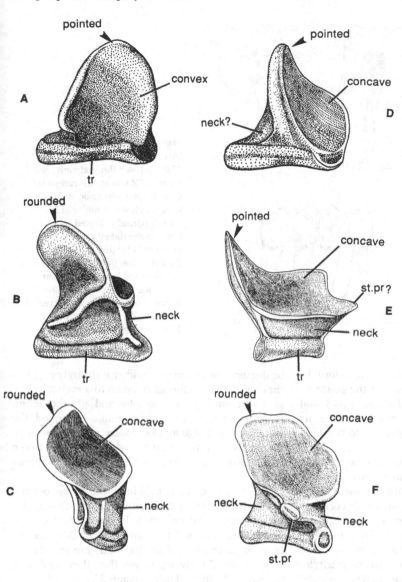

Figure 6.5. Structure of the left quadrate in some cynodonts and *Morganucodon*. (A) *Thrinaxodon* (posterior view, based on an unpublished specimen). (B) *Pachygenelus* (anterior view, based on a camera lucida drawing provided by Professor A. W. Crompton). (C) *Pachygenelus* (medial view). (D) *Probainognathus* (posterior view, based on several MCZ specimens). (E) *Oligokyphus* (posterior view, modified from Kühne, 1956). (F) *Morganucodon* (posteromedial view, modified from Kermack et al., 1981, on the basis of personal observation).

forms a pointed angle ("dorsal process" of Kühne, 1956). Thus the shape of the dorsal margin and the orientation of the quadrate dorsal plate support a tritheledontid-mammal relationship rather than a tritylodontid-mammal relationship.

The presence of a "stapedial process" on the quadrate is the only character in which tritylodontids are more similar to *Morganucodon* than to *Pachygenelus* (Kühne, 1956; Hopson, 1966; Sues, 1986; Rowe, 1988; Sun and Cui, 1989). Rowe (1988) cited the presence of the "stapedial process" in the quadrate of tritylodontids to support the tritylodontid-mammal hypothesis. This is reasonable, because tritylodontids are the only group among nonmammalian cynodonts in which the quadrate articulates with the stapes via a projecting process. Nevertheless, Sues (1985) noted that the stapedial processes of tritylodontids and that of *Morganucodon*

differ in their orientation relative to the trochlea: The process is nearly parallel to the trochlear axis in tritylodontids (Figure 6.5F), but perpendicular to the trochlear axis in *Morganucodon* (Figure 6.5D). Luo and Crompton (in press) note that the stapedial process develops as a projection from the edge of the dorsal plate (instead of the neck) in most tritylodontids, including *Oligokyphus* (Kühne, 1956), *Bienotherium* (Hopson, 1966), and *Kayentatherium* (Sues, 1986). By contrast, the stapedial process in *Morganucodon* is developed from the neck. The issue of homology of the stapedial process between tritylodontids and *Morganucodon* is further complicated by *Bienotheroides* (Sun and Cui, 1989; Luo and Crompton, in press). *Bienotheroides* differs both from other tritylodontids and from *Morganucodon* in that the stapedial process develops from the quadrate neck, rather than from the

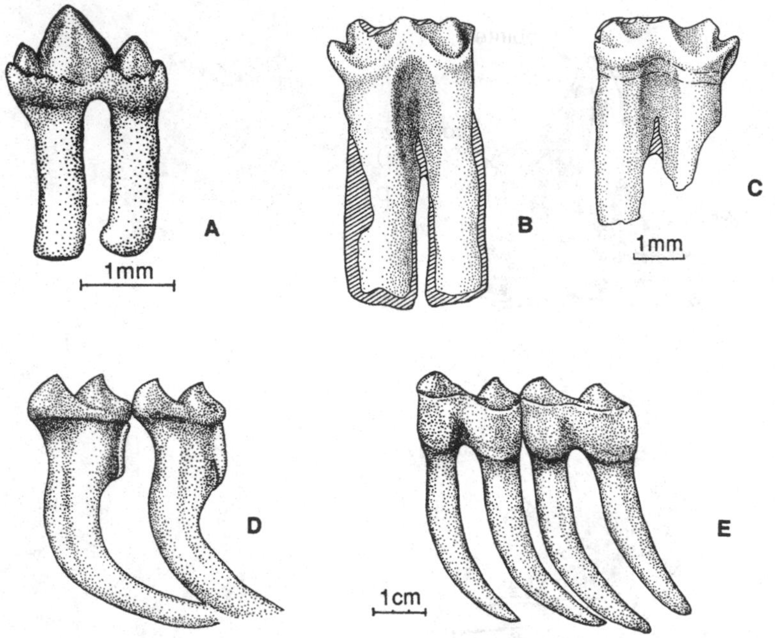

Figure 6.6. Postcanine roots of tritylodontids and mammals. (A) *Morganucodon* (lingual view, based on an MCZ specimen): completely divided roots. (B) *Sinoconodon* (lingual view of a molar of IVPP 8691): partially divided root. (C) *Sinoconodon* (labial view of the same molar of IVPP 8691). (D) *Bienotheroides* (labial view, reversed and modified from Cui and Sun, 1987, fig. 2): single root. (E) *Bienotherium* (labial view; reversed and modified from Cui and Sun, 1987): completely divided roots.

dorsal plate, thus differing from other tritylodontids, but similar to that of *Morganucodon*. Yet the quadrate of *Bienotheroides* differs from that of *Morganucodon* and resembles that of other tritylodontids, because its stapedial process projects nearly parallel to the long axis of the trochlea.

The character distribution of the quadrate suspension among early mammals is ambiguous with regard to the tritylodontid-mammal hypothesis. Rowe (1988) and Wible (1991; Wible and Hopson, 1993) correctly noted that a direct contact between the quadrate and the anterior paroccipital process of the petrosal is present in both tritylodontids and *Morganucodon*. The anterior paroccipital process of the petrosal in tritylodontids is massive and bulbous (Figure 6.3D). However, the anterior paroccipital process in *Morganucodon* is proportionately much smaller and has a low crest. On the lateral side of this crest is a shallow fossa (Figure 6.3). Crompton and Luo (1993) suggested that the crest and its associated fossa are the homologues of the crista parotica and the fossa incudis, respectively, of *Ornithorhynchus* and triconodontids (Crompton and Sun, 1985; Luo, 1989). The anterior paroccipital process in tritylodontids lacks these characters. *Sinoconodon* also poses a difficult problem with respect to direct comparison of the anterior paroccipital process between tritylodontids and *Morganucodon*. In *Sinoconodon*, the anterior paroccipital process appears to be covered, at least in part, by the squamosal (IVPP 4729). The squamosal in *Sinoconodon* is similar to that in *Pachygenelus* (Luo and Crompton, in press) in that it contacts the anterior paroccipital process, unlike the conditions in tritylodontids and *Morganucodon*. As will

be discussed later, current evidence strongly suggests that *Sinoconodon* is the sister-taxon to a monophyletic group comprising *Morganucodon* and all other mammals. Thus it may not be tenable to regard the condition of *Morganucodon* to be representative of mammals and to identify the similarity between *Morganucodon* and tritylodontids as a single character state.

The postcanine roots held in alveoli by periodontal ligaments in tritylodontids probably are the only unambiguous dental similarity between tritylodontids and mammals (Sues, 1985; Rowe, 1988). Other specialized dental features of tritylodontids are so different from those of early mammals that they lend little support to the tritylodontid-mammal hypothesis (Sues, 1985). For example, the primitive dental features in *Sinoconodon* (Crompton and Luo, 1993) would be very difficult to interpret in the framework of the tritylodontid-mammal hypothesis. The canines are lost in tritylodontids; the distinct canines in *Sinoconodon* had multiple replacements. The occlusal pattern between postcanine cusps is established from the moment of eruption in tritylodontids, while such wear facets are absent in *Sinoconodon* and develop only by a substantial amount of occlusal wear after the eruption in other Liassic mammals. The divided postcanine roots have been cited as a shared apomorphy of tritylodontids and mammals (Kemp, 1983; Rowe, 1988; Wible, 1991). However, the two postcanine roots in *Sinoconodon* are only partially divided (Figure 6.6). Incomplete division of the roots also occurs in some (although not all) postcanines of *Morganucodon* and *Kuehneotherium* (Parrington, 1971).

Direct comparison of the multiple roots of the postcanines between morganucodontids and tritylodontids is also complicated by the great range of variation of postcanine roots in tritylodontids (Cui and Sun, 1987). These multiple roots can be divided to (or even above) the level of the gum line, as in *Bienotherium*, or connected by dentine blades below the gum line along transverse rows, as in *Oligokyphus* and *Lufengia* (Kühne, 1956; Cui and Sun, 1987). Cui and Sun (1987) also noted that the lower postcanine roots coalesce into a single root in the derived tritylodontid *Bienotheroides*. The complete division between the anterior and posterior transverse rows of roots and the incomplete division among the roots within each transverse row create a problem in defining the so-called multiple roots of tritylodontids. Docodonts have multiple roots, with complete division in the upper molars but not in the lower molars (Lillegraven and Krusat, 1991). The "multiple postcanine roots" of tritylodontids should be treated as a character state separate from the condition with two partially divided roots in *Sinoconodon*, which I consider the primitive condition of mammals.

It is noteworthy that different technical treatments of the highly specialized dental character states of tritylodontids can affect the results of parsimony analyses. The presence and absence of divided roots were treated as binary character states in some previous analyses (Rowe, 1988; Wible, 1991). I designate the incipient division of postcanine roots in *Pachygenelus* (Shubin et al., 1991) and the incomplete root division in *Sinoconodon* (Figure 6.6) as two additional intermediate character states in my parsimony analysis (character 9 in Table 6.1 and Appendix 6.1). Rowe (1988) and Wible (1991) designated the absence and presence of postcanine wear facets as binary character states. In this analysis, two additional character states are recognized for the preexisting wear facets of tritylodontids and for the wear facets of morganucodontids, triconodontids, and kuehneotheriids that are developed through mastication after eruption (character 16 in Table 6.1). If these multistate characters are treated as unordered data in parsimony analysis to avoid any a priori interpretation of character evolution (Gauthier, Kluge, and Rowe, 1988; Rowe, 1988), they do not unambiguously support a tritylodontid-mammal relationship.

The hypothesis of a tritylodontid-mammal sister-taxon relationship is contradicted by several dental apomorphies shared by tritylodontids, diademodontids, and traversodontids (Watson, 1942; Young, 1947; Kühne, 1956; Crompton and Ellenberger, 1957; Crompton, 1972b; Sues, 1985). Crompton (1972b), Sues (1985) and Hopson and Barghusen (1986) suggested that tritylodontids and traversodontids form a monophyletic group, which in turn is related to diademodontids. These three families form a clade diagnosed by the apomorphic crown pattern of the postcanines, which have transversely expanded crowns with multiple cusps (Romer, 1967). Crompton (1972b) showed that in traversodontids and tritylodontids, the rows of cusps of the upper postcanines interlock with the rows of cusps on the lower. The upper and lower cusps already match at the time of tooth eruption. The postcanines of all tritylodontids and most traversodontids lack cingula, which are present in other cynodonts and many early mammals. The rate of tooth replacement was reduced. The occlusal movement of the lower jaw was directed posteriorly. The postcanine roots were held in alveoli by periodontal ligaments, which was cited in support of tritylodontid-mammal affinity but, is also present in traversodontids (J. A. Hopson, pers. commun.). These three herbivorous groups also lack foramina for the internal carotid artery in the basisphenoid (Hopson and Barghusen, 1986; Rowe, 1988; Wible, 1991). Presumably the internal carotid artery entered the braincase through the ventral opening of the cavum epitericum. By contrast, in *Probainognathus* (MCZ 4015 and 4017) and *Pachygenelus* (Allin and Hopson, 1991; Wible and Hopson, 1993), internal carotid foramina are present in the basisphenoid, as in other nonmammalian cynodonts and mammals. The tritylodontid-mammal sister-group hypothesis would entail a difficult (if not insurmountable) postulation that these shared apomorphies, especially the occlusal specializations in these herbivorous groups, arose repetitively or that the condition in tritylodontids represents a secondary reversal from the putative common ancestor shared with mammals.

Tritheledontid-mammal hypothesis

This hypothesis gains the strongest support from the characters of the temporomandibular joint, and from the occlusal movement of the lower jaw inferred from the dental wear and symphysis in tritheledontids. The posterior end of the lateral ridge on the dentary in *Diarthrognathus* and *Pachygenelus* is considerably enlarged (Crompton, 1963, plate 1; Crompton and Luo, 1993). Although a dentary condyle is not fully developed in tritheledontids, the enlarged posterior end of the lateral ridge closely resembles the mammalian dentary condyle. The ventromedial surface of the zygomatic moiety of the squamosal of *Pachygenelus* has a concave area (area gl in Figure 6.7). The structure and orientation of this concave area are very similar to (if not identical with) those of the glenoid on the squamosal of *Sinoconodon* (IVPP 4729) (Figure 6.7). Notwithstanding Gow's (1981) questioning of the presence of the dentary–squamosal articulation in *Diarthrognathus*, it appears that tritheledontids are most similar to *Sinoconodon* in the structure of the temporomandibular joint among nonmammalian cynodonts. Another interesting feature of tritheledontids is

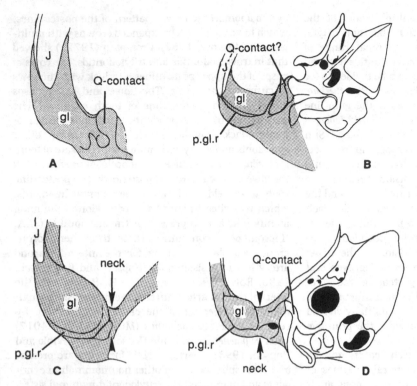

Figure 6.7. Stylized illustration of the glenoid region of the squamosal. (A) *Pachygenelus* (based on Allin and Hopson, 1991, and a camera lucida drawing provided by Professor A. W. Crompton). (B) *Sinoconodon* (squamosal based on IVPP 4729; petrosal based on several specimens). (C) *Haldanodon* (modified from Lillegraven and Krusat, 1991). (D) A morganucodontid (squamosal based on MCZ 20988; petrosal modified from Crompton and Luo, 1993). Abbreviations: gl, glenoid area for mandibular articulation; J, jugal; p.gl.r, postglenoid ridge; Q-contact, contact area between quadrate and basicranium. Arrows point to the constricted neck medial to the glenoid in *Haldanodon* and morganucodontids. Stippled area represents the squamosal.

that the posterior end of the lateral ridge of the dentary is raised above the level of the alveoli, a condition comparable to that of *Morganucodon* and triconodontids, but not *Sinoconodon*.

A unilateral occlusion and a dorsomedial movement of the mandible are important in the feeding mechanism in early mammals (Crompton and Hylander, 1986; Crompton, 1989). The mandibular symphysis in tritheledontids is unfused and probably was mobile (Gow, 1980; Shubin et al., 1991). This may have permitted unilateral occlusion of the lower jaw (Shubin et al., 1991). The wear facets on the lingual side of the upper postcanines and the labial side of the lower postcanines indicate that during occlusion the lower jaw moved dorsomedially. The dorsomedial movement and the unilateral occlusion of the lower jaw are correlated to the load-bearing temporomandibular joint formed by the incipient dentary–squamosal contact (Crompton and Hylander, 1986).

Additional evidence supporting a tritheledontid–mammal sister-group relationship comes from other areas of the skull. The gracile zygomatic arch of tritheledontids is most similar to those of *Sinoconodon*

and *Morganucodon* (Crompton, 1958; Hopson and Barghusen, 1986; Crompton and Luo, 1993). The maximum vertical depth of the zygomatic arch relative to the length of the skull is about 8 percent for *Diarthrognathus* (Crompton, 1958, 1963), 7 percent for *Pachygenelus* (Allin and Hopson, 1991), and 5–7 percent for *Sinoconodon* (IVPP 4729, 8687), *Morganucodon* (Kermack et al., 1981), and the unidentified Cloverly Formation triconodontid. The same ratio is about 9 percent in *Haldanodon* (Lillegraven and Krusat, 1991). By contrast, in the tritylodontids this ratio ranges from 20 to 23 percent. The same ratio is about 17–18 percent in *Cynognathus, Probelesodon, Massetognathus, Probainognathus,* and chiniquodontids, and about 12 percent in *Thrinaxodon* (Fourie, 1974). The thinning of the zygomatic arch is perhaps correlated to the incipient dentary–squamosal contact in tritheledontids, but its functional implications have yet to be explored.

The palate in tritheledontids is much broader posteriorly than that in tritylodontids. The narrowest distance across the palate between the lateral edges of the pterygoids is much greater in *Diarthrognathus* (Crompton,

1958) and *Pachygenelus* (Allin and Hopson, 1991) than in tritylodontids. The intermediate pterygopalatal ridges on the plate posteriorly extend to the anterior border of the basisphenoid in tritheledontids, whereas the pterygo-palatal ridges terminate much farther anteriorly in *Bienotherium, Yunnanodon,* and other tritylodontids (Z. Luo, personal observation). Separation of the round window and jugular foramen is better developed than in tritylodontids. The hypoglossal foramen is at the margin of the jugular foramen in *Diarthrognathus* (Crompton, 1958, fig. 6). In all these features, tritheledontids resemble *Sinoconodon* and *Morganucodon* more closely than tritylodontids.

Tritheledontids show a mosaic of primitive and derived characters. Although they are very "mammal-like" in the temporomandibular joint, the lower jaw, and some features of the basicranium, they are very primitive in other areas of the skull when compared with tritylodontids. The orbital vacuity in *Diarthrognathus* is large. The structure of the braincase is similar to that in *Thrinaxodon* and *Probainognathus* and thus primitive (Crompton, 1958). The anterior paroccipital process in *Pachygenelus* is covered by the squamosal, a primitive condition characteristic of other nonmammalian cynodonts (Allin and Hopson, 1991; Luo and Crompton, in press). It lacks the hyoid ("stapedial") muscle fossa and the distal bifurcation of the paroccipital process that are shared by tritylodontids, *Sinoconodon, Dinnetherium,* and *Morganucodon.* Many areas of the tritheledontid cranium, such as the braincase, orbit, and petrosal, simply have no characters that could be interpreted as apomorphies in favor of the tritheledontid-mammal sister-group relationship.

The tritheledontid dentition neither strongly supports nor contradicts the tritheledontid-mammal hypothesis. The postcanines of tritheledontids are ankylosed to the jaws and have alternating replacement, both being primitive characters (Crompton, 1958; Gow, 1980). But not all of the dental characters are primitive by comparison with tritylodontids and other cynodonts. Shubin et al. (1991) reported that the roots of the postcanines have two grooves along the length of the root, indicating incipient division, an apomorphic condition when compared with other nonmammalian cynodonts (excluding tritylodontids). Gow (1980) and Shubin et al. (1991) noted that extensive wear facets are present on the lingual sides of the upper postcanines and the labial sides of the lower postcanines. Conspicuous wear facets are also developed on the incisors (Gow, 1980) and the canines (Shubin et al., 1991). Because the teeth must have functioned long enough for the wear to have developed, these well-developed wear facets of *Pachygenelus* suggest that the functional life of the individual teeth must have been quite long relative to the life span of the animal. This suggests that the tooth replacement rates would have been slow by comparison with other cynodonts (excluding tritylodontids), which lack any evidence for slow replacement.

Solution to the sister-taxon problem

Both hypotheses of mammalian relationships are supported by a number of apomorphies, but neither hypothesis is without problems. The main difficulty with the tritylodontid-mammal hypothesis is that too many apomorphic features of tritylodontids are more derived than the corresponding features in primitive mammals such as *Sinoconodon* and *Adelobasileus* (e.g., greatly reduced rate of tooth replacement and the hypertrophied paroccipital process). By contrast, the main weakness of the tritheledontid-mammal hypothesis is that too many tritheledontid characters are primitive (e.g., the orbit and braincase). For a solution to this sister-taxon controversy, we shall have to obtain more evidence on the critical taxa, such as *Sinoconodon, Adelobasileus,* and *Haldanodon,* and more anatomical information through more detailed studies, such as serial sectioning, on the existing taxa.

Based on my own definition of characters and interpretation of character distributions among relevant taxa (Table 6.1, Appendix 6.1), and using the minimal number of transformational steps as the criterion for selecting among competing hypotheses (Farris, 1982; Swofford, 1990), I favor the hypothesis of a tritheledontid-mammal sister-group relationship over the tritylodontid-mammal hypothesis. The cladogram in Figure 6.1A is the preferred phylogenetic scheme of this analysis.

The comparison of the two hypotheses is shown in Figure 6.8. Based on the character matrix (Table 6.1, Appendix 6.1), the tritheledontid-mammal hypothesis has consistently fewer transformational steps than the tritylodontid-mammal hypothesis, regardless of different permutations in the arrangement of the out-groups. Nevertheless, the difference between the two hypotheses is quite small. Given the same out-group arrangement proposed by Rowe (1988), there are only six more steps required for the cladogram in Figure 6.8A (tritylodontid-mammal hypothesis) than for the cladogram in Figure 6.8B (tritheledontid-mammal hypothesis). Given the out-group arrangement proposed by Kemp (1983), the former hypothesis (Figure 6.8C) involves only two steps more than the latter hypothesis (Figure 6.8D). Given the out-group arrangement proposed by Wible (1991) (Figure 6.8E,F), the tritheledontid-mammal hypothesis is also favored by only two steps less. The tritylodontid-mammal hypothesis remains a viable alternative hypothesis, despite criticisms (Wible, 1991) regarding the interpretation and distribution of many characters that were initially cited in its support (Kemp, 1983; Rowe, 1986, 1988; Wible, 1991).

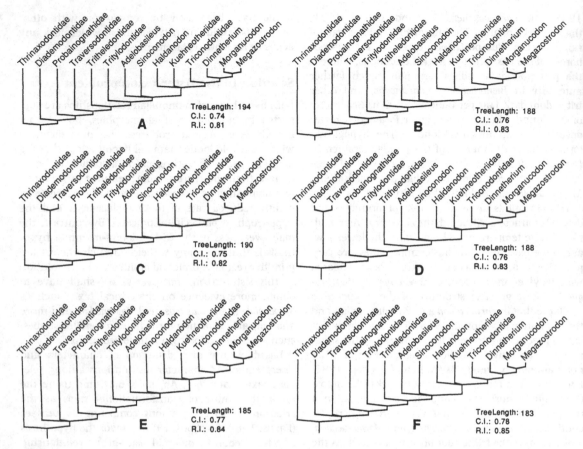

Figure 6.8. Influence of permutation in the arrangement of cynodont out-groups on the mammalian sister-taxon hypotheses. The left row (A, C, E) list the tritylodontid-mammal hypothesis with different permutations of out-group taxa. The right row (B, D, F) lists the tritheledontid-mammal hypothesis with different permutations of out-group taxa. (A and B) Permutation in out-group arrangement according to Rowe (1988). (C and D) Permutation in out-group arrangement according to Kemp (1983). (E and F) Permutation in out-group arrangement according to Wible (1991). Based on the character distributions listed in Table 6.1 and Appendix 6.1, a tritheledontid-mammal group is consistently favored over the tritylodontid-mammal group by a small difference, regardless of various permutations in the arrangement of more distantly related cynodonts.

Alternative predictions on functional evolution

Parsimony analysis is perhaps the most effective technical approach for selecting the preferred hypothesis, given ubiquitous and rampant convergence in the evolution of advanced nonmammalian cynodonts and early mammals. However, to force the selection of one hypothesis over its alternative implies that many bona fide similarities supporting the alternative hypothesis must be ruled out as convergences. Without understanding the anatomical characters that form the apomorphies, selecting between these two alternatives can be quite arbitrary, especially when the difference in the number of apomorphies is so small. To understand what kinds of characters have made up the

apomorphies is as meaningful as the number of apomorphies, if not more so. The full consequence of either hypothesis can be appreciated only if we can have a clear understanding of what kinds of characters are considered as synapomorphies versus what characters are discarded as convergences in either case.

Previously published evidence for the tritylodontid-mammal relationship concentrated heavily on the petrosal, quadrate, and dentition. However, more recent studies (Kemp, 1983; Crompton and Luo, 1993; Luo and Crompton, in press) indicate that the mammalian apomorphic characters from these three areas have nearly equal distributions between tritheledontids and tritylodontids. None of these character systems provides unambiguous support for either hypothesis (Table 6.2). Both hypotheses would have to assume almost equal numbers of convergences in the petrosal, quadrate, and dentition.

The most clear-cut differences between the two hypotheses are in the orbit, palate, temporomandibular joint, and lower jaw (Table 6.2). The apomorphic characters of the orbital region predominantly support the tritylodontid-mammal hypothesis. By contrast, the apomorphic characters in the temporomandibular joint, lower jaw, and palate predominantly support the tritheledontid-mammal hypothesis. To accept the latter hypothesis implies that the orbital apomorphies shared by tritylodontids and early mammals are convergent. By the same token, the tritylodontid-mammal hypothesis would have to assume that the apomorphies shared by tritheledontids and mammals in the temporomandibular joint, mode of occlusion, and palate are convergences. Because important changes in feeding mechanisms are reflected by the temporomandibular joint and lower jaw at the transition to mammals, clear-cut differences in the apomorphies in these critical areas between the two competing hypotheses would lead to different understandings of the evolution of the masticatory apparatus. The tritheledontid-mammal hypothesis differs significantly in a posteriori predictions regarding the functional evolution of the mammalian feeding mechanism.

Within the framework of the tritheledontid-mammal hypothesis, evolution of the cranial structure through cynodont–mammal transition can be best characterized as evolution of more efficient mastication in a carnivorous and/or insectivorous context. Tritheledontids, *Sinoconodon*, morganucodontids, and kuehneotheriids all have either an incipient or well-developed dentary–squamosal articulation, as well as a mobile symphysis. Tritheledontids and all early mammals (except haramiyids) appear to be adapted to carnivory/insectivory. Thus the common ancestor of mammals and their sister-taxon must have been a generalized carnivore and/or insectivore, and the transition from nonmammalian cynodonts to mammals is characterized by changes in carnivorous/insectivorous adaptations. The tritheledontid-mammal hypothesis would lead to a scenario in which the major features of the functional evolution in the origin of mammals were the development of an increasingly stronger temporomandibular articulation, a more flexible symphysis (which permitted fine control of the mandibular movement), and increasingly precise matching of the opposing postcanines leading to the differentiation of wear facets for shearing. All these occurred without any major disruption of the basic adaptation to carnivory and/or insectivory.

While this hypothesis results in a straightforward scenario of evolution of feeding functions, the tritylodontid-mammal hypothesis offers no simple explanation for the evolution of feeding adaptations. Within the framework proposed by Rowe (1988) and Wible

(1991), the dentary–squamosal articulation either is secondarily lost in tritylodontids or is derived in parallel in tritheledontids and mammals. The common ancestor could be either a specialized herbivore or a carnivore. The transition from advanced nonmammalian cynodonts to mammals would be characterized by alternating changes between carnivory and herbivory adaptations among major clades. Both the carnivorous/insectivorous adaptation and herbivorous adaptation arose repetitively. Within the framework of the tritylodontid-mammal hypothesis, interpretation of the evolution of feeding function is enormously complicated.

In the following discussion, I shall assume that tritheledontids and mammals are sister-taxa.

Monophyly of Mammalia

Monophyly of Mammalia (Mammaliaformes of Rowe, 1988) is supported by numerous unequivocal apomorphies in the temporomandibular joint, petrosal, and palate. The temporomandibular joint is characterized by a fully developed squamosal glenoid with a strong postglenoid ridge and by the dentary condyle. As mentioned earlier, in *Diarthrognathus* and *Pachygenelus* (Crompton, 1963; Crompton and Luo, 1993), a concave area on the zygomatic moiety of the squamosal is very similar to the glenoid of *Sinoconodon* in its orientation and topography. *Sinoconodon* is more advanced in having a postglenoid ridge that buttresses the temporomandibular articulation (Figure 6.7). This postglenoid ridge is also present in *Haldanodon* (Lillegraven and Krusat, 1991), *Morganucodon oehleri* (IVPP 8684), and the Cloverly Formation triconodont (MCZ 19973). The dentary condyle in mammals is derived from the posterior end of the dentary in nonmammalian cynodonts (Crompton, 1963, 1972a; Crompton and Luo, 1993).

Many petrosal apomorphies support the monophyly of Mammalia, the most important of which is the cochlear housing (Figure 6.9). In non-mammalian cynodonts, the basisphenoid wing ("parasphenoid ala") is very prominent on the basicranium and forms a part of the cochlear housing. It extends posterolaterally to participate in the periphery of the oval window in *Thrinaxodon* (Figure 6.9A), but it has withdrawn from the oval window in derived cynodonts such as *Massetognathus* (MCZ 3807) and *Probainognathus* (Figure 6.9B). The cochlear housing is formed by two bones, the basisphenoid and prootic, in cynodonts (Olson, 1944; Fourie, 1974; Quiroga, 1979; Gow, 1985, 1986a; Allin, 1986). By contrast, the cochlear housing is formed exclusively by the petrosal in most mammals (Gow, 1986a; Graybeal et al., 1989; Luo and Ketten, 1991). The part of the petrosal with the cochlea forms the promontorium in *Sinoconodon*, *Megazostrodon*, and *Morganucodon*, and the basisphenoid wing is lost in these taxa. *Adelobasileus* (Lucas and Hunt, 1990;

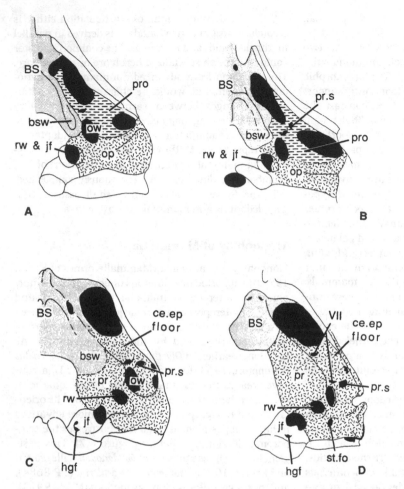

Figure 6.9. Schematic of the transformation of the basicranium and the cochlear housing through the cynodont–mammal transition. (A) *Thrinaxodon*: Separate prootic and opisthotic; basisphenoid wing reaches the oval window and contributes to the cochlear housing; the round window and the jugular foramen are confluent. (B) *Probainognathus* (based on MCZ 4015 and 4017): The basisphenoid wing has receded from the oval window; otherwise similar to *Thrinaxodon*. (C) *Adelobasileus* (based on NMMNH 12721, Lucas and Hunt, 1990): Prootic and opisthotic fused; the basisphenoid wing is reduced; an incipient promontorium is present; the jugular foramen is separate from the round window, and the hypoglossal foramina are separate from the jugular foramen; the cavum epitericum is ventrally floored anterior to the tympanic opening for the prootic sinus. (D) Morganucodontid (based on IVPP 8684, IVPP 4727; *Dinnetherium*, from Crompton and Luo, 1993): The basisphenoid wing is lost, and the promontorium is fully developed; two facial nerve foramina are present; the "stapedial fossa" is present; otherwise similar to *Adelobasileus*. Abbreviations: BS, basisphenoid; bsw, basisphenoid wing; ce.ep floor, petrosal floor for the cavum epitericum; hgf, hypoglossal foramen (XII); jf, jugular foramen; op, opisthotic; ow, oval window; pr, promontorium; pro, prootic; pr.s, tympanic opening for the prootic sinus vein; rw, round window; st.fo, putative "stapedial muscle fossa."

Lucas and Luo, 1993) may represent an intermediate stage of the transformation, where the promontorium develops at the expense of the basisphenoid wing (Figure 6.9C). The basisphenoid wing is greatly reduced in this taxon, and a petrosal promontorium is developed. It is still uncertain (although possible) that the vestigial basisphenoid wing contributes to the cochlear housing. The semicircular canals of the inner ear in mammals are entirely enclosed by the petrosal (Miao, 1988; Lillegraven and Krusat, 1991; Luo and Ketten, 1991;

Z. Luo and A. W. Crompton, unpublished data), whereas in nonmammalian cynodonts the semicircular canals are enclosed by both the petrosal and the exoccipital (Fourie, 1974; Quiroga, 1979).

The periphery of the oval window is modified in mammals. The oval window is formed by a thickened ring of bone in *Probainognathus* (MCZ 4015, 4017), *Luangwa* (Kemp, 1980), *Massetognathus* (MCZ 3807), *Yunnanodon* (Z. Luo, personal observation), *Kayentatherium* (Sues, 1986), and *Diarthrognathus*

(Crompton, 1958). Allin (1975, 1986) and Kemp (1979) believed that this thickened ring probably functioned to stop the proximal end of the stapes from jamming into the inner ear. This thickened bony ring of the oval window is absent in *Adelobasileus* (Lucas and Hunt, 1990; Lucas and Luo, 1993), *Sinoconodon*, and all other early mammals.

The anterior lamina of the petrosal forms a large portion of the lateral wall of the braincase in mammals, as compared with nonmammalian cynodonts. In *Sinoconodon, Adelobasileus, Morganucodon, Dinnetherium, Trioracodon*, and the Cloverly triconodontid, the exits for the maxillary and mandibular branches of the trigeminal nerve (V) are encircled by the anterior lamina of the petrosal, instead of being positioned at the anterior border of the anterior lamina as in nonmammalian cynodonts (Kermack and Kielan-Jaworowska, 1971; Crompton and Jenkins, 1979; Crompton and Sun, 1985). Two mammals show variation in this feature. In both *Megazostrodon* (Gow, 1986b) and *Haldanodon* (Lillegraven and Krusat, 1991), a large trigeminal foramen is positioned between the ascending process of the epipterygoid and the anterior lamina of the petrosal. By the congruence with the distribution of other characters (Figure 6.1, Table 6.1, Appendix 6.1), the character state represented by *Sinoconodon* and *Adelobasileus* is regarded as the diagnostic character of all mammals, whereas the condition in *Megazostrodon* and *Haldanodon* is interpreted as a secondary reversal. On the petrosal in mammals, the foramen for the prootic vein is present in the lateral trough; the floor to the cavum epiptericum is formed anterior to the tympanic opening of the prootic sinus (Crompton and Sun, 1985; Luo, 1989). Both these features are absent from most nonmammalian cynodonts. However, both features are present in *Probainognathus* and *Massetognathus* (Rougier et al., 1992; Crompton and Luo, 1993), and they may be equivocal as mammalian apomorphies.

Monophyly of Mammalia is further supported by several apomorphies in the occipital and palatal regions. Mammals and nonmammalian cynodonts clearly differ in their positions for the exits of the hypoglossal nerve (XII) (Figure 6.9). In the exoccipital in mammals, the condylar foramen for the hypoglossal nerve (XII) is well separated from the jugular foramen. The number of openings for the exit of the hypoglossal nerve (XII) may vary among mammals. *Sinoconodon* and *Morganucodon* possess a single condylar foramen, whereas *Dinnetherium* and *Haldanodon* have two openings for the hypoglossal nerve. In *Haldanodon* (Lillegraven and Krusat, 1991), one of the hypoglossal foramina is separated from the jugular foramen, whereas the other seems to be connected to the jugular foramen by a narrow gap. In view of the variation in *Haldanodon*, this mammalian apomorphy is defined as at least one condylar foramen for the hypoglossal nerve being

separated from the jugular foramen. By contrast, the condylar (hypoglossal) and jugular foramina are confluent in most nonmammalian cynodonts. Although Kühne (1956) illustrated separate hypoglossal foramina in *Oligokyphus*, his observation cannot be confirmed in *Tritylodon* (Rowe, 1986), *Kayentatherium* (MCZ 8812), *Yunnanodon* (IVPP 5071, 8694), and other tritylodontids (Sun, 1984; Clark and Hopson, 1985). The hypoglossal foramen seems to be positioned at the margin of the jugular foramen in *Diarthrognathus* (Crompton, 1958, fig. 2).

The occipital condyle is more cylindrical in *Adelobasileus* (Lucas and Hunt, 1990) and *Morganucodon* than in nonmammalian cynodonts. The two occipital condyles are separated by a deep odontoid notch for the dens of the axis in *Adelobasileus* (Lucas and Luo, 1993), *Morganucodon* (Kermack et al., 1981), and *Sinoconodon* (Zhang and Cui, 1983). The only exception is *Haldanodon*, which lacks this notch (Lillegraven and Krusat, 1991). The odontoid notch is absent from all nonmammalian cynodonts except *Exaeretodon* (MCZ 4781). The notch between the occipital condyles in *Exaeretodon* is interpreted here as a convergence to mammals. The palate is considerably wider at the level anterior to the basisphenoid in *Adelobasileus* (Lucas and Luo, 1993) and *Sinoconodon* than in tritylodontids (IVPP 5071, 8694) and other nonmammalian cynodonts (excluding tritheledontids).

Several additional diagnostic characters of early mammals may be equivocal in view of the controversy on the relationships of out-group taxa. The palatal region in early mammals is characterized by a conspicuous expansion of the palatine, as a result of which the transverse flange of the pterygoid is positioned more posteriorly than in nonmammalian cynodonts, relative to the postcanine row. The ascending (orbital) process of the palatine and the orbitosphenoid bone form the medial wall of the orbit. The greater palatine nerve and the lesser palatine nerve have separate openings in the orbital wall. If a tritheledontid-mammal sister-taxon relationship is assumed, these mammalian apomorphies can be interpreted as convergent to tritylodontids.

Assuming that tritheledontids are the sister-taxon of mammals, two equivocal dental characters are diagnostic of Mammalia: The postcanines have two partially divided roots, and the mode of postcanine replacement is changed from continuous and alternating replacement to partial replacement of the more posterior teeth in the postcanine row (Crompton and Luo, 1993).

Relationships of *Sinoconodon*

Contra Patterson and Olson (1961), *Sinoconodon* retains many plesiomorphies in the dentition (Crompton and Sun, 1985; Crompton and Luo, 1993). Both the incisors and canines show multiple replacements. In

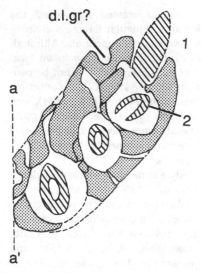

d.l.gr?

a

a'

Figure 6.10. Schematic transverse section through the mandibular symphysis of *Sinoconodon* (based on serial sections 166–168 of the anterior part of the skull, IVPP 8689). Abbreviations: aa', mandibular symphysis; d.l.gr, dental lamina groove for tooth replacement; 1 and 2, putative generations of replacement for the same incisor alveolus.

several specimens of *Sinoconodon* (IVPP 8683, 8688, 8690), small incisors (or their alveoli) alternate with larger incisors (or alveoli), a tooth size pattern characteristic of many therapsids with continuous and alternating replacement of teeth. The incisors are replaced alternating, probably at least three times. In transverse sections of the symphyseal region of the mandible in one of the largest specimens (probably an older individual), a functioning incisor and a preerupting incisor are present at a single incisor site on the mandible (Figure 6.10). This may be interpreted as evidence of continuing replacement of incisors even in older individuals. As far as can be determined in the different ontogenetic stages in the available sample of skulls, the canine was replaced more than four times (Figure 6.11) (Crompton and Luo, 1993).

Crompton and Luo (1993) noted that the replacement of postcanines is very similar to the pattern of gomphodont cynodonts, as observed by Crompton (1963) and Hopson (1971). The anterior postcanines are lost without replacement, resulting in an increasingly larger postcanine diastema during ontogeny. At least two postcanines are added to the posterior end of the tooth row in older individuals. The small ultimate postcanine in younger ontogenetic stages probably was replaced by a larger tooth in older specimens. Thus at least one "molariform" postcanine was replaced. The postcanine roots are divided only along their distal parts. In *Morganucodon* and *Kuehneotherium*, although some postcanines have incompletely divided roots,

most postcanines have completely divided roots (Mills, 1971; Parrington, 1971, 1973).

The postcanines lack differentiation into "premolariform" and "molariform" teeth (Crompton and Sun, 1985). Cingula are absent from all postcanines except on one upper tooth of one skull, which has a very faint labial cingulum. The adjacent postcanines of *Sinoconodon* do not interlock with one another as they do in triconodontids and morganucodontids (Crompton and Luo, 1993). In at least two skulls of *Sinoconodon*, the number of upper postcanines does not match the number of the lowers. Thus, the upper and lower postcanines in *Sinoconodon* lack the one-to-one opposition that is characteristic of all other known Early Jurassic mammals (Crompton and Sun, 1985; Crompton and Luo, 1993). The crown of each upper postcanine has a principal cusp A, with a posterior cusp C slightly larger than an anterior cusp B. Two small accessory cusps, the posterior cusp D and the anterior cusp E, are also present. None of these cusps match precisely with the cusps of the lower postcanines. Consequently, the postcanine crowns lack wear facets, which probably is correlated with the absence of precise one-to-one alignment of the corresponding upper and lower post-canines. The one-to-one alignment is a crucial precondition for the development of wear facets in early mammals (Kermack et al., 1968; Clemens and Mills, 1971; Crompton, 1974).

There is some evidence that the modes of skull growth are different between *Sinoconodon* and *Morganucodon*. The available skulls of *Sinoconodon* with fully functioning adult postcanines show wide ontogenetic variation in size: The smallest skull (IVPP 8683) is less than 3 cm long, and the largest skull is more than 6 cm. The alternating replacement of the incisors and canines and partial replacement of the postcanines continued as the skull increased in size. Based on this information, I interpret that the skull of *Sinoconodon* experienced continuous slow growth, as in modern diapsid reptiles and in nonmammalian cynodonts. By contrast, the lengths of the four complete adult skulls of *Morganucodon* discovered thus far (Kermack et al., 1981; Z. Luo, personal observation) range from 2.8 to 3.1 cm. The size range of the skulls represented in this sample is much smaller than that of *Sinoconodon*, suggesting that *Morganucodon* probably lacked the continuous slow growth of *Sinoconodon*. Gow (1985) also argued that *Morganucodon* had acquired the rapid growth of mammals. He notes that the maxillary bones and mandibles from the Welsh fissure deposits, as reported by Kermack et al. (1973, 1981), have disproportionately large numbers of adult specimens and relatively few juvenile specimens. Gow (1985) implied that the small number of juvenile specimens in the sample indicates that the juvenile stage of *Morganucodon* was quite short.

The petrosal and the squamosal display several

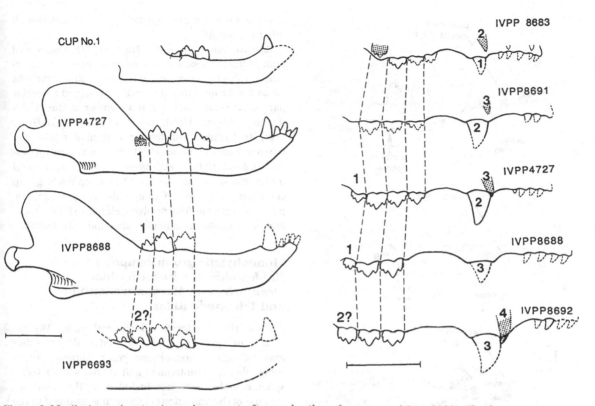

Figure 6.11. Canine and postcanine replacement in *Sinoconodon* (from Crompton and Luo, 1993). The dentaries are based on (left side, from top to bottom): CUP 1, IVPP 4727, IVPP 8688, IVPP 6693. The upper jaws are based on (right side, from top to bottom): IVPP 8683, IVPP 8691, IVPP 4727, IVPP 8688, IVPP 8692. The larger specimens toward the bottom represent older individuals. Numbers represent the inferred generations of tooth replacement at an alveolus. The erupting teeth are stippled. Scale bars each represent 1 cm.

intermediate character states that are more primitive than the conditions of morganucodontids, but more derived than those of nonmammalian cynodonts. *Sino-conodon* resembles *Morganucodon* and *Haldanodon* in that the prootic sinus vein has an independent opening in the lateral trough of the petrosal (Crompton and Sun, 1985; Crompton and Luo, 1993); it differs from *Morganucodon*, but resembles nonmammalian cyno-donts, in that the vessel passed through the cavum epiptericum (Luo, 1989; Crompton and Luo, 1993). The anterior paroccipital process is bulbous; its lateral end is at least partially covered by a lamina of the squamosal, in an arrangement almost identical with the paroccipital region in tritheledontids (Figure 6.7). The lateral flange of the petrosal forms a very promi-nent lateral shelf perforated by two vascular foramina, as in tritylodontids and *Luangwa* (Kemp, 1980; Z. Luo, personal observation).

Sinoconodon differs from *Morganucodon* and *Haldanodon* in that the anterior part of the Meckelian groove is separated from and is nearly parallel to the ventral border of the horizontal ramus of the mandible (Figure 6.12). By contrast, in morganucodontids and tricon-odontids, the Meckelian groove intersects the ventral border of the mandible (Figure 6.12). The groove converges toward the ventral border of the mandible in docodontids (Kron, 1979; Lillegraven and Krusat, 1991) and dryolestids (Krebs, 1971). The character state of *Sinoconodon* is clearly more primitive, because it is also present in *Diarthrognathus* and all nonmam-malian cynodonts. The dentary condyle is at the same level as the postcanine alveoli in *Sinoconodon* (Crompton and Sun, 1985), whereas the condyle is raised above the level of the alveoli in all other mammals.

However, *Sinoconodon* is more derived than mor-ganucodontids (but similar to triconodontids) in that the posterior part of the Meckelian groove on the mandible is very reduced. The medial ridge of the dentary overhanging the groove is also reduced. The separation of the round window and the jugular fora-men is wider than in *Morganucodon* and *Dinnetherium*. The canines are larger than those of any early mammals and advanced nonmammalian cynodonts.

The overwhelming numbers of apomorphies in

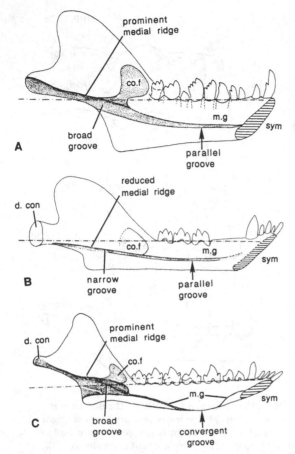

Figure 6.12. Schematic of the transformation of the Meckelian groove through cynodont–mammal transition (medial view of the left mandible). (A) A generalized tritheledontid. Sources: Crompton (1964), Shubin et al. (1991), Crompton and Luo (1993). (B) *Sinoconodon* (based on IVPP 4727 and CUP 1). (C) *Morganucodon* (after Kermack et al., 1973). Abbreviations: d. con, dentary condyle; sym, mandibular symphysis; co.f, fossa (depression) for coronoid bone; m.g, Meckelian groove. Stippled areas represent the depressions and grooves for the accessory jaw bones. Hatched areas indicate mandibular symphysis. The dashed lines represent the level of the alveoli of the postcanines.

the palate, temporomandibular joint, and petrosal strongly suggest that *Sinoconodon* and all other mammals form a robust monophyletic group. But within this monophyletic group, *Sinoconodon* differs from morganucodontids and other mammals in many dental and skull features. In most of these differences, *Sinoconodon* retains more primitive character states than other mammals. In a few cases, *Sinoconodon* is more derived than other mammals (such as reduction of the posterior part of the Meckelian groove and the medial ridge of the dentary), and these are regarded as autapomorphies. Given this mosaic of primitive and derived characters, *Sinoconodon* is identified as the

sister-taxon to a monophyletic taxon that includes all other mammals.

Adelobasileus (Lucas and Hunt, 1990; Lucas and Luo, 1993) possesses numerous mammalian apomorphies in the basicranium, braincase, and occiput. As far as can be determined from the preserved posterior part of the skull, this taxon is a member of Mammalia (Lucas and Hunt, 1990; Lucas and Luo, 1993). However, its dentition and temporomandibular joint are missing (Lucas and Luo, 1993). Because of the incomplete information, *Adelobasileus* is placed at a basal trichotomy with *Sinoconodon* and a monophyletic group comprising the rest of mammals. Its phylogenetic position cannot be fully resolved relative to *Sinoconodon* and other Liassic mammals (Lucas and Luo, 1993).

Monophyletic group comprising *Haldanodon*, Kuehneotheriidae, *Megazostrodon*, Morganucodontidae, and Triconodontidae

I argue that *Haldanodon*, kuehneotheriids, *Megazostrodon*, morganucodontids, triconodontids, and other mammals form a monophyletic group. Although multituberculates, monotremes, and therians with tribosphenic molars are not included in this analysis, because of the limitations of space, it is assumed in the following discussion that these clades also belong to this monophyletic group. The diagnosis of this group is based mostly on dental apomorphies. Although no information is available on the skull of kuehneotheriids, the petrosal and squamosal in *Megazostrodon*, *Morganucodon*, *Dinnetherium*, triconodontids, and *Haldanodon* (Lillegraven and Krusat, 1991) show that this grouping is also supported by many skull apomorphies (Crompton and Luo, 1993).

The following dental characters are diagnostic of this group (Crompton, 1974): differentiation of premolariform and molariform postcanines; a one-to-one occlusal relationship between the opposing upper and lower molars; interlocking of the adjacent molars by the cingular cusps; cusps of the opposing upper and lower molars maintaining consistent positional relationships; and consistent development of wear facets. Most permanent postcanines have two roots that are completely divided (Mills, 1971; Parrington, 1971, 1973), although *Haldanodon* has multiple roots in the upper molars (Lillegraven and Krusat, 1991). None of these dental characters is present in tritheledontids and *Sinoconodon*.

Tooth replacement is greatly reduced in this group. As demonstrated by Mills (1971), and Parrington (1978), *Morganucodon* and *Kuehneotherium* have the diphyodont dental replacement characteristic of mammals. This pattern is present in *Dinnetherium* (Jenkins, Crompton, and Downs, 1983; Jenkins, 1984; Crompton and Luo, 1993) and probably also in the unidentified

Cloverly Formation triconodont (Crompton and Luo, 1993). Gow (1986b) reports that the second molar may have been replaced in *Megazostrodon*, based on evidence of the degree of molar wear. If this can be confirmed by study of more nearly complete growth series for the taxon, the tooth replacement of *Megazostrodon* will prove to be an exception. In any event, *Megazostrodon* has lost the multiple replacement of the canines and incisors and the primitive postcanine replacement found in *Sinoconodon*.

The petrosals of *Megazostrodon*, *Morganucodon*, *Dinnetherium*, triconodontids, and *Haldanodon* share several apomorphies that are absent from *Sinoconodon* and other nonmammalian cynodonts. The lateral flange, which forms a broad shelf with two vascular foramina in *Luangwa*, tritylodontids, and *Sinoconodon*, is greatly reduced in *Megazostrodon*, morganucodontids, and triconodontids. These latter groups and *Adelobasileus* are reduced to one vascular foramen in the lateral flange. In *Haldanodon* (Lillegraven and Krusat, 1991), both the shelf formed by the lateral flange and the vascular foramen are absent. The prootic vein is enclosed by a canal in the petrosal in all these taxa, including *Megazostrodon*. By inference, the prootic vein was separated from the trigeminal ganglion contained in the trigeminal recess (Luo, 1989). It should be noted that the prootic canal is shorter in *Megazostrodon* (A. W. Crompton, pers. commun.), *Trioracodon* (Kermack, 1963; J. Wible, pers. commun.), and the Cloverly Formation triconodontid (Crompton and Jenkins, 1979; Z. Luo, personal observation on MCZ 19969) than in *Morganucodon* and *Dinnetherium*. Nevertheless, the course of this vein in *Megazostrodon* and *Trioracodon* is separated from the trigeminal recess as in morganucodontids and the Cloverly Formation triconodontid (Luo, 1989). The ventral transverse shelf of the septomaxilla, which is present in nonmammalian cynodonts (Sues, 1986; Wible, Miao, and Hopson, 1990) and *Sinoconodon* (Wible et al., 1990; Crompton and Luo, 1993), has been lost in *Morganucodon* (IVPP 4729) and triconodontids (MCZ 19973).

In this monophyletic group, the floor of the cavum epiptericum has enclosed the palatine branch and the geniculate ganglion of the facial nerve into the trigeminal recess. Thus, in tympanic view, there are two openings for the facial nerve branches (VII), with the anterior opening for the palatine branch and the posterior opening for the hyomandibular branch (Figures 6.5 and 6.9).

The squamosal glenoid in *Haldanodon*, morganucodontids, and triconodontids is formed by a platform that is separated from the cranial moiety of the squamosal by a narrow neck (Figure 6.7). The articular area of the glenoid faces ventrally, instead of anteroventrally and medially as in *Pachygenelus* and *Sinoconodon*. Crompton and Sun (1985) pointed out that the ventrally facing glenoid and the dorsoventrally

compressed dentary condyle would facilitate the rotation of the lower jaw, in correlation with the increased mobility of the symphysis of the mandible.

The dentary condyle is raised slightly above the level of the postcanine alveoli in *Haldanodon* (Lillegraven and Krusat, 1991), and much more so in morganucodontids and triconodontids. By contrast, in *Sinoconodon*, the condyle is level with the alveoli (Figure 6.12). *Haldanodon*, kuehneotheriids, morganucodontids, and triconodontids share an apomorphy: The anterior segment of the Meckelian groove converges toward or intersects with the mandibular border.

Relationships of *Haldanodon*

Haldanodon is a member of a monophyletic group that includes kuehneotheriids, *Megazostrodon*, morganucodontids, and triconodontids, but it does not include *Sinoconodon* and *Adelobasileus*. Some cranial specializations of *Haldanodon* may be correlated to its fossorial adaptation (Lillegraven and Krusat, 1991). The dentary condyle in *Haldanodon* is large and bulbous (Lillegraven and Krusat, 1991), a feature also present in *Sinoconodon* (Crompton and Sun, 1985; Crompton and Luo, 1993). By contrast, the dentary condyle in *Morganucodon*, *Dinnetherium*, and triconodontids is slender and dorsoventrally compressed (Kermack et al., 1973). The size and the outline of the condyle (in posterior view) in *Dinnetherium* (Jenkins, 1984) show no significant differences from those in *Morganucodon* (Kermack et al., 1973; Crompton and Hylander, 1986). The main distinction between *Morganucodon* and *Dinnetherium* is that in the latter, the angular process of the mandible is expanded in such a way that it becomes continuous with the ventral side of the condyle (Jenkins, 1984).

Haldanodon also differs from morganucodontids and triconodontids in some petrosal features. The promontorium in *Sinoconodon* and *Haldanodon* is less inflated than that in *Morganucodon*. Both taxa have a longitudinal ridge that is topographically similar to the basisphenoid wings in nonmammalian cynodonts. The cochlea in *Sinoconodon* is exclusively housed by the promontorium, but it does not extend the full length of the promontorium (IVPP 8689). Lillegraven and Krusat (1991) indicated that this is also the case in *Haldanodon*.

Relationships of Morganucodontidae and Triconodontidae

As a group, morganucodontids and triconodontids are diagnosed by several apomorphies in molar wear facets and in the cranium. In both taxa, lower cusp *a* occludes between cusp *A* and cusp *B* of the upper molars (Crompton and Jenkins, 1968, 1979; Crompton, 1974). The promontorium is much more rounded than in *Sinoconodon* and *Adelobasileus* and lacks the flat medial face found in *Sinoconodon*, *Haldanodon*, and *Megazostrodon*. The ventral opening of the cavum epiptericum is greatly reduced (Crompton and Luo, 1993). In *Morganucodon* (IVPP 8682) and *Dinnetherium* (MCZ 20991), the quadrate remains attached to the petrosal, despite the postmortem damage and disarticulation of the squamosal. This indicates that the squamosal did not contribute to the suspension of the quadrate and that the petrosal was the only bone in the cranium supporting the quadrate (Luo, 1989). The dorsal plate of the quadrate fits the fossa incudis of the anterior paroccipital process, while the concave contacting facet of the quadrate straddles the crest of the crista parotica (Figure 6.3). The posterior paroccipital process possesses a large and ventrally projecting process (*Morganucodon*, IVPP 8684; *Dinnetherium*, MCZ 20872; triconodont, MCZ 19969, 19973) (Luo, 1988, 1989).

Another apomorphy diagnostic of both morganucodontids and triconodontids is the withdrawal of the jugal bone from the glenoid. In *Probainognathus* (Romer, 1970; Crompton, 1972a), *Diarthrognathus* (Crompton, 1958), *Sinoconodon* (Crompton and Sun, 1985; Crompton and Luo, 1993), and *Haldanodon* (Lillegraven and Krusat, 1991), the jugal extends posteriorly to the glenoid region. But it is withdrawn from the glenoid in *Morganucodon* (Z. Luo, personal observation on IVPP 8684; Kermack et al., 1981) and triconodonts (MCZ 19973). The withdrawal of the jugal from the glenoid may also be diagnostic of a larger group that includes monotremes (van Bemmelen, 1901; Kuhn, 1971; Zeller, 1989) and multituberculates (Miao, 1988; Hopson, Kielan-Jaworowska, and Allin, 1989).

Relationships of *Megazostrodon* and Kuehneotheriidae

It is hypothesized here that *Megazostrodon*, morganucodontids, and triconodontids belong to a monophyletic group. *Megazostrodon* differs from *Morganucodon* and triconodontids in several cranial and dental characters. The differences in the occlusal relationships of the upper and lower principal cusps between *Megazostrodon* and *Morganucodon* have long been noted (Crompton and Jenkins, 1968; Mills, 1971; Crompton, 1974). In *Megazostrodon*, the principal cusp *a* of the lower molars occludes between cusp *B* of the opposing upper molars and cusp *C* of the preceding upper molars (Crompton, 1974; Crompton and Jenkins, 1979; Gow, 1986b). By contrast, in *Morganucodon*, *Dinnetherium*, and triconodontines (though not in amphilestines), the lower cusp *a* occludes between cusp *A* and cusp *B* (Crompton and Luo, 1993).

The anterior and posterior paroccipital processes in the petrosal in *Megazostrodon* are strongly divergent (Gow, 1986b; Rowe, 1986), at least more so than

in *Morganucodon*, *Dinnetherium*, and triconodontids (Crompton and Luo, 1993). Rowe (1986, fig. 39) illustrated the lateral end of the anterior paroccipital process as partially covered by the squamosal, a condition present in *Sinoconodon* and *Pachygenelus*, but not in *Morganucodon* (Figure 6.5), *Dinnetherium*, and triconodontids (Crompton and Luo, 1993, fig. 12). The pterygoparoccipital foramen is enclosed by bone (Gow, 1986b, fig. 3), which is similar to *Adelobasileus* (Lucas and Hunt, 1990). A flat facet on the ventromedial side of the promontorium is very similar to that of *Sinoconodon* (Gow, 1986b, fig. 4). Gow (1986b) suggested that *Megazostrodon* is similar to *Dinnetherium* but differs from *Morganucodon watsoni* in possessing a massive dentary condyle. However, the condyles of *Morganucodon* and *Dinnetherium* show no significant differences in size and structure and thus should be assigned to the same character state, whereas the massive condyle of *Megazostrodon*, as reported by Gow (1986b), is designated as a separate character state herein.

It is noteworthy that many of these presumably distinctive character states of *Megazostrodon* cannot be traced to the successive out-groups *Sinoconodon* and *Pachygenelus* and thus cannot be identified as plesiomorphic character states for the Early Jurassic mammals (Maddison et al., 1984; Gauthier et al., 1988). While the interpretation of several distinctive characters of *Megazostrodon* is based on fragmentary materials, information on many critical features of the basicranium is currently unavailable. With these caveats, I hypothesize that *Megazostrodon* and *Morganucodon* are sister-taxa. The presumed distinctive features of *Megazostrodon* are treated as autapomorphies.

The phylogenetic analysis suggests that kuehneotheriids belong to the monophyletic group including *Haldanodon*, *Megazostrodon*, morganucodontids, and triconodontids. However, it fails to solve the issue of how kuehneotheriids are related to other mammalian taxa within this monophyletic group. The fossil record of kuehneotheriids is based mostly on their dentition. As pointed out by Rowe (1988), including the incomplete taxa in a phylogenetic analysis decreases the resolution of the analysis and increases the number of unresolvable polychotomies. PAUP analysis of the character matrix (Table 6.1 and Appendix 6.1) shows that it would be equally parsimonious to place kuehneotheriids on almost every node of the Mammalia (Figure 6.1), because of its incompleteness.

Stepwise transformation of major mammalian apomorphies

Gow (1985) observed that the character-state changes in the cynodont–mammal transition can be classified in two categories: the long-term trends of cumulative directional change versus the sudden profound changes associated with the emergence of the higher taxon. He argues that only "sudden and profound" changes should be used as the apomorphies of Mammalia. He identifies four major mammalian apomorphies as "sudden and profound" changes: small size, determinate growth, presence of promontorium, and diphyodonty (Gow, 1985, p. 560). This distinction between the long-term, cumulative changes and the sudden emergence of the mammalian characters invites a basic question: What is the predominant mode of character transformation in the origin of mammals?

An understanding of the mode of character transformation depends on the overall phylogenetic framework. The phylogenetic relationship of the transitional clades provide the foundation for interpreting the evolution of the diagnostic characters of mammals. The new information on *Sinoconodon* (Crompton and Sun, 1985; Crompton and Luo, 1993), *Adelobasileus* (Lucas and Hunt, 1990; Lucas and Luo, 1993), *Haldanodon* (Lillegraven and Krusat, 1991), and *Megazostrodon* (Gow, 1986; Rowe, 1986) has refined the phylogenetic positions of these taxa. Based on the phylogenetic framework proposed in this study, (Figure 6.1A), I suggest that the transformation of the "diagnostic" mammalian characters occurred stepwise in such transitional clades as tritheledontids, *Sinoconodon*, and *Adelobasileus*. The major mode in the transformation of mammalian apomorphies is stepwise change at successively less inclusive hierarchical levels in cynodont phylogeny, not the sudden emergence of mammalian apomorphies without precursors. In the following discussion, I shall support this hypothesis of stepwise transformation by discussing the evolution of several character complexes: the temporomandibular joint, mammalian dentition, the petrosal promontorium, and the quadrate-petrosal suspension. Historically, these characters have played important roles in the diagnosis of Mammalia.

Temporomandibular joint

The presence of the dentary–squamosal articulation as the temporomandibular joint has been consistently used in all published diagnoses of Mammalia; see Rowe (1988) for a review of the literature. The prevalent view was that the dentary–squamosal articulation had multiple origins in cynodont phylogeny (Barghusen and Hopson, 1970; Gow, 1985). The main evidence cited in support of this view was that the dentary–squamosal jaw joint allegedly evolved in both tritheledontids (Crompton, 1958) and *Probainognathus* (Romer, 1970) among nonmammalian cynodonts. The dentary–squamosal joints in these taxa and mammals were believed to have evolved in parallel.

The hypothesis of multiple origins of the dentary–squamosal jaw joint was considerably weakened by reinterpretation of *Probainognathus*, which showed

that the temporomandibular joint in this form is formed by the surangular and squamosal (Crompton and Jenkins, 1979; Crompton and Hylander, 1986), rather than by the dentary and squamosal as suggested by Romer (1970).

Among all nonmammalian cynodonts, tritheledontids resemble *Sinoconodon* most closely in the structure of the squamosal. This similarity is congruent with a large suite of other apomorphies shared by the two taxa. The most parsimonious phylogenetic framework of this study (Figure 6.1) postulates that the dentary–squamosal articulations in tritheledontids and mammals are homologous (synapomorphic). The dentary–squamosal articulation went through at least two intermediate character states (Figure 6.7). In the common ancestor of tritheledontids and mammals, the enlarged posterior end of the lateral ridge of the dentary and the concave area in the ventral side of the zygomatic ramus of the squamosal presumably were present. Although the dentary lacked a distinctive condyle, and the squamosal a well-defined glenoid, the basic design of a full dentary–squamosal joint was already developed. In the common ancestor of *Sinoconodon* and other mammals, the squamosal glenoid articulated with a spherical dentary condyle, and this articulation was buttressed by a thickened postglenoid ridge (Figure 6.7B). The glenoid, which faced medially and anteroventrally, had the same orientation as that in tritheledontids. The dentary–squamosal joint is further modified in *Morganucodon watsoni*. The dentary condyle is dorsoventrally compressed, and the glenoid is formed by a ventrally facing platform on the squamosal (Kermack et al., 1981; Crompton and Sun, 1985). This modified jaw articulation, together with the greatly reduced symphysis, is correlated with the greater degree of mandibular rotation during occlusion than was possible in *Sinoconodon* and tritheledontids (Crompton and Sun, 1985; Crompton and Luo, 1993). The squamosal glenoid in *Haldanodon* is very similar to that in *Morganucodon oehleri*.

Dentition

Traditional views hold that because the dentary–squamosal joint originated repeatedly, the diagnosis of mammals should be supplemented by several dental characters, such as (1) differentiation of premolars and molars, (2) precise occlusion between opposing molars capable of shearing food, (3) diphyodont dentition indicating determinate growth, and (4) dorsomedial movement of the lower postcanines relative to the uppers during mastication (Hopson, 1969, 1973; Barghusen and Hopson, 1970; Crompton, 1974; Kemp, 1982; Kermack and Kermack, 1984). For example, Kermack et al. (1973, p. 163) argued that the dentary–squamosal articulation evolved in parallel in several lineages, because it developed as a secondary consequence of the complex shearing function of the cheek teeth.

These views need to be amended in the light of new information on tritheledontids and *Sinoconodon*. *Sinoconodon* lacks the differentiation of premolars and molars, and its postcanines have no wear facets. This indicates that precise occlusion was not developed in this taxon even though it had a fully formed dentary–squamosal jaw articulation. The multiple replacement of incisors and canines and the replacement of the more posterior postcanines in *Sinoconodon* indicate that it did not have a diphyodont dentition. Given the phylogenetic framework (Figure 6.1), it is most parsimonious to postulate that the dentary–squamosal articulation was fully developed prior to the evolution of the diphyodont dentition, the differentiation of the premolar and molar, and the shearing functions of the molars. The wear facets of the postcanines in tritheledontids extend the length of the tooth and are not homologous to molar wear facets in mammals (Gow, 1980; Shubin et al., 1991). Nonetheless, the wear facets suggest that some degree of dorsomedial movement of the lower teeth relative to the uppers occurred in this group. If tritheledontids are regarded as the sister-taxon to mammals (Hopson and Barghusen, 1986; Shubin et al., 1991; Crompton and Luo, 1993), the dorsomedial movement of the lower jaw must have been derived from the common ancestor of *Sinoconodon* and tritheledontids. It was only after the divergence of *Sinoconodon* and other mammals that several new dental and mandibular functions were developed. These included the increased mandibular rotation during occlusion, as indicated by the reduced symphysis and more flexible jaw joint. The precise matching of the wear facets developed in correlation with the reduced rates of tooth replacement. In summary, these dental characters, which were regarded as generally diagnostic of all mammals, must now be attributed to several different levels of cynodont phylogeny.

The evolution of the divided postcanine roots in most Early Jurassic mammals took place in two steps, according to this proposed phylogenetic scheme. Shubin et al. (1991) reported that a postcanine of *Pachygenelus* has an incipient division of the root. The two portions of the root are still connected by a thin sheet of dentine, so that the transverse section of the root forms a figure 8 (H.-D. Sues, pers. commun.). The division of the postcanine roots in *Sinoconodon* is not complete (Figure 6.6). The dentine connection between the two roots extends below the level of the gum line for nearly half of the height of the entire roots. In *Morganucodon* (Parrington, 1971, 1978) and *Dinnetherium* (Jenkins, Crompton, and Downs, 1983), the roots of most postcanines are divided above the level of the gum line. Traditionally, the presence versus absence of divided postcanine roots has been viewed as two discrete

character states (Kemp, 1983; Rowe, 1988; Wible, 1991). The transformation of the divided postcanine roots should include two more intermediate character states, which are represented by tritheledontids and *Sinoconodon*, respectively.

Petrosal promontorium

The promontorium is traditionally considered a major diagnostic feature of mammals (Crompton and Sun, 1985; Gow, 1985; Rowe, 1988; Luo and Ketten 1991). Currently, no transformational sequence is postulated for the evolution of the promontorium. Gow (1986a) initially proposed that the promontorium in mammals was formed by fusion of the parasphenoid ala with the periotic (petrosal) bone, but later revised that (Gow, 1986a, addendum). I argue that the petrosal promontorium with an elongated cochlear canal was also developed through three intermediate transformational steps. In primitive cynodonts, such as *Thrinaxodon* (Figure 6.9), the basisphenoid wing ("parasphenoid ala" of Gow, 1986a; Rowe, 1988) forms part of the rim of the oval window. The basisphenoid contributes extensively to the housing of the cochlea (Fourie, 1974). In *Probainognathus* (MCZ 4015, 4017) and *Tritylodon* (Z. Luo, personal observation; Gow, 1986a, addendum), the basisphenoid wing has slightly receded from the rim of the oval window. No nonmammalian cynodont has any indication of a promontorium.

The first step in transformation of the mammalian promontorium is represented by *Adelobasileus* (Lucas and Hunt, 1990). *Adelobasileus* has an incipient promontorium. The basisphenoid wing is smaller and is displaced anteromedially by the incipient promontorium (Figure 6.9). The second step in transformation can be characterized as the development of a flat promontorium with a short cochlear canal. This step is represented by *Sinoconodon* and probably *Haldanodon*. In the common ancestry of *Sinoconodon*, *Haldanodon*, and other mammals, the basisphenoid wing is lost, and the promontorium is fully developed. The promontorium in *Sinoconodon* (Crompton and Luo, 1993), *Haldanodon* (Lillegraven and Krusat, 1991), and *Megazostrodon* (Gow, 1986b; Rowe, 1986) is characterized by a longitudinal crest and a flat medial facet. The cochlea did not extend the whole length of the promontorium in *Sinoconodon*. The third step in transformation is represented by *Morganucodon*, *Dinnetherium*, and the Cloverly Formation triconodontid. These taxa possess a fully expanded, rounded promontorium. The cochlea extends the full length of the promontorium (Crompton and Sun, 1985; Graybeal et al., 1989; Luo and Ketten, 1991). Therefore, the fully developed promontorium and the elongated cochlea in most Liassic mammals are derived through three separate steps in cynodont phylogeny.

Quadrate suspension

The transformation of the quadrate (incus) played a critical role in the origin of the mammalian middle ear. Major transformations in cynodont phylogeny occurred in the dorsal plate of the quadrate that articulates with the cranium (Figures 6.3 and 6.4). They were accomplished in several steps. The first step is represented by the development of a concave articular facet of the dorsal plate in the common ancestor of *Probainognathus*, traversodontids, tritylodontids, tritheledontids, and mammals. In noncynodont therapsids and in primitive cynodonts such as *Thrinaxodon*, the articular facet of the dorsal plate is flat to slightly convex. The dorsal plate is oriented at a slight angle to the axis of the trochlea. By contrast, the facet is concave in *Probainognathus*, tritylodontids, tritheledontids, and *Morganucodon*. The second step is represented by the rotation of the dorsal plate relative to the axis of the trochlea in the common ancestor of *Probainognathus*, tritylodontids, tritheledontids, and mammals. The dorsal plate is rotated at an increasing angle to the trochlea in *Probainognathus*, tritylodontids, *Morganucodon*, and tritheledontids. The third step is characterized by the development of a neck between the dorsal plate and the trochlea of the quadrate in tritylodontids, tritheledontids, and *Morganucodon*. The dorsal plate is separated by this neck from the trochlea. The fourth step is the introduction of the stapedial process in *Morganucodon*, but the interpretation of the stapedial process as an apomorphy of mammals may be equivocal, because the feature is also present in tritylodontids (Kemp, 1983; Sues, 1985; Rowe, 1988; Wible, 1991).

Acknowledgments

My greatest gratitude goes to my postdoctoral adviser, Professor A. W. Crompton, who not only stimulated and supported me throughout this work but also gave me full access to the collections of cynodonts and mammals he is studying. I am grateful to Professor A. L. Sun, Mr. G. Cui, and Dr. X.-c. Wu, who generously provided skull specimens of *Yunnanodon*, *Sinoconodon*, and *Morganucodon* from Lufeng. Professor Farish A. Jenkins, Jr., graciously allowed me to use the MCZ collections of early mammals as comparative material. Dr. S. G. Lucas allowed me to study the holotype of *Adelobasileus cromptoni*. I also thank Mr. Charles Schaff for his help and support and Mr. William A. Amaral for skillfully preparing many specimens of *Sinoconodon* and *Morganucodon* used in this study. Mr. Laszlo Meszoly prepared two of the illustrations. During the course of this study, I benefitted from discussion with Drs. A. W. Crompton, W. A. Clemens, J. A. Hopson, T. Rowe, H.-D. Sues, N. H. Shubin, A. Sun, J. R. Wible, D. Miao, and J. A. Lillegraven. Drs. Lillegraven and Hopson made numerous comments and suggestions for improvement of an earlier draft, and Drs. N. C. Fraser and H.-D. Sues provided editorial help. The research was initiated with support from NSF grants BSR-8818098 and BSR-9020034 (to A. W. Crompton) and was completed with financial support from the College of Charleston.

References

Allin, E. F. 1975. Evolution of the mammalian middle ear. *Journal of Morphology* 147: 403–436.

1986. Auditory apparatus of advanced mammal-like reptiles and early mammals. Pp. 283–294 in N. Hotton III, P. D. MacLean, J. J. Roth, and E. C. Roth (eds.). *The Ecology and Biology of Mammal-like Reptiles*. Washington, D. C.: Smithsonian Institution Press.

Allin, E. F., and J. A. Hopson. 1991. Evolution of the auditory system in Synapsida ("mammal-like reptiles" and primitive mammals) as seen in the fossil record. Pp. 587–614 in D. B. Webster, R. R. Fay, and A. N. Popper (eds.), *The Evolutionary Biology of Hearing*. New York: Springer-Verlag.

Barghusen, H. R. 1968. The lower jaw of the cynodonts (Reptilia, Therapsida) and the evolutionary origin of the mammalian adductor jaw musculature. *Postilla* 116: 1–49.

1986. On the evolutionary origin of the therian tensor veli palatini and tensor tympani muscles. Pp. 253–262 in N. Hotton III, P. D. MacLean, J. J. Roth, and E. C. Roth (eds.), *The Ecology and Biology of Mammal-like Reptiles*. Washington, D.C.: Smithsonian Institution Press.

Barghusen, H. R., and J. A. Hopson. 1970. Dentary–squamosal joint and the origin of mammals. *Science* 168: 573–575.

Bemmelen, J. F. van. 1901. Der Schädelbau der Monotremen. *Denkschriften der Medicinisch-Naturwissenschaftlichen Gesellschaft, Jena* 6: 730–798.

Bonaparte, J. F. 1962. Descripción del cráneo y mandíbula de *Exaeretodon frenguellii*, Cabrera y su comparacion con Diademodontidae, Tritylodontidae y los cinodontos sudamericanos. *Publicaciones del Museo Municipal de Ciencias Naturales y Tradicional de Mar del Plata* 1: 135–202.

1966. Sobre las cavidades cerebral, nasal y otras estructuras del cráneo de *Exaeretodon* sp. (Cynodontia-Traversodontidae). *Acta Geologica Lilloana* 8: 5–31.

Brink, A. S. 1955. A study on the skeleton of *Diademodon*. *Palaeontologia Africana* 3: 3–39.

1963. Notes on new *Diademodon* specimens in the collection of the Bernard Price Institute. *Palaeontologia Africana* 8: 97–111.

Clark, J. M., and J. A. Hopson. 1985. Distinctive mammal-like reptile from Mexico and its bearings on the phylogeny of Tritylodontidae. *Nature* 315: 398–400.

Clemens, W. A. 1979. A problem in morganucodontid taxonomy (Mammalia). *Zoological Journal of the Linnean Society* 66: 1–14.

1986. On the Triassic and Jurassic mammals. Pp. 237–246 in K. Padian (ed.), *The Beginning of the Age of Dinosaurs: Faunal Change across the Triassic-Jurassic Boundary*. Cambridge University Press.

Clemens, W. A., and J. R. E. Mills. 1971. Review of *Peramus tenuirostris* Owen (Eupantotheria, Mammalia). *Bulletin of the British Museum (Natural History), Geology* 20: 89–113.

Crompton, A. W. 1958. The cranial morphology of a new genus and species of ictidosaurian. *Proceedings of the Zoological Society of London* 140: 697–750.

——— 1963. Tooth replacement in the cynodont *Thrinaxodon liorhinus* Seeley. *Annals of the South African Museum* 46: 479–521.

——— 1964. On the skull of *Oligokyphus*. *Bulletin of the British Museum (Natural History), Geology* 9: 70–82.

——— 1972a. Evolution of the jaw articulation in cynodonts. Pp. 231–253 in K. A. Joysey and T. S. Kemp (eds.), *Studies in Vertebrate Evolution*. Edinburgh: Oliver and Boyd.

——— 1972b. Postcanine occlusion in cynodonts and tritylodonts. *Bulletin of the British Museum (Natural History), Geology* 21: 27–71.

——— 1974. The dentitions and relationships of the Southern African Triassic mammals, *Erythrotherium parringtoni* and *Megazostrodon rudnerae*. *Bulletin of the British Museum (Natural History), Geology* 24: 399–437.

——— 1989. The evolution of mammalian mastication. Pp. 23–40 in D. B. Wake and G. Roth (eds.), *Complex Organismal Functions: Integration and Evolution in Vertebrates*. New York: Wiley.

Crompton, A. W., and F. Ellenberger. 1957. On a new cynodont from the Molteno Beds and the origin of the tritylodontids. *Annals of the South African Museum* 44: 1–13.

Crompton, A. W., and W. L. Hylander. 1986. Changes in mandibular function following the acquisition of a dentary–squamosal jaw articulation. Pp. 263–282 in N. Hotton III, P. D. MacLean, J. J. Roth, and E. C. Roth (eds.), *The Ecology and Biology of Mammal-like Reptiles*. Washington, D. C.: Smithsonian Institution Press.

Crompton, A. W., and F. A. Jenkins, Jr. 1968. Molar occlusion in Late Triassic mammals. *Biological Reviews* 43: 427–458.

——— 1979. Origin of mammals. Pp. 59–72 in J. A. Lillegraven, Z. Kielan-Jaworowska, and W. A. Clemens (eds.), *Mesozoic Mammals: The First Two-thirds of Mammalian History*. Berkeley: University of California Press.

Crompton, A. W., and Z. Luo. 1993. Relationships of the Liassic mammals, *Sinoconodon, Morganucodon oehleri*, and *Dinnetherium*. Pp. 30–44 in F. S. Szalay, M. J. Novacek, and M. C. McKenna (eds.), *Mammal Phylogeny*. New York: Springer-Verlag.

Crompton, A. W., and A. Sun. 1985. Cranial structure and relationships of the Liassic mammal *Sinoconodon*. *Zoological Journal of the Linnean Society* 85: 99–119.

Cui, G. 1976. [*Yunnania*, a new tritylodontid from Lufeng, Yunnan.] *Vertebrata Palasiatica* 25: 1–7. [in Chinese]

Cui, G., and A. Sun. 1987. [Postcanine root system of tritylodonts.] *Vertebrata Palasiatica* 25: 245–259. [in Chinese]

Edgeworth, F. H. 1935. *Cranial Muscles of Vertebrates*. Cambridge University Press.

Farris, J. S. 1982. Outgroups and parsimony. *Systematic Zoology* 31: 328–334.

Fourie, S. 1974. The cranial morphology of *Thrinaxodon liorhinus* Seeley. *Annals of the South African Museum* 65: 337–400.

Fraser, N. C., G. M. Walkden, and V. Stewart. 1985. The first pre-Rhaetic therian mammal. *Nature* 314: 161–163.

Gauthier, J., A. G. Kluge, and T. Rowe. 1988. Amniote phylogeny and the importance of fossils. *Cladistics* 4: 105–209.

Gow, C. E. 1980. The dentitions of the Tritheledontidae (Therapsida: Cynodontia). *Proceedings of the Royal Society of London* B208: 461–481.

——— 1981. *Pachygenelus, Diarthrognathus* and the double articulation. *Palaeontologia Africana* 24: 15.

——— 1985. Apomorphies of the Mammalia. *South African Journal of Science* 81: 558–560.

——— 1986a. The side wall of the braincase in cynodont therapsids and a note on the homology of the mammalian promontorium. *South African Journal of Zoology* 21: 136–148.

——— 1986b. A new skull of *Megazostrodon* (Mammalia: Triconodonta) from the Elliot Formation (Lower Jurassic) of southern Africa. *Palaeontologia Africana* 26: 13–23.

Graybeal, A., J. Rosowski, D. R. Ketten, and A. W. Crompton. 1989. Inner ear structure in *Morganucodon*, an early Jurassic mammal. *Zoological Journal of the Linnean Society* 96: 107–117.

Hall, A. V. 1991. A unifying theory for methods of systematic analysis. *Biological Journal of the Linnean Society* 42: 425–456.

Hopson, J. A. 1964. The braincase of the advanced mammal-like reptile *Bienotherium*. *Postilla* 87: 1–30.

——— 1966. The origin of the mammalian middle ear. *American Zoologist* 6: 437–450.

——— 1969. The origin and adaptive radiation of mammal-like reptiles and nontherian mammals. *Annals of New York Academy of Sciences* 167: 199–216.

——— 1971. Postcanine replacement in the gomphodont cynodont *Diademodon*. Pp. 1–21 in D. M. Kermack and K. A. Kermack (eds.), *Early Mammals*. London: Academic Press.

——— 1973. Endothermy, small size, and the origin of mammalian reproduction. *American Naturalist* 107: 446–452.

Hopson, J. A., and H. R. Barghusen. 1986. An analysis of therapsid relationships. Pp. 83–106 in N. Hotton III, P.D. MacLean, J. J. Roth, and E. C. Roth (eds.), *The Ecology and Biology of Mammal-like Reptiles*. Washington, D. C.: Smithsonian Institution Press.

Hopson, J. A., and A. W. Crompton. 1969. Origin of mammals. *Evolutionary Biology* 3: 15–72.

Hopson, J. A., Z. Kielan-Jaworowska, and E. F. Allin. 1989. The cryptic jugal of multituberculates. *Journal of Vertebrate Paleontology* 9: 201–209.

Jenkins, F. A., Jr. 1984. A survey of mammalian origins. Pp. 32–47 in P. D. Gingerich and C. E. Badgley (eds.), *Mammals: Notes for a Short Course*. University of Tennessee Department of Geological Sciences, Studies in Geology no. 8.

Jenkins, F. A., Jr., A. W. Crompton, and W. R. Downs. 1983. Mesozoic mammals from Arizona: new evidence on mammalian evolution. *Science* 222: 1233–1235.

Kemp, T. S. 1979. The primitive cynodont *Procynosuchus*: functional anatomy of the skull and relationships. *Philosophical Transactions of Royal Society of London* B285: 73–122.

———— 1980. Aspects of the structure and functional anatomy of the Middle Triassic cynodont *Luangwa*. *Journal of Zoology (London)* 191: 193–239.

———— 1982. *Mammal-like Reptiles and the Origin of Mammals*. London: Academic Press.

———— 1983. The interrelationships of mammals. *Zoological Journal of the Linnean Society* 77: 353–384.

Kermack, D. M., and K. A. Kermack. 1984. *Evolution of Mammalian Characters*. London: Croom Helm.

Kermack, D. M., K. A. Kermack, and F. Mussett. 1968. The Welsh pantothere *Kuehneotherium praecursoris*. *Zoological Journal of the Linnean Society* 47: 407–423.

Kermack, K. A. 1963. The cranial structure of the triconodontids. *Philosophical Transactions of the Royal Society of London* B246: 83–103.

Kermack, K. A., and Z. Kielan-Jaworowska. 1971. Therian and non-therian mammals. Pp. 103–115 in D. M. Kermack and K. A. Kermack (eds.), *Early Mammals*. London: Academic Press.

Kermack, K. A., F. Mussett, and H. W. Rigney. 1973. The lower jaw of *Morganucodon*. *Zoological Journal of the Linnean Society* 53: 87–175.

———— 1981. The skull of *Morganucodon*. *Zoological Journal of the Linnean Society* 71: 1–158.

Krebs, B. 1971. Evolution of the mandible and lower dentition in dryolestids (Pantotheria, Mammalia). Pp. 89–102 in D. M. Kermack and K. A. Kermack (eds.), *Early Mammals*. London: Academic Press.

Kron, D. G. 1979. Docodonta. Pp. 91–98 in J. A. Lillegraven, Z. Kielan-Jaworowska, and W. A. Clemens (eds.), *Mesozoic Mammals: The First Two-thirds of Mammalian History*. Berkeley: University of California Press.

Kuhn, H.-J. 1971. Die Entwicklung und Morphologie des Schädels von *Tachyglossus aculeatus*. *Abhandlungen der Senckenbergischen Naturforschenden Gesellschaft* 528:1–224.

Kühne, W. G. 1956. *The Liassic Therapsid Oligokyphus*. London: Trustees of the British Museum (Natural History).

Lillegraven, J. A., and G. Krusat. 1991. Cranio-mandibular anatomy of *Haldanodon exspectatus* (Docodontia; Mammalia) from the Late Jurassic of Portugal and its implications to the evolution of mammalian characters. *Contributions to Geology, University of Wyoming* 28: 39–138.

Lucas, S. G., and A. Hunt. 1990. The oldest mammal. *New Mexico Journal of Science* 30: 41–49.

Lucas, S. G., and Z. Luo. 1993. *Adelobasileus* from the Upper Triassic of West Texas: the oldest mammal. *Journal of Vertebrate Paleontology* 13: 309–334.

Luo, Z. 1988. Two distinct patterns of apomorphous petrosal character among major mammalian groups and their phylogenetic implications. *Journal of Vertebrate Paleontology* 8 (suppl. to 3): 20A.

———— 1989. The petrosal structures of Multituberculata (Mammalia) and the molar morphology of the early arctocyonids (Condylarthra: Mammalia). Ph. D. dissertation, University of California, Berkeley.

Luo, Z., and A. W. Crompton. In press. Transformations of the quadrate (incus) in the origin of mammals. *Journal of Vertebrate Paleontology*.

Luo, Z., and D. R. Ketten. 1991. CT scanning and computerized reconstructions of the inner ear of multituberculate mammals. *Journal of Vertebrate Paleontology* 11: 220–228.

Maddison, W. P., M. J. Donaghue, and D. R. Maddison. 1984. Outgroup analysis and parsimony. *Systematic Zoology* 33:83–103.

Maddison, W. P., and D. R. Maddison. 1986. MacClade Program, Version 2.1 (privately distributed).

Miao, D. 1988. Skull morphology of *Lambdopsalis bulla* (Mammalia: Multituberculata) and its phylogenetic implications to mammalian evolution. *Contributions to Geology, University of Wyoming, Special Paper* 4: 1–104.

Miller, M. E., G. C. Christensen, and H. E. Evans. 1964. *Anatomy of the Dog*. Philadelphia: Saunders.

Mills, J. R. E. 1971. The dentition of *Morganucodon*. Pp. 26–63 in D. M. Kermack and K. A. Kermack (eds.), *Early Mammals*. London: Academic Press.

Novacek, M. J. 1986. The skull of leptictid insectivorans and the higher-level classification of eutherian mammals. *Bulletin of the American Museum of Natural History* 183: 1–112.

Olson, E. C. 1944. Origin of mammals based upon cranial morphology of the therapsid suborders. *Special Papers of the Geological Society of America* 55: 1–122.

Parrington, F. R. 1971. On the Upper Triassic mammals. *Philosophical Transactions of the Royal Society of London* B261: 231–272.

———— 1973. The dentition of the earliest mammals. *Zoological Journal of the Linnean Society* 52: 85–95.

———— 1978. A further account of the Triassic mammals. *Philosophical Transactions of the Royal Society of London* B282: 177–204.

Patterson, B., and E. C. Olson. 1961. A triconodontid mammal from the Triassic of Yunnan. Pp. 129–191 in *International Colloquium on the Evolution of Lower and Nonspecialized Mammals*. Brussels: Koninklijke Vlaamse Academie voor Wetenschapen, Letteren en Schone Kunsten van Belgie.

Patterson, C. 1982. Morphological characters and homology. Pp. 21–74 in K. A. Joysey and A. E. Friday (eds.), *Problems of Phylogenetic Reconstruction*. Systematics Association Special Volume 21. London: Academic Press.

Quiroga, J. C. 1979. The inner ear of two cynodonts (Reptilia–Therapsida) and some comments on the evolution of the inner ear from pelycosaurs to mammals. *Gegenbaurs Morphologisches Jahrbuch* 125:178–190.

Romer, A. S. 1967. The Chañares (Argentina) Triassic reptile fauna. III. Two new gomphodonts, *Massetognathus pascuali* and *M. teruggii*. *Breviora* 264: 1–25.

———— 1970. The Chañares (Argentina) Triassic reptile fauna. VI. A chiniquodontid cynodont with an incipient squamosal–dentary jaw articulation. *Breviora* 344: 1–18.

Rougier, G. W., J. R. Wible and J. A. Hopson 1992. Reconstruction of the cranial vessels in the Early Cretaceous mammal *Vincelestes neuquenianus*: implications for the evolution of the mammalian cranial vascular system, *Journal of Vertebrate Paleontology* 12: 188–216.

Rowe, T. 1986. Osteological diagnosis of Mammalia L. 1758, and its relationship to extinct Synapsida. Ph.D. dissertation, Museum of Paleontology, University of California at Berkeley.

1987. Definition and diagnosis of the phylogenetic system. *Systematic Zoology* 36: 208–211.

1988. Definition, diagnosis, and origin of Mammalia. *Journal of Vertebrate Paleontology* 8: 241–264.

Shubin, N. H., A. W. Crompton, H.-D. Sues, and P. E. Olsen. 1991. New fossil evidence on the sister-group of mammals and early Mesozoic faunal distribution. *Science* 251: 1063–1065.

Story, G. 1951. The carotid arteries in the Procyonidae. *Fieldiana, Zoology* 32: 477–557.

Sues, H.-D. 1985. The relationships of the Tritylodontidae (Synapsida). *Zoological Journal of the Linnean Society* 85: 205–217.

1986. The skull and dentition of two tritylodontid synapsids from the Lower Jurassic of western North America. *Bulletin of the Museum of Comparative Zoology, Harvard University* 151: 217–268.

Sun, A. L. 1984. Skull morphology of the tritylodont genus *Bienotheroides* of Sichuan. *Scientia Sinica* B27: 270–284.

Sun, A. L., and G. Cui. 1986. A brief introduction to the Lower Lufeng saurischian fauna (Lower Jurassic: Lufeng, Yunnan, People's Republic of China). Pp. 275–278 in K. Padian (ed.), *The Beginning of the Age of Dinosaurs: Faunal Change across the Triassic–Jurassic Boundary.* Cambridge University Press.

1987. [Otic region in tritylodont *Yunnanodon*.] *Vertebrata Palasiatica* 25: 1–7. [in Chinese]

1989. [The discovery of a tritylodont from the Xinjiang Autonomous Region.] *Vertebrata Palasiatica* 27: 1–8. [in Chinese]

Swofford, D. L. 1990. PAUP 3.0. Phylogenetic Analysis Using Parsimony (distributed by D. L. Swofford, Illinois Natural History Survey, Champaign, Ill.).

Wahlert, J. H. 1974. The cranial foramina of protrogomorphous rodents: an anatomical and phylogenetic study. *Bulletin of the Museum of Comparative Zoology, Harvard University* 146: 363–410.

Watson, D. M. S. 1942. On Permian and Triassic tetrapods. *Geological Magazine* 79: 81–116.

Wible, J. R. 1991. Origin of Mammalia: the craniodental evidence re-examined. *Journal of Vertebrate Paleontology* 11: 1–28.

Wible, J. R., and J. A. Hopson 1993. Basicranial evidence for early mammal phylogeny. Pp. 45–62 in F. S. Szalay, M. J. Novacek, and M. C. McKenna (eds.), *Mammal Phylogeny.* New York: Springer-Verlag.

Wible, J. R., D. Miao, and J. A. Hopson. 1990. The septomaxilla of fossil and recent synapsids and the problem of the septomaxilla of monotremes and

armadillos. *Zoological Journal of the Linnean Society* 98: 203–228.

Wiley, E. O. 1979. *Phylogenetics: The Theory and Practice of Phylogenetic Systematics.* New York: Wiley.

Young, C. C. 1947. Mammal-like reptiles from Lufeng, Yunnan, China. *Proceedings of Zoological Society of London* 117: 537–597.

Zeller, U. 1989. Die Entwicklung und Morphologie des Schädels von *Ornithorhynchus anatinus* (Mammalia: Prototheria: Monotremata). *Abhandlungen der Senckenbergischen Naturforschenden Gesellschaft* 545: 1–188.

Zhang, F., and G. Cui. 1983. [New material and new understanding of *Sinoconodon*.] *Vertebrata Palasiatica* 21: 32–41. [in Chinese]

Appendix 6.1: Dental and cranial characters

The phylogenetic hypotheses of Figures 6.1 and 6.8 were based on the distribution of 82 morphological characters among 14 taxa of advanced nonmammalian cynodonts and early mammals (Table 6.2). The following list is to provide short descriptions of the characters and definitions of the character-states, followed by the numerical coding of the character-states. The main focus of this analysis is on the relationships of Tritylodontidae, Tritheledontidae, and Mammalia, which are treated as the in-group taxa. The out-group taxa used in this study are chosen from the phylogenetic studies of Kemp (1982, 1983), Hopson and Barghusen (1986), Rowe (1986, 1988), and Wible (1991). These include (in successively more distant order): Traversodontidae, Diademodontidae, Probainognathidae, and Thrinaxodontidae. Many characters listed here were also discussed in those previous studies. The successively more derived characters are coded 0, 1, 2, 3, 4; but in the phylogenetic algorithms these are treated as unordered data. Table 6.1 presents the summary of the character-state coding for all 82 characters and all 14 taxa. The character matrix of Table 6.1 was analyzed using MacClade (Maddison and Maddison, 1986) and PAUP (Swofford, 1990).

Dentition

1. Mode of occlusion: bilateral (0), unilateral (1)
2. Direction of mandibular movement during occlusion: orthal (0), posterodorsal (1), dorsomedial (2)
3. Rotation of the mandible during occlusion (as suggested by inclination of wear facets on lower molars): absent (0), present (1), increased rotation (2)
4. Replacement of incisors and canines: alternating (0), diphyodont (1)
5. Number of incisors: more than four (0), reduced to three (1), reduced to two (2)
6. Canine: present (0), enlarged (1), absent (2)
7. Pattern of postcanine replacement: alternating (0), partial (1), diphyodont (2), sequential addition posteriorly (3)
8. Differentiation of postcanine crowns: undifferentiated (0), differentiated into premolariform and molariform (1)
9. Roots of postcanines: single (0), incipient root division as indicated by vertical grooves on the root (1), incomplete separation of roots (2), complete separation of roots (3), multiple roots (4)

10. Positional correspondence of upper and lower postcanines: absent (0), present (1)
11. Interlocking of adjacent lower postcanines: absent (0), cusp *d* of anterior molar fits into embayment between cusp *b* and cusp *e* of the succeeding molar (1), cusp *d* fits into embayment between cusp *e* and cusp *f* (2), cusp *d* applied to medial side of cusp *b* (3)
12. Cingula on lower postcanines: present (0), present, with well-developed cingular cusps (1), present, with well-developed labial cingula (2), absent (3)
13. Arrangement of main cusps of upper postcanines: in single longitudinal row (0), multiple cusps that may form multiple longitudinal rows or transverse rows (or both) (1), in reversed triangle (2)
14. Occlusion of principal cusps of upper and lower molariform postcanines: principal cusps of upper and lower molariforms lack consistent contact relationship (0), principal cusp *a* is positioned anterior to cusp *A* but posterior to cusp *B* of same tooth (1), principal cusp *a* anterior to cusp *B* but posterior to cusp *C* of preceding tooth (2), interdigitating occlusion between multiple cusps (3)
15. Relative size of longitudinally arranged molar cusps: *b* larger than *c* (0), *c* larger than *b* (1)
16. Functional development of wear facets on molars: absent (0), absent at eruption but developed later by wear (1), wear facets present at eruption (2)
17. Relationships of wear facets to main cusps: wear facet absent (0), simple longitudinal facet that extends entire length of crown (1), principal cusp bears two longitudinal wear facets (*a* of lower tooth bears two facets, which either contact facets of cusp *B* of opposing tooth and cusp *C* of the preceding tooth or contact facets of cusps *A* and *B* of opposing tooth) (2), single facet supported by two cusps (*a* and *c* of lower tooth contact the facet supported by cusps *A* and *B* of opposing tooth; single facet supported by cusps *a* and *b* contacts facet supported by *A* and *C* of preceding tooth) (3), multiple cusps, with each cusp bearing one or two transverse facets (4)
18. Angle of longitudinal wear facets to vertical plane: wear facet absent (0), less inclined (1), more inclined (2)

Mandible

19. Symphysis: fused (0), unfused (1), unfused and further reduced (2)
20. Groove for dental lamina: present (0), vestigial or absent (1)
21. Medial dentary ridge overhanging posterior segment of Meckelian groove: broad groove with prominent ridge (0), intermediate (1), shallow groove and low medial dentary ridge (2)
22. Middle segment of Meckelian groove: parallel (or nearly parallel) to and separated from ventral edge of mandible (0), converges toward ventral edge of mandible (1), intersects ventral edge of mandible (2)

Temporomandibular joint

23. Surangular: participates in temporomandibular joint (TMJ) (0), does not participate in TMJ (1)
24. Dentary: does not participate in TMJ (0), contacting

squamosal by enlarged posterior end of lateral ridge (1), presence of slender and dorsoventrally compressed condyle (2), presence of bulbous condyle (3)
25. Position of dentary–squamosal articulation relative to level of postcanine alveoli: below alveolar level (0), about same level (1), above alveolar level (2)
26. Squamosal glenoid that articulates with mandible: absent (0), formed by small and medially facing facet (1), formed by broad and anteroventrally facing glenoid (2), glenoid buttressed by postglenoid ridge (3), glenoid facing ventrally and separated from cranial moiety by neck (4)
27. Cranial moiety of squamosal: narrow (0), broad (1)
28. Posterior extension of jugal along zygomatic arch: extending back near quadratojugal notch of squamosal (0), extending back near squamosal glenoid for surangular (1), extending back to glenoid for dentary (2), reduced and receding from glenoid (3)
29. Dorsal contacting facet of quadrate: flat to slightly convex (0), concave (1)
30. Rotation of dorsal plate relative to trochlear axis: less than 10 degrees (0), about 45 degrees (1), over 90 degrees (2), parallel to trochlear axis (3)
31. Dorsal margin of dorsal plate of quadrate: retains pointed angle (0), has rounded margin (1)
32. Neck of quadrate: absent (0), incipient, represented by lateral emargination (1), well developed (2)
33. Stapedial contact of quadrate: by recess on proximal end of trochlea (0), by stapedial process extended from margin of quadrate contact facet (1), by stapedial process (crus longus) from neck of quadrate (2)

Petrosal

34. Prootic and opisthotic: separated (0), fused in adult (1)
35. Petrosal promontorium: absent (0), incipiently developed (1), fully developed, with longitudinal crest and flat medial facet (2), fully rounded (3)
36. Housing of cochlea formed by both basisphenoid wing and prootic (0), or by petrosal (prootic) only (1)
37. Length of cochlea: short (0), elongated without extending full length of promontorium (1), elongated and extending full length of promontorium (2)
38. Semicircular canals enclosed by both exoccipital and opisthotic (0), or by petrosal (opisthotic) only (1)
39. Internal acoustic meatus: without separate foramina for cochlear and vestibular nerves (0), with separate foramina (1)
40. Hyoid (stapedial) muscle fossa: absent (0), present (1)
41. Fenestra ovalis: with thickened ring (0), without thickened ring (1)
42. Separation of foramen rotundum and jugular foramen: confluent (0), narrowly separated (1), widely separated (2)
43. Relative proportion of ventral opening of cavum epiptericum to size of cavum epiptericum floor anterior to tympanic opening of prootic sinus canal: large ventral opening with no floor (0), large opening with incipient floor (1), reduced opening with extensive floor (2), very reduced opening (3), opening completely closed (4)
44. Width of lateral trough anterior to tympanic opening of facial nerve (VII): broad (0), narrow (1)

45. Foramen and passage of prootic sinus: separate tympanic foramen for prootic sinus absent from lateral trough of petrosal (0), foramen present in lateral trough, but vessel passed through cavum epiptericum (1), vessel separated by petrosal canal from cavum epiptericum (2)

46. Bifurcation of paroccipital process: absent (0), present (1)

47. Anterior paroccipital process: laterally covered by squamosal (0), bulbous, without squamosal cover (1), crest of crista parotica and fossa incudis differentiated, without squamosal cover (2)

48. Posterior paroccipital process: ventrally flat (0), ventrally projecting (1)

49. Relationship of geniculate ganglion and greater superficial petrosal nerve to floor of cavum epiptericum: floor dorsal to nerve and ganglion (0), floor of cavum enclosing geniculate ganglion and greater superficial petrosal nerve (palatine nerve) (1)

50. Foramina of maxillary and mandibular branches of trigeminal nerves: at suture of anterior lamina of petrosal and ascending process of alisphenoid (0), enclosed by anterior lamina of petrosal (1)

51. Pterygo-paroccipital foramen: enclosed by lateral flange of petrosal and/or rami of pterygoid and epipterygoid (0), squamosal contributes to enclosure of foramen (1), open (2)

52. Lateral flange of petrosal (prootic): forming lateral shelf perpendicular to anterior lamina (0), lateral flange forming broad shelf with vertical component (L-shaped) anterior to pterygo-paroccipital foramen (1), lateral shelf reduced (2)

53. Vascular foramina perforating lateral flange of petrosal (anterior to pterygo-paroccipital foramen): absent (0), one foramen present (1), two foramina present (2)

Orbitotemporal region

54. Maximum vertical depth of zygomatic arch relative to length of skull: 10–12% (0), 15–20% (1), over 20% (2), below 9% (3) (*Thrinaxodon* 12%, *Massetognathus* 17%, *Probainognathus* 18%, *Cynognathus* 18%, *Kayentatherium* 22%, *Diarthrognathus* 8%, *Sinoconodon*, *Morganucodon*, and triconodontid 5–7%, *Haldanodon* ~9%)

55. Postorbital bar: present (0), absent (1)

56. Presence of ascending process of palatine: absent (0), present but small (1), large and contributing to enclosure of orbital vacuity (2)

57. Separation of greater and lesser palatine foramina in orbit: absent (0), present (1)

58. Palatine: excluded from subtemporal border of orbit (0), or participates in subtemporal border by displacing pterygoid posteriorly (1)

59. Orbitosphenoid: unossified (0), ossified to form anterior floor of braincase but does not contact ascending process of palatine (1), contributing to medial orbital wall by contacting ascending process of palatine (2)

60. Frontal: without ventral vertical process contributing to medial wall of orbit (0), with process (1)

61. Contact of frontal with ascending process of alisphenoid: by anterior corner of ascending process (0), over half of dorsal margin of ascending process (1)

62. Maxilla: excluded from subtemporal margin of orbit in ventral view (0), participating in rounded subtemporal margin of orbit (1), forming well-defined edge along subtemporal margin (2)

63. Anterior ascending branch of the anteria diploetica: transmitted in open groove in temporal region (0), partially enclosed in canal (1), completely enclosed in canal (2)

Skull roof

64. Pineal foramen: present (0), absent (1)

65. Parietals: separate (0), fused in adults (1)

66. Interparietal in dorsal view of skull: separate bone in adult (0), fused with other bones (1)

67. Expansion of braincase in parietal region: absent (0), expanded (1), very expanded (2)

Palate and basicranium

68. Posterior edge of bony secondary palate: anterior to posterior end of postcanine row (0), level with posterior end of postcanine row (1), posterior to posterior end of postcanine row (2)

69. Width of palate anterior to level of basisphenoid: narrow (0), broad (1)

70. Transverse process of pterygoid: massive (0), reduced (1), absent (2)

71. Pterygopalatal crest: present but does not reach basisphenoid (0), reaches basisphenoid (1), absent (2)

72. Internal carotid artery: entered the braincase through foramina in basisphenoid (0), through ventral opening of cavum epiptericum (1)

73. Curvature of basisphenoid area of basicranium: flat (0), sigmoidal (1)

74. Basisphenoid wing ("parasphenoid ala"): long and borders on oval window (0), slightly reduced and excluded from oval window (1), greatly reduced (2), absent (3)

Occipital region

75. Separation of hypoglossal foramina from jugular foramen: absent (0), hypoglossal foramen at margin of jugular foramen (1), at least one foramen separated from jugular foramen (2), two foramina separated from jugular foramen (3)

76. Presence of odontoid notch: absent (0), present (1)

77. Shape of occipital condyle: spherical (0), cylindrical (1)

78. Size of occipital condyle: small (encircles foramen magnum for ventral one-third of its circumference) (0), large (encircles foramen magnum for more than half of its circumference) (1)

79. Posttemporal canal flanked dorsolaterally: by tabular (0), by squamosal (1)

80. Tabular: present (0), absent (1)

Nasal region

81. Transverse shelf of septomaxilla: present (0), absent (1)

82. Facial process of premaxilla: small (0), large but not reaching nasal (1), reaching nasal (2)

PART II

Faunal assemblages

These chapters examining the different Triassic and Jurassic assemblages are not intended as exhaustive reviews of all early Mesozoic tetrapod-bearing localities. Rather, they reflect areas where the greatest advances have been made in recent years. Other early Mesozoic tetrapod assemblages that include a number of small forms are known from southern Africa (e.g., Kitching and Raath, 1984), Argentina (Bonaparte, 1978), Brazil (Barberena, Araujo, and Lavina, 1985), and India (Jain, 1980), but much of that material remains unpublished or has been described only in a rather preliminary fashion.

Although the contribution by Benton et al. (Chapter 7) on Middle Triassic vertebrates of England obviously describes a local assemblage and does not necessarily reflect faunal compositions globally, we include it because it serves to provide some limited background for events in the Late Triassic. By the same token, the profound changes that took place in the terrestrial realm at the transition from the Triassic to the Jurassic are not always fully apparent until later on, and the contributions on Middle Jurassic small tetrapods from Britain by Evans and Milner (Chapter 18) and by Metcalf and Walker (Chapter 19) provide some insight into what has traditionally been regarded as a barren period in the continental fossil record. Although the British Middle Jurassic assemblages are from a limited geographic area, it is to be hoped that the surprising diversity of small and medium-sized tetrapods recorded here will encourage reconnaissance for Middle Jurassic vertebrate-bearing continental strata elsewhere.

Recent discoveries of several diverse tetrapod assemblages in the rift basins of the Triassic–Jurassic Newark Supergroup in eastern North America have provided many new insights into the spatial and temporal distribution of early Mesozoic tetrapods. On the basis of material from a new site of early to middle Carnian age in the Richmond basin in Virginia, Sues and Olsen (1990) questioned previous distinctions of Gondwanan and Laurasian tetrapod communities during the Late Triassic. As discussed by Sues et al. in Chapter 8, this locality has produced a typically "Gondwanan" type of fauna that is numerically dominated by traversodont eucynodonts, implying that the supposed distinction between Late Triassic tetrapod assemblages from Laurasia and Gondwana may well reflect differences in stratigraphic age rather than geographic separation. Another attribute of the Newark Supergroup strata is the cyclicity displayed by their lacustrine sedimentary rocks, representing repeated fluctuations in lake levels. These cyclical sequences are thought to reflect major climatic cycles driven by variations in the earth's orbit, similar to the Milankovitch cycles seen in Pleistocene and Holocene lake sediments (Olsen, 1986). Together with the distributions of vertebrate and plant fossils, analysis of these cycles offers the potential for high-resolution correlation between the deposits from the various Newark Supergroup rift basins. In the northern part of the Newark Supergroup, the age of the bone-bearing earliest Jurassic McCoy Brook Formation in the Fundy basin of Nova Scotia is well constrained by several separate lines of evidence. Significantly, conventional potassium-argon and potassium-argon isochron dates of an associated extrusive tholeiitic basalt (North Mountain Basalt) allow an absolute estimate of the age of these assemblages. Shubin et al. (Chapter 13) present a review of this material and its implications for faunal change across the Triassic–Jurassic boundary. Although sedimentary sequences in the Newark Supergroup extend from the Middle Triassic through to the Early Jurassic, the Norian strata apparently are poorly fossiliferous, and the Newark Supergroup does not currently present a continuous sequence of Late Triassic–Early Jurassic tetrapod and plant assemblages that could be used for a definitive test of the various recent hypotheses concerning faunal and floral changes across the Triassic–Jurassic boundary.

The contribution by Kaye and Padian (Chapter 9) surveys a range of intriguing bones and teeth from the Upper Triassic of Arizona, many of which currently defy taxonomic identification. Such fossils are frequently excluded from published reports for this very reason. One of the main conclusions resulting from the Front Royal workshop was the importance of publishing records of such enigmatic remains. In fact, we came to realize just how widespread some of these problematica appear to be in early Mesozoic assemblages worldwide. The inability to place a particular fossil in any established taxonomic group is no justification for ignoring it. Conodonts presented a taxonomic enigma for a long time; yet their tremendous value in biostratigraphic correlation was undisputed.

Generally speaking, new taxa founded on very fragmentary remains with few diagnostic characters are not desirable. Discovery of more complete fossil material and particularly of abundant skeletal remains documenting ontogenetic series may render such taxa invalid. At the same time, adopting an overly conservative approach is perhaps also not always warranted. In this volume, Hunt and Lucas (Chapter 12) proposed several new taxa of Late Triassic ornithischian dinosaurs on the basis of isolated tooth crowns. While recognizing the validity of arguments against this decision, they believe that the value of describing and naming tooth taxa outweighs the arguments against this procedure, because such taxa are of great potential use in biostratigraphic and paleobiogeographic studies.

One problem in dealing with the Triassic–Jurassic boundary is the use of the Rhaetian stage of the Upper Triassic. In recent years there has been a school of thought that advocates abandonment of this stage altogether, considering it part of the Norian (Tozer, 1979, 1984; Palmer, 1983; Buffetaut and Wouters, 1986). The 'Rhaetian' is of very limited duration when compared with the other Triassic stages (Harland et al., 1990). Furthermore, the so-called Rhaetian ammonoid fauna cannot be clearly distinguished from that of the underlying Norian stage (Tozer, 1984). Attempts to define the Rhaetian on palynological criteria (Schuurman, 1979) proved to be of limited value (Fisher and Dunay, 1981). The rationale to abandon the concept of a Rhaetian stage is thus convincing to us and avoids the confusion that frequently arises from the use of the chronostratigraphic stage name "Rhaetian" and the "Rhaetic," which is a lithostratigraphic unit of limited geographic extent. Nevertheless, the abrupt change to marine conditions in parts of western Europe during the latest Norian was

so marked that it provides a localized reference point that is difficult to ignore. Consequently, different opinions persist, even in this volume, where Hunt and Lucas (Chapter 12) continue to recognize the Rhaetian as a valid stage, whereas the editors and the majority of contributors have chosen to consider it part of the Norian stage. Further work at finer stratigraphic levels is needed to resolve these differences.

References

Barberena, M. C., D. C. Araujo, and E. L. Lavina. 1985. Late Permian and Triassic tetrapods from southern Brazil. *National Geographic Research* 1: 5–20.

Bonaparte, J. F. 1978. El Mesozoico de América del Sur y sus tetrápodos. *Opera Lilloana* 26: 5–596.

Buffetaut, E., and G. Wouters. 1986. Amphibian and reptile remains from the Upper Triassic of Saint-Nicolas-de-Port (eastern France) and their biostratigraphic significance. *Modern Geology* 10: 133–145.

Fischer, M. J., and R. E. Dunay. 1981. Palynology and the Triassic–Jurassic boundary. *Review of Palaeobotany and Palynology* 34: 129–135.

Harland, W. B., R. L. Armstrong, A. V. Cox, L. E. Craig, A. G. Smith, and D. G. Smith. 1990. *A Geologic Time Scale 1989.* Cambridge University Press.

Jain, S. L. 1980. The continental Lower Jurassic fauna from the Kota Formation, India. Pp. 99–123 in L. L. Jacobs (ed.), *Aspects of Vertebrate History: Essays in Honor of Edwin Harris Colbert.* Flagstaff: Museum of Northern Arizona Press.

Kitching, J. W., and M. A. Raath. 1984. Fossils from the Elliot and Clarens formations (Karoo sequence) of the northeastern Cape, Orange Free State and Lesotho, and a suggested biozonation based on tetrapods. *Palaeontologia Africana* 25: 111–125.

Olsen, P. E. 1986. A 40-million-year lake record of early Mesozoic orbital climatic forcing. *Science* 234: 842–848.

Palmer, A. R. 1983. Decade of North American Geology (DNAG) Geologic Time Scale. *Geology* 11: 503–504.

Schuurman, W. M. L. 1979. Aspects of Late Triassic palynology. 3. Palynology of latest Triassic and earliest Jurassic deposits of the Northern Limestone Alps in Austria and southern Germany, with special reference to a palynological characterization of the Rhaetian Stage in Europe. *Review of Palaeobotany and Palynology* 27: 53–75.

Sues, H.-D., and P. E. Olsen. 1990. Triassic vertebrates of Gondwanan aspect from the Richmond basin of Virginia. *Science* 249: 1020–1023.

Tozer, E. T. 1979. Latest Triassic ammonoid faunas and biochronology, western Canada. *Geological Survey of Canada, Professional Paper* 79-1B: 127–135.

1984. The Trias and its ammonoids: The evolution of a time scale. *Geological Survey of Canada, Miscellaneous Report* 35: 1–171.

7

A review of the British Middle Triassic tetrapod assemblages

MICHAEL J. BENTON, GEOFFREY WARRINGTON,
ANDREW J. NEWELL, AND PATRICK S. SPENCER

Introduction

Remains of fossil tetrapods were first recovered from Middle Triassic deposits in England in 1823. Buckland (1837) noted that "part of a jaw and other bones of a Saurian, found in the sandstone at Guy's Cliffe near Warwick, were presented to the Oxford Museum, by the late Butic Greathead, Esq." in 1823. He identified these as the remains of a phytosaur by comparison with specimens in Germany. The original specimen has been lost, but it probably comprised the jaws of the temnospondyl amphibian *Mastodonsaurus*. At about the same time, tracks of then unidentified footprints were exposed at Storeton Hill Quarry, Cheshire, in Middle Triassic rocks (Tresise, 1989). About 1824, footprints were discovered in sandstone near Tarporley, Cheshire, but their significance was not realized immediately (Egerton, 1838).

Further fossil bones, "apparently of Phytosaurus [sic], were found at Warwick by Dr Lloyd of Leamington" in October 1836 (Buckland, 1837), and in June 1838 superbly preserved handlike footprints were identified in the Storeton Quarries near Birkenhead, Merseyside. These immediately attracted wide attention and were recognized as very like footprints from the German Triassic that had been named *Chirotherium* in 1835 (Swinton, 1960; Sarjeant, 1974; Tresise, 1989).

Specimens collected by Dr. Lloyd from Coton End Quarry, Warwick, and from Leamington were identified by Murchison and Strickland (1840, p. 344) as teeth of *Megalosaurus* and of a "Saurian," as well as an unidentified vertebra; these were the first British Middle Triassic skeletal remains to be figured (Murchison and Strickland, 1840, pl. 28, figs. 6–10). Other finds probably had been made at Coton End Quarry, because Murchison and Strickland (1840, p. 343) stated that it "has been most productive of the remains of *Vertebrata*." One of these specimens

(Murchison and Strickland, 1840, pl. 28, fig. 9) was reidentified as "a smooth curved tooth" and was named *Anisodon gracilis* by Owen (1841b, pl. 62A, fig. 3). Later, Owen (1842a, p. 535) suggested that this specimen was a terminal claw-bearing phalanx of the amphibian *Labyrinthodon pachygnathus*. Owen (1842a, pp. 523–524) identified a second specimen as a vertebra of *Labyrinthodon leptognathus*, and Owen (1841b) identified some of the teeth as *Cladeiodon lloydi*, which was later regarded as a dinosaur.

Collections of new material not seen by Murchison and Strickland were received by Owen during 1840–1841; these came from quarries in and around Warwick (Figure 7.1), from Dr. Lloyd, and from quarries at Grinshill (Figure 7.1), north of Shrewsbury, from T. Ogier Ward, a Shrewsbury physician. In a paper presented to the Geological Society of London on February 24, 1841, Owen clearly viewed most of this material as representing various species of *Labyrinthodon* [i.e., *Mastodonsaurus*, which had been described by Jaeger (1828) from the German Upper Triassic]. In an abstract of that paper, Owen (1841a) included in *Labyrinthodon* a great range of different amphibian and reptile bones and, tentatively, the producer of the *Chirotherium* footprints. However, before that paper was published in full (Owen, 1842a), and before the British Association meeting in a August 1841 (Owen, 1842b), he had received a new cranium from Dr. Ward, which enabled him to separate the Grinshill animal from *Labyrinthodon*. He described it as *Rhynchosaurus articeps*, a new genus and species of reptile (Owen, 1842b,c). He regarded it as a "lacertian" (i.e., a lizard), but did not connect it with the Warwick material he was studying, which he retained in *Labyrinthodon* (Owen, 1842a). It has subsequently been realized (e.g., Walker, 1969; Benton, 1990) that all the Grinshill material, and much of that from Warwick, pertains to *Rhynchosaurus*. This includes the "tooth"

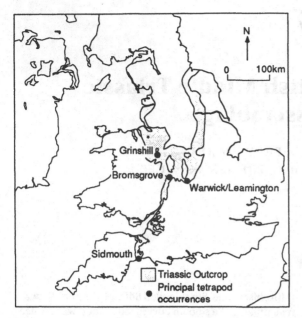

Figure 7.1. The main British Triassic outcrop (stippled), with localities mentioned in the text.

(actually a premaxilla) and the vertebra figured by Murchison and Strickland (1840), which are thus the first figured rhynchosaur fossils. The teeth referred to *Cladeiodon* by Owen (1841b), together with some other fossils from Warwick, were regarded as dinosaurian by Huxley (1870) and ascribed by Huene (1908a) to *Teratosaurus*, a rauisuchian (Galton, 1985; Benton, 1986). However, such generalized archosaur teeth probably are unidentifiable.

Subsequent descriptive work by Owen (1845, 1859, 1863), Huxley (1887), Woodward (1907), Watson (1910), Huene (1929), Hughes (1968), and Benton (1990) has shown that the Grinshill fauna consists exclusively of *Rhynchosaurus articeps*. Further tetrapod fossils, comprising temnospondyl amphibians, a rhynchosaur, a prolacertiform, and various archosaurs, were described from the Warwick area by Huxley (1859, 1869, 1870, 1887), Miall (1874), Burckhardt (1900), Wills (1916), Huene (1908a, 1929), Walker (1969), Paton (1974), Galton (1985), and Benton (1990).

New English sources of Middle Triassic fossil tetrapods were announced after Owen's time. Huxley (1869) described a rhynchosaur jaw bone from the south Devon coast near Sidmouth (Figure 7.1), and subsequent collecting there (Seeley, 1876; Metcalfe, 1884; Carter, 1888; Spencer and Isaac, 1983; Benton, 1990; Milner et al., 1990) has yielded an extensive fauna of temnospondyl amphibians, procolophonids, a rhynchosaur, archosaurs, and other unidentified animals. A solitary, well-preserved temnospondyl skull was

discovered at Stanton, Staffordshire (Ward, 1900), and temnospondyl amphibians, a rhynchosaur, archosaurs, a prolacertiform(?), and a nothosaur were recovered from Bromsgrove, Worcestershire (Figure 7.1) (Wills, 1907, 1910, 1916; Walker, 1969; Paton, 1974; Benton, 1990). Footprints, mainly *Chirotherium*, produced by a rauisuchian archosaur, and rhynchosauroid prints were reported from Middle Triassic deposits at numerous localities in the Cheshire basin (Thompson, 1970a, figs. 4 and 5) and the Midlands (Sarjeant, 1974).

The following account of the British Middle Triassic tetrapod faunas is based upon skeletal remains; vertebrate ichnofaunas are noted only where found in association with such remains. The stratigraphy and sedimentology of the host deposits, and the occurrence, composition, taphonomy, and paleoecology of the faunas, have been reviewed by three of us (MJB, AJN, and PSS) and GW has compiled independent evidence of age and has contributed to the assessment of the faunas.

Abbreviations utilized in the text are as follows:
BATGM, Bath Geology Museum
BGS (GSM), British Geological Survey (Geological Survey Museum), Keyworth, Nottingham
BIRUG, Birmingham University, Geology Department collections
BMNH, British Museum (Natural History), London
CAMSM, Cambridge University, Sedgwick Museum
EXEMS, Royal Albert Memorial Museum, Exeter
SHRBM, Shrewsbury Borough Museum
WARMS, Warwickshire Museum, Warwick

Stratigraphic and depositional setting

Terrestrial Middle Triassic tetrapod faunas are less well-known globally than those of Late Triassic age. The best-known European Middle Triassic deposit, the Muschelkalk, extends over parts of Germany, Switzerland, and Poland. This facies is famous for its diverse fauna of nothosaurs, placodonts, and ichthyosaurs and is relatively well dated by ammonoids and other marine fossils, but it does not occur in Britain.

Sedgwick (1829) recognized the British New Red Sandstone as equivalent, in part, to the German Triassic and considered some units equivalent to the German Buntsandstein and Keuper. Hull (1869) equated the English Bunter Sandstone with the German Buntsandstein (broadly Early Triassic in age) and the Lower Keuper Sandstone with the German Lettenkohle (latest Middle Triassic to early Late Triassic in age). He argued that a major unconformity in the British sequence corresponded to most of the Middle Triassic and represented the Muschelkalk (Figure 7.2). Warrington et al. (1980) advocated abandonment of the terms "Bunter" and "Keuper" as applied in Britain and established a lithostratigraphic nomenclature with

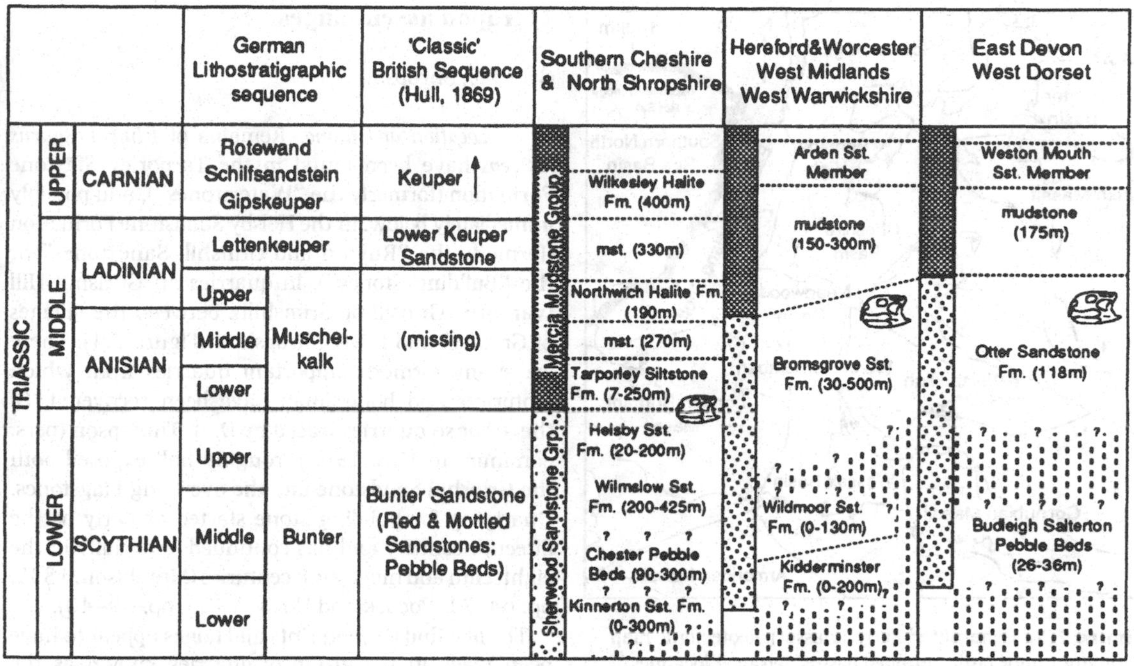

		German Lithostratigraphic sequence		'Classic' British Sequence (Hull, 1869)	Southern Cheshire & North Shropshire	Hereford&Worcester West Midlands West Warwickshire	East Devon West Dorset
TRIASSIC — UPPER	CARNIAN	Rotewand Schilfsandstein Gipskeuper		Keuper ↑	Wilkesley Halite Fm. (400m)	Arden Sst. Member mudstone (150-300m)	Weston Mouth Sst. Member mudstone (175m)
MIDDLE	LADINIAN	Lettenkeuper		Lower Keuper Sandstone	mst. (330m)		
		Upper	Muschel-kalk	(missing)	Northwich Halite Fm. (190m)		Otter Sandstone Fm. (118m)
	ANISIAN	Middle			mst. (270m)	Bromsgrove Sst. Fm. (30-500m)	
		Lower			Tarporley Siltstone Fm. (7-250m)		
LOWER	SCYTHIAN	Upper		Bunter Sandstone (Red & Mottled Sandstones; Pebble Beds)	Helsby Sst. Fm. (20-200m) / Wilmslow Sst. Fm. (200-425m)	Wildmoor Sst. Fm. (0-130m)	Budleigh Salterton Pebble Beds (26-36m)
		Middle	Bunter		Chester Pebble Beds (90-300m)	Kidderminster Fm. (0-200m)	
		Lower			Kinnerton Sst. Fm. (0-300m)		

(Mercia Mudstone Group; Sherwood Sandstone Grp.)

Figure 7.2. Stratigraphic setting of the British Middle Triassic tetrapod faunas. Correlations of the standard Triassic divisions and the German Triassic sequence with the British Triassic, as proposed by Hull (1869) for the "classical" British succession, and by Warrington et al. (1980, modified after the account in this chapter) for currently recognized lithostratigraphic units. Skulls indicate levels of main tetrapod faunas.

correlations based on palynomorphs and other fossils, where possible (Figure 7.2).

Palynological work (Warrington, 1967, 1970b; Geiger and Hopping, 1968) has shown that deposits of Middle Triassic age are present in Britain, where correlatives of the Muschelkalk, including brackish-water to littoral marine facies, occur in the upper part of the Sherwood Sandstone Group and lower parts of the Mercia Mudstone Group in central and northern parts of England (Geiger and Hopping, 1968; Warrington, 1974a; Ireland et al., 1978; Warrington et al., 1980).

The Sherwood Sandstone Group includes the former "Bunter Sandstone" and the arenaceous (lower) parts of the former British "Keuper." Its boundaries are diachronous, the lower ranging from Late Permian to Early Triassic and the upper from Early to Middle Triassic in age (Warrington et al., 1980). The Sherwood Sandstone Group comprises up to 1,500 m of arenaceous deposits that form the lower part of British Triassic successions. The sandstones are red, yellow, or brown in color, and pebbly units occur, especially in the Midlands. Most of the deposits are of fluvial origin, but there are many eolian units (Thompson, 1970a,b), and marine influences are evident toward the top.

The Mercia Mudstone Group corresponds broadly with the former "Keuper Marl" and encompasses the dominantly argillaceous and evaporitic units that overlie the Sherwood Sandstone Group throughout much of Britain. Its lower boundary may be sharp, but there is commonly a passage upward from predominantly sandy to predominantly silty and muddy facies at a diachronous interface that varies regionally from Early to Middle Triassic in age. The upper boundary, associated with a marine transgression that apparently occurred approximately contemporaneously throughout much of Europe, lies within the Rhaetian stage (sensu Richter-Bernburg, 1979). The Mercia Mudstone Group comprises dominantly red mudstones with subordinate siltstones. Extensive developments of halite and of sulfate evaporite minerals suggest deposition in hypersaline epeiric seas, connected to marine environments, in associated sabkhas, and in playas (Warrington, 1974b).

Triassic deposits have a broad U-shaped outcrop in the English Midlands, with a continuation south westward to South Wales and Devon. Smaller outcrops occur in northwest England, in Northern Ireland, and in Scotland (Warrington et al., 1980, figs. 2 and 3). The tectonic and sedimentary regimes established during Permian times continued into the Triassic, with deposition in fault-bounded basins in southern and western Britain and on the more regionally subsiding Eastern England Shelf, which formed the onshore

Figure 7.3. Generalized Early Triassic paleotectonic map showing major fault-bounded basins; tetrapod localities correspond with lowland areas in the Wessex basin and the central Midlands (Worcester Graben and adjacent areas).

marginal part of the Southern North Sea basin (Audley-Charles, 1970; Holloway, 1985) (Figure 7.3). In the Late Permian and Early Triassic, renewed and extensional subsidence in the Wessex basin, Worcester graben, and Needwood and Cheshire basins resulted in the establishment of an axial drainage system that flowed northward from the Variscan Highlands (Holloway, 1985). The south-to-north regional paleoslope and the proximal-to-distal depositional pattern that developed are reflected in the diachronous nature of the Sherwood Sandstone–Mercia Mudstone boundary (Figure 7.2), with coarse clastics being deposited in the south, whereas mudstones and evaporites accumulated farther north (Warrington, 1970a,b; Warrington et al., 1980; Warrington and Ivimey-Cook, 1992). This general sedimentary pattern was complicated locally by the introduction of coarse-grained deposits along basin margins and the deposition of marine intertidal sediments during Middle Triassic marine incursions. The widespread occurrence of transgressive intertidal facies of Middle Triassic age indicates extremely low relief in central England and suggests that the contemporary vertebrates were disporting themselves in lowland areas close to sea level, a suggestion first offered in 1839 by Buckland, who proposed (1844) a paleoenvironment of intertidal sandbanks.

Tetrapod assemblages

Grinshill, Shropshire

Location and fauna. Remains of *Rhynchosaurus articeps* have been found in the Tarporley Siltstone Formation (formerly the "Waterstones"), and possibly immediately below, in the Helsby Sandstone Formation (formerly the "Ruyton and Grinshill Sandstones" or the "Building Stones"), in quarries on Grinshill Hill (variously Grinsill or Grimshill), between the villages of Grinshill and Clive, Shropshire (Figure 7.4). There are many formerly important quarries from which footprints and bones may have been recovered. Of the 40 or so quarries traced by D. B. Thompson (pers. commun. to MJB, 1992), roughly half exposed both the Grinshill Sandstone and the overlying Flagstones. Quarrying for building stone started as early as the fifteenth century and has continued, especially in the eighteenth and nineteenth centuries (Murchison, 1839, pp. 64, 73; Pocock and Wray, 1925, pp. 39–40).

The first finds of footprints and bones appear to have been made in the group of quarries known as the Bridge Quarries, located between SJ 517329 and SJ 519327, the central Bridge Quarry (at about SJ 518328) being the main source, according to manuscript sources (D. B. Thompson, pers. commun. to MJB, 1992). Only one working quarry (Figure 7.5A,B), now owned by English China Clays (ECC) Quarries Division, remains (being centered on SJ 527237). It yields footprints very commonly, but bones are recovered only episodically [e.g., in 1971, 1984 (twice), and 1991].

The only tetrapod species reported thus far is *Rhynchosaurus articeps* Owen, 1842 (Figure 7.6). About 17 individuals have been collected since 1840, at least three of which were found recently in the active quarry. This species, redescribed in detail by Benton (1990), is a small rhynchosaur with a skull length of 60–80 mm (mean: 70 mm) and a total body length of 360–540 mm (mean: 470 mm). It shows the characteristic rhynchosaurian adaptations for feeding on tough vegetation (Benton, 1983, 1984) and differs from its larger relatives mainly in having slender bones, presumably a scaling effect of its small size.

The absence of other tetrapod body fossils from Grinshill is unusual and may be related to environmental or paleogeographic factors, although it is not clear what these may have been. Rauisuchians were also present there, as indicated by rare *Chirotherium* footprints (discussed later), but no bones of these large reptiles have yet come to light.

Host deposits. The Grinshill Sandstone comprises some 20 m of buff and yellow, medium-grained, well-sorted sandstones. These are well cemented and contain many small occurrences of manganese hydroxide. Large-scale cross-beds, at times reactivated, suggesting aeolian deposition, are visible in vertical quarry faces

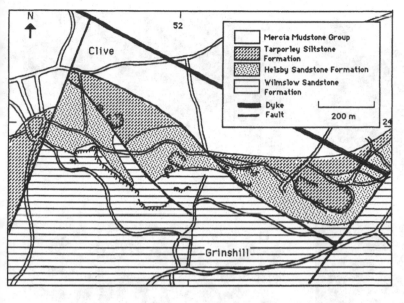

Figure 7.4. Grinshill localities. The map is based on published maps of the British Geological Survey (BGS 1:63360 scale geological sheet 138, Wem) and on field observations by MJB.

(Thompson, 1985). The formation may be equivalent to part of the Helsby Sandstone Formation (Sherwood Sandstone Group) of central and northern parts of the Cheshire basin (Warrington et al., 1980).

The Grinshill Sandstone is separated sharply from the Tarporley Siltstone Formation (Mercia Mudstone Group) by about 0.3 m of loose sand, speckled with manganese oxide flakes and barite nodules, termed the Esk Bed (Pocock and Wray, 1925, pp. 39–40; Thompson, 1985).

The Tarporley Siltstone Formation, which varies from about 10 to 270 m in thickness in the Cheshire basin (Warrington et al. 1980, table 4), is only about 6–10 m thick at Grinshill. Thompson (1985) identified two facies, interpreted by him as "tropical arid belt fluvial and marine-marginal hypersaline lagoon (salina) deposits."

Facies A comprises rippled cross-laminated fine- to medium-grained sandstones that occupy trough-shaped erosion channels. The sandstones bear transverse and linguoid ripples, which reflect northward paleocurrents. They are green-gray in color when freshly exposed. Rippled surfaces bear rhynchosauroid footprints (discussed later), invertebrate trace fossils, and raindrop impressions (Figure 7.5C) that were noted at Grinshill (Ward, 1840; Buckland, 1844) soon after the first-ever report of such structures from the Middle Triassic of Storeton, Cheshire, in 1838 (Tresise, 1991). Rhynchosaur bones were recovered in 1984 from the lowest of these units in the active quarry (D. B. Thompson, pers. commun. to MJB, 1992). This facies is interpreted as representing low-energy fluvial environments with rivers that occasionally dried up; exposed semisolidified muds were pitted during sporadic rain showers (Thompson, 1985). These evi-

dently moist conditions may have supported a seasonal vegetation sought by the herbivorous rhynchosaurs.

Facies B comprises interbedded fine sandstones (to 0.1 m), siltstones, and mudstones (10–20 mm thick), with primary current lineation, asymmetric current and wave ripple marks, and under-surfaces with load casts, flutes, and prod marks. Adhesion ripples have been observed on flat and current ripple surfaces (D. B. Thompson, pers. commun. to MJB, 1992). Mud cracks and pseudomorphs after halite, rhynchosauroid footprints, and invertebrate trace fossils, including a meandering "worm" trail about 15 mm wide, have been observed in situ by MJB. The mudstones and siltstones are stained red in parts, particularly near the base of the Tarporley Siltstone Formation. This facies, with evidence for current activity (?rivers) and wave activity (?lagoons), appears to represent "fluvial-intertidal rather than lake marginal" environments (Thompson, 1985). Brackish pools occasionally dried out, leaving salt crystals and mud-cracked surfaces. Half-damp sand flats developed adhesion ripple features. Rhynchosaurs and other reptiles (discussed later) walked across the muds.

Occurrences of reptiles. As noted by Owen (1842b, p. 146) and by Ward (in litt., BMNH), specimens of *Rhynchosaurus articeps* occur in two sediment types: in a fine-grained sandstone and in a coarser pinkish gray sandstone, termed "burr-stone" by Owen. The fine sandstone is gray to beige in color and has subrounded sand grains, greenish mud flakes, and specks of mica and manganese minerals. Slabs show fine parallel lamination, and some bedding surfaces show ripple marks and irregular clasts up to 10 mm in diameter.

Walker (1969, p. 470) noted that the specimens of

Figure 7.5. Tetrapod-bearing Middle Triassic deposits and localities in central England. Grinshill Stone Quarry prior to becoming owned by ECC Quarries Ltd., looking northward (A) and eastward (B), and a rain-printed slab from that quarry (C) (coin 30 mm in diameter). In the quarry (A, B), the eolian Helsby Sandstone Formation forms clear, manually squared faces, and the overlying Tarporley Siltstone Formation occupies the slope above. (D) Coten End Quarry as it is now, largely overgrown, looking eastward. (E) Section in the banks of the River Avon at Guy's Cliffe, showing the cross-bedding and contorted strata noted by Huene (1908b). (F) Site of the main fossiliferous quarry at Bromsgrove, now filled and forming part of the grounds of a hospital. (All photographs by AJN.)

R. articeps came from the Tarporley Siltstone Formation (the fine-grained gray sandstone) and from the top of the Grinshill Sandstone Formation (the coarser sandstone). This latter rock type was identified in the Building Stones in Aikin and Murchison's succession (Murchison, 1839, p. 40), and both rock types were described in Pocock and Wray's (1925, pp. 39–40)

section, in which the top of the Grinshill Sandstone is described as "Hard Burr: Hard yellowish-white sandstone, 2ft. 6in." Thompson (1985) doubted that any bones had been found in the Grinshill Sandstone, but a find in 1991 by Philip Page, a quarryman of ECC Quarries Ltd., appears to confirm the likelihood of the original evaluation.

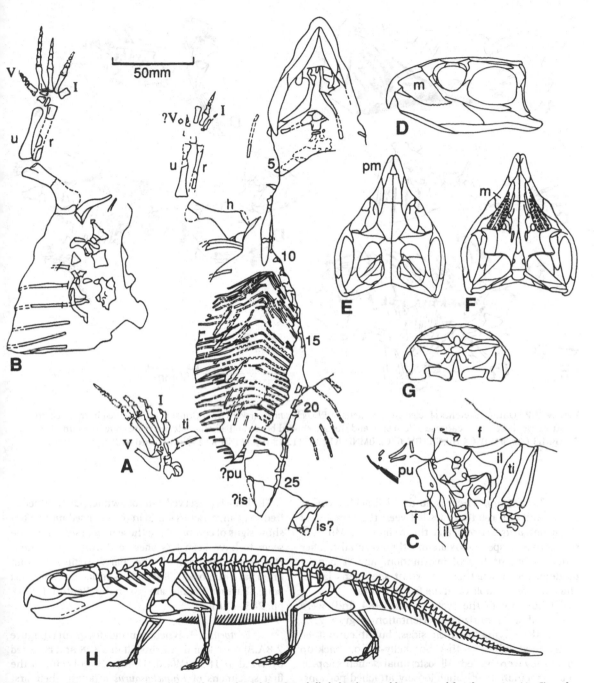

Figure 7.6. *Rhynchosaurus articeps*, the only member of the Grinshill skeletal assemblage: typical fossil remains (A–C) and restorations (D–H). (A) Partial skeleton lacking the tail and the limbs of the left side, in ventral view (BMNH R1237, R1238). (B) Dorsal vertebrae, ribs, and right forelimb in posteroventral view (SHRBM 6). (C) Pelvic region, right leg with ankle bones, presacral vertebrae 22–25, sacral vertebrae 1 and 2, and caudal vertebrae 1–8 (BATGM M20a/b). Restoration of the skull, based on SHRBM G132/1982 and 3 and BMNH R1236, in lateral (D), dorsal (E), ventral (F), and occipital (G) views. (H) Restoration of the skeleton in lateral view in walking pose. Abbreviations: f, femur; h, humerus; il, ilium; is, ischium; m, maxilla; pm, premaxilla; pu, pubis; r, radius; ti, tibia; u, ulna. Vertebrae and digits are numbered. (All based on Benton, 1990.)

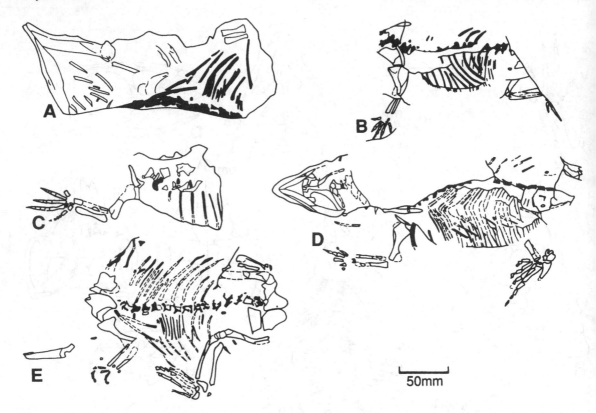

Figure 7.7. Outline sketches of selected specimens of *Rhynchosaurus articeps* from Grinshill as preserved in the rock, to illustrate the pose of the carcasses. Vertebrae and ribs are shaded black, and all other elements are shown in outline. SHRBM 3 (A), SHRBM 4 (B), SHRBM 6 (C), BMNH R1237/R1238 (D), BMNH R1239 (E). (Based on Benton, 1990.)

Taphonomy. There is no detailed field information about the relationships between the remains of *Rhynchosaurus articeps* and the sediments. Museum slabs bearing specimens are mostly too small to offer much sedimentological information, and it cannot be determined whether the extant specimens lay in channels or in pools or at the feet of dunes.

All but one of the reptiles (the 1984 find) were preserved in a horizontal orientation (Figure 7.7), rather than lying on their sides, but it cannot be determined whether they lay belly-up or back-up when they were buried. All postcranial elements appear to be articulated, although loosely attached portions, such as gastralia and scapulae, may have moved slightly out of position. Most skull specimens show slight distortion and disarticulation of loosely sutured elements. This is probably the result of collapse during burial or during postdepositional sediment compaction. The remains evidently were buried rapidly; there is no evidence that parts of the skeleton were removed by water currents, wind, or scavengers before fossilization. One skull (BMNH R1236) shows tectonic damage; a small fault offsets the posterior parts of the parietal, braincase, squamosal, and mandible by about 5 mm.

The bone is preserved as a soft, white, partly mineralized substance. Bones found in the finer sediment often show signs of compression; those in coarser sandstone seem to have been less affected during fossilization. However, bones found in the coarser sediment sometimes have iron-oxide-filled hollows replacing cancellous bone. Further details are given by Benton (1990, pp. 283–286).

Footprints. Rhynchosauroid footprints (Figure 7.8A,B) were found at Grinshill in 1838 and reported by Ward in 1839 (Ward, 1840), the collector of the first specimens of *Rhynchosaurus articeps*. Their first appearance in print was in an addendum by Murchison (1839, p. 734). They were found beneath the rubbly red sandstone called "Fee," on ripple-marked surfaces in a finely laminated, buff-colored sandstone presumably equivalent to part of Facies B of Thompson (1985). Ward (1840, p. 75) described the prints as differing "from those of *Chirotherium* in having only three toes, armed with long nails, directed forwards, not spreading out, and one hind toe on the same side as the longest fore toe, pointing backwards, and having a very long claw. No impression of the ball of the foot

Figure 7.8. Footprints from Grinshill Quarries. *Rhynchosauroides* prints on a ripple-marked surface, SHRBM "2" (A, B), and *Chirotherium*-type prints, SHRBM "1" (C), SHRBM "3" (D). All specimens are from the Tarporley Siltstone Formation of unknown quarries in the Clive-Grinshill area. Magnifications are × 0.2 (A), × 0.6 (B), × 0.27 (C), × 0.22 (D). Photographs by B. Bennission (SHRBM).

in this example; but in another there are three toes and a depression for a ball not unlike that of a dog." He later repeated and stressed the existence of a least one backward-pointing hind toe (Ward, 1874).

Ward (1841) attributed the footprints to the amphibian "*Inosopus scutulatus* [sic] (Owen MS.)," a form of *Labyrinthodon*. The name may be the same as *Labyrinthodon* (*Anisopus*) *scutulatus*, given by Owen (1842a, pp. 538–541, pl. 46, figs. 1–5) to a collection of small bones from Leamington, which he interpreted as amphibian in origin, but which are now regarded as those of a prolacertiform such as *Macrocnemus* (discussed later). Owen (1842b, p. 146; 1842c, pp. 355–356) recognized the footprints as probably those of *Rhynchosaurus articeps*, following a written communication by Ward.

Buckland described these finds graphically: "impressions of small drops of rain and footsteps have ... been found by Mr Ward.... On the same slabs are also very distinct small ripple marks produced by water, the undulations of which shew the direction in which the water ran, while the impressions of the rain, being in an oblique direction, shew in what direction the wind blew at the moment when this shower fell. The footsteps on the same slab shew the direction in which the animal was running" [Buckland, 1840, pp. 246–247; 1844 (for 1839), p. 5].

Beasley (1896, 1898, 1902, 1904, 1905, 1906), described further specimens of footprints from Grinshill and concluded that they probably were varieties of his rhynchosauroid print, type D1. This is a four-toed print, about 30–40 mm in total length, of which often only the three longest digits, presumably representing digits II–IV, are preserved (Figure 7.8A,B). The digit impressions taper from their roots to a pointed tip. The longest digit can reach 35 mm, with a width halfway along of 5 mm. The digits decrease in length from IV to II, and digit I can be very short, often less than a quarter the size of II, but occasionally reaching half its length. An impression of a short digit V is occasionally preserved, but set back from the others, and diverging slightly, but not in the backward-pointing position emphasized by Ward (1840, 1874). The digits often appear to curve to one side, and the claws, in particular, are bent that way. The palm of the print is not preserved. Rhynchosauroid footprints from Grinshill are preserved in the museums in Manchester, Shrewsbury, Ludlow, and Warwick, and in the British Geological Survey collections (Sarjeant, 1983).

Maidwell (1911, p. 142) named Beasley's rhynchosauroid D1 prints *Rhynchosauroides articeps*, but without designating a type specimen. He referred to Beasley's 1896 and 1905 papers, in which a specimen from Weston Quarries, Runcorn, north Cheshire, is illustrated. These authors refer to the original specimens from Grinshill, but these appear never to have been figured, probably because the prints are less well defined than examples from neighboring localities.

Thompson (1985, pp. 119–121) recorded rhynchosauroid footprints in both Facies A and, especially, Facies B at Grinshill. He also noted footprints of *Chirotherium* type (in the museums in Shrewsbury and Ludlow), but could not say whether they came from Facies A or Facies B (Figure 7.8C,D) (Beasley, 1904, p. 225; Sarjeant, 1983, p. 553). Prints collected at Grinshill by J. Stanley have been documented by Delair and Sarjeant (1985).

Warwick and Leamington, Warwickshire

Location and fauna. Fossil amphibian and reptile remains from this area came from the Bromsgrove Sandstone Formation, between about 1840 and 1870, and include the specimens described by Owen (1842a), Huxley (1859, 1869, 1870, 1887), and Miall (1874). The main productive site, Coton End Quarry (SP 289655; Figure 7.5D), Warwick, was worked for building stone in the early nineteenth century, but it was long abandoned when Beasley (1890b, p. 148) reported a visit to the site. The faunal list from the area is based on fossils from this quarry (Figure 7.9). Unless stated otherwise data are from Walker (1969), Paton (1974), Shishkin (1980), Kamphausen (1983), Galton (1985), and Benton (1990).

1. Stenotosaurinae incertae sedis: *Stenotosaurus leptognathus* (Owen, 1842), nomen dubium, jaws and other skull fragments. A small capitosauroid temnospondyl with a flattened crocodile-shaped skull, 210 mm long. The capitosauroids had been assigned to two species of *Cyclotosaurus* by Paton (1974), but Shishkin (1980) assigned *C. leptognathus* to *Stenotosaurus*, and Kamphausen (1983) then argued that the type specimen of this species was indeterminate, thus making the taxon a nomen dubium.

2. Cyclotosaurinae incertae sedis: *Cyclotosaurus pachygnathus* (Owen, 1842), nomen dubium, jaws and other skull fragments. A moderate-sized capitosauroid temnospondyl with a lower skull than *S. leptognathus*.

3. *Mastodonsaurus* sp. [including material referred to as *Mastodonsaurus jaegeri* (Owen, 1842) and *Mastodonsaurus lavisi* (Seeley, 1876)]: a single jaw from Coton End, and one from Guy's Cliffe (the 1823 find), as well as assorted skull fragments. A large mastodonsaurid temnospondyl; estimated skull length 500–600 mm.

4. Cf. *Macrocnemus* (includes *Rhombopholis scutulata* Owen, 1842): partial skeleton (from Old Leamington Quarry) and isolated limb elements. A small lizardlike animal, 50–80 mm long.

5. *Rhynchosaurus brodiei* Benton, 1990: skull and mandible remains and isolated postcranial elements. A moderate-sized rhynchosaur with a skull 90–140 mm long and an estimated body length of 0.5–1.0 m.

6. *Bromsgroveia walkeri* Galton, 1985: vertebrae, sacrum, ilium, ischium, and ?femur. A moderate- to large-sized rauisuchian.

7. "Large thecodontian": an ilium.

8. Archosaur indet. (*Cladeiodon lloydi* Owen, 1841): about ten isolated teeth, which could belong to *Bromsgroveia*, to the "large thecodontian," or to another, as yet unidentified carnivorous archosaur.

9. "Prosauropod dinosaur": a cervical vertebra. If a true dinosaur, this could be the oldest known.

Three other localities in the Warwick-Leamington area also yielded vertebrate remains in the nineteenth century. Old Leamington Quarry (?SP 325666) produced remains of the fish *Gyrolepis* (Walker, 1969, p. 472) and of *Mastodonsaurus*, "*Stenotosaurus*," cf. *Macrocnemus* (type specimen of Owen's *Rhombopholis scutulata*) (Owen, 1842a, pp. 538–541, pl. 46, figs. 1–5), *Rhynchosaurus brodiei*, and a "prosauropod" tooth (Murchison and Strickland, 1840, pl. 28, fig. 7a; Huene, 1908a, fig. 265). Cubbington Heath Quarry (SP 332694) yielded *Mastodonsaurus* and "*Stenotosaurus*" (Huxley, 1859; Woodward, 1908; Wills, 1916, pp. 9–11, pl. 3). Guy's Cliff (SP 294668) (Figure 7.5E) produced remains of the jaws of *Mastodonsaurus* (Owen, 1842a, pp. 537–538, pl. 44, figs. 4–6, pl. 37, figs. 1–3; Miall, 1874, p. 433).

Host deposits. The Bromsgrove Sandstone Formation is 20–35 m thick in the Warwick district (Old, Sumbler, and Ambrose, 1987, p. 20). Coton End Quarry (Figure 7.5D) exposes 10–11 m of channeled and cross-bedded water-laid buff and red sandstone in units 1–3 m thick, with occasional impersistent marl and clay bands 0.1–0.5 m thick.

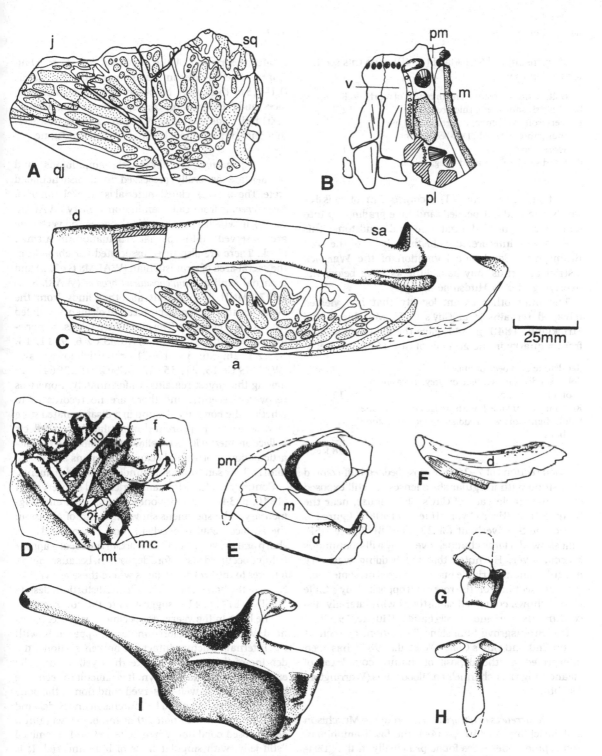

Figure 7.9. Typical elements of the Warwick fauna. (A) Left posteroexternal corner of the skull of "*Cyclotosaurus pachygnathus*" (Cyclotosaurinae incertae sedis) in lateral view (WARMAS Gz 13). (B) part of the snout of "*Stenotosaurus leptognathus*" (Stenotosaurinae incertae sedis) in palatal view (WARMS Gz 38). (C) Posterior portion of a left lower jaw of "*Stenotosaurus leptognathus*" in lateral view (WARMS Gz 35). (D) Scattered bones of cf. *Macrocnemus* (*Rhombopholis scutulata*) (WARMS Gz 10). (E–H) Assorted remains of *Rhynchosaurus brodiei*: anterior part of the skull in lateral view (WARMS Gz 6097/BMNH R8495) (E), anterior part of a dentary in medial view (WARMS Gz 950) (F), middorsal vertebra in right lateral view (WARMS Gz 17) (G), and interclavicle in ventral view (WARMS Gz 34) (H). (I) Right ilium of *Bromsgroveia walkeri* in lateral view (WARMS Gz 3). Abbreviations: a, angular; d, dentary; f, femur; j, jugal; m, maxilla; mt, metatarsal; pl, palatine; pm, premaxilla; qj, quadratojugal; sa, surangular; sq, squamosal. (A–C, after Paton, 1974; D and I after Owen, 1842a; E–H, after Benton, 1990.)

Murchison and Strickland (1840) gave this section for the quarry:

a. Soft, white sandstone and thin beds of marl   8 feet
b. Whitish sandstone, thick-bedded   12
c. Very soft sandstone, colored brown by
   manganese; called "Dirt-bed" by the
   workmen   1
d. Hard sandstone, called "Rag," about   2
                                                  ‾‾‾‾‾‾
                                                  23 feet.

Old et al. (1987, fig. 11) recorded 7 m of massive sandstone and flat-bedded sandstone grading up into 4 m of cross-bedded sandstone and mudstone; this section was interpreted as in the middle of the thin Bromsgrove Sandstone Formation of the Warwick district and thus may be as little as 10 m below the overlying Mercia Mudstone Group.

The only other extant locality that has yielded tetrapod remains is at Guy's Cliffe. Murchison and Strickland (1840, p. 344) gave the following section from a quarry in the grounds of Guy's Cliffe House:

Sandstone and beds of marl   8 feet
Solid sandstone, whitish or grey, occasionally
   of a reddish tint   12
Red, micaceous marl, with wedges of sandstone   8
Solid, light-colored, reddish tinted sandstone
   about   20
                                                  ‾‾‾‾‾‾
                                                  48 feet.

Good sections of 7–10 m of cross-bedded buff-colored sandstone with irregular shale lenses are still exposed in an old stable yard of Guy's Cliffe House, near the bank of the River Avon. Huene (1908b) figured a section on the riverbank (SP 29376678) (Figure 7.5E) that showed a large channel covered by a discontinuous breccia layer. He noted that the bedding was very irregular and that ripple marks occurred on some beds. Another section "on the rocky cliff opposite Guy's Cliffe House shows contorted sandstones with laterally discontinuous marl and breccia bands" (Huene, 1908b).

The Bromsgrove Sandstone Formation, as seen at Coton End and Guy's Cliffe (Old et al., 1987), has been interpreted as the deposit of mature complexes of meandering river channels and floodplains (Warrington, 1970b).

Occurrences of tetrapods. According to Murchison and Strickland (1840, p. 344), the fossil amphibian and reptile bones were found principally in the "Dirt-bed" in Coton End Quarry. Hull (1869, pp. 88–89) stated that the amphibian remains occurred in the "Water Stones," but Walker (1969) noted that the reptiles came from the upper part of the "Building Stones" (i.e., the Bromsgrove Sandstone Formation). Most of the Warwick specimens have been cleared of matrix, but some show a yellow or greenish-colored fine- to medium-grained sandstone with coarse patches that might accord with Murchison and Strickland's

(1840) description of the "Dirt-bed." In Coton End Quarry, laterally discontinuous marl and clay bands 0.1–0.5 m thick may be observed in the weathered faces, and these probably correspond to the fossiliferous "Dirt-bed." One specimen (WARMS Gz 34) is, however, in a hard, fine-grained, laminated gray sandstone.

Taphonomy. The Warwick amphibians and reptiles are generally preserved in a disarticulated state. The only associated material is the skeleton of cf. *Macrocnemus* from Old Leamington Quarry (WARMS Gz 10), in which a number of limb bones and vertebrae are preserved, still in partial articulation, in a small block. There are also three associated vertebrae from the sacral area of a rauisuchian (WARMS Gz 1, 2) and a partial skull of *Rhynchosaurus brodiei* (WARMS Gz 6097/BMNH R8495), but all other finds from the Warwick area are single postcranial elements, isolated teeth, or jaw elements. The amphibian fossils, comprising parts of the skull roof (WARMS Gz 6, 9, 11, 13, 14, 20, 26, 36, 38, 1057) or partial lower jaws [WARMS Gz 15, 27, 35, 37; BGS(GSM) 27964], are among the largest remains. Unfortunately, none was recovered recently, and there are no records as to whether the bones were found in a disarticulated state or whether their apparent disarticulated state reflects collection methods. We believe that most of the disarticulation is original, because specimens still in the matrix show sandstone surrounding isolated elements.

Contrary to Murchison and Strickland's observation (1840, p. 344) that the bones were "rolled and fragmentary," the specimens show little sign of abrasion; the surface detail is excellent, and broken ends are sharp and unworn. Some, at least, of the breaks appear to have occurred just before deposition, because matrix adheres to the broken surfaces where these are visible. None of the bones shows significant distortion, despite Miall's (1874, p. 417) suggestion to the contrary.

The bone in the Warwick specimens is preserved as hard, white to buff-colored material, apparently with the internal structure intact. In broken sections, the dentine of the rhynchosaur teeth is yellow, and the enamel is stained dark brown. It is difficult to reconcile the current hard, well-preserved condition of the bone with the description given by Murchison and Strickland (1840, p. 344), who noted that the bones were in a decomposed condition when collected and resembled "stiff jelly, with singular hues of blue and red. It is necessary to remove them with a solution of gum arabic, as the best means of preserving them."

Most of the Warwick bones show evidence of transport, but insufficient to cause abrasion. Carcasses, both large and small, were generally broken up (with the exceptions noted earlier), and some bones were broken through sharply. Skulls of amphibians and of the rhynchosaur seem to have survived with less disarticulation. It is not known whether the bones were deposited in channel lag deposits, on cross-bedded point bars, or in finer- grained overbank deposits. The few specimens on which the matrix survives show no sign of clasts indicative of a channel lag.

Footprints. Brodie and Kirshaw (1873) record *Rhynchosaurus* footprints from Warwick, now on display in the Warwick Museum; the locality is not recorded, but Sarjeant (1974, p. 315) suggested that it was Coton End. Beasley (1890a) reported finding "labyrinthodont" and "*Rynchosaurus* [*sic*]" footprints in Coton End Quarry, and later (Beasley; 1898, p. 236; 1906, p. 162) referred to a large slab with footprints from Coton End. He also noted (Beasley, 1906, p. 164) a "slab of Keuper sandstone with impressions of plants upon it, from Coton End Quarry, 1872." The "plants" are "longitudinally ribbed" markings that terminate "in narrow rods which look like continuations of the longitudinal ribs" and hence are probably groove marks produced by objects (plant stems, pebbles, bones, etc.) transported by water (Cummins, 1958). This slab lacks "distinct footprints," although a photograph in Beasley's collection shows prints of *Rhynchosauroides* (Sarjeant, 1984, p. 142, no. 46).

Bromsgrove, Worcestershire

Location and fauna. Collections of isolated amphibian and reptile bones were made by L. J. Wills early in this century from the Bromsgrove Sandstone Formation in quarries on Rock Hill (SO 948698), Bromsgrove, near Birmingham; the specimens are labeled as having come from "Wilcox S. Quarry." The Bromsgrove Sandstone Formation (Sherwood Sandstone Group) (Warrington et al., 1980) comprises, in ascending order, the Burcot, Finstall, and Sugarbrook members (Old et al., 1991). The Rock Hill Quarries (Wills, 1907, 1910, pp. 254–256) worked beds in the middle (Finstall) member (formerly the "Building Stones") (Wills, 1970, p. 250; Old et al., 1991), but are now filled in (Figure 7.5F).

The fossil amphibian and reptile remains, together with plants and invertebrates from the Bromsgrove area (Old et al., 1991) (Figure 7.10), were collected almost exclusively by Wills (1907, 1908, 1910). The vertebrate faunal list is based on these publications and on the Wills collections in the CAMSM and BIRUG. Other fossils not listed here include plants, annelids, bivalves, arthropods, and fish (Old et al., 1991).

1. *Mastodonsaurus* sp.: skull plate, ? vertebra, ? tooth. A large mastodonsaurid temnospondyl; estimated skull length 500–600 mm.
2. Cyclotosaurinae incertae sedis: *Cyclotosaurus pachygnathus* (Owen, 1842), nomen dubium, lower jaw piece. A moderately sized capitosauroid temnospondyl; estimated skull length 300 mm.

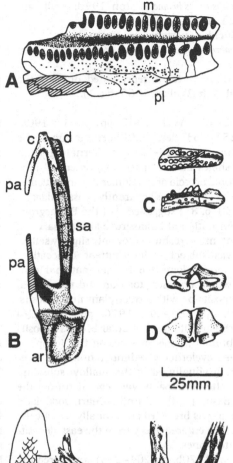

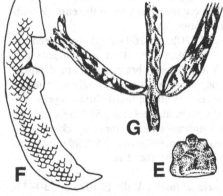

Figure 7.10. Elements of the Bromsgrove fauna and flora. (A) Part of the right upper jaw of *Mastodonsaurus* sp. in ventral view (BIRUG 1867). (B) Posterior end of a right lower jaw of "*Cyclotosaurus pachygnathus*" in dorsal view (BIRUG 52). (C) Left maxilla of *Rhynchosaurus brodiei* in ventral (top) and lingual (bottom) views (CAMSM G336). (D) Neural arch of a nothosaur dorsal vertebra in anterior (top) and dorsal (bottom) views (CAMSM G351). (E) Headshield of the scorpion *Willsiscorpio bromsgroviensis* (CAMSM G1). (F) Male cone of the conifer *Voltzia heterophylla* showing external surface (CAMSM K1001). (G) Leafy stem of the horsetail *Schizoneura paradoxa* (CAMSM K1079). Abbreviations: ar, articular; c, coronoid; d, dentary; m, maxilla; pa, prearticular; pl, palatine; sa, surangular. (A and B after Paton, 1974; C after Benton, 1990; D after Walker, 1969; E–G after Wills, 1910.)

3.   *Rhynchosaurus brodiei* Benton, 1990: small right and left maxilla.
4.   Rauisuchian: vertebra and teeth.
5.   cf. *Macrocnemus*: a dorsal vertebra (Walker, 1969).
6.   Nothosaur vertebra, represented only by the neural arch (Walker, 1969, fig. 1).

Host deposits. Wills (1907, pp. 29–32; 1908; 1910, pp. 257–263) described the succession in the Rock Hill Quarries as 15–20 m of alternating sandstones and shales and a "marl conglomerate" (i.e., intraformational breccia or conglomerate), in lenticular units, with the sandstones apparently cross-bedded.

Wills (1950, p. 85) suggested that the Bromsgrove Sandstone Formation at Bromsgrove formed part of a delta built out into a freshwater, or only slightly saline, lake that was subject to intermittent desiccation. In modern terminology, the intraformational conglomerates would represent torn-up and redeposited overbank deposits or within-river-plain fine deposits. Wills (1970, pp. 263–266; 1976, pp. 107–126) interpreted the fossiliferous lenticular beds as deposits formed in channels, pools, or lakes on the floodplain and envisaged cyclothemic sedimentation from temporary rivers gradually filling the shallow subsiding Midland Cuvette. This basin was separated from the North Sea Basin by the Pennine-Charnwood land barrier, which was breached occasionally so that the Muschelkalk Sea entered briefly from the east, depositing the "Waterstones."

Warrington (1970b, pp. 204–205) considered the Bromsgrove Sandstone Formation as comprising a sequence of low-sinuosity, braided stream deposits, followed by deposits representing higher-sinuosity, meandering rivers; fining-upward fluvial cycles are well developed (Old et al., 1991, fig. 5). The floral and faunal evidence indicates freshwater or brackish conditions at the time of deposition of the fossiliferous units (Wills, 1910; Ball, 1980; Old et al., 1991).

Occurrences of fossils. Wills (1907, pp. 30–31; 1908, p. 312) noted that the majority of the fossils came from "lenticular beds of marl and shale, while some appear in the sandstone." Some horizons were very carbonaceous, and those contained abundant fragmentary arachnid remains. The red marl and red sandstone were barren of fossils, and plants occurred in the gray sandstone.

Wills (1907, p. 33) noted that "the Labyrinthodont remains, next to the plants, are the most abundant fossils, but are apparently confined to the marl conglomerate." This unit was the source of most of the bones. Wills (1907, p. 31) believed that the marl conglomerate formed "a definite horizon in all four quarries." It was known locally as "Cat-brain" and consisted "of small pieces of marl, mostly gray in color, cemented, along with bits of bone or wood and sand, into a compact rock. This hardens to a very tough stone, though one only fit for rough work.... They are associated with one or more laminae, covered with fragments of carbonized wood. Further, it is in, or close to, these marl-conglomerates that most of the teeth and bones of the vertebrates and pieces of stems of plants are found – a significant fact when we consider how many bone-beds are conglomeratic, especially in the Trias...in some cases [the conglomerates] appear to have decayed *in situ*; they are then reduced to a friable and crumbly state, while their color is in parts ochreous and others brown, instead of the usual green" (Wills, 1910, p. 260–261). In a cored borehole at Sugarbrook (SO 9621 6818), some 3 km southeast of Rock Hill, plant remains and crustaceans occurred in similar beds in the upper half of the Finstall Member and at the base of the Sugarbrook Member of the Bromsgrove Sandstone Formation (Old et al., 1991, fig. 5).

Taphonomy. The bones from Bromsgrove are all isolated pieces: jaw fragments of temnospondyl amphibians, vertebrae of a rauisuchian, a tooth of an archosaur, a vertebra of a macrocnemid, partial maxillae of *Rhynchosaurus*, and a damaged neural arch of a nothosaur. Fine details, such as the sculpture on the temnospondyl bones (Paton, 1974) and sharp posterior teeth in maxillae of *Rhynchosaurus*, are often preserved. However, the specimens are all single elements, and some transport by water seems evident, as at Warwick. Most of the specimens have been removed from the matrix, so that little can be determined about their original state. Wills (1910, pp. 260–261) implied that the bones were found in a fragmentary condition and that disarticulation and damage were predepositional.

The bone is now in a hard and well-preserved state, with all internal structure of bone and tooth intact. However, Wills (1910, p. 261) noted that the bones suffered some damage when they were found in parts of the marl conglomerate that had decayed: "we find bones in this decayed rock which are of the consistency of hard soap when first extracted, but quickly harden on exposure to the atmosphere." This is reminiscent of the description given by Murchison and Strickland (1840) of the initial state of the bones from Warwick described earlier.

Sidmouth to Budleigh Salterton, Devon

Location and fauna. Bones of amphibians and reptiles have been found at localities on the coast between Budleigh Salterton and Sidmouth (Figure 7.11). Several specimens collected in the nineteenth century came from below High Peak (SY 104858), 2 km west of Sidmouth, and from the mouth of the River Otter,

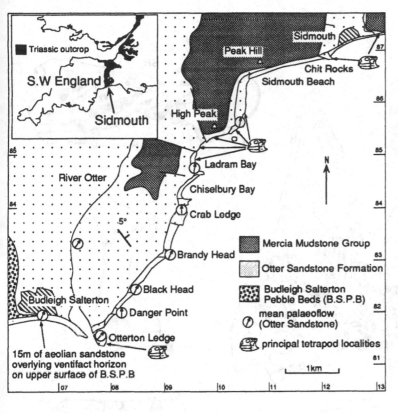

Figure 7.11. Map of the coastal outcrop of the Otter Sandstone Formation between Sidmouth and Budleigh Salterton, Devon. The major Triassic formations are indicated, together with mean fluvial paleoflow directions and principal tetrapod localities. Based on fieldwork by AJN and collecting by PSS.

on its east bank (SY 077820), just east of Budleigh Salterton. Since 1982, extensive collections have been made from at least 15 localities between Budleigh Salterton and Sidmouth (SY 0807 8212 to SY 1066 8639) and just east of Sidmouth (SY 1297 8730), with specimens being collected from fallen blocks of red sandstone and in situ, from horizons at the base of the cliff and on the foreshore (Spencer and Isaac, 1983; Benton, 1990; Milner et al., 1990). Some of the best recent finds of tetrapods have come from Ladram Bay and Chiselbury Bay, parts of the coast not noted as fossiliferous by the Victorian authors.

The fossils all occur in the Otter Sandstone Formation, formerly the "Upper Sandstone" (Ussher, 1876; Woodward and Ussher, 1911). At the western end of the outcrop, at Otterton Point, 12 m of dark red sandstones near the base of the Otter Sandstone Formation are exposed. The cliffs rise to a height of 155 m at High Peak, near the eastern end of the outcrop, where the sequence was described by Whitaker (1869), Lavis (1876), Ussher (1876), and Irving (1888).

The faunal list of amphibians and reptiles is based on Benton (1990), Milner et al. (1990), and more recent unpublished work (MJB and PSS, unpublished data). Other fossils, not listed here, include plants, invertebrates, and fish (Figures 7.12, 7.13, and 7.17).

1. "*Mastodonsaurus lavisi*" (Seeley, 1876), nomen dubium: skull fragments, part of lower jaw, and elements of the pectoral girdle of a large capitosaurid temnospondyl; estimated skull length 500–600 mm. Milner et al. (1990) argued that the type specimen of *M. lavisi* was indeterminate, and the taxon is a nomen dubium.

2. *Eocyclotosaurus* sp.: remains of a skull, about 150 mm long, and other fragments.

3. Capitosauridae inc. sed.: posterior part of a mandible.

4. *Rhynchosaurus spenceri* Benton, 1990: skull and mandible remains, isolated maxillae, and post-cranial elements, as well as a partial skeleton collected in 1990. A moderate-sized rhynchosaur with a skull length of 40–175 mm (mean 116 mm) and an estimated mean body length 0.8 m (range 0.4–1.3 m).

5. *Tanystropheus* sp.: a small tooth.

6. Procolophonidae inc. sed.: three small dentaries, a maxilla, and an interclavicle.

7. Rauisuchian: numerous teeth and a few vertebrae.

8. ?Ctenosauriscid archosaur: a long neural spine, possibly part of the dorsal "sail"; this identification is provisional (Milner et al., 1990).

Host deposits. The Otter Sandstone Formation comprises about 118 m of medium- to fine-grained red sandstones. These dip gently eastward in the coast

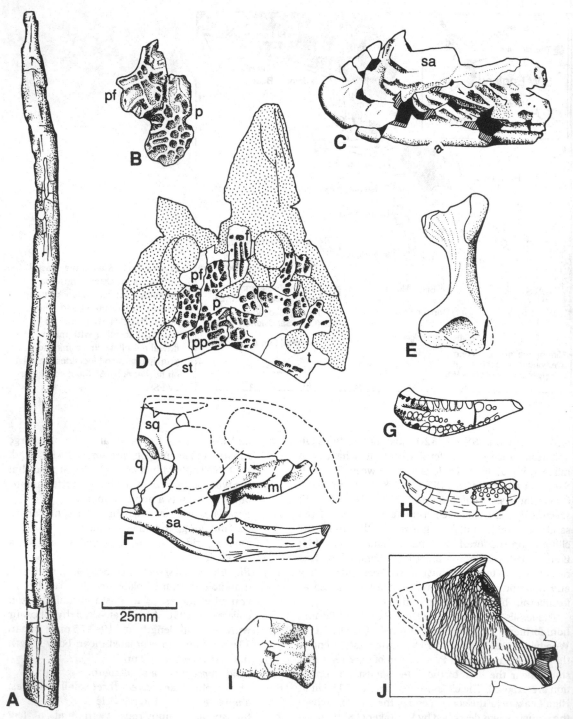

Figure 7.12. Larger elements of the Otter Sandstone Formation fauna from Devon. (A) Spine of an unknown vertebrate, possibly a dorsal neural spine of a ctenosauriscid archosaur (EXEMS 60/1985.88). (B) Fragment of the skull roof of *Mastodonsaurus lavisi* in dorsal view (EXEMS 60/1985.287). (C) Posterior portion of a right mandibular ramus of an unidentified capitosaurid, in lateral view (EXEMS 60/1985.78). (D) Incomplete skull roof of *Eocyclotosaurus* sp., in dorsal view (EXEMS 60/1985.72). (E–H) Remains of *Rhynchosaurus spenceri*: left humerus in ventral view (EXEMS 60/1985.282) (E), restored skull in right lateral view (EXEMS 60/1985.292) (F), right maxilla in ventral view (EXEMS 60/1985.292) (G), and right dentary in lingual view (BMNH R9190) (H). (I) Vertebra of an archosaur (Bristol University, unnumbered). (J) Neopterygian fish *Dipteronotus cyphus* (EXEMS 60/1985.293). Abbreviations: as in Figures 7.6, 7.9, and 7.10, and f, frontal; j, jugal; p, parietal; pf, postfrontal; pp, postparietal; t, tabular. (A and I, original; B–D and J after Milner et al., 1990; E–H after Benton, 1990.)

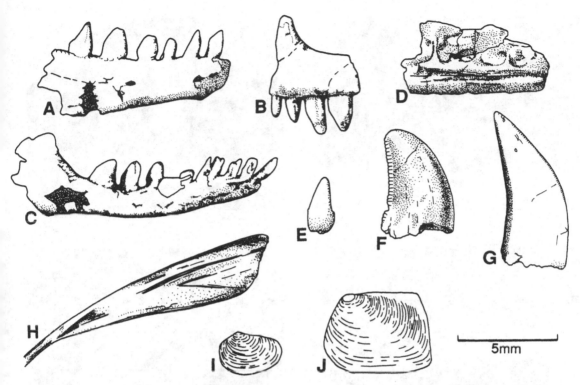

Figure 7.13. Smaller elements of the Otter Sandstone Formation fauna of Devon. Right dentaries (A, C) and a left maxilla (B) of a procolophonid, all in lateral view (EXEMS 60/1985.311, 3, and 154). (D) Dentary fragment of an unknown small pleurodont reptile, showing pits for teeth, in lingual view. (E) Tooth of *?Tanystropheus*, showing small accessory cusps (EXEMS 60/1985.143). (F, G) Recurved teeth of two kinds of unknown archosaurs (Bristol University, unnumbered). (H) Unidentified insect wing (Bristol University, unnumbered). I, J Carapaces of the conchostracan *Euestheria* (Bristol University, unnumbered). (A–C and E after Milner et al., 1990; D and F–J, original.)

section, and the formation continues northward to Somerset and eastward as far as Hampshire and the Isle of Wight beneath younger Triassic sediments, reducing in thickness to 15 m in the east, and 30–60 m beneath Somerset. It rests unconformably on the Budleigh Salterton Pebble Beds, a unit of fluvial conglomerates 20–30 m thick derived from the south and west (Henson, 1970; Smith, 1990; Smith and Edwards, 1991). The contact, visible just west of Budleigh Salterton (Figure 7.14A) (SY 057815), is marked by an extensive ventifact horizon (Leonard, Moore, and Selwood, 1982) that represents a nonsequence of unknown duration and is interpreted by Wright, Marriott, and Vanstone (1991) as a desert pavement associated with a shift from a semiarid to an arid climate. This contact and the layer containing ventifacts have been noted also at inland exposures (Smith and Edwards, 1991, p. 74).

The lowest beds of the Otter Sandstone Formation, exposed west of Budleigh Salterton and in the middle of the foreshore there (SY 064817), are red, rather structureless, well-sorted sandstones (Henson, 1970; Selwood et al., 1984).

At Otterton Point (SY 078819), hard, calcite-cemented, cross-bedded sandstone units (less than 0.5 m thick) contain calcite-cemented rhizoliths, up to 1 m deep, and other calcrete formations (Figure 7.14D) (Mader, 1990; Purvis and Wright, 1991). Purvis and Wright (1991) attributed the large vertical rhizoliths to deep-rooted phreatophytic plants that colonized bars and abandoned channels on a large braidplain. Ussher (1876, p. 380) observed that the sandstones here "contain two or three conglomerate beds, and few pebbles in false-bedded lines." Irving (1888, p. 153) described "an irregular band of breccia . . . intercalated with the sandstones, just above high-water mark," and containing fragments of slate, granite, sandstone, and quartzite. Woodward and Ussher (1911, pp. 10–11) traced this "brecciated horizon" as far as Ladram Bay, 3.5 km to the northeast of Otterton Point.

Farther east, calcretes occur more sporadically, and the formation is dominated by sandstones in large and small channels (Figure 7.14C), with occasional siltstone lenses. The sandstones occur in cycles, often with conglomeratic bases, and fine upward through cross-bedded sandstones to ripple-marked sandstones

Figure 7.14. Sedimentology and stratigraphy of the Otter Sandstone Formation. (A) The lower contact of the Otter Sandstone Formation with the Budleigh Salterton Pebble Beds (contact marked with an arrow), seen on the coast just west of Budleigh Salterton; note the person for scale. (B) The upper contact of the Otter Sandstone Formation with the Mercia Mudstone Group (contact marked with an arrow), seen below Peak Hill, just west of Sidmouth. (C) Mudclast-lined erosion surface and cross-bedded sandstones in the Otter Sandstone Formation in Chiselbury Bay; note the meter-pole for scale. (D) Calcrete nodules exposed in a vertical cliff section in the Otter Sandstone Formation; the camera lens cap is 50 mm in diameter. (All photographs by AJN.)

(Figure 7.15). The Otter Sandstone Formation is capped by water-laid siltstones and mudstones of the Mercia Mudstone Group (Figure 7.14B).

Henson (1970), Laming (1982, pp. 165, 167, 169), and Mader and Laming (1985) interpreted the Otter Sandstone Formation as comprising fluvial and eolian deposits. Sandstones near the base of the formation are eolian and were accumulated in dunes produced by easterly winds (Henson, 1970), these being transverse barchanoid dune ridges in modern terminology. The middle and upper parts of the formation are of fluvial origin; sandstones were deposited by ephemeral braided streams flowing from the south and southwest (Selwood et al., 1984). The comparatively thin mudstones are interpreted as the deposits of temporary lakes on the floodplain, with impersistent rivers fed from reservoirs in breccia outwash fans elsewhere, in turn recharged by flash floods and episodic rainfall. Numerous calcrete horizons occur and indicate subaerial soil and sub-surface calcrete formation in semiarid conditions (Mader and Laming, 1985; Lorsong, Clarey, and Atkinson, 1990; Mader, 1990; Purvis and Wright, 1991).

The climate was semiarid, with long dry periods when riverbeds dried out, and seasonal or occasional rains leading to violent river action and flash floods. However, there is little evidence of complete aridity; desiccation cracks and pseudomorphs after halite are uncommon in the Otter Sandstone Formation (Lavis, 1876; Woodward and Ussher, 1911; Henson, 1970). The relative scarcity of plant fossils may reflect oxidizing conditions in an arid climate (Laming, 1982, p. 170).

Occurrences of fossils. Recent collections of amphibian and reptile bones have come from the top 40 m or so of the Otter Sandstone Formation and occur in all lithologies, but most commonly in intraformational conglomerates and breccias (Spencer and Isaac,

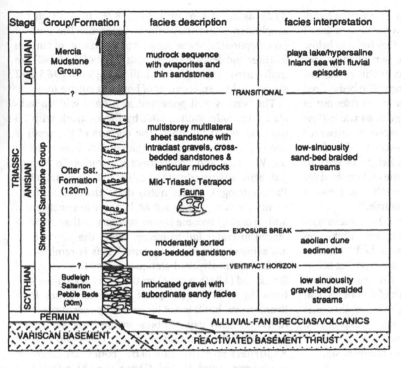

Stage		Group/Formation	facies description	facies interpretation
TRIASSIC / LADINIAN		Mercia Mudstone Group	mudrock sequence with evaporites and thin sandstones	playa lake/hypersaline inland sea with fluvial episodes
		? ————————————————— TRANSITIONAL ———————————————		
TRIASSIC / ANISIAN	Sherwood Sandstone Group	Otter Sst. Formation (120m)	multistorey multilateral sheet sandstone with intraclast gravels, cross-bedded sandstones & lenticular mudrocks Mid-Triassic Tetrapod Fauna	low-sinuousity sand-bed braided streams
			————————— EXPOSURE BREAK —————————	
			moderately sorted cross-bedded sandstone	aeolian dune sediments
		? ————————————————— VENTIFACT HORIZON ———————————		
SCYTHIAN		Budleigh Salterton Pebble Beds (30m)	imbricated gravel with subordinate sandy facies	low sinuosity gravel-bed braided streams
	PERMIAN			
VARISCAN BASEMENT			ALLUVIAL-FAN BRECCIAS/VOLCANICS REACTIVATED BASEMENT THRUST	

Figure 7.15. Sedimentology and interpreted depositional environments of the Otter Sandstone Formation and adjacent units, based on fieldwork by AJN.

1983). In breccias exposed west of Chiselbury Bay (Figure 7.11), the abundance of tetrapod finds declines significantly lower in the sequence. The bones are generally in a fine- to medium-grained reddish sandstone that often contains clasts of pinkish, greenish, or ochreous calcrete and mudflakes up to 20 mm in diameter. The more complete fish specimens are, however, preserved in dark red siltstone, sometimes in association with plants and conchostracan crustaceans. Plant remains are preserved in iron oxide in all the lower-energy deposits, and their preservation appears to be controlled by the sedimentation regime.

The only specimens found in situ by Spencer and Isaac (1983, p. 268) came from "the lowest of three intraformational conglomerates," but these were "indeterminate bone fragments." Since 1983, four rhynchosaur specimens (EXEMS 60/1985.284, 285, 292, and 7/1986.3) have been collected in situ from a single horizon at beach level, and a partial rhynchosaur skeleton was found at the top of the foreshore exposures in Ladram Bay in 1990. It is likely that fossils occur at numerous levels throughout the Otter Sandstone Formation, but most have been found in fallen blocks on the shore, and locating the original horizons in the cliffs is difficult.

The most clearly localized of the older finds is a jaw of *Rhynchosaurus* [BGS(GSM) 90494] recovered from a large displaced block on the east bank of the River Otter (SY 0775 8196) "where the sandstone is somewhat brecciiform" (Whitaker, 1869, p. 156). Metcalfe (1884) reported white fragments, which he identified

as bone, in the "harder parts of the sandstones, at numerous points near Budleigh Salterton and Otterton Point."

The Victorian authors believed that one or more discrete bonebeds occurred at the eastern end of the outcrop. Lavis (1876) and Metcalfe (1884) placed it "about 10 feet from the top of the sandstone." Hutchinson (1906) and Woodward and Ussher (1911) placed it "about 50 feet below the base of the Keuper Marls," some 40 feet (13 m) lower in the section.

Lavis (1876) made his finds in fallen blocks from an "ossiferous zone" consisting of up to four beds, "characterized by lithological differences, inasmuch as the matrix is composed of much coarser sandstone, containing here and there masses of marl varying in size from that of a pea to that of a hen's egg. . . . In these beds ripple-marks are very plentiful. The fragments of bone which are found in this zone seem to be very slightly waterworn." Metcalfe (1884) gave further details of this locality at High Peak, stating that bones were found in fallen blocks of sandstone from a light-colored band in the cliff close below the base of the "Upper Marls" (Mercia Mudstone Group). Carter (1888) recovered bone material and coprolites from this locality.

Hutchinson (1879, p. 384) gave the most detailed account of the fossiliferous horizons. He found equisetalean stems in a bed at the top of the sandstone and "about eight or ten feet above" two or three "White bands" that appeared as clear horizons in the cliff face. Then, "one or two steps below" the White bands "is what I venture to call the Saurian or Batrachian band, in which Mr Lavis found his Labyrinthodon; but I cannot exactly say how many feet this band is below the white bands, because the fall down of the under cliff has concealed the stratification at this place; but it may be fifty feet below, and amongst the beds of red rock. Be that as it may, the Saurian band rises out of the beach somewhere under Windgate, as the hollow between the two hills is called, and ascends westwards into High Peak Hill, and having proceeded for about half-a-mile, and having attained a height of sixty or seventy feet above the sea, a fall of the cliff enabled Mr Lavis to find his specimens on the beach, and I was so fortunate as to see them soon afterwards."

Woodward and Ussher (1911, pp. 12–13) summarized an unpublished section drawn up by Hutchinson in 1878 in which he located the bone bed "100 feet above the talus on the beach, and about 50 feet below the base of the Keuper Marls." No trace of any tetrapod-bearing horizon in the form of a bonebed can be seen today, and there is no evidence that one existed. The Victorian geologists evidently expected to find bones at discrete levels and had no concept of restricted lenticular deposits, such as channel lags.

Taphonomy. The tetrapod fossils are generally isolated elements – jaws, teeth, partial skulls, or single postcranial bones. Exceptions are the partial articulated skull and lower jaws of *Rhynchosaurus spenceri* (EXEMS 60/1985.292), the associated humerus, radius, and ulna of that species (EXEMS 60/1985.282), two sets of vertebrae (EXEMS 60/1985.15, 57), and a recently collected partial rhynchosaur skeleton that comprises much of the trunk region, the pelvis, and the hindlimbs, with the bones in close association, but mostly slightly disarticulated.

About half of the identifiable tetrapod bones found are rhynchosaur remains, and most of these are parts of the skull, especially the jaw elements. This is a phenomenon of preservation, rather than selective collecting, and probably reflects the good preservation potential of teeth and jawbones. This is especially true for rhynchosaurs in which the maxilla and dentary are composed of unusually dense bone, the teeth are firmly ankylosed, and the bone is virtually indestructable. The amphibians are represented mainly by skull and pectoral girdle elements, all relatively dense and with characteristic sculpture. The small reptiles are represented by teeth and small segments of jaw, and the larger archosaur(s) by teeth and vertebrae. These selective features of preservation are comparable with the situation at Warwick.

The incompleteness of most specimens is largely the result of predepositional disarticulation and breakage, as is shown by their context in the sediment. Some specimens show signs of possible abrasion during transport (e.g., EXEMS 60/1985.37–45, 56, 284,

312), as noted also by Lavis (1876, p. 277) on his amphibian bones, but others, especially the jaws of procolophonids, show detailed preservation of surface features and delicate sharp teeth. Most of the fossils are undistorted, although the skull EXEMS 60/1985.292 shows slight displacement of bones at suture lines.

The bone is well preserved as a hard whitish substance (usually stained pink by the matrix), with all internal structure intact. The dentine of the teeth is yellow, and the enamel is stained dark brown, as in the Warwick and Bromsgrove specimens. In an unpublished manuscript (1882, BMNH, Department of Palaeontology), Carter noted that "the smaller fragments are more or less soft and cheesy in consistence, and on drying, become almost powdery so that on being placed in water fall to pieces, while the larger ones are somewhat more competent." This is reminiscent of the previously cited statements by Murchison and Strickland (1840, p. 344) on the Warwick bones, and those by Wills (1910, p. 261) on the bones from Bromsgrove, but it is not borne out by recent observations on freshly exposed bone in the Otter Sandstone Formation. Examination of Carter's collection (BMNH R330) has shown that most of his "bones" are coprolites or calcretes, and indeed Carter (1888) referred to coprolites containing fish scales from the Otter Sandstone Formation.

Other sites

Isolated tetrapod remains have been collected from Middle Triassic deposits at other sites in the English Midlands, but none has shown the potential of the sites discussed here. Those other sites include the following:

1. Hollington, Staffordshire (SK 060388), source of some articulated gastralia of an unknown reptile (BMNH R3227), described by Woodward (1905) as "*Hyperodapedon* sp.," but probably not rhynchosaurian.
2. Stanton, Staffordshire (SK 126462), source of the best-known English Triassic temnospondyl skull (Ward, 1900) (BMNH R3174), the holotype of *Stenotosaurus stantonensis*, described by Woodward (1904) as *Capitosaurus stantonensis*, synonymized with *Cyclotosaurus leptognathus* (Owen, 1842) by Paton (1974), and transferred to *Stenotosaurus* by Shishkin (1980).
3. Emscote, near Warwick (?SP 298658), source of a bone fragment labeled "*Rhombopholis scutulatus*" (WARMS Gz 126), collected by J. W. Kirshaw in 1868.

Paleoecology

It is difficult to determine the ecology of these faunas in detail because of the limited material available. However, it is possible to estimate the relative impor-

Table 7.1. *Numbers of individuals in the Warwick (Wa.), Bromsgrove (Br.), and Devon (De.) assemblages, calculated as minimum number of individuals (MNI) and nonredundant maximum number (NRMAX), and converted to percentages*

Taxon	MNI			NRMAX		
	Wa.	Br.	De.	Wa.	Br.	De.
Amphibians	6 (33%)	2 (25%)	6 (25%)	14 (29%)	7 (41%)	15 (25%)
Rhynchosaurus	4 (22%)	1 (12%)	9 (38%)	15 (31%)	2 (12%)	29 (48%)
Archosaur(s)	5 (28%)	3 (38%)	5 (21%)	15 (31%)	5 (29%)	11 (18%)
Procolophonid	0	0	3 (12%)	0	0	5 (8%)
Nothosaur	0	1 (12%)	0	0	2 (12%)	0
Prolacertiform	3 (17%)	1 (12%)	1 (4%)	5 (10%)	1 (6%)	1 (2%)
Totals	18	8	24	49	17	61

Sources: Data obtained from Paton (1974), Benton (1990), Milner et al. (1990), and from inspection of collections.

tance of the different tetrapod-body fossil types as represented in existing collections and to apply broader knowledge of the habits of the different amphibians and reptiles.

The numbers of individuals of each taxon have been estimated in two ways. The minimum number of individuals (MNI) has been obtained by counting up the most-represented elements in the collections (e.g., complete skulls, right maxillae, left femora) as an unequivocal figure for the absolute minimum number of individuals required to produce all of the known fossils. The second measure, the nonredundant maximum number of individuals (NRMAX), is based on the initial assumption that each bone found separately represents a different individual. This assumed maximum number is reduced by an examination of historical sources and by collecting details that may provide information on original associations of material in the same, or different, repositories. An attempt is made to discover all isolated bones that can be fitted together or that can be associated as being probably the remains of a single individual.

It is likely that the MNI underestimates the number of individuals present in an ancient fauna, and the NRMAX probably overestimates the number. In cases such as the Triassic faunas, where isolated elements dominate, the NRMAX probably gives a better estimate of true numbers, especially where detailed collecting data are available. For example, the Devon tetrapods have been collected from numerous sites along a 7-km coastal section, and site information is taken into account in associating material: A left maxilla and a right maxilla found 4 km apart are unlikely to have come from the same individual, although the MNI figure would imply that! Hence, in the following discussion, the NRMAX figures are used, although MNI figures are also given.

It is not clear, however, which of these measures should give the best *proportional* estimates. There is no reason, for example, that an MNI value based on skulls for one species should be in proportion with that based on right femora for another; indeed, it is more likely that such figures, based of necessity on different skeletal elements, will not be in proportion to true ancient diversities. The NRMAX could be a better proportional estimator for preservation of ancient faunas if all taxa and finds are equally well documented. Of course, both the MNI and NRMAX are influenced by preservation factors, and neither can give a good estimate of the original *living* faunal compositions; both MNI and NRMAX figures are given here (Table 7.1, Figure 7.16, data in Appendix 7.1).

The Grinshill fauna stands out from the others paleoecologically in being monospecific (Figure 7.16). About seventeen individuals of *Rhynchosaurus articeps* have been collected (NRMAX), an MNI of seven, based on skulls, all of which are well preserved. No bones of any other animal have been found in association, though tracks signify the coexistence of rauisuchians.

The Warwick, Bromsgrove, and Devon faunas (Figure 7.16) show similar diversities of taxa (Table 7.1). Comparisons with the Bromsgrove fauna are difficult because fossils are so sparse and so incomplete that many are not clearly identifiable. In both the Warwick and the Devon faunas, the rhynchosaur *Rhynchosaurus* dominates (31% and 48% of all individuals, respectively). Amphibians are more abundant at Warwick (29%) than in Devon (25%), and archosaurs likewise (31% at Warwick, 18% in Devon).

It is possible to infer likely food chains (Figure 7.17). The top carnivores at Warwick and in Devon were archosaurs, probably a rauisuchian (*Bromsgroveia*), and some others, which preyed on the temnospondyl amphibians, rhynchosaurs, and smaller reptiles. The

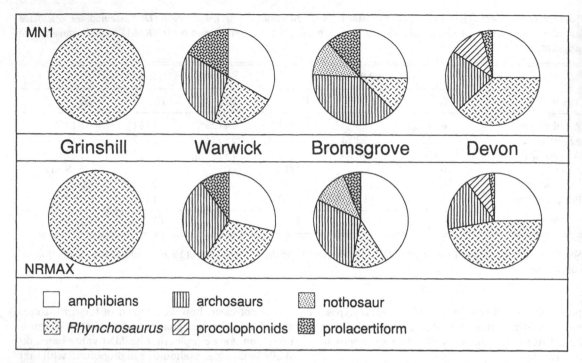

Figure 7.16. Pie charts showing the relative abundances of major tetrapod groups at the four English Middle Triassic localities. The percentages are calculated from the data in Appendix 7.1 and summarized in Table 7.1. MNI figures are minimum numbers of individuals, based on the single maximally represented element; NRMAX figures are based on evidence of localities and dates of finds. MNI figures are minimum estimates and NRMAX figures are maximum estimates of faunal diversity.

amphibians presumably fed almost exclusively on fish. The rhynchosaurs were herbivores, feeding on elements of the equisetalean and coniferalean flora found in association at Bromsgrove. Smaller (100–150 mm long) procolophonid reptiles may also have had herbivorous diets, feeding on low herbaceous plants thus far unrepresented in Middle Triassic collections. The macrocnemid may have fed on arthropods, represented by scorpions from Bromsgrove (Wills, 1910, 1947; Kjellesvig-Waering, 1986), or it may have been a shallow-water predator.

Fish are known from these sites. Only a tooth plate of a ceratodontid lungfish (*Ceratodus laevissimus*) has been found in Coton End Quarry, Warwick (Woodward, 1893), but a more diverse ichthyofauna, comprising the remains of *Acrodus*, a selachian, the dipnoan *Ceratodus*, and the holotype, an almost complete specimen, of the deep-bodied perleidid "palaeonisciform" *Dipteronotus cyphus*, has been found at Bromsgrove (Wills, 1910; Old et al., 1991). The Otter Sandstone of Devon has yielded *Dipteronotus*, about 60–70 mm long, represented by several entire specimens (Milner et al., 1990), and coprolites containing fish scales (Carter, 1888).

Other fossils are represented best at Bromsgrove (Old et al., 1991) and, to a lesser extent, in Devon. Plant remains from the Finstall Member at Bromsgrove

comprise *Equisetites arenaceus*(?), *Schizoneura paradoxa* (roots, pith casts, leaves), *Chiropteris digitata*(?), *Yuccites vogesiacus* (leaves and stems), *Voltzia*(?) (pith casts and decorticated stems), *Aethophyllum*, and cones (*Strobilites, Willsiostrobus bromsgrovensis, W. willsi*). This macrofossil association comprises equisetalean pteridophytes and coniferalean gymnosperms. Microfloras from the Finstall Member reflect a more diverse source flora that included lycopsids, sphenopsids, pteropsids, and gymnosperms; the last, principally conifers, but including cycadalean types, dominated the flora. Invertebrates include the annelid *Spirorbis* (Ball, 1980), a bivalve, *Mytilus*(?) (Wills, 1910), the branchiopod crustacean *Euestheria*, and the scorpions *Bromsgroviscorpio, Mesophonus, Spongiophonus,* and *Willsiscorpio* (Wills, 1910, 1947; Kjellesvig-Waering, 1986).

A similar, but less diverse, association of plant remains and invertebrates is present in the Otter Sandstone Formation of Devon. Plant remains there comprise rhizomes and stems referable to *Schizoneura*, arthropod cuticle, an insect wing, and branchiopod crustaceans (*Euestheria, Lioestheria*) (P. S. Spencer, unpublished data). Rhizoliths indicate the existence of a contemporary indigenous vegetation interpreted as comprising conifers (Mader, 1990) and phreatophytic plants (Purvis and Wright, 1991).

Figure 7.17. Reconstructed scene during the Middle Triassic in Devon, based on specimens from the Otter Sandstone Formation between Sidmouth and Budleigh Salterton. A scorpion (mid-foreground) contemplates a pair of procolophonids on the rocks. Opposite them, a temnospondyl amphibian has spotted some paleonisciform fish, *Dipteronotus*, in the water. Two *Rhynchosaurus* stand in the middle distance, and behind them a pair of rauisuchians lurk. The plants include *Equisetites* (horsetails) around the waterside and *Voltzia*, a conifer tree. (Drawn by Pam Baldaro, based on her color painting.)

Pteridophytes in the associations from Worcestershire and Devon probably populated damp tracts bordering river channels or in floodplain areas; the gymnosperm components may reflect drier habitats. Scorpions signify dry terrestrial habitats, but the crustaceans indicate the existence of seasonal pools of fresh to brackish water. The records of lungfishes from Worcestershire suggest that rivers there were prone to seasonal drought, but were penetrated intermittently from the north by marine-sourced waters that introduced those and other fish (e.g., a selachian), bivalves, and semi-aquatic reptiles (? prolacertiform and nothosaur) with marine affinity. Temnospondyl amphibians indicate the existence of at least seasonal bodies of fresh water, necessary for breeding, on a floodplain that was inhabited also by herbivorous rhynchosaurs and procolophonids and the carnivorous rauisuchians.

Ages of the tetrapod assemblages

The tetrapod faunas reviewed here have generally been regarded as Middle Triassic, though the views

based upon these remains and the associated macro-fossils have varied with regard to the assignment of a more precise, Anisian or Ladinian, age. Walker, on the basis of reptiles, favored (1969) an early to middle Ladinian age or (1970) a late Anisian or, preferably, Ladinian age and suggested that the Devon fauna is older than those from the Midlands; this has been construed as indicating an Anisian age for the Devon assemblage. Paton (1974) also proposed an early Ladinian age, on the basis of amphibian remains, but regarded the Devon and Midlands faunas as similar in age. Milner et al. (1990) advocated an Anisian age for the Devon vertebrate fauna.

There is generally no direct independent evidence for the ages of the vertebrate assemblages. Associated footprints, at Grinshill, and macrofossils (plants and invertebrates), known from the Bromsgrove and Devon localities, afford no satisfactory biostratigraphic infor-mation. Wills (1970, pp. 260–261), for example, noted that the Bromsgrove plants indicated late Scythian to late Ladinian ages, by comparison with German material; the scorpions suggested an Anisian or Ladinian age; and the conchostracan *Euestheria* suggested a late Ladinian to Norian age. Rhynchosauroid footprints indicate merely a Triassic age, while footprints of *Chirotherium* are usually Early or Middle Triassic in age.

Pollen and spores provide an independent means of correlating British Triassic deposits with the stages based upon marine faunas in the Tethyan realm. They were first recorded from the Bromsgrove Sandstone Formation by Wills (1910), who recovered spores and pollen from plant remains from the vertebrate-bearing locality at Bromsgrove. Within the past three decades, formations in the Sherwood Sandstone Group and Mercia Mudstone Group in many parts of Britain have been assigned ages on the basis of comparison of palynoflorules with those documented from independently dated Triassic sequences elsewhere in Europe.

Palynological information available from northwest and central England indicates that the tetrapod faunas from Grinshill and localities elsewhere in the Midlands are pre-Ladinian, probably Anisian, in age. The Anisian–Ladinian boundary is characterized palynologically by the last occurrence of *Stellapollenites thiergartii* at the boundary (Visscher and Brugman, 1981; van der Eem, 1983) or in the basal Ladinian (Brugman, 1986) and by the appearance of *Ovalipollis pseudoalatus* at the boundary (van der Eem, 1983) or in the basal Ladinian (Visscher and Brugman, 1981). Slightly above the base of the Ladinian, *Camerosporites secatus* and *Duplicisporites* spp. appear, followed by *Echinitosporites iliacoides*, signifying an early Ladinian (Fassan substage) age (Visscher and Brugman, 1981; van der Eem, 1983; Brugman, 1986). The base of the Anisian stage is marked by the lowest occurrences of *Stellapollenites thiergartii* and *Angustisulcites* spp. (Visscher and Brugman, 1981; Brugman, 1986).

Palynoflorules from the Sherwood Sandstone Group and Mercia Mudstone Group in the Cheshire basin and the contiguous West Lancashire–East Irish Sea basin (Warrington, 1970b, 1974c; Fisher, 1972a,b; Warrington, in Earp and Taylor, 1986; Warrington, in Wilson and Evans, 1990) indicate, by reference to the foregoing criteria, that the succession from the basal Helsby Sandstone Formation to a level above the Northwich Halite Formation is of Anisian age. *Angustisulcites klausi* was recorded from the basal Helsby Sandstone in northeast Cheshire (Warrington, 1970b), and assemblages from the Tarporley Siltstone Formation at Liverpool (Fisher, 1972a,b) include *Angustisulcites* spp. together with *Stellapollenites thiergartii* and other taxa, such as *Perotrilites minor*, indicative of an Anisian age; evidence of Anisian age was also obtained from this formation in the Chester district (Warrington, in Earp and Taylor, 1986). Few datable palynoflorules have been recorded from the Mercia Mudstone Group in the Cheshire basin. However, assemblages from mudstones overlying the Northwich

Halite Formation indicate a correlation of that unit with the Preesall Salt of west Lancashire, which is assigned a late Anisian age (Warrington, in Wilson and Evans, 1990). Thus, the base of the Ladinian stage in the Cheshire basin is placed within the Mercia Mudstone Group, above the Northwich Halite Formation, and the tetrapod-bearing sequence at Grinshill in the southern part of the basin is therefore pre-Ladinian in age. The reassessment of pollen and spores from the basal Helsby Sandstone Formation as Anisian, rather than late Scythian as previously suggested (Warrington, 1970b), indicates that the Grinshill vertebrates probably are not older than Anisian.

In the central Midlands, palynoflorules recovered from the Bromsgrove Sandstone Formation in the Bromsgrove (Clarke, 1965; Warrington, 1970b), Stratford-upon-Avon (Warrington, in Williams and Whittaker, 1974), and Banbury (Warrington, 1978) areas include *Angustisulcites* spp. and *Stellapollenites thiergartii*, thus indicating an Anisian age. Though the "Waterstones" at Bromsgrove were regarded as early Ladinian (Warrington, 1970b), this is now considered unlikely, as none of the pollen and spore taxa now used to indicate the base of the Ladinian has been recorded in the Bromsgrove succession. Furthermore, the presence of *Tsugaepollenites oriens* in the Bromsgrove Sandstone Formation at Stratford-upon-Avon (Warrington, in Williams and Whittaker, 1974), and possibly near Banbury (Warrington, 1978), implies a correlation with beds in the Kirkham Mudstone Formation (Mercia Mudstone Group) in west Lancashire that are dated as Anisian (Warrington, in Wilson and Evans, 1990). North of the Warwick-Leamington area, the lower part of the Mercia Mudstone Group has yielded *Stellapollenites thiergartii* and is assessed as Anisian; the presence of *Echinitosporites iliacoides* at a higher level in the group there indicates a Ladinian to earliest Carnian age (Warrington in Worssam and Old, 1988). As in the Cheshire basin, therefore, the Anisian–Ladinian boundary in the central Midlands is placed above the stratigraphic level of the tetrapod faunas. Palynoflorules from the Bromsgrove succession afford direct evidence of Anisian age for the tetrapod-bearing sequence there; the lowest pollen and spore assemblages from Bromsgrove were once regarded as late Scythian (Warrington, 1970b), but on the basis of the presence of *Angustisulcites klausi*, they are now reassessed as Anisian.

No pollen and spores have been recovered from the Otter Sandstone Formation in Devon. This formation is poorly constrained palynologically by occurrences of Late Permian palynoflorules in the lower part of the Permo-Triassic succession near Exeter (Warrington and Scrivener, 1988, 1990) and Carnian (Late Triassic) taxa in the Mercia Mudstone Group, 135 m above the Otter Sandstone Formation (Warrington, 1971; Holloway et al., 1989).

The vertebrate evidence can now be reviewed in light of the palynological evidence, which indicates that the tetrapod assemblages in the Midlands are pre-Ladinian, but not older than Anisian; though palynological evidence is lacking in Devon, these findings support the interpretation of the comparable Devon tetrapod assemblage as Anisian in age (Milner et al., 1990).

Gardiner (in Milner et al., 1990) considered that the fish assemblage indicates a Ladinian age for the Bromsgrove and Devon localities, and presumably also Warwick. *Dipteronotus cyphus* is known from Devon and Bromsgrove, indicating that the assemblages are coeval, and the shark *Palaeobates keuperinus* from Bromsgrove is closely similar to *P. angustissimus*, of Ladinian age, from the upper Muschelkalk and Lettenkohle of Germany, Poland, and France. Further, *Gyrolepis albertii* from the Bromsgrove Sandstone Formation is also known from the upper Muschelkalk of various parts of Germany, as well as from younger horizons.

Paton (1974) considered the temnospondyl amphibians from Warwick, Bromsgrove, and Devon to be essentially the same and to indicate a mean age of early Ladinian, based on comparisons with German material. Milner et al. (1990) noted, however, that the British specimens of *Mastodonsaurus* from Warwick, Devon, and Bromsgrove indicate an Anisian to Carnian age, based on comparisons with German material. The other amphibian genus, *Eocyclotosaurus*, is more useful, being known from the *Voltzia* Sandstone of France and the lower Röt of Germany, both dated as latest Scythian or early Anisian in age, and from the Holbrook Member of the Moenkopi Formation of Arizona, dated as early Anisian (Morales, 1983). Hence, the amphibians would appear to indicate an Anisian age for the Warwick, Bromsgrove, and Devon assemblages.

The reptiles generally point to a Middle Triassic, possibly Anisian, age for all the formations (Milner et al., 1990). The procolophonids, macrocnemid, tanystropheid, nothosaur, and rauisuchian archosaurs could all be Anisian or Ladinian in age, although Milner et al. (1990) prefer an Anisian age on the basis of the primitive nature of the Devon procolophonid. The three species of *Rhynchosaurus* fall in the cladogram (Benton, 1990, p. 298) between *Stenaulorhynchus* from the Manda Formation of Tanzania (generally dated as Anisian) and the Hyperodapedontinae (*Hyperodapedon*, *Scaphonyx*), which all are Carnian in age. Hence, the rhynchosaurs, present in all four regions, might indicate a Ladinian age, but the order of branching in a cladogram need not match stratigraphic order.

Part of the problem in dating is, as Milner et al. (1990) noted, the fact that the four English faunal associations may date from the time of the marine Muschelkalk (Anisian/Ladinian) in central Europe. Huene (1908c) and Wills (1910) equated the English faunas with the German Lettenkohle (late Ladinian), which immediately followed the Muschelkalk. Wills (1948) and Milner et al. (1990), on the other hand, recognized more similarities with the Scythian/Anisian late Buntsandstein and *Voltzia* Sandstone faunas of Germany and France, which immediately preceded the Muschelkalk. It is likely, of course, that the English localities fall somewhere between, which may be suggested by the absence of their commonest element, *Rhynchosaurus*, in the central European pre- and post-Muschelkalk terrestrial faunas.

Acknowledgments

We thank Nick Fraser (Virginia Museum of Natural History) and Hans-Dieter Sues (Royal Ontario Museum) for the opportunity to attend the Front Royal meeting and for their encouragement to write this chapter. David Thompson (Keele University) supplied a great deal of sedimentological information and added greatly to the account of Grinshill. We thank Pam Baldaro (Bristol University) for drafting Figures 7.4, 7.6, 7.10, 7.12, 7.13, 7.16, and 7.17, especially for Figure 7.17, which is taken from her original color painting of the Otter Sandstone scene. Bruce Bennisson (Shrewsbury Museum) kindly supplied the photographs in Figure 7.8. We thank Andrew Milner for helpful advice. MJB was funded by the Royal Society, MJB and PSS by NERC grant GR3/7040, and AJN by the Department of Education (Northern Ireland). GW publishes with the approval of the Director, British Geological Survey (N.E.R.C.).

References

Audley-Charles, M. G. 1970. Stratigraphical correlation of the Triassic rocks of the British Isles. *Quarterly Journal of the Geological Society of London* 126: 19–47.

Ball, H. W. 1980. *Spirorbis* from the Triassic Bromsgrove Sandstone Formation (Sherwood Sandstone Group) of Bromsgrove, Warwickshire. *Proceedings of the Geologists' Association* 91: 149–154.

Beasley, H. C. 1890a. A visit to Warwick. *Transactions of the Liverpool Geological Association* 10: 27–30.

1890b. The life of the English Trias. *Proceedings of the Liverpool Geological Society* 6: 145–165.

1896. An attempt to classify the footprints in the New Red Sandstone of this district. *Proceedings of the Liverpool Geological Society* 7: 391–409.

1898. Notes on examples of footprints, &c., from the Trias in some provincial museums. *Proceedings of the Liverpool Geological Society* 8: 233–237.

1902. The fauna indicated in the Lower Keuper Sandstone of the neighbourhood of Liverpool. *Transactions of the Liverpool Biological Society* 16: 3–26.

1904. Report on footprints from the Trias – Part I. *Report of the British Association for the Advancement of Science* 1904(1903): 219–231.

1905. Report on footprints from the Trias – Part II. *Report of the British Association for the Advancement of Science* 1905(1904): 275–282.

1906. Notes on footprints from the Trias in the museum of the Warwickshire Natural History and

Archaeological Society at Warwick. *Report of the British Association for the Advancement of Science* 1906(1905): 162–165.

Benton, M. J. 1983. The Triassic reptile *Hyperodapedon* from Elgin: functional morphology and relationships. *Philosophical Transactions of the Royal Society of London* B302: 605–720.

——— 1984. Tooth form, growth, and function in Triassic rhynchosaurs (Reptilia, Diapsida). *Palaeontology* 27: 737–776.

——— 1986. The late Triassic reptile *Teratosaurus* – a rauisuchian, not a dinosaur. *Palaeontology* 29: 293–301.

——— 1990. The species of *Rhynchosaurus*, a rhynchosaur (Reptilia, Diapsida) from the Middle Triassic of England. *Philosophical Transactions of the Royal Society of London* B328: 213–306.

Brodie, P. B., and J. W. Kirshaw. 1873. Excursion to Warwickshire, July 10th and 11th, 1871. *Proceedings of the Geologists' Association* 2: 284–287.

Brugman, W. A. 1986. Late Scythian and Middle Triassic palynostratigraphy in the Alpine realm. *Albertiana* 5: 19–20.

Buckland, W. 1837. On the occurrence of Keuper-Sandstone in the upper region of the New Red Sandstone formation or Poikilitic system in England and Wales. *Proceedings of the Geological Society of London* 2: 453–454.

——— 1840. Ichnology. *Proceedings of the Geological Society of London* 3: 245–247.

——— 1844. [President's address, 1839]. *Proceedings of the Ashmolean Society* 16: 5–7.

Burckhardt, R. 1900. On *Hyperodapedon gordoni*. *Geological Magazine* (4) 7: 486–492, 529–535.

Carter, H. J. 1888. On some vertebrate remains in the Triassic strata of the south coast of Devonshire between Budleigh Salterton and Sidmouth. *Quarterly Journal of the Geological Society of London* 44: 318–319.

Clarke, R. F. A. 1965. Keuper miospores from Worcestershire, England. *Palaeontology* 8: 294–321.

Cummins, W. A. 1958. Some sedimentary structures from the Lower Keuper Sandstones. *Liverpool and Manchester Geological Journal* 2: 37–43.

Delair, J. B., and W. A. S. Sarjeant. 1985. History and bibliography of the study of fossil vertebrate footprints in the British Isles: supplement 1973–1983. *Palaeogeography, Palaeoclimatology, Palaeoecology* 49: 123–160.

Earp, J. R., and B. J. Taylor. 1986. Geology of the country around Chester and Winsford. *Memoirs of the British Geological Survey* (sheet 109), London: HMSO.

Egerton, P. G. 1838. On two casts in sandstone of the impressions of the hind foot of a gigantic *Chirotherium*, from the New Red Sandstone of Cheshire. *Proceedings of the Geological Society of London* 3: 14–15.

Fisher, M. J. 1972a. A record of palynomorphs from the Waterstones (Triassic) of Liverpool. *Geological Journal* 8: 17–22.

——— 1972b. The Triassic palynofloral succession in England. *Geoscience and Man* 4: 101–109.

Forster, S. C., and G. Warrington. 1985. Geochronology of the Carboniferous, Permian and Triassic. *Geological Society of London, Memoir* 10: 99–113.

Galton, P. M. 1985. The poposaurid thecodontian *Teratosaurus suevicus* v. Meyer, plus referred specimens mostly based on prosauropod specimens from the Middle Stubensandstein (Upper Triassic) of Nordwürttemberg. *Stuttgarter Beiträge zur Naturkunde* B116: 1–29.

Geiger, M. E., and C. A. Hopping. 1968. Triassic stratigraphy of the southern North Sea Basin. *Philosophical Transactions of the Royal Society of London* B254: 1–36.

Henson, M. R. 1970. The Triassic rocks of south Devon. *Proceedings of the Ussher Society* 2: 172–177.

Holloway, S. 1985. Triassic Sherwood Sandstone Group (excluding the Kinnerton Sandstone Formation and the Lenton Sandstone Formation). Pp. 31–34 in A. Whittaker (ed.), *Atlas of Onshore Sedimentary Basins in England and Wales: Post-Carboniferous Tectonics and Stratigraphy*. London: Geological Society.

Holloway, S., A. E. Milodowski, G. E. Strong, and G. Warrington. 1989. The Sherwood Sandstone Group (Triassic) of the Wessex Basin, southern England. *Proceedings of the Geologists' Association* 100: 383–394.

Huene, F. von 1908a. Die Dinosaurier der europäischen Triasformation. *Geologische und Paläontologische Abhandlungen, Supplement-Band* 1: 1–419.

——— 1908b. Note on two sections in the Lower Keuper Sandstone of Guy's Cliff, Warwick. *Geological Magazine* (5) 5: 100–102.

——— 1908c. Eine Zusammenstellung über die englische Trias und das Alter ihrer Fossilien. *Centralblatt für Mineralogie, Geologie, und Paläontologie* 1908: 9–17.

——— 1929. Ueber Rhynchosaurier und andere Reptilien aus den Gondwana-Ablagerungen Südamerikas. *Geologische und Paläontologische Abhandlungen, Neue Folge* 17: 1–62.

Hughes, B. 1968. The tarsus of rhynchocephalian reptiles. *Journal of Zoology (London)* 156: 457–481.

Hull, E. 1869. The Triassic and Permian rocks of the Midland counties of England. *Memoirs of the Geological Survey of the United Kingdom* 1869: 1–127.

Hutchinson, P. O. 1879. Fossil plant, discovered near Sidmouth. *Transactions of the Devonshire Association for the Advancement of Science, Literature, and Art* 11: 383–385.

——— 1906. Geological section of the cliffs to the west and east of Sidmouth, Devon. *Report of the British Association for the Advancement of Science* 1906 (1905): 168–170.

Huxley, T. H. 1859. On a fragment of a lower jaw of a large labyrinthodont from Cubbington. *Memoirs of the Geological Survey of the United Kingdom* 1859: 56–57.

——— 1869. On *Hyperodapedon*. *Quarterly Journal of the Geological Society of London* 25: 138–152.

——— 1870. On the classification of the Dinosauria, with observations on the Dinosauria of the Trias. *Quarterly Journal of the Geological Society of London* 26: 32–50.

——— 1887. Further observations of *Hyperodapedon gordoni*.

Quarterly Journal of the Geological Society of London 43: 675–694.

Ireland, R. J., J. E. Pollard, R. S. Steel, and D. B. Thompson. 1978. Intertidal sediments and trace fossils from the Waterstones (Scythian-Anisian?) at Daresbury, Cheshire. *Proceedings of the Yorkshire Geological Society* 41: 399–436.

Irving, A. 1888. The red-rock series of the Devon coast-section. *Quarterly Journal of the Geological Society of London* 44: 149–163.

Jaeger, G. F. 1828. *Ueber die fossile Reptilien, welche in Württemberg aufgefunden worden sind.* Stuttgart: Metzler.

Kamphausen, D. 1983. *Stenotosaurus gracilis,* ein neuer Capitosauride (Stegocephalia) aus den Unteren Röttonen Oberfrankens. *Neues Jahrbuch für Geologie und Paläontologie, Monatshefte* 1983: 119–128.

Kjellesvig-Waering, E. N. 1986. A restudy of the fossil Scorpionida of the world. *Palaeontographica Americana* 55: 1–287.

Laming, D. J. C. 1982. The New Red Sandstone. Pp. 148–178 in E. M. Durrance and D. J. C. Laming (eds.), *The Geology of Devon.* Exeter: Exeter University Press.

Lavis, H. J. 1876. On the Triassic strata exposed in the cliff sections near Sidmouth, and a note on the occurrence of an ossiferous zone containing bones of a *Labyrinthodon. Quarterly Journal of the Geological Society of London* 32: 274–277.

Leonard, A. J., A. G. Moore, and E. B. Selwood. 1982. Ventifacts from a deflation surface marking the top of the Budleigh Salterton Pebble Beds, east Devon. *Proceedings of the Ussher Society* 5: 333–339.

Lorsong, J. A., T. J. Clarey, and C. D. Atkinson. 1990. Lithofacies architecture of sandy braided stream deposits in the Otter Sandstone, U.K. P. 139 in *Sediments 1990, Nottingham, England, Abstracts of Posters.* Utrecht: International Sedimentological Union.

Mader, D. 1990. *Palaeoecology of the Flora in Buntsandstein and Keuper in the Triassic of Middle Europe. Vol. 1. Buntsandstein.* Stuttgart: Gustav Fischer Verlag.

Mader, D., and D. J. C. Laming. 1985. Braidplain and alluvia-fan environmental history and climatological evolution controlling origin and destruction of aeolian dune fields and governing overprinting of sand seas and river plains by calcrete pedogenesis in the Permian and Triassic of south Devon (England). Pp. 519–528 in D. Mader (ed.), *Aspects of Fluvial Sedimentation in the Lower Triassic Buntsandstein.* Berlin: Springer-Verlag.

Maidwell, F. T. 1911. Notes on footprints from the Keuper of Runcorn Hill. *Proceedings of the Liverpool Geological Society* 11: 140–152.

Metcalfe, A. T. 1884. On further discoveries of vertebrate remains in the Triassic strata of the south coast of Devon between Budleigh Salterton and Sidmouth. *Quarterly Journal of the Geological Society of London* 40: 257–262.

Miall, L. C. 1874. On the remains of Labyrinthodontia from the Keuper Sandstone of Warwick. *Quarterly Journal of the Geological Society of London* 30: 417–435.

Milner, A. R., B. G. Gardiner, N. C. Fraser, and M. A. Taylor. 1990. Vertebrates from the Middle Triassic Otter Sandstone Formation of Devon. *Palaeontology* 33: 873–892.

Morales, M. 1983. A preliminary report on the terrestrial paleoecology of the Triassic Moenkopi Formation. *Geological Society of America, Abstracts with Programs* (5) 15: 284.

Murchison, R. I. 1839. *The Silurian System.* London: John Murray.

Murchison, R. I., and H. E. Strickland. 1840. On the upper formations of the New Red Sandstone System in Gloucestershire, Worcestershire, and Warwickshire; etc. *Transactions of the Geological Society of London* (2) 5: 331–348.

Old, R. A., R. J. O. Hamblin, K. Ambrose, and G. Warrington. 1991. *Geology of the Country around Redditch.* London: HMSO.

Old, R. A., M. G. Sumbler, and K. Ambrose. 1987. *Geology of the Country around Warwick.* London: British Geological Survey.

Owen, R. 1841a. [The skeleton of three species of *Labyrinthodon*]. *Athenaeum* 1841: 581–582.

——— 1841b. *Odontography.* London: Hippolyte Bailliere.

——— 1842a. Description of parts of the skeleton and teeth of five species of the genus *Labyrinthodon* (*Lab. leptognathus, Lab. pachygnathus,* and *Lab. ventricosus,* from the Coton-end and Cubbington Quarries of the Lower Warwick Sandstone; *Lab. Jaegeri,* from Guy's Cliff, Warwick; and *Lab. scutulatus,* from Leamington); with remarks on the probable identity of the *Cheirotherium* with this genus of extinct batrachians. *Transactions of the Geological Society of London* (2) 6: 515–543.

——— 1842b. Report on British fossil reptiles. Part II. *Report of the British Association for the Advancement of Science* 1842(1841): 60–204.

——— 1842c. Description of an extinct lacertian, *Rhynchosaurus articeps,* Owen, of which the bones and foot-prints characterize the upper New Red Sandstone at Grinshill, near Shrewsbury. *Transactions of the Cambridge Philosophical Society* (2) 7: 355–369.

——— 1845. Description of certain fossil crania discovered by A. G. Bain, esq., in the sandstone rocks of the southeastern extremity of Africa, referable to different species of an extinct genus of Reptilia (*Dicynodon*), and indicative of a new tribe or suborder of Sauria. *Transactions of the Geological Society of London* (2) 7: 59–84.

——— 1859. Note on the affinities of *Rhynchosaurus. Annals and Magazine of Natural History* (3) 4: 237–238.

——— 1863. Notice of a skull and parts of the skeleton of *Rhynchosaurus articeps. Philosophical Transactions of the Royal Society of London* 152: 466–467.

Paton, R. 1974. Capitosauroid labyrinthodonts from the Trias of England. *Palaeontology* 17: 253–289.

Pocock, R. W., and D. A. Wray. 1925. *The Geology of the Country around Wem.* London: Geological Survey of the United Kingdom.

Purvis, K., and V. P. Wright. 1991. Calcretes related to phreatophytic vegetation from the Middle Otter Sandstone of South West England. *Sedimentology* 38: 539–551.

Purvis, K., V. P. Wright and A. Leonard. 1990. Calcretes related to phreatophytic vegetation from the Upper Triassic Otter Sandstone (Sherwood Sandstone Group) of S. W. England. Pp. 441–442 in *Sediments 1990, Nottingham, England, Abstracts of Papers.* Utrecht: International Sedimentological Union.

Richter-Bernburg, G. 1979. Nachwort. Bemerkungen zum Begriff Rhaet. Pp. 151–152 in J. Wiedmann, F. Fabricius, L. Krystyn, J. Reitner, and M. Urlichs, *Ueber Umfang und Stellung des Rhaet. Newsletters in Stratigraphy* 8: 133–152.

Sarjeant, W. A. S. 1974. A history and bibliography of the study of fossil vertebrate footprints in the British Isles. *Palaeogeography, Palaeoclimatology, Palaeoecology* 16: 265–378.

——— 1983. British fossil footprints in the collections of some principal British museums. *Geological Curator* 3: 541–560.

——— 1984. The Beasley collection of photographs and drawings of fossil footprints and bones, and of fossil and recent sedimentary structures. *Geological Curator* 4: 133–163.

Sedgwick, A. 1829. On the geological relations and internal structure of the Magnesian Limestone, and the lower portions of the New Red Sandstone Series in their range through Nottinghamshire, Derbyshire, Yorkshire, and Durham, to the southern extremity of Northumberland. *Transactions of the Geological Society of London* (2) 3: 37–124.

Seeley, H. G. 1876. On the posterior portion of a lower jaw of a *Labyrinthodon* (*L. lavisi*). *Quarterly Journal of the Geological Society of London* 32: 278–284.

Selwood E. B., R. A. Edwards, S. Simpson, J. A. Chesher, R. J. O. Hamblin, M. R. Henson, B. W. Riddolls, and R. A. Waters. 1984. *Geology of the Country around Newton Abbott.* London: British Geological Survey.

Shishkin, M. A. 1980. The Luzocephalidae, a new Triassic labyrinthodont family. *Paleontological Journal* 14 (1): 88–101.

Smith, S. A. 1990. The sedimentology and accretionary styles of an ancient gravel-bed stream: the Budleigh Salterton Pebble Beds (Lower Triassic), southwest England. *Sedimentary Geology* 67: 199–219.

Smith, S. A., and R. A. Edwards. 1991. Regional sedimentological variations in Lower Triassic fluvial conglomerates (Budleigh Salterton Pebble Beds), southwest England: some implications for palaeogeography and basin evolution. *Geological Journal* 26: 65–83.

Spencer, P. S., and K. P. Isaac. 1983. Triassic vertebrates from the Otter Sandstone Formation of Devon, England. *Proceedings of the Geologists' Association* 94: 267–269.

Swinton, W. E. 1960. The history of *Chirotherium. Liverpool and Manchester Geological Journal* 2: 443–473.

Thompson, D. B. 1970a. The stratigraphy of the so-called Keuper Sandstone Formation (Scythian–?Anisian) in the Permo-Triassic Cheshire Basin. *Quarterly Journal of the Geological Society of London* 126: 151–181.

——— 1970b. Sedimentation of the Triassic (Scythian) Red Pebbly Sandstones in the Cheshire Basin and its margins. *Geological Journal* 7: 183–216.

——— 1985. *Field Excursions to the Cheshire, Irish Sea, Stafford, and Needwood Basins.* Chester: Poroperm Ltd.

Tresise, G. 1989. *The Invisible Dinosaur.* Liverpool: National Museums and Galleries on Merseyside.

——— 1991. The Storeton Quarry discoveries of Triassic vertebrate footprints, 1838: John Cunningham's account. *Geological Curator* 5: 225–229.

Ussher, W. A. E. 1876. On the Triassic rocks of Somerset and Devon. *Quarterly Journal of the Geological Society of London* 32: 367–394.

Van der Eem, J. G. L. A. 1983. Aspects of Middle and Late Triassic palynology. 6. Palynological investigations in the Ladinian and Lower Karnian of the western Dolomites. *Review of Palaeobotany and Palynology* 39: 189–300.

Visscher, H., and W. A. Brugman. 1981. Ranges of selected palynomorphs in the Alpine Triassic of Europe. *Review of Palaeobotany and Palynology* 34: 115–128.

Walker, A. D. 1969. The reptile fauna of the 'Lower Keuper' Sandstone. *Geological Magazine* 10: 470–476.

——— 1970. Discussion contributions. *Journal of the Geological Society of London* 126: 217–218.

Ward, J. 1900. On the occurrence of labyrinthodont remains in the Keuper Sandstone of Stanton. *Transactions of the North Staffordshire Field Club* 34: 108–112.

Ward, T. O. 1840. On the foot-prints and ripple-marks of the New Red Sandstone of Grinshill Hill, Shropshire. *Report of the British Association for the Advancement of Science* 1840 (1839): 75–76.

——— 1841. The *Labyrinthodon. Salopian Journal* 28 April 1841: 2.

——— 1874. Note on the *Rhynchosaurus articeps* Owen. *Nature (London)* 11: 8.

Warrington, G. 1967. Correlation of the Keuper Series of the Triassic by miospores. *Nature (London)* 214: 1323–1324.

——— 1970a. The 'Keuper' Series of the British Trias in the northern Irish Sea and neighbouring areas. *Nature (London)* 226: 254–256.

——— 1970b. The stratigraphy and palaeontology of the 'Keuper' Series of the Central Midlands of England. *Journal of the Geological Society of London* 126: 183–223.

——— 1971. Palynology of the New Red Sandstone of the south Devon coast. *Proceedings of the Ussher Society* 2: 307–314.

——— 1974a. Triassic. Pp. 145–160 in D. H. Rayner and J. E. Hemingway (eds.), *The Geology and Mineral Resources of Yorkshire.* York: Yorkshire Geological Society.

——— 1974b. Les évaporites du Trias britannique. *Bulletin de la Société Géologique de France, Série 7* 16: 708–723.

——— 1974c. Studies in the palynological biostratigraphy of the British Trias. 1. Reference sections in west Lancashire and north Somerset. *Review of Palaeobotany and Palynology* 17: 133–147.

——— 1978. Palynology of the Keuper, Westbury and Cotham beds and the White Lias of the Withycombe Farm Borehole. *Bulletin of the Geological Survey of Great Britain* 68: 22–28.

Warrington, G., M. G. Audley-Charles, R. E. Elliott, W. B. Evans, H. C. Ivimey-Cook, P. E. Kent, P. L. Robinson, F. W. Shotton, and F. M. Taylor. 1980. A correlation of Triassic rocks in the British Isles.

Special Report of the Geological Society of London 13: 1–78.

Warrington, G., and H. C. Ivimey-Cook. 1992. Triassic. *Atlas of Palaeogeography and Lithofacies*. Bath: Geological Society of London.

Warrington, G., and R. C. Scrivener. 1988. Late Permian fossils from Devon: regional geological implications. *Proceedings of the Ussher Society* 7: 95–96.

——— 1990. The Permian of Devon, England. *Review of Palaeobotany and Palynology* 66: 263–272.

Watson, D. M. S. 1910. On a skull of *Rhynchosaurus* in the Manchester Museum. *Report of the British Association for the Advancement of Science* 1910(1909): 155–158.

Whitaker, W. 1869. On the succession of beds in the "New Red" on the south coast of Devon, and on the locality of a new specimen of *Hyperodapedon*. *Quarterly Journal of the Geologial Society of London* 25: 152–158.

Williams, B. J., and A. Whittaker. 1974. Geology of the country around Stratford-upon-Avon and Evesham. *Memoirs of the Geological Survey of Great Britain*. London: HMSO.

Wills, L. J. 1907. On some fossiliferous Keuper rocks at Bromsgrove, Worcestershire. *Geological Magazine* (5) 4: 28–34.

——— 1908. Note on the fossils from the Lower Keuper of Bromsgrove. *Report of the British Association for the Advancement of Science* 1908(1907): 312–313.

——— 1910. On the fossiliferous Lower Keuper rocks of Worcestershire. *Proceedings of the Geologists' Association* 21: 249–331.

——— 1916. The structure of the jaw of Triassic labyrinthodonts. *Proceedings of the Birmingham Natural History Society* 14: 1–16.

——— 1947. British Triassic scorpions. *Monographs of the Palaeontographical Society* 100–101: 1–137.

——— 1948. *The Palaeogeography of the Midlands*. Liverpool University Press.

——— 1950. *The Palaeogeography of the Midlands*, 2nd ed. Liverpool University Press.

——— 1970. The Triassic succession in the central Midlands in its regional setting. *Journal of the Geological Society of London* 126: 225–283.

——— 1976. The Trias of Worcestershire and Warwickshire. *Report of the Institute of Geological Sciences* 76(2): 1–209.

Wilson, A. A., and W. B. Evans. 1990. Geology of the country around Blackpool. *Memoirs of the British Geological Survey*. London: HMSO.

Woodward, A. S. 1893. Palaeoichthyological notes. *Annals and Magazine of Natural History* (6)12: 281–287.

——— 1904. On two new labyrinthodont skulls of the genera *Capitosaurus* and *Aphaneramma*. *Proceedings of the Zoological of London* 1904: 170–176.

——— 1905. On some abdominal ribs of *Hyperodapedon* from the Keuper Sandstone of Hollington. *Report and Transactions of the North Staffordshire Field Club* 34: 115–117.

——— 1907. On *Rhynchosaurus articeps* (Owen). *Report of the British Association for the Advancement of Science* 1907(1906): 293–299.

——— 1908. On a mandible of *Labyrinthodon leptognathus*. *Report of the British Association for the Advancement of Science* 1908(1907): 298–300.

Woodward, H. B., and W. A. E. Ussher. 1911. The geology of the country near Sidmouth and Lyme Regis. *Memoirs of the Geological Survey of the United Kingdom*. London: HMSO.

Worssam, B. C., and R. A. Old. 1988. Geology of the country around Coalville. *Memoirs of the British Geological Survey*. London: HMSO.

Wright, V. P., S. B. Marriott, and S. D. Vanstone. 1991. A 'reg' palaeosol from the Lower Triassic of south Devon: stratigraphic and palaeoclimatic implications. *Geological Magazine* 128: 517–523.

Appendix 7.1

Documentation of the identifiable specimens of Middle Triassic tetrapods from England. This list is based on published works, examination of collections in museums, and (for Devon) recent collecting. The MNI estimates, where greater than 1, are justified in terms of the maximally represented skeletal part.

	MNI	NRMAX
Grinshill		
Rhynchosaurus articeps (Benton, 1990, pp. 219–20); also, foot-prints, *Rhynchosauroides*, *Chirotherium* (i.e., rauisuchian)	7 (skulls)	17
Warwick (all Coton End, unless otherwise stated)		
"*Stenotosaurus leptognathus*" (Paton, 1974; WARMS Gz 6, 11, 35, 38)	2 (left squamosals)	4
"*Cyclotosaurus pachygnathus*" (Paton, 1974; WARMS Gz 13, 14, 26, 36)	2 (right tabulars)	4
Mastodonsaurus sp. (Paton, 1974; WARMS Gz 9, 15, 20, 37, 1075),	2 (posterior right mandibular rami)	5
Amphibian indet. (Paton, 1974; WARMS Gz 27)	0	1
cf. *Macrocnemus* (Walker, 1969; WARMS Gz 19, 21, 3787 [= 4714])	1	3
WARMS Gz 10, Leamington	2 (two sizes on slab)	2
Rhynchosaurus brodiei (listed in Benton, 1990, p. 220)	3 (left dentaries)	14
plus one from Leamington	1	1
Bromsgroveia walkeri (Galton, 1975; WARMS Gz 1/2, 3, 5, 121, 128, 970, 1036)	1	7
"Large thecodontian" (Walker, 1969; WARMS Gz 4713)	1	1

"Prosauropod"   1   3
  (Walker, 1969;
  WARMS Gz 982;
  BMNH R2628; BGS
  [GSM] 4873)
"Cladeiodon"   1   3
  (Walker, 1969;
  WARMS Gz 7, 8, 954,
  957, 969)
  WARMS Gz 956, Leek   1   1
  Wootton

Bromsgrove
"Cyclotosaurus pachgnathus" 1   1
  (Paton, 1974; BIRUG
  52)
Mastodonsaurus sp.   1   2
  (Paton, 1974; BIRUG
  1867; CAMSM G369)
"Amphibian indet."
  (Paton, 1974; BIRUG
  51; CAMSM G332,
  333, 334, 335)
cf. Macrocnemus   1   1
  (Walker, 1969;
  CAMSM G343)
Rhynchosaurus brodiei   1   2
  (Benton, 1990, p. 220)
Bromsgroveia walkeri   1   3
  (Walker, 1969; Galton,
  1985; BIRUG 768;
  CAMSM G353, 357)
"Large thecodontian"   1   1
  (Walker, 1969;
  CAMSM G344–349
  [= 344a–f]
"Cladeiodon"   1   1
  (Walker, 1969;
  CAMSM G352, and
  others)
Nothosaur   1   1
  (Walker, 1969;
  CAMSM G351, ? G354)

Devon

Eocyclotosaurus sp.   2 (left tabulars)   3
  (Milner et al., 1990;
  EXEMS 60/1985.72, ?
  75, 310)
Mastodonsaurus lavisi   2 (right mandibles)   4
  (Milner et al., 1990;
  BMNH R331, R4215;
  EXEMS 60/1985.287,
  309)
Capitosaurid inc. sed.   1   1
  (Milner et al., 1990;
  EXEMS 60/1985.78)
Amphibian indet.   1   7
  (Milner et al., 1990;
  EXEMS 60/1985.2, 4,
  79, 96, 148, 183,
  308)
Procolophonids   3 (right dentaries)   5
  (Milner et al., 1990;
  EXEMS 60/1985.3, 9,
  87, 154, 311)
Rhynchosaurus spenceri   9   29
  (Benton, 1990,
  pp. 221–2; plus new
  specimens in Bristol)
Tanystropheus sp.   1   1
  (Milner et al., 1990;
  EXEMS 60/1985.143)
?Ctenosauriscid   1   1
  (Milner et al., 1990;
  EXEMS 60/1985.88)
"Thecodontians"   4 (?)   10
  (Teeth: EXEMS 60/
  1985.6, 8, 25, 27, 28,
  51, 133, 140, 148,
  155, 165, 176, 180,
  Others: EXEMS 60/
  1985.1, 7, 20, 22,
  24, 53, 54, 64, 73,
  84, 97, 150)

8

Small tetrapods from the Upper Triassic of the Richmond basin (Newark Supergroup), Virginia

HANS-DIETER SUES, PAUL E. OLSEN, AND
PETER A. KROEHLER

Introduction

During the Triassic period, profound changes took place in the composition of continental biotas (Cox, 1967; Benton, 1986; Olsen and Sues, 1986). Late Permian communities of continental tetrapods were dominated by nonmammalian synapsids. During the Triassic, these assemblages gave way to communities that were dominated by archosaurian reptiles, especially dinosaurs, but also included other important elements of later continental biotas, such as lepidosaurs, mammaliaform synapsids, and turtles. Despite the obvious importance of this faunal transition, the pattern of change among Triassic continental tetrapods has yet to be fully documented. Much of the currently available fossil record has been recovered from stratigraphically poorly constrained strata of Middle and Late Triassic age in the southern continents, especially in Argentina, Brazil, and Tanzania (Romer, 1966; Cox, 1973). These assemblages appear to differ from the classic Late Triassic tetrapod assemblages from Europe and the American Southwest in their taxonomic composition, which is puzzling in view of the fact that the Triassic was the only period in tetrapod history during which a single landmass existed for the entire length of the period (Cox, 1973; Parrish, Parrish, and Ziegler, 1986).

The early Mesozoic Newark Supergroup comprises the remnants of the sedimentary and igneous fill of an extensive series of partially fault-bounded basins that formed in continental crust along the eastern margin of North America in response to extensional forces during the initial phase of the breakup of Laurasia (Olsen, Schlische, and Gore, 1989). Its sedimentary rocks, ranging in age from the Middle Triassic to Early Jurassic, have traditionally been regarded as virtually devoid of tetrapod bones, although tracks and trackways representing a considerable variety of tetrapods are locally very abundant. Starting in the 1950s, renewed collecting efforts, first by Donald Baird (formerly of Princeton University) and subsequently by Paul Olsen and his associates, have resulted in the discovery of several stratigraphically well constrained tetrapod-bearing localities in strata of the Newark Supergroup (Olsen, 1988; Olsen et al., 1989).

We report here on an early Late Triassic tetrapod assemblage from the Tomahawk Creek Member of the Turkey Branch Formation in the Richmond basin of east-central Virginia. It is quite unlike any other known from North America and raises a number of important questions regarding early Mesozoic biogeography and faunal change. A brief report on the Tomahawk assemblage has already been published (Sues and Olsen, 1990). This chapter presents a preliminary review of the small tetrapods and discusses the general significance of the Tomahawk tetrapod assemblage.

Geological setting

The Richmond basin (Figure 8.1) is located in east-central Virginia close to the eastern edge of the Piedmont, some 19 km west of Richmond. It is surrounded by igneous and metamorphic rocks of the Piedmont Province and is bounded on the western side by a series of normal faults. As exposed today, the basin is 53 km long and about 15 km wide at its widest point, covering an area of only 273 km^2 (Cornet and Olsen, 1990). Its small size, compared with many other basins of the Newark Supergroup, however, is deceptive, because close similarities in facies development suggest that it was once continuous with the Taylorsville basin, which lies just 11 km to the north (Ressetar and Taylor, 1988; Cornet and Olsen, 1990). A large portion of the latter basin underlies the Atlantic Coastal Plain to the northeast, where it is probably

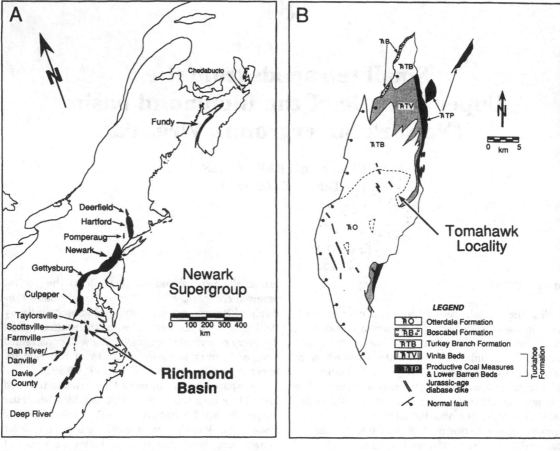

Figure 8.1. (A) Distribution of the principal rift basins of the Newark Supergroup in eastern North America. (B) Geological map of the Richmond basin. (Modified from Olsen et al., 1989.)

continuous with the recently discovered Queen Anne basin of the Delmarva Peninsula in Maryland (Hansen, 1988) and possibly a basin in southern New Jersey (Sheridan, Olsson, and Miller, 1991). The resulting structure would represent one of the largest rifts of the Newark Supergroup, one with a very distinctive stratigraphic sequence and sedimentary fill.

The Richmond basin occupies the southwestern side of an elongated rift valley region and may preserve only a portion of the original rift valley south of the Taylorsville and Queen Anne basins. Despite a long history of intermittent study of the geological structure of the basin, dating back to the pioneering work of Lyell (1847), it is still rather poorly understood because suitable exposures and outcrops are scarce. Extensive drilling for oil and gas and seismic-reflection work in the 1980s have clarified some aspects of the stratigraphic sequence and structure of the basin (Cornet, 1989; Cornet and Olsen, 1990). The Richmond basin contains some of the stratigraphically oldest sedimentary rocks of the Newark Supergroup currently

recognized south of Nova Scotia (Olsen et al., 1989). Based on palynological data, the ages of its strata may range from late Ladinian to early Carnian (Ediger, 1986) or, more likely, from early to middle Carnian (Cornet, 1989; Cornet, in Olsen et al., 1989). A few dikes of Jurassic diabase intrude both the rocks in the Richmond basin and the surrounding Piedmont. The Carnian sedimentary sequence rests unconformably upon igneous and metamorphic basement. The Richmond and Taylorsville basins contain gray and black sedimentary rocks of lacustrine to paludal origin, rather than the red and brown playa and fluvial sedimentary rocks characteristic of other Newark Supergroup basins (Olsen et al., 1989). The Richmond basin contains relatively extensive coals and highly bioturbated shallow-water lacustrine and fluvial sequences, which suggest persistently humid conditions (Cornet and Olsen, 1990). Cornet and Olsen divide the main basin sequence into three formations (from oldest to youngest): Tuckahoe Formation, Turkey Branch Formation, and Otterdale Sandstone.

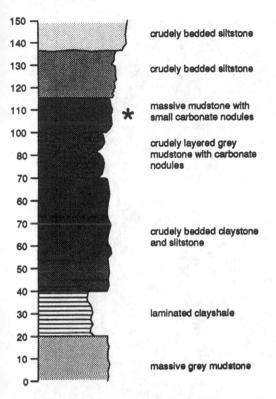

crudely bedded siltstone

crudely bedded siltstone

massive mudstone with small carbonate nodules

crudely layered grey mudstone with carbonate nodules

crudely bedded claystone and siltstone

laminated clayshale

massive grey mudstone

Figure 8.2. Stratigraphic section of the Turkey Branch Formation at the Tomahawk Locality and in its immediate surroundings. Asterisk denotes tetrapod-bearing mudstone horizon.

Grammer (1818) first reported the occurrence of fossil fishes in pits and mine shafts in the famous Coal Measures of the Richmond basin. In 1845, Charles Lyell visited the Blackheath mines, situated south of the James River and northeast of the town of Midlothian in Chesterfield County, and obtained a specimen of the distinctive redfieldiid 'holostean' *Dictyopyge macrura*, which he subsequently illustrated in his description of the coal field in the Richmond basin (Lyell, 1847). *Dictyopyge* is endemic to the Richmond, Taylorsville, and smaller associated basins and occurs in large numbers throughout much of the stratigraphic sequence (Schaeffer and McDonald, 1978). The fish specimens generally are incomplete and disarticulated and commonly are associated with conchostracans, ostracods, plant remains, coprolites, and occasional gastropods and reptilian teeth.

Olsen discovered the vertebrate-bearing Tomahawk Locality in July 1981, during geological reconnaissance in the Richmond basin. The first geological map of the Richmond basin by Shaler and Woodworth (1899) shows a dip symbol at this location, indicating that an outcrop already existed there at the time of their survey. The site (USNM locality 39981) is located along the northeastern bank of the old course of Old

Hundred Road (VA 652), 0.1 mile (0.16 km) east of the eastern branch of Tomahawk Creek, near Midlothian, Chesterfield County. We refer to the site as the Tomahawk Locality because of its proximity to Little Tomahawk Creek on the old Tomahawk Plantation. The tetrapod bones occur in a 15–20-cm-thick stratum of massive calcareous mudstone, together with small (1–10 mm) carbonate nodules, poorly preserved root traces, and countless fish scales and bones (Figure 8.2). At least two irregular layers of fissile, more silty mudstone extend within the massive mudstone and contain vast quantities of fish scales and bones along with isolated tetrapod bones and teeth. The fossiliferous strata form part of the middle Tomahawk Creek Member of the Turkey Branch Formation (B. Cornet, pers. commun.), an extensive sequence of sedimentary rocks of shallow-water lacustrine origin. The poorly exposed stratigraphic sequence in the area of the excavation consists of laminated dark gray claystone with conchostracans, that grades upward into massive mudstone and nodular limestone. The Turkey Branch Formation is unconformably overlain by the coarse-grained sandstones of the possibly late Carnian Otterdale Formation, which appears to represent a braided-stream deposit and thus far has yielded only petrified wood.

Most tetrapod remains occur as dissociated bones (or fragments of bones) and teeth. For this reason, we have generally adopted a conservative approach in taxonomic identification of the material now at hand. In a few instances, skulls and partial skeletons are disarticulated, but the component elements still remain in close association. The superbly preserved bones and teeth show relatively few signs of crushing and distortion.

Most of the associated fish material can be assigned to the ubiquitous *Dictyopyge*. A few isolated teeth document the presence of small hybodont sharks referable to *Lissodus* (A. K. Johansson, pers. commun.). The only invertebrates preserved in the tetrapod-bearing mudstone are as yet unidentified gastropods, which are documented by shell fragments and rare complete specimens in steinkern preservation. Macroscopic plant remains comprise a single poorly preserved fern pinnule (B. Cornet, pers. commun.), poorly preserved root traces, and carbonized wood scraps.

Field collection and preparation

The locality was initially quarried by means of hand tools. As the area of excavation was expanded, a backhoe was used on several occasions to remove several meters of overlying clay and weathered rock, but care was taken not to uncover the actual fossiliferous horizon during that phase of site preparation.

The bones and teeth can be readily separated from the enclosing mudstone matrix using needles ground from rods of tungsten carbide and mounted in pin vises. The fossils usually are stained black and display even minute structural details. Many teeth have light gray or bluish enamel and black dentine. Most of the small bones and teeth are penetrated by minute fractures, and often they rapidly disintegrate as the enclosing mudstone dries. Application of cyanoacrylate glues was used to retard disintegration during collecting; the adhesives can be removed with acetone or by careful peeling with a needle during preparation in the laboratory.

Bulk samples of fossiliferous mudstone were disintegrated in hot water or treated using kerosene and water. The resulting residue was screened, dried, and manually sorted under a dissecting microscope. This procedure yielded numerous skeletal remains of a variety of small tetrapods, as well as abundant scales referable to *Dictyopyge*.

Systematic paleontology

Synapsida

Cynodontia

"Traversodontidae"

Boreogomphodon jeffersoni Sues and Olsen, 1990

This small traversodont cynodont is by far the most abundant identifiable tetrapod taxon at the Tomahawk Locality. [We use only the informal term "traversodont" because the family Traversodontidae Huene, 1936 probably constitutes a paraphyletic grouping (Hopson, 1984).] It is represented by three excellently preserved partial skulls (including one with a natural endocast), scattered remains of a fragmentary skeleton, a number of isolated dentaries, maxillae, and premaxillae, and many isolated teeth. In addition, a few isolated cynodont limb-bones (humerus, femur) may prove referable to this taxon.

B. *jeffersoni* most closely resembles *Traversodon stahleckeri* Huene, 1936 from the Carnian Santa Maria Formation of southern Brazil and *Luangwa drysdalli* Brink, 1963 from the Anisian Ntawere Formation of Zambia (Kemp, 1980) in the structure of its upper postcanine teeth. The upper postcanines have buccolingually expanded crowns, with three principal cusps posteriorly and a large anterior basin (Figure 8.3). Only a single accessory cusp is developed anterior to the buccal principal cusp. A posterior cingulum is present. The enamel typically shows coarse vertical wrinkling. The lower postcanine teeth differ from those of *Luangwa*, *Traversodon*, and most other known traversodont taxa in the presence of three, rather than two, anterior cusps. This character state is shared only by *Arctotraversodon* from the Wolfville Formation (middle to upper Carnian) of Nova Scotia and an undescribed traver-

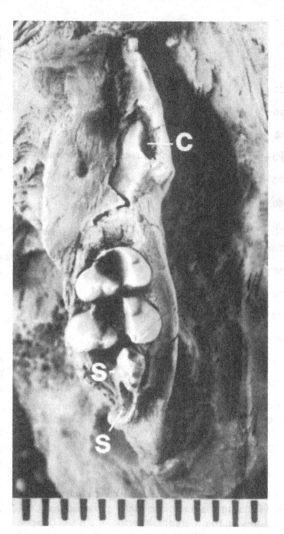

Figure 8.3. Left maxilla of *Boreogomphodon jeffersoni* Sues and Olsen, 1990, USNM 437632 (holotype), in occlusal view. Specimen coated with ammonium chloride. Divisions of scale bar each equal 1 mm. Abbreviations: c, canine alveolus; s, sectorial tooth and alveolus for sectorial tooth.

sodont from the Lettenkeuper (Ladinian) of southern Germany (Sues, Hopson, and Shubin, 1992).

The buccal surface of the maxilla bears a pronounced longitudinal ridge that overhangs the tooth row buccally. The dentary has a large posterior mental foramen. These features, in conjunction with the inset lower and upper rows of postcanine teeth, provide suggestive evidence for a flexible cheek and a buccal oral vestibule, as inferred for other traversodont cynodonts (Hopson, 1984).

Most of the jaws referable to *Boreogomphodon* found to date represent juvenile individuals. This assessment is based on the small overall size of the specimens, the

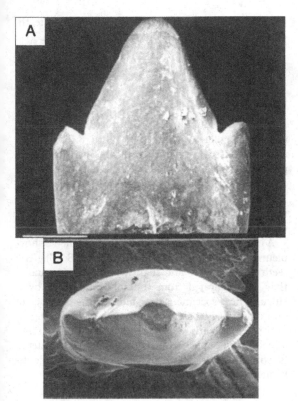

Figure 8.4. Isolated postcanine tooth of *Microconodon tenuirostris* Osborn, 1886, USNM 448600, in buccal (A) and occlusal (B) views. Scale bar equals 400 µm.

low number (four or five) of postcanine teeth, the proportional shortness of the snout, and the proportional depth of the dentary (Hopson, 1984). It is further borne out by the presence of multicuspid "sectorial" teeth in the two posterior alveoli of several maxillae (Figure 8.3). Such teeth have previously been recorded only in juvenile specimens of other gomphodont cynodonts, such as *Diademodon* (Hopson, 1971) and *Massetognathus* (J. A. Hopson, pers. commun.).

Cynodontia incertae sedis

Microconodon tenuirostris Osborn, 1886

Three dentaries and several isolated postcanine teeth document the presence of this very mammal-like cynodont, which previously was known only from a right dentary from the upper Carnian Cumnock Formation of North Carolina (Osborn, 1886; Simpson, 1926). The teeth are also very similar to the isolated teeth of *Pseudotriconodon* from the middle Norian Steinmergel-Gruppe of Luxembourg (Hahn, Lepage, and Wouters, 1984; Sigogneau-Russell and Hahn, Chapter 10).

The postcanine teeth typically bear three pointed cusps that are anteroposteriorly aligned behind one another (Figure 8.4). The tall principal cusp is anteroposteriorly long and has sharp anterior and posterior cutting edges. It is symmetrically flanked by much smaller cusps anteriorly and posteriorly, Cingula are absent. The root of an isolated upper(?) postcanine tooth shows a pronounced anteroposterior constriction. The dentary has a long, low tooth-bearing ramus and lacks a distinct angular process. Its robust symphyseal region holds alveoli for one canine and three slightly procumbent incisors. In the smallest known dentary from the Tomahawk Locality, the postcanine dentition comprises simple anterior and more posterior tricuspid teeth; furthermore, no diastema is present. In the largest dentary found to date, the anterior postcanines have been lost, and a prominent diastema separates the canine and the exclusively tricuspid postcanines.

To date, we have found no differences to justify taxonomic distinction between the material from the Tomahawk Locality and the holotype of *Microconodon tenuirostris* Osborn, 1886 (Academy of Natural Sciences, Philadelphia, no. 10248). The phylogenetic relationships of *Microconodon* are beyond the scope of this chapter and will be discussed elsewhere.

Diapsida

Archosauria

Crurotarsi

To date, only phytosaurs (Parasuchia) and a new suchian archosaur with diagnostic dorsal dermal armor, *Euscolosuchus olseni* Sues, 1992, have been confidently identified.

Although phytosaurs are rather large reptiles, their isolated tooth crowns, representing a wide spectrum of sizes, are very common in both the quarried and screen-washed material and hence merit inclusion in this review. Long, slender, and conical tooth crowns, with a round cross-section and smooth cutting edges, probably represent teeth from the anterior end of the snout, whereas the more robust and labiolingually compressed crowns with finely serrated cutting edges presumably are from the posterior regions of the jaws, much as in other phytosaurs. Many of the presumed anterior teeth occasionally show very pronounced vertical fluting of the enamel and closely resemble certain isolated phytosaurian tooth crowns from the Upper Triassic of Pennsylvania illustrated by Huene (1921). The material is of little diagnostic value. Traditionally, the mostly very fragmentary skeletal remains of phytosaurs from the Upper Triassic of eastern North America have all been referred to *Rutiodon* Emmons, 1856, but in most cases the available evidence neither supports nor contradicts this assumption.

Figure 8.5. *Uatchitodon kroehleri* Sues, 1991, USNM 448624, nearly complete tooth in side view. Scale bar equals 1 mm.

?Archosauriformes incertae sedis

Uatchitodon kroehleri Sues, 1991

This taxon, of uncertain, possibly archosauriform, affinities, is known only from its highly distinctive teeth, which represent the earliest instance of a presumed oral venom-delivery system in reptiles recorded to date (Sues, 1991). The labiolingually strongly compressed, recurved, bladelike tooth crowns bear deeply infolded, enamel-lined median grooves on both their labial and lingual surfaces (Figure 8.5). Judging from the close structural similarity to the venom grooves on the teeth of extant poisonous snakes and lizards of the genus *Heloderma*, it seems likely that these features functioned in venom conduction. The grooves become narrow and shallow toward the tip of the tooth and disappear before reaching it. The anterior and posterior cutting edges of the crowns are serrated on all but two of the teeth recovered to date, with typically six or seven denticles per millimeter (Figure 8.6A). Inspection at higher magnification shows that the sharp cutting edge of each individual denticle is denticulated as well (Figure 8.6B). The tooth crowns of *Uatchitodon kroehleri* have an average height of about 10 mm, but

several fragments indicate the presence of larger teeth. The root indicates thecodont tooth implantation.

Diapsida

Lepidosauria

A new sphenodontian lepidosaur is represented by isolated maxillae with teeth (Figure 8.7A,B). The acrodont teeth have bluntly conical crowns that bear prominent radial ridges on the enamel. An enlarged anterior tooth is followed by three or four smaller teeth. The sphenodontian from the Tomahawk Locality appears to provide the stratigraphically oldest record of this group known to date.

A number of lizardlike jaws and jaw fragments show pleurodont tooth implantation (Figure 8.7C). The teeth are columnar and slightly recurved. Lingually, they are separated from the subdental ridge by a sulcus. These specimens may indicate the presence of an unidentified lepidosaur. Although of great interest as a potential early Late Triassic record of Squamata (see Rieppel, Chapter 2), the currently available material is too fragmentary to permit definite taxonomic identification.

Parareptilia sensu Gauthier et al., 1988

Procolophonia

Two distinctive new taxa are each documented by a jaw fragment with diagnostic teeth. One is clearly referable to the Procolophonidae, and the other may be related to certain advanced taxa of that group. This material has been described and compared in detail elsewhere (Sues and Olsen, 1993).

Importance of the Tomahawk tetrapod assemblage

The preliminary faunal inventory presented here underscores the unusual importance of the Tomahawk tetrapod assemblage for the study of the early Mesozoic history of continental biotas. First, it represents a very diverse assemblage of continental vertebrates of early Late Triassic age from North America and contains a number of previously unknown taxa. Second, the Tomahawk tetrapod assemblage is unlike other known North American assemblages (from the Chinle and Dockum formations in the American Southwest and the rest of the Newark Supergroup) in the numerical predominance of traversodont cynodonts, which indicates close faunal ties to the Gondwanan realm (Argentina, Brazil, and Tanzania) (Romer, 1966; Cox, 1973). The slightly younger tetrapod assemblages from very similar strata of the Pekin and Cumnock formations of North Carolina (Olsen et al., 1989)

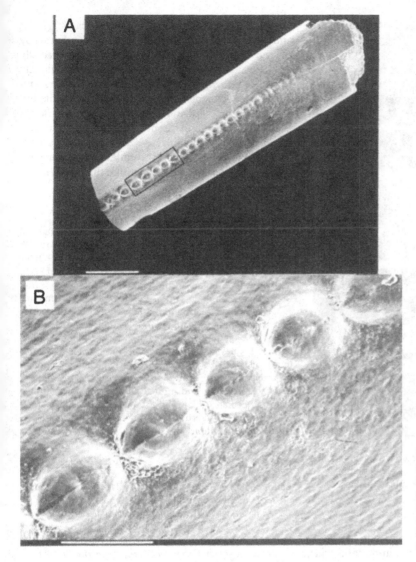

Figure 8.6. *Uatchitodon kroehleri* Sues, 1991. (A) Tooth fragment in posterior view, showing denticles along cutting edge (carina). Scale bar equals 1 mm. (B) Closeup of denticles delineated by rectangular box in A. Scale bar equals 200 µm.

include taxa known from the Chinle and Dockum formations of the American Southwest and demonstrate that the distinctive faunal composition of the Tomahawk assemblage probably is not related to differences in depositional environments. The Tomahawk tetrapod assemblage shares the presence of phytosaurs with other Laurasian assemblages, but it apparently lacks metoposaurid temnospondyl amphibians, which otherwise were widely distributed throughout Europe (Fraas, 1889), North America (Colbert and Imbrie, 1956; Hunt, 1989), Morocco (Dutuit, 1976), and India (Roy-Chowdhury, 1965) during the Late Triassic.

The Tomahawk tetrapod assemblage is demonstrably slightly older than other well-documented Laurasian assemblages, and Sues and Olsen (1990) have previously suggested that the traditionally recognized differences between Carnian tetrapod assemblages from Laurasia and Gondwana reflect differences in stratigraphic age, rather than geographic separation. The apparent faunal provinciality might thus reflect poor stratigraphic sampling of the transition from the Middle to the Late Triassic in Laurasia. Floral provinciality during the early part of the Late Triassic, however, is well established, with Laurasian plant assemblages dominated by cycadophytes and conifers and Gondwanan macrofloras dominated by the seed-fern *Dicroidium* and palynofloras of the Ipswich-Onslow type (Cornet and Olsen, 1985). This floral provinciality does not correspond geographically and temporally in a simple way to the distribution of continental tetrapods. In India, typical Laurasian tetrapod assemblages with abundant phytosaurs and metoposaurs (Roy Chowdhury, 1965; Chatterjee and Roy-Chowdhury, 1974) occur in association with *Dicroidium*-dominated

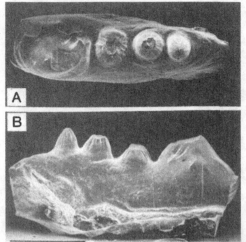

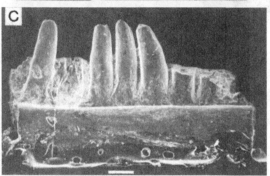

Figure 8.7. Maxilla fragment of undescribed sphenodontian in occlusal view (A) and side view (B). Scale bar equals 2 mm. (C) Dentary fragment of indeterminate lepidosaur in lingual view. Scale bar equals 400 μm.

assemblages of plant macrofossils and with Ipswich-Onslow-type palynofloras (Kumaran and Maheswari, 1980). The tetrapod material from the Richmond basin described here is associated with diverse, typically Laurasian florules (Fontaine, 1883; Cornet and Olsen, 1990) although B. Cornet (pers. commun.) has recently identified two new plant taxa with Gondwanan affinities.

Romer (1966) informally distinguished three successive stages in the historical development of continental Triassic tetrapod assemblages worldwide. He did not attempt to fit these stages into a chronostratigraphic framework, but they broadly correlate with the Triassic standard sequence (Shubin and Sues, 1991). The Early Triassic A-type assemblages are composed predominantly of nonmammalian therapsids and are best known from the Beaufort Group (Karoo Supergroup) of South Africa. B-type assemblages are characterized by the abundance of traversodont cynodonts and, in most (but not all) cases, rhynchosaurian reptiles and are well documented from the Middle and Late Triassic of Argentina, Brazil, and

Tanzania. The Late Triassic (and Early Jurassic) (Olsen and Sues, 1986) C-type assemblages are dominated by a great diversity of archosaurian reptiles, especially dinosaurs. Together with tetrapod material (mostly still undescribed), including both rhynchosaurs and traversodont cynodonts, from the Wolfville Formation of Nova Scotia (Baird and Take, 1959; Baird, in Carroll et al., 1972; Baird and Olsen, 1983; Sues et al., 1992), the Tomahawk assemblage clearly establishes both the existence of B-type communities in the Northern Hemisphere and their persistence into the Late Triassic.

Acknowledgments

We gratefully acknowledge financial support from the National Geographic Society (grants 3952-88 and 4232-89 to H-DS and PEO), the National Science Foundation (NSF BSR-8717707 to PEO and NSF EAR-9016677 to H-DS and PEO), and the Smithsonian Institution (Walcott and Research Opportunities funds to H-DS). J. Adams of J. K. Timmons Associates (Midlothian) and the Department of Transportation of the Commonwealth of Virginia kindly granted permission to develop the Tomahawk Locality. Special thanks must go to E. B. Sues for her enthusiastic and capable assistance in the field. W. W. Amaral meticulously prepared a number of vertebrate fossils. M. A. Parrish drew Figures 8.2 and 8.5 and D. Breger took and printed all SEM photographs. We thank K. Padian (University of California at Berkeley) and J. M. Parrish (Northern Illinois University) for helpful comments on the manuscript.

References

Baird, D., and P. E. Olsen. 1983. Late Triassic herpetofauna from the Wolfville Fm. of the Minas Basin (Fundy Basin) Nova Scotia, Can. *Geological Society of America, Abstracts with Program* 15: 122.

Baird, D., and W. F. Take. 1959. Triassic reptiles from Nova Scotia. *Geological Society of America Bulletin* 70: 1565–1566.

Benton, M. J. 1986. The Late Triassic tetrapod extinction events. Pp. 303–320 in K. Padian (ed.), *The Beginning of the Age of Dinosaurs: Faunal Change across the Triassic–Jurassic Boundary.* Cambridge University Press.

Brink, A. S. 1963. Two new cynodonts from the Ntawere Formation of the Luangwa Valley of Northern Rhodesia. *Palaeontologia Africana* 8: 77–96.

Carroll, R. L., E. S. Belt, D. L. Dineley, D. Baird, and D. C. McGregor. 1972. *Vertebrate Paleontology of Eastern Canada.* Guidebook, Field Excursion A59. Montreal: 24th International Geological Congress.

Chatterjee, S., and T. Roy-Chowdhury. 1974. Triassic Gondwana vertebrates from India. *Indian Journal of Earth Sciences* 1: 96–112.

Colbert, E. H., and J. Imbrie. 1956. Triassic metoposaurid amphibians. *Bulletin of the American Museum of Natural History* 110: 403–452.

Cornet, B. 1989. Late Triassic angiosperm-like pollen from the Richmond rift-basin of Virginia, U.S.A. *Palaeontographica* B213: 37–87.

Cornet, B., and P.E. Olsen. 1985. A summary of the biostratigraphy of the Newark Supergroup of eastern North America, with comments on early Mesozoic provinciality. Pp. 67–81 in R. Weber (ed.), *Simposio sobre Floras del Triásico Tardío, su Fitografía y Paleoecología. Memoria, III Congreso Latinoamericano de Paleontologia*. Universidad Nacional Autonoma de Mexico, Mexico D.F.

——— 1990. *Early to Middle Carnian (Triassic) Flora and Fauna of the Richmond and Taylorsville Basins, Virginia and Maryland, U.S.A.* Guidebook no.1, Martinsville: Virginia Museum of Natural History.

Cox, C. B. 1967. Changes in terrestrial vertebrate faunas during the Mesozoic. Pp. 77–89 in W. B. Harland (ed.), *The Fossil Record*. Geological Society of London.

——— 1973. Triassic tetrapods. Pp. 213–233 in A. Hallam (ed.), *Atlas of Palaeobiogeography*. Amsterdam: Elsevier Scientific.

Dutuit, J.-M. 1976. Introduction a l'étude paléontologique du Trias continental Marocain: déscription des premiers Stégocephales recueillis dans le couloir d'Argana (Atlas occidental). *Mémoires du Muséum National d'Histoire Naturelle, sér. C* 36: 1–253.

Ediger, V. S. 1986. Paleopalynological biostratigraphy, organic matter deposition, and basin analysis of the Triassic-Jurassic(?), Richmond rift basin, VA. Ph.D. dissertation, Pennsylvania State University.

Emmons, E. 1856. *Geological Report on the Midland Counties of North Carolina*. New York: George P. Putnam.

Fontaine, W. M. 1883. *Contributions to the Knowledge of the Older Mesozoic Flora of Virginia*. Monographs of the United States Geological Survey, Vol. 6. Washington: Government Printing Office.

Fraas, E. 1889. Die Labyrinthodonten der schwäbischen Trias. *Palaeontographica* 36: 1–158.

Gauthier, J. A., A. G. Kluge, and T. Rowe. 1988. The early evolution of the Amniota. Pp. 103–155 in M. J. Benton (ed.), *The Phylogeny and Classification of the Tetrapods. Vol. 1: Amphibians, Reptiles, Birds*. Oxford: Clarendon Press.

Grammer, J., Jr. 1818. Account of the coal mines in the vicinity of Richmond, Virginia. *American Journal of Science* (1) 1: 125–130.

Hahn, G., J. C. Lepage, and G. Wouters. 1984. Cynodontier-Zähne aus der Ober-Trias von Medernach, Grossherzogtum Luxemburg. *Bulletin de la Société belge de Géologie* 93: 357–373.

Hansen, J. J. 1988. Buried rift basin underlying coastal plain sediments, central Delmarva Peninsula, Maryland. *Geology* 16: 779–782.

Hopson, J. A. 1971. Postcanine replacement in the gomphodont cynodont *Diademodon*. Pp. 1–21 in D. M. Kermack and K. A. Kermack (eds.), *Early Mammals*. London: Academic Press.

——— 1984. Late Triassic traversodont cynodonts from Nova Scotia and southern Africa. *Palaeontologia Africana* 25: 181–201.

Huene, F. von. 1921. Reptilian and stegocephalian remains from the Triassic of Pennsylvania in the Cope collection. *Bulletin of the American Museum of Natural History* 44: 561–574.

——— 1936. *Die fossilen Reptilien des südamerikanischen Gondwanalandes. Ergebnisse der Sauriergrabungen in Südbrasilien 1928/29, Lieferung 2*. Tübingen: Verlag Franz F. Heine.

Hunt, A. P. 1989. Comments on the taxonomy of North American metoposaurs and a preliminary phylogenetic analysis of the family Metoposauridae. Pp. 293–300 in S. G. Lucas and A. P. Hunt (eds.), *The Dawn of the Age of Dinosaurs in the American Southwest*. Albuquerque: New Mexico Museum of Natural History.

Kemp, T. S. 1980. Aspects of the structure and functional anatomy of the Middle Triassic cynodont *Luangwa*. *Journal of Zoology (London)* 191: 193–239.

Kumaran, K. P. N., and H. K. Maheswari. 1980. Upper Triassic sporae dispersae from the Tiki Formation. 2: Miospores from the Janar Nala section, South Gondwana Basin, India. *Palaeontographica* B173: 26–84.

Lyell. C. 1847. On the structure and probable age of the coal-field of the James River, near Richmond, Virginia. *Quarterly Journal of the Geological Society of London* 3: 261–280.

Olsen, P.E. 1988. Paleontology and paleoecology of the Newark Supergroup (early Mesozoic, eastern North America). Pp. 185–230 in W. Manspeizer (ed.), *Triassic-Jurassic Rifting: Continental Breakup and the Origin of the Atlantic Ocean and Passive Margins, Part A*. Amsterdam: Elsevier.

Olsen, P. E., R. W. Schlische, and P. J. W. Gore (eds.). 1989. *Tectonic, Depositional, and Paleoecological History of early Mesozoic Rift Basins, Eastern North America*. Guidebook for Field Trip T351, 28th International Geological Congress. Washington, D.C.: American Geophysical Union.

Olsen, P. E., and H.-D. Sues. 1986. Correlation of continental Late Triassic and Early Jurassic sediments, and the Triassic-Jurassic tetrapod transition. Pp. 321–351 in K. Padian (ed.), *The Beginning of the Age of Dinosaurs: Faunal Change across the Triassic-Jurassic Boundary*. Cambridge University Press.

Osborn, H. F. 1886. Observations on the Triassic mammals *Dromatherium* and *Microconodon*. *Proceedings of the Academy of Natural Sciences* 37: 359–363.

Parrish, J. M., J. T. Parrish, and A. M. Ziegler. 1986. Permian-Triassic paleogeography and paleoclimatology and implications for therapsid distribution. Pp. 109–131 in N. Hotton III, P. D. MacLean, J. J. Roth, and C. E. Roth (eds.), *The Ecology and Biology of Mammal-like Reptiles*. Washington, D. C.: Smithsonian Institution Press.

Ressetar, R., and G. K. Taylor. 1988. Late Triassic depositional history of the Richmond and Taylorsville basins, eastern Virginia. Pp. 423–443 in W. Manspeizer (ed.), *Triassic-Jurassic Rifting: Continental Breakup and the Origin of the Atlantic Ocean and Passive Margins, Part A*. Amsterdam: Elsevier.

Romer, A. S. 1966. The Chañares (Argentina) Triassic reptile fauna. I. Introduction. *Breviora* 247: 1–14.

Roy Chowdhury, T. 1965. A new metoposaurid amphibian from the Upper Triassic Maleri Formation of central India. *Philosophical Transactions of the Royal Society of London* B250: 1–52.

Schaeffer, B., and N. G. McDonald. 1978. Redfieldiid fishes

from the Triassic–Jurassic Newark Supergroup of eastern North America. *Bulletin of the American Museum of Natural History* 159: 129–174.

Shaler, N. S., and J. B. Woodworth. 1899. Geology of the Richmond basin, Virginia. *United States Geological Survey, Annual Report* (2)19: 385–519.

Sheridan, R. E., R. K. Olsson, and J. J. Miller. 1991. Seismic reflection and gravity study of proposed Taconic suture under the New Jersey Coastal Plain: implications for continental growth. *Geological Society of America Bulletin* 103: 402–414.

Shubin, N. H., and H.-D. Sues. 1991. Early Mesozoic tetrapod distributions: patterns and implications. *Paleobiology* 17: 214–230.

Simpson, G. G. 1926. Mesozoic Mammalia. V. *Dromatherium* and *Microconodon. American Journal of Science* (5) 12: 87–108.

Sues, H.-D. 1991. Venom-conducting teeth in a Triassic reptile. *Nature* 351: 141–143.

1992. A remarkable new armored archosaur from the Upper Triassic of Virginia. *Journal of Vertebrate Paleontology* 12: 142–149.

Sues, H.-D., J. A. Hopson, and N. H. Shubin. 1992. Affinities of ?*Scalenodontoides plemmyridon* Hopson, 1984 (Synapsida: Cynodontia) from the Upper Triassic of Nova Scotia. *Journal of Vertebrate Paleontology* 12: 168–171.

Sues, H.-D., and P. E. Olsen. 1990. Triassic vertebrates of Gondwanan aspect from the Richmond basin of Virginia. *Science* 249: 1020–1023.

1993. A new procolophonid and an unusual tetrapod of uncertain, possibly procolophonian affinities from the Upper Triassic of Virginia. *Journal of Vertebrate Paleontology* 13: 282–286.

9

Microvertebrates from the Placerias Quarry: a window on Late Triassic vertebrate diversity in the American Southwest

FRAN TANNENBAUM KAYE AND KEVIN PADIAN

Introduction

More bones have been recovered from the famous Placerias Quarry (UCMP locality A269) (Camp and Welles, 1956; Jacobs and Murry, 1980; Murry and Long, 1989) than from almost any Triassic locality in the Western Hemisphere, with the exception of the famous Ghost Ranch Quarry in New Mexico (Colbert, 1947, 1989). Nearly 3,000 fossil bones from the Placerias Quarry are catalogued in the collections of the University of California Museum of Paleontology (UCMP), yet only those of the large dicynodont *Placerias* itself have been thoroughly described (Camp and Welles, 1956). From Ghost Ranch, only the small theropod traditionally referred to *Coelophysis* by Colbert (1947, 1989; Padian, 1986; pace Hunt and Lucas, 1991) has been described, though the faunal diversity is hardly monospecific. The small theropod is by far the major faunal element at Ghost Ranch, much as *Plateosaurus* is in the famous quarry at Trossingen in southern Germany (Weishampel, 1984; Weishampel and Westphal, 1986). Although *Placerias* is only one of many large vertebrates, including various phytosaurs, aetosaurs, poposaurs, and metoposaurs, common at the Placerias Quarry, its skeletal remains represent about two-thirds of the collected macrofaunal material.

Frequently, microvertebrates tell an entirely different story about diversity, taphonomy, and paleoecology than the macrovertebrates can tell alone. Generally, our perception of the existing diversity is greatly increased by assessing microvertebrates, and often entire sectors of Mesozoic faunas are captured that otherwise would be missing or poorly represented (Estes, 1964; Jacobs and Murry, 1980; Tannenbaum, 1983; Curtis, 1989; Jensen and Padian, 1989; Murry, 1989). Nowhere is this more true than in the Placerias Quarry, as this study and an earlier one by Jacobs and Murry (1980) show. In this chapter we describe the diversity of the Placerias Quarry microfauna, compare it to the macrofaunal diversity, and discuss what it shows us about paleoecology, biostratigraphy, and faunal change in the Late Triassic and across the Triassic–Jurassic boundary.

Historical and paleoenvironmental setting

As Dr. S. P. Welles tells it, Professor C. L. Camp of the University of California stopped for a haircut in the small Arizona town of St. Johns in 1927. Camp learned from the local barber that some fossil bones had been recovered by his brother the cobbler a few miles to the southwest of town. Following the barber's directions, Camp indeed came across a deposit of fossil bones, Triassic in age, and collected surface float for the remainder of the day, without considering the area very promising (C. L. Camp, 1927 field notes, UCMP: 947–948). In 1930 Camp returned to the area and went out once again under directions of the cobbler and the barber toward the previous fossil site, where his crew began to excavate (C. L. Camp, 1930 field notes, UCMP: 1099–1102). He was surprised some time later by the arrival of the barber, who asked him what he thought he was doing; the previously discovered locality was 1.5 miles (2.4 km) to the north. The locality that Camp had prospected in 1927 became the "Old A269" of the University of California Museum of Paleontology (Camp and Welles, 1956); the one discovered by him in 1930 became the true A269, the Placerias Quarry, one of the richest and most diverse deposits of Triassic fossil bones in the hemisphere.

The Placerias Quarry is located in a grassy valley near Romero Spring, an area of meadows cut by abrupt washes and flanked by rolling hills of Triassic silts overlain by thin Quaternary alluvia. Badlands are not common in this area, as they are in the Blue Hills just northeast of St. Johns and in the Petrified Forest National Park 50 miles (80 km) to the northwest (Cooley, 1957; Harshbarger, Repenning, and Haskell, 1957; Akers, 1964; Repenning, Cooley, and Akers 1969; Stewart, Poole, and Wilson, 1972; Long and Padian, 1986; Colbert, 1989). Fossiliferous Triassic deposits are mainly discovered through surface float, which is still common on the juniper- and scrub-studded point on which the Placerias Quarry is situated. The original quarry horizon was a bed of limestone concretions from which many broken or poorly preserved bones were recovered; this gives way to typical reddish- brown Chinle mudstones and silts, with occasional sandstone lenses, in which "two fossiliferous levels [were] separated by 2 feet of barren sediments" (Camp and Welles, 1956, p. 259). The upper level contained mostly coprolites, with some scattered phytosaurid and metoposaurid bones; the lower level preserved all the bones of Placerias, along with those of aetosaurs, poposaurs, phytosaurs, various smaller reptiles, and metoposaurs. Calcareous nodules, gypsum lenses, carbonaceous particles, and yellow-brown sulfurous deposits are common, and this suggested to Camp and Welles that the original environment had been a soft, vegetated pond bottom or other quiet water deposit (Green, 1956; Dubiel et al., 1991). Many bones show evidence of trampling, and some were collected in an evidently disturbed, vertical position; some show tooth marks, and virtually no bones are associated. Coprolites, mostly about the size of a man's thumb, are still the most abundant fossils in the quarry, and it is within these coprolitic deposits that many of the discovered microvertebrates have been concentrated. Many microfossils show wear, pitting, and surface etching characteristic of passage through a gut.

Camp and his crews worked the Placerias Quarry from 1931 through 1934 and collected nearly 5,000 kg of fossil bone and nodules, representing some 3,000 individual elements. The quarry was not extensively worked again until 1978, when field crews from the Museum of Northern Arizona (MNA) reopened the site and another excavation about 72 m to the east, which they called the Downs Quarry (MNA locality 207–10) (Jacobs and Murry, 1980), several meters stratigraphically above the original site. Approximately 5.8 metric tons of sediment were collected and screen-washed, and the microvertebrates picked and sorted, through the efforts of Louis Jacobs and Will Downs at the MNA. Jacobs and Murry (1980) published an account of the excavation, a review of the paleoecology and taphonomy, and a comprehensive faunal list of the taxonomic components of both quarries. In 1989 and 1990, UCMP field crews under the direction of the second author reopened the Placerias Quarry and explored the Blue Hills northeast of the town of St. Johns, recovering several hundred additional bones and several skeletons.

Diversity of the microfauna of the Placerias and Downs quarries

Jacobs and Murry (1980) concluded that the Placerias and Downs quarries sample the same fauna, and that is true but for the local restriction of Placerias to the floor of the original quarry. However, the microvertebrate fauna, which shows no significant taxonomic differences between the two sites (Jacobs and Murry, 1980; Tannenbaum, 1983), greatly deepens our understanding of the diversity and paleoecology of the deposits. Jacobs and Murry (1980) estimated that the microfaunal diversity doubles that known previously from the macrofauna alone, and subsequent work has supported that estimate (Tannenbaum, 1983; Murry, 1987; Murry and Long, 1989). The study reported here suggests an even greater diversity of microvertebrates, though much of the material is not identifiable to specific, generic, or even familial rank of known vertebrate taxa. Names and tentative higher taxonomic assignments have been given to some of these micromorphs (Murry, 1987) [see also Sues (1991) on a similar tooth morph found in the Newark Supergroup of eastern North America], but we have chosen not to name each distinct micromorph in this fauna, for three reasons. First, often the specimen cannot be classified below the traditional Linnean ranking of class. Second, the association of parts is impossible to assess, as is the variability due to ontogeny, sexual dimorphism, position in the dentition or vertebral column, and populational variation. Third, some specimens cannot be assigned to specific parts of the skeleton, whether because of poor or partial preservation or because parts like these apparently have never been described in the literature. All these factors could artificially, though not without probable cause, inflate the apparent taxonomic diversity of the Placerias Quarry and Downs Quarry microfauna. Therefore, we take the conservative course of illustrating and describing problematica as such, and assigning other specimens to the lowest diagnostic level possible.

The MNA collection of microvertebrates from the Placerias and Downs quarries was brought to Berkeley in the fall of 1980 by Louis Jacobs and Will Downs and made available to us for study. The collection consists of approximately 4,000 specimens that had been wet-screened, washed in acetic acid, and sorted in a preliminary fashion into about 300 half-dram vials. The first author undertook the project to sort and identify the specimens further, based on similarities and differences of morphotypes, into approximately 1,000 half-dram vials containing from one to about 50 specimens each, thus establishing a reference collection of 200 "vouchers" or representative specimens that could be described more explicitly. Each vial was assigned its MNA locality and specimen number; these are described and illustrated in this chapter and in earlier work by Tannenbaum (1983), where data on precise stratigraphic and geographic positions of specimens within the quarry are also reported. Taxonomic identifications and systematic positions were further revised by the second author in the course of preparing this report, based on more recently published studies of Chinle vertebrates.

The preservation states of the specimens vary from fair to poor, and most specimens are worn or fragmentary. Many, especially bone fragments, show pitting and damaged surface textures, suggesting passage through a gut with strong digestive acids. Many specimens were recovered from coprolites, and virtually all were dissociated (Jacobs and Murry, 1980; W. R. Downs, pers. commun.). The most numerous and most readily identifiable elements are teeth, scales, vertebral centra and neural arches, proximal and distal ends of limb bones, and phalanges. Many of the limb bones bear unfused epiphyses, suggesting a high frequency of juveniles in the collections. This is not surprising in view of the tiny size of most of the microvertebrates (note scales on photographs). There are few identifiable and diagnostic cranial and limb girdle bones, which might be expected, given the delicacy of these elements, especially in juveniles.

In our taxonomic approach we have tried to follow cladistic methods as far as practicable. Wherever possible, we have tried to identify diagnostic characters (synapomorphies) visible in the microvertebrates that have been documented in the published literature to the restrictive taxonomic level. The hierarchical specificity of our identifications generally correlates with the availability of diagnostic characters. More general identifications reflect probable assignments based on the known presence of the taxon in other Late Triassic assemblages. In some cases, identifications reflect virtual guesswork based on overall form, but we classify these as such in order to stimulate further study and discussion. Most specimens in the MNA reference collection of some 200 specimen vials are described and figured in this chapter (Figures 9.1–9.128; see also Tannenbaum, 1983), but it should be remembered that although the reference specimens are fairly representative of the microvertebrate diversity, they represent only about 25 percent of the entire sample numerically.

Diversity of the nontetrapod microvertebrates

The focus of this volume is on Triassic and Jurassic fossils that pertain to small tetrapods, but the microfaunas often do not separate so easily. Many taxa are not clearly identifiable, and fish parts may be easily mistaken for those of reptiles (Murry and Long, 1989). Moreover, nontetrapod microvertebrates are the ones that contribute most significantly to the increase in known Chinle diversity, compared with macrofaunal diversity alone (Jacobs and Murry, 1980; Tannenbaum, 1983). And finally, knowledge of the entire microvertebrate fauna contributes to an understanding of the paleoecology and taphonomy of the original environment. Therefore, we shall summarize briefly the nontetrapod diversity of the Placerias and Downs quarries (Figures 9.1–9.47); details can be found in Tannenbaum (1983), and further information about the Chinle nontetrapod microfauna in Jacobs and Murry (1980) and Murry (1986, 1987, 1989).

As pointed out by Camp and Welles (1956), and supported by all later workers, the Placerias Quarry (and the Downs Quarry, which, although slightly higher in the section, samples the same depositional environment and fauna) (Jacobs and Murry, 1980; Tannenbaum, 1983) is the site of an ancient pond or marsh, an environment of low-energy water flow that may have been seasonally low or dry (as indicated by numerous evaporites). Tetrapods came to drink there, some lived there, and more than a few (including at least 39 individuals of *Placerias*) died there and were preserved. The nontetrapods included various sharks, actinopterygians, and at least one lungfish and one coelacanth. Only the lungfish was previously known from the macrofauna, so the other nontetrapods increase the known diversity by at least ten recognizable taxa, plus three possible new shark taxa and as many as 26 new osteichthyans (Table 9.1, *sed caveat lector*), though each new incertae sedis morph may not represent a new taxon.

Jacobs and Murry (1980) noted the presence of *Xenacanthus moorei* (the most common shark by far) and *Lissodus* ("*Lonchidion*") *humblei* (a hybodont), to which we add the possible presence of the ctenacanth shark *Phoebodus* (Figure 9.1) (Woodward, 1893), known elsewhere from the upper Keuper of Germany, and the hybodont shark *Acrodus* (Figure 9.6) (Murry, 1989). Two other micromorphs represent possible additional shark taxa; they include a tricuspid, shiny black tooth (Figure 9.4) and a worn tooth (Figure 9.5) superficially similar to the crushing teeth of the Cretaceous genus *Ptychodus* (Case, 1967).

Among the actinopterygians are various palaeonisciforms. Jacobs and Murry (1980) identified the redfieldiids *Cionichthys* and *Lasalichthys* or *Synorichthys*, all common components of Triassic freshwater faunas of the Chinle. In all probability these genera were present, but we could not diagnose them below the family level. The palaeoniscid *Turseodus* was reported by Jacobs and Murry (1980), and to this we add *Gyrolepis* (Figure 9.17), represented by a ventral trunk scale with a distinctive sculpturing pattern (Guerin, 1957), as well as a number of teeth, scales, and jaw fragments of palaeoniscids (Figure 9.18) (Schaeffer, 1952). The pholidopleurid *Australosomus* (Figure 9.15) (Schaeffer, 1967a) is a new record for the Placerias

Figures 9.1–9.6. Chondrichthyes. **Figures 9.7–9.22.** Osteichthyes.

Figure 9.1. *Phoebodus* sp., V3500, tricuspid tooth, ?labial view. "V" numbers denote specimens of the Museum of Northern Arizona, localities MNA 207–9 and 207–10 (equivalent to UCMP locality A269). For details and explanation of nontetrapod specimens (Figures 9.1–9.47), see Tannenbaum (1983). Scale bars in all figures equal 1 mm. **Figure 9.2.** *Xenacanthus moorei*, V3501, tricuspid tooth: (a) occlusal view, (b) basal view, (c) anterolateral view. **Figure 9.3.** *Lissodus* cf. *humblei*, V3525, partial dorsal fin spine, lateral view. **Figure 9.4.** Type A, V3526, tricuspid tooth, ?lingual view. **Figure 9.5.** Type B, V3542, skatelike crushing tooth: (a) occlusal view, (b) basal view. **Figure 9.6.** *Acrodus* sp., V3566, dense rectangular plates in matrix: (a) ?dorsal view, (b) side view. **Figure 9.7.** Ceratodontid, V3527, marginal tooth plate, occlusal view. **Figure 9.8.** Ceratodontid, V3528, marginal tooth plate, occlusal view. **Figure 9.9.** Ceratodontid, V3558, marginal tooth plate: (a) occlusal

Quarry microfauna, represented by distinct fused vertebral basiventral and interventral elements. The possible presence of colobodontids was noted by Jacobs and Murry (1980), and we confirm this (Figure 9.14), based on comparison to figures in Murry (1989). Among Semionotiformes, we identified several scales and teeth (Figures 9.19–9.22) of semionotids. The teeth are generally conical, with blunt opaque-to-translucent tips, and are often preserved in jaw fragments. The scales are smooth and rhombic, often with an anterodorsal extension for peg-and-socket articulations (Schaeffer and Dunkle, 1950).

In addition to the foregoing taxa, we separated a variety of micromorph types that appear to pertain to actinopterygians (Figures 9.23–9.47), but have escaped more precise identification. These include jaw fragments, often with teeth, isolated teeth and scales, spines, skull and dermal elements, erectile fin structures, vertebral fragments, and unidentifiable yet distinctive skeletal parts (Tannenbaum, 1983, pp. 28–39).

Aquatic sarcopterygians were previously known from ceratodontid lungfish tooth plates recovered from the macrofauna, and Jacobs and Murry (1980) reported their presence in the microfauna. Our reference specimens include several dozen tooth plates of assorted structure (V3527, V3528, V3558, V3531–3541; Figures 9.7–9.9), a vomerine tooth (V3529), and two jaw fragments (V3530). Ceratodontid tooth plates are highly characteristic (Jain, 1968; Martin et al., 1981). No vertebral centra were found, which is in accord with Schaeffer's (1967b, p. 193) observation that with one nonapplicable exception, "centra are absent in all known post-Devonian dipnoans." Dipnoan bones, including the nonenameloid areas on the tooth plates, display a distinctive pitted structure, and this trait was observed on most specimens. The relatively high abundance and morphologic diversity of the tooth plates suggest ontogenetic and taxonomic variation, and it is likely that several taxa, such as *Ceratodus* and *Arganodus*, are represented.

In addition to these fairly common dipnoans, we report for the first time the presence in this quarry of coelacanths, comparable to *Moenkopia* (Schaeffer and Gregory, 1961) and *Diplurus* (Schaeffer, 1952). The group is represented by several scale fragments (V3591; Figure 9.10), a cleithrum (V3604; Figure 9.11), and a right quadrate with a small piece of the pterygoid

attached (V3577; Figure 9.12). The quadrate appears to be the smallest known for the Chinle Formation. The scale fragments are more or less rounded in outline, with various numbers of thin, elongate, translucent ridges on the non-embedded scale surface, and very fine striae on the presumably embedded parts of the scale (Schaeffer, 1952, pp. 51–52). Some of the ridges are parallel, and others are varied in their orientations. It is not possible to determine from which part of the trunk the scales are derived, because they are too fragmentary.

Diversity of the small tetrapods

As noted earlier, nearly all the macrovertebrate remains from the Placerias Quarry are from tetrapods, with only occasional coelacanth teeth representing fish. The microvertebrates are about evenly divided between tetrapods and nontetrapods, but within tetrapods the microvertebrate diversity differs substantially from that of the macrovertebrates. Before discussing proportional representation, however, we shall describe the tetrapod taxa recognized from the quarry, as well as a range of micromorphs that are more difficult to identify to taxonomic level or anatomical provenience.

Systematic paleontology

Monophyletic hierarchy: Amphibia: Temnospondyli: Stereospondyli: Metoposauridae

Reference specimens. V3613, atlas; V3614, atlas or dorsal centrum (Figure 9.48); V3615, left clavicle (Figure 9.49); V3616, tooth (Figure 9.50); V3617, tooth. The reference specimens representing the Metoposauridae include two opisthocoelous centra, one left clavicle, and two isolated teeth. The two centra, V3613 and V3614, are circular in outline and have distinct notochordal pits on their anterior faces. Neither centrum has the neural arch preserved. The centra are notable because most centra belonging to temnospondyls are amphicoelous, but the two specimens described here are opisthocoelous. It is possible that at least one of the centra is part of the axis, as described by Branson (1905), or from the anterior thoracic series (Branson and Mehl, 1929). Sawin (1945; p. 385)

Figures (*Continued*)

view, (b) basal view. **Figure 9.10.** Coelacanthid, V3591, scale fragments, external view. **Figure 9.11.** Coelacanthid, V3604, cleithrum: (a) ?lateral view, (b) ?medial view. **Figure 9.12.** Coelacanthid, V3577, right quadrate, posteromedial view. **Figure 9.13.** Redfieldiid, V3543, dermal bone with enameloid implanted plates. **Figure 9.14.** Colobodontid, V3544, rostral bone with tubercles: (a) dorsal view, (b) basal view. **Figure 9.15.** cf. *Australosomus* sp., V3597, fused basiventral and interventral, anterior view. **Figure 9.16.** cf. *Australosomus* sp., V3600, same as Figure 9.15. **Figure 9.17.** *Gyrolepis* sp., V3567, ventral trunk scale, external view. **Figure 9.18.** Palaeoniscid incertae sedis, V3545, jaw fragment, lingual view. **Figure 9.19.** Semionotid, V3554, two teeth, side view. **Figure 9.20.** Semionotid, V3555, scale, internal view. **Figure 9.21.** Semionotid, V3556, three scales, external view. **Figure 9.22.** Semionotid, V3564, jaw fragment, ?labial view.

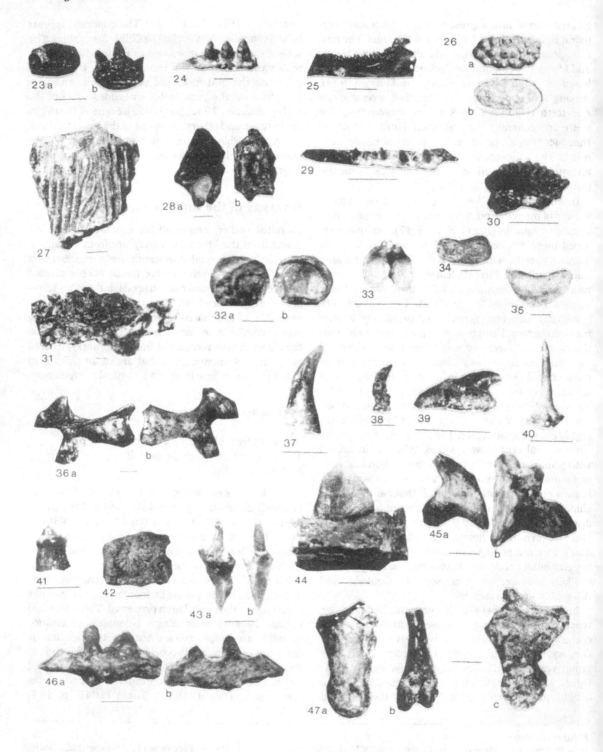

Figures 9.23–9.47. Osteichthyes.
Figure 9.23. Type A, V3565, tooth-bearing (possibly palatal) element: (a) occlusal view, (b) basal view. **Figure 9.24.** Type B, V3559, jaw fragment with teeth, ?lingual view. **Figure 9.25.** Type C, V3576, tooth-bearing ectopterygoid, ? lingual view. **Figure 9.26.** Type D, V3579, pustulate ovoid dermal element: (a) external view, (b) internal view. **Figure 9.27.** Type E, V3582, dermal ?cranial element, external view. **Figure 9.28.** Type F, V3589, proximal fin spine: (a) ?posterior view, (b) ?anterior view. **Figure 9.29.** Type G, V3586, jaw element, occlusal view. **Figure 9.30.** Type H, V3587, palatal or vomerine element, side view. **Figure 9.31.** Type I, V3588, jaw fragment, lingual view. **Figure 9.32.** Type J, V3602, circular dermal ?scale or armor: (a) external view, (b) internal view. **Figure 9.33.** Type K, V3603, fin erectile element, proximal view. **Figure 9.34.** Type L, V3598, intercentrum, dorsal view. **Figure 9.35.** Type M, V3599, intercentrum, side view.

noted that "various centra in association with ilia are opisthocoelous with the outline of the intercentrum changing from a bilaterally compressed circle to a roughly rectangular form as seen in outline." Therefore, V3614 may be from a proximal caudal, with its "slightly rectangular outline," and V3613 may represent a more anterior centrum, but it is not possible to determine the precise vertebral position.

The left clavicle, V3615, is highly sculptured and roughly triangular in outline. It is approximately 12.25 mm in length along the lateral scapular ridge, and approximately 7.0 mm in length from the scapular ridge to the mesial part of the element. Sawin's (1945, pp. 387–388) description of the clavicle of *Buettneria howardensis* is applicable, though not diagnostic of the taxon. A thickened ridge providing a greater area for attachment to the interclavicle corresponds to the indentation of the clavicular border on the interclavicles (Case, 1932).

The two isolated teeth appear to be labyrinthodont on the basis of gross microscopic examination. The external enamel surfaces are radially striated, and each tooth is asymmetrically curved from base to tip. V3616 is 3.5 mm from base to tip, and V3617 is 4.25 mm from base to broken tip.

Amphibia incertae sedis, Type A

Reference specimens. V3593, assorted centra (Figure 9.51); V3594, single centrum. V3593 and V3594 include numerous amphicoelous pleurocentra. Each specimen consists of an asymmetrical ring of bone around a large notochordal opening. The heaviest concentration of bone is present ventrally, and the notochordal opening is open on its dorsal border. The texture of the bone varies from relatively dense and solid to pitted and porous, the latter perhaps from the effects of digestive acids dissolving the cartilaginous sheath and penetrating the underlying bone. The centra appear quadrangular in outline and vary in size from about 1.0 to 3.5 mm in height and 1.0 to 2.5 mm in anteroposterior length. None of the specimens bears transverse or neural processes.

Type B

Reference specimens. V3595 (Figure 9.52) includes three quadrangular vertebral centra similar to those of V3593 and V3594, except that a ring of bone completely and symmetrically surrounds the notochordal opening. These specimens also bear lateral grooves, ostensibly for the attachment of lateral processes. The specimens are pitted; they vary in height from about 2.0 to 4.5 mm and vary in anteroposterior length from 2.0 to 3.0 mm. These specimens are more suggestive of amphibians than of fishes because of the relatively dense texture of the bone and their definite quadrangular outline, which is more typical of the temnospondyls than of the Late Triassic fishes.

Type C

Reference specimen. V3596 (Figure 9.53) is a single, subcircular amphicoelous centrum, symmetrically constructed around a large notochordal opening. The bone appears dense, but there are no distinct lateral structures visible, again perhaps because the bone has been partially digested by a predator. The specimen is aproximately 3.75 mm in height and about 2.50 mm in anteroposterior length.

Type D

Reference specimen. V3667 (Figure 9.54) is a single large labyrinthodont tooth, approximately 11.0 mm in height and 4.5 mm in maximum width across its base. There are two distinct areas of the tooth: the bottom third, which has strong lateral ridges externally, and which in basal view displays a prominent infolding of the enamel, and the upper two-thirds of the tooth (above the jawline in life), where the enamel is completely smooth. The tooth is suggestive of a temnospondyl amphibian rather than a fish, because fishes, such as *Megalichthys*, with this tooth type are otherwise unknown from the fauna of the Placerias Quarry and contemporaneous faunas, whereas temnospondyls are common.

Monophyletic hierarchy: Amniota: Reptilia: Diapsida: Lepidosauria: Sphenodontia

Reference specimens. V3605, single jaw fragment; V3606, three jaw fragments (Figure 9.55); V3607, single jaw fragment; V3608, three jaw fragments; V3609, single jaw fragment (Figure 9.56); V3610, several jaw fragments; V3612, numerous jaw frag-

---

Figures (*Continued*)
Figure 9.36. Type N, V3592, unidentified element: (a) side view, (b) opposite side view. **Figure 9.37.** Type P, V3574, fish spine, side view. **Figure 9.38.** Type Q, V3571, ?denticulated fish spine, side view. **Figure 9.39.** Type R, V3572, dermal denticle, dorsolateral view. **Figure 9.40.** Type S, V3573, ?fish spine, side view. **Figure. 9.41.** Type T, V3568, tooth with lingual and labial collar, side view. **Figure 9.42.** Type U, V3569, unidentified dermal element. **Figure 9.43.** Type V, V3570, tooth with long, laterally compressed root: (a) lingual view (note apparent pleurodont implantation), (b) labial view. **Figure 9.44.** Type W, V3560, asymmetrical, laterally compressed tooth and jaw fragment, ?labial view. **Figure 9.45.** Type X, V3575, distal quadrate: (a) anterior view, (b) posterior view. **Figure 9.46.** Type Y, V3561, jawlike fragment: (a) ?labial view, (b) ?lingual view. **Figure 9.47.** Type Z, V3708, median element, possibly fin erectile element: (a) side view, (b) posterior view, (c) V3710, side view (note somewhat expanded, laterally compressed proximal "plate").

Table 9.1. *Diversities of Placerias/Downs Quarry macrovertebrates and microvertebrates*

Taxon	Macro	Micro
Vertebrata: Gnathostomata		
Chondrichthyes		
Ctenacanthidae: *Phoebodus* sp. +		1
Xenacanthidae: *Xenacanthus moorei*		24
Hybodontidae: *Lissodus (Lonchidion) humblei*		1
Hybodontidae: *Acrodus* sp. +		4
Incertae sedis: Type A		1
Type B		1
Osteichthyes		
Actinopterygii		
Redfieldiidae		
Cionichthys sp.	b	
Lasalichthys sp./*Synorichthys* sp.	b	
Palaeoniscidae		
Gyrolepis sp. +		1
cf. *Turseodus* sp.		1
Pholidopleuridae		
cf. *Australosomus* sp. +		2
Colobodontidae	b	
Semionotidae		4
Incertae sedis: Types A–Z		35
Sarcopterygii		
Dipnoi: Ceratodontidae	b	16
Crossopterygii: Coelacanthidae +		3
Amphibia: Temnospondyli		5
Metoposaurus sp.	a, b	
"*Anaschisma*" sp.	a, c	
Incertae sedis: Type A		2
Type B		1
Type C		1
Type D		1
Amniota		
Synapsida: Therapsida		
Cynodontia +		9
Dicynodontia: *Placerias gigas*	a	
Reptilia: Parareptilia: Procolophonidae	b, d	
Reptilia: Diapsida		
Lepidosauromorpha		
?Kuehneosauridae	c	
Sphenodontidae		8
?Lepidosauromorpha incertae sedis	c, e	
Archosauromorpha		
Prolacertiformes: *Tanytrachelos* sp.	b	1
Trilophosauridae: *Trilophosaurus jacobsi*	c	
Trilophosaurus sp.	b	1
?Proterochampsid	d	
Archosauria: Pseudosuchia		
Phytosauridae: *Rutiodon* sp.	a	1
Paleorhinus sp.	d	
Stagonolepididae		
Stagonolepis ("Calyptosuchus") wellesi	a, d, g	

Table 9.1. (*Continued*)

Taxon	Macro	Micro
Desmatosuchus haplocerus	a	
Desmatosuchus sp.	a	3
"Acaenasuchus geoffreyi"	f	2
Poposauridae: Postosuchus kirkpatricki	a	
Poposaurus gracilis	a	
"Chatterjeea elegans"	f	?1
Crocodylomorpha: Hesperosuchus agilis	b	
"large sphenosuchian"	d	
Archosauria: Dinosauria		
"Chindesaurus bryansmalli"	f	
Ornithischia: Revueltosaurus callenderi +		1
Ornithischia incertae sedis		1
Saurischia		
Sauropodomorpha: ?Anchisauridae	b, d	1
Theropoda: ?Coelophysis sp.	b, d	?
Archosauria: Incertae sedis Type A		2
Type B		1
Type C		1
Type D		1
Type E		1
Reptilia incertae sedis: cf. Uatchitodon +		1
Other Reptilia indet.		55

Note: Taxa are listed according to monophyletic hierarchies adapted from Gauthier (1986), Benton (1988), and other sources. The presence of a taxon in the macrofauna (Macro) is given with a cited reference; its presence in the microfauna (Micro) is given along with the number of MNA reference-collection vials segregated in this study from the entire collection, or with cited references of other authors if our study did not confirm the presence of a given taxon. Note that a single reference-collection vial may contain from one to 50 specimens, and there may be more examples in the full MNA Placerias/Downs microvertebrate collection; therefore, analyses of relative diversity and abundance should not be undertaken from this table alone. For more information about the reference-specimen collection and horizontal and vertical placement of specimens in the quarries, see Tannenbaum (1983). A plus sign indicates taxa reported here for the first time in the Placerias microfauna.

[a]Camp and Welles (1956).

[b]Jacobs and Murry (1980).

[c]Murry (1987).

[d]Murry and Long (1989).

[e]Murry (1987, p. 783, fig. 9) astutely compared a jaw fragment with teeth to *Fulengia*, once generally thought to be a lepidosauromorph; however, Evans and Milner (1989) showed that *Fulengia* was probably a juvenile sauropodomorph, and as Murry (1987) notes, the tooth implantation differs between *Fulengia* and the Placerias Quarry specimen. Murry also suggested the presence of "eolacertilians" or "eosuchians" based on the presence of tooth types with subpleurodont and acrodont implantation similar to those of *Daedalosaurus* and *Coelurosauravus*, as described by Carroll (1978). However, the structures first identified by Carroll as acrodont dentition now seem to be from the fringed edge of the supratemporals in *Coelurosauravus* (Evans and Haubold, 1987), and as yet no independent evidence of these Permo-Triassic diapsid groups has been reported in the Chinle Formation. Nevertheless, Murry is quite probably correct that other lepidosauromorphs besides sphenodontians, which he first recognized in 1987, are present in the Placerias Quarry.

[f]*Nomina nuda* of Murry and Long (1989). These taxa have not yet been published or diagnosed, but it is possible that some of the intended names will match some of the unnamed "incertae sedis types" that we list and describe.

[g]Murry and Long (1989) synonymized without explanation *Calyptosuchus* Long and Ballew, 1985, with *Stagonolepis* Agassiz, 1844, and apparently referred to it the species *C. wellesi*.

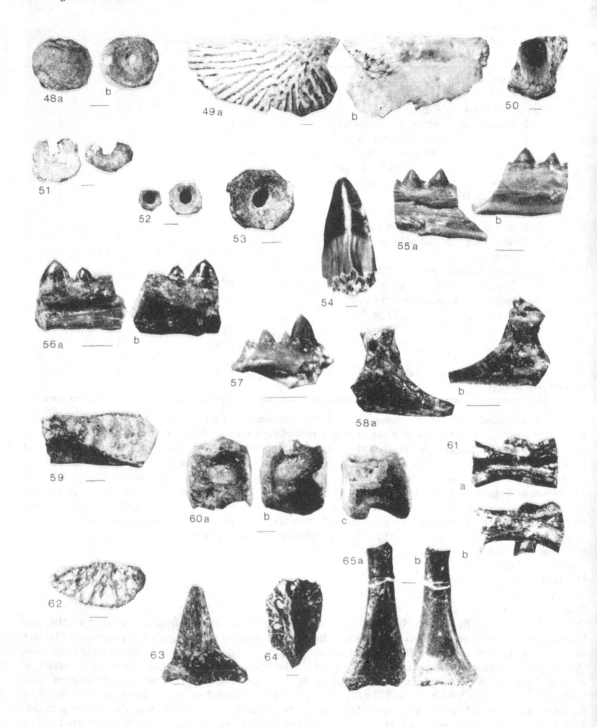

Figures 9.48–9.54. Amphibia. **Figures 9.55–9.65.** Reptilia.
Figure 9.48. Metoposaurid, V3614, atlas or dorsal centrum: (a) anterior view, (b) posterior view.
Figure 9.49. Metoposaurid, V3615, left clavicle: (a) ventral view, (b) dorsal view. **Figure 9.50.** Metoposaurid, V3616.
tooth, lingual view. **Figure 9.51.** Type A, V3593, two centra, anterior or posterior views. **Figure 9.52.** Type B,
V3595, two centra, anterior or posterior views. **Figure 9.53.** Type C, V3596, centrum, anterior or posterior view.
Figure 9.54. Type D, V3687, labyrinthodont tooth, side view. **Figure 9.55.** Sphenodontid, V3606, jaw fragment with
teeth: (a) lingual view, (b) labial view. **Figure 9.56.** Sphenodontid, V3609, jaw fragment with teeth: (a) lingual view,
(b) labial view. **Figure 9.57.** Sphenodontid, V3611, jaw fragment with teeth, labial view. **Figure 9.58.** *Tanytrachelos* sp.,
V3627, plowshare-shaped posterior cervical rib: (a) lateral view, (b) medial view. **Figure 9.59.** Trilophosaurid,
V3730, jaw fragment, occlusal view. **Figure 9.60.** *Rutiodon* sp., V3626, midcervical centrum: (a) dorsal view, (b) ventral
view, (c) lateral view. **Figure 9.61.** cf. *Desmatosuchus* sp., V3667, anterior caudal centrum: (a) dorsal view, (b) ventral

ments; V3611, single jaw fragment (Figure 9.57). All the reference-collection specimens assigned to the Sphenodontia are jaw fragments containing small, laterally compressed, triangular teeth with acrodont implantation. Most of the specimens have been compared to specimens of *Clevosaurus* (Robinson, 1973), *Planocephalosaurus* (Fraser, 1982, 1986, 1988), and UCMP specimens V6844/82468 and V6844/82469, to which the MNA specimens appear quite similar. The MNA specimens generally conform to the diagnosis of sphenodontians given by Robinson (1973). She noted, however, that in *Clevosaurus* the anterior maxillary and dentary teeth tend to alternate in size, which is not observable in most of the rather fragmentary material in the MNA collection. Robinson (1973, p. 469) also noted that posterior to the anterior maxillary and dentary teeth, there are "four very much larger robust conical acrodont teeth with broad bases ... each bear[ing] a relatively large flat flange posteriorly." However, none of these flanged teeth have been seen on the MNA specimens. This absence does not mean that the MNA specimens are not sphenodontian, because Robinson (1973) also noted that "there may be as many as three smaller unflanged conical teeth behind the flanged series" on the maxilla, which may be seen in V3611. The overall appearance of the jaw fragments and teeth, with their triangular, laterally compressed teeth and acrodont implantation, is highly suggestive of sphenodontians. In particular, the MNA specimens are reminiscent of *Planocephalosaurus*, because the teeth are simpler and more laterally compressed than in other contemporaneous forms (Fraser, 1986, 1988). Murry (1987) has identified numerous sphenodontians from the Placerias Quarry microfauna.

Monophyletic hierarchy: Diapsida: Archosauromorpha: Prolacertiformes: Tanystropheidae

Tanytrachelos sp.

Reference specimen. V3627 (Figure 9.58) is a single plowshare-shaped posterior cervical rib and is one of the diagnostic elements of *Tanytrachelos*, according to Olsen (1979). It is approximately 4 mm in maximum anteroposterior length along the ventral blade, and about 3 mm in maximum dorsoventral height. The ventral blade is thin, gradually thickening dorsally toward the ascending shaft. The blade is slightly concave medially. The shaft has double articular processes dorsally, with paired nutrient foramina located on the posterior end of the dorsal shaft.

Monophyletic hierarchy: Archosauromorpha: Trilophosauria

Trilophosaurus sp.

Reference specimen. V3730 (Figure 9.59) is a possible trilophosaurid jaw fragment identified by indistinct traces of possibly three or four transversely elongated tricuspid teeth. The fragment is about 6.0 mm in anteroposterior length and 3.0 mm in mediolateral width. The presence of trilophosaurids would not be unexpected in the collection, as they have been found in the Dockum Formation of western Texas in beds of supposed Chinle age (Case, 1928; Gregory, 1945), and Murry (1987) has already described them from the Placerias Quarry. However, they do not appear in the macrofauna. Consequently the identification is provisional, and as Fraser (1986, 1988) notes, laterally broadened tricuspid teeth could equally pertain to procolophonids, sphenodontians, or other unknown forms, depending on the mode of tooth implantation and other jaw features.

Monophyletic hierarchy: Archosauria: Pseudosuchia: Phytosauridae

cf. *Rutiodon* sp.

Reference specimen. V3626 (Figure 9.60) is a midcervical centrum, similar in appearance to those of *Rutiodon* (Camp, 1930). It is amphicoelous and has a notochordal pit. The specimen has distinctive parapophyseal facets located laterally, and articulations for cervical ribs located ventrally. The centrum has a hexagonal appearance when viewed anteriorly, and a heart-shaped appearance when viewed posteriorly. There are no attached neural processes. The centrum is approximately 3 mm in height and about 4 mm in width across the dorsal surface, including the transverse process.

Monophyletic hierarchy: Pseudosuchia: Stagonolepididae

Desmatosuchus sp.

Reference specimens. V3667, anterior caudal centrum (Figure 9.61); V3633, lateral dermal plate fragment (Figure 9.62). Long and Ballew (1985) provide

---

Figures (*Continued*)
view. **Figure 9.62.** cf. *Desmatosuchus* sp., V3633, lateral dermal plate fragment, caudal region, external view.
Figure 9.63. Stagonolepidid, type A, V3631, lateral "horned" dermal plate, external view. **Figure 9.64.** Stagonolepidid, type A, V3632, paramedial dermal plate, external view. **Figure 9.65.** *Postosuchus* Chatterjee, 1985, V3719, distal humerus: (a) anterior view, (b) posterior view.

good comparative figures of stagonolepidid dermal armor; the genus is already well represented in the macrofauna (Camp and Welles, 1956).

Type A

Reference specimens. V3631, lateral "horned" dermal plate (Figure 9.63); V3632, paramedial dermal plate (Figure 9.64). This aetosaurian scute type is identified by distinctive dermal armor with diagnostic sculpturing (Walker, 1961; Long and Ballew, 1985). Some small stagonolepidid specimens pertain to an as yet unconstituted genus and species ("*Acaenasuchus geoffreyi* Long") alluded to by Murry and Long (1989; P.A. Murry and R.A. Long, unpublished data).

Monophyletic hierarchy: Pseudosuchia: Poposauridae

Postosuchus sp.

Reference specimen. V3719 (Figure 9.65) is the distal part of a humerus with part of the shaft intact. The medial and lateral condyles are relatively widely spaced and flattened anteroposteriorly. The texture of the bone at the condyles is spongy. The humerus lacks a well-defined epiphysis, suggesting that the specimen is probably juvenile. The broken specimen is approximately 6.75 mm in proximodistal length.

Archosauria incertae sedis, Type A

Reference specimens. V3669, single anterior dorsal centrum (Figure 9.66); V3670, single dorsal centrum. V3669 and V3670 are both slightly amphicoelous, nonnotochordal dorsal centra, referred to a new, as yet undescribed archosaur (P. A. Murry and R. A. Long, unpublished data). Dorsally, each centrum has a wide, shallow groove with well-developed dorsolateral shelf-like transverse processes. The ventral surface is rounded and pinched at midlength. The anterior and posterior faces have a heart-shaped outline. The larger specimen, V3669, is about 11 mm in anteroposterior length, and the smaller specimen, V3670, is about 8.5 mm in anteroposterior length.

Type B

Reference specimen. V3634 (Figure 9.67) is a right dorsal presacral dermal plate, about 3.25 mm in length. Its shape and ornamentation do not show affinities to either phytosaurs or aetosaurs, but the specimen is presumed archosaurian because dermal plates are not known in nonarchosaurian reptiles of this period.

Type C

Reference specimen. V3635 (Figure 9.68) is a fragment of dermal armor with a recurved clawlike process. It also does not show affinities to phytosaurs or aetosaurs.

Type D

Reference specimen. V3711 (Figure 9.69) is a proximal coracoid fragment, presumed archosaurian by its elongated shape.

Type E

Reference specimen. V3689 (Figure 9.70). Included here are five generalized archosaurian teeth, serrated anteriorly and posteriorly. The largest tooth is about 18.0 mm in height, and the smallest is about 4.0 mm in height. Such teeth are common in archosauriforms and vary taxonomically, ontogenetically, and according to their placement in the jaw in some taxa, so finer taxonomic resolution is extremely difficult.

Monophyletic hierarchy: Archosauria: Dinosauria: Ornithischia

Revueltosaurus sp.

Reference specimen. V3690 (Figure 9.71) is a single tooth, which at first Tannenbaum (1983) tentatively referred to the Prosauropoda; see Galton (1973) and Galton and Cluver (1976) for comparison. It is laterally flattened on one side and swollen on the other side, with large serrations on the anterior and posterior cutting edges. The swollen side of the tooth has a shallow ridge located medially, which fades out at both the anterior and posterior ends of the tooth. The complete tooth is approximately 1.0 mm in exposed height. In these respects the tooth matches the ornithischian form genus *Revueltosaurus* (Hunt, 1989; Padian, 1990), and given the polymorphism of the form genus reported by those authors, other teeth in this collection (e.g., Figures 9.104 and 9.108) may pertain to it too. However, as Padian (1990) and Sereno (1991) point out, caution is warranted in the taxonomic assignment of isolated teeth with multicusped or serrated denticular forms, because these have so often evolved among reptiles.

Monophyletic hierarchy: Amniota: Synapsida: Therapsida: Cynodontia

Reference specimens. V3618, left ischium (Figure 9.72); V3619, left ischium fragment (Figure 9.73); V3620, single tooth (tricuspid) (Figure 9.74); V3621, single tooth (tricuspid); V3622, three teeth (tricuspid); V3623, single tooth (tricuspid); V3624, numerous teeth (tricuspid) (Figure 9.75); V3563, single tooth (bicuspid); V3625, single canine tooth. The presence of cynodonts in the MNA collection is most definitively shown by the two ischia, V3618 and V3619. These specimens, especially V3618, which is almost complete,

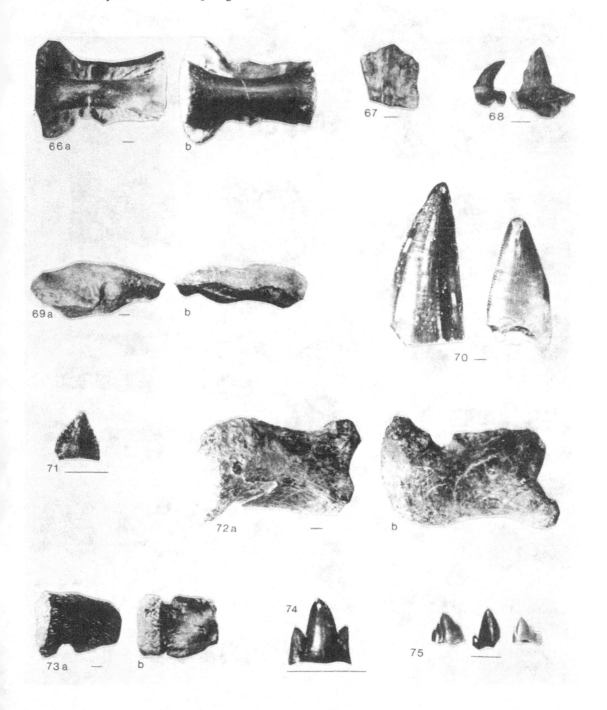

Figures 9.66–9.71. Archosauria. **Figures 9.72–9.75.** Cynodontia.
Figure 9.66. Type A, V3669, anterior dorsal centrum: (a) dorsal view, (b) ventral view. **Figure 9.67.** Type B, V3634, right dorsal presacral dermal plate, external view. **Figure 9.68.** Type C, V3635, two "horned" dermal plates, external view.
Figure 9.69. Type D, V3711, proximal portion of coracoid: (a) anterior view, (b) proximal view. **Figure 9.70.** Type E, V3689, two generalized anteriorly and posteriorly serrated archosaurian teeth. **Figure 9.71.** *Revueltosaurus* sp., V3690, tooth, labial or lingual view. **Figure 9.72.** Cynodont, V3618, ischium: (a) lateral view, (b) medial view.
Figure 9.73. Cynodont, V3619, anterior ischium fragment: (a) lateral view, (b) medial view. **Figure 9.74.** Cynodont, V3620, postcanine tooth, ?labial view. **Figure 9.75.** Cynodont, V3624, three postcanine teeth, ?labial view.

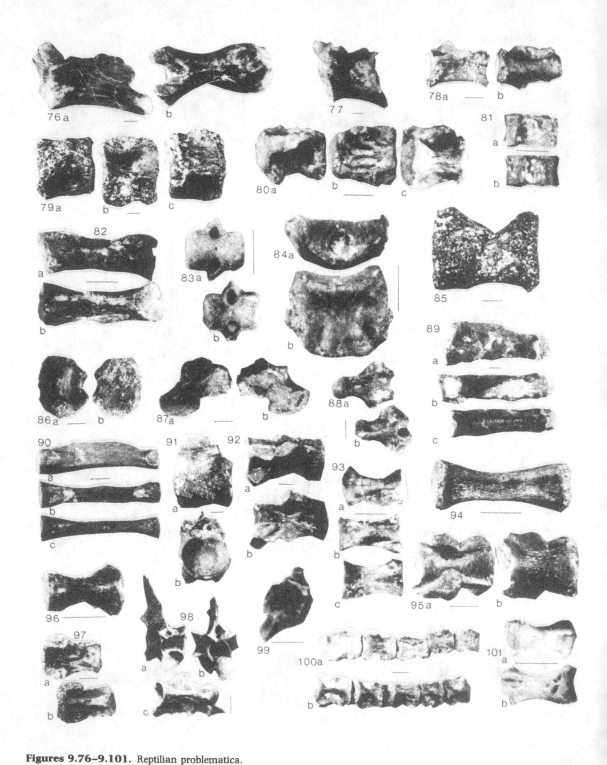

Figures 9.76–9.101. Reptilian problematica.
Figure 9.76. V3637, procoelous posterior caudal centrum: (a) lateral view, (b) dorsal view. **Figure 9.77.** V3638, amphicoelous sacral centrum, lateral view. **Figure 9.78.** V3639, amphicoelous caudal centrum: (a) lateral view, (b) ventral view. **Figure 9.79.** V3640, amphicoelous, possibly cervical centrum: (a) lateral view, (b) dorsal view, (c) ventral view. **Figure 9.80.** V3641, amphicoelous, possibly posterior cervical centrum: (a) lateral view, (b) dorsal view, (c) ventral view. **Figure 9.81.** V3643, ?intercentrum: (a) side view, (b) opposite side view. **Figure 9.82.** V3645, caudal centrum: (a) lateral view, (b) dorsal view. **Figure 9.83.** V3646, possible vertebral element: (a) side view, (b) opposite side view. **Figure 9.84.** V3647, amphicoelous, dorsoventrally flattened centrum: (a) anterior view, (b) ventral view. **Figure 9.85.** V3648, worn, pitted centrum, ventral view. **Figure 9.86.** V3650, ?intercentrum: (a) side view, (b) opposite side view. **Figure 9.87.** V3651, dorsoventrally flattened procoelous centrum: (a) dorsal view, (b) ventral view.

correspond very closely to drawings and descriptions of the postcranial skeleton of *Thrinaxodon* and other South African cynodonts given by Jenkins (1971). V3618 is approximately 14.75 mm in anteroposterior length, and about 9 mm in greatest length across the head of the ischium. This almost complete specimen appears to be missing the ventromedially directed plate. Jenkins (1971) noted, however, that an entire ischial plate has yet to be discovered for cynodonts, which is understandable given how thin the bone appears to have been in that area of the ischium. Medially, the iliac facet on the proximal end of the ischium V3618 may be seen, and there is a relatively long posterodorsal ischial tuberosity. Immediately anterior to the broken edge where the ischial plate would have been, there is a pronounced shallow groove that faces posterolaterally. Laterally, a ridge extends from the middle of the lateral acetabular rim to the posterodorsal corner of the broken ischial plate, becoming less pronounced posteriorly. An acetabular facet is present on the head of the ischium for articulation with the femur, as are symphyseal surfaces for the pubis and ischium (Jenkins, 1971).

V3619 is also a left ischium, although it is only a small fragment representing the head of the ischium with the shallow posterolaterally directed groove. This specimen, if complete, would have been approximately one-half the size of V3618. The acetabular facet and symphyseal surfaces are not distinct on this specimen. Most of the teeth in the reference collection referred to the Cynodontia appear to be postcanine teeth. These have three anteroposteriorly linearly arranged cusps, with the central cusp higher by varying degrees than the two lateral cusps. There is a general tendency for the middle cusp of the more anterior postcanine teeth to be relatively higher and narrower than the more posterior postcanines in adults, though in juveniles (e.g., V3563) the morphology of the postcanines may include unicuspid and bicuspid teeth (Estes, 1961). Comparison with an adult specimen of *Thrinaxodon liorhinus* (UCMP V36115/40466) reveals that the teeth from the MNA collection are more laterally compressed and smaller. V3622 might be referable to *Pachygenelus*, because it appears to have an additional small lateral fourth cusp. An examination of Watson's

(1913) original description of *Pachygenelus*, however, suggests that this reference is not correct. The fifth molar of *Pachygenelus* does have an additional small fourth cusp, but V3622 lacks the prominent cingulum in both drawings and description and also appears much more laterally compressed than the fifth molar of *Pachygenelus*. Only one possible canine (V3625) has been identified out of scores of cynodont teeth. Fourie (1974, p. 394), however, noted that "during replacement of the canines the crown of the functional canine is shed, but . . . the root is gradually resorbed." One would therefore not expect to see rooted canines, which may make it difficult to distinguish the caniniform teeth of cynodonts from those of reptiles. Also lacking in the collection are any recognizable "simple conical" or gomphodont teeth described by Hopson (1971) for juvenile as well as adult specimens of *Diademodon* sp., nor do any of the ostensibly cynodont teeth in the MNA collection have roots. These absences are puzzling, unless the teeth described as belonging to cynodonts in fact belong to another taxon or are the shed crowns of juveniles. The specimen pictured in Figure 9.74 resembles the upper postcanine teeth of *Pachygenelus monus* (Gow, 1980), though without a well-defined external cingulum (Shubin et al., 1991, fig. 1B); the specimens in Figure 9.75 are more similar to the lower postcanine teeth of *Pachygenelus monus* (Gow, 1980). Both Pterosauria and *Tanystropheus* have heterodont dentitions that include tricuspid teeth, but such teeth have not yet been unambiguously identified at either the Placerias Quarry or Downs Quarry.

The apparent lack of cynodont vertebrae is perhaps surprising, but comparative material and adequate reference drawings are scarce, and the specimens under study are dissociated, so it is difficult to distinguish cynodont from reptilian centra using clearly diagnostic characters.

Problematica

The remainder of the descriptive section of this chapter is devoted to specimens that appear to be reptilian but cannot be adequately diagnosed. Some of these specimens are recognizable skeletal elements, but others

---

Figures (*Continued*)
Figure 9.88. V3652, procoelous centrum with flaring transverse processes: (a) dorsal view, (b) ventral view. **Figure 9.89.** V3653, procoelous posterior caudal centrum: (a) side view, (b) dorsal view, (c) ventral view. **Figure 9.90.** V3654, procoelous posterior caudal centrum: (a) side view, (b) dorsal view, (c) ventral view. **Figure 9.91.** V3655, anterior third of caudal centrum: (a) side view, (b) anterior view. **Figure 9.92.** V3656, anterior caudal centrum: (a) side view, (b) dorsal view. **Figure 9.93.** V3659, posterior caudal centrum: (a) side view, (b) dorsal view, (c) ventral view. **Figure 9.94.** V3660, caudal centrum, ventral view. **Figure 9.95.** V3661, caudal centrum: (a) dorsal view, (b) ventral view. **Figure 9.96.** V3662, caudal centrum, ventral view. **Figure 9.97.** V3663, two caudal centra: (a) dorsal view, (b) ventral view. **Figure 9.98.** V3671, anterior caudal vertebra: (a) anterodorsal view, (b) posterior view, (c) ventral view. **Figure 9.99.** V3676, possible sacral centrum, dorsal view. **Figure 9.100.** V3677, five caudals similar to that in Figure 9.99: (a) dorsal view, (b) ventral view. **Figure 9.101.** V3678, possible caudal centrum: (a) dorsal view, (b) ventral view.

are not. The figures must therefore supplement the necessarily brief verbal descriptions.

Reptilian vertebral elements. The following probable reptilian vertebral elements are classified, where possible on the basis of their locations in the vertebral column, but their taxonomic ranks are uncertain. These distinctive structures are presented here primarily to document the existence of unusual and previously unrecognized types.

V3637 (Figure 9.76): A procoelous posterior caudal centrum with a thin ridge on the ventral surface.

V3638 (Figure 9.77): An amphicoelous sacral centrum with a broken sacral rib and a smooth ventral surface.

V3639 (Figure 9.78): An amphicoelous caudal centrum with broken transverse processes. The centrum is deeply notochordal, with a thickened border around the notochordal pit. It is pinched ventrally and laterally and has a narrow but flat ventral surface.

V3640 (Figure 9.79): This is a slightly amphicoelous, nonnotochordal, possible cervical centrum. It has a deep pit on the dorsal surface; its ventral surface has a narrow ridge that does not extend to the anterior and posterior borders, which have a beveled appearance.

V3641 (Figure 9.80): This is also a slightly amphicoelous, nonnotochordal centrum, possibly from a posterior cervical. Its dorsal surface has two deep oval pits separated by a thin ridge; its ventral surface is pinched along the midline, with a weak thin ridge extending across its surface. The anterior face of the centrum is an ovoid polygon; the posterior face is relatively larger and heart-shaped. The anterior face has facets, possibly for rib attachment, extending obliquely upward from the ventral surface.

V3642: Another slightly amphicoelous, notochordal cervical centrum superficially similar in appearance to V3641. Dorsally there is a shallow ovoid pit, further divided into two separate pits by a deeply set ridge. The ventral surface is beveled, with a prominent anteroposterior ridge extending the length of the centrum. Anterior and posterior faces of the centrum are polygonally heart-shaped in outline; parapophyseal and diapophyseal facets are only faintly visible.

V3643 (Figure 9.81): A small, saddle-shaped possible intercentrum, with an approximate figure-eight outline in side view. It is more concave on one surface than the other and has no obvious articular surfaces other than concavities.

V3644: A slightly amphicoelous caudal, missing its posterior third. The anterior face is rectangular in outline and nonnotochordal. The ventral surface has a moderately expanded keel extending the anteroposterior length of the specimen. No transverse processes are present.

V3645 (Figure 9.82): These are two amphicoelous caudal centra, one complete, one broken. They are nonnotochordal, but their anterior and posterior faces have shallow depressions. The ventral surface has a shallow anteroposterior groove; the dorsal surface has a narrow hourglass appearance. No transverse processes are present.

V3646 (Figure 9.83): A small, possible vertebral element, this specimen has an hourglass outline with handlelike processes connecting the two attached mirror-image halves of the "hourglass." These halves are distinctly convex on one surface and concave on the other.

V3647 (Figure 9.84): The anterior and posterior faces of these amphicoelous, notochordal centra have a definite half-moon outline. Dorsally the centra are smooth and flat, with upward-curving transverse processes fused dorsolaterally. Ventrally there is a faint, broad, flattened ridge extending anteroposteriorly.

V3648 (Figure 9.85): This is a very worn and pitted centrum, pinched at midlength. It seems to be a good example of the digestive effects of a predator such as a phytosaur.

V3649: This very worn and pitted vertebral element has a semicircular outline; it is another possible example of the effects of digestive action.

V3650 (Figure 9.86): This possible intercentrum has a flat oval outline; it is completely flat on one surface and slightly convex on the other; halfway along the flattened surface there is a distinct semilunar notch.

V3651 (Figure 9.87): This flattened, procoelous, nonnotochordal centrum has a wide, fan-shaped posterior condyle. The centrum narrows midway along its anteroposterior length. Broken anterolateral processes give added width to the appearance of the anterior of the specimen. Its dorsal surface has a narrow-bore pit within the anterior region. The ventral surface has a median ridge extending anteroposteriorly, which is broad and flat anteriorly, highest at midlength, and indistinct posteriorly.

V3652 (Figure 9.88): This procoelous, nonnotochordal centrum has a ventrally directed posterior condyle and is roughly triangular in outline. Transverse processes are present anterolaterally, and a small neural spine dorsally; the ventral aspect is smooth and slightly concave. This specimen is possibly a basioccipital fragment.

V3653 (Figure 9.89): This relatively large, procoelous, nonnotochordal posterior caudal centrum has a well-developed posterior condyle. The anterior face of the specimen, as well as the entire body of the centrum, has a quadrangular outline. The neural arch is broken, but the base is completely fused to the body of the centrum. Ventrally there is a deep, narrow, anteroposterior groove; no transverse processes or chevron facets are present.

V3654 (Figure 9.90): Another procoelous, nonnotochordal centrum, possibly from a posterior caudal, with a weak posterior condyle. The anterior face of the centrum is quadrangular in outline; a fused, narrow neural arch is present dorsally. Weak facets are present posteroventrally for chevron attachment; no transverse processes are present.

V3655 (Figure 9.91): This is the anterior third of a nonnotochordal caudal centrum. The anterior face of the centrum has a circular outline; the neural arch is present and completely fused to the body of the centrum, with a short neural spine. The ventral surface has a shallow, narrow groove.

V3656 (Figure 9.92): This slightly amphicoelous, nonnotochordal anterior caudal centrum has well-developed articular surfaces for the the transverse processes, which are located dorsolaterally on the anterior half of the centrum. Dorsally, there is a shallow groove divided by a shallow, narrow ridge; ventrally, there is a sharp, narrow keel extending anteroposteriorly, and chevron facets.

V3659 (Figure 9.93): Two small, amphicoelous, notochordal posterior caudal centra bear thickened bone around the anterior and posterior faces of the centra. The ventral surface is smooth and flat, with well-developed chevron facets; the dorsal surface is a shallow, wide groove, bordered by thin edges.

V3660 (Figure 9.94): This amphicoelous, nonnotochordal caudal centrum lacks transverse processes. Dorsally, the specimen appears pinched at its midlength and is flat; ventrally, there is a wide, shallow groove that appears pinched at midlength.

V3661 (Figure 9.95): Another amphicoelous centrum, with a narrow notochordal pit on its posterior face. Dorsally, there is a pronounced hourglass outline created by the dorsolateral areas for the transverse processes; ventrally, there are no grooves or ridges or chevron facets visible, though they may have been worn away at the anterior and posterior ventral surfaces of the centrum.

V3662 (Figure 9.96): Two small, amphicoelous, nonnotochordal caudal centra have a rounded ventral surface, with no transverse processes or chevron facets.

V3663 (Figure 9.97): Two amphicoelous, nonnotochordal caudal centra may represent different parts of the caudal region of the same specimen or taxon. There is a ventral keel on one of the specimens, and a rounded ventral surface on the other; the specimen with the ventral keel also has a shallow ridge dividing the dorsal surface longitudinally, while the specimen without the ventral keel has a shallow undivided depression at midlength on its dorsal surface.

V3664: This is a slightly amphicoelous, nonnotochordal caudal centrum with a rectangular, blocky appearance. Its dorsal surface has a narrow-bore deep pit located medially; the ventral surface is keeled. The anterior and posterior faces are worn, and chevron facets cannot be seen.

V3671 (Figure 9.98). This procoelous, nonnotochordal anterior caudal vertebra is completely preserved. It has a high, rectangular neural arch directly fused to the centrum, with a thin, high neural spine bearing a distinct midlateral ridge that culminates in an anteriorly projecting nub. The narrow but rounded ventral surface has posteroventral chevron facets.

V3676 (Figure 9.99): A small, flat, amphicoelous notochordal centrum, possibly sacral, with ascending dorsolateral transverse processes and a bluntly ridged ventral surface.

V3677 (Figure 9.100): Numerous anterior caudals are similar to those of V3676; they are flat, amphicoelous, and mostly notochordal, with very wide transverse processes.

V3678 (Figure 9.101): This is a small procoelous, nonnotochordal centrum, possibly caudal, and dorsoventrally flattened. The anterior face is rectangular in outline, while the posterior face consists of a flattened condyle, slightly flared laterally. No transverse processes are present. A deep, narrow groove extends anteroposteriorly on the ventral surface, and notches extend from the groove on the anterior and posterior faces of the centrum. No chevron facets are present.

Reptilian teeth. Thirteen presumably reptilian tooth morphs are found in the collection.

cf. *Uatchitodon* Sues, 1991

Reference specimen. V3680 (Figure 9.102): Two teeth are faintly serrated anteriorly and posteriorly and laterally compressed. Each has a thin groove lingually and labially that extends from the base of the tooth to just below the tip, terminating in foramina just below the tip. The base of each tooth is lenticular in cross section, with two lateral canals separated by a thin median groove running perpendicular to the lingual and labial grooves. The morphology corresponds in many details to those reported by Sues (1991) for the type and referred specimens of *Uatchitodon kroehleri* from the early middle Carnian of the Newark Supergroup of Virginia. These presumably venom-conducting teeth were noted previously in the Placerias Quarry microfauna by Jacobs and Murry (1980) and Tannenbaum (1983).

The remaining dental micromorphs are not assignable to known taxa as far as we know, and for reasons noted earlier we choose not to name them here.

V3681 (Figure 9.103): This incisiform, laterally compressed tooth is slightly recurved posteriorly; faint serrations are on the posterior edge; faint dorsolateral striations are in the external enamel.

V3682 (Figure 9.104): This small, asymmetrical, somewhat laterally compressed tooth is roughly triangular in outline; it is denticulated on the presumed anterior edge, and serrated on the other edge. The

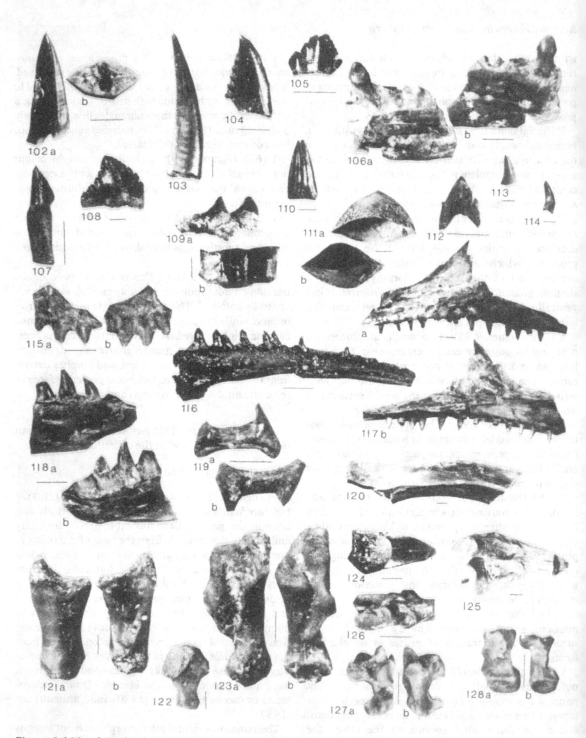

Figure 9.102. cf. *Uatchitodon* sp., V3680, serrated tooth with lingual and labial grooves and foramina: (a) side view, (b) basal view. **Figure 9.103.** V3681, laterally compressed, slightly recurved tooth, side view. **Figure 9.104.** V3682, denticulated and finely serrated tooth, side view. Possibly referable to the form genus *Revueltosaurus* (Hunt, 1989). **Figure 9.105.** V3683, broad, leaf-shaped denticulated tooth, side view. **Figure 9.106.** V3701, jaw fragments with spatulate teeth: (a) lingual view, (b) labial view. **Figure 9.107.** V3693, incisiform tooth with long conical root and laterally compressed, asymmetrical cusp, side view. **Figure 9.108.** V3697, serrated and denticulated triangular tooth crown, side view. Possibly referable to the form genus *Revueltosaurus* (Hunt, 1989). **Figure 9.109.** V3684, two connected, transversely broadened, bicuspid teeth: (a) lateral view, (b) occlusal view. **Figure 9.110.** V3685, unserrated tooth with vertical striations and bicuspid tip, side view. **Figure 9.111.** V3686, two serrated, hollow tooth buds: (a) ventro lateral view, (b) occlusal view.

presumed lingual side of the tooth is more laterally compressed than the other, and this edge has a dorsoventral ridge along the occlusal third of the tooth near the tip. In many features it is comparable to the ornithischian tooth form genus *Revueltosaurus* (Hunt, 1989; Padian, 1990). Murry and Long (1989) reported the presence of "prosauropod" or anchisaurid sauropodomorph remains in the Placerias Quarry, and as both Hunt (1989) and Padian (1990) noted, it is often very difficult to separate isolated tooth crowns of early ornithischians and "prosauropods," which were contemporaneous in the Late Triassic and Early Jurassic.

V3683 (Figure 9.105): A single tooth with broad denticles on the presumed anterior edge. These denticles are serrated, and somewhat finer serrations appear on the edge. The tooth has a broad median ridge on both lateral surfaces, lending a swollen appearance to the tooth.

V3701 (Figure 9.106): V3701 consists of two maxilla fragments, each containing a single spatulate tooth and one to three tooth alveoli. The tooth is concave lingually, convex labially, and has pleurodont implantation.

V3693 (Figure 9.107): A single incisiform, unserrated tooth, with an asymmetrical laterally flattened triangular cusp and a long, conical root, superficially resembling an ornithischian incisor (Charig and Crompton, 1974, p. 180, fig. 1).

V3697 (Figure 9.108): Several broadly triangular, serrated and/or denticulated isolated tooth cusps, some with serrations on the denticulations. Superficially, these teeth are ornithischian-like (Hunt, 1989; Padian, 1990) and may pertain to *Revueltosaurus*. The teeth are also similar in overall appearance to the leaf-shaped serrated teeth found in many genera of the Iguanidae and may represent a stem-group relative of the Squamata (Estes, 1964).

V3684 (Figure 9.109): These two large, transversely broadened teeth are bicuspid and swollen at the base. The cusps are connected by a strong loph, and the teeth are apparently directly attached to the jaw by means of acrodont implantation. The teeth do not appear to be procolophonid because they are too broad transversely (Watson, 1914; Case, 1928; Colbert,

1960; Gow, 1977a,b), nor do they seem to be trilophosaurid because such teeth are tricuspidate on both uppers and lowers (Gregory, 1945).

V3685 (Figure 9.110): A single long, isolated tooth that is neither compressed nor serrated. It has a small two-pointed tip, and an oval basal cross-section with a single median pit.

V3686 (Figure 9.111): V3686 consists of two serrated and hollow tooth buds that are broadly and symmetrically triangular in outline.

V3688 (Figure 9.112): A small tooth, very compressed laterally, with one main triangular center cusp flanked by two symmetrically lateral incipient cusps.

V3691 (Figure 9.113): V3691 consists of several nonspecific reptilian teeth that are serrated on one edge.

V3692 (Figure 9.114): V3692 consists of several nonspecific reptilian teeth that are unserrated and often display microstriations in the enamel.

Reptilian jaw elements. Some jaw fragments still contain teeth, but most of the isolated tooth crowns found in the quarry cannot be associated with jaws, so even their mode of implantation is unknown. Murry (1987) described several jaw fragments from this quarry and other Late Triassic sites that showed pleurodont or subpleurodont implantation, suggesting lepidosauromorph affinities or perhaps even a relict of Paleozoic diapsid groups outside the lepidosauromorph–archosauromorph dichotomy. Sphenodontids are known from this quarry (Tannenbaum, 1983; Murry, 1987), and it is strange that squamates or their stem-group representatives have not been positively identified in Triassic strata thus far.

V3562 (Figure 9.115): V3562 may be a premaxillary fragment, but its affinities are unknown. It has three teeth with apparent subpleurodont implantation and a blunt, undifferentiated appearance. An ascending process on the jaw fragment is broken but seems to consist of an asymmetrical narrowing of the entire jaw fragment, creating an articular surface. Lingually(?) there is a shelf at the jawline, and labially(?) there is a single nutrient foramen.

V3585 (Figure 9.116): V3585 consists of a single

Figures (*Continued*)
Figure 9.112. V3688, laterally compressed, incipiently tricuspid tooth, side view. **Figure 9.113.** V3691, tooth serrated on one edge only, side view. **Figure 9.114.** V3692, unserrated tooth, side view. **Figure 9.115.** V3562, ?premaxilla: (a) labial view, (b) lingual view. **Figure 9.116.** V3585, dentary of unknown affinity, lingual view. **Figure 9.117.** V3629, almost complete maxilla: (a) labial view, (b) lingual view. **Figure 9.118.** V3630, possible anterior dentary fragment: (a) labial view, (b) lingual view. **Figure 9.119.** V3706, stapes: (a) side view, (b) opposite side view. **Figure 9.120.** V3583, sigmoid rib, side view. **Figure 9.121.** V3712, possible left ulna: (a) posterior view, (b) anterior view. **Figure 9.122.** V3636, calcaneal tuber, side view. **Figure 9.123.** V3628, "hooked" fifth metatarsal: (a) side view, (b) another view. **Figure 9.124.** V3702, superficially ungual-like element, ventral view. **Figure 9.125.** V3703, superficially ungual-like element, ventral view. **Figure 9.126.** V3704, true ungual phalanx, ventral view. **Figure 9.127.** V3584, unknown asymmetrical element: (a) side view, (b) opposite side view. **Figure 9.128.** V3590, unknown element with an S-shaped ridge extending from end to end: (a) side view, (b) opposite side view.

dentary, broken at both anterior and posterior ends, and tapering toward the anterior end. The dentary fragment contains eight teeth and eleven alveoli. The implantation appears to be subpleurodont, and the smallest teeth are anterior. The teeth have sharp tips differentiated in form but not texture. There is a thin groove extending anteroposteriorly along the lower lingual surface of the dentary.

V3629 (Figure 9.117): V3629 is a single, almost complete maxilla with a broken facial process. There are 21 conical teeth with blunt translucent tips and pleurodont implantation. The groove for articulation with the jugal is seen lingually, and numerous nutrient foramina are visible labially.

V3630 (Figure 9.118): V3630 consists of a jaw fragment, possibly representing the most anterior section of a dentary. It bears three unserrated, slightly recurved teeth and a fourth partial tooth, all displaying pleurodont implantation.

Reptilian skeletal elements. Nine reference-collection vials contain elements that can be identified anatomically but not taxonomically.

V3706 (Figure 9.119): Single isolated stapes, apparently complete.

V3707: Single isolated stapes, somewhat shorter and stouter in appearance than V3706.

V3583 (Figure 9.120): A long, slightly sigmoid, completely ossified rib, broken at the proximal and distal ends. (Compared with UCMP V6148/65419, identified as a rib belonging to a small reptile. The locality is in Quay County, New Mexico, Redondo Formation, just below the Petrified Forest Member of the Chinle.)

V3712 (Figure 9.121): This is a laterally compressed long bone, possibly a left ulna. At the proximal end is an olecranonlike process and a radial notchlike articular surface on the medial side. The epiphyseal surface is not well defined, and the proximal and distal ends of the bone are spongy in appearance, suggesting that this element belonged to a juvenile.

V3636 (Figure 9.122): V3636 was kindly identified by the late Richard Estes as a calcaneal tuber of an unknown small reptile.

V3628 (Figure 9.123): V3628 is a single, isolated "hooked" fifth metatarsal, with robust tuberosities at its proximal end and a pronounced lateral groove for strong tendon adduction. It does not belong to any known lepidosaurian, crocodiloid, pterosaur, or chelonian, as far as the available literature and comparative specimens indicate.

V3702 (Figure 9.124): V3702 superficially resembles an ungual phalanx because of its hooked, tapered outline. But it is not like that of any known ungual, and it is perfectly bilaterally symmetrical. In the presumed plantar view there is a knoblike process located in the proximal half of the element; the distal half tapers to a point. On the presumed dorsal surface there is a median ridge that has its maximum height at midlength.

V3703 (Figure 9.125): V3703 consists of two elements that also resemble ungual phalanges, but also display perfect bilateral symmetry. Proximally, each element has two separate and distinctive saddle-shaped articular surfaces. In plantar view, the specimens show two shallow but separate and distinct impressions. Lateral to the main articular surfaces are laterally extended processes that are also visible in dorsal view. Dorsally, the bone is finely sculptured, with a thin, deep groove extending along a proximodistal ridge. Both elements also taper to a point at their distal ends, suggesting that, like V3702, they are median elements.

V3704 (Figure 9.126): V3704 consists of several true ungual phalanges, each having double articular surfaces at the proximal end. Each specimen also has a well-developed ridge extending proximodistally along the dorsal surface. All the elements taper to a point at their distal ends.

Probably reptilian elements of unknown identity. It is not clear to which parts of the skeleton these elements pertain.

V3584 (Figure 9.127): This specimen is at first glance suggestive of a vertebra, but its asymmetry precludes that identification. It has several processes extending from what may be its proximal and distal ends. It may be a tarsal or carpal element, but there are no definite landmarks upon which to base an identification.

V3590 (Figure 9.128): V3590 consists of two specimens that are virtual mirror images. They are small, rodlike elements, each having an S-shaped ridge extending lengthwise from end to end on one surface, and a relatively large knoblike process on the opposite surface. These specimens may represent tarsal or carpal elements, or perhaps even otic elements, but they are not identifiable at this time.

Discussion

Distribution and relative abundance of microvertebrate fossils

Table 9.1 summarizes the taxonomic diversity of the microfauna of the Placerias and Downs quarries; it should be noted that it lists only the relative numbers represented in the reference collection (see explanation of table). Some contrasts are immediately apparent (Jacobs and Murry, 1980; Tannenbaum, 1983). Among nontetrapods, only coelacanth teeth have been recorded from the macrofauna. The microfauna, most specimens of which do not exceed a few millimeters in size, includes about 50 percent nontetrapods. Of these, about half are chondrichthyans, and half osteichthyans, a great many of which cannot be identified to a lower taxonomic rank. Some have escaped even anatomical identification thus far and are relegated to the uncertain category of "ichthyolith," following Murry (1986). Whereas Murry in his fine study listed five ichthyolith types in the roughly contemporaneous Dockum strata of Texas, we find 26 in the Placerias Quarry; however, some could represent different parts of the same taxon. New records of identifiable nontetrapod taxa from the quarry include the ctenacanth shark *Phoebodus*, the hybodont shark *Acrodus*, the palaeoniscid actinopterygian *Gyrolepis*, the pholidopleurid actinopterygian *Australosomus*, and a coelacanth.

Our study of the small tetrapod remains generally confirms the previous results of Jacobs and Murry (1980), adding to that list considerable cynodont material, as well as sphenodontians, which were also recognized by Murry (1987). All other taxa that appear in the macrofauna are found in the microfauna, with the exception of *Placerias* itself, which makes up two-thirds of the macrovertebrate abundance (Camp and Welles, 1956). In addition, many taxa are represented in the microfauna but not the macrofauna; these include sphenodontians, cynodonts, prolacertiforms, trilophosaurs, a small aetosaur, and a host of indeterminate forms, as well as two taxa (the ornithischian *Revueltosaurus* and cf. *Uatchitodon*, of uncertain affinities) known only from teeth.

Of the small tetrapods, about 20 percent by number are amphibian, and about 80 percent are amniote remains. The figure for the amphibians is strange, for several reasons. Amphibians and phytosaurs are relatively rare in the Placerias Quarry macrofauna, yet they dominate the macrofauna in typical assemblages of the Petrified Forest Member (Long and Padian, 1986), which are primarily those of lowland stream and floodplain environments. Though amphibians are more abundant in the Placerias Quarry microfauna than they are in the macrofauna, they do not approach their representation in Murry's (1986, fig. 9.16) Kalgary locality in the Dockum Formation, though they greatly exceed their representation in the Rotten Hill and Otis Chalk localities. Assuming similar collection techniques, which seems reasonable, it appears that the various localities are sampling somewhat different environments.

Among the amniote microvertebrate remains, about 20 percent are cynodont synapsids, 5 percent each trilophosaurs (or procolophonids?) and sphenodontians, and about 30 percent archosaurs. The remaining 40 percent cannot be identified to lower taxonomic levels, and many are anatomically problematic. This representation is also unusual in that the taxa are largely terrestrial, not aquatic. Phytosaurs are not as numerous as one might expect in an assemblage so heavily dominated by nontetrapods. However, this accords with the relatively low percentage of metoposaurid

amphibians, which in the macrofauna are normally highly associated with phytosaurs. As just noted, this association tends to dominate typical Petrified Forest Member assemblages, and synapsids, procolophonids(?) or trilophosaurs, sphenodontians, aetosaurs, and poposaurs typically are rarer.

The data discussed in the preceding paragraphs and in Table 9.1 are summarized in Figure 9.129.

Paleoenvironment of the Placerias Quarry

The Placerias Quarry is a very unusual assemblage for the Chinle, as Camp and Welles (1956) recognized. The domination of the macrofauna by the dicynodont *Placerias* is known nowhere else in the Chinle or Dockum formations, nor in any other contemporaneous Late Triassic beds from North America. It is as unusual in this respect as the domination of *Coelophysis* at Ghost Ranch (Colbert, 1947, 1989) and that of *Plateosaurus* at Trossingen (Weishampel, 1984; Weishampel and Westphal, 1986). Similar concentrations of Late Triassic tetrapods include those of the small stagonolepidid *Aetosaurus ferratus* in the Upper Triassic of Germany (Fraas, 1877) and the large metoposaurid amphibians found in the Chinle Formation near Lamy, New Mexico, excavated by crews from Harvard University (Romer, 1939) and elsewhere.

The Placerias Quarry macrofauna contained a minimum of 39 individuals of *Placerias*, and probably many more were represented. As Camp and Welles (1956) noted, juveniles of *Placerias* are not strongly represented. Taphonomic effects include nearly complete disarticulation of skeletons (even cranial bones), tooth marks, breakage, wear, and repositioning. (Camp and Welles noted that some bones were standing on edge in the sediment.) Trampling, disturbance, and scavenging are obvious processes that would cause these effects. Camp and Welles (1956) suggested that rooting around in the sediments was a main factor in disturbance and repositioning, and this is consistent with both scavenging and herbivory, assuming that aquatic plants were abundant. Dubiel et al. (1991) confirm the environmental interpretation of "an organic-rich, possibly partially anoxic pond," with "periodic, possibly seasonal fluctuations in water table." According to them, the water table was relatively high throughout the times of deposition of the Petrified Forest Member, and precipitation was both abundant and seasonal, with characteristic "megamonsoons" providing influxes of water and sediment that contributed to the preservation of many lowland environments that include a wide diversity of tetrapods (Parrish, 1991). However, these authors note that phytosaurs and metoposaurs are the most commonly preserved tetrapods (Long and Padian, 1986), though *Placerias*, aetosaurs, and poposaurs are most common in the pondlike or marshlike habitat of the Placerias Quarry.

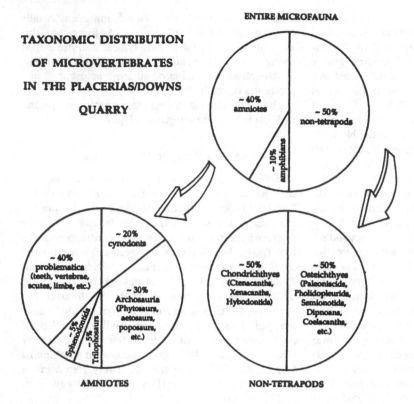

TAXONOMIC DISTRIBUTION
OF MICROVERTEBRATES
IN THE PLACERIAS/DOWNS
QUARRY

ENTIRE MICROFAUNA

~ 40% amniotes

~ 50% non-tetrapods

~ 10% amphibians

~ 20% cynodonts

~ 40% problematica (teeth, vertebrae, scutes, limbs, etc.)

~ 30% Archosauria (Phytosaurs, aetosaurs, poposaurs, etc.)

~5% Sphenodontids

~5% Trilophosaurs

~ 50% Chondrichthyes (Ctenacanths, Xenacanths, Hybodontids)

~ 50% Osteichthyes (Paleoniscids, Pholidopleurids, Semionotids, Dipnoans, Coelacanths, etc.)

AMNIOTES   **NON-TETRAPODS**

Figure 9.129. Taxonomic distributions of microvertebrates in the Placerias and Downs quarries.

Our analysis of the Placerias Quarry microvertebrates accords with the observations of Camp and Welles (1956) that several depositional levels are reflected in the quarry sites. The reddish brown mudstones contain numerous sandstone stringers, lenses of limestone in which blocky, fragmented bones are found, and layers of gypsum and calcite evaporites, as well as pockets of yellow-brown sulfurous material; carbonized plant remains are also abundant in places, and some horizons are nearly solidly coprolitic. Camp and Welles noted that the concentration of *Placerias* itself, which they called "a Golgotha of dicynodonts," was limited to the quarry floor, and "in the next few feet of supervening sediments only phytosaurs and amphibians were found" (1956, p. 262), and these are not abundant. The Placerias and Downs quarries sample essentially the same microfauna, as Jacobs and Murry (1980) concluded, though with somewhat differing representation. Amniotes are less abundant in the Downs Quarry (about 23%) than in the Placerias Quarry (about 52%), but fish are more abundant. The Placerias Quarry was sampled far more extensively than was the Downs Quarry, but the overall diversities were not significantly different. Nor were there any clear differences in diversity among stratigraphic levels or geographic locations within the quarry area (Tannenbaum, 1983). The sandstone stringers were particularly rich in microvertebrates (Jacobs and Murry, 1980). Hence,

because diversity was fairly uniform among sampled subsections, yet relative abundances differed, it would seem fair to say that local differences in mode of deposition, environment, water depth, and chemical composition were the primary factors affecting relative abundances, and this accords with expectations for a seasonal, variable water table driven by monsoonal precipitation (Dubiel et al., 1991).

Comparisons with local microfaunas of the Late Triassic and Early Jurassic

Murry (1989) noted an important transition in patterns of microvertebrate diversity through the Petrified Forest Member of the Chinle Formation in Arizona. The "Lower Unit" and "Upper Unit" of this member, though not lithologically distinct, do have some general faunal differences that have appeared to many to reflect a general discontinuity in time (Camp, 1930; Long and Padian, 1986; Hunt and Lucas, 1990; Parrish, 1991); they are separated in the Petrified Forest National Park [located about 50 miles (80 km) northwest of St. Johns] by the Sonsela Sandstone, a white, laterally extensive bed differing substantially in chemical and petrographic composition from the tan-brown sandstones found elsewhere in the park (Elzea, 1983). Murry (1989) found microfaunal differences between the two units as well. No xenacanth shark or *Lissodus* specimens were recovered from the Upper Unit, though they are present in the Lower Unit localities, as are redfieldiids, colobodontids, other fishes, lungfishes, and trilophosaurs. Low-level taxonomic changes in *Turseodus, Acrodus,* metoposaurs, and sphenodontians were also recorded between the Lower Unit and the Upper Unit. The overall faunal comparison shows that the Placerias Quarry microfauna is completely similar to the Lower Unit microfauna of the park, which is also true for the macrofauna, and this observation is consistent with the low placement of the Placerias Quarry in the Petrified Forest Member, as noted by previous workers (Camp and Welles, 1956; Jacobs and Murry, 1980; Long and Padian, 1986).

Curtis (1989) studied several hundred microvertebrate remains from four localities within the Kayenta Formation (Lower Jurassic) of northern Arizona, collected by field crews from Harvard University and UCMP during the early 1980s. He noted further differences between the Late Triassic and Early Jurassic microfaunas. Xenacanth and ctenacanth sharks are absent from the Kayenta; hybodonts are present, but *Acrodus* replaced *Lissodus,* as Murry (1989) already noted in Upper Petrified Forest strata. *Turseodus* and the pholidopleurid actinopterygians were not found in the Kayenta microfaunas, as expected, but osteichthyans were generally absent except for some doubtful unidentifiable fragments; this is unusual, because they are abundant elsewhere in the Jurassic, so their absence must be for preservational or paleoenvironmental reasons (Clark and Fastovsky, 1986). Metoposaurid amphibians were replaced by lissamphibians, including anurans and caecilians. Sphenodontians, crocodylomorphs, and cynodonts (including Mammaliaformes) (Jenkins, Crompton, and Downs, 1983) were the only amniotes represented in the microfauna, though turtles, dinosaurs, and pterosaurs are among the amniotes known from the macrofauna. Significantly, no "typical" Triassic amniote taxa were found in the microvertebrate sample, nor are they known from the macrofauna, suggesting the extinction of these groups by the time the Glen Canyon Group was deposited (Padian, 1989a,b; Curtis, 1989; pace Colbert, 1981; Clark and Fastovsky, 1986; Sues et al., Chapter 16).

Conclusions

Microvertebrates from the Placerias and Downs quarries greatly expand the known diversity of taxa from these localities. Jacobs and Murry (1980) estimated that the microvertebrates doubled the known diversity based previously on macrovertebrates. This seems reasonable, but is probably conservative; the factor could be closer to 5 than to 2, or even greater. We simply do not know, because the remains are so incomplete and dissociated. It is significant that such a large part of the micromorph diversity comes from forms that are so difficult to classify or even to identify anatomically. It is clear that paleontologists are just scratching the surface of what is to be known about the diversity of Triassic microvertebrates, as other authors (Jacobs and Murry, 1980; Murry, 1986, 1987, 1989) have noted.

New taxa reported since Jacobs and Murry's (1980) initial survey of the microfauna include the ctenacanth shark *Phoebodus,* the hybodont shark *Acrodus,* the palaeoniscid *Gyrolepis,* the pholidopleurid *Australosomus,* a coelacanth, various nonmammalian cynodont synapsids, several sphenodontian reptiles (also reported by Murry, 1987), the ornithischian dinosaur *Revueltosaurus,* and possibly the problematic reptile *Uatchitodon,* as well as dozens of micromorphs representing unknown taxa. Though reptiles are better represented in the Placerias than in the Downs Quarry, their diversities are comparable, and no apparent horizontal or vertical differences in diversity were discovered.

The microvertebrates support previous assessments that the Placerias Quarry represents a quiet pond or marshlike depositional setting with abundant but seasonally variable precipitation, in contrast to the floodplains and braided streams typical of Chinle environments of that time. Microvertebrate taxa are about equally divided in their representation between tetrapods and nontetrapods, and the ecological profiles of the various taxa suggest that whereas some members were aquatic denizens, many were terrestrials that probably came to the water to drink and feed. The macrovertebrates show a high terrestrial component and unusual taxonomic representation as well, dominated as they are by the dicynodont *Placerias* and various aetosaurs and poposaurs, rather than the phytosaurs and metoposaurs typical of those times in lowland environments. The large, hippopotamus-like *Placerias* may have habitually wallowed in the marshlike muds of this site, as its broad, spatulate unguals and stout form might suggest. However, if these unguals were sheathed by a horny hoof, as seems to have been the case for some hadrosaurian dinosaurs (Norman, 1985), their suitability for marsh wallowing may have been less likely than suitability for a more terrestrial habit. In either case, they and their young faced strong predation pressures from the aquatic phytosaurs, which are unusually rare in this quarry, and the terrestrial poposaurs, which are unusually common here.

Acknowledgments

For valuable discussions and assistance with identification of these specimens we thank J. A. Gauthier, R. A. Long, Donald Baird, Amy McCune, Paul Olsen, J. T. Gregory, the late Richard Estes, Will Downs, Phil Murry, John Bolt, Howard Hutchison, Mike Parrish, and Kent Curtis. Louis Jacobs, Mike Parrish, and Hans-Dieter Sues provided very helpful reviews of the manuscript, but must be considered blameless for any errors we introduced. Judy Bacskai and George Shkurkin of the Bibliography of Fossil Vertebrates were very helpful with tracking down obscure references, and photographic and logistic assistance was provided by Howard Schorn, UC Berkeley Scientific Photo, and Reuben Andrade. Figure 9.129 was drawn by Phyllis Spowart. We are especially grateful to Louis Jacobs and Will Downs for providing us this material for study and to the Museum of Northern Arizona for its extended loan. We particularly want to thank Mike and Jeannie Udall, Mrs. Dorothalene Tuckness, and Mr. Gordon ("Sonny") Cowley for access to their lands. This study was based in part on a master's degree thesis submitted to the UC Berkeley Department of Paleontology by the first author in 1983, and we thank Professors W. A. Clemens and Marvalee Wake for serving on the thesis committee. This research was supported by American Chemical Society grant 13577-G2 and National Geographic Society grants 2327-81 and 2484-82 to the second author, and by the UCMP. This is contribution no. 1560 from the University of California Museum of Paleontology.

References

Akers, J. P. 1964. Geology and ground water in the central part of Apache County, Arizona. *United States Geological Survey Water Supply Paper* 1771: 1–107.

Benton, M. J. (ed.). 1988. *Phylogeny and Classification of the Tetrapods, Vol. 1.* Oxford: Clarendon Press.

Branson, E. B. 1905. Structure and relationships of American Labyrinthodontidae. *Journal of Geology* 13: 568–610.

Branson, E. B., and M. G. Mehl. 1929. Triassic amphibians from the Rocky Mountain region. *University of Missouri Studies* 4: 155–239.

Camp, C. L. 1930. A study of the phytosaurs, with description of new material from western North America. *Memoirs of the University of California* 10: 1–161.

Camp, C. L., and S. P. Welles. 1956. Triassic dicynodont reptiles. Part 1. *Memoirs of the University of California* 13: 255–341.

Carroll, R. L. 1978. Permo-Triassic "lizards" from the Karroo System. Part 2. A gliding reptile from the Upper Permian of Madagascar. *Palaeontologia Africana* 21: 143–159.

Case, E. C. 1928. Indications of a cotylosaur and of a new form of fish from the Triassic beds of Texas, with remarks on the Shinarump Conglomerate. *University of Michigan Contributions from the Museum of Paleontology* 3: 1–14.

  1932. On the caudal region of *Coelophysis* sp. and on some new or little known forms from the Upper Triassic of western Texas. *University of Michigan Contributions from the Museum of Paleontology* 3: 89–91.

Case, G. R. 1967. *Fossil Shark and Fish Remains of North America.* New York: Grafco Press.

Charig, A. J., and A. W. Crompton. 1974. The alleged synonymy of *Lycorhinus* and *Heterodontosaurus. Annals of the South African Museum* 64: 167–189.

Chatterjee, S. 1985. *Postosuchus*, a new thecodontian reptile from the Triassic of Texas, and the origin of tyrannosaurs. *Philosophical Transactions of the Royal Society of London* B309: 395–460.

Clark, J. M., and D. E. Fastovsky. 1986. Vertebrate biostratigraphy of the Glen Canyon Group in northern Arizona. Pp. 285–301 in K. Padian (ed.), *The Beginning of the Age of Dinosaurs: Faunal Change across the Triassic–Jurassic Boundary.* Cambridge University Press.

Colbert, E. H. 1947. Little dinosaurs of Ghost Ranch. *Natural History* 56: 392–399, 427–428.

  1960. A new Triassic procolophonid from Pennsylvania. *American Museum Novitates* 2022: 1–19.

  1981. A primitive ornithischian dinosaur from the Kayenta Formation of Arizona. *Museum of Northern Arizona Bulletin* 53: 1–61.

  1989. The Triassic dinosaur *Coelophysis. Museum of Northern Arizona Bulletin* 57: 1–160.

Cooley, M. E. 1957. Geology of the Chinle Formation in the Upper Little Colorado drainage area, Arizona and New Mexico. M. A. thesis, University of Arizona.

Curtis, K. M. 1989. A taxonomic analysis of a microvertebrate fauna from the Kayenta Formation (Early Jurassic) of Arizona and its comparison to an Upper Triassic microvertebrate fauna from the Chinle Formation. M. A. thesis, University of California, Berkeley.

Dubiel, R. F., J. T. Parrish, J. M. Parrish, and S. C. Good. 1991. The Pangaean megamonsoon – evidence from the Upper Triassic Chinle Formation, Colorado Plateau. *Palaios* 6: 347–370.

Elzea, J. 1983. A petrographic and stratigraphic analysis of the Petrified Forest Member (Triassic Chinle Formation) sandstones, Petrified Forest National Park, Arizona. B. A. honors thesis, Department of Paleontology, University of California, Berkeley.

Estes, R. 1961. Cranial anatomy of the cynodont reptile *Thrinaxodon liorhinus. Bulletin of the Museum of Comparative Zoology, Harvard University* 125: 165–180.

  1964. Fossil vertebrates from the Late Cretaceous Lance Formation, eastern Wyoming. *University of California Publications in Geological Sciences* 49: 1–180.

Evans, S. E., and H. Haubold. 1987. A review of the Upper Permian genera *Coelurosauravus, Weigeltisaurus* and *Gracilisaurus* (Reptilia: Diapsida). *Zoological Journal of the Linnean Society* 90: 275–303.

Evans, S. E., and A. R. Milner. 1989. *Fulengia*, a supposed early lizard reinterpreted as a prosauropod dinosaur. *Palaeontology* 32: 223–230.

Fourie, S. 1974. The cranial morphology of *Thrinaxodon liorhinus* Seeley. *Annals of the South African Museum* 65: 337–400.

Fraas, O. 1877. Die gepanzerte Vogel-Echse aus dem Stubensandstein bei Stuttgart. *Württembergische naturwissenschaftliche Jahreshefte* 33: 1–21.

Fraser, N. C. 1982. A new rhynchocephalian from the British Upper Trias. *Palaeontology* 25: 709–725.

  1986. Terrestrial vertebrates at the Triassic–Jurassic boundary in south west Britain. *Modern Geology* 10: 147–157.

  1988. The osteology and relationships of *Clevosaurus*

(Reptilia: Sphenodontida). *Philosophical Transactions of the Royal Society of London* B321: 125–178.

Galton, P. 1973. On the anatomy and relationships of *Efraasia diagnostica* (Huene) n. gen., a prosauropod dinosaur (Reptilia: Saurischia) from the Upper Triassic of Germany. *Paläontologische Zeitschrift* 47: 229–255.

Galton, P., and M. A. Cluver. 1976. *Anchisaurus capensis* (Broom) and a revision of the Anchisauridae (Reptilia: Saurischia). *Annals of the South African Museum* 69: 121–159.

Gauthier, J. A. 1986. Saurischian monophyly and the origin of birds. *Memoirs of the California Academy of Sciences* 8: 1–55.

Gow, C. E. 1977a. New procolophonids from the Triassic *Cynognathus* Zone of South Africa. *Annals of the South African Museum* 72: 109–124.

1977b. Tooth function and succession in *Procolophon trigoniceps*. *Palaeontology* 20: 695–704.

1980. The dentitions of the Tritheledontidae (Therapsida: Cynodontia). *Proceedings of the Royal Society of London* B208: 461–481.

Green, T. E. 1956. Disturbed Chinle beds, St. Johns, Apache County, Arizona. M.A. thesis, University of Texas.

Gregory, J. T. 1945. Osteology and relationships of *Trilophosaurus*. *University of Texas Bulletin* 4401: 275–359.

Guerin, S. 1957. Contribution à l'étude géologique et paléontologique du Trias supérieur et du Lias inférieur de la région de Saint-Rambert-en-Bugey (Jura méridional). *Sciences de la Terre* 5: 13–51.

Harshbarger, J. W., C. A. Repenning, and I. J. Haskell. 1957. Stratigraphy of the uppermost Triassic and the Jurassic rocks of the Navajo country. *United States Geological Survey Professional Paper* 291: 1–74.

Hopson, J. A. 1971. Postcanine replacement in the gomphodont cynodont *Diademodon*. Pp. 1–21 in D. M. Kermack and K. A. Kermack (eds.), *Early Mammals*. London: Academic Press.

Hunt, A. P. 1989. A new ?ornithischian dinosaur from the Bull Canyon Formation (Upper Triassic) of east-central New Mexico. Pp. 355–358 in S. G. Lucas and A. P. Hunt (eds.), *Dawn of the Age of Dinosaurs in the American Southwest*. Albuquerque: New Mexico Museum of Natural History.

Hunt, A. P., and S. G. Lucas. 1990. Re-evaluation of "*Typothorax*" *meadei*, a Late Triassic aetosaur from the United States. *Paläontologische Zeitschrift* 64: 317–328.

1991. *Rioarribasaurus*, a new name for a Late Triassic dinosaur from New Mexico (USA). *Paläontologische Zeitschrift* 65: 191–198.

Jacobs, L. L., and P. A. Murry. 1980. The vertebrate community of the Triassic Chinle Formation near St. Johns, Arizona. Pp. 55–72 in L. L. Jacobs (ed.), *Aspects of Vertebrate History*. Flagstaff: Museum of Northern Arizona Press.

Jain, S. L. 1968. Vomerine teeth of *Ceratodus* from the Maleri Formation (Upper Triassic, Deccan, India). *Journal of Paleontology* 42: 96–99.

Jenkins, F. A., Jr. 1971. The postcranial skeleton of African cynodonts. *Bulletin of the Peabody Museum of Natural History, Yale University* 36: 1–216.

Jenkins, F. A., Jr., A. W. Crompton, and W. R. Downs. 1983. Mesozoic mammals from Arizona: new evidence on mammalian evolution. *Science* 222: 1233–1235.

Jensen, J. A., and K. Padian. 1989. Small pterosaurs and dinosaurs from the Uncompahgre fauna (Brushy Basin Member, Morrison Formation: ?Tithonian), Late Jurassic, western Colorado. *Journal of Paleontology* 63: 364–373.

Long, R. A., and K. L. Ballew. 1985. Aetosaur dermal armor from the Late Triassic of southwestern North America, with special reference to material from the Chinle Formation of Petrified Forest National Park. *Museum of Northern Arizona Bulletin* 54: 45–68.

Long, R. A., and K. Padian. 1986. Vertebrate biostratigraphy of the Late Triassic Chinle Formation, Petrified Forest National Park, Arizona: preliminary results. Pp. 161–170 in K. Padian (ed.), *The Beginning of the Age of Dinosaurs: Faunal Change across the Triassic–Jurassic Boundary*. Cambridge University Press.

Martin, M., D. Sigogneau-Russell, P. Coupatez, and G. Wouters. 1981. Les Cératodontidés (Dipnoi) du Rhétien de Saint-Nicolas-de-Port (Meurthe-et-Moselle). *Géobios* 14: 773–791.

Murry, P. A. 1986. Vertebrate paleontology of the Dockum Group, western Texas and eastern New Mexico. Pp. 109–138 in K. Padian (ed.), *The Beginning of the Age of Dinosaurs: Faunal Change across the Triassic–Jurassic Boundary*. Cambridge University Press.

1987. New reptiles from the Upper Triassic Chinle Formation of Arizona. *Journal of Paleontology* 61: 773–786.

1989. Microvertebrate fossils from the Petrified Forest and Owl Rock members (Chinle Formation) in Petrified Forest National Park and vicinity, Arizona. Pp. 249–278 in S. G. Lucas and A. P. Hunt (eds.), *Dawn of the Age of Dinosaurs in the American Southwest*. Albuquerque: New Mexico Museum of Natural History.

Murry, P. A., and R. A. Long. 1989. Geology and paleontology of the Chinle Formation, Petrified Forest National Park and vicinity, Arizona and a discussion of vertebrate fossils of the southwestern Upper Triassic. Pp. 29–64 in S. G. Lucas and A. P. Hunt (eds.), *Dawn of the Age of Dinosaurs in the American Southwest*. Albuquerque: New Mexico Museum of Natural History.

Norman, D. 1985. *The Illustrated Encyclopedia of Dinosaurs*. New York: Crescent Books.

Olsen, P. E. 1979. A new aquatic eosuchian from the Newark Supergroup (Late Triassic–Early Jurassic) of North Carolina and Virginia. *Postilla* 176: 1–14.

Padian, K. 1986. On the type material of *Coelophysis* Cope (Saurischia: Theropoda), and a new specimen from the Petrified Forest of Arizona (Late Triassic: Chinle Formation). Pp. 45–60 in K. Padian (ed.), *The Beginning of the Age of Dinosaurs: Faunal Change across the Triassic–Jurassic Boundary*. Cambridge University Press.

1989a. Presence of the dinosaur *Scelidosaurus* indicates Jurassic age for the Kayenta Formation (Glen

196

Canyon Group, northern Arizona). *Geology* 17: 438–441.

1989b. Did "thecodontians" survive the Triassic? Pp. 401–414 in S. G. Lucas and A. P. Hunt (eds.), *Dawn of the Age of Dinosaurs in the American Southwest*. Albuquerque: New Mexico Museum of Natural History.

1990. The ornithischian form genus *Revueltosaurus* from the Petrified Forest of Arizona (Late Triassic: Norian; Chinle Formation). *Journal of Vertebrate Paleontology* 10: 268–269.

Parrish J. M. 1991. Vertebrate paleoecology of the Chinle Formation (Late Triassic) of the southwestern United States. *Palaeogeography, Palaeoclimatology, Palaeoecology* 72: 227–247.

Repenning, C. A., M. E. Cooley, and J. P. Akers. 1969. Stratigraphy of the Chinle and Moenkopi formations, Navajo and Hopi Indian Reservations, Arizona, New Mexico, and Utah. *United States Geological Survey Professional Paper* 521-B: 1–34.

Robinson, P. 1973. A problematic reptile from the British Upper Trias. *Journal of the Geological Society of London* 129: 457–479.

Romer, A. S. 1939. An amphibian graveyard. *Scientific Monthly* October 1939: 337–339.

Sawin, H. J. 1945. Amphibians from the Dockum Triassic of Howard Country, Texas. *University of Texas Bulletin* 4401: 361–399.

Schaeffer, B. 1952. The Triassic coelacanth fish *Diplurus*, with observations on the evolution of the Coelacanthini. *Bulletin of the American Museum of Natural History* 99: 25–78.

1967a. Late Triassic fishes from the western United States. *Bulletin of the American Museum of Natural History* 135: 285–342.

1967b. Osteichthyan vertebrae. *Zoological Journal of the Linnean Society* 47: 185–195.

Schaeffer, B., and D. H. Dunkle. 1950. A semionotid fish from the Chinle Formation, with considerations of its relationships. *American Museum Novitates* 1457: 1–29.

Schaeffer, B., and J. T. Gregory. 1961. Coelacanth fishes from the continental Triassic of the western United States. *American Museum Novitates* 2036: 1–18.

Sereno, P. C. 1991. *Lesothosaurus*, "fabrosaurids," and the early evolution of Ornithischia. *Journal of Vertebrate Paleontology* 11: 168–197.

Shubin, N. H., A. W. Crompton, H.-D. Sues, and P. E. Olsen. 1991. New fossil evidence on the sister-group of mammals and early Mesozoic faunal distributions. *Science* 251: 1063–1065.

Stewart, J. H., F. G. Poole, and R. F. Wilson, 1972. Stratigraphy and origin of the Chinle Formation and related Upper Triassic strata in the Colorado Plateau region. *United States Geological Survey Professional Paper* 690: 1–336.

Sues, H.-D. 1991. Venom-conducting teeth in a Triassic reptile. *Nature* 351: 141–143.

Tannenbaum, F. A. 1983. The microvertebrate fauna of the Placerias and Downs Quarries, Chinle Formation (Upper Triassic), near St. Johns, Arizona. M. A. thesis, Department of Paleontology, University of California, Berkeley.

Walker, A. D. 1961. Triassic reptiles from the Elgin area: *Stagonolepis, Dasygnathus* and their allies. *Philosophical Transactions of the Royal Society of London* B244: 103–204.

Watson, D. M. S. 1913. On a new cynodont from the Stormberg. *Geological Magazine* 10: 145–148.

1914. *Procolophon trigoniceps*, a cotylosaurian reptile from South Africa. *Proceedings of the Zoological Society of London* 1914: 735–747.

Weishampel, D. B. 1984. Trossingen: E. Fraas, F. von Huene, R. Seemann, and the "Schwäbische Lindwurm" *Plateosaurus*. Pp. 249–253 in W.-E. Reif and F. Westphal (eds.), *Third Symposium on Mesozoic Terrestrial Ecosystems, Short Papers*. Tübingen: ATTEMPTO Verlag.

Weishampel, D. B., and F. Westphal. 1986. Die Plateosaurier von Trossingen. *Ausstellungskataloge der Universität Tübingen* 19: 1–27.

Woodward, A. S. 1893. Palaeichthyological notes. *Annals and Magazine of Natural History* (6) 12: 281–287.

10

Late Triassic microvertebrates from central Europe

DENISE SIGOGNEAU-RUSSELL AND GERHARD HAHN

Introduction

The type area for the Triassic is the Germanic Basin, where three broad lithostratigraphic units have traditionally been recognized: the mostly continental Buntsandstein, the marine Muschelkalk, and the more continental Keuper. The stages of the Triassic period were subsequently defined in marine sedimentary rocks exposed in the Alps. Correlation between the chronostratigraphic and lithostratigraphic units has proved difficult. We are concerned here only with the Keuper, which broadly corresponds to the late Ladinian, Carnian, and Norian. (The "Rhaetian" is considered here upper Keuper and latest Norian.)

In the Germanic and Paris basins, the Keuper, representing some 24 million years, is thin, reaching a maximum thickness of 500 m in southwestern Germany. There apparently were localized areas of nondeposition during that time interval; consequently, fossils from a given lithostratigraphic level need not necessarily be contemporaneous.

Paleogeography

During the Triassic, variations in the rate of sea-floor spreading (Ziegler, 1982) caused changes in the volume of the mid-oceanic ridges, which in turn resulted in fluctuations in sea level.

At the beginning of the Triassic (Figure 10.1), the incipient Paris Basin was already a vast depression surrounded by the east–west-trending Ardenno-Rhenanian Massif to the north and the "éperon bourguignon" to the south, and it was linked to the Germanic Basin. In western Europe, during Muschelkalk deposition, the Tethys Sea transgressed northward and westward.

At the end of the Middle Triassic, during the onset of the Keuper deposition, the Tethys partly regressed

from central Europe, but as a result of increasing peneplanation the sea extended farther into the Paris Basin (Figure 10.2). Shallow seas, deltas, tidal flats, and lagoons occupied both the Germanic and Paris basins. At the edges of the lagoons, detrital and evaporitic regimes existed. In Germany, the detrital input came from the east and south (with the uplifting of the East European craton and the Bohemian Massif), whereas in the Paris Basin it came from the west (Massif Armoricain) and south (Massif Central) (Figure 10.1). In conjunction with this oscillation in sea level, conditions of aridity alternated with more humid episodes (in the Paris Basin: middle Lettenkohle, "grès à roseaux," and "Rhaetian" sandstones). During the periods of desiccation, salts accumulated in areas of active subsidence (Courel et al., 1973).

At the end of the Triassic ("Rhaetian"), a second transgression of the Tethys (by then connected with the Arctic Sea) inundated the rapidly subsiding basins.

Sedimentology

The Keuper beds are thin in the southern Germanic Basin and even thinner in the Paris Basin. In the latter basin, the Triassic outcrops form the most external of the concentric exposures; they have been eroded to the north, except in the area of the "Golfe du Luxembourg" (Figures 10.1 and 10.3).

Keuper deposition (Figure 10.4) began with the regressive episode of the Lettenkohle ("charbon des argiles" with fish, plants, and bivalves). It continued with the Bunte Mergel (in Germany) and the Marnes irisées (in France), corresponding to a subsidence in a relatively calm and supersaturated environment. Evaporites are common and indicate a temporary increase in aridity. However, there are fluvial intercalations (only one in the Paris Basin sensu stricto) with cross-bedded stratification and rare fossil plants

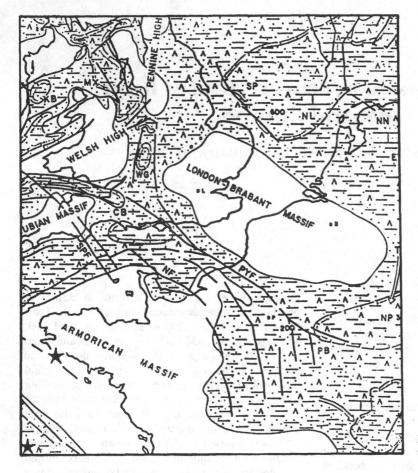

Figure 10.1. Paleogeography of the Paris Basin in the Triassic.

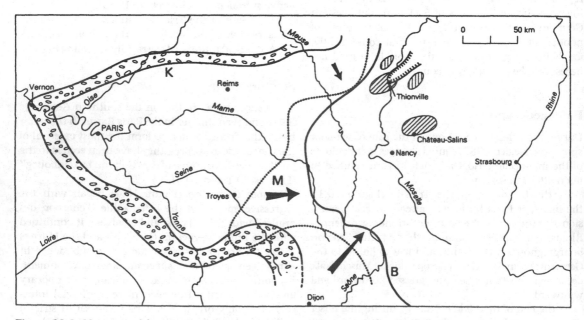

Figure 10.2. Maximum of the Triassic transgression in the Paris Basin: M, in the Muschelkalk; K, in the Keuper.

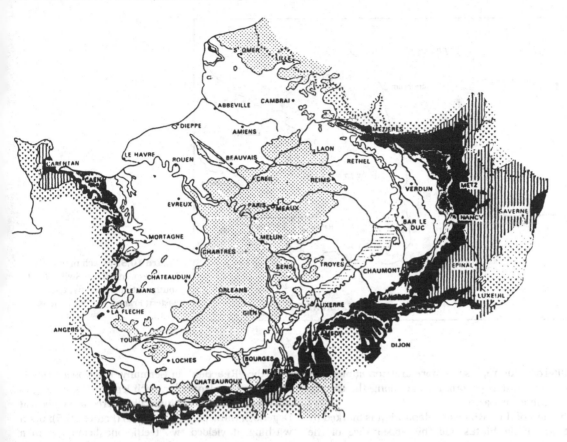

Figure 10.3. Mesozoic outcrops in the Paris Basin; the most external (in black) represents the Triassic.

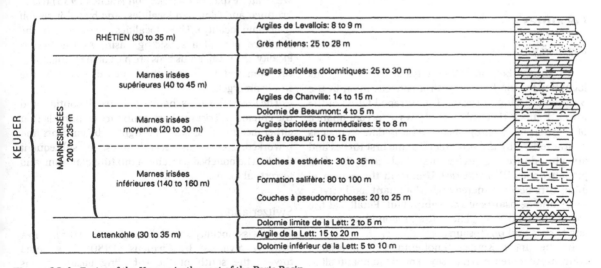

Figure 10.4. Facies of the Keuper in the east of the Paris Basin.

and bivalves. The series ends with the "Rhaetian" facies, conformable on older Keuper sediments (except on the edges of the basins, where one finds the Cimmerian unconformity) consisting of cross-stratified sandstones, conglomerates, and mudstones, representing deposits in very shallow water (such as deltas, tidal flats, channels, or sandy bars along the continental margins). In the Paris Basin, the basal sandstones are usually fossiliferous. They do not exceed 25 m in thickness: they are fine- to medium-grained and contain intercalations of green clay. They are overlain by the lagoonal "Argiles de Levallois," which is poorly

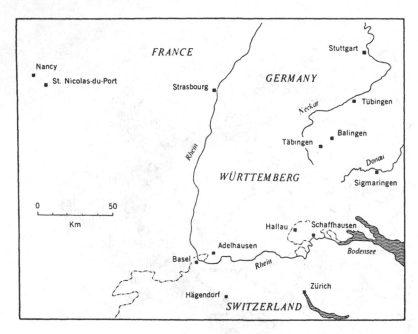

Figure 10.5. Sketch map with localities in northern Switzerland, southwestern Germany, and eastern France. (After Clemens, 1980.)

fossiliferous and represents more arid conditions. The "Rhaetian" outcrops form a ridge along the eastern part of the Paris Basin.

The end of the "Rhaetian" deposition is marked by a stratigraphic hiatus and the appearance of the ammonite *Psiloceras planorbis*.

Localities

Germany

In Germany, Benton (1986) cites seven fossiliferous localities for the Keuper, with a concentration in the Stuttgart-Tübingen region. The tetrapod fauna of the lower Keuper (Lettenkohle) shows a predominance of temnospondyl amphibians. The middle Keuper is characterized by the occurrence of the first turtles and an abundance of archosaurs, including the first prosauropods (*Plateosaurus*). Higher in the sequence, dinosaurs become increasingly dominant, and phytosaurs and aetosaurs are also abundant. Finally, at the top of the sequence, plateosaurs predominate, with a few turtles and phytosaurs.

In the upper Keuper (Knollenmergel), dinosaurs continue to increase in diversity. The oldest mammalian tooth known to date was found in the Knollenmergel near Halberstadt (Sachsen-Anhalt).

It is interesting to note that during the Norian, dinosaurs markedly increased in abundance (from 7 percent of the fauna to 95 percent – mostly made up of *Plateosaurus* – even if preservational biases exaggerate these proportions). In addition, it is remarkable for the apparent absence of nonmammalian synapsids.

In the "Rhaetian" of Germany, the Upper Triassic sandstones are lenticular, each lens corresponding to a delta. In 1847, Plieninger discovered a bonebed at Degerloch (now part of Stuttgart) (Figure 10.5); upon washing, it yielded two teeth, one belonging to a haramiyid, the other to a tritylodontid. Similar beds were later explored by Erika von Huene (1933) (Gaisbrunnen and Olgahain localities) and by Schindewolf (Olgahain locality). These localities are situated in the northern part of a subsiding basin. At the former locality, the fossiliferous beds are part of the "Rhaetian" sandstone; at the latter, the bonebed occurs in the Knollenmergel.

The fossils (bivalves, haramiyids, tritylodontids, and the enigmatic *Tricuspes*) are a mixture of animals that either died locally or were brought in by currents and reworked, sometimes several times, by subsequent floods. The bonebed contains a mixture of marine and terrestrial forms.

Switzerland

The Swiss locality of Hallau (Figure 10.5), most recently reviewed by Clemens (1980), is situated toward the south of the subsiding basin. It was discovered by Schalch in 1873. Peyer screen-washed material from the site in 1942 and even used the heavy-liquid flotation process. He published a monograph on the synapsid remains in 1956. The fossils, including reworked material, represent selachians, dipnoans, amphibians, sauropterygians, dinosaurs, pterosaurs, and mammals (haramiyids, *Tricuspes*, and morganucodontids). The age of the locality is some-

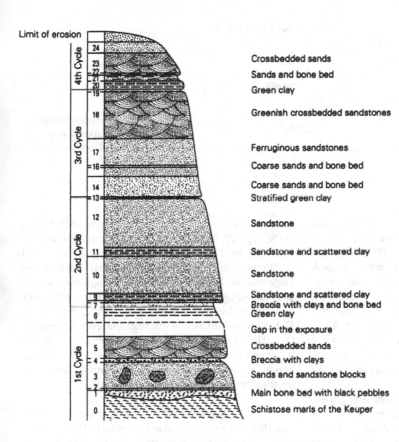

Bed Number

1.10 m: Oolitic green clay	6
0.80 m: Gap in the exposure	
1.30 m: Crossbedded sands	5
0.15 m: Breccia with clays	4
1.50 m: Sand and sandstone blocks	3
0.10 m: Ferruginous sands	2
0.10/0.50/1.05 m: Main bone bed with black pebbles	1
Schistose marls of the Keuper	0

Limit of erosion

Crossbedded sands

Sands and bone bed

Green clay

Greenish crossbedded sandstones

Ferruginous sandstones

Coarse sands and bone bed

Coarse sands and bone bed

Stratified green clay

Sandstone

Sandstone and scattered clay

Sandstone

Sandstone and scattered clay
Breccia with clays and bone bed
Green clay

Gap in the exposure

Crossbedded sands

Breccia with clays

Sands and sandstone blocks

Main bone bed with black pebbles

Schistose marls of the Keuper

Figure 10.6. Section of the Saint-Nicolas-de-Port Quarry. (From Laugier, 1971.)

where between the top of the middle Keuper and the early Hettangian (Clemens, 1980).

France

In the middle Keuper of the Paris Basin (Marnes irisées) there is little to compare with the Germanic Basin succession and faunal material (Corroy, 1928). However, several productive localities have recently been explored in the upper Keuper.

The Saint-Nicolas-de-Port (or Rosières-aux-Salines) locality has been known since 1851 (Levallois, 1862). A stratigraphic section was first published in 1928 by Corroy, when a quarry was opened for the construction of a railway. Laugier (1971) further described the section and documented the diversity and abundance of the faunal remains.

The first mammal-like tooth was found in 1975 by

G. Wouters, a Belgian collector who was interested in fishes at the time. In 1976, D. E. Russell and D. Sigogneau-Russell started large-scale screen-washing; several tons of sediment were processed, and more than one ton of concentrate remains to be sorted. G. Wouters worked independently, assembling a large collection, which he gave to the Muséum National d'Histoire Naturelle (Paris) in 1984. Later, he began collecting for the Institut Royal des Sciences Naturelles of Bruxelles. At the same time, several other amateur collectors made their own collections. Unfortunately, the quarry is now being used as a city dump and is progressively being reduced in size. Furthermore, there are plans for transforming the entire area into a golf course.

A conglomerate layer (10–110 cm) at the base of the Saint-Nicolas section (Figure 10.6) corresponds to the base of the "Rhaetian." Above this lies a succession

Figure 10.7. Geographic situation of the Belgian and Luxembourg localities mentioned in the text.

of sands and clays, with intercalated bonebeds interpreted as recurrent invasions of the sea after a more brackish period (Al Khatib, 1976). The sands are white to greenish in color, alternating with a reddish carbonaceous sandstone, and contain a few concentrations of green clay; unfortunately, they contain no characteristic pollen. From granulometric analyses of the sediments, Al Khatib (1976) interpreted the paleoenvironment as a very shallow sea close to the continent. The fine sands have characteristics of beach and river deposits, and the very fine sands correspond to tidal zones.

The age of the locality has been disputed. Sigogneau-Russell (1983a) placed the fossiliferous beds at the base of the lower "Rhaetian." Buffetaut and Wouters (1986) advocated a slightly older age. They considered the fossils to be typical of the uppermost middle Keuper (i.e., immediately below the "Rhaetian"). However, Buffetaut and Wouters (1986) pointed out that the faunas of the uppermost middle Keuper could be very similar to those of the upper Keuper.

Other vertebrates at Saint-Nicolas-de-Port include actinopterygians, selachians, dipnoans, amphibians (plagiosaurids and capitosaurids), reptiles (sphenodontians, phytosaurs, prosauropod and small theropod dinosaurs), therapsids (cynodonts), and mammals. At present, none of these groups permits a more precise biostratigraphic assessment.

Belgium and Luxembourg

In Belgium and Luxembourg, the efforts of amateur collectors have led to the discovery or rediscovery of several sites (Figure 10.7).

Habay-la-Vieille. Fossils were found at Habay as early as 1907, but new outcrops were discovered and exploited in 1982 and 1983 by J. C. Lepage and G. Wouters in the course of the construction of the Liège–Arlon freeway (Wouters, Sigogneau-Russell, and Lepage, 1984). Several fossiliferous units have been recognized: HLV 1 (marine), HLV 2 (estuarine), HLV 3 (estuarine), HLV 4 (marine), and HLV 5 (marine). Only HLV 2 and 3 have yielded mammals. The age is believed to be lower "Rhaetian." The fossils include cynodonts and a few mammals (among them the multituberculate *Mojo* and one therian).

Hachy (Sagnette). Three marine units (HS1, HS2, and HS3) were discovered at Hachy by G. Wouters and J. C. Lepage in 1980 and 1981. The age is thought to be middle "Rhaetian." Only mammals of uncertain relationships have been documented.

Attert. Discovered by J. C. Lepage in 1983 (Wouters, Lepage, and Delsate, 1985). This site has yielded actinopterygian fish and rare cynodonts and mammals (haramiyids). It may be Norian in age.

Medernach. Vertebrate fossils at Medernach, Luxembourg, were first discovered by J. C. Lepage in 1982. This site is equivalent in age to the upper Bunte Mergel (middle Norian) (Hary and Muller, 1967) and contains selachians, actinopterygians, phytosaurs, and cynodonts.

Unter der Kirchen. This includes two localities discovered by G. Wouters in 1979 and J. C. Lepage in 1980. The deposits are marine in origin and apparently

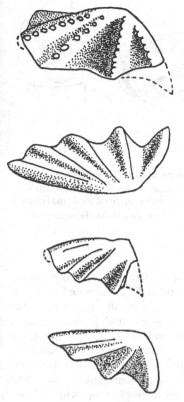

Figure 10.8. Tooth plates of *Ceratodus phillipsi* from Saint-Nicolas-de-Port. (After Martin et al., 1981.)

cover the "Rhaetian"–Hettangian boundary. Details on the fauna are not yet available.

"Rhaetian" vertebrates

At Saint-Nicolas, fish are very abundant and were the first vertebrates to be described from this locality. The following identifications were made by C. J. Duffin (pers. commun.) and J. Herman (in Al Khatib, 1976). Selachians are represented by *Polyacrodus*, *Hybodus minor*, *Nemacanthus* sp., *Orthacodus*, and *Lonchidion*. Chondrosteans include *Birgeria acuminata*, *Colobodus* sp., *Gyrolepis albertii*, and *Saurichthys longidens*. *Sargodon tomicus* is the sole holostean recorded. According to Herman (Al Khatib, 1976) these fish have been reworked from the underlying beds. From the Belgian localities, Duffin et al. (1983) recorded the following additional selachians: *Acrodus minimus*, *Pseudodalatias barstonensis*, and *Vallisia coppi*. The fish from Hallau have not been revised since the study by Peyer (1956).

Dipnoans. The Ceratodontidae from Saint-Nicolas were studied by Martin et al. (1981), based on about 200 tooth plates, of which only a quarter are complete, and most are separated from their bony support. Most plates are worn and represent both juveniles and

adults. Martin et al. (1981) conducted a thorough morphological study of these tooth plates and were able to clarify the very confused systematics. They recognized two species at Saint-Nicolas: *Ceratodus kaupi* and *C. phillipsi* (Figure 10.8). *C. kaupi* is also known from reworked fossils from the continental beds of the German Muschelkalk and from marine Triassic beds of Spitzbergen; *C. phillipsi* also occurs at Hallau. Several other species have been cited from the Germanic Keuper, but they are in need of revision, as at least some are apparently synonymous with *C. phillipsi*.

Amphibians. At Saint-Nicolas, temnospondyl amphibians (Buffetaut and Wouters, 1986) are represented by abundant vertebrae of Plagiosauridae (with lengths varying from 5 to 16 mm), which appear similar to those of certain forms from the German Muschelkalk and the Keuper of Lorraine. A clavicle and dermal plates have also been attributed to this group. Capitosaurids are documented by a palatine. The amphibian remains from Hallau have been attributed to temnospondyls, but they have not been identified more precisely.

Reptiles. At the Belgian localities, remains of ichthyosaurs and plesiosaurs have been recognized (Buffetaut and Wouters, 1986). The Saint-Nicolas reptile assemblage is dominated by the phytosaur *Rutiodon ruetimeyeri*, represented mostly by isolated tooth crowns. Three types of teeth have been attributed to this taxon: large, laterally compressed teeth with serrated cutting edges and smooth enamel; large, recurved teeth that are circular in transverse section and have serrated edges and fluted enamel; long and slender teeth that are circular in transverse section and have fluted enamel, but lack any serrations. Such teeth are characteristic of the heterodont phytosaurian dentition. Some osteoderms may belong to the same species. Another phytosaur, *Belodon*, has been identified at Medernach. *Rutiodon ruetimeyeri* is also well known from the famous German locality of Halberstadt (top of middle Keuper) (Kuhn, 1939).

The prosauropod *Plateosaurus* is less abundant at Saint-Nicolas than are the phytosaurs. Its teeth are coarsely serrated and slightly asymmetrical. This genus has also been recorded at Sagnette (Duffin et al., 1983) and Hallau. A *Procompsognathus*-like theropod may be represented by a few small teeth that are compressed and recurved and have serrated edges. *Procompsognathus* is also known from the German Keuper.

Cynodonts. The precise constitution of the Mammalia is equivocal. McIntyre (1972) restricts mammals to placentals and marsupials and their last common ancestor. Rowe (1988, fig. 3) adds the multituberculates to this assemblage, but excludes Kuehneotheriidae and Haramiyidae from it. Wible

Table 10.1. *Occurrences of Cynodontia at localities in western and central Europe*

	Species	Provenance
Tritylodontidae	—	—
Traversodontidae	*Microscalenodon nanus*	Gaume
Chiniquodontoidea	*Pseudotriconodon wildi*	Medernach
	Pseudotriconodon sp.	Gaume
	Lepagia gaumensis	Gaume
	Gaumia longiradicata	Gaume
	Gaumia? incisa	Gaume
	Meurthodon gallicus	Rosières-aux-Salines

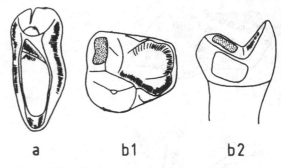

a   b1   b2

Figure 10.9. *Microscalenodon nanus* Hahn, Lepage, and Wouters, 1988: (a) ?left upper postcanine tooth, occlusal view; (b1) ?left lower postcanine, occlusal view; (b2) same tooth, anterior view. (Modified from Hahn, Lepage, and Wouters, 1988.)

(1991) excludes Multituberculata, but adds Monotremata to Mammalia. On the other hand, Ax (1984, figs. 69 and 70) includes all Synapsida within Mammalia. If we were to follow McIntyre and Rowe, all of the teeth discussed in this section would be regarded as non-mammalian, but following Ax, they would all be considered mammals. We continue to use the classical definition of Mammalia, regarding the presence of three middle ear bones and a single bone in the lower jaw as the diagnostic mammalian characteristics.

Three groups of Cynodontia are present in the Upper Triassic of western and central Europe (Table 10.1): Tritylodontidae, Traversodontidae, and Chiniquodontoidea. The Tritylodontidae are well known from England and from Germany, but they are not found in Belgium, France, and Luxembourg. The reasons for this disjunct distribution are unknown, but competition with haramiyids, which shared a herbivorous diet, might at least be partly responsible.

Traversodontidae are known only from two tiny teeth from Gaume (Belgium), interpreted as an upper and a lower postcanine. *Microscalenodon nanus* Hahn, Lepage, and Wouters, 1988 (Figure 10.9) is the stratigraphically youngest and, in its dimensions, smallest taxon of the Traversodontidae. The upper postcanine is very narrow anteroposteriorly and bears only one labial cusp. Cuspules on the base of the cusp are missing. The masticatory surface is well worn and flat and rises somewhat lingually. The anterior and posterior borders are similar. The lower postcanine is subrectangular in shape; it bears two large anterior cusps and a deeply basined "talonid." The anterolabial cusp is very high, and longer than wide; a sharp crest crosses this cusp anteroposteriorly. The anterolingual cusp is broader than long and is crossed by a transverse

crest. This crest and the anterior slope of the cusps are hardly worn. The "talonid" is surrounded by an elevated rim that bears a small cusp behind the anterolabial cusp. An interdental pressure facet is visible on the anterior wall of the crown. *M. nanus* is characterized by its minute size (tooth length ca. 1 mm) and by the presence of only one labial cusp on the upper postcanines (a plesiomorphic feature). The latter feature indicates a relationship with African taxa such as *Scalenodon*. The South American traversodontids show a more or less distinct second labial cusp, and they are, in this respect, more advanced than their African relatives. *M. nanus* was a small animal, probably not larger than a mouse, and perhaps fed on insects and soft plant matter.

Members of the Chiniquodontoidea are the most common cynodonts in all "Rhaetian" localities, but their teeth have been described in detail only from Belgium and Luxembourg. Three problems complicate their treatment. First, the teeth are quite distinct from those of other therapsids and mammals. Second, the dentition in the better-known genera of the Chiniquodontoidea has been poorly studied. (If the whole skull is preserved, the analysis of the teeth is often considered unimportant in systematic descriptions.) Third, the problem of subdivision of the Chiniquodontoidea into families has yet to be resolved. These three problems are briefly discussed next.

The postcanine teeth are characterized by a "triconodont" crown with three to five laterally compressed cusps arranged in a longitudinal row. This type of tooth is adapted to a cutting function, and it evolved independently in reptiles, therapsids, and mammals. Therefore, the systematic position of the Chiniquodontoidea cannot be deduced from the structure of the crown alone. The root is perhaps the more critical part of the tooth. It may be completely undivided or may bear a longitudinal furrow, or it may be divided either at its tip or along its entire length.

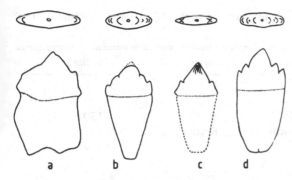

Figure 10.10. Postcanine teeth of different chiniquodontid taxa from the Upper Triassic of Belgium, France, and Luxembourg: (a) *Lepagia gaumensis* Hahn, Wild, and Wouters, 1987; (b) *Gaumia longiradicata* Hahn, Wild, and Wouters, 1987; (c) *Gaumia? incisa* Hahn, Wild, and Wouters, 1987; (d) *Pseudotriconodon wildi* Hahn, Lepage, and Wouters, 1984. Top row, occlusal views; bottom row, lingual views.

In the Triconodonta, the root in premolars and molars is completely divided, and cingula accompany the row of main cusps. This combination of characters is not found among the "Rhaetian" teeth considered in this chapter.

"Triconodont" teeth with tricuspid to pentacuspid crowns and undivided roots are present in young individuals of the Triassic reptiles *Tanystropheus* and *Macrocnemus* (in adults, the teeth have single cusps). However, in this instance, they are broadly ovoid in occlusal view, whereas the teeth described here are very narrow (Figure 10.10). Moreover, the roots of these reptilian teeth are, at most, only slightly tapered distally. There is nothing to suggest that teeth of these forms occur in our "Rhaetian" collections.

It is more difficult to distinguish between teeth of cynodonts and those of the pterosaur *Eudimorphodon*; both are found together, at least at Medernach. Nevertheless, in *Eudimorphodon*, the crown is much shorter and higher than in cynodonts, and the enamel is strongly wrinkled, with numerous ridges (Hahn, Lepage, and Wouters, 1984, pl. 1, figs. 3 and 4). In cynodonts, the enamel is generally smooth and may be wrinkled only at the tips of the main cusps (Figure 10.10).

The structure of the root is very important in the identification of "triconodont" teeth (Figure 10.10). Therefore, it is most unfortunate that in the treatment of many cynodont skulls, the dentition is described only in a general way, and the roots frequently not at all. Only when the teeth of cynodonts are studied in the same detailed manner as those of mammals might it prove possible to carry out detailed taxonomic studies of the isolated small teeth from the "Rhaetian" bonebeds.

We distinguish four different morphological groups

of teeth (= genera; Table 10.2) in the "Rhaetian" of Belgium, France, and Luxembourg that may belong to the superfamily Chiniquodontoidea (excluding Trithelodontidae).

The subdivision of the Chiniquodontoidea is equivocal. Originally, three families were established:

1. Dromatheriidae Gill, 1872, with *Dromatherium* Emmons, 1857, and *Microconodon* Osborn, 1886; both genera were erected on the basis of single dentaries from the Upper Triassic of North Carolina.

2. Chiniquodontidae Huene, 1936, with *Chiniquodon* Huene, 1936, and *Belesodon* Huene, 1936, from the Middle to Upper Triassic of Argentina and Brazil. Other genera, such as *Probelesodon* Romer, 1969 (of which the whole skeleton is known), from the Middle Triassic of Argentina, and *Aleodon* Crompton, 1955, from the Middle Triassic of Tanzania, were later referred to this family.

3. Therioherpetidae Bonaparte and Barberena, 1975, with the sole genus *Therioherpeton* Bonaparte and Barberena, 1975, from the Upper Triassic of Brazil.

Hopson and Kitching (1972) restricted the Chiniquodontoidea to the Chiniquodontidae. *Therioherpeton* was undescribed at the time and they regarded *Dromatherium* and *Microconodon* as "Cynodontia incertae sedis." Hopson and Barghusen (1986, fig. 12), in their cladistic analysis of the Cynodontia, restricted the Chiniquodontoidea to the Chiniquodontidae and Probainognathidae. The Dromatheriidae and Therioherpetidae were not mentioned. Hahn, Lepage, and Wouters (1987) separated the Dromatheriidae and Chiniquodontidae, but incorporated the Therioherpetidae into the Chiniquodontidae. Following Hopson and Kitching (1972), they grouped *Probainognathus* with the Chiniquodontidae. Lucas and Oakes (1988) discussed the Dromatheriidae in detail, and, following Simpson (1926), they considered *Dromatherium* and *Microconodon* to be poorly known representatives of two different families and argued that *Therioherpeton* probably was not closely related to them. Carroll (1988) referred all these genera to the Chiniquodontidae.

This complex situation cannot be resolved by studies of isolated teeth. If three families are recognized, then the isolated teeth from the "Rhaetian" can be distinguished on the basis of root structure. In *Therioherpeton*, the crown and root are separated by a sharp constriction, and the root shows a broad longitudinal groove (Bonaparte and Barberena, 1975, fig. 5). In the "Dromatheriidae," the crown and root are not separated by constriction, and the incipient division extends along the full length of the root (Simpson, 1926, fig. 2B). In the Chiniquodontidae excluding *Therioherpeton*, the roots show not even a longitudinal groove (Bonaparte and Barberena, 1975). Following this scheme, the "Rhaetian" genera of Chiniquodontoidea can be grouped as follows (Tables 10.1 and 10.2):

Table 10.2. *Characters of the different types of chiniquodontoid teeth from the Late Triassic ("Rhaetian") of Belgium, France, and Luxembourg*

Location	Attribute	Pseudotriconodon	Gaumia	Lepagia	Meurthodon
Root	Number	1(+)	1	1	2
	Height (% of crown)	150	200	100	~150
	Shape	Semielliptical	Conical	Rectangular	Narrow, conical
Crown	Number of cusps	3–5	3–5	3–5	4
	Arrangement of cusps	Symmetrical	Symmetrical	More or less symmetrical	Not symmetrical
	Separation of cusps	Distinct	Distinct	Less distinct	Very distinct
	Base of crown	Not incised	Distinctly incised	Separated by shallow groove	Incised
	Length (mm)	0.7–1.6	1.1–1.6	1.2–3.0	4.0

1. *Pseudotriconodon* Hahn, Lepage, and Wouters, 1984 (Figure 10.10d). Crown is tricuspid to pentacuspid. In most specimens, it is more or less symmetrical, with the central cusp being the highest. The cusps are clearly separated from each other. The enamel is smooth. The base of the crown is not constricted. The root is semielliptical in outline, about 1.5 times as high as the crown. Its distal tip is divided (Hahn et al., 1984, pl. 2, fig. 6a–c) and shows two pulp canals. This demonstrates incipient division of the root. The length of the teeth ranges from 0.7 mm to 1.6 mm. *Pseudotriconodon* is known from Medernach (*P. wildi* Hahn, Lepage, and Wouters, 1984, type species), from Gaume (*P.? sp.* Hahn, Wild, and Wouters, 1987), and from the Norian of New Mexico (*P. chatterjeei* Lucas and Oakes, 1988). But it is apparently absent from Hallau (Switzerland). The genus occurs at Saint-Nicolas, but it is yet to be described.

In the shape of the crown, separation of cusps, and absence of a constriction between crown and root, the teeth of *Pseudotriconodon* most closely resemble those of *Microconodon tenuirostris* Osborn, 1886. In *Microconodon*, however, the root is apparently divided along its whole height by a slight furrow (Simpson, 1926, fig. 2b), whereas in *Pseudotriconodon* the furrow is either confined to the tip of the root or, more often, completely missing. Therefore, on this basis, *Pseudotriconodon* can be grouped either with the Dromatheriidae (if *Microconodon* and *Dromatherium* are united in one family) or with *Microconodon* (if separate families are established for both genera).

2. *Gaumia* Hahn, Wild, and Wouters, 1987 (Figure 10.10b,c). *Gaumia* resembles *Pseudotriconodon* in the shape of its crown. However, the root is longer (about twice as high as the crown), conical, and tapering at its distal end; the pulp canal is narrow. The base of the crown is distinctly constricted. The root is apparently undivided. The length of the postcanines in *Gaumia* ranges from 1.1 mm to 1.6 mm.

Three species can be distinguished. *G. longiradicata* Hahn, Wild, and Wouters, 1987, is characterized by smooth enamel on all cusps. A second, as yet undescribed species is known from Hallau (Peyer, 1956, pl. 9, figs. 18, 27, 34, 43, and 44). In this form, the main cusp is characterized by the presence of a few indistinct enamel ridges at its tip. The third species that probably can be attributed to *Gaumia* is *G.? incisa* Hahn, Wild, and Wouters, 1987. In this form, the enamel ridges are more numerous and better defined, but they are generally restricted to the lingual face of the main cusp. Unfortunately, the crown of this species is not complete, and therefore referral to *Gaumia* is still tentative.

The systematic position of *Gaumia* among the Chiniquodontoidea is not clear. The undivided root suggests that *Gaumia* is not closely related to *Dromatherium*, *Microconodon*, and *Therioherpeton*. It may be a member of the Chiniquodontidae, but the postcanines of the chiniquodontid genera are not sufficiently well known to allow detailed comparison with *Gaumia*.

3. *Lepagia* Hahn, Wild, and Wouters, 1987 (Figure 10.10a). In *Lepagia gaumensis* Hahn, Wild, and Wouters, 1987, the only known species, the crown is also tricuspid to pentacuspid, but in contrast to *Pseudotriconodon* and *Gaumia*, the cusps are less symmetrical and less distinct. The root is nearly rectangular in side

Figure 10.11. *Meurthodon gallicus* gen. et sp. nov., postcanine tooth in occlusal view (top) and lateral view (bottom). (From Russell et al., 1976.)

view, is subequal in height to the crown, and does not taper distally. It is depressed laterally, and the pulp canal is restricted to a long and narrow slit. The crown and root are separated by a shallow constriction. The teeth are between 1.2 mm and 3.0 mm long. A portion of the lower jaw is preserved in the holotype (Hahn, Wild, and Wouters, 1987, pl. 1, fig. 1a–e, and text-fig. 2a,b). There is a thin splenial, which extends halfway up on the lingual face of the dentary. *L. gaumensis* is known from Gaume and from Hallau (Switzerland) (Peyer, 1956, pl. 5, fig. 66a–d, and pl. 10, fig. 68a–d; Kindlimann, 1984, fig. 4a, b).

The splenial of *Lepagia* resembles that of *Probainognathus* Romer, 1970, from the Middle Triassic of Argentina (Kermack and Kermach, 1984, fig. 3-1E; Hahn, Wild, and Wouters, 1987, figs. 1 and 2). The somewhat irregular shapes of the tooth cusps are similar in the two genera (Carroll, 1988, fig. 17-34c–f), but in *Probainognathus* a cuspidate cingulum is present on the postcanines, whereas it is missing from *Lepagia*. The shape of the root is unknown in *Probainognathus*, the roots being hidden in the jaw. If *Lepagia* is related to *Probainognathus*, it is the more primitive of the two, despite its younger geological age. We tend to group *Lepagia* with the Probainognathidae (or with the Chiniquodontidae, if *Probainognathus* is regarded as a member of this family).

4. *Meurthodon* gen. nov. (Figure 10.11). The crown is tetracuspid and, in contrast to all other discussed genera, asymmetrical in the arrangement of the cusps, with the second cusp being the highest and the first cusp the smallest. The apex of the tooth is displaced anteriorly. The root is completely divided in a mammal-like fashion. The crown is constricted at its base. Only one species is known, *M. gallicus* sp. nov. from Saint-Nicolas-de-Port. The structure and the systematic position of the holotypic tooth were discussed in detail by Russell, Russell, and Wouters (1976). These authors

came to the conclusion that *Therioherpeton cargnini* Bonaparte and Barberena, 1975, would be the most closely related form, and we endorse that opinion here. In *T. cargnini*, the postcanines also bear four cusps (Bonaparte and Barberena, 1975, fig. 5), and the second cusp is also the largest. Furthermore, the structure of the root in *T. cargnini* is a precursor of that seen in *Meurthodon*. The root in this form is almost completely divided; only a thin sheet of dentine connects the two portions. If this sheet were lost, the condition seen in *Meurthodon* would be realized. *Meurthodon* is the first chiniquodontid known with completely separated roots. Parallel evolution is seen in the Tritylodontidae, where as many as five roots may be present in the upper postcanines (Cui and Sun, 1987, fig. 31). *Meurthodon* is interpreted by us as a younger and (in the structure of its root) more advanced relative of *Therioherpeton*, and it is included the Therioherpetidae.

There apparently were several small chiniquodontoids present in the latest Triassic of Europe that were the size of a rat or a mouse. Although their dentitions had a cutting function, we consider them to have been too small to be carnivorous. Perhaps they fed on insects with hard elytrae. Unfortunately, too little is known about these tiny therapsids, which were contemporaneous with, and perhaps competitors of, the first mammals. In the "Rhaetian," the two groups co-existed, whereas in the Jurassic, mammals began to flourish, and these small cynodonts fell into decline.

Mammals. Few mammalian teeth have been documented from the Tübingen-Stuttgart area, but they include the holotypes of *Thomasia antiqua* (Plieninger, 1847) and *Tricuspes tuebingensis* E. Huene, 1933. At Hallau, Peyer (1956) recovered some 150 mammalian teeth, which, according to Clemens (1980), constitute only half of the collection available. The mammalian assemblage from Saint-Nicolas-de-Port consists of more than 1,000 complete or partial teeth. These are all isolated teeth, sometimes rolled, sometimes very well preserved, but often cracked. The three known mammalian subclasses are represented: Allotheria, Eotheria, and Theria. In other localities of the Paris Basin, mammals remain rare and, with one exception, consist of haramiyids.

Allotheria. These are by far the most abundant, the most widely represented, and perhaps the most varied "Rhaetian" mammals. They belong to three families: Haramiyidae, Theroteinidae, and Paulchoffatiidae.

The Paulchoffatiidae are known from a single tooth, *Mojo usuratus*, from the Lorraine locality (Figure 10.12) (Hahn, Lepage, and Wouters, 1987). *Mojo* is separated by an enormous chronological gap from later paulchoffatiids, but the two longitudinal rows of cusps and the nature of the wear facets (with cusps completely

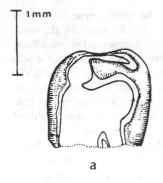

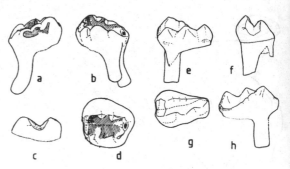

Figure 10.13. *Haramiya* I (a–d) and II (e–h) in different views. (From Sigogneau-Russell, 1989.)

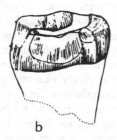

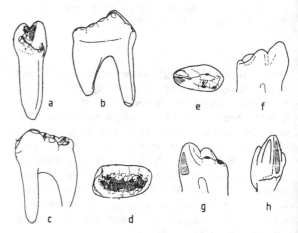

Figure 10.12. *Mojo usuratus*: (a) occlusal view; (b) anterior view. (From Hahn et al., 1989.)

Figure 10.14. *Thomasia* I (a–d) and II (e–h) in different views. (From Sigogneau-Russell, 1989.)

eroded and confluent at their base) are consistent with reference to this group.

The Haramiyidae of the Germanic and Paris basins, discussed by Hahn (1973) and Clemens (1980), have been reviewed in detail (Sigogneau-Russell, 1989). There is a single genus with perhaps three species. The upper teeth (Figure 10.13) correspond to what is called *Haramiya*; they bear two rows of cusps, one consisting of three subequal cusps, the other of up to five unequal cusps. The lower teeth, which are relatively narrow, correspond to what used to be called *Thomasia* (Figure 10.14); they also bear two rows of cusps, one with two unequal cusps, the other with three to five, which are also unequal in size. A longitudinal depression is closed at one end by a low wall and at the other end by the main cusps of the two rows. The roots are quite different in the two types of teeth: subequal in size and transversely compressed in *Haramiya*, unequal in size and direction of compression in *Thomasia*.

Sigogneau-Russell has described additional haramiyid teeth as *Haramiya* II (Figure 10.13e–h) and *Thomasia* II (Figure 10.14e–h), interpreted as premolars, and other teeth considered incisors. In the so-called premolars, the *B* row undergoes a reduction at the anterior end in the upper jaw, and the posterior end in the lower jaw. There is an interesting similarity between the tooth identified as an I2 (Figure 10.15) and the incisors of the Late Jurassic Paulchoffatiidae. Sigogneau-Russell

(1990) formally distinguished the most common species of *Haramiya* from Saint-Nicolas from those of the British fissures by naming a new species, *H. butleri*, for the French specimens.

Based on tooth structure and wear facets, the following conclusions concerning the dentition of haramiyids can be drawn: There must have been an anteroposterior inversion of the upper and lower postcanines, with the largest cusps positioned posteriorly in the upper jaw and anteriorly in the lower jaw. Such an inversion was first suggested by Parrington (1946) and later adopted by Hahn (1973).

Furthermore, it has been concluded that the masticatory movement was intermediate between the strictly orthal movement of most cynodonts and the clearly anteroposterior movement of the multituberculates, the amplitude of the anteroposterior component being limited in the haramiyids. We have not been able to demonstrate the direction of this movement, because we cannot distinguish between leading and trailing edges on worn cusps. However, a movement of the mandible from the rear to the front of the upper teeth

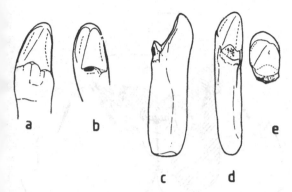

Figure 10.15. Two types (a, b and c–e) of incisors attributed to Haramiyidae. (From Sigogneau-Russell, 1989.)

would best explain the relative extent of the various wear facets. It should be added that several problems persist in this interpretation of the haramiyid dentition, although ongoing work by P. Butler may resolve some of the problems.

We consider the enigmatic Haramiyidae to be related to the Multituberculata, and they may not be as closely related to the paulchoffatiids as we first thought (e.g., Hahn, 1973). On the basis of the orientation of the main axis of the teeth, haramiyids and traversodontids do not have a close relationship (Hahn, 1973; Sigogneau-Russell, 1989). The origin of the haramiyids remains uncertain, and it is not known if they share a common origin with other mammals. The very poorly known tritheledontid cynodonts or their precursors (because the known forms are younger than the haramiyids) seem to be the only potential link with other mammals.

Sigogneau-Russell, Frank, and Hemmerlé (1986) erected the Theroteinidae on the basis of isolated teeth from Saint-Nicolas. None of the other localities of the Germanic and Paris basins has yielded theroteinid teeth. The subcircular upper teeth (Figure 10.16,1) are composed of three rows of low cusps (*A*, *B*, and *C*) separated by a narrow valley. The enamel is wrinkled. In row *A*, the middle cusp is by far the largest, its median flank extending into the adjoining valley. Row *B* consists of a large cusp followed by two smaller ones. Row *C* is composed of three or four small cusps. The three rows are confluent at one end; at the other end, the tooth is bounded by two cusps positioned below the level of the three main rows of cusps. Wear is complicated but apparently constant. The lower molars (Figure 10.16,2) (Sigogneau-Russell et al., 1986; Hahn, Sigogneau-Russell, and Wouters, 1989) are ovoid in outline, composed of two longitudinal, very unequal, broad rows referred to as *a* and *b*, which are confluent at one end. Each consists of two main cusps and additional cuspules, and they are separated by a valley. Again, wear was a complicated process for which the various specimens show different stages.

The lack of completely preserved roots raises a question regarding their mammalian affinities, but in one or two specimens described by Hahn et al. (1989) there is the indication of divided roots. In addition, the nature of the tooth crown and the wear facets and the ultrastructure of the enamel, which is similar to that described for haramiyids and *Kuehneotherium* (Frank, Sigogneau-Russell, and Voegel, 1984), are considered as mammalian characteristics.

All of these allotherian groups probably had different dietary specializations. The low cusps and delicate roots of the theroteinid teeth perhaps suggest a diet of rather soft fruit. The high cusps, extensive wear, and stout roots of the paulchoffatiid molars may be indicative of a diet of hard seeds, whereas the haramiyid teeth, which are rarely strongly worn, may be indicative of a diet of soft plant matter.

Eotheria. These constitute the second most abundant group in the Swiss, German, and French localities, where they are represented by several taxa. From Saint-Nicolas, only the relatively large *Brachyzostrodon* has been described to date (Figure 10.17). It is known from at least two species (Sigogneau-Russell, 1983b; Hahn, Sigogneau-Russell, and Godefroit, 1991), and it is characterized by its very stocky teeth, with wrinkled enamel. On the lower molars, cusps *a* and *c* are of nearly equal size; the talonid is short and carries three cuspules. There is a strong kuehneocone, similar to that in *Megazostrodon* (Crompton, 1974). Indeed, in general tooth structure, *Brachyzostrodon* is very similar to *Megazostrodon*, but the occlusal mode of *Brachyzostrodon*, as deduced from the wear facets, seems to indicate a closer relationship to *Morganucodon*. However, the heavy abrasion of the teeth of *Brachyzostrodon* and the orientation of the abraded surfaces suggest a considerable transverse component in the

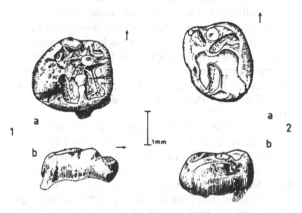

Figure 10.16. (1) Theroteinid upper molar in occlusal (a) and lateral (b) views; (2) theroteinid lower molar in occlusal (a) and anterior (b) views. (From Hahn et al., 1989.)

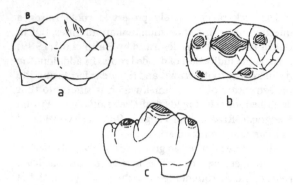

Figure 10.17. *Brachyzostrodon coupatezi*: right lower molar in lingual (a), occlusal (b), and labial (c) views. (From Sigogneau-Russell, 1983b.)

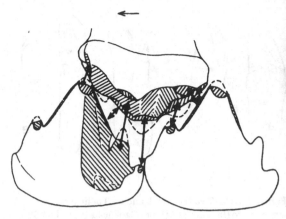

Figure 10.19. *Woutersia mirabilis*, suggested occlusion between upper and lower molars. (From Sigogneau-Russell, 1983b.)

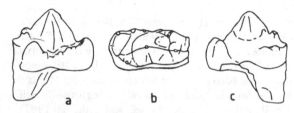

Figure 10.18. *Helvetiodon schuetzi*: right upper molar in lingual (a), occusal (b), and labial (c) views. (From Clemens, 1980.)

masticatory cycle, which suggests a greater crushing component than inferred for either *Megazostrodon* or *Morganucodon*. The British genus *Wareolestes* (Freeman, 1979) is insufficiently known to make detailed comparison. *Morganucodon* is also present at Saint-Nicolas, as is another morganucodontid yet to be described. In addition to the new species, *Morganucodon peyeri*, Clemens (1980) described two new genera referable to the Morganucodontidae, *Helvetiodon* and *Hallautherium*, from Hallau. *M. peyeri* is a small species with narrow teeth, showing weak cingula on the upper teeth but a large anterior cingular cusp on the lower molars.

Helvetiodon schuetzi (Figure 10.18) is characterized by larger upper molariform teeth, with well-developed labial and lingual cingula and cusps. Clemens referred several premolariform and incisiform teeth to this taxon.

Hallautherium schalchi is known from small lower teeth. They lack a lingual cingulum and hence a kuehneocone, but they exhibit a posterior labial basin.

Tricuspes occurs in Germany, at Hallau, and at Saint-Nicolas. With regard to this genus, Clemens (1980) noted that "the complexity of the morphology of the crown, particularly the slight angulation in alignment of the main cusps," and the "partially divided root suggest but do not demonstrate a mammalian affinity." Nothing further can be added.

Theria. Therians are rare, and to date they have been recorded only from Saint-Nicolas and Habay-la-Vieille. *Woutersia* (Figure 10.19) is a relatively large and stocky kuehneotheriid diagnosed on the basis of the lower molars, characterized by low cusps, a large kuehneocone, a large cusp *e*, a short labial cingulum, and a small talonid (Sigogneau-Russell, 1983c). Upper molars with a very strong cingulum, which thickens labially to form a ledge, have been attributed to *Woutersia*. *Kuehneotherium* seems to be slightly more primitive than *Woutersia*. Two species of *Kuehneotherium* occur at Saint-Nicolas: *K. praecursoris* and a second species yet to be described.

The Late Triassic mammals of western Europe are the oldest known, but their diversity indicates an even earlier origin for mammals. It is also interesting to note that, like the eutherians at the Cretaceous–Tertiary boundary, these early mammalian taxa survived the Triassic–Jurassic extinctions with no apparent catastrophic decline.

Conclusion

As this review demonstrates, our knowledge of the small tetrapod fauna of the "Rhaetian" of central and western Europe is scanty and is confined almost entirely to isolated teeth. Nevertheless, at least two interesting trends have become apparent as a result of the research of recent decades: the abundance of small cynodonts and the great diversity among the mammals. Both trends are surprising considering the lack of a record of cynodonts and mammals in the underlying strata of the Upper Triassic.

This abrupt change in the faunal composition between middle and upper Norian is surely not the result of a sudden invasion of Europe by cynodonts at the beginning of the "Rhaetian" coupled with the abrupt origin of mammals at the same time. It is more likely that the stark differences in the composition of tetrapod assemblages result from differences in preservation and collecting methods. In the middle Keuper, only large vertebrates have traditionally attracted the attention of collectors, and the small cynodonts and mammals have gone unnoticed, whereas in the "Rhaetian" the fossils mostly represent microvertebrates. The contrast between the middle and upper Norian faunas probably would be decreased if modern methods of screen-washing were to be applied to sedimentary rocks from the lower and middle parts of the Keuper.

The study of nonsynapsid microvertebrates is still in its infancy, especially since only isolated skeletal elements are present. As a result, the faunal assemblages of the Upper Triassic of the Germanic and Paris basins are still very poorly known. However, no valid synthesis can be attempted without taking into account all the components of these assemblages. Moreover, obvious differences from other faunas cannot yet be discussed in terms of time, geography, and ecology and must await more comprehensive studies of the localities discussed earlier.

Acknowledgments

We thank Drs. N. C. Fraser and H.-D Sues for inviting us to the stimulating workshop on early Mesozoic vertebrates. We are extremely grateful to Dr. M. Durand, Vandoeuvres-les-Nancy, for revision and improvement of our geological introduction.

References

Al Khatib, R. 1976. Le Rhétien de la bordure orientale du bassin de Paris et le "calcaire à gryphées" de la région de Nancy. Etude pétrographique et sédimentologique. Thesis, Université Nancy.

Ax, P. 1984. *Das Phylogenetische System*. Stuttgart: Gustav Fischer Verlag.

Benton, M. J. 1986. The Late Triassic tetrapod extinction events. Pp. 303–320 in K. Padian (ed.), *The Beginning of the Age of Dinosaurs: Faunal Change across the Triassic–Jurassic Boundary*. Cambridge University Press.

Bonaparte, J. F., and M. C. Barberena. 1975. A possible mammalian ancestor from the Middle Triassic of Brazil (Therapsida-Cynodontia). *Journal of Paleontology* 49: 931–936.

Buffetaut, E., and G. Wouters. 1986. Amphibian and reptile remains from the Upper Triassic of Saint-Nicolas-de-Port (eastern France) and their biostratigraphic significance. *Modern Geology* 10: 133–145.

Carroll, R. L. 1988. *Vertebrate Paleontology and Evolution*. San Francisco: Freeman.

Clemens, W. A. 1980. Rhaeto-Liassic mammals from Switzerland and West Germany. *Zitteliana* 5: 51–92.

Corroy, G. 1928. Les vertébrés du Trias de Lorraine et le Trias lorrain. *Annales de Paléontologie* 17: 11–56.

Courel, L., M. Durand, J.C. Gall, and G. Jurain. 1973. Quelques aspects de la transgression triasique dans le nordest de la France. Influence d'un éperon bourguignon. *Revue de Géologie physique et Géologie dynamique* 15: 547–553.

Crompton, A. W. 1955. On some Triassic cynodonts from Tanganyika. *Proceedings of the Zoological Society of London* 125: 617–669.

1974. The dentitions and relationships of the southern African Triassic mammals, *Erythrotherium parringtoni* and *Megazostrodon rudnerae*. *Bulletin of the British Museum (Natural History), Geology* 24: 397–437.

Cui G., and Sun A. 1987. [Postcanine root system in tritylodonts.] *Vertebrata Palasiatica* 25: 256–280. [in Chinese]

Duffin, C. J., P. Coupatez, J. C. Lepage, and G. Wouters. 1983. Rhaetian (Upper Triassic) marine faunas from "Le golfe du Luxembourg" in Belgium (Preliminary note). *Bulletin de la Société belge de Géologie* 92: 311–315.

Emmons, E. 1857. *American Geology*, Pt. 6. Albany: Sprague.

Frank, R. M., D. Sigogneau-Russell, and J. C. Voegel. 1984. Tooth ultrastructure of Late Triassic Haramiyidae. *Journal of Dental Research* 63: 661–664.

Freeman, E. 1979. A middle Jurassic mammal bed from Oxfordshire. *Palaeontology* 22: 135–166.

Gill, T. 1872. Arrangement of the families of mammals. With analytical tables. *Smithsonian Miscellaneous Collections* 230(I–IV): 1–98.

Hahn, G. 1973. Neue Zähne von Haramiyiden aus der deutschen Ober-Trias und ihre Beziehungen zu den Multituberculaten. *Palaeontographica* A141: 1–15.

Hahn, G., J. C. Lepage, and G. Wouters. 1984. Cynodontier-Zähne aus der Ober-Trias von Medernach, Grossherzogtum Luxembourg. *Bulletin de la Société belge de Géologie* 93: 357–373.

1987. Ein Multituberculaten-Zahn aus der Ober-Trias von Gaume (S-Belgien). *Bulletin de la Société belge de Géologie* 96: 39–47.

1988. Traversodontiden-Zähne (Cynodontia) aus der Ober-Trias von Gaume (Süd-Belgien). *Bulletin de Institut royal des Sciences naturelles Belge* 58: 171–186.

Hahn, G., D. Sigogneau-Russell, and P. Godefroit. 1991. New data on *Brachyzostrodon* (Mammalia; Upper Triassic). *Geologica et Palaeontologica* 25: 237–249.

Hahn, G., D. Sigogneau-Russell. and G. Wouters. 1989. New data on Theroteinidae – their relations with Paulchoffatiidae and Haramiyidae. *Geologica et Palaeontologica* 23: 205–215.

Hahn, G., R. Wild, and G. Wouters. 1987. Cynodontier-Zähne aus der Ober-Trias von Gaume (S-Belgien). *Mémoires pour servir à l'Explication des Cartes Géologiques et Minières de la Belgique* 24: 1–33.

Hary, A., and A. Muller. 1967. Zur stratigraphischen Stellung des Bonebeds von Medernach (Luxembourg). *Neues Jahrbuch für Geologie und Paläontologie, Monatshefte* 1967: 333–341.

Hopson, J. A., and H. R. Barghusen. 1986. An analysis of therapsid relationships. Pp. 83–106 in N. Hotton III, P. D. MacLean, J. J. Roth, and E. C. Roth (eds.), *The Ecology and Biology of Mammal-like Reptiles.* Washington, D. C.: Smithsonian Institution Press.

Hopson, J. A., and J. W. Kitching. 1972. A revised classification of cynodonts (Reptilia, Therapsida). *Palaeontologia Africana* 14: 71–85.

Huene, E. von. 1933. Zur Kenntnis des württembergischen Rhätbonebeds mit Zahnfunden neuer Säuger und säugerähnlicher Reptilien. *Jahreshefte des Vereins für vaterländische Naturkunde Württemberg* 84: 65–128.

Huene, F. von. 1936. *Die fossilen Reptilien des südamerikanischen Gondwanalandes an der Zeitenwende. Ergebnisse der Sauriergrabungen in Südbrasilien 1928/29.* Lieferung 2. Tübingen: F. F. Heine.

Kermack, D. M., and K. A. Kermack. 1984. *The Evolution of Mammalian Characters.* London: Croom Helm.

Kindlimann, R. 1984. Ein bisher unbekannt gebliebener Zahn eines synapsiden Reptils aus dem Rät von Hallau (Kanton Schaffhausen, Schweiz). *Mitteilungen der naturforschenden Gesellschaft Schaffhausen* 32: 3–11.

Kuhn, O. 1939. Beiträge zur Keuperfauna von Halberstadt. *Paläontologische Zeitschrift* 21: 258–286.

Laugier, R. 1971. Le Lias inférieur et moyen du Nord-Est de la France. *Science de la Terre* 21: 1–291.

Levallois, J. 1862. Aperçu de la constitution géologique du département de la Meurthe. *Mémoires de l'Académie Stanislas* 1862: 246–301.

Lucas, S. G., and W. Oakes. 1988. A Late Triassic cynodont from the American South-West. *Palaeontology* 31: 445–449.

McIntyre, G. T. 1972. The trisulcate petrosal pattern of mammals. *Evolutionary Biology* 6: 275–303.

Martin, M., D. Sigogneau-Russell, P. Coupatez, and G. Wouters. 1981. Les Cératodontidés (Dipnoi) du Rhétien de Saint-Nicolas-de-Port (Meurthe-et-Moselle). *Géobios* 14: 773–791.

Osborn, H. F. 1886. A new mammal from the American Triassic. *Science* 8: 540.

Parrington, F. R. 1946. On a collection of Rhaetic mammalian teeth. *Proceedings of the Zoological Society of London* 116: 707–728.

Peyer, B. 1956. Ueber Zähne von Haramiyden, von Triconodonten und von wahrscheinlich synapsiden Reptilien aus dem Rhät von Hallau Kt. Schaffhausen, Schweiz. *Schweizerische Paläontologische Abhandlungen* 72: 1–72.

Plieninger, W. 1847. *Microlestes antiquus* und *Sargodon tomicus* in der Grenzbreccie von Degerloch. *Jahreshefte des Vereins für vaterländische Naturkunde Württemberg* 3: 164–167.

Romer, A. S. 1969. The Chañares (Argentina) Triassic reptile fauna. 5. A new chiniquodontid cynodont, *Probelesodon lewisi* – cynodont ancestry. *Breviora* 333: 1–24.

Rowe, T. 1988. Definition, diagnosis, and origin of Mammalia. *Journal of Vertebrate Paleontology* 8: 241–264.

Russell, D., D. Russell, and G. Wouters. 1976. Une dent d'aspect mammalien en provenance du Rhétien français. *Géobios* 9: 377–392.

Schalch, F. 1873. Beiträge zur Kenntniss der Trias am südöstlichen Schwarzwalde. Doctoral dissertation, Universität Würzburg.

Sigogneau-Russell, D. 1983a. Caractéristiques de la faune mammalienne du Rhétien de Saint-Nicolas-de-Port. *Bulletin Informatique de la Géologie du Bassin de Paris* 20: 5–53.

    1983b. Nouveaux taxons de Mammifères rhétiens. *Acta Palaeontologica Polonica* 28: 233–249.

    1983c. A new therian mammal from the Rhaetic locality of Saint-Nicolas-de-Port (France). *Zoological Journal of the Linnean Society* 78: 175–186.

    1989. Haramiyidae (Mammalia, Allotheria) en provenance du Trias supérieur de Lorraine (France). *Palaeontographica* A206: 137–198.

    1990. Reconnaissance formelle d'une nouvelle espèce d'*Haramiya* dans l'hypodigme français des Haramiyidae (Mammalia, Allotheria). *Bulletin du Muséum National d'Histoire Naturelle, Paris, sér. 4* (12), C, 1: 85–88.

Sigogneau-Russell, D., R. M. Frank, and J. Hemmerlé. 1986. A new family of mammals from the lower part of the French Rhaetic. Pp. 99–108 in K. Padian (ed.), *The Beginning of the Age of Dinosaurs: Faunal Change across the Triassic–Jurassic Boundary.* Cambridge University Press.

Simpson, G. G. 1926. Mesozoic Mammalia. V. *Dromatherium* and *Microconodon. American Journal of Science* (5) 12: 87–108.

Wible, J. R. 1991. Origin of Mammalia: the craniodental evidence reexamined. *Journal of Vertebrate Paleontology* 11: 1–28.

Wouters, G., J. C. Lepage, and D. Delsate. 1985. Nouveau gisement de thérapsides et mammifères dans le Trias supérieur d'Attert, en Lorraine belge. *Bulletin de la Société belge de Géologie* 94: 1– 253.

Wouters, G., D. Sigogneau-Russell, and J. C. Lepage. 1984. Découverte d'une dent d'Haramiyidé (Mammalia) dans des niveaux rhétiens de la Gaume (en Lorraine belge). *Bulletin de la Société belge de Géologie* 93: 351–355.

Ziegler, P. A. 1982. *Geological Atlas of Western and Central Europe.* Amsterdam: Shell International.

Appendix 10.1: Cynodontia: Therioherpetidae Bonaparte and Barberena, 1975

Meurthodon gen. nov.

Derivation of name. After the river Meurthe (Sigogneau-Russell, 1983).

Type species. Meurthodon gallicus sp. nov. (by monotypy).

Distribution. Upper Triassic ("Rhaetian") of Saint-Nicolas-de-Port (= Rosières-aux-Salines), France.

Diagnosis. Crown of the postcanines tetracuspid; all cusps laterally compressed, arranged in a longitudinal row, very distinctly separated from each other and with a cutting function. The first cusp is the smallest, and the second cusp is the largest; cusps three and four are somewhat smaller than the second cusp, but distinctly larger than the first cusp. No cingula. Crown distinctly separated from the root. Root completely subdivided in a mammal-like manner.

Discussion. *Meurthodon* differs from *Therioherpeton* mainly in the structure of the postcanine root. In *Therioherpeton* the root is also divided, but both parts are connected by a thin sheet of dentine. In *Meurthodon* this sheet is reduced, and the two roots are completely separated. The cusps are higher and better defined in *Meurthodon* than in *Therioherpeton*, but this may be a result of stronger wear in *Therioherpeton*. For further comparisons, see Russell et al. (1976).

Meurthodon gallicus sp. nov.

Derivation of name. *gallicus* (Latin), Gallic, referring to "Gallia," the Roman name for France.

Holotype. Postcanine tooth SNP-1-W (Figure 10.11), housed in the Institut de Paléontologie, Muséum National d'Histoire Naturelle, Paris (Russell et al., 1976, pl. 1, figs. 1–3).

Type locality. Quarry at Rosières-aux-Salines, region of Saint-Nicolas-de-Port, near Nancy, Lorraine, France.

Type stratum. "Rhaetian," Upper Triassic.

Diagnosis. Only known species of *Meurthodon*, as diagnosed earlier.

Description. See Russell et al. (1976, pp. 379–380).

11

Assemblages of small tetrapods from British Late Triassic fissure deposits

NICHOLAS C. FRASER

Introduction

Ancient fissure and cave systems are important sources of fossil terrestrial vertebrates. Pleistocene cave deposits are particularly notable sources of fossil vertebrates, and there is a considerable literature documenting widespread accumulations resulting from roosting raptor pellets, middens, and fluvial agents, as reviewed by Andrews (1990). Older systems are equally important, and the early Mesozoic cave and fissure systems in Britain are well known for their faunas of reptiles and early mammals (Robinson, 1957a; Kermack, Mussett, and Rigney, 1973; Fraser, 1985), but details of the sedimentology and taphonomy are limited. Predators have been cited as possible agents of bone accumulation for some of the Mesozoic fissure assemblages (Kühne, 1956; Fraser and Walkden, 1983; Evans and Kermack, Chapter 15).

For Pleistocene cave systems, accurate ages for the contained sediments frequently can be determined from uranium-series dating of calcite speleothems, but there is much more difficulty in dating the Mesozoic cave and fissure deposits. In the Mesozoic systems the consolidated sediments usually are isolated from normally bedded sequences; pollen and spores either are absent or, where present, do not provide an age beyond a very coarse resolution. Dating of the sediments has largely been based on macrofossil evidence – principally vertebrates – and that has generated much controversy.

The Triassic of Britain

Detailed interpretation of the depositional environments of the Triassic in Britain is difficult because of a general lack of macrofossils and diagnostic sedimentary structures. Nevertheless, much of the period is represented by continental red beds, and sediment deposition is thought to have taken place mostly under arid or semiarid conditions, with the sediments possibly representing giant playa or desert plain deposits (Tucker, 1977, 1978; Jeans, 1978). Toward the end of the period, a major change was effected by a progressive marine transgression (Warrington, 1981). The red Keuper marls were replaced by green and gray mudstones that are comparable to sequences from modern sabkha environments. In southwest Britain, the transition to full marine conditions was abrupt, and the top of the marl, which marks the base of the Penarth Group (Westbury Formation), is intensely burrowed. Pebble beds and bonebeds also typify the transition to marine sequences in some areas. Both marine and freshwater species are represented in the bonebeds. Above the Westbury Formation lies the Cotham Member of the Lilstock Formation (Figure 11.1), which includes nonmarine gray and green marls, with local Keuper-like red marls. These indicate fluctuating conditions and periodic temporary retreats of the sea, with local reestablishment of subaerial conditions (Hamilton, 1961; Poole, 1978; Warrington, 1978), before renewed onsets of the marine transgression, which continued into the Jurassic and eventually inundated the whole of southwest Britain. Extensive occurrences of fissure and cave deposits offer us the potential to gain a clear and detailed picture of the nature of terrestrial faunas at that time period. The fissures generally regarded as Triassic in age will be discussed here, and the Jurassic systems are reviewed by Evans and Kermack in Chapter 15.

The local base of the Jurassic is recognized by the appearance of the ammonite *Psiloceras* (Cope et al., 1980; Warrington et al., 1980). Clearly, this is of no value in determining the boundary in terrestrial deposits, but the assemblages from the western localities of the Vale of Glamorgan and Windsor Hill (Figure 11.2) are normally regarded as Jurassic in age, and

CHRONOSTRATIGRAPHY			LITHOSTRATIGRAPHY	
Warrington *et al.* 1980	Benton (This Volume)			
HETTANGIAN	HETTANGIAN		*Planorbis* Zone	Blue Lias
			Pre-*Planorbis* Zone	
RHAETIAN	LATE NORIAN 2	SEVATIAN	Lilstock Formation	Penarth Group
			Westbury Formation	
	LATE NORIAN 1		Blue Anchor Formation	Mercia Mudstone Group
			Keuper Marl	
NORIAN	MIDDLE NORIAN 2	ALAUNIAN		
	MIDDLE NORIAN 1	LACIAN		
	EARLY NORIAN			
CARNIAN	CARNIAN	TUVALIAN		

Figure 11.1. Chronostratigraphy and local lithostratigraphy of southwest Britain.

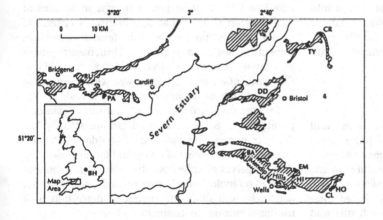

Figure 11.2. The Bristol Channel area showing the locations of the main Triassic vertebrate-bearing karst systems: BA, Batscombe; BH, Breedon-on-the-Hill; CL, Cloford; CR, Cromhall; DD, Durdham Down; EM, Emborough; HI, Highcroft; HO, Holwell; PA, Pant-y-ffynon; RU, Ruthin; TY, Tytherington.

they will thus be omitted from this discussion. The remaining fissure localities (Figure 11.2) are here regarded as mostly Triassic in age (although the Mesozoic sediments at Holwell extend into the Jurassic), and they comprise both terrestrial and marine assemblages.

The broad paleoenvironmental picture of southwest Britain in Late Triassic and Early Jurassic times is apparently one of change from essentially playa lakes to a shallow sea dotted with small limestone islands. However, the precise nature and timing of this change, and its effects on the terrestrial faunas, remain unresolved, and there have been several rather different interpretations of the fissure faunas.

Some authors consider the majority of Triassic assemblages lacking any apparent marine influences to predate the transgression (Fraser and Walkden, 1983), whereas others argue that these assemblages actually postdate the initial transgression and are part of the Cotham Member (Whiteside and Robinson, 1983) and are therefore representative of periodic minor regressions and perhaps localized temporary returns to sabkha-like conditions. Whatever the outcome of this controversy, the profound environmental change at the close of the Triassic is undisputed, and consequently any perceived changes in local assemblages around the Triassic–Jurassic boundary may at least partly reflect paleoenvironmental changes and may be independent of various patterns of global mass extinctions that have been postulated for the close of the Triassic (Hallam, 1981, 1990; Olsen and Sues, 1986; Benton, 1991).

Before the distinction between Triassic and Jurassic fissure assemblages can be discussed further, the use of the Rhaetian stage needs to be addressed. Gümbel (1861) introduced the Rhaetian stage and designated it as the uppermost part of the Triassic. However, as Buffetaut and Wouters (1986) indicated, many stratigraphers have placed it at the base of the Jurassic, and the validity of the Rhaetian has been seriously questioned in recent years. Tozer (1979, 1984) noted that there is no clear distinction between the upper Norian and Rhaetian ammonoid faunas. Fisher and Dunay (1981) were unable to provide any palynological evidence to support a well-defined Rhaetian stage. In Britain, the marine incursion that established characteristic "Rhaetic" sediments affected wide areas almost simultaneously (Kent, 1970). Consequently, it has been used as a stratigraphic marker in British sequences. Nevertheless, in terms of correlating these same sequences with deposits worldwide, the lithostratigraphic term "Rhaetic" is of limited value, particularly because it does not even correspond exactly to the chronostratigraphic stage name "Rhaetian," which generally extends somewhat lower. Benton (Chapter 22) has attempted to split the terrestrial Norian into five units, namely, early, middle 1, middle 2, late 1, and late 2, the last being equivalent to the Rhaetian. Following this scheme, late Sevatian (uppermost Norian) will be substituted for Rhaetian.

History of research

Published work on the British Mesozoic fissure and cave localities is mainly restricted to those areas neighboring the Bristol Channel, but vertebrate material has been reported from Triassic karst features in the Carboniferous limestone of the English Midlands (A.R.I. Cruickshank, pers. commun.). I have collected fragments of bone from Mesozoic sediments in fissure and cave systems at Breedon-on-the-Hill Quarry, Leicestershire (Figure 11.2), but they are poorly preserved and not diagnostic. Although it is probably not as rich as the Bristol Channel region, the region of the Midlands should not be overlooked in future studies.

The history of research into the Bristol Channel Mesozoic fissure systems goes back to the mid-1800s (Moore, 1867, 1881), and work has continued intermittently on the vertebrate assemblages since then; for details, see Evans, and Kermack (Chapter 15). Although many of the classic Triassic localities, such as Gurney Slade, Highcroft, and Batscombe (Figure 11.2), are either completely overgrown or fully worked through quarrying operations, other sites are still actively

quarried. In the latter, including Holwell, some of the earliest recognized fissure systems are still exposed, but continued quarrying operations have progressively exposed and then quarried away further Mesozoic karst features.

For a long time it was generally accepted that the vertebrate assemblages of the British Mesozoic fissure deposits could be broadly separated into "Norian" (essentially early Sevatian) and "Rhaeto-Liassic" (late Sevatian and younger) types. Robinson (1957a) regarded the sediments contained within mature solution-etched features, and bearing archosaurs, lepidosaurs, and procolophonids (but no palynomorphs), to be early Sevatian in age, and the assemblages as representing an upland fauna. These have been referred to as the "Complex-A" assemblages (Shubin and Sues, 1991). By contrast, the apparently late Sevatian and younger ("Rhaeto-Liassic") deposits supposedly occur in immature slot-type fissures, and Robinson presumed that these reflected an island archipelago paleoenvironment, with mammals and nonmammalian synapsids, in addition to archosaurs and lepidosaurs. Palynomorphs of cycads, ferns, and conifers are also usually present. These have been referred to as the "Complex-B" assemblages.

This broad separation of the British fissures was challenged on the basis of two separate lines of evidence. First, Marshall and Whiteside (1980) identified characteristic late Sevatian palynomorphs in fissures at Tytherington Quarry that by Robinson's criteria of fissure type and the nature of the faunal assemblage were of the classic Complex-A type. Thus, fissure systems and faunal assemblages that Robinson regarded as diagnostic of an early Sevatian age were shown also to occur in late Sevatian sediments. Whiteside (1983, 1986) even suggested that many (if not all) of Robinson's putative early Sevatian assemblages are actually late Sevatian in age. Marshall and Whiteside (1980) also questioned the concept of an upland fauna, indicating that the palynomorphs from the Tytherington fissures showed no indication of being upland forms and that acritarchs and dinocysts suggested deposition in a marginal marine environment.

With the discovery of a therian mammal tooth at a second fissure Complex-A locality, Emborough Quarry (Fraser, Walkden, and Stewart, 1985), Robinson's twofold division was once more questioned. However, in that instance, the age of the sediments was not called into question. Robinson relied heavily on a topographic study as evidence that the sediments were deposited prior to the late Sevatian. These arguments were emphasized by Fraser et al. (1985), and they need not be repeated here. Accepting the Emborough deposits and fossils as early Sevatian or older, the presence or absence of synapsids has no direct bearing on the age of these fissure assemblages. So although both Triassic and Jurassic fissure deposits are still recognized by all current workers, the previously accepted broad distinctions in the constituent faunas are invalid.

The western fissure localities in the Vale of Glamorgan are situated on what was once one of the highest limestone plateaus in southwest Britain. During Early Jurassic times this plateau formed an island, known as St. Bride's Island (Evans and Kermack, Chapter 15). Robinson (1971) and Cope et al. (1980) have presented good evidence indicating that St. Bride's Island was completely submerged by early Sinemurian times, placing an upper age constraint on the terrestrial faunas. The lower age limit is less well constrained. Whiteside and co-workers prefer a range extending upward from the late Sevatian, whereas Fraser and co-workers consider that they extend down at least to the early Sevatian. More recently, Simms (1990) and Benton (1991) have suggested that some of the assemblages may even date back to the middle or late Carnian.

As a result of the present blurring between Triassic and Jurassic fissure assemblages, there is inevitably some overlap in this discussion of the Late Triassic localities (Complex-A) and the review of the Early Jurassic sites (Complex-B) by Evans and Kermack (Chapter 15), particularly because certain localities apparently contain both Triassic and Jurassic infills (e.g., Holwell). Evidence will be presented here suggesting that subtle differences in the fissure assemblages may be age-correlated and that it might ultimately prove possible to place at least some of the Triassic fissure deposits into an accepted age sequence based on a few widespread genera and species that were of limited distribution in time.

British Triassic vertebrate-bearing fissure localities

The main localities regarded here to be Triassic in age are as follows:

England

Holwell (in part) U.K. National Grid Reference (ST 727 452)
Cromhall (ST 704 916)
Tytherington (ST 660 890)
Emborough (ST 623 505)
Batscombe (ST 460 550)
Cloford (ST 702 475)
Highcroft (Gurney Slade) (ST 623 499)
Durdham Down (ST 565 748)?

Wales

Ruthin (SS 975 796)
Pant-y-ffynon (ST 046 743)

Figure 11.3. Triassic fissure system at Cromhall Quarry showing the Mesozoic sediments against the Carboniferous limestone host rock, and the surface opening.

The Triassic paleokarst features of Britain consist of a range of different types. At Cromhall Quarry, many of the features have been considered to consist of separate fissures and chambers developed underneath dolines (Figure 11.3). Elsewhere there are larger caverns that are lined by thick bands of calcite with large dog-tooth crystals projecting into the infilling sediments. The large calcite crystals are clearly zoned, and the surfaces show evidence of erosion. Simms (1990) compared these features to crystal-lined passages in the Jewel Cave (South Dakota) and made the interesting suggestion that, like the Jewel Cave, the Cromhall calcite-lined caverns and passages form parts of thermal-spring conduits. Although other solution-etched fissures at Cromhall lack the calcite lining, the sediment infills contain detrital fragments of the calcite, indicating that these sediments postdate those filling the calcite-lined features. Sometimes these Triassic water-courses are exposed as very large caverns, such as the one at Pant-y-ffynon.

Neptunian dykes, open submarine slots infilled by direct sedimentation, are also a feature of some of the Mesozoic deposits in the Bristol Channel area. Holwell Quarry exhibits some good examples of Neptunian

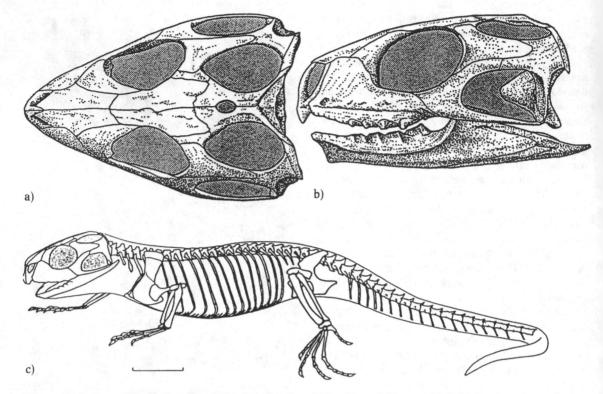

a)

b)

c)

Figure 11.4. Restoration of *Clevosaurus hudsoni*: skull in dorsal (a) and lateral (b) views; (c) skeleton in lateral view. Scale bar represents 2.0 cm. (After Fraser, 1988a.)

dykes together with injection slots. In the case of injection fissures, they opened up beneath overlying Mesozoic sediments, which at the time were in various stages of lithification. Sediments were sucked into the deep parallel-sided slots, and the resulting infill, characterized by vertically aligned clasts, contrasts with the horizontal lamination of the sediments in the Neptunian dykes. Injection fissures may represent single or multiple phases of tectonic movement. Tectonically generated fissures also opened subaerially, and these filled with windblown debris and broken wall rock.

The major English localities

Holwell Quarry

As mentioned earlier, the relatively rapid marine transgression toward the close of the Triassic provides a useful local stratigraphic marker in southwest Britain. The so-called Rhaetic bonebeds, occurring in the basal part of the Penarth Group, contain abundant teeth, scales, spines, and bones. These bonebeds presumably reflect successive reworked strandline deposits. Typical of the smaller elements are teeth of the shark genera *Polyacrodus*, *Lissodus*, and *Palaeospinax* and scales of *Gyrolepis*. Certain fissure deposits at Holwell Quarry have been widely regarded as basal Penarth Group on the basis of the occurrence of a similar faunal suite

(Kühne, 1946; Savage, 1977). Recent studies of material originally collected from Holwell Quarry by Charles Moore in the mid-1800s (now housed in Bath Museum) and by Walter Kühne in 1939 (housed in the University Museum of Zoology, Cambridge) have also revealed sphenodontid remains along with the rich fish assemblage (Fraser and Duffin, in prep.). The majority of the jawbones are attributed to the genus *Diphydontosaurus*, but a single premaxilla has been identified as *Clevosaurus* sp. aff. *C. hudsoni*.

Cromhall Quarry

Cromhall Quarry is the most extensively studied of all the Triassic fissure localities. Vertebrate remains were first identified there in 1938 by F. G. Hudson and described by Swinton (1939) as a new species of sphenodontian, *Clevosaurus hudsoni* (Figure 11.4). Since then, a number of different authors have documented further new tetrapods from that locality (Robinson, Kermack, and Joysey, 1952; Robinson, 1957a; Halstead and Nicoll, 1971; Fraser, 1982, 1986a; Fraser and Walkden, 1984; Fraser and Unwin, 1990). At present, the total number of tetrapod taxa identified exceeds 15, but the list continues to grow as more matrix is prepared. Furthermore, new fissures continue to appear, and these sometimes contain vertebrate material. Consequently, while work continues at this locality, the faunal list is expected to increase slowly.

Many different karstic features have been exposed at different times in Cromhall Quarry, and many of these are still at least partly preserved. However, the formation and subsequent filling of the paleokarstic features at Cromhall are not clearly understood. The majority of vertebrate remains recovered from Cromhall have been found from a line of fissure/cave systems on the western side of the quarry (Fraser, 1985). Sphenodontians are the most common group, but a number of different archosaurian and procolophonid remains have been collected. More recently, the southeastern part of the quarry has been worked, revealing a sequence of normally bedded Mesozoic sediments that includes marls overlying a dolomitic conglomerate. The marls contain abundant fish fragments and remains of the sphenodontians *Clevosaurus hudsoni* and *Diphydontosaurus* sp. Deep narrow fissures filled with red, green, and yellow clays also occur in the southeastern corner. They probably represent grikes or small invasion vadose caves (Simms, 1990). Some of these slot features extend directly from the cover sequence and contain similar vertebrate remains.

Based on the sedimentology and geomorphology, Walkden and Fraser (1993) have outlined a time scale (Figure 11.5) for the filling of each of the different sites at Cromhall relative to the cover sequence. Fraser and Walkden (1983) argued that in most cases the individual fills and associated assemblages at Cromhall reflected once-living communities and are not greatly biased by particular environmental or taphonomic factors. The distribution of vertebrate remains within the different karstic infills and the cover sequences are relatively constant. Consequently, vertebrates are potential fine-scale biostratigraphic indicators.

The assemblages of fish and tetrapod remains contained within the sequences of marls (and associated slot fissures) overlying the dolomitic conglomerate are remarkably similar to the basal Penarth Group assemblages from the Holwell fissures and to Penarth Group sequences at Aust Cliff, Blue Anchor Point, Garden Cliff, and elsewhere (Fraser and Duffin, in prep.). Thus the youngest sediments at Cromhall can be assigned with some certainty to the basal Penarth Group (i.e., Westbury Formation). Walkden and Fraser (1993) have argued that the fissure/cave systems on the western side of Cromhall Quarry all predate the cover sequence. Furthermore, they suggest that the order in which the systems were filled can be deduced, which theoretically allows us to date all the Mesozoic assemblages at Cromhall relative to the base of the Penarth Group and assess the range of the various taxa (Figure 11.6). Although the upper age limit of the Cromhall sediments is now fairly well constrained, the lower limit remains uncertain. Simms (1990) considers formation of the fissures and caves to have been initiated during middle Carnian times; on that assumption, the oldest sediments must be at least slightly younger than

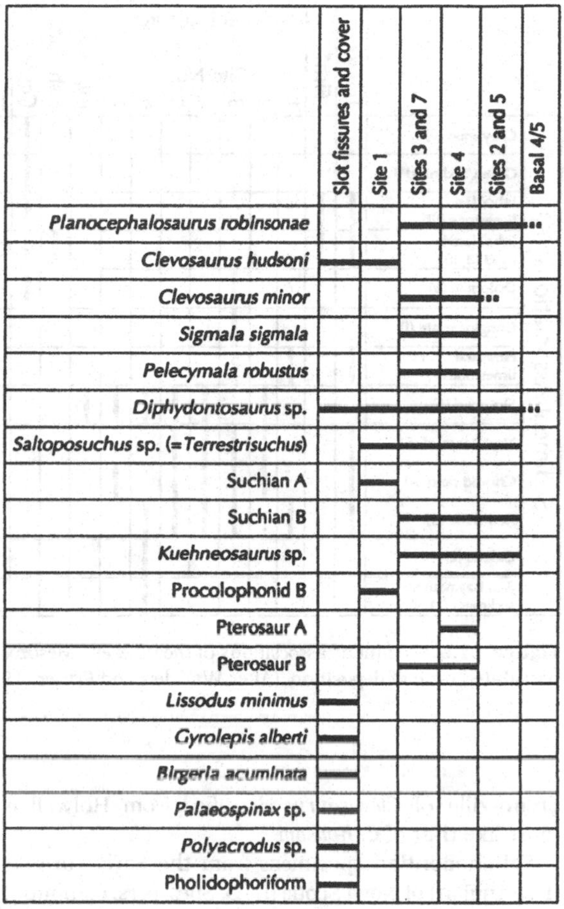

	Slot fissures and cover	Site 1	Sites 3 and 7	Site 4	Sites 2 and 5	Basal 4/5
Planocephalosaurus robinsonae			■	■	■	•••
Clevosaurus hudsoni	■	■				
Clevosaurus minor			■	■	•••	
Sigmala sigmala			■	■		
Pelecymala robustus			■	■		
Diphydontosaurus sp.	■	■	■	■	■	•••
Saltoposuchus sp. (= *Terrestrisuchus*)		■	■	■	■	
Suchian A		■				
Suchian B					■	
Kuehneosaurus sp.			■			
Procolophonid B		■				
Pterosaur A					■	
Pterosaur B					■	
Lissodus minimus	■					
Gyrolepis alberti	■					
Birgeria acuminata	■					
Palaeospinax sp.	■					
Polyacrodus sp.	■					
Pholidophoriform	■					

Figure 11.5. Distribution of vertebrates at Cromhall Quarry, based on sedimentological interpretations by Walkden and Fraser (1993); age of sediments increases from left to right.

middle Carnian. Benton (1991) noted that certain elements of the German Stubensandstein (lower to middle Norian) and the Lossiemouth Sandstone Formation (upper Carnian) (e.g., *Leptopleuron*, *Brachyrhinodon*, and *Scleromochlus*) are closely comparable to forms from Cromhall.

Although these theories remain to be tested, it is interesting to note certain points concerning the distribution of the various genera and species. First, on the basis of this study, the sphenodontian *Planocephalosaurus* may be a relatively abundant and possibly long-ranging genus. There is no indication, however, that it extends beyond the Blue Anchor Formation. Second, although *Clevosaurus* is present in most Mesozoic vertebrate-bearing sediments, the distribution at the species level seems to be quite restricted, with *C. hudsoni* occurring only in the cover sequence and in fissure infills that Walkden and Fraser (1993) regard as the youngest. It is perhaps significant that the

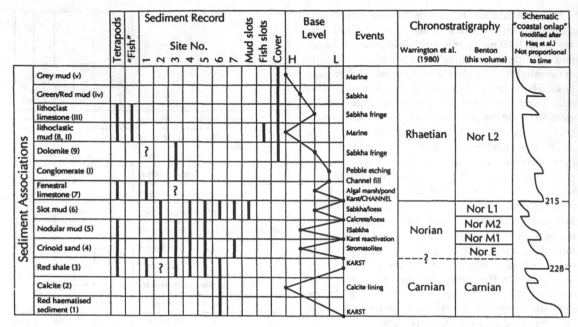

Figure 11.6. Sediment associations of the different Mesozoic fissure infills at Cromhall Quarry, together with the postulated time of deposition. (After Walkden and Fraser, 1993.)

premaxilla of *Clevosaurus* identified from Holwell is most like that of *C. hudsoni*.

Sphenodontian specimens from the Early Jurassic (Hettangian) of Nova Scotia (H.-D. Sues, pers. commun.) and from the Early Jurassic of the Lufeng basin (Wu, Chapter 3) have been attributed to *Clevosaurus*. However, they are somewhat different from the species of *Clevosaurus* described from Cromhall Quarry. On the other hand, fragmentary sphenodontian material, currently being studied by D. Pacey, from a fissure at Pant Quarry designated as Pant 4 (Evans and Kermack, Chapter 15) seems to more closely resemble the Canadian and Chinese clevosaurs. The Pant 4 fissure probably is early Sinemurian in age (Evans and Kermack, Chapter 15).

Although the sphenodontians are the most widespread, most abundant, and most fully described group of tetrapods in the Cromhall fissure deposits, archosaurs and procolophonids also form important components of the Triassic tetrapod assemblages, and these may also be of some stratigraphic significance. Preliminary work on the procolophonids from Cromhall Quarry indicates that there are at least two species that closely resemble *Leptopleuron* from the Carnian of Scotland, but the Cromhall specimens possess a third small quadratojugal spine in addition to the two prominent spines present in *Leptopieuron*. It has also been suggested that *Tricuspisaurus*, originally described by Robinson (1957b) from Ruthin, is a procolophonid (Fraser, 1986b). If it can be shown more conclusively, as has

been suggested (Olsen, 1986), that procolophonids did not survive the Triassic–Jurassic boundary, then their occurrence in certain of the fissure deposits implies a Triassic age. However, at present, Late Triassic procolophonids are insufficiently described to be of value as fine-scale biostratigraphic indicators (e.g., they cannot be used to distinguish between the Mercia Mudstone Group and Penarth Group).

There are very few descriptions of archosaurs from the Triassic fissure assemblages, and these are largely confined to the articulated remains of *Terrestrisuchus* (= *Saltoposuchus*) (Figure 11.7) and *Thecodontosaurus* (Figure 11.8) from Pant-y-ffynon. However, there is a greater abundance and diversity of archosaurs than the literature suggests. A sphenosuchian crocodylomorph, *Saltoposuchus* (= *Terrestrisuchus*), has been documented from Cromhall Quarry (Fraser, 1985), and there may be more than one species. Perhaps of more importance are two completely new archosaurian taxa that are well represented by disassociated elements. One exhibits a cranial structure that resembles that of sphenosuchian crocodylomorphs: The squamosal is kidney-shaped and lacks a ventral process, and the maxilla has the same proportions as that in *Saltoposuchus*, but the teeth are subconical and lack serrations. However, in the structure of the braincase, it has a curious mosaic of characteristics: The parabasisphenoid is primitive, with two large carotid foramina positioned close to the midline, reminiscent of sphenodontians, but the opisthotic-prootic region

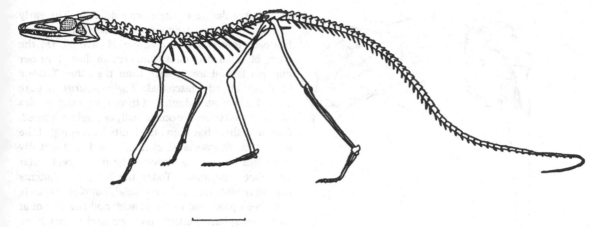

Figure 11.7. Restoration of *Saltoposuchus gracilis*. Scale bar represents 4.0 cm. (After Crush, 1984.)

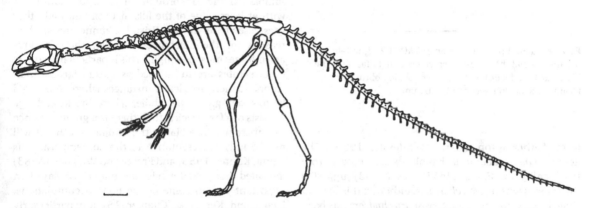

Figure 11.8. Restoration of *Thecodontosaurus*. Scale bar represents 4.0 cm. (After Kermack, 1984.)

appears advanced, with a well-developed metotic fissure bounded by a loop of the opisthotic. From what is known of the postcranial skeleton, it exhibits no crocodylomorph characteristics: The coracoid does not possess an elongated posterior process, and there is no indication of the pubis being excluded from the acetabulum. The ulnare and radiale are unknown. At least superficially the skull shows some resemblance to *Erpetosuchus*. It also shows a similarity to *Gracilisuchus*, which has also previously been compared to crocodylomorphs (Brinkman, 1981).

The second of the two archosaurs, which does not occur together with the first, is a much more gracile form, with a very lightly built skull. Although this archosaur is known only from completely disassociated material, elements are relatively abundant, and significant amounts of the cranial and postcranial skeleton are known. Superficially it resembles the enigmatic genus *Scleromochlus*. The antorbital fenestrae are greatly enlarged. The configuration of the temporal region is reminiscent of aetosaurs, and it is thought that the temporal fenestrae are reduced. Of particular interest is the structure of the pelvic girdle, where the pubis bears two obturator foramina (Figure 11.9). Benton and Clark (1988) cite this as a synapomorphy of the Stagonolepididae. However, the distribution of this character among basal archosaurs is not well known, and the feature may occur in forms such as *Euparkeria*.

Additional archosaurian fragments include remains of a rhamphorhynchoid pterosaur (Fraser and Unwin, 1990) and possible stagonolepidid bones, including an osteoderm (Fraser, 1988b). There is a variety of isolated laterally compressed teeth bearing serrations that are not all consistent with those of *Saltoposuchus*, indicating the presence of at least one further archosaur.

Tytherington Quarry

Detailed descriptions of sphenodontian material from other Mesozoic fissure localities in the Bristol Channel region are limited. The cranial morphology of *Diphydontosaurus avonis* has been described in some detail

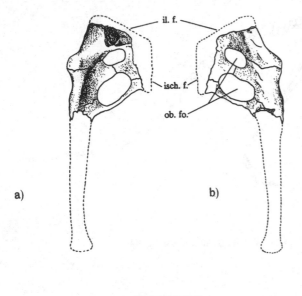

a)   b)

Figure 11.9. Pubis of suchian B, VMNH 203, in lateral (a) and medial (b) views. Abbreviations: il. f., facet for the ilium; isch. f., facet for the ischium; ob. fo., obturator foramen. Scale bar represents 5.0 mm.

from Tytherington Quarry (Whiteside, 1986). The genus *Planocephalosaurus* has also been recorded from this locality, but Fraser and Walkden (1983) suggested the species was not *P. robinsonae*, although this has not since been substantiated. *Clevosaurus hudsoni* has been documented from Tytherington, and additional species may be present. In addition to *Thecodontosaurus* and *Saltoposuchus*, D. I. Whiteside (pers. commun.) has identified at least one other archosaur at Tytherington Quarry.

Like many of the other localities in the region, the vertebrates from Tytherington have originated from a number of different fissure and cave systems. Largely on the basis of palynological evidence, Whiteside and colleagues (Marshall and Whiteside, 1980; Whiteside and Robinson, 1983; Whiteside, 1986) have advocated a late Sevatian age (Westbury Formation or Cotham member of the Lilstock Formation) for the whole suite of Tytherington fissures. Whereas this is undoubtedly true for some of the vertebrate-bearing Mesozoic sediments, not all the deposits contain palynomorphs. Furthermore, like Cromhall, not all species listed for the locality occur together in the same deposit, but a detailed faunal profile for each fissure has not been published to date. Therefore, the fissure fills at Tytherington are not necessarily all contemporaneous.

Emborough Quarry
Emborough Quarry has been extensively studied and has been central to much of the previous research on

the relative dating of the fissure deposits. The early Mesozoic sediments are confined to a single pocket that has been interpreted as a collapsed cave fill. On the basis of published data, Emborough has a rather different faunal assemblage than the other Triassic localities already discussed. *Kuehneosaurus* (Figure 11.10) is very abundant, and three additional reptiles and *Kuehneotherium* are occasionally or rarely recovered. However, there has been only limited screening of the matrix for disassociated elements, and undoubtedly there has been a bias toward the more spectacular articulated specimens. Today the Triassic sediments are not readily accessible, but small samples of matrix have been processed in acetic acid, and remains of at least one sphenodontian and two archosaurs have been identified. At the same time, *Kuehneosaurus* was still found to be the most common form in these samples. If the Emborough fill is contemporaneous with at least some of the fills at Cromhall and other sites elsewhere in the Bristol Channel area, then reasons for the differences between these various assemblages must be sought. This is particularly critical if vertebrates are to be used as biostratigrphic indicators, as there are clearly instances where they could be misleading. On the whole, where the assemblage consists of a few species, then there is a greater chance that there will be a bias in the sample and that it will not be fully representative of the contemporaneous fauna. Kühne (1956) and Fraser and Walkden (1983) indicated that predator selection may have played an important role in some of the bone accumulations. Evans and Kermack (Chapter 15) are particularly concerned that some of the observed faunal differences have resulted from artifacts of preservation and sampling. Certainly, low-diversity assemblages are more likely to be unreliable for biostratigraphic purposes, but in relatively high diversity assemblages (as seen at Cromhall) biased samples are not considered to be a significant factor (Fraser and Walkden, 1983, p. 345).

Other English localities (nonworking quarries)

Other localities on the English side of the Bristol Channel area that have been considered to bear Triassic assemblages include Batscombe, Cloford, Highcroft, Durdham Down, and Barnhill quarries.

Batscombe
On the basis of the abundant remains of *Kuehneosaurus*, Robinson considered the Mesozoic sediments at Batscombe Quarry to be contemporaneous with Emborough. No other taxa were recorded from Batscombe, and unfortunately the site is now completely overgrown. Robinson (1967) erected a new genus, *Kuehneosuchus*, for the Batscombe gliding reptile, but the supposed differences between the Emborough and Batscombe forms have not since been confirmed.

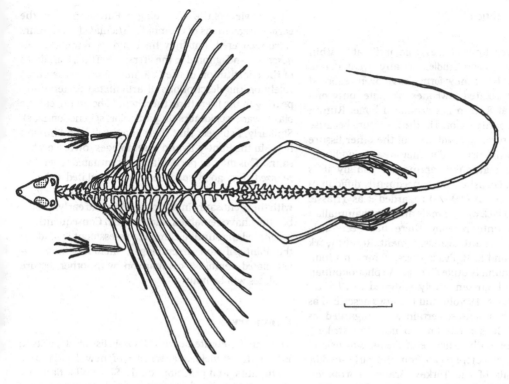

Figure 11.10. Restoration of *Kuehneosaurus*. Scale bar represents 4.0 cm. (After Robinson, in Romer, 1966.)

Cloford

To date, sediments collected from Cloford Quarry have yielded little terrestrial material, but the assemblage of abundant teeth of *Palaeospinax, Polyacrodus, Birgeria, Hybodus minor,* and *Lissodus minimus* resembles that of the Holwell fissures, which are considered to be late Sevatian (=Rhaetian) in age. Preliminary analysis of archosaur material recovered from Cloford Quarry indicates the presence of *Rysosteus.* Halstead and Nicoll (1971) document the occurrence of *Rysosteus* (=*Ryzosteus*), a small terrestrial alleged archosaur from the pebble beds and bonebed at Aust Cliff. Relatively abundant bones of this form were recovered in 1963 and are now housed in Bristol Museum, although they have not been described in detail. In addition, the genus has been documented from the Westbury Formation at Newark, Nottinghamshire (Martill and Dawn, 1986).

Highcroft

A few isolated fragments of jaws have been collected from Highcroft Quarry, Gurney Slade. These specimens, which are housed in the Museum of Natural History (London), most closely resemble *Clevosaurus.* Once again, the deposits have been worked away, and no further material has been collected from this locality in recent years.

Durdham Down

Although Robinson (1957a) did not regard Durdham Down as a fissure deposit, Halstead and Nicoll (1971) maintained that it was a collapsed cave. This site is best known for the remains of *Thecodontosaurus* (Riley and Stutchbury, 1840), but among the bone-bearing matrix are remains of other taxa, including an articulated limb and isolated jaw of *Diphydontosaurus.*

The remains that have been identified from these old localities are not inconsistent with a Triassic age. In particular, the pronounced flanges on the *Clevosaurus* tooth fragments from Highcroft more closely resemble the presumed Triassic form *C. hudsoni* than the Early Jurassic species of *Clevosaurus* from Canada and China. Furthermore, although *Diphydontosaurus* extends into the Penarth Group (Holwell Quarry and probably Tytherington and Cromhall), it has not been recorded from within the Early Jurassic sequences. Its presence at Durdham Down suggests a Triassic age, but the evidence is inconclusive. In the absence of extensive assemblages and the improbability of excavating further material from any of these localities, it is not possible to comment further on their age.

Barnhill Quarries

Prospecting Mesozoic infills at Barnhill Quarries in 1980 yielded a few scraps of bone, including a small fragment of a maxilla of *Planocephalosaurus.*

The Welsh localities

Ruthin

The assemblages from a Mesozoic infill at Ruthin Quarry have not been studied in any great detail, although a number of new forms have been recorded on the basis of isolated jawbones. To date, only one fissure system at Ruthin has produced bone. Ruthin may have been partly ignored in the literature because it does not tie in readily with any of the other fissure assemblages in the Bristol Channel area. In addition, although bone material is abundant, usually it is highly worn and polished and consequently difficult to work with. Robinson (1957a) regarded it as Triassic on the basis of a lack of mammals and nonmammalian synapsids, but until recently there has been little evidence to support any age assignment. Recent work by the author and S. E. Evans (pers. commun.) indicates that the fauna is quite diverse. A sphenodontian premaxilla has been tentatively assigned to *Planocephalosaurus* (Fraser, 1986b), and the jaws described as *Tricuspisaurus* can almost certainly be regarded as procolophonid. It is interesting to note the striking similarity in the tooth structure of *Tricuspisaurus* and an undescribed procolophonian from the early–middle Carnian deposits of the Turkey Branch Formation (H.-D. Sues and P. E. Olsen, unpublished data). Of perhaps more significance is an isolated tooth in the collections of the National Museum of Wales that was collected by T. M. Thomas some time before 1952. This tooth is almost certainly referable to *Clevosaurus*. It lacks the extensive flange of additional teeth of *C. hudsoni* and is more consistent with an additional tooth from the mandible of *C. minor*.

Pant-y-ffynon

K. A. Kermack (1956) first reported the occurrence of a crocodile from Pant-y-ffynon that was described in more detail by Crush (1984) as a new genus, *Terrestrisuchus*. However, Benton and Clark (1988) have suggested that *Terrestrisuchus* is synonymous with *Saltoposuchus*. D. Kermack (1984) described well-preserved material of the prosauropod *Thecodontosaurus* and also reported the presence of a "thecodont" and a coelurosaur.

Pant-y-ffynon is more readily comparable to the rest of Robinson's "sauropsid" localities than is Ruthin, but it has been partially separated from them because archosaurs apparently are more predominant than lepidosaurs. However, the descriptions of the vertebrates from Pant-y-ffynon are almost entirely based on articulated specimens, with no indication of the extent of disassociated remains. (There has been only limited screening of matrix for disassociated material.) Likewise, some of the earlier descriptions of material from Cromhall were based solely on articulated specimens of sphenodontians, and the disassociated specimens were largely ignored at that time, thereby producing a biased view of the assemblages. Furthermore, in the earlier excavations of Cromhall, articulated archosaurs were recovered, but they have mostly remained undescribed and ignored in the literature. Thus, if analyses of the constitution of the Cromhall fauna were based solely on the descriptions of articulated material appearing in the literature, there would be an unrealistic bias toward lepidosaurs (in particular, sphenodontians). Similarly, I suspect that there may well be a "human bias" in the Pant-y-ffynon assemblages toward archosaurs. This may be particularly accentuated when it is realized that good specimens of articulated sphenodontians (probably *Clevosaurus*) were recovered along with the archosaurs in the late 1950s and early 1960s, but they have remained undescribed. Consequently, it is believed that until a complete assessment is made of the Pant-y-ffynon assemblages, including any disassociated remains, comparison with other fissure localities will be limited.

Summary

The rich vertebrate-bearing Triassic fissure deposits in Britain have yielded a wealth of new lepidosaurs, archosaurs, and procolophonids. Sphenodontians are the most abundant group represented, and many of them have been described in detail. Other Late Triassic localities worldwide are also beginning to produce sphenodontians, indicating their widespread distribution and importance in early Mesozoic terrestrial assemblages. Although the procolophonids and some of the archosaurs remain to be described formally, initial examination suggests that they will be valuable in future phylogenetic studies of these groups.

The fissures of the Bristol Channel area did not all fill synchronously, but the age constraints are unclear. Previous distinctions were made between those predating the local marine transgression and those postdating the transgression. These were at least partly based on the presence or absence of taxa, essentially at the ordinal level and above. Such distinctions are not at a sufficiently high resolution to be meaningful in studies of putative end-Triassic extinctions, and furthermore they have also been shown to be unreliable.

The more recent studies of the Triassic fissure and cave systems have been more detailed, and the distribution of taxa has been recorded at the generic and specific levels. In addition, parallel sedimentological and geological studies indicate that it might be possible to arrange the fissures in stratigraphic sequence. These initial studies suggest that certain of the small vertebrates are potential zone fossils (although, at the same time, the effects of factors such as predator selection must not be overlooked, particularly in low-diversity assemblages). If the fissures can be confidently arranged in stratigraphic sequence, they could provide useful tests for postulated end-Triassic extinctions.

Acknowledgments

Amey Roadstone Corporation generously continues to allow me free access to their quarrying operations in southwest Britain. M. J. Benton and H.-D. Sues critically read the manuscript and offered many helpful suggestions. I have profited greatly from discussions with M. J. Benton, M. J. Simms, A. R. I. Cruickshank, S. E. Evans, K. A. Joysey, G. M. Walkden, and D. I. Whiteside, but all errors are entirely my responsibility.

References

Andrews, P. 1990. *Owls, Caves, and Fossils.* University of Chicago Press.

Benton, M. J. 1991. What really happened in the Late Triassic? *Historical Biology* 5: 263–278.

Benton, M. J., and J. M. Clark. 1988. Archosaur phylogeny and the relationships of the Crocodylia. Pp. 295–338 in M. J. Benton (ed.), *The Phylogeny and Classification of the Tetrapods. Vol. 1: Amphibians, Reptiles and Birds.* Oxford: Clarendon Press.

Brinkman, D. 1981. The origin of the crocodiloid tarsi and the interrelationships of thecodontian archosaurs. *Breviora* 464: 1–23.

Buffetaut, E., and G. Wouters. 1986. Amphibian and reptile remains from the Upper Triassic of Saint-Nicolas-de-Port (eastern France) and their biostratigraphic significance. *Modern Geology* 10: 133–145.

Cope, J. C. W., K. L. Duff, C. E. Parsons, H. S. Torrens, W. A. Wimbledon, and J. Wright. 1980. A correlation of Jurassic rocks in the British Isles. Pt. 1: Introduction and Lower Jurassic. *Geological Society Special Report* 14: 1–73.

Crush, P. J. 1984. A late Triassic sphenosuchid crocodilian from Wales. *Palaeontology* 27:131–157.

Fisher, M. J., and R. E. Dunay. 1981. Palynology and the Triassic/Jurassic boundary. *Review of Palaeobotany and Palynology* 34: 129–135.

Fraser, N. C. 1982. A new rhynchocephalian from the British Upper Trias. *Palaeontology* 25: 709–725.

1985. Vertebrate faunas from Mesozoic fissure deposits of Southwest Britain. *Modern Geology* 9: 273–300.

1986a. New Triassic sphenodontids from Southwest England and a review of their classification. *Palaeontology* 29: 165–186.

1986b. Terrestrial vertebrates at the Triassic–Jurassic boundary in Southwest Britain. *Modern Geology* 10: 147–157.

1988a. The osteology and relationships of *Clevosaurus* (Reptilia: Sphenodontida). *Philosophical Transactions of the Royal Society of London* B321: 125–178.

1988b. Latest Triassic terrestrial vertebrates and their biostratigraphy. *Modern Geology* 13: 125–140.

Fraser, N. C., and C. J. Duffin. In prep. The vertebrate fauna of Holwell Quarry. *Proceedings of the Geological Association.*

Fraser, N. C., and D. M. Unwin. 1990. Pterosaur remains from the Upper Triassic of Britain. *Neues Jahrbuch für Geologie und Paläontologie, Monatshefte* 1990: 272–282.

Fraser, N. C., and G. M. Walkden. 1983. The ecology of a Late Triassic reptile assemblage from Gloucestershire, England. *Palaeogeography, Palaeoclimatology, Palaeoecology* 42: 341–365.

1984. The postcranial skeleton of *Planocephalosaurus robinsonae.* *Palaeontology* 27: 575–595.

Fraser, N. C., G. M. Walkden, and V. Stewart. 1985. The first pre-Rhaetic therian mammal. *Nature* 314: 161–163.

Gümbel, C. W. 1861. *Geognostische Beschreibung des bayerischen Alpengebirges und seines Vorlandes.* Gotha: Perthes.

Hallam, A. 1981. The end-Triassic bivalve extinction event. *Palaeogeography, Palaeoclimatology, Palaeoecology,* 35: 1–44.

1990. The end-Triassic mass extinction event. *Geological Society of America Special Paper* 247: 577–583.

Halstead, L. B., and P. G. Nicoll. 1971. Fossilized caves of Mendip. *Studies in Speleology* 2: 93–102.

Hamilton, D. 1961. Algal growths in the Rhaetic Cotham Marble of southern England. *Palaeontology* 4: 324–333.

Jeans, C. V. 1978. The origin of the Triassic clay assemblages of Europe with special reference to the Keuper Marl and Rhaetic of parts of England. *Philosophical Transactions of the Royal Society of London* A289: 549–639.

Kent, P. E. 1970. Problems of the Rhaetic on the East Midlands. *Mercian Geologist* 3: 361–372.

Kermack, D. 1984. New prosauropod material from South Wales. *Zoological Journal of the Linnean Society* 82: 101–117.

Kermack, K. A. 1956. An ancestral crocodile from South Wales. *Proceedings of the Linnean Society* 166: 1–2.

Kermack, K. A., F. Mussett, and H. W. Rigney. 1973. The lower jaw of *Morganucodon. Zoological Journal of the Linnean Society* 53: 87–175.

Kühne, W. G. 1946. The geology of the fissure-filling "Holwell 2;" the age determination of the mammalian teeth therein; and a report on the technique employed when collecting the teeth of *Eozostrodon* and Microleptidae. *Proceedings of the Zoological Society of London* 116: 729–733.

1956. *The Liassic Therapsid Oligokyphus.* London: Trustees of the British Museum.

Marshall, J. E. A., and D. I. Whiteside. 1980. Marine influences in the Triassic "uplands." *Nature* 287: 627–628.

Martill, D. M., and A. Dawn. 1986. Fossil vertebrates from new exposures of the Westbury Formation (Upper Triassic) at Newark, Nottinghamshire. *Mercian Geologist* 10: 127–133.

Moore, C. 1867. On abnormal conditions of secondary deposits. *Quarterly Journal of the Geological Society of London* 23: 449–568.

1881. On abnormal geological deposits in the Bristol district. *Quarterly Journal of the Geological Society of London* 37: 67–82.

Olsen, P. E. 1986. Discovery of earliest Jurassic reptile assemblages from Nova Scotia implies catastrophic end to the Triassic. *Lamont-Doherty Newsletter* 12: 1–3.

Olsen, P. E., and H.-D. Sues. 1986. Correlation of continental Late Triassic and Early Jurassic sediments, and the Triassic–Jurassic tetrapod transition. Pp. 321–351 in K. Padian (ed.), *The Beginning of the Age of Dinosaurs: Faunal Change across the Triassic–Jurassic Boundary.* Cambridge University Press.

Poole, E. G. 1978. Stratigraphy of the Withycombe Farm Borehole, Oxfordshire. *Bulletin of the Geological Survey of Great Britain* 57: 1–85.

Riley, H., and S. Stutchbury. 1840. A description of various fossil remains of three distinct saurian animals, recently discovered in the Magnesian Conglomerate near Bristol. *Transactions of the Geological Society of London* (2) 5: 349–357.

Robinson, P. L. 1957a. The Mesozoic fissures of the Bristol Channel area and their vertebrate faunas. *Zoological Journal of the Linnean Society* 43: 260–282.

——— 1957b. An unusual sauropsid dentition. *Zoological Journal of the Linnean Society* 43: 283–293.

——— 1967. Triassic vertebrates from lowland and upland. *Science and Culture* 33: 169–173.

——— 1971. A problem of faunal replacement on Permo-Triassic continents. *Palaeontology* 14: 131–153.

Robinson, P. L., K. A. Kermack, and K. A. Joysey. 1952. Exhibition of a new Upper Triassic land fauna from Slickstone Quarry, Gloucestershire. *Proceedings of the Geological Society of London* 108: 86–87.

Romer, A. S. 1966. *Vertebrate Paleontology*; 3rd ed. University of Chicago Press.

Savage, R. J. G. 1977. The Mesozoic strata of the Mendip Hills. In R. J. G. Savage (ed.), *Geological Excursions in the Bristol District.* University of Bristol.

Shubin, N. H., and H.-D. Sues. 1991. Biogeography of early Mesozoic continental tetrapods: patterns and implications. *Paleobiology* 17: 214–230.

Simms, M. J. 1990. Triassic palaeokarst in Britain. *Cave Science* 17: 93–101.

Swinton, W. E. 1939. A new Triassic rhynchocephalian from Gloucestershire. *Annals and Magazine of Natural History* (11)4: 591–594.

Tozer, E. T. 1979. Latest Triassic ammonoid faunas and biochronology, western Canada. *Geological Survey of Canada Reports* 79-1B: 127–135.

——— 1984. The Trias and its ammonoids: The evolution of a time scale. *Geological Survey of Canada Miscellaneous Report* 35: 1–171.

Tucker, M. E. 1977. The marginal Triassic deposits of South Wales: continental facies and palaeogeography. *Geological Journal* 12: 169–199.

——— 1978. Triassic lacustrine sediments from South Wales: shore-zone clastics, evaporites and carbonates. Pp. 205–244 in A. Matter and M. E. Tucker (eds.), *Modern and Ancient Lake Sediments.* Oxford: Blackwell.

Walkden, G. M., and N. C. Fraser. 1993. Late Triassic fissure sediments and vertebrate faunas: environmental change and faunal succession at Cromhall, South West Britain. *Modern Geology* 18.

Warrington, G. 1978. Palynology of the Keuper, Westbury and Cotham Beds and the White Lias of the Withycombe Farm Borehole. *Bulletin of the Geological Survey of Great Britain* 68: 22–28.

——— 1981. The indigenous micropalaeontology of British Triassic shelf sea deposits. Pp. 61–70 in J. W. Neale and M. D. Brasier (eds.), *Microfossils from Recent and Fossil Shelf Seas.* Chichester: Horwood.

Warrington, G., M. G. Audley-Charles, R. E. Elliot, W. B. Evans, H. C. Ivimey-Cook, P. Kent, P. L. Robinson, F. W. Shotton, and F. M. Taylor. 1980. A correlation of Triassic rocks in the British Isles. *Geological Society Special Report* 13: 1–78.

Whiteside, D. I. 1983. A fissure fauna from Avon. Ph.D. thesis, Bristol University.

——— 1986. The head skeleton of the Rhaetian sphenodontid *Diphydontosaurus avonis* gen. et sp. nov. and the modernizing of a living fossil. *Philosophical Transactions of the Royal Society of London* B312: 379–430.

Whiteside, D. I., and D. Robinson. 1983. A glauconitic clay-mineral from a speleological deposit of Late Triassic age. *Palaeogeography, Palaeoclimatology, Palaeoecology* 41: 81–85.

12

Ornithischian dinosaurs from the Upper Triassic of the United States

ADRIAN P. HUNT AND SPENCER G. LUCAS

Introduction

Ornithischian dinosaurs are rare components of Late Triassic vertebrate assemblages, and the described taxa comprise *Pisanosaurus mertii* Casamiquela, 1967 from Argentina (Bonaparte, 1976), *"Thecodontosaurus" gibbidens* Cope, 1878 from eastern North America (Huene, 1921; Olsen, 1980; Galton, 1983), and *Technosaurus smalli* Chatterjee, 1984 and *Revueltosaurus callenderi* Hunt, 1989 from western North America (Figure 12.1). In addition, there have been several reports of isolated ornithischian teeth that usually are described as "fabrosaur-like" from Morocco and the western United States (Figure 12.1) (Dutuit, 1972; Jacobs and Murry, 1980; Murry, 1982, 1986; Galton, 1984, 1986; Lucas, Oakes, and Froehlich, 1985; Hunt and Lucas, 1989; Murry and Long, 1989). The main purpose of this chapter is to review all the skeletal evidence pertaining to the presence of ornithischian dinosaurs in Upper Triassic strata of the United States. This necessitates a review of the utility of dinosaurian taxa based on isolated teeth. Several new taxa are recognized, and tooth function in Triassic ornithischian dinosaurs is briefly reviewed.

The ichnogenus *Atreipus* occurs in late Carnian–middle Norian strata of the Newark Supergroup in eastern North America and has been attributed to an ornithischian dinosaur (Olsen and Baird, 1986). Because this chapter concerns the skeletal records of ornithischians, we shall not consider this ichnotaxon in detail. However, we do note that the comparatively elongate pedal digit III of *Atreipus* (Olsen and Baird, 1986, figs. 6.6–6.10) is most comparable to that of saurischians. Were it not for the presence of manus impressions, *Atreipus* undoubtedly would be considered saurischian in origin. We consider it most parsimonious to consider *Atreipus* a saurischian, as the pedal structure suggests, rather than an ornithischian, based on the presence of a manus imprint but ignoring the structure of the pedal imprint.

Abbreviations utilized in the text are as follows:
AMNH, American Museum of Natural History, New York
MNA, Museum of Northern Arizona
MNHN, Museum National d'Histoire Naturelle, Paris
NMMNH, New Mexico Museum of Natural History, Albuquerque
TTUP, Texas Tech University Paleontology collection, Lubbock
YPM, Yale Peabody Museum, New Haven

Triassic ornithischians based on teeth

In the nineteenth century, many taxa of dinosaurs (and other tetrapods) were erected on the basis of isolated teeth. The majority of these taxa have subsequently been shown to be nomina dubia (e.g., Coombs, 1988; Coombs and Galton, 1988). As a result, few attempts have been made to identify isolated teeth at either the generic or specific level (Currie, Rigby, and Sloan, 1990). However, in recent years there has been increasing interest in isolated dinosaurian teeth (Galton, 1986; Hunt, 1989; Bakker et al., 1990; Coombs, 1990; Currie et al., 1990; Padian, 1990), with the realization that some taxa can be identified on this basis (Bakker et al., 1990; Currie et al., 1990; Sereno, 1991).

A major problem when considering the taxonomic status of isolated teeth of poorly known animals is to assess the polarity of characters. We test polarities against the phylogenetic scheme of Ornithischia (with some modifications) proposed by Sereno (1986). Thus, the dental synapomorphies of the Ornithischia apparent in isolated teeth are as follows: (1) low, triangular tooth crown in lateral view (Sereno, 1986); (2) recurvature absent from maxillary and dentary teeth (Sereno, 1986); (3) well-developed neck separat-

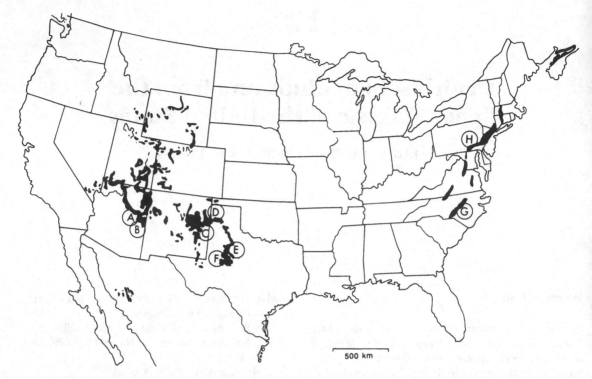

Figure 12.1. Map of North America showing outcrops of Late Triassic strata (in black) and localities that yield ornithischian fossils: A, Petrified Forest National Park, Arizona; B, Downs Quarry, Apache County, Arizona; C, Bull Canyon, Quay County, New Mexico; D, Revuelto Creek, Quay County, New Mexico; E, Kalgary, Crosby County, Texas; F, Post Quarry, Garza County, Texas; G, Pekin, North Carolina; H, Emiggsville, York County, Pennsylvania.

ing crown from root (Sereno, 1986); (4) prominent large denticles arranged at 45° or greater to the mesial and distal edges; (5) premaxillary teeth distinct from dentary/maxillary teeth; (6) maxillary and dentary teeth asymmetrical in mesial and distal views. Note that the absence of recurvature in maxillary and dentary teeth is also observed in other archosaurs (e.g., in the posterior portions of the dentition in heterodont phytosaurs). We consider *Euparkeria* an out-group, as it has a dentition typical of primitive archosaurs. *Euparkeria* has dentary and maxillary teeth that are laterally compressed and recurved and have fine serrations along the mesial and distal margins at about 90° to the tooth margin, and they lack a constriction between the crown and the root (Ewer, 1965, pp. 304–305, fig. 4). The premaxillary teeth are only slightly different from the dentary and maxillary teeth (Ewer, 1965, pp. 304–305).

Coombs (1990) listed five sources of tooth variation that need to be considered before new taxa are created for isolated teeth: (1) positional, (2) ontogenetic, (3) intraspecific, (4) taxonomic, and (5) chimeric. If these factors are taken into consideration, and the teeth under consideration remain distinct from those of other taxa (i.e., they possess apomorphies), then they

can be identified at the generic and specific levels and can provide the basis for taxonomic names. If the teeth possess apomorphies, then a taxon based on them is no more a form taxon than one based on nondental skeletal remains (contra Padian, 1990). As with all taxonomy, the value of classifying isolated teeth is dependent on how usable and useful the classification is. A case in point concerns the tooth taxon *Revueltosaurus callenderi*.

Isolated ornithischian teeth have been known for several years from the Upper Triassic strata of the southwestern United States (e.g., Jacobs and Murry, 1980; Murry, 1982, 1986, 1987, 1989; Lucas et al., 1985; Murry and Long, 1989). However, until recently, none of these teeth had been adequately described or illustrated, and hence they could not be utilized for biochronological, paleoecological, or paleobiogeographic purposes, because teeth from different localities could not be compared. In 1989, Hunt named and described a new ornithischian, *Revueltosaurus callenderi*, on the basis of isolated teeth from eastern New Mexico. Within months, teeth that had been accessioned into the Field Museum of Natural History collection as unidentified ornithischian teeth were recognized as pertaining to *Revueltosaurus* (Padian, 1990). These additional specimens came from northern Arizona, from strata of the same age as those that yielded *Revueltosaurus* in New Mexico (lower *Typothorax* biochron of Hunt and Lucas, 1990, and lower *Pseudopalatus* biochron of Hunt, 1991). Thus, the tooth taxon *Revueltosaurus* has proved to be of both biochronological and paleobiogeographic utility, as it supports existing biochronologies and demonstrates the presence of the same dinosaurian taxon in localities 250 km apart.

In addition, it should be noted that in the majority of strata of Late Triassic age, ornithischian dinosaurs are known only from isolated teeth (Cope, 1878; Dutuit, 1972; Jacobs and Murry, 1980; Murry, 1982, 1986, 1989; Galton, 1984, 1985; Lucas et al., 1985; Hunt, 1989; Hunt and Lucas, 1989; Murry and Long, 1989). To facilitate study of the early evolutionary diversification of herbivorous dinosaurs, these teeth should be named and described, when this is justifiable, so that the geographic and stratigraphic ranges of taxa can be established. A comparable case is the application of a nomenclature by Estes (1964) to fragmentary but diagnostic remains of reptiles and amphibians from Late Cretaceous microvertebrate localities. This work sparked a revolution in our understanding of the biochronology, paleoecology, and evolution of small amphibians and reptiles during the Late Cretaceous.

We thus argue that further description of diagnostic isolated teeth will permit better biochronological and paleobiogeographic resolution of the evolutionary diversification of Late Triassic herbivorous dinosaurs (Hunt, 1991). However, we are acutely aware of the problems associated with naming isolated teeth. We are unable to investigate variation in size and shape within a single dentition, nor can we study variation due to ontogeny. Nevertheless, we are able to recognize apomorphies of each of the taxa that we erect and to present differential diagnoses for them. This approach is necessarily typological, but that is the case in the study of most microvertebrate remains.

History of study

Cope (1878, p. 231) named a new species of *Thecodontosaurus*, *T. gibbidens*, based on two teeth from strata that are now assigned to the Newark Supergroup in Pennsylvania (Figure 12.2). Among the diagnostic features of this species, Cope (1878, p. 231) noted "the greater convexity of the external face" of the teeth compared with the type species of *Thecodontosaurus*. These teeth were later illustrated and further described by Huene (1921, p. 571, figs. 14 and 15), who agreed with Cope's generic assignment, although he later expressed some doubt, referring to this taxon as ?*Thecodontosaurus* (Huene, 1932). Thus, by implication, both Cope and Huene considered these teeth to represent a prosauropod, as *Thecodontosaurus* Riley and Stutch-

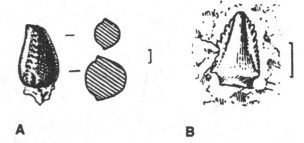

A   **B**

Figure 12.2. *"Thecodontosaurus" gibbidens* from the Upper Triassic of Pennsylvania. (A) Premaxillary tooth (AMNH 2339) in distal view, with cross sections at two levels (From Huene, 1921, fig. 14.) (B) Premaxillary tooth (AMNH 2339) in lingual view (From Huene, 1921, fig. 15.)

bury, 1840 is an undoubted prosauropod from the Upper Triassic of southwestern England (Galton and Cluver, 1976; Galton, 1984).

Galton (1976, p. 76) first suggested that *"Thecodontosaurus" gibbidens* represents an ornithischian dinosaur, and later authors concurred (Olsen, 1980; Galton, 1983; Padian, 1990). Prosauropod teeth are straight and narrow in mesial and distal views (Figure 12.3A,B) (Galton, 1984, p. 13, fig. 3M,N), whereas ornithischian teeth are more bulbous and inclined (Figure 12.3C,D) (Galton, 1984, p. 13, fig. 3K,L). [Note that we use terms from dentistry to describe the orientations of ornithischian teeth, following some recent authors such as Thulborn (1971).] By these criteria it is obvious that *"Thecodontosaurus" gibbidens* is not a prosauropod, but an ornithischian, and it requires a new generic name.

P. E. Olsen collected ornithischian teeth from the Pekin Formation (Newark Supergroup) of North Carolina. These four teeth (two premaxillary and two dentary/maxillary teeth) were briefly described by Galton (1983, p. 122). On the basis of the structure of the premaxillary teeth, Galton considered these specimens to represent *"Thecodontosaurus" gibbidens*. However, these teeth differ from those of *"T." gibbidens* in a number of features, as detailed later, and we believe that they deserve separate generic status.

Additional ornithischian teeth have been collected during the past decade from various strata of Late Triassic age in Arizona (Jacobs and Murry, 1980; Tannenbaum, 1983; Kirby, 1989; Murry and Long 1989; Kaye and Padian, Chapter 9) and New Mexico (Lucas et al., 1985; Hunt, 1989; Hunt and Lucas, 1989), and bones and teeth have been reported from Texas (Murry, 1982, 1986, 1987, 1989; Chatterjee, 1984, 1986). *Revueltosaurus callenderi* occurs at four localities, three in the Bull Canyon Formation of east-central New Mexico and one in the Painted Desert Member of the Petrified Forest Formation in northeastern Arizona (Hunt, 1989; Hunt and Lucas, 1989; Padian, 1990; A. P. Hunt, unpublished data). *Techno-*

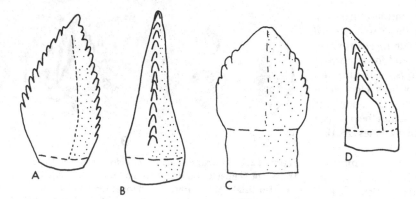

Figure 12.3. Prosauropod (A, B) and ornithischian (C, D) teeth in buccal (A, C) and distal (B, D) views.

saurus smalli is known only from the Cooper Member of the Dockum Formation in Texas (Chatterjee, 1984).

Study of additional specimens, as described later, indicates that at least three unnamed taxa are also present in Upper Triassic strata in North America. One of these occurs in different stratigraphic units in different states and thus may be of biochronological value. These taxa are described later. Another primitive ornithischian may be present in the Owl Rock Formation of northeastern Arizona (Kirby, 1989), and it is currently under study by Kirby.

Systematic paleontology

Dinosauria Owen, 1842

Ornithischia Seeley, 1888

Family incertae sedis

Genus *Galtonia* gen. nov.

Type species. Galtonia gibbidens.

Included species. Known only from the type species.

Etymology. For P. M. Galton, who recognized these teeth as ornithischian, and for his many contributions to the study of ornithischian dinosaurs.

Distribution. Late Triassic of Pennsylvania.

Diagnosis. Ornithischian that differs from other members of the order, except *Fabrosaurus australis*, in possessing narrow, elongate premaxillary teeth (width/height = 2.5/6.5) with denticulated margins that are nearly symmetrical in lingual view. Distinguished from *Fabrosaurus australis* in lacking thinner mesial than distal margins (Sereno, 1991, p. 187) and in being less recurved (Thulborn, 1970a, fig. 6).

Discussion. This genus is currently known only from premaxillary teeth (Figures 12.2 and 12.4A–F),

but these are distinct from the premaxillary teeth of other ornithischians. *Tecovasaurus* and *Lucianosaurus* are also currently known only from dentary/maxillary teeth, but they are consistently much smaller in size than those of *Galtonia* (lectotype is 6.5 mm high). Although Galton (1983, p. 122) used the name *Thecodontosaurus* without quotation marks, he did recognize that these teeth do not represent a prosauropod.

Galtonia gibbidens (Cope, 1878)

Thecodontosaurus gibbidens Cope, 1878, p. 231
Thecodontosaurus gibbidens, Huene, 1921, p. 571, figs. 14 and 15
"*Thecodontosaurus*" *gibbidens*, Galton, 1976, p. 76
"*Thecodontosaurus gibbidens*," Olsen, 1980, p. 40
Thecodontosaurus gibbidens, Galton, 1983, p. 122

Lectotype. AMNH 2339, nearly complete premaxillary tooth (Huene, 1921, fig. 15) (Figures 12.2A, 12.4A–C, 12.8A), plus complete tooth (Huene, 1921, fig. 14).

Referred specimens. AMNH 2339, broken premaxillary tooth; AMNH 2327, complete premaxillary tooth (Figure 12.4D–F).

Type locality. Near Emiggsville, York County, Pennsylvania.

Type horizon. New Oxford Formation (Upper Triassic: upper Carnian).

Diagnosis. As for genus.

Description. The lectotype is a tall, narrow premaxillary tooth with a height up to 6.5 mm, a basal-crown width of 2.5 mm, and a basal-crown length of 5 mm (Figures 12.4A–C, 12.8A). In lingual view the tooth is nearly symmetrical, and the denticulated margin is offset from the medial ridge of the

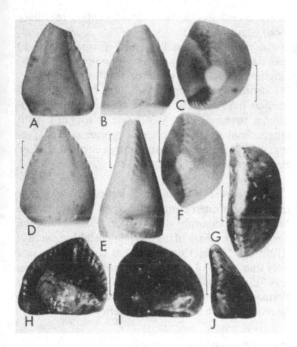

Figure 12.4. Late Triassic ornithischian taxa from the Newark Supergroup of the eastern North America. All scale bars are 2 mm. Cast of holotype of *Galtonia gibbidens* (AMNH 2339) in lingual (A), buccal (B), and occlusal (C) views, and cast of referred premaxillary tooth (AMNH 2327) in buccal (D), mesial or distal (E), and occlusal (F) views. Holotype of *Pekinosaurus olseni* (YPM 8545) in occlusal (G), lingual (H), buccal (I), and mesial or distal (J) views.

tooth, which is triangular and narrows to the apex. In mesial or distal view the tooth is markedly asymmetrical, with the denticle row being closer to the lingual margin. The labial margin of the tooth is rounded, and there is slight development of a median ridge near the apex. The tooth has a constriction below the crown. The tip of the lectotype is damaged, but AMNH 2327 has a terminal wear facet (Figure 12.4D–F).

Discussion. The syntypes of "*Thecodontosaurus*" *gibbidens* were collected by C. M. Wheatley from near Emiggsville, York County, Pennsylvania, from strata now referred to the New Oxford Formation (Cope, 1885, p. 403; Olsen, 1980, p. 40). Huene (1921, p. 574) wrongly implied that these teeth came from the Lockatong Formation at Phoenixville, Chester County, Pennsylvania.

Cope (1878) did not identify a holotype, and Huene (1921) did not designate a lectotype for "*Thecodontosaurus*" *gibbidens*. Therefore, AMNH 2339, the more

nearly complete tooth (Huene, 1921, fig. 14) (Figures 12.2B, 12.4A–C, 12.8A), which is one of the syntypes, is here designated as the lectotype. It should be noted that two teeth are included under catalogue number AMNH 2339. The other tooth (AMNH 2339), which was illustrated by Huene (1921, fig. 15) (Figure 12.2A) but is not refigured here, now lacks most of the upper half of the crown. A third premaxillary tooth is present in the AMNH collection (AMNH 2327; Figure 12.4D–F), but its tip is too nearly complete to be the other specimen illustrated by Huene.

Teeth of *Galtonia* differ from those of *Thecodontosaurus antiquus*, the type species of the genus (Galton, 1984, fig. 4F), in being bulbous and asymmetrical in mesial or distal view and in possessing a broad, ovoid basal cross-section. In contrast, teeth of *Thecodontosaurus* are narrow and symmetrical in mesial or distal view and have a narrow basal cross-section.

Huene's illustration (1921, fig. 15) of the lectotype is misleading in showing a proportionally thinner (mesial-distal) tooth that is symmetrical in lingual view (Figure 12.2B). The lectotype is actually slightly asymmetrical, and the median ridge on the lingual side is slightly concave posteriorly.

Genus *Pekinosaurus* gen. nov.

Type species. *Pekinosaurus olseni* sp. nov.

Included species. Known only from the type species.

Etymology. From Pekin, North Carolina, which is near where the holotype was collected, and for the Pekin Formation, which is the stratigraphic unit that yielded the holotype.

Distribution. Late Triassic of North Carolina.

Diagnosis. Ornithischian distinguished by the possession of very wide and low dentary/maxillary teeth (basal-crown height/length > 5/5.8), with no cingulum and relatively short, broad premaxillary teeth (basal-crown width/height > 2.6/4.0).

Pekinosaurus olseni sp. nov.

Holotype. YPM 8545, dentary/maxillary tooth (Figures 12.4G–J, 12.8D).

Referred specimens. YPM 8545, one dentary/maxillary tooth (Figure 12.5A–C) and two premaxillary teeth (Figure 12.5D–F).

Etymology. For P. E. Olsen, who found the holotype, and for his diverse contributions to the study of the Newark Supergroup.

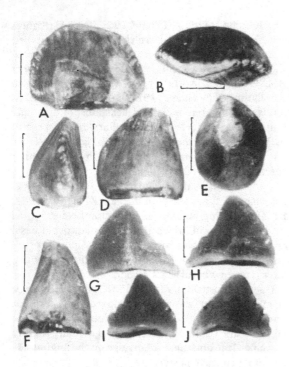

Figure 12.5. Late Triassic ornithischian taxa from North America. Scale bars are 2 mm (A–F) and 1 mm (G–J). Referred dentary/maxillary tooth of *Pekinosaurus olseni* (YPM 8545) in lingual (A), occlusal (B), and mesial or distal (C) views. Referred premaxillary tooth of *P. olseni* (YPM 8545) in lingual (D), occlusal (E), and lateral (F) views. Holotype of *Tecovasaurus murryi* (NMMNH P-18192) in buccal (G) and lingual (H) views. Referred dentary/maxillary tooth of *T. murryi* (NMMNH P-18193) in buccal (I) and lingual (J) views.

Type locality. East of Pekin, North Carolina.

Type horizon. Pekin Formation (Upper Triassic: upper Carnian).

Diagnosis. As for genus.

Description. The holotype is a mesiodistally long, labiolingually broad (5.8 mm) and low (5 mm) dentary/maxillary tooth (Figures 12.4G–J, 12.8D) that is asymmetrical in lingual view (Figures 12.4H, 12.8D). The denticles are close to the plane of the lingual surface. The tooth appears rounded, but this is due to its overall shape. Lingually, there is a small, shallow medial ridge developed near the apex. There are vertical, longitudinal ridges in the medial portion of the apical half of the tooth that are best seen on the referred dentary/maxillary tooth. In mesial and distal views the tooth is markedly asymmetrical, with the denticulated margin on the lingual edge. The labial side of the tooth is rounded. The constriction of the tooth below the

crown is best seen in the referred dentary/maxillary tooth (Figure 12.5A).

The referred premaxillary teeth (Figure 12.5D–F) are rounded in cross-section (basal-crown width/length = 2.6/3.2) and recurved. On their lingual side, the denticulated margins are offset from the broad median ridge, which, on one specimen, has longitudinal striations. The labial side is rounded, with longitudinal striations on one specimen. The premaxillary teeth are at least slightly constricted below the crown (Figures 12.5D, 12.8B).

Discussion. *Pekinosaurus* occurs in rocks that often are considered middle Carnian in age, on the basis of palynomorphs (Gore, 1989). However, on the basis of the presence of the stagonolepidid *Longosuchus* (Hunt and Lucas, 1990), which is elsewhere known only from strata of late Carnian age (Hunt and Lucas, 1990), we consider it more likely that the Pekin Formation is of late Carnian age as well. More data will be required to resolve this discrepancy.

Genus *Tecovasaurus* gen. nov.

Type species. *Tecovasaurus murryi* sp. nov.

Included species. Known only from the type species.

Etymology. For the Tecovas Formation, which yielded the holotype of the type species.

Distribution. Late Triassic of Texas and Arizona.

Diagnosis. Ornithischian with dentary/maxillary crowns that are low and mesiodistally long and are markedly asymmetrical, with up to five large denticles on the distal margin and up to twelve denticles on the steeper-sloping mesial margin; the mesial denticles do not reach the base of the crown, and there is no cingula.

Discussion. These teeth undoubtedly represent an ornithischian, as the crowns are subtriangular in lateral view (Sereno, 1986) and are separated from the root by a neck (Sereno, 1986), which is most apparent in Figure 12.6D,E.

Tecovasaurus murryi sp. nov.

Holotype. NMMNH P-18192, dentary/maxillary tooth (Figures 12.5G, H, 12.7A, 12.8F).

Referred specimens. NMMNH P-18193 (Figures 12.5I,J, 12.8E), NMMNH P-18196 (Figure 12.6C), Tecovas Member of Dockum Formation, Crosby County, Texas; MNA Pl. 1704; four dentary/maxillary

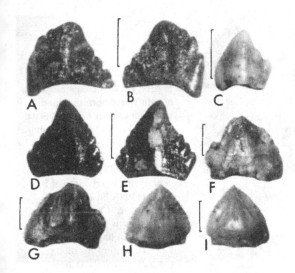

Figure 12.6. Late Triassic ornithischian teeth from east-central New Mexico and northeastern Arizona. All scale bars are 1 mm. Referred dentary/maxillary tooth of *Tecovasaurus murryi* (MNA Pl. 1704) in lingual (A) and buccal (B) views. (C) Referred dentary/maxillary tooth of *T. murryi* (NMMNH P-18196) in lingual view. Referred dentary/maxillary tooth of *T. murryi* in buccal (D) and lingual (E) views. Holotype of *Lucianosaurus wildi* (NMMNH P-18194) in lingual (F) and buccal (G) views. Referred dentary/maxillary tooth of *L. wildi* (NMMNH P-18195) in lingual (H) and buccal (I) views.

teeth (Figure 12.6A,B,D,E); MNA Pl. 1699, dentary/maxillary tooth, Placerias Quarry, Blue Mesa Member of Petrified Forest Formation, Arizona.

Etymology. For Phillip A. Murry, to honor his many contributions to the study of Late Triassic microvertebrates.

Type locality. NMMNH locality 1430, Crosby County, Texas.

Type horizon. Tecovas Member of the Dockum Formation, Chinle Group (Upper Triassic: upper Carnian).

Diagnosis. As for genus.

Description. *Tecovasaurus murryi* has small (up to 2 mm high) and quite asymmetrical (Figure 12.5G–J, 12.8E,F) teeth. The distal portion of the tooth is longer, relative to the axis of maximum height, than the mesial one and has up to six large denticles. The mesial margin has about twice as many smaller denticles. The

labial surface of the tooth is slightly more rounded than the lingual side. Small teeth (Figure 12.6C) have fewer large denticles.

Discussion. *Tecovasaurus murryi* is represented by a range of tooth sizes. The smallest tooth (Figure 12.6C) has only two denticles on the distal margin, which is broken and thus was not as truncated as it now appears.

One of the teeth included in MNA Pl. 1704 is only tentatively assigned to *Tecovasaurus murryi* (Figure 12.6A,B). This tooth differs from other teeth in that the number of denticles on the anterior margin is only four. However, this tooth is similar to teeth of *T. murryi* in that it is asymmetrical and recurved and has a steeper mesial face than distal; also, it has mesial denticles that do not extend to the base of the tooth crown, and distal denticles that do, and it has less than five denticles on the distal margin. Thus, this tooth is tentatively assigned to *T. murryi*.

Tecovasaurus occurs in both the Tecovas Formation of Crosby County, Texas, and the lower Blue Mesa Member of the Petrified Forest Formation (sensu Lucas, 1992) of Arizona. Both of these stratigraphic units contain a diverse tetrapod fauna, including *Rutiodon* and *Stagonolepis*, which are indicative of latest Carnian age (Hunt and Lucas, 1990, 1991a,c; Lucas, 1993). This age is supported by palynological evidence (Litwin, Traverse, and Ash, 1991).

Genus *Lucianosaurus* gen. nov.

Fabrosaurid, Lucas et al., 1985, p. 205, fig. 5
Fabrosaur?, Hunt and Lucas, 1989, p. 82

Type species. *Lucianosaurus wildi* sp. nov.

Included species. Known only from the type species.

Etymology. For Luciano Mesa, which is close to the locality that yielded the holotype.

Distribution. Late Triassic of New Mexico.

Diagnosis. Ornithischian distinguished by the possession of dentary/maxillary teeth without cingula and with asymmetrical basal crowns and, in some teeth, one accessory cusp.

Lucianosaurus wildi sp. nov.

Holotype. NMMNH P-18194, dentary/maxillary tooth (Figures 12.6F,G, 12.7C, 12.8G).

Referred specimen. NMMNH P-18195, dentary/maxillary tooth (Figures 12.6H,I, 12.7D).

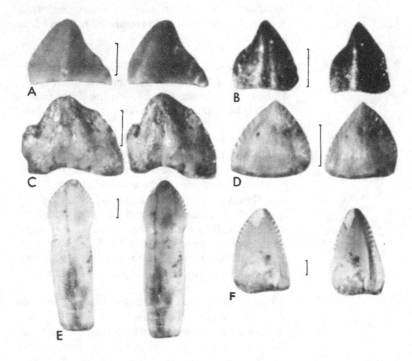

Figure 12.7. Stereophotographs of Late Triassic ornithischian teeth from North America. Scale bars are 1 mm (A–D) and 2 mm (E–F). (A) Holotype of *Tecovasaurus murryi* (NMMNH P-18192) in buccal view. (B) Referred dentary/maxillary tooth of *T. murryi* (MNA Pl. 1704) in buccal view. (C) Holotype of *Lucianosaurus wildi* (NMMNH P-18194) in lingual view. (D) Referred dentary/maxillary tooth of *L. wildi* (NMMNH P-18195) in lingual view. (E) Dentary/maxillary tooth of *Revueltosaurus callenderi* (NMMNH P-4958). (F) Premaxillary tooth of *R. callenderi* (NMMNH P-4959) in lingual view.

Etymology. For Dr. Rupert Wild, in recognition of his diverse contributions to Triassic vertebrate paleontology.

Type locality. NMMNH locality 110, Guadalupe County, New Mexico.

Type horizon. Upper portion of Bull Canyon Formation (Upper Triassic: ?middle Norian).

Diagnosis. As for genus.

Description. Lucianosaurus wildi has dentary/maxillary teeth that are constricted below the crown, but the level of the base of the crown varies for the mesial and distal margins of the tooth (Figures 12.6F–I, 12.8G). In the holotype, the side of the crown that is most elevated (?distal) bears a small accessory cusp (Figures 12.6F,G, 12.8G). The referred specimen also has an asymmetrical base to the crown, but no accessory cusp (Figure 12.6H,I). On the lingual side, the denticulated margin is set slightly back from the face of the tooth. On both labial and lingual sides there are fine longitudinal striations. In mesial and distal views the teeth are asymmetrical, with more rounded labial margins.

Discussion. We refer NMMNH 18195 to *Lucianosaurus wildi*, even though it lacks an accessory cusp, because it shares the asymmetrical basal crown with

the holotype of the taxon. We have not observed this feature in any other ornithischian teeth. Given the presence of this distinct feature in two teeth from the same locality, we consider it most parsimonious to refer both teeth to the same taxon and to consider the presence or absence of the accessory denticle as being a feature related to placement within the dentition. *Lucianosaurus* is the only Triassic ornithischian that has accessory cusps, except for *Technosaurus smalli*, which has both mesial and distal accessory cusps (Figure 12.8I).

Revueltosaurus callenderi Hunt, 1989

Revised diagnosis. Large ornithischian distinguished by the possession of tall premaxillary teeth (length/height = 0.72), with denticulated margins that are slightly recurved and are twice the height of dentary/maxillary teeth, as well as having relatively high dentary/maxillary teeth (length/height = 0.88) that lack accessory cusps.

Discussion. Sereno (1991) suggested that the holotype, paratypes, and referred specimens of this taxon lack (1) clear association, (2) clear positional information, and (3) distinctive characters. The teeth in question were all collected from one quarry in the lower Bull Canyon Formation of east-central New Mexico (Hunt, 1989). These teeth are much larger

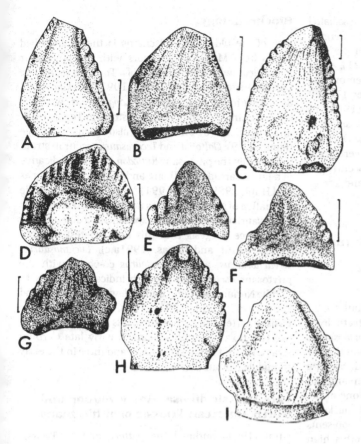

Figure 12.8. Late Triassic ornithischian teeth from North America. Scale bars are 1 mm (E–G) and 2 mm (A–D, H, I). (A) Holotype of *Galtonia gibbidens* (AMNH 2339) in lingual view. (B) Referred premaxillary tooth of *Pekinosaurus olseni* (YPM 8545) in lingual view. (C) Premaxillary tooth of *Revueltosaurus callenderi* (NMMNH P-4959) in lingual view. (D) Holotype of *P. olseni* (YPM 8545) in lingual view. (E) Referred dentary/maxillary tooth of *Tecovasaurus murryi* (NMMNH P-18193) in buccal view. (F) Holotype of *T. murryi* (NMMNH P-18192) in lingual view. (G) Holotype of *Lucianosaurus wildi* (NMMNH P-18194) in buccal view. (H) Dentary/maxillary tooth of *R. callenderi* (NMMNH P-4958) in lingual view. (I) Dentary/maxillary tooth in holotypic jaw of *Technosaurus smalli* (TTUP 9021) in buccal view.

than those of other Triassic ornithischians. In a collection of over 2,000 specimens of fossil vertebrates in NMMNH from this formation, all teeth of ornithischians are similar in size and structure to the specimens described by Hunt (1989). Given the large size of these specimens, the lack of evidence for the presence of another ornithischian in this formation, and the fact that all the specimens described by Hunt (1989) came from the same quarry, it appears to be most parsimonious to assume that these specimens represent one taxon.

It is evident that the holotype, paratype, and referred specimens of *Revueltosaurus callenderi* represent dentary/maxillary teeth (Figures 12.7E, 12.8H; compare Hunt, 1989, pl. 9A–D, and Sereno, 1991, figs. 4A,B, 5A,C) and premaxillary teeth (Figures 12.7F, 12.8C; compare Hunt, 1989, pl. 8E–H, and Sereno, 1991, fig. 6C). Thus, there is no lack of clarity regarding the positions of the teeth. Given the clear association of the specimens referred to *Revueltosaurus callenderi* by Hunt (1989), and given that positional information can be inferred, the taxon can be diagnosed by the characters listed earlier. Thus, *Revueltosaurus callenderi* is not a nomen dubium.

Revueltosaurus callenderi occurs in the lower part of the Bull Canyon Formation of eastern New Mexico (Hunt, 1989) and the lower part of the Painted Desert Member of the Petrified Forest Formation (sensu Lucas, 1992) in Arizona (Padian, 1990). Both of these units contain *Pseudopalatus* and *Paratypothorax*, which indicate an early Norian age (Hunt and Lucas, 1990; Lucas, 1992), and this is supported by palynological evidence (Litwin et al., 1991).

Technosaurus smalli Chatterjee, 1984

Lectotype. TTUP P 9021, right dentary (Chatterjee, 1984, fig. 1d,e,j) (Figure 12.8I).

Revised diagnosis. Ornithischian distinguished by the possession of dentary/maxillary teeth that have anterior and posterior accessory cusps and longitudinal striations in the lower constricted area of the crown.

Discussion. Chatterjee (1984) named *Technosaurus smalli* on the basis of cranial and postcranial material from the Cooper Member of the Dockum Formation of Texas. Sereno (1991) has demonstrated that the holotype of this taxon includes the right dentary of an ornithischian and part of the lower jaw and a premaxilla of a small prosauropod. The associated putative astragalus is not recognizable as such (Sereno, 1991). Thus, we designate a lectotype from the syntypes of *Technosaurus smalli* (TTUP P 9021), right dentary (Chatterjee, 1984, fig. 1d,e,j). We remove the other specimens included under the number TTUP P 9021 from *T. smalli* and follow Sereno (1991) in identifying them as Prosauropoda indet. *T. smalli* is known only from the Cooper Member of the Dockum Formation in West Texas, which, based on vertebrate fossils, is early Norian in age based on the occurrence of *Pseudopalatus* and *Paratypothorax* (Hunt and Lucas, 1990, 1991c).

Late Triassic ornithischians outside the United States

Morocco

Dutuit (1972) described a fragmentary, but tooth-bearing, jaw fragment and two isolated teeth from the Argana Formation of Morocco as *Azendohsaurus laarousi*. Thulborn (1974) first pointed out that the jaw fragment represents, in fact, a prosauropod, because the teeth are symmetrical and straight in anterior view. Galton (1984), however, noted that one of the isolated teeth (MNHN XVI) is curved and markedly asymmetrical in anterior view and thus represents an ornithischian dinosaur. Additional ornithischian specimens from the Upper Triassic of Morocco include isolated teeth (Galton, 1984, fig. 3k,l, 1985, fig. 5N,O, 1986, fig. 16.3L,M) and maxillae with teeth that will be described by J.-M. Dutuit (pers. commun.). These leaf-shaped teeth lack cingula and represent a new genus of ornithischian.

Canada

Galton (1983) mentioned a partial maxilla from the Wolfville Formation of Nova Scotia. This bone has a preserved replacement tooth similar to teeth of *Pekinosaurus*. The tooth lacks cingula (Galton, 1983), and the jaw fragment represents another ornithischian occurrence.

Argentina

Pisanosaurus mertii from the upper Carnian Ischigualasto Formation is based on a fragmentary skeleton (Casamiquela, 1967). The teeth have never been well described or illustrated, but undoubtedly possess cingula (Sereno, 1991). The structures of the distal end of the tibia and of the astragalus indicate that it is more primitive than *Fabrosaurus* (*Lesothosaurus*), but the maxilla and dentary appear more advanced in their emargination of the cheek (Sereno, 1991).

Biochronology

One of the oldest ornithischians is the undescribed genus from Morocco. It occurs with the phytosaur *Paleorhinus*, which is indicative of a Tuvalian (early late Carnian) age (Hunt and Lucas, 1991c). *Pekinosaurus* occurs with the stagonolepidid *Longosuchus*, which is known from *Paleorhinus*-bearing strata in Texas (Hunt and Lucas, 1990) and thus is also Tuvalian in age (Figure 12.9). *Galtonia* and *Tecovasaurus* occur in strata that contain the phytosaur *Rutiodon*, which is indicative of a latest Carnian age (Long and Ballew, 1985; Lucas and Hunt, 1989; Hunt, 1991). Although the Wolfville Formation of Nova Scotia has not yet yielded diagnostic phytosaurian or aetosaurian specimens, a diverse body of evidence suggests that this unit is latest Carnian in age (Hunt and Lucas, 1991a,c). *Lucianosaurus*, *Revueltosaurus*, and *Technosaurus* co-occur with the phytosaur *Pseudopalatus*, which indicates an early to middle Norian age (Long and Ballew, 1985; Lucas and Hunt, 1989; Hunt, 1991). Thus, principally on the basis of vertebrate correlations, it is evident that there are two taxa of ornithischians in early late Carnian strata, two in the latest Carnian, and three in the early to middle Norian (Figure 12.9).

Late Triassic dinosaurian evolution and North American Triassic ornithischians

Hunt (1991) outlined the pattern of Late Triassic dinosaurian evolution. The earliest dinosaurs were late Carnian (Tuvalian) in age and included ornithischians. In Europe, South America, and South Africa, prosauropod dinosaurs became very common in the early to middle Norian and were the dominant terrestrial herbivores. Herbivorous dinosaurs other than prosauropods were apparently either rare (South America, Morocco) or absent (South Africa, Europe).

The North American Triassic ornithischians were consistent with that pattern. The earliest of them were Tuvalian in age (*Pekinosaurus*). These dinosaurs were always rare, and apparently there was never more than one taxon present in any fauna. Thus, it seems that nonprosauropod dinosaurs were minor elements of the terrestrial herbivore fauna in the Late Triassic. Extinction of other low-browsing herbivores at the end of the Triassic, notably aetosaurs, may have been related to the diversity of small Early Jurassic herbivorous dinosaurs. During the Early Jurassic, ornithischians were relatively common and diverse (e.g., the stem ornithischian *Fabrosaurus*, basal thyreophorans such as *Scelidosaurus* and *Emausaurus*, *Tatisaurus*, and diverse heterodontosaurids).

Tooth wear in Triassic ornithischians

Teeth of *Revueltosaurus*, *Galtonia*, and *Pekinosaurus* exhibit similar tooth wear. Maxillary/dentary teeth

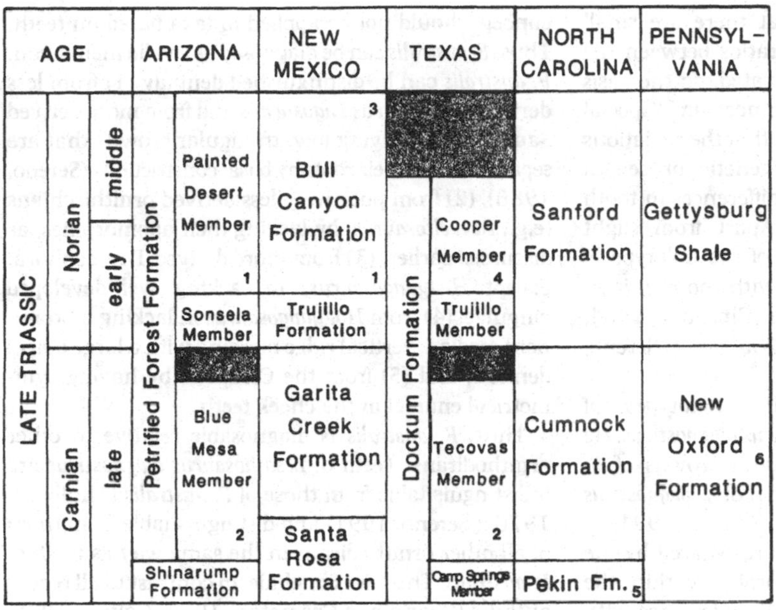

AGE			ARIZONA	NEW MEXICO	TEXAS		NORTH CAROLINA	PENNSYL-VANIA
LATE TRIASSIC	Norian	middle	Petrified Forest Formation	Painted Desert Member	Bull Canyon Formation 3	Dockum Formation	Sanford Formation	Gettysburg Shale
		early				Cooper Member 4		
				1	1			
		early		Sonsela Member	Trujillo Formation	Trujillo Member		
	Carnian	late		Blue Mesa Member	Garita Creek Formation	Tecovas Member	Cumnock Formation	New Oxford Formation 6
				2	Santa Rosa Formation	2		
				Shinarump Formation		Camp Springs Member	Pekin Fm. 5	

Figure 12.9. Correlation of Late Triassic ornithischian-bearing strata in North America. Ornithischian taxa: 1, *Revueltosaurus callenderi*; 2, *Tecovasaurus murryi*; 3, *Lucianosaurus wildi*; 4, *Technosaurus smalli*; 5, *Pekinosaurus olseni*; 6, *Galtonia gibbidens*.

have double wear facets on the mesially and distally inclined occlusal surfaces. These wear facets are poorly developed and generally represent a planing-off of the denticles, although the enamel is sometimes breached. Such wear is common on teeth of *Fabrosaurus* (Thulborn, 1971; Weishampel, 1984; Sereno, 1991). This pattern of wear suggests that maxillary and dentary tooth positions were staggered (Thulborn, 1971; Weishampel, 1984) and that through isognathous occlusion the teeth of the dentary sheared past those of the maxilla along inclined wear surfaces (Weishampel, 1984). However, Sereno (1991) demonstrated that opposing maxillary and dentary teeth could not be aligned to occlude with opposing crowns in alternation, and he therefore concluded that tooth-to-tooth occlusion occurred only locally within the jaws of *Fabrosaurus*. None of the American Triassic taxa exhibit tooth wear as extreme as that seen on some teeth of *Fabrosaurus* (e.g., Thulborn, 1971, fig. 4; Crompton and Attridge, 1986, fig. 17.8).

Premaxillary teeth of *Revueltosaurus*, *Galtonia*, and *Pekinosaurus* have terminal wear facets, which, in more severely worn teeth, extend down the posterior margins. Terminal tooth wear is not known on premaxillary teeth of *Fabrosaurus* (Thulborn, 1971), although the premaxillary teeth of the heterodontosaurid *Heterodontosaurus* have wear facets on their lingual sides (Thulborn, 1970b). The wear facets on the premaxillary teeth of the American Triassic ornithischians could have formed as a result of occlusion with a predentary beak or through contact with tough vegetation, although such wear is unknown in *Fabrosaurus*,

which has premaxillary teeth with similar structure. However, the extension of wear down the posterior margins of the teeth suggests tooth-to-tooth occlusion. This evidence suggests that these Triassic ornithischians had prominent teeth on the anterior dentary to occlude with premaxillary teeth. Thus, it is likely that these dinosaurs had little in the way of a horny beak unless there was a very small edentulous predentary.

This admittedly somewhat tenuous line of reasoning suggests that these Triassic ornithischians were less derived than *Fabrosaurus*, which shows small anterior dentary teeth and no evidence of wear on premaxillary teeth. The North American Triassic taxa also apparently were less derived than *Fabrosaurus* in having premaxillary teeth with denticulated margins.

The taxonomic status of *Fabrosaurus*

The Early Jurassic ornithischian *Fabrosaurus australis* Ginsburg, 1964 was named on the basis of a fragmentary dentary with cingulated leaf-shaped teeth from the Stormberg Group of Lesotho. Subsequently, Thulborn (1970a, 1971, 1972) referred complete skeletal material to this taxon. Charig and Crompton (1974) argued that the teeth of *Fabrosaurus* are not diagnostic. Galton (1978, p. 139) thought that Thulborn's (1970a) material of *Fabrosaurus* actually represents a new genus, *Lesothosaurus*. He noted that the holotype of *Fabrosaurus australis* has a broader dentary, that it possesses special foramina not present in Thulborn's specimens, and that there are small differences in tooth height/width ratios between the two forms. Gow (1981, p. 43) indicated, on the basis of new material, that the differences in "special foramina" were ontogenetic effects, that the variations in dentary width could be ontogenetic or sexual dimorphic effects, and that minor differences in tooth proportions are not diagnostic. Apart from slight proportional differences, the teeth of the holotype of *Fabrosaurus australis* are identical with those in jaws referred by Thulborn to that taxon (Ginsburg, 1964, unnumbered figure; Thulborn, 1970a, fig. 2; Sereno, 1991, figs. 1 and 4).

Sereno (1991) redescribed the holotypes of *Fabrosaurus australis* and *Lesothosaurus diagnosticus*. He noted that the structure of the tooth crowns of *F. australis* "corresponds in detail" to that of *L. diagnosticus* (Sereno, 1991, p. 173). However, Sereno (1991, p. 173) then concluded that the features shared by the teeth of these taxa are plesiomorphic within the Ornithischia and that they share these features with other ornithischians such as *Scutellosaurus*, *Scelidosaurus*, *Hypsilophodon*, *Psittacosaurus*, and *Stegoceras*. Further, Sereno (1991) stated that he could not identify any autapomorphies of the teeth of *F. australis*, and thus he regarded this taxon as a nomen dubium.

It is clear that no other ornithischian taxa have teeth identical with those of *F. australis*. All the genera listed by Sereno (1991) as sharing the plesiomorphic characters of *F. australis* are distinct and possess autapomorphies or can otherwise be distinguished from this taxon. Thus, for example, the teeth of *Scelidosaurus* (Galton, 1986, fig. 16.6U) are distinguished from those of *F. australis* by being proportionally almost twice as tall, by having a less pinched root relative to the crown, by having denticles that do not extend to the base of the crown, and by possessing prominent mesial and distal denticles that extend ventrally below the other denticles and are separated from the median portion of the tooth by deep furrows extending almost to the base of the crown. Thus, it is highly misleading to note only the plesiomorphic characters that unite *F. australis* and *Scelidosaurus*.

The basic cladistic problem is that *F. australis* lacks autapomorphies. However, this need not be a problem in a cladistic analysis. As Gauthier (1984, 1986) and Padian (1986) have argued, a taxon without apomorphies can be diagnosed by a combination of character states plesiomorphic relative to more derived taxa and synapomorphic relative to less derived taxa. Gauthier (1984, 1986) called such taxa "metataxa," and examples include *Archaeopteryx* (Gauthier, 1984, 1986) and *Rioarribasaurus* ("*Coelophysis*," Hunt and Lucas, 1991b). This concept has previously been applied only to skeletal material. However, if the polarity of characters is clearly established, as discussed earlier, then there is no reason that the metataxon

concept should not be applied to taxa based on teeth. Thus, *F. australis* can be diagnosed as a valid metataxon. *F. australis* can be distinguished dentally (1) from less derived taxa such as *Lagosuchus* and from more derived Saurischia by having low, triangular crowns that are separated from their roots by basal constrictions (Sereno, 1986), (2) from putatively less derived ornithischians (e.g., *Revueltosaurus*) by lacking their apomorphies, as discussed earlier, (3) from more derived Thyreophora, except *Huayangosaurus*, in lacking well-developed cingulae, (4) from *Huayangosaurus* in lacking a prominent median, vertical ridge merging with a large apical denticle, and (5) from the Cerapoda by having symmetrical enamel in the cheek teeth.

Thus, *F. australis* is diagnosable relative to other ornithodirans. Teeth of *Lesothosaurus diagnosticus* are indistinguishable from those of *F. australis* (Thulborn, 1970a; Sereno, 1991) and distinguishable from those of all other ornithodirans in the same way as teeth of *F. australis*. Thus, we conclude, in contrast to all recent authors (Charig and Crompton, 1974; Galton, 1978, 1986; Crompton and Attridge, 1986; Weishampel and Witmer, 1990; Sereno, 1991), that *Lesothosaurus diagnosticus* is an invalid binomen and that all material from Stormberg assigned to that taxon should be referred to *F. australis*.

Relationships of Triassic ornithischians from North America

Galtonia, *Pekinosaurus*, *Tecovasaurus*, *Lucianosaurus*, *Revueltosaurus*, and *Technosaurus* are members of the Ornithischia, on the basis of the following dental characters: (1) teeth low and triangular in lateral view, (2) lack of recurvature in maxillary and dentary teeth, (3) crown and root separated by a constriction or "neck," (4) prominent large denticles arranged at 45° or greater to the mesial and distal edges, (5) distinct differences between premaxillary teeth and dentary/maxillary teeth, and (6) maxillary and dentary teeth asymmetrical in mesial and distal views. These taxa are, in addition, less derived than *Pisanosaurus*, *Fabrosaurus*, and the Thyreophora because they lack cingula, and they are distinguished from the Cerapoda by the presence of symmetrical enamel on the "cheek" teeth. The presence of a cingulum is derived relative to *Euparkeria* and all primitive archosaurs. It is undoubtedly a character associated with a herbivorous diet. Thus, it appears probable that the ornithischian taxa from North America are the least derived members of the order.

Acknowledgments

We thank Eugene Gaffney, Howard Hutchison, and Michael Morales for permission to study specimens under their care.

Randy Pence for illustrations used in Figure 12.6, and Nick Fraser and two anonymous reviewers for their helpful comments.

References

Bakker, R. T., P. M. Galton, J. Siegwarth, and J. Filla. 1990. A new latest Jurassic vertebrate fauna, from the highest levels of the Morrison Formation at Como Bluff, Wyoming, with comments on Morrison biochronology. Part IV. The dinosaurs: A new *Othnielia*-like hypsilophodontoid. *Hunteria* 2(6): 8–13.

Bonaparte, J. F. 1976. *Pisanosaurus mertii* Casamiquela and the origin of the Ornithischia. *Journal of Paleontology* 50: 808–820.

Casamiquela, R. M. 1967. Un nuevo dinosaurio ornitisquio triásico (*Pisanosaurus mertii*) de la Formación Ischigualasto, Argentina. *Ameghiniana* 5: 47–64.

Charig, A. J., and A. W. Crompton. 1974. The alleged synonymy of *Lycorhinus* and *Heterodontosaurus*. *Annals of the South African Museum* 64: 167–189.

Chatterjee, S. 1984. A new ornithischian dinosaur from the Triassic of North America. *Naturwissenschaften* 71: 630–631.

1986. The Late Triassic Dockum vertebrates: their stratigraphic and paleobiogeographic significance. Pp. 139–150 in K. Padian (ed.), *The Beginning of the Age of Dinosaurs: Faunal Change across the Triassic–Jurassic Boundary.* Cambridge University Press.

Coombs, W. P., Jr. 1988. The status of the dinosaurian genus *Diclonius* and the taxonomic utility of hadrosaurian teeth. *Journal of Paleontology* 62: 812–817.

1990. Teeth and taxonomy in ankylosaurs. Pp. 269–279 in K. Carpenter and P. J. Currie (eds.), *Dinosaur Systematics: Perspectives and Approaches.* Cambridge University Press.

Coombs, W. P., Jr., and P. M. Galton. 1988. *Dysganus*, an indeterminate ceratopsian dinosaur. *Journal of Paleontology* 62: 818–821.

Cope, E. D. 1878. On some saurians found in the Triassic of Pennsylvania, by C. M. Wheatley. *Proceedings of the American Philosophical Society* 17: 177.

1885. [Untitled.] *Proceedings of the American Philosophical Society* 1886: 403–404.

Crompton, A. W., and J. Attridge. 1986. Masticatory apparatus of the larger herbivores during the Late Triassic and Early Jurassic times. Pp. 223–236 in K. Padian (ed.), *The Beginning of the Age of Dinosaurs: Faunal Change across the Triassic–Jurassic Boundary.* Cambridge University Press.

Currie, P. J., J. K. Rigby, Jr., and R. E. Sloan. 1990. Theropod teeth from the Judith River Formation of southern Alberta, Canada. Pp. 107–125 in K. Carpenter and P. J. Currie (eds.), *Dinosaur Systematics: Perspectives and Approaches.* Cambridge University Press.

Dutuit, J.-M. 1972. Decouverte d'un Dinosaure ornithischien dans le Trias supérieur de l'Atlas occidental marocain. *Comptes Rendus de l'Académie des Sciences de Paris, sér. D* 275: 2841–2844.

Estes, R. 1964. Lower vertebrates from the Late Cretaceous Lance Formation, eastern Wyoming. *University of California Publications in the Geological Sciences* 49: 1–180.

Ewer, R. F. 1965. The anatomy of the thecodont reptile *Euparkeria capensis* Broom. *Philosophical Transactions of the Royal Society of London* B248: 379–435.

Galton, P. M. 1976. Prosauropod dinosaurs (Reptilia: Saurischia) of North America. *Postilla* 169: 1–98.

1978. Fabrosauridae, the basal family of ornithischian dinosaurs (Reptilia: Ornithopoda). *Paläontologische Zeitschrift* 52: 138–159.

1983. The oldest ornithischian dinosaurs in North America from the Late Triassic of Nova Scotia, N. C. and PA. *Geological Society of America, Abstracts with Programs* 15: 122.

1984. An early prosauropod dinosaur from the Upper Triassic of Nordwürttemberg, West Germany. *Stuttgarter Beiträge zur Naturkunde, Ser. B,* 106: 1–25.

1985. Diet of prosauropod dinosaurs from the late Triassic and early Jurassic. *Lethaia* 18: 105–123.

1986. Herbivorous adaptations of Late Triassic and Early Jurassic dinosaurs. Pp. 203–221 in K. Padian (ed.), *The Beginning of the Age of Dinosaurs: Faunal Change across the Triassic–Jurassic Boundary.* Cambridge University Press.

Galton, P. M., and M. A. Cluver. 1976. *Anchisaurus capensis* (Broom) and a revision of the Anchisauridae (Reptilia: Saurischia). *Annals of the South African Museum* 69: 121–159.

Gauthier, J. A. 1984. A cladistic analysis of the higher categories of the Diapsida. Ph.D. dissertation, University of California, Berkeley.

1986. Saurischian monophyly and the origin of birds. *Memoirs of the California Academy of Science* 8: 1–55.

Ginsburg, L. 1964. Decouverte d'un Scélidosaurien (Dinosaure ornithischien) dans le Trias supérieur du Basutoland. *Comptes Rendus de l'Académie des Sciences de Paris* 258: 2366–2368.

Gore, P. J. W. 1989. Boren Clay Products Quarry. Pp. 23–25 in P. E. Olsen, R. W. Schlische, and P. J. W. Gore (eds.), *Tectonic, Depositional and Paleoecological History of Early Mesozoic Rift Basins, Eastern North America.* Washington, D.C.: American Geophysical Union.

Gow, C. E. 1981. Taxonomy of the Fabrosauridae (Reptilia: Ornithischia) and the *Lesothosaurus* myth. *South African Journal of Science* 77: 43.

Huene, F. von. 1921. Reptilian and stegocephalian remains from the Triassic of Pennsylvania in the Cope collection. *Bulletin of the American Museum of Natural History* 44: 561–574.

1932. Die fossile Reptil-Ordnung Saurischia, ihre Entwicklung und Geschichte. *Monographien zur Geologie und Paläontologie* 1 (4): 1–361.

Hunt, A. P. 1989. A new ?ornithischian dinosaur from the Bull Canyon Formation (Upper Triassic) of east-central New Mexico. Pp. 355–358 in S. G. Lucas and A. P. Hunt (eds.), *The Dawn of the Age of Dinosaurs in the American Southwest.* Albuquerque: New Mexico Museum of Natural History.

1991. The early diversification pattern of dinosaurs in the Late Triassic. *Modern Geology* 16: 43–59.

Hunt, A. P., and S. G. Lucas. 1989. Late Triassic vertebrate localities in New Mexico. Pp. 72–101 in S. G. Lucas and A. P. Hunt (eds.), *The Dawn of the Age of Dinosaurs in the American Southwest.* Albuquerque: New Mexico Museum of Natural History.

1990. Re-evaluation of "*Typothorax meadei*," a Late Triassic aetosaur from the United States. *Paläontologische Zeitschrift* 64: 317–328.

1991a. A new rhynchosaur from the Upper Triassic of West Texas and the biochronology of Late Triassic rhynchosaurs. *Palaeontology* 34: 927–938.

1991b. *Rioarribasaurus*, a new name for a Late Triassic dinosaur from New Mexico (USA). *Paläontologische Zeitschrift* 65: 191–198.

1991c. The *Paleorhinus* biochron and the correlation of the nonmarine Upper Triassic of Pangaea. *Palaeontology* 34: 487–501.

Jacobs, L. L., and Murry, P. A. 1980. The vertebrate community of the Triassic Chinle Formation near St. Johns, Arizona. Pp. 55–71 in L. L. Jacobs (ed.), *Aspects of Vertebrate History: Essays in Honor of Edwin Harris Colbert.* Flagstaff: Museum of Northern Arizona Press.

Kirby, R. E. 1989. Late Triassic vertebrate localities of the Owl Rock Member (Chinle Formation) in the Ward Terrace area of northern Arizona. Pp. 12–28 in S. G. Lucas and A. P. Hunt (eds.), *The Dawn of the Age of Dinosaurs in the American Southwest.* Albuquerque: New Mexico Museum of Natural History.

Litwin, R. J., A. Traverse, and S. R. Ash. 1991. Preliminary palynological zonation of the Chinle Formation, southwestern U.S.A., and its correlation to the Newark Supergroup (eastern U.S.A.). *Review of Palaeobotany and Palynology* 68: 269–287.

Long, R. A., and K. A. Ballew. 1985. Aetosaur dermal armor from the Late Triassic of southwestern North America with special reference to material from the Chinle Formation of Petrified Forest National Park. *Bulletin of the Museum of Northern Arizona* 54: 45–68.

Lucas, S. G. 1993. The Chinle Group: revised stratigraphy and chronology of Upper Triassic nonmarine strata in the western United States. *Bulletin of the Museum of Northern Arizona*

Lucas, S. G., and A. P. Hunt. 1989. Vertebrate biochronology of the Late Triassic. *28th International Geological Congress, Abstracts* 2: 335–336.

Lucas, S. G., W. Oakes, and J. W. Froehlich. 1985. Triassic microvertebrate locality, Chinle Formation, east-central New Mexico. Pp. 205–212 in S. G. Lucas and J. Zidek (eds.), *Santa Rosa–Tucumcari Region.* Socorro: New Mexico Geological Society Guidebook 36.

Murry, P. A. 1982. Biostratigraphy and paleoecology of the Dockum Group, Upper Triassic of Texas. Ph.D. dissertation. Southern Methodist University.

1986. Vertebrate paleontology of the Dockum Group, western Texas and eastern New Mexico. Pp. 109–137 in K. Padian (ed.), *The Beginning of the Age of Dinosaurs: Faunal Change across the Triassic–Jurassic Boundary.* Cambridge University Press.

1987. Notes on the stratigraphy and paleontology of the Upper Triassic Dockum Group. *Journal of the Arizona-Nevada Academy of Science* 22: 73–84.

1989. Geology and paleontology of the Dockum Formation (Upper Triassic), West Texas and eastern New Mexico. Pp. 102–144 in S. G. Lucas and A. P. Hunt (eds.), *The Dawn of the Age of Dinosaurs in the American Southwest.* Albuquerque: New Mexico Museum of Natural History.

Murry, P. A., and R. A. Long. 1989. Geology and paleontology of the Chinle Formation, Petrified Forest National Park and vicinity, Arizona, and a discussion of vertebrate fossils from the southwestern Upper Triassic. Pp. 29–64 in S. G. Lucas and A. P. Hunt (eds.), *The Dawn of the Age of Dinosaurs in the American Southwest.* Albuquerque: New Mexico Museum of Natural History.

Olsen, P. E. 1980. A comparison of the vertebrate assemblages from the Newark and Hartford basins (Early Mesozoic, Newark Supergroup) of eastern North America. Pp. 35–53 in L. L. Jacobs (ed.), *Aspects of Vertebrate History: Essays in Honor of Edwin Harris Colbert.* Flagstaff: Museum of Northern Arizona Press.

Olsen, P. E., and D. Baird. 1986. The ichnogenus *Atreipus* and its significance for Triassic biostratigraphy. Pp. 61–87 in K. Padian (ed.), *The Beginning of the Age of Dinosaurs: Faunal Change across the Triassic–Jurassic Boundary.* Cambridge University Press.

Padian, K. 1986. On the type material of *Coelophysis* Cope (Saurischia: Theropoda) and a new specimen from the Petrified Forest of Arizona (Late Triassic: Chinle Formation). Pp. 45–60 in K. Padian (ed.), *The Beginning of the Age of Dinosaurs: Faunal Change across the Triassic–Jurassic Boundary.* Cambridge University Press.

1990. The ornithischian form genus *Revueltosaurus* from the Petrified Forest of Arizona (Late Triassic: Norian; Chinle Formation). *Journal of Vertebrate Paleontology* 10: 268–269.

Sereno, P. C. 1986. Phylogeny of the bird-hipped dinosaurs (Order Ornithischia). *National Geographic Research* 2: 234–256.

1991. *Lesothosaurus*, "fabrosaurids," and the early evolution of Ornithischia. *Journal of Vertebrate Paleontology* 11: 168–197.

Tannenbaum, F. 1983. The microvertebrate fauna of the *Placerias* and Downs quarries, Chinle Formation (Upper Triassic) near St. Johns, Arizona. M.S. thesis. University of California, Berkeley.

Thulborn, R. A. 1970a. The skull of *Fabrosaurus australis*, a Triassic ornithischian dinosaur. *Palaeontology* 13: 414–432.

1970b. The systematic position of the Triassic ornithischian dinosaur *Lycorhinus angustidens*. *Zoological Journal of the Linnean Society* 49: 235–245.

1971. Tooth wear and jaw action in the Triassic ornithischian dinosaur *Fabrosaurus*. *Journal of Zoology* (London) 164: 165–179.

1972. The post-cranial skeleton of the Triassic

ornithischian dinosaur *Fabrosaurus australis.* *Palaeontology* 15: 29–60.

——— 1974. A new heterodontosaurid dinosaur (Reptilia: Ornithischia) from the Upper Triassic Red Beds of Lesotho. *Zoological Journal of the Linnean Society* 55: 151–175.

Weishampel, D. B. 1984. Evolution of jaw mechanics in ornithopod dinosaurs. *Advances in Anatomy, Embryology and Cell Biology* 87: 1–110.

Weishampel, D. B., and L. M. Witmer. 1990. *Lesothosaurus, Pisanosaurus,* and *Technosaurus.* Pp. 416–425 in D. B. Weishampel, P. Dodson, and H. Osmolska (eds.), *The Dinosauria.* Berkeley: University of California Press.

13

Early Jurassic small tetrapods from the McCoy Brook Formation of Nova Scotia, Canada

NEIL H. SHUBIN, PAUL E. OLSEN, AND HANS-DIETER SUES

Introduction

The biotic transition at the Triassic–Jurassic boundary was a key event in the history of life. Major tetrapod groups such as mammaliaforms, crocodylomorph archosaurs, dinosaurs, pterosaurs, and turtles first appeared in the Late Triassic and coexisted with many taxa that had had Paleozoic origins. Many of the latter disappeared during a major extinction at the end of the Triassic (Olsen and Sues, 1986; Olsen, Shubin, and Anders, 1987), and as a consequence, continental tetrapod communities acquired a distinctly modern composition. The temporal placement, taxonomic composition, and duration of that interval of faunal change have always been poorly understood, because global correlations of continental tetrapod assemblages are difficult, and definitely Early Jurassic (Hettangian) occurrences of such assemblages were unknown.

The traditional view of early Mesozoic faunal change postulated a competitive replacement of various groups of early archosaurs, procolophonians, metoposaurid temnospondyls, and nonmammalian therapsids with a new assemblage composed of dinosaurs, mammaliaforms, crocodylomorph archosaurs, turtles, and other taxa (Colbert, 1958). The stratigraphic underpinnings of this view changed significantly during the 1970s. A number of important vertebrate-bearing assemblages of allegedly Late Triassic age were redated as Early Jurassic (Olsen and Galton, 1977, 1984), extending the range of many taxa into the Jurassic and thereby reducing the number of groups thought to have become extinct at the Triassic–Jurassic boundary (Olsen and Sues, 1986). Furthermore, it became apparent that the Late Triassic was an interval of high faunal diversity, whereas the Early Jurassic apparently was one of much decreased diversity (Olsen and Sues, 1986; Olsen et al., 1987). Major difficulties in stratigraphic correlations still hinder studies of the Triassic–Jurassic transition.

The Lower Jurassic standard sections in Europe comprise sedimentary rocks of marine origin; hence, stratigraphic correlations with continental Lower Jurassic strata elsewhere have to rely on indirect criteria, such as pollen and spores and radiometric dates.

The recent discovery of a diversified assemblage of tetrapods in the lower part of the McCoy Brook Formation (Fundy Group, Newark Supergroup; Lower Jurassic) in the Fundy basin in Nova Scotia, Canada, permits a more refined temporal calibration of the faunal turnover at the Triassic–Jurassic boundary (Olsen et al., 1987; Olsen, Schlische, and Gore, 1989). Numerous biostratigraphic and chronostratigraphic controls, a range of depositional environments, and the great abundance of well-preserved skeletal remains of fossil vertebrates in the lower part of the McCoy Brook Formation provide a unique opportunity for the study of Early Jurassic continental tetrapod communities. The tetrapod assemblage from the McCoy Brook Formation is characterized by taxa known from the Upper Triassic of eastern North America and elsewhere, but lacks many characteristic Late Triassic forms such as procolophonians and parasuchian and rauisuchian archosaurs. It includes tritheledontid cynodonts, ornithischian and prosauropod dinosaurs, crocodylomorph archosaurs, and sphenodontian lepidosaurs. In this chapter, we review the small tetrapods from the McCoy Brook Formation of Nova Scotia and discuss their geographic and stratigraphic significance.

Geological setting and fossil distribution

The early Mesozoic sedimentary and igneous rocks of the Newark Supergroup of eastern North America were deposited in a series of rift basins that formed during the initial phase of the breakup of the supercontinent Pangaea. The Fundy basin of Nova Scotia and New Brunswick, Canada, is the largest of all exposed

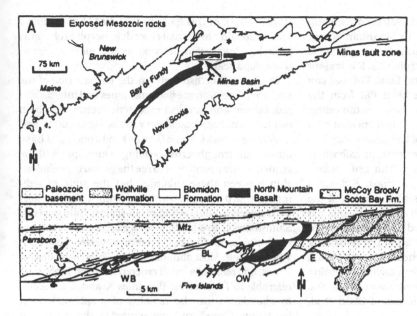

Figure 13.1. Fundy basin, showing the exposed surfaces of early Mesozoic continental sediments. The area marked by an asterisk represents the major localities of Late Triassic and Early Jurassic age. (B) Five Islands–Wasson Bluff region, showing the extensional strike-slip duplexes; BL, Blue Sac; E, Lower Economy; Mfz, Minas Fracture Zone; OW, Old Wife Point; WB, Wasson Bluff. (Adapted from Olsen and Schlische, in Olsen et al., 1989.)

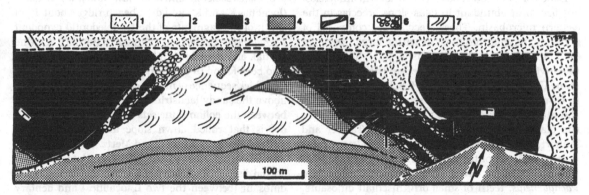

Figure 13.2. Geological map of Wasson Bluff, with subbasin developed within the faulted upper surface of the North Mountain Basalt. Key to lithologies: 1, Carboniferous basement rocks; 2, Blomidon Formation; 3, North Mountain Basalt; 4, fluviolacustrine strata of McCoy Brook Formation; 5, "fish bed"; 6, talus-slope deposits of McCoy Brook Formation; 7, eolian sandstones of McCoy Brook Formation. (Adapted from Olsen and Schlische, in Olsen et al., 1989.)

rift basins of the Newark Supergroup (Figure 13.1). The Fundy Group comprises a sequence of predominantly red clastic rocks and tholeiitic basalts more than 1,000 m thick. It is divided into five formations, ranging in age from Middle Triassic to Early Jurassic, of which the McCoy Brook Formation is the youngest (Figures 13.1 and 13.2).

Outcrops of the lower part of the McCoy Brook Formation at Wasson Bluff near Parrsboro, Cumberland County, have exposed the richest fossil-bearing deposits of the Fundy Group known to date. The structure of the deposits reflects synrift sedimentation and deformation in two microbasins (Olsen and Schlische, 1990) (Figure 13.2) adjacent to the Minas fault zone. The Minas fault zone was a major left-lateral strike-slip

fault in the early Mesozoic (Figure 13.1) (Olsen and Schlische, 1990). Strata of the McCoy Brook Formation represent fluvial, lacustrine, playa, eolian, and alluvial-fan depositional settings and are underlain by the tholeiitic basalt flows of the North Mountain Basalt. A series of microbasins was generated by faulting on the underlying North Mountain Basalt, two of which will be described here (Figure 13.2). Each was filled with the various sediments of the basal McCoy Brook Formation listed earlier, containing tetrapod remains that vary in preservational quality and abundance. Vertebrate remains are known from all major lithofacies of the McCoy Brook Formation at Wasson Bluff; Table 13.1 lists the vertebrate taxa and their distribution. The "fish bed" (Figure 13.2) consists of purple and greenish siltstone, limestone, and basalt detritus; along with the upper surface of the North Mountain Basalt, it forms a useful marker horizon for stratigraphic correlations between the two microbasins. It is present in all of the microbasins at Wasson Bluff. This bed can be traced in outcrop along the tidal flat from the eastern edge of the western microbasin to the center of the eastern microbasin, where it is truncated by a fault that forms the eastern side of the latter basin. The step-faulted southern rim of the eastern microbasin is exposed in the tidal zone, and the "fish bed" onlaps only the southernmost basalt outcrops. Sediments onlapping the small fault blocks to the north of the "fish bed" are continuous with the gray and red sandstones below it. The "fish bed" dips to the south, and its projection extends above all of the remaining fill of the eastern microbasin. All these observations are consistent with the fill in the eastern microbasin being older than the fill in the western microbasin. Site designations follow the nomenclature of Olsen et al. (1989, fig. 11.6).

Bones of small tetrapods occur in several lithofacies, but are most abundant in talus-slope breccias in the eastern microbasin. These breccias are composed of generally angular basalt clasts with a large size range (< 1 cm to > 3 m), with extensive interstitial spaces filled with bedded red-to-orange sandstone and mudstone (Tanner and Hubert, 1988; Schlische and Olsen, in Olsen et al., 1989). The presence of bedding within the matrix shows that the matrix accumulated after the clasts were more or less in place (Tanner and Hubert, 1988). Sandy mudstone and sandstone talus on the east side of the eastern microbasin (locality E) has produced, in order of abundance, Protosuchus, Pachygenelus, teeth of small ornithischian dinosaurs, jaw fragments of an additional taxon of crocodylomorph, jaw fragments of Clevosaurus, and coprolites. Orange sandstone with less abundant basalt clasts occurs at the westernmost edge of the eastern microbasin (locality F) and has produced bones of Clevosaurus.

The western microbasin at Wasson Bluff is mostly infilled by finer-grained material than that in the eastern microbasin. The "fish bed" outcrops at the eastern margin of the westen microbasin (locality J), where it contains locally abundant remains of semionotid and paleonisciform fish, hybodont sharks, occasional isolated teeth and bones of small ornithischian dinosaurs, and coprolites. Bones and scales are most abundant between large clasts of basalt, where the "fish bed" onlaps a possible wave-cut terrace on the upper surface of the flow. The "fish bed" dips to the west below most of the fill of the western microbasin, but a small piece of it is exposed on the western side of the microbasin (about 12 m west of locality N). This outcrop occurs in a notch in a very large block of basalt, apparently within the broad paleosurface expression of the fault zone bounding the western edge of this microbasin

(Figure 13.2). At this spot, the "fish bed" has produced teeth belonging to a small crocodylomorph archosaur, probably Protosuchus, as well as the more common semionotid scales.

Overlying the "fish bed" on the eastern side of the western microbasin are fine sandstones and interbedded red, brown, and purplish mudstones containing both unidirectional and oscillatory cross-laminations and desiccation cracks. Some thicker sandstone units have dune-scale trough cross-bedding. Outcrops of these sequences are insufficient to see the geometric relationship between these subunits, but the suite of sedimentary structures is consistent with a shallow lake with associated fluvial and small-scale deltaic systems. Sandstones in this shallow-water lacustrine sequence have produced the best-preserved vertebrates from Wasson Bluff. Most abundant are isolated osteoderms, bones, and teeth of a small crocodylomorph archosaur referable to Protosuchus (localities K and K'). Second in abundance are the remains of a sphenodontian lepidosaur (Clevosaurus), represented by skulls, a partial articulated skeleton, and many jaws and jaw fragments (localities K and K', and on a beach ridge about 10 m south of J). Fragments of the tritheledontid cynodont Pachygenelus are next in abundance (locality K'). One partial, articulated juvenile prosauropod skeleton has also been found in these units (locality K).

A thick sequence of mostly eolian dune sands occurs above the sandy lacustrine units. First-order surfaces between the eolian dune sets are covered with basalt debris that rolled down slope and was buried by advancing dunes (Hubert and Mertz, 1984; Schlische and Olsen, in Olsen et al., 1989). To date, only one partial dissociated skeleton of a fairly large prosauropod dinosaur (between the two L localities) and dentary fragments of Clevosaurus have been recovered from the interdune deposits (between locality L and locality K'). Along the western and northern edges of the western microbasin, the dune sands are bound laterally by paleocliffs of basalt composed of the previously mentioned blocks of basalt (> 10 m in length) from the fault zone (between O and M), as well as interbedded basalt talus deposits (Figure 13.2). A fragmentary and badly tectonically deformed prosauropod skeleton has been found in a sandstone tongue below a preserved talus cone (locality N). Osteoderms and bones of Protosuchus have been found in talus-slope deposits along the northern edge of the subbasin.

Overlying the dune deposits, and exposed only in the adjacent tidal flats, are decimeter-scale, laterally continuous sandstone beds with oscillatory ripples; they appear to be composed of the recycled eolian sand (about 200 m south of the stretch of cliffs between localities N and K'). These beds are interbedded with very thin mudstone layers and contain footprints of Grallator sp. (probably small theropod) and Batrachopus sp. (small crocodylomorph). The ripples and recycled dune sand are consistent with deposition in a very shallow lake or playa.

Age relations

The Middle Triassic to Early Jurassic sedimentary rocks of the Newark Supergroup of eastern North America possess a variety of chronostratigraphic and biostratigraphic controls. The ages of continental tetrapod assemblages can be constrained by means of radiometric dates, palynoflorules, tetrapod footprint assemblages, and tetrapod bone assemblages. Each of these criteria alone presents its own particular difficulties, and only several lines of evidence in conjunction permit correlation of the Late Triassic (Norian) and Early Jurassic (Hettangian) strata in the Newark Supergroup with the European type sections (Olsen and Galton, 1977; Cornet, 1977).

Newark Supergroup deposits of Early Jurassic age are typified by an abundance of conifer (cheirolepidiaceous) pollen of the form-genus *Corollina* (particularly *C. meyeriana*), by complete absence of typical Late Triassic pollen taxa such as *Patinasporites* and *Vallisporites*, by the presence of characteristic Connecticut Valley–type tetrapod footprints (Olsen and Galton, 1977; Olsen and Baird, 1986), by the presence of flood basalts and associated intrusives with radiometric dates clustering around 202 Ma, and by a total absence of typically Late Triassic tetrapod taxa such as procolophonids and phytosaurs (Olsen et al., 1987).

The age of the lower part of the McCoy Brook Formation is constrained by several independent lines of evidence. Palynoflorules from the Scots Bay and McCoy Brook formations are dominated by *Corollina meyeriana*, which indicates a Hettangian or younger age. The uppermost Blomidon Formation has produced palynomorph assemblages at a number of localities and in subsurface samples. All are dominated by *Corollina*; however, two outcrop localities produce fairly abundant bisaccates, and one (from Partridge Island) contains *Patinasporites*. The latter occurrence indicates a Jurassic age for at least part of the uppermost Blomidon Formation and suggests that the Triassic–Jurassic boundary occurs within a few meters of the North Mountain Basalt. The North Mountain Basalt is geochemically very similar to the older flows in the rest of the Newark Supergroup (Puffer and Philpotts, 1988). U-Pb ages for zircons from the basalt (202 ± 1 Ma) (Hodych and Dunning, 1992) are indistinguishable from ages determined in more southern intrusions and associated flows (Sutter, 1988; Dunning and Hodych, 1990). These similarities, together with the broad outlines of the cyclostratigraphy of the upper Blomidon and lower Scots Bay formations, are consistent with the age of the tetrapod fossils from the basal McCoy Brook Formation being maximally 100,000 to 200,000 years younger than the palynologically

identified Triassic–Jurassic boundary (Olsen et al., 1987).

Tetrapod diversity

Skeletal remains of small tetrapods recovered from the lower portion of the McCoy Brook Formation comprise two different taxa of crocodylomorph archosaurs, prosauropod and small ornithischian dinosaurs, a sphenodontian lepidosaur, and a tritheledontid cynodont (Table 13.1). Although well preserved and very abundant, the tetrapod bones recovered to date have nearly always been dissociated. This fact makes identification and comparisons difficult, particularly for many postcranial bones. Exceptions include an articulated hindlimb of a small, presumably juvenile anchisaurid prosauropod and an articulated partial skeleton and two incomplete skulls of a sphenodontian lepidosaur. Shubin et al. (1991) have recently published a preliminary description of the tritheledontid cynodont. Detailed anatomical and phylogenetic studies of the sphenodontian and of the crocodylomorph archosaurs from the McCoy Brook Formation are currently being prepared by the authors. Many specimens have yet to be prepared, and the material discussed next represents only an initial selection of specimens of particular anatomical and taxonomic interest. All the fossils described here, and in forthcoming papers, will be housed in the collections of the Nova Scotia Museum (NSM), Halifax, and the Museum of Comparative Zoology (MCZ), Harvard University.

Diapsida

Archosauria

Crocodylomorpha

Crocodyliformes

Protosuchus sp. nov.

Occurrence. Basalt talus-slope breccias, sites E and F in the eastern microbasin and the banks of a creek draining onto the beach about 50 m east of site M in the western microbasin; lake-margin and fluvial sandstones in the eastern microbasin, sites K and K', in the banks of a small brook draining onto the beach about 15 m east of site L, and about 12 m west of locality N.

Discussion. Skeletal remains of this small crocodylomorph referable to the genus *Protosuchus*, especially isolated dorsal osteoderms, are the most common tetrapod fossils from the basalt talus-slope breccias. The new protosuchid is closely related to *Protosuchus richardsoni* (Brown, 1933) from the Lower Jurassic

Table 13.1. *Faunal list of the vertebrate taxa of the McCoy Brook Formation*

Chondrichthyes
  Elasmobranchii
    Hybodontiformes
      cf. *Hybodus* sp. (Ll, La)[a]
Osteichthyes
  Actinopterygii
    ?Redfieldiidae
      scales and skull bones (Ll, La)
    Semionotidae
      aff. *Semionotus* sp. (Ll, La, Lc)
Synapsida
  Therapsida
    Cynodontia
      Tritheledontidae
        Pachygenelus cf. *P. monus* (Fl, A)
Reptilia
  Lepidosauria
    Sphenodontia
      Sphenodontidae
        Clevosaurus sp. nov. (Fl, A)
  Archosauria
    Crocodylomorpha
      Protosuchidae
        Protosuchus sp. nov. (Fl, A)
      Sphenosuchidae gen. indet. (Fl, A)
  Dinosauria
    Sauropodomorpha
      cf. *Ammosaurus* sp. (Fl, A, E)
    Ornithischia
      "Fabrosauridae" indet. (La, A)

Ichnotaxa
  Otozoum moodi (Fl)
  Batrachopus sp. (Fl)
  Grallator sp. (Fl)
  Anomoepus scambus (Fl)

[a]Abbreviations for lithofacies: A, basalt agglomerate with orange or red sandstone or mudstone matrix; E, eolian dune sandstones; Fl, fluviolacustrine sandstone and mudstone; La, lacustrine basalt agglomerate; Lc, lacustrine red mudstones; Ll, lacustrine limestones. (La, Lc, and Ll are collectively referred to as the "fish bed" in the text.)

Moenave Formation of Arizona and to *P. haughtoni* (Busbey and Gow, 1984) from the upper Stormberg Group of southern Africa (Clark, in Benton and Clark, 1988). It is distinguished from other species of *Protosuchus* mainly by the presence of two greatly enlarged teeth in the dentary and the less spatulate development

of the symphysis. The derived presence of two caniniform teeth in the dentary is shared by the problematical protosuchid *Platyognathus hsui* Young, 1944, from the Lower Jurassic Lower Lufeng Formation of Yunnan, China, but the latter is too poorly known for meaningful comparisons. A diagnosis and detailed anatomical description of the cranial material of the new species of *Protosuchus* from the McCoy Brook Formation will be presented elsewhere (H.-D. Sues, N. H. Shubin, and P. E. Olsen, unpublished data).

Archosauria

Crocodylomorpha

Sphenosuchia

Occurrence. Basalt talus-slope breccias, sites E and F.

Discussion. A maxilla with six teeth (NSM 988GF9.1) resembles the corresponding bone in *Terrestrisuchus* (Crush, 1984, fig. 2). MCZ 9112 probably represents a larger specimen of the same taxon. Some of the numerous isolated postcranial bones of crocodylomorph archosaurs from the basalt talus-slope breccias at Wasson Bluff probably are also referable to this gracile crocodylomorph.

The mediolaterally compressed tooth crowns are recurved and have distinctly serrated, gently convex or nearly straight posterior cutting edges; the anterior cutting edges bear serrations only near the tip. A moderately deep antorbital fossa is developed anteroventral to the large antorbital fenestra. Interdental plates are present. The shapes of the tooth crowns, the absence of sculpturing on the maxilla, and the large size of the antorbital fossa in the maxilla preclude reference to *Protosuchus* from the same strata, but the currently available material is insufficient for more precise taxonomic identification.

Archosauria

Dinosauria

Saurischia

cf. *Ammosaurus* sp.

Occurrence. Sandstones between the two L localities and locality K; below basalt talus, locality N.

Discussion. A poorly preserved partial skeleton, a dissociated skeleton of a larger specimen, and an incomplete pelvic girdle and hindlimb of a juvenile indicate the presence of anchisaurid prosauropods in

A

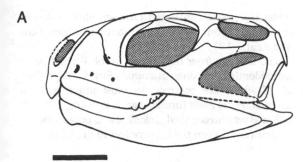

B

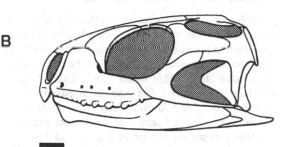

Figure 13.3. Reconstructed skulls of (A) *Clevosaurus* sp. nov. (based on MCZ 9105 and NSM 988GF1.1) and (B) *Clevosaurus hudsoni* from the Upper Triassic of southwest Britain (redrawn from Fraser, 1988.) Scales equal 5 mm.

the McCoy Brook Formation. An articulated pelvic girdle and hindlimb found at site K are referable to a small, presumably juvenile prosauropod. This identification is based on the ischium; more precise designation must await further preparation of the specimen. Numerous worn gastroliths are present within the rib cage of one of the larger specimens. These gastroliths consist of smoothly oblong pebbles that range from 1 to 3 cm in maximum diameter. A badly worn partial maxilla of *Clevosaurus* was found in direct association with the gastroliths.

Archosauria

Dinosauria

Ornithischia

"Fabrosauridae"

Occurrence. Basalt talus-slope breccias, site E; "fish bed," site J.

Discussion. Numerous isolated teeth, especially from site J, indicate the presence of small ornithischian dinosaurs. At present, it is impossible to ascertain the taxonomic diversity of this material. One very well preserved tooth, MCZ 9119, resembles maxillary teeth of *Lesothosaurus* illustrated by Sereno (1991, fig. 4). Its

tooth crown is symmetrical and labiolingually compressed. The anterior and posterior cutting edges of the tooth crown each bear seven denticles. These denticles represent the apices of subparallel, vertical ridges that extend along the labial face of the crown. The median ridge terminates in a blunt apical denticle.

Diapsida

Lepidosauria

Sphenodontia

Clevosaurus sp. nov.

Occurrence. Lake-margin and fluvial sandstones, localities K and K'; basalt talus-slope breccias, site F; interdune sandstones, between localities L and K', and associated with the gastroliths of the prosauropod at locality N.

Discussion. A new sphenodontian lepidosaur is one of the most common fossil vertebrates in the McCoy Brook Formation and is most abundant in the lake-margin and fluvial sandstones at sites K and K'. The known material includes two well-preserved partial skulls (Figure 13.3) and an incomplete, partially articulated postcranial skeleton (MCZ 9106, from beach ridge about 10 m south of site J). The new form is referable to *Clevosaurus* Swinton, 1939 (which was originally known only from Late Triassic fissure fillings in southwestern England) (Fraser, 1988), based especially on the shared presence of a long posterior process of the premaxilla that excludes the maxilla from the posterior margin of the external narial opening. It differs from the English species of this genus in a number of cranial and dental features and very closely resembles a new clevosaur from the Lower Lufeng Formation of Yunnan, China (Wu, Chapter 3). A detailed description of the available material and diagnosis of the new species will be provided elsewhere (H.-D. Sues, N. H. Shubin, and P. E. Olson, in press).

Synapsida

Cynodontia

Tritheledontidae

Pachygenelus cf. *P. monus* Watson, 1913

Occurrence. Basalt talus-slope breccias, site E; lake-margin and fluvial sandstones, sites K and K'.

Discussion. The currently available tritheledontid specimens from the McCoy Brook Formation are dentally indistinguishable from *Pachygenelus monus* Watson, 1913, from the middle and upper Elliot Formation of southern Africa. Shubin et al. (1991) refer to this material as *Pachygenelus* cf. *P. monus* only because of the considerable geographic distance that separated the southern African and Canadian occurrences even during Early Jurassic times.

There are two procumbent incisors in each dentary and in each premaxilla. Wear is confined to the posterolingual surfaces of the lower incisors and the anterior surfaces of the upper ones. The small canine is separated from the first postcanine tooth by a short diastema.

Each of the postcanine teeth has a large central cusp flanked by an accessory cusp anteriorly and posteriorly. Occasionally the distinct buccal cingulum carries cuspules. This is more pronounced on posterior teeth than on anterior ones. A single wear facet is developed along the lingual aspect of the large central cusp; in cases of extreme wear, the facet extends onto the anterior accessory cusp. Each of the lower postcanine teeth bears a large, recurved principal cusp, followed by two cusps that decrease in height posteriorly. A prominent, cuspidate lingual cingulum extends along the entire length of the tooth. The cingulum becomes more prominent on the posterior postcanine teeth. Occlusal wear occurs on the buccal side of the main cusp and occasionally extends onto the posterior accessory cusps. The roots of the lower postcanines are very large.

Paleobiogeography

The tetrapods from the McCoy Brook Formation support the hypothesis of very close similarities in taxonomic composition between continental tetrapod communities in the Early Jurassic (Shubin and Sues, 1991). The presence of the tritheledontid *Pachygenelus* cf. *P. monus* is particularly significant in this context. *Pachygenelus monus* is known only from the middle to upper Elliot Formation (upper Stormberg Group) of southern Africa (Kitching and Raath, 1984). The sphenodontian and archosaurian reptiles all show close phylogenetic affinities to taxa from the upper Stormberg Group (middle to upper Elliott Formation and Clarens Formation) (Kitching and Raath, 1984), the early Mesozoic fissure fillings of southwestern England (Fraser, 1986), and the Lower Lufeng Formation of Yunnan, China (Simmons, 1965; Luo and Wu, Chapter 14). These affinities extend to the family level and often even to the generic level. The sphenodontian lepidosaur *Clevosaurus* is common in Upper Triassic fissure fillings in southwestern England (Fraser, 1988), and it has been recorded from the Lower Jurassic Lower Lufeng Formation of Yunnan (Wu, Chapter 3) and possibly from the Lower Jurassic Forest Sandstone of Zimbabwe,

which is generally considered a stratigraphic equivalent of the upper Stormberg Group in southern Africa (Gow and Raath, 1977).

The absence from the McCoy Brook Formation of tritylodontid cynodonts, mammaliaforms, and turtles, all of which are common in most other tetrapod assemblages of Early Jurassic age, is puzzling and may reflect as yet unrecognized ecological and/or taphonomic differences between the occurrences in question.

Implications for faunal change

Early Jurassic assemblages of continental tetrapods worldwide are characterized by the absence of many taxa, such as procolophonians, phytosaurs, rauisuchian archosaurs, that were common during the Late Triassic (Olsen and Sues, 1986; Olsen et al., 1987; Shubin and Sues, 1991). This fact and the apparent global homogeneity of the known Early Jurassic tetrapod assemblages are consistent with the hypothesis of an extinction event during or at the end of the Late Triassic. Analyses of Triassic turnover have resulted in proposals of at least two major extinction events in the Late Triassic, one at the end of the Carnian (not supported by the currently available evidence from the Newark Supergroup) and another at the end of the Norian (Benton, 1986, and Chapter 22; Olsen and Sues, 1986; Olsen et al., 1987). The relative timing of these extinction events awaits further stratigraphic study. Particularly vexing in this regard is the poor sampling of much of the Norian stage, which is well-represented only by assemblages from the Germanic Basin of Europe (Benton, 1986).

The presence of the Manicouagan bolide impact site in Québec and evidence of an impact in the form of shocked quartz in strata of latest Triassic age elsewhere (Bice et al., 1992) suggest that the Late Triassic may prove to be an important testing ground for the role of impacts in mass extinctions. Olsen et al. (1987) suggested a link between the Manicouagan impact and the apparent mass extinctions at the end of the Triassic. However, the Manicouagan crater has recently yielded dates of 212.9 ± 0.4 Ma (^{40}Ar/^{39}Ar) (Shepard, 1986) and 214 ± 1 Ma (U-Pb) (Hodych and Dunning, 1992). These dates are internally very consistent, but they are incompatible with dates on the lava flows and intrusions that directly overlie the palynologically identified Triassic–Jurassic boundary. ^{40}Ar/^{39}Ar and U-Pb dates from these basalts and associated intrusions from Virginia to Nova Scotia cluster around 201–202 Ma (Sutter, 1988; Dunning and Hodych, 1990; Hodych and Dunning, 1992), and that is probably the best available date for the Triassic–Jurassic boundary. If the Triassic–Jurassic boundary is correctly identified in the Newark Supergroup, and if the forgoing dates are correct, Manicouagan could have had nothing to do with the extinctions at the boundary. A search for a large-scale extinction event at around 212–216 Ma in the Milankovitch-cycle-calibrated section of the Newark basin has not been successful.

The Manicouagan impact probably was not the cause of the supposed mass extinctions in the late Carnian (Benton, 1991, and Chapter 22). The Carnian–Norian boundary has not been directly dated in marine sections; based on Milankovitch-cycle calibration in the Newark basin, the age of the palynologically dated Carnian–Norian boundary is significantly older than 214 Ma (contra Olsen et al., 1987) and closer to 220 Ma. Despite the gigantic size of the Manicouagan impact, that event apparently was insufficient to generate an obvious global extinction.

Taken at face value, however, the discovery by Bice et al. (1992) of shocked-quartz horizons at the Triassic–Jurassic boundary in marine deposits in Italy provides very strong support for an impact origin of the Triassic–Jurassic mass extinctions. Faunal and floral data (Hallam, 1981; Olsen and Sues, 1986; Olsen et al., 1987; Bice et al., 1992) are consistent with a catastrophic event as well. The structure or structures made by the impacts of bolide(s) responsible for the Triassic–Jurassic boundary extinctions have yet to be identified, however, now that Manicouagan has been excluded.

The early Mesozoic thus apparently provides at least two cases of giant impacts: one that produced a mass extinction and one that did not. In addition, Bice et al. (1992) report two additional horizons with shocked quartz just below that coincident with the mass extinctions, suggesting two more impacts without massive faunal change. A search is under way for impact ejecta within the appropriate parts of the Newark sequences. With high-resolution chronostratigraphy now becoming available for the Late Triassic and Early Jurassic, and strong evidence for multiple giant impacts and massive faunal and floral changes, the early Mesozoic becomes an unparalleled laboratory for the study of impacts and mass extinctions.

Acknowledgments

Donald Baird's pioneering studies on the fossil vertebrates of the Fundy Group laid the foundation for our work. Our research was aided by the Museum of Comparative Zoology, Harvard University, particularly by F. A. Jenkins, Jr., W. W. Amaral, and C. R. Schaff. Logistic support and field permits were obtained through the Nova Scotia Museum, and the assistance of R. Grantham and R. Ogilvie is greatly appreciated. This work was supported by grants from the National Geographic Society (to NHS and PEO), the Sloan Foundation (to PEO), the National Science Foundation (BSR 8717707 to PEO), the American Chemical Society (to PEO), the Miller Institute for Basic Research in Science, Berkeley (to NHS and PEO), and the Natural Sciences and Engineering Research Council of Canada (to H-DS).

References

Benton, M. J. 1986. The Late Triassic tetrapod extinction events. Pp. 303–320 in K. Padian (ed.), *The Beginning of the Age of Dinosaurs: Faunal Change across the Triassic–Jurassic Boundary.* Cambridge University Press.

1991. What really happened in the Late Triassic? *Historical Biology* 5: 263–278.

Benton, M. J., and J. M. Clark. 1988. Archosaur phylogeny and the relationships of the Crocodylia. Pp. 295–338 in M. J. Benton (ed.), *The Phylogeny and Classification of the Tetrapods. Vol. 1: Amphibians, Reptiles and Birds.* Oxford: Clarendon Press.

Bice, D., C. R. Newton, S. McCauley, P. W. Reiners, and C. A. McRoberts. 1992. Shocked quartz at the Triassic–Jurassic boundary in Italy. *Science* 255: 443–446.

Brown, B. 1933. An ancestral crocodile. *American Museum Novitates* 638: 1–4.

Busbey, A. B., III, and C. E. Gow. 1984. A new protosuchian crocodile from the Upper Triassic Elliot Formation of South Africa. *Palaeontologia Africana* 25: 127–149.

Colbert, E. H. 1958. Tetrapod extinctions at the end of the Triassic period. *Proceedings of the National Academy of Sciences USA* 44: 973–977.

Cornet, B. 1977. The palynostratigraphy and age of the Newark Supergroup. Ph.D. dissertation, Pennsylvania State University.

Crush, P. J. 1984. A late upper Triassic sphenosuchid crocodilian from Wales. *Palaeontology* 27: 131–157.

Dunning, G., and J. P. Hodych. 1990. U/Pb zircon and baddelyte ages for the Palisades and Gettysburg sills of the northeastern United States: implications for the age of the Triassic/Jurassic boundary. *Geology* 18: 795–798.

Fraser, N. C. 1986. Terrestrial vertebrates at the Triassic–Jurassic boundary in South West Britain. *Modern Geology* 10: 147–157.

1988. The osteology and relationships of *Clevosaurus* (Reptilia: Sphenodontida). *Philosophical Transactions of the Royal Society of London* B321: 125–178.

Gow, C. E., and M. A. Raath. 1977. Fossil vertebrate studies in Rhodesia: Sphenodontid remains from the Upper Triassic of Rhodesia. *Palaeontologia Africana* 20: 121–122.

Hallam, A. 1981. The end-Triassic bivalve extinction event. *Palaeogeography, Palaeoclimatology, Palaeoecology* 35: 1–44.

Hodych, J. P., and G. R. Dunning. 1992. Did the Manicouagan impact trigger end-of-Triassic mass extinction? *Geology* 20: 51–54.

Hubert, J. F., and K. A. Mertz. 1984. Eolian sandstones in Upper Triassic–Lower Jurassic red beds of the Fundy Basin, Nova Scotia. *Journal of Sedimentary Petrology* 54: 798–810.

Johnson, A. L. A., and M. J. Simms. 1989. The timing and cause of Late Triassic marine extinctions: evidence from scallops and crinoids. Pp. 174–194 in S. F. Donovan (ed.), *Mass Extinctions: Processes and Evidence.* London: Belhaven.

Kitching, J. W., and M. A. Raath. 1984. Fossils from the Elliot and Clarens formations (Karoo sequence) of the northeastern Cape, Orange Free State and Lesotho, and a suggested biozonation based on tetrapods. *Palaeontologia Africana* 25: 111–125.

Olsen, P. E., and D. Baird. 1986. The ichnogenus *Atreipus* and its significance for Triassic biostratigraphy. Pp. 61–87 in K. Padian (ed.), *The Beginning of the Age of Dinosaurs: Faunal Change across the Triassic–Jurassic Boundary.* Cambridge University Press.

Olsen, P. E., S. J. Fowell, and B. Cornet. 1990. The Triassic–Jurassic boundary in continental rocks of eastern North America: a progress report. Pp. 585–593 in V. L. Sharpton and P. D. Ward (eds.), *Global Catastrophes in Earth History: An Interdisciplinary Conference on Impacts, Volcanism, and Mass Mortality.* Geological Society of America Special Paper 247.

Olsen, P. E., and P. M. Galton. 1977. Triassic–Jurassic extinctions: are they real? *Science* 197: 983–986.

1984. A review of the reptile and amphibian assemblages from the Stormberg of southern Africa, with special emphasis on the footprints and the age of the Stormberg. *Palaeontologia Africana* 25: 87–110.

Olsen, P. E., and R. W. Schlische. 1990. Transtensional arm of the early Mesozoic Fundy rift basin: penecontemporaneous faulting and sedimentation. *Geology* 18: 695–698.

Olsen, P. E., R. W. Schlische, and P. J. W. Gore (eds.). 1989. *Tectonic, Depositional, and Paleoecological History of Early Mesozoic Rift Basins, Eastern North America.* Field Trip Guidebook T351. 28th International Geological Congress. Washington, D. C.: American Geophysical Union.

Olsen, P. E., N. H. Shubin, and M. H. Anders. 1987. New Early Jurassic tetrapod assemblages constrain Triassic–Jurassic extinction event. *Science* 237: 1025–1029.

Olsen, P. E., and H.-D. Sues. 1986. Correlation of continental Late Triassic and Early Jurassic sediments, and patterns of the Triassic–Jurassic tetrapod transition. Pp. 321–351 in K. Padian (ed.), *The Beginning of the Age of Dinosaurs: Faunal Change across the Triassic–Jurassic Boundary.* Cambridge University Press.

Puffer, J. H., and A. R. Philpotts. 1988. Eastern North American quartz tholeiites: geochemistry and petrology. Pp. 579–605 in W. Manspeizer (ed.),

Triassic–Jurassic Rifting: Continental Breakup and the Origin of the Atlantic Ocean and Passive Margins. Amsterdam: Elsevier.

Sereno, P. C. 1991. *Lesothosaurus,* "fabrosaurids," and the early evolution of Ornithischia. *Journal of Vertebrate Paleontology* 11: 168–197.

Shepard, J. B. 1986. The Triassic/Jurassic boundary and the Manicouagan impact: implications of ^{40}Ar/^{39}Ar dates on periodic extinction models. Senior thesis, Princeton University.

Shubin, N. H., A. W. Crompton, H.-D. Sues, and P. E. Olsen. 1991. New fossil evidence on the sister-group of mammals and early Mesozoic faunal distributions. *Science* 251: 1063–1065.

Shubin, N. H., and H.-D. Sues. 1991. Biogeography of early Mesozoic continental tetrapods: patterns and implications. *Paleobiology* 17: 214–230.

Simmons, D. J. 1965. The non-therapsid reptiles from the Lufeng basin, Yunnan, China. *Fieldiana, Geology* 15: 1–93.

Sues, H.-D., N. H. Shubin, and P. E. Olsen. In press. A new sphenodontian (Lepidosauria: Rhynchocephalia) from the McCoy Brook Formation (Lower Jurassic) of Nova Scotia, Canada. *Journal of Vertebrate Paleontology.*

Sutter, J. E. 1988. Innovative approaches to dating igneous events in the early Mesozoic basins of the eastern United States. Pp. 194–200 in A. J. Froelich and G. R. Robinson, Jr. (eds.), *Studies on the Early Mesozoic Basins of the Eastern United States.* United States Geological Survey Bulletin 1776.

Swinton, W. 1939. A new Triassic rhynchocephalian from Gloucestershire. *Annals and Magazine of Natural History* (11) 4: 591–594.

Tanner, L. H., and J. F. Hubert. 1988. Debris-flow and talus-slope conglomerates in the Lower Jurassic McCoy Brook Formation, Fundy Basin, Nova Scotia. *Geological Society of America, Abstracts with Program* 20: 74.

Watson, D. M. S. 1913. On a new cynodont from the Stormberg. *Geological Magazine* (5) 10: 145–148.

Young, C. C. 1944. On a supposed new pseudosuchian from Upper Triassic saurischian-bearing beds of Lufeng, Yunnan, China. *American Museum Novitates* 1264: 1–4.

14

The small tetrapods of the Lower Lufeng Formation, Yunnan, China

ZHEXI LUO AND XIAO-CHUN WU

Introduction

Since the pioneering exploration by Young and Bien in the late 1930s, the diverse fossil vertebrates of the Lufeng Basin in Yunnan, China, have been an important source of information for our understanding of Triassic–Jurassic tetrapod assemblages (Figures 14.1 and 14.2). The Lower Lufeng Formation yielded one of the most diverse faunas of early Mesozoic terrestrial tetrapods (Young, 1940a,b, 1944, 1946, 1947a,b, 1951; Bien, 1941; Patterson and Olson, 1961; Rigney, 1963; Simmons, 1965). Subsequent fieldwork by the Institute of Vertebrate Paleontology and Paleoanthropology (IVPP) has produced additional, better-preserved specimens and additional new taxa (Chow and Hu, 1959; Chow, 1962; Young, 1974, 1978, 1982; Cui, 1976, 1981; Wu, 1986). Phylogenetic studies of the Lufeng taxa have provided important data on the evolutionary history of several major tetrapod groups (Young, 1951; Patterson and Olson, 1961; Kermack, Mussett, and Rigney, 1973, 1981; Carroll and Galton, 1977; Jenkins, 1984; Crompton and Sun, 1985; Clark, 1986; Clemens, 1986; Sues, 1986b; Rowe, 1988; Crompton and Luo, 1993; Wu and Chatterjee, 1993; Wu, Chapter 3).

Many fossil specimens collected from the Lower Lufeng Formation by the early expeditions lack precise stratigraphic details (e.g., Rigney, 1963), and some biostratigraphic problems remain unsolved. Fieldwork to establish the stratigraphic ranges of the taxa did not begin in earnest until the early 1970s. Based on more detailed stratigraphic work, Sun, Cui, Li, and Wu (1985) recognized two successive but discrete faunal assemblages within the Lower Lufeng Formation. The lower faunal assemblage from the Dull Purplish Beds is dominated by prosauropod dinosaurs and the tritylodontid *Bienotherium*. The upper assemblage from the Dark Red Beds has a much greater taxonomic diversity. It is characterized by the appearance of early mammals, crocodylomorphs, and ornithischian dinosaurs. More recently, fieldwork by personnel from the IVPP and a joint expedition of the IVPP and Texas Tech University have clarified the stratigraphic ranges of several taxa within the Lower Lufeng Formation.

Much of what we know about the Lufeng tetrapods is based on studies by C. C. Young (e.g., 1951, 1982). Following the prevailing scheme of systematics of his time, Young assigned many diapsids in the Lufeng fauna to "primitive thecodontians." He further used this as faunal evidence to argue for a Late Triassic (Rhaetic) age for the Lower Lufeng Formation. A persisting problem in faunal studies of the Lufeng concerns the systematic relationships of the diapsid taxa assigned by Young (e.g., 1951, 1982) and Simmons (1965) to "thecodontians," which are a paraphyletic group (Benton and Clark, 1988). Recent studies show that some of the "thecodontians" are sphenosuchians (Wu, 1991), and some should be reassigned to crocodyliforms, based on new phylogenetic studies (Crush, 1984; Clark, 1986; Wu, 1986; Wu and Chatterjee, 1993). A "protorosaur" established by Young (1951) is a misidentified sphenodontian. A systematic review of the Lower Lufeng fauna has been long overdue in light of this new information on the systematic relationships of the diapsids.

The first goal of this chapter is to provide updated information on the diversity of the small, nondinosaurian tetrapods, in the wake of the recent studies, which have considerably changed the taxonomy of some vertebrate groups from the Lower Lufeng Formation (Crompton and Sun, 1985; Evans and Milner, 1989; Wu, 1991; Crompton and Luo, 1993; Wu and Chatterjee, 1993). The second goal is to describe the available stratigraphic ranges of the known Lufeng tetrapods, based on specimens collected since the early 1970s with verifiable records of relative stratigraphic

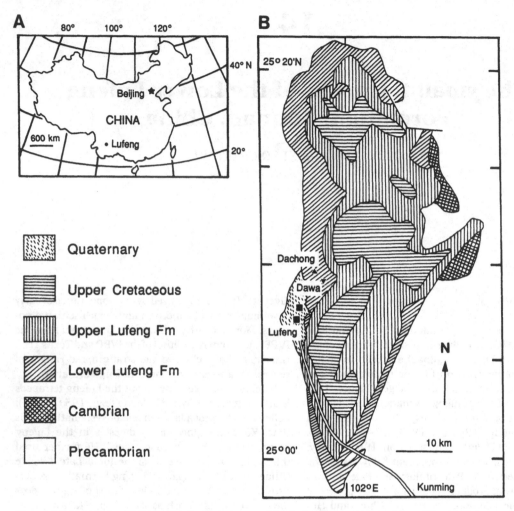

A

80°   100°   120°

40° N

Beijing ★

CHINA

600 km

• Lufeng

20°

Quaternary

Upper Cretaceous

Upper Lufeng Fm

Lower Lufeng Fm

Cambrian

Precambrian

B

25° 20'N

Dachong

Dawa

Lufeng

N

25° 00'

102°E   Kunming

10 km

Figure 14.1. (A) Map of China. (B) Geological map of the Lufeng Basin. (Adapted from Wu, 1991.)

positions (Cui, 1976, 1981; Zhang and Cui, 1983; Sun et al., 1985; Wu, 1986, 1991; Z. Luo and X. Wu, personal observations). Abbreviations of fossil collections: CUP, former Catholic University of Peking (specimens housed in the Field Museum of Natural History, Chicago); IVPP, Institute of Vertebrate Paleontology and Paleoanthropology, Academia Sinica, Beijing.

Preservation and stratigraphy of tetrapod fossils

The early Mesozoic strata of the Lufeng Basin have traditionally been divided into the Lower and the Upper Lufeng formations (Bien, 1941). They consist of a mosaic of lacustrine, fluvial, and overbank deposits, most of which are mudstones and siltstones containing some calcareous nodules. The Lower Lufeng Formation has yielded very abundant vertebrate fossils, whereas only fragmentary remains of fish, turtles, and sauro-

pod dinosaurs are known from the Upper Lufeng Formation. The most fossiliferous outcrops of the Lower Lufeng Formation known to date are in the Dachong-Dawa area (Figure 14.1). Most outcrops are slightly eroded, and in some areas they are surrounded by vegetation and fields. Thus it is difficult to find an uninterrupted stratigraphic section that contains all vertebrate-bearing strata. Based on two composite sections (AA' and BB', Figure 14.2), we estimate that the total thickness of the Lower Lufeng Formation in the Dachong-Dawa area as about 400 m (Figure 14.3). Most of the small nondinosaurian vertebrates are found in calcareous nodules, in which skulls or parts of skulls are preserved. In some cases, postcranial bones may be associated with the skull. Isolated teeth and postcranial elements of small vertebrates are relatively rare.

The early stratigraphic studies of the Lower Lufeng Formation by Young (1940a,b, 1944, 1946, 1947a,b)

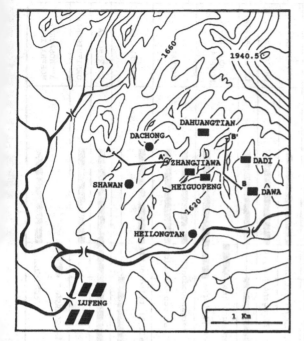

Figure 14.2. Simplified topographic map of the Dachong-Dawa area, Lufeng County, Yunnan Province, China. Major localities for fossil vertebrates in the Dull Purplish Beds are shown by circles; those in the Dark Red Beds are shown by rectangles. The stratigraphic sections of the Dull Purplish Beds and Dark Red Beds in Figure 14.3 are measured from A to A' and from B to B', respectively. (Adapted from Wu, 1991.)

and Bien (1941) were summarized by Young (1951). Bien and Young recognized that most of the vertebrate fossils occurred in two major stratigraphic horizons in the Lower Lufeng Formation: the lower Dull Purplish Beds and the upper Dark Red Beds (Table 14.1). But they did not recognize these as two separate faunal assemblages. Since the early work by Bien and Young, many more fossils have been collected in the area. The collection at the Field Museum of Natural History, assembled by Father E. Oehler of the former Catholic University of Peking in 1948–1949, is very well known (Rigney, 1963; Hopson, 1964, 1965; Simmons, 1965; Kermack et al., 1973, 1981; Clark, 1986). Researchers at the Institute of Vertebrate Paleontology and Paleoanthropology in Beijing, especially Professor Sun Ailin and her associates, have greatly expanded the collection of vertebrate fossils from the Lower Lufeng Formation. As a result, Sun et al. (1985) now recognize two distinctive faunal assemblages in the Lower Lufeng Formation: a lower faunal assemblage from the Dull Purplish Beds, dominated by saurischian dinosaurs and the tritylodontid *Bienotherium*, and an upper assemblage from the Dark Red Beds, which is domi-

nated by ornithischian dinosaurs, early mammals, and more diverse tritylodontids (Table 14.1).

Because of several recent studies of the diapsids from the Lufeng Basin (Clark, 1986; Wu, 1986, 1991, and Chapter 3; Wu and Chatterjee, 1993), it is clear that many diapsids formerly regarded as "primitive thecodontians" are either crocodylomorphs or sphenodontians. Crompton and Luo (1993) show that the several supposedly diagnostic dental features of species of *Sinoconodon* are based on ontogenetic variation. Thus, it is necessary to revise the taxonomy of the faunal assemblages initially proposed by Sun et al. (1985) and Sun and Cui (1986). The new systematic information suggests that the lower faunal assemblage is characterized by the first appearance of the sphenodontian *Clevosaurus*, in addition to *Bienotherium* and saurischians. The upper faunal assemblage is characterized by the first appearance of mammals and ornithischians, as well as more derived and diverse crocodyliforms, sphenodontians, and tritylodontids. Table 14.2 and Figure 14.3 summarize the stratigraphic positions of these taxa, based on our firsthand field records. The justification and evidence for this revision of the Lufeng faunal assemblages are presented in the following sections.

Systematic paleontology

Synapsida

Tritylodontidae

Bienotherium Young, 1940

Bienotherium yunnanense Young, 1940, is documented by abundant dental and cranial material (Young, 1947a; Hopson, 1964, 1965) More recently, Cui and Sun (1987) have described the postcanine roots. Specimens of *B. yunnanense* collected in recent years are all from the Dull Purplish Beds. This has led Sun et al. (1985; Cui and Sun, 1987) to suggest that *B. yunnanense* is restricted to the Dull Purplish Beds (strata 3 and 4). Most specimens collected in earlier years by Young, Bien, and Oehler lacked verifiable stratigraphic data, although the specimens were said to have been collected from the Dull Purplish Beds. The only possible occurrence of *B. yunnanense* in the Dark Red Beds (stratum 6, Table 14.2 and Figure 14.3) was reported by Young (1947a, p. 540). He described two limb bones of *B. yunnanense* from the "Huangchiatian" locality (= Dahuangtian in Figure 14.2) in the Dark Red Beds. However, postcranial bones are not diagnostic of *B. yunnanense*, and no character of the postcranial skeleton has been established to distinguish *Bienotherium* from other tritylodontids. Later fieldwork has failed to confirm the presence of *B. yunnanense* in the Dark Red Beds. There is now substantial evidence that *B.*

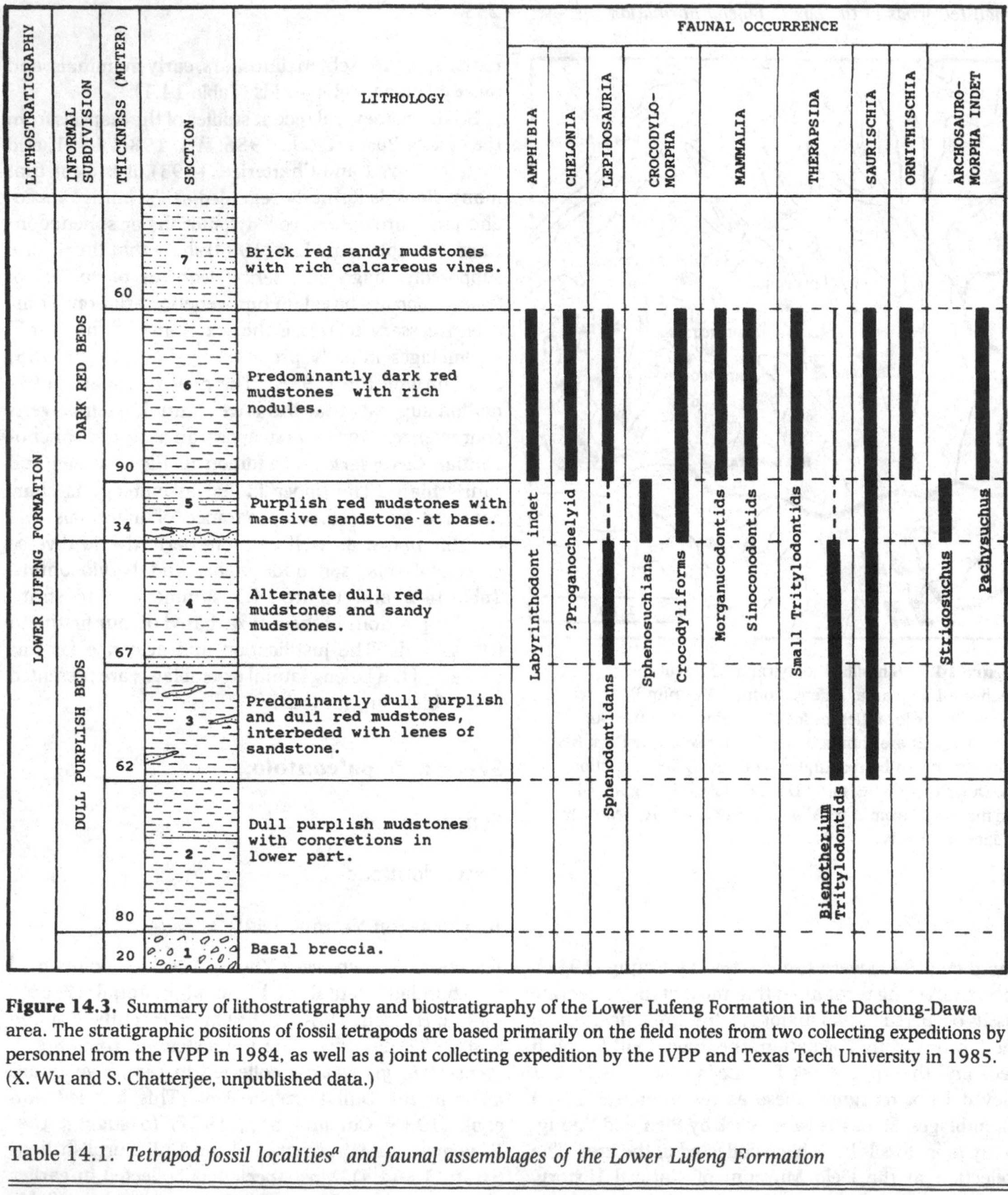

Figure 14.3. Summary of lithostratigraphy and biostratigraphy of the Lower Lufeng Formation in the Dachong-Dawa area. The stratigraphic positions of fossil tetrapods are based primarily on the field notes from two collecting expeditions by personnel from the IVPP in 1984, as well as a joint collecting expedition by the IVPP and Texas Tech University in 1985. (X. Wu and S. Chatterjee, unpublished data.)

Table 14.1. *Tetrapod fossil localities[a] and faunal assemblages of the Lower Lufeng Formation*

Major tetrapod fossil localities	Stratigraphic horizons[b]	Faunal assemblages[c]	Index taxa (this study)
Dawa Dadi	Dark Red Beds	Upper faunal assemblage	*Morganucodon* *Yunnanodon*
Dahuangtian Heiguopeng Zhangjiawa	Dark Red Beds		Ornithischia Crocodylomorpha
Dachong	Dull Purplish Beds	Lower faunal assemblage	*Bienotherium*
Heilongtan Shawan	Dull Purplish Beds	Lower faunal assemblage	Saurischia

[a] All major fossil localities are shown in Figure 14.2.
[b] Bien (1941); Young (1951).
[c] Sun et al. (1985); Sun and Cui (1986).

Table 14.2. *Tetrapod taxa from the Lower Lufeng Formation*

Taxon	Stratum 6	Stratum 5	Strata 3 & 4
Amphibia			
Labyrinthodontia indet.	×		
Testudines			
Proganochelyidae indet.	×		
Diapsida			
Lepidosauria			
Sphenodontia			
Clevosaurus ("Dianosaurus") petilus[a,b]			×
Clevosaurus wangi[b]	×		
Clevosaurus mcgilli[b]	×		
Crocodylomorpha			
Sphenosuchia			
Dibothrosuchus elaphros Simmons, 1965		×	
Crocodyliformes			
Platyognathus hsui Young, 1944	×		
Microchampsa scutata Young, 1951 (nomen dubium)	×		
Dianosuchus changchiawaensis Young, 1982	×		
Unnamed new taxon[c]		×	
Archosauria indet.			
Strigosuchus licinus Simmons, 1965		×	
Pachysuchus imperfectus Young, 1951	×		
Saurischia			
Lufengosaurus huenei Young, 1941	×	×	×
Lufengosaurus magnus Young, 1941	×	×	×
Yunnanosaurus huangi Young, 1941	×	×	×
Yunnanosaurus robustus Young, 1941	×	×	×
Gyposaurus sinensis Young, 1941	×	×	×
Sinosaurus triassica Young, 1948	×	×	×
Lukousaurus yini Young, 1948	×	×	×
Ornithischia			
Tatisaurus oehleri Simmons, 1965	×		
Tawasaurus minor Young, 1982	×		
Dianchungosaurus lufengensis Young, 1982	×		
Synapsida			
Tritylodontidae			
Bienotherium yunnanense Young, 1940			×
Bienotherium minor Young, 1947			×
Bienotherium magnum Chow, 1962	×		
Lufengia delicata Chow and Hu, 1959	×		
Yunnanodon brevirostre Cui, 1976	×		
Dianzhongia longistrata Cui, 1981	×		
Mammalia			
Sinoconodon rigneyi Patterson and Olson, 1961	×		
Morganucodon oehleri Rigney, 1963	×		
Morganucodon heikuopengensis[d]	×		
Unnamed new taxon[e]	×		
Synapsida indet.			
Kunminia minima Young, 1947	×		

Note: This table lists all tetrapod taxa and their respective stratigraphic positions.

[a] Young (1982).

[b] Wu (Chapter 3, this volume).

[c] Wu and Chatterjee (1993).

[d] Young (1978).

[e] Z. Luo and A. W. Crompton (unpublished data).

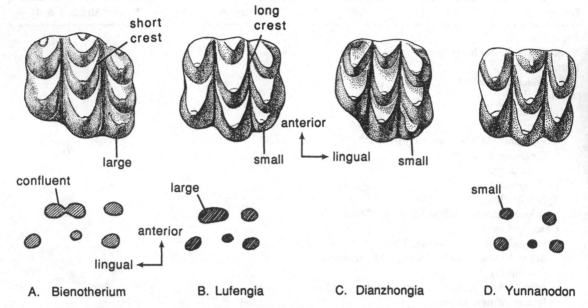

short crest

long crest

anterior

lingual

small

large

small

confluent

large

anterior

small

lingual

A. **Bienotherium**   B. **Lufengia**   C. **Dianzhongia**   D. **Yunnanodon**

Figure 14.4. Right upper postcanines of tritylodontids from the Lower Lufeng Formation. Top row: cusp patterns. Bottom row: cross-sections near the bases of the roots. (A) *Bienotherium yunnanense.* (Adapted from Cui and Sun, 1987, and Hopson, 1965.) (B) *Lufengia delicata.* (Adapted from Cui and Sun, 1987, and Hopson, 1965.) (C) *Dianzhongia longirostrata.* (Adapted from Cui, 1981, on the basis of IVPP 8694.) (D) *Yunnanodon brevirostre.* (Adapted from Cui and Sun, 1987.)

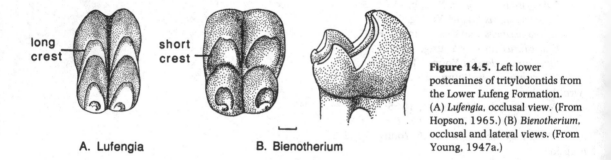

long crest

short crest

Figure 14.5. Left lower postcanines of tritylodontids from the Lower Lufeng Formation. (A) *Lufengia*, occlusal view. (From Hopson, 1965.) (B) *Bienotherium*, occlusal and lateral views. (From Young, 1947a.)

A. **Lufengia**   B. **Bienotherium**

yunnanense is restricted to the Dull Purplish Beds (Sun et al., 1985; Z. Luo and X. Wu, personal observation).

Adult individuals of *B. yunnanense* have two upper incisors and one lower incisor on each side (Young, 1947a; Hopson, 1965; Cui, 1981; Clark and Hopson, 1985). Hopson (1965) recognized that I^3 and I_2 are present in subadult individuals, but have been lost by the more mature individuals. Thus the full incisor count should be I^3/I_2 (Hopson, 1965).

The upper postcanines have two cusps in the buccal row, three cusps in the middle row, and three cusps in the lingual row. Additional small cuspules may be present on the anterior end of the cusp row in some teeth (Figure 14.4). The cusps of the middle row have the crescentic crests typical of tritylodontids. By contrast, the crests are not as well developed in the buccal and lingual cusps, and consequently the buccal and lingual cusps appear to be more conical (Chow and Hu,

1959; Hopson, 1965; Cui 1981; Cui and Sun, 1987). *Bienotherium* differs from *Lufengia* (Chow and Hu, 1959) and *Dianzhongia* (Cui, 1981) in the following features: the cusps in *Bienotherium* have shorter crests than in the latter two forms (Figure 14.4) (Chow and Hu, 1959; Hopson, 1965; Cui and Sun, 1987); a much larger posterior cusp is developed in the lingual row (Chow and Hu, 1959; Cui and Sun, 1987); and more conical (and less crescentic) cusps are present in both the lingual and buccal rows. The lower postcanines have two crescentic cusps in both the lingual and buccal rows (Figure 14.5), and the lower anterior cusps are much larger than the posterior cusps in *Bienotherium*. A third cuspule may be present at the anterior end of the cusp row in some teeth.

Cui and Sun (1987) show that the upper postcanines of *Bienotherium* have six long roots arranged in two transverse rows. Three roots in the posterior row are

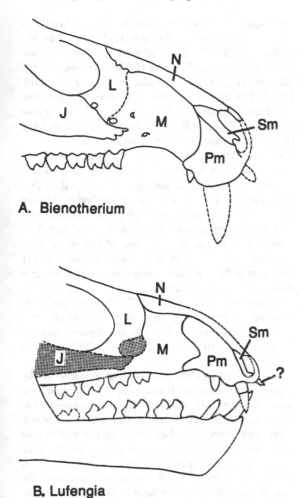

A. Bienotherium

B. Lufengia

Figure 14.6. Lateral views of the rostral region in
Lufengia and *Bienotherium*. (A) *Bienotherium*. (From Clark
and Hopson, 1985; reversed.) (B) *Lufengia* (IVPP 8685,
partially prepared). Abbreviations: J, jugal; L, lacrimal;
M, maxilla; N, nasal; Pm, premaxilla; Sm, septomaxilla.
Stippling represents damaged areas of the specimens.
Dashed lines represent reconstruction.

equally spaced. *Bienotherium* can be distinguished from
other tritylodontids from the Lower Lufeng Formation
by several features of its postcanine roots (Cui and Sun,
1987). In the upper postcanines, the anterior medial
root is fused with the anterior lingual root at the base
(Figure 14.4). The lower postcanines have two roots.
Each lower root is transversely broadened and anteroposteriorly flattened. Two grooves are present along
the length of each root, such that the root appears to
have a figure-eight shape in transverse section (Cui
and Sun, 1987). Each root can be interpreted as being
fused from two formerly divided roots (Cui and Sun,
1987) or roots with incipient division. A similar
condition is found in the lower postcanine roots of

Oligokyphus (Kühne, 1956). The lower postcanine
roots are distinct from the columnar and rounded
lower postcanine roots of *Lufengia* and *Yunnanodon* (Cui
and Sun, 1987).

The cranial structure of *Bienotherium yunnanense* is
primitive for tritylodontids. Hopson (1964) reported
that the quadrate has a posterodorsal process, which
is primitive for tritylodontids. The rostrum is elongated
(Hopson, 1965; Clark and Hopson, 1985). The premaxilla is not as expanded as those in *Lufengia*,
Bienotheroides, and *Kayentatherium* (Sun, 1984; Clark
and Hopson, 1985; Sues, 1986a; Sun and Cui, 1986),
whereas the maxilla is very large, a primitive character
for tritylodontids (Clark and Hopson, 1985). On the
basis of the relative proportions of the premaxilla and
the maxilla, *Bienotherium* can be distinguished from
Lufengia and probably also from *Dianzhongia* (Figure 14.6).

Young (1974) established *Oligokyphus sinensis* from
the Lower Lufeng Formation on a poorly prepared
partial dentary. Quoting a personal communication
from Hopson, Sues (1985a) suggested that *O. sinensis*
might be an immature specimen of *Bienotherium*
yunnanense. Sun and Cui (1986) pointed out that "*O.*
sinensis" possessed only two cusps in each row of the
lower postcanines. This differs from *Oligokyphus* (Kühne,
1956; Sues, 1985a) from the British fissure deposits
and the Kayenta Formation of Arizona, which has
three or four cusps in each row on the lower postcanines. We therefore treat "*O. sinensis*" as a junior
synonym of *B. yunnanense*.

Bienotherium magnum Chow, 1962 is represented by
a single specimen from the Heiguopeng ("Heikopeng")
Locality in the Dark Red Beds (stratum 6). More
specimens have been found, but not described. *B.*
magnum is much larger than any specimen of *B.*
yunnanense. However, the range of size variations for
Bienotherium is still unknown. *Kayentatherium* shows
a very broad range in the size and proportions of the
skull (Sues, 1986a). If *Bienotherium* has a comparable
range in size, then *B. magnum* may simply be a large
individual of *B. yunnanense* (Sues, 1986a). With that
caveat in mind, we tentatively follow most authors in
regarding *B. magnum* as a valid species (Chow, 1962;
Hopson and Kitching, 1972; Sun et al., 1985).

The classification of two other species, *Bienotherium*
minor and *B. elegans* (Young, 1947a), is a matter
of debate. Hopson and Kitching (1972) suggested that
B. elegans was based on an immature juvenile of
B. yunnanense. By contrast, Sun and Cui (1986, table
21.1) retained *B. elegans* as a valid species. Study of a
large sample of *Bienotherium* from the Dull Purplish
Beds will be necessary to clarify this problem. Hopson
and Kitching (1972) also suggested that *B. minor*
pertained to *Lufengia delicata* (Chow and Hu, 1959).
However, Cui and Sun (1987) regarded *B. minor* and
L. delicata as two separate species. Several major
collecting expeditions by IVPP have shown that *B.*

minor is restricted to the Dull Purplish Beds (stratum 4), whereas *Lufengia* is found only in the Dark Red Beds (stratum 6). On the basis of the stratigraphic separation of *B. minor* and *L. delicata* and their differences in postcanine structure (see later discussion of *Lufengia*), we retain *B. minor* and *L. delicata* as separate taxa, pending further studies of juvenile specimens of *Bienotherium*.

Lufengia Chow and Hu, 1959

All specimens of *Lufengia delicata* (Chow and Hu, 1959; Cui and Sun, 1987) with reliable field stratigraphic data are from the Dark Red Beds (stratum 6) (Cui and Sun, 1987; X. Wu, unpublished field notes). Sun et al. (1985, p. 7) speculated that the single specimen of *Lufengia* reported by Young (1982) from the Dachong locality in the Dull Purplish Beds might be a small, perhaps juvenile, specimen of *Bienotherium yunnanense*. We agree with Sun et al. (1985) and consider *Lufengia* to be restricted to the Dark Red Beds (stratum 6).

Lufengia is the smallest of the known tritylodontids from the Lower Lufeng Formation. The overall length of a complete skull is about 4 cm. The premaxilla is much larger than that of *Bienotherium*, but the maxilla is much smaller than that of *Bienotherium* (Figure 14.6). Clark and Hopson (1985, fig. 4) suggested that in *Lufengia*, the premaxilla contacts the nasal, and the facial process of the maxilla is much reduced in comparison with *Bienotherium*. We have been able to confirm both features in a partial skull specimen (IVPP 8685, Figure 14.6B). The rostrum of *Lufengia* is much shorter than those of *Bienotherium* and *Dianzhongia* (but not *Yunnanodon*). Young (1974) pointed out that the short diastema in *Lufengia* is correlated to the short rostrum of the cranium. Other aspects of the skull are poorly known.

Hopson (1965) reported that *Lufengia* had three upper and two lower incisors. However, Young (1974) suggested that it had three lower incisors, based on the number of preserved alveoli. In the specimen available to us, at least two upper and two lower incisors are present, but because of the damage to the rostrum and the mandibular symphysis of the specimen, the presence of the third incisor can neither be confirmed nor denied.

Young (1974) reported that in one specimen of *Lufengia* there were four postcanines and the alveoli for two additional postcanines; thus the maximum number of postcanines in *Lufengia* would be six. The actual number of postcanines may be dependent on the growth stage (Chow and Hu, 1959; Hopson, 1965). The postcanine cusps have crests that are better developed than in *Bienotherium* (Chow and Hu, 1959; Hopson, 1965). The middle row of the upper postcanines has three cusps. Among these, the anterior cusp is the smallest, and it is separated farther from the middle cusp than the middle cusp is from the posterior cusp.

The lingual row has three cusps. In this row the middle and posterior cusps are closer to each other than the middle cusp is to the anterior cusp. The posterior cusp is considerably smaller and tightly appressed to the adjacent cusp (Chow and Hu, 1959; Cui and Sun, 1987). The buccal row has two cusps. The first two lower postcanines may have three cusps in each row (Young, 1974), and the other lower postcanines have two cusps in each row (Cui and Sun, 1987).

Cui and Sun (1987) showed that the upper postcanines of *Lufengia* have five roots, with the anterolingual root larger and more flattened than the other four (Figure 14.4). By contrast, *Bienotherium* has six roots in the upper postcanines, the anterolingual root and anteromedian root fused to one another. The median root in the posterior row of *Lufengia* is very reduced, similar to that in *Yunnanodon*, but differing from that in *Bienotherium*. *Lufengia* possesses transverse dentine sheets that connect the bases of the roots in each row (Cui and Sun, 1987). This is similar to *Oligokyphus*, but differs from *Bienotherium* (Kühne, 1956; Cui and Sun, 1987). The lower postcanines of *Lufengia* have two columnar roots (Cui and Sun, 1987).

Lufengia can be distinguished from *Bienotherium* by several characters (Cui and Sun, 1987). First, the posterolingual cusp is smaller and more appressed to the middle lingual cusp than in *Bienotherium*. Second, each upper postcanine has five roots; the middle root of the posterior row is very reduced. By contrast, upper postcanines of *Bienotherium* have six roots, with fusion of the anterolingual and anteromedian roots; the middle root of the posterior row is well developed. Third, the transverse dentine sheets that connect the bases of the roots of the upper postcanines are better developed in *Lufengia* than in *Bienotherium*. Fourth, the two lower postcanine roots in *Bienotherium* are transversely broad and are marked by a median groove, suggesting that each root was formed by fusion of two originally separate roots; the lower postcanine roots are more columnar in *Lufengia*.

The juvenile specimens of tritylodontids usually show a size gradient along the tooth row – the postcanine teeth become increasingly larger posteriorly. In juvenile specimens of *Kayentatherium* (Sues, 1986a), the size of the postcanines increases posteriorly; the anteriormost postcanine is only half the size of the posteriormost tooth. However, in adults, this size gradient of postcanines is diminished. A similar size gradient is also present in immature specimens of *Oligokyphus*. Yet in adults, the size of the anteriormost postcanine almost equals that of the more posterior postcanines (Kühne, 1956, fig. 26). According to Cui and Sun (1987), the size gradient for postcanines in most specimens of *Lufengia* is not as pronounced as in the juvenile individuals of *Oligokyphus* (Kühne, 1956; Sues, 1985b) and *Kayentatherium* (Sues, 1986a). These features have led Cui and Sun (1987) to argue that *Lufengia* does not represent juvenile individuals of *Bienotherium yunnanense*, as suggested by other workers.

There is also substantial biostratigraphic evidence for separating *Lufengia* from *Bienotherium yunnanense*. All specimens of *B. yunnanense* are from the Dull Purplish Beds, whereas all specimens of *Lufengia* (including the holotype) are from the Dark Red Beds (Cui and Sun, 1987; X. Wu, unpublished data). Hopson and Kitching (1972) suggested that *Bienotherium minor* was synonymous with *Lufengia delicata*, but Cui and Sun (1987) maintained the two as separate species. This taxonomic issue is partly related to the lack of well-documented distinction between *B. minor* and *B. yunnanense*. In light of the more recent data (Sun et al., 1985; Cui and Sun, 1987), we believe that *Lufengia* is a valid taxon. However, it remains uncertain whether or not *B. minor* can be separated from *B. yunnanense*.

Yunnanodon Cui, 1986

Yunnanodon brevirostre ("*Yunnania brevirostre*" of Cui, 1976; *Yunnanodon brevirostre* of Cui, 1986) is known from the Dark Red Beds (stratum 6). The most diagnostic feature in the upper postcanines of *Yunnanodon* is the presence of only two lingual cusps; all other tritylodontids from the Lower Lufeng Formation have three cusps (Cui and Sun, 1987). *Yunnanodon* resembles *Lufengia* but differs from *Bienotherium* in having strongly crescentic crests of the cusps (Figure 14.4). The buccal row has two cusps in most postcanines, although a very small and indistinct cusp is present at the anterior end of the buccal rows in the right anterior two postcanines of the holotype (Cui, 1976). Three cusps are present in the middle row. The crescentic crests of the central cusp of the middle row extend anteriorly to flank the anterior small cusp, which is separated farther from the middle cusp than the middle cusp is from the posterior one. The upper postcanines have five gracile roots, lacking connecting dentine sheets, and the posterior middle root is vestigial. This contrasts with *Lufengia*, where a transverse dentine sheet connects the roots (Cui and Sun, 1987). The lower postcanines mostly have two rows with two cusps in each row. Occasionally a small third cusp may be present at the posterior end. The lower postcanine has two columnar roots. Two incisors are preserved in the type specimen, and these probably are homologous to I^2 and I^3 of *Bienotherium*. It is unclear whether or not I^1 is present, because the tip of the rostrum is slightly damaged. The lower incisors of *Yunnanodon* are unknown.

The skull of *Yunnanodon* has a short rostrum and very high vault in the frontal region (Cui, 1976) (Figure 14.7). The dorsal flexion of the basisphenoid–parasphenoid is strongly developed. The transverse

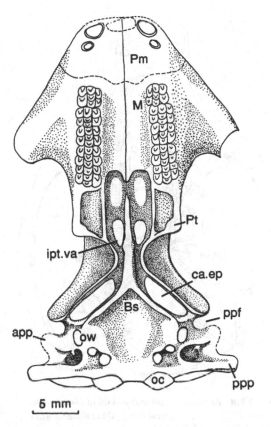

Figure 14.7. Reconstruction of the skull of *Yunnanodon* (based on IVPP 5071), ventral view. Abbreviations: app, anterior paroccipital process of the petrosal; Bs, basisphenoid; ca.ep, ventral opening of the cavum epiptericum; ipt.va, interpterygoid vacuity; oc, occipital condyle; ow, fenestra ovalis; M, maxilla; Pm, premaxilla; ppf, pterygoparoccipital foramen; ppp, posterior paroccipital process of petrosal; Pt, pterygoid.

flange of the pterygoid is much reduced. The lateral flange of the petrosal strongly flares ventrally. The median palatal crest is very pronounced, more so than in *Dianzhongia*. Two interpterygoid vacuities are present in the palate, a character also present in juvenile *Kayentatherium* (Sues, 1986a) and *Dianzhongia* (Figure 14.8).

Sues (1986b) raised the possibility that "*Yunnania*" Cui, 1976 might be synonymous with *Lufengia*, because they share long crests on the cusps of the upper postcanines. We consider *Yunnanodon* and *Lufengia* distinct taxa. These two forms differ in the pattern of the upper postcanine cusps. *Lufengia* has three cusps in the lingual row, with the posterior cusp reduced in size; *Yunnanodon* has only two lingual cusps. The transverse dentine sheet connecting the upper postcanine roots in the same row is much better developed in *Lufengia* than in *Yunnanodon* (Cui and Sun, 1987). In *Lufengia*, the anterolingual root of the upper postcanine is large

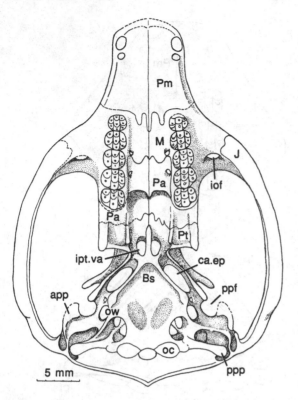

Figure 14.8. Reconstruction of the skull of *Dianzhongia* (based on IVPP 8694), ventral view. Abbreviations as in Figure 14.7; also, iof, infraorbital foramen (posterior opening); J, jugal; Pa, palatine.

and ellipsoid in cross-section. By contrast, this root is small and rounded in transverse section in *Yunnanodon* (Cui and Sun, 1987).

Dianzhongia Cui, 1981

Dianzhongia longirostrata Cui, 1981, has a long rostrum with a slight constriction in the postincisive diastema (Figure 14.8). The premaxilla is expanded, and the maxilla is much reduced, in comparison with *Bienotherium* (Clark and Hopson, 1985). The palatal suture between the premaxilla and the maxilla is positioned between the first upper postcanines. The anterior transverse suture between the palatine and the maxilla is positioned at the level of the fourth postcanines. These rostral and palatal structures of *Dianzhongia* are very similar to those of *Kayentatherium* (Sues, 1986a). The anterior end of the jugal is not very expanded. The transverse process of the pterygoid is more developed than that of *Yunnanodon*.

Cui (1981) stated that only one large upper incisor was present on each side in the type specimen. However, in a smaller specimen (IVPP 8694, Figure 14.8) there are two upper incisors. As described by Hopson

(1965), older individuals of *Bienotherium* lost I[3]. It is possible that the type specimen of *Dianzhongia*, which apparently is an older individual, had lost a third incisor.

The postcanine crowns of *Dianzhongia* are very similar to those of *Lufengia* described by Chow and Hu (1959) and Cui and Sun (1987). The postcanines of *Dianzhongia* have two cusps in the buccal row and three cusps each in the middle and lingual rows. The posterior cusp in the lingual row is much smaller than the adjacent cusps and is appressed to the posterolateral side of the middle cusp of the lingual. This is very similar to the postcanines of *Lufengia*. *Dianzhongia* is also indistinguishable from *Lufengia* in the size and proportions of the cusps in the middle row. The anterior cusp is separated farther from the middle cusp than the middle cusp is from the posterior cusp. Both features are found in *Lufengia* (Chow and Hu, 1959; Cui and Sun, 1987).

There are two possible interpretations of the similarities between *Dianzhongia* and *Lufengia*, which occur in the same stratigraphic level of the Lower Lufeng Formation (stratum 6). The first is that *Lufengia delicata* and *Dianzhongia longirostrata* are sympatric but related species, and the similarities between them are synapomorphies. The second possibility is that *L. delicata* and *D. longirostrata* represent different ontogenetic stages of the same taxon.

Cui (1981) listed several features to distinguish *Dianzhongia* from *Lufengia*: the large size of the skull, only one upper incisor in each premaxilla, the long and slightly constricted rostrum, and long postincisive diastema. The total skull length of the holotype of *Dianzhongia longirostrata* is 76 mm, whereas the lengths of several skulls of *Lufengia* are around 40 mm (Cui, 1981). The difference in skull size is equivocal at best. A new skull referable to *Dianzhongia* is about 55 mm long, much smaller than that of the holotype. Sues (1986a) and Kühne (1956) have documented the wide range of variation in skull size in *Kayentatherium* and *Oligokyphus*, respectively. Hopson (1965) documented the loss of the posterior incisors in older *Bienotherium*, which would result in a longer postincisive diastema. Sues (1986a,b) indicated that the rostral constriction was slightly different between smaller and larger individuals of *Kayentatherium*. Thus, it is conceivable that *Dianzhongia* represents larger individuals of *Lufengia* and that the difference between the two in the rostrum represents a proportional change during growth. However, lacking a reasonably large sample of these skulls, we tentatively recognize *Dianzhongia* as a separate taxon, pending further study.

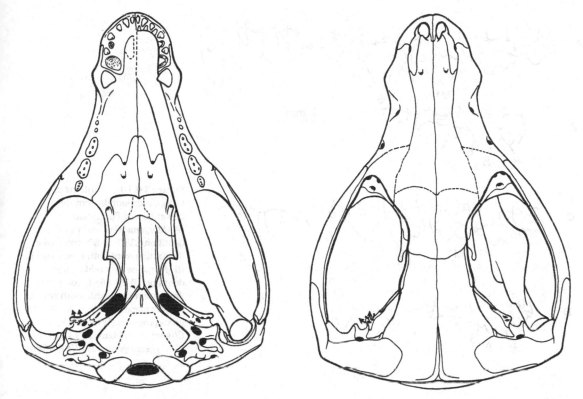

Figure 14.9. Reconstruction of the skull of *Sinoconodon*, ventral (left) and dorsal (right) views. (From Crompton and Luo, 1993.)

Mammalia

Sinoconodon Patterson and Olsen, 1961

All specimens referred to *Sinoconodon rigneyi* Patterson and Olson, 1961, have been collected from the Dark Red Beds (stratum 6). The cranium of *Sinoconodon* has many diagnostic mammalian features (Figure 14.9). It has a complete medial orbital wall, a fully developed promontorium, an expanded parietal region of the braincase, and a robust dentary condyle articulating with the glenoid fossa on the squamosal (Patterson and Olson, 1961; Kermack et al., 1981; Crompton and Sun, 1985; Crompton and Luo, 1993). However, the dentition of *Sinoconodon* is quite different from that in other Early Jurassic mammals. The postcanines are not differentiated into premolariform and molariform teeth. They also lack cingula and occlusal wear facets that are characteristic of other mammals (Crompton and Sun, 1985; Crompton and Luo, 1993). *Sinoconodon* is also distinctive in that the dentary condyle lies at the same level as the postcanine alveoli, rather than above the alveolar level as in *Morganucodon* (Crompton and Sun, 1985). The petrosal of *Sinoconodon* shows some differences from that of *Morganucodon* (Crompton and Luo, 1993; Luo, Chapter 6).

Crompton and Luo (1993) showed that *Sinoconodon* has a very wide range of ontogenetic variation. The length of the skull can range from about 30 mm to over 60 mm (Luo, Chapter 6). The anterior postcanines seen in small individuals have been lost in the larger individuals, resulting in an increasingly larger postcanine diastema in larger skulls. Additional postcanines have been added at the posterior end of the postcanine row. The posterior postcanine in smaller, presumably younger individuals probably was replaced in larger, presumably older individuals. Replacement of postcanines in *Sinoconodon* is similar to that in gomphodont cynodonts (Crompton and Luo, 1993; Luo, Chapter 6). The canine of *Sinoconodon* was replaced at least four times. In some specimens, the incisors alternate in size. This alternating arrangement of small and large teeth is typical of nonmammalian cynodonts with continuous and alternating replacement of teeth, indicating that the incisors in *Sinoconodon* retained an alternating replacement (Crompton and Luo, 1993). *Sinoconodon* probably had multiple generations of incisors.

Young (1982) separated *Sinoconodon parringtoni* from *S. rigneyi* on the basis of three characters: a short and robust rostrum, a larger canine, and a vertically deep skull. This diagnosis is no longer valid in view of the new information from the much larger sample of

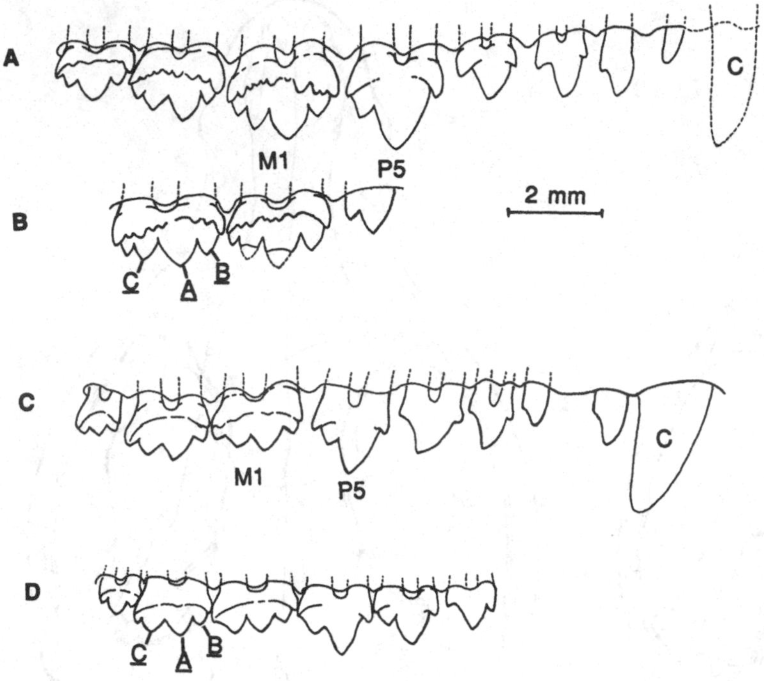

Figure 14.10. Dentitions of morganucodontids from the Lower Lufeng Formation. (A) *Morganucodon oehleri*, upper dentition (IVPP 8682; composite illustration from both tooth rows). (B) *Morganucodon oehleri*, upper dentition (IVPP 8684; composite illustration from both tooth rows). (C) *Morganucodon* ("*Eozostrodon*") *heikuopengensis* (IVPP 4729). (From Crompton and Luo, 1993.) (D) *M.* ("*E.*") *heikuopengensis* (IVPP 8686).

Sinoconodon mentioned earlier. For example, the difference in canine size between the holotype of *S.* "*parringtoni*" and that of *S. rigneyi* can be interpreted as an artifact of continuous replacement of canines. The shorter rostrum of *S.* "*parringtoni*" in comparison with that of *S. rigneyi* can be interpreted as a difference between small and large individuals of the same taxon.

Sinoconodon yangi Zhang and Cui, 1983 was diagnosed primarily by the difference in number of incisors. Zhang and Cui (1983) noted that the holotype of *S. yangi* has four incisors, rather than three as in *S. rigneyi*. Because the incisors were continuously replaced, the number of incisors would not remain constant through various growth stages. This difference can be interpreted as a consequence of continuing incisor replacement. Similarly, *Sinoconodon* "*changchiawaensis*" ("*Lufengoconodon changchiawaensis*" Young, 1982) represents just another ontogenetic stage in the entire growth series of *Sinoconodon* (Crompton and Luo, 1993). It should be regarded as a junior synonym of *S. rigneyi*.

Morganucodon Kühne, 1949

So far, four complete skulls (CUP 2320, IVPP 4729, IVPP 8682, IVPP 8684) and several partial skulls and mandibles of this mammal have been collected from the Dark Red Beds (stratum 6). Two species of *Morganucodon* can be recognized among these specimens, based on differences in the dentition (Figure 14.10). More detailed studies of the skulls and dentitions are being carried out by Luo and Crompton.

The holotype of *Morganucodon oehleri* Rigney, 1963 is a nearly complete skull (CUP 2320); the dentition and skull were described by Mills (1971) and Kermack et al. (1973, 1981). *M. oehleri* has a much better developed labial cingulum in the upper molars than do *M. heikuopengensis* and *M. watsoni*. The notch between cusp A and cusp B is much deeper in *M. oehleri* than in *M. heikuopengensis*. The difference in height between cusp A and cusp B is much more pronounced in the former than in the latter. The number of premolars may be dependent on the growth stage. In the holotype of *M. oehleri* (CUP 2320), only four premolars are present (Kermack et al., 1973), whereas there are five premolars in IVPP 8684. The difference in the number of premolars is related to the loss of the anterior premolars in larger (older) individuals, a dental replacement pattern well documented in *M. watsoni* and other early mammals (Mills, 1971; Parrington, 1973; Crompton, 1974). Four upper incisors and four lower incisors are present.

Morganucodon ("*Eozostrodon*") *heikuopengensis* Young, 1978 (IVPP 4729) was described in two papers (Young, 1978; Crompton and Luo, 1993). The putative synonymy of *Morganucodon* and *Eozostrodon* was a controversy in the taxonomy of morganucodontids (Kühne,

1949; Mills, 1971; Parrington, 1971, 1973, 1978; Kermack et al., 1973; Clemens, 1979). Clemens (1979) pointed out that the type material for *Eozostrodon parvus* is very limited. Thus we regard *Eozostrodon* as a valid taxon from Holwell Quarry in England, distinct from *Morganucodon*. *Morganucodon oehleri* has been broadly accepted as a valid taxon. The holotype of *Eozostrodon heikuopengensis* (Young, 1978) shows many similarities to *M. oehleri*. Thus, we believe that they are congeneric, and *E. heikuopengensis* Young, 1978 should be called *Morganucodon heikuopengensis*.

As discussed earlier, the main evidence supporting the separation of *M. oehleri* from *M. heikuopengensis* concerns the sizes of the main cusps and the cingular cusps. *M. oehleri* (Figure 14.10) has much better developed labial cingular cusps, and its main cusp *A* is more pronounced than the adjacent cusps *B* and *C*. In addition, the largest molar is M1 in *M. heikuopengensis*, but M2 in *M. oehleri*. M3 is much smaller than M2 in *Morganucodon heikuopengensis*, but less so in *M. oehleri* (Figure 14.10).

Young (1978) noted that *M. heikuopengensis* has five upper premolars, whereas *M. oehleri* has only four. *M. heikuopengensis* lacks the postcanine diastema present in *M. oehleri*. Young argued that this is a major distinction between *M. heikuopengensis* and *M. oehleri*. Both Mills (1971) and Parrington (1971) showed that the anterior premolars are progressively lost in older individuals of *M. watsoni*. Thus the difference in the number of premolars between *M. oehleri* and *M. heikuopengensis* represents ontogenetic variation, rather than a taxonomic difference, and should not be used to diagnose either species.

Diapsida

Lepidosauria

Sphenodontia

Sphenodontians are well represented in the Lufeng fauna (Figure 14.11). Wu (1991, and Chapter 3) reported two new species of *Clevosaurus*: *C. wangi* and *C. mcgilli*. *Clevosaurus* ("*Dianosaurus*") *petilus*, initially referred to the Protorosauria by Young (1982), is also a sphenodontian (Wu, 1991, and Chapter 3). *Fulengia youngi*, described by Carroll and Galton (1977) as a primitive lacertilian, based on an almost complete skull from stratum 5 of the Dark Red Beds of the Dadi locality, has been shown to be a juvenile of the prosauropod dinosaur *Lufengosaurus huenei* (Evans and Milner, 1989). Nevertheless, the sphenodontians listed in the following sections (Wu, 1991, and Chapter 3) indicate the presence of lepidosaurs in the Lufeng Basin.

Clevosaurus Swinton, 1939

Clevosaurus wangi Wu (Chapter 3) is based on an almost complete skull (Figure 14.11A) from stratum 6 near Dawa. The skull possesses several diagnostic features of *Clevosaurus*: The paroccipital process fits into a fossa on the squamosal and separates the supratemporal from the quadrate head; the squamosal sits in a depression on the cephalic head of the quadrate; the epipterygoid has an expanded dorsal end; and the posterior end of the dentary is elongate and tapers abruptly. The well-preserved middle ear and braincase indicate that *C. wangi* had an impedance-matching ear functionally comparable to those of lizards.

Clevosaurus mcgilli Wu (Chapter 3) is the smallest of the Lufeng sphenodontians, with a skull length ranging from 21 to 23 mm. It is based on two incomplete skulls from stratum 6 near Dawa (Figure 14.11C). Most of its diagnostic features are in the orbital region: The jugal is characterized by an expanded dorsal process and by an elongate anterior process that meets the ventral process of the prefrontal at the anteroventral corner of the orbit. The postorbital is T-shaped, with a strongly curved ventral margin for the broad dorsal extremity of the jugal. The palatine is broad and free posteriorly, resulting in an L-shaped suborbital fenestra. The enlarged palatal tooth row extends nearly parallel to the marginal tooth row of the maxilla. The snout is very short, as in *Brachyrhinodon* (Fraser and Benton, 1989) and *Polysphenodon* (Carroll, 1985; Fraser and Benton, 1989) from the Upper Triassic of Europe.

Clevosaurus ("*Dianosaurus*") *petilus* (Young, 1982; Wu, 1991, and Chapter 3) is represented by an incomplete skull (Figure 14.11B) from stratum 4 at Dachong. It has the following diagnostic characters of sphenodontians: The dentary has an elongate posterior and a coronoid process. The palatine possesses a lateral row of enlarged teeth. The teeth in the upper and lower jaws are acrodont. *C. petilus* differs from other species of *Clevosaurus* in that the large supratemporal fenestra is diagonally oriented, the basipterygoid process of the basipterygoid is slender, and the prearticular extends anteriorly beyond the anteroventral border of the coronoid.

The three species of *Clevosaurus* from Lufeng share many derived features with *C. hudsoni* from Britain (Robinson, 1973; Fraser, 1988). In all these forms, the maxilla is separated from the external naris by the posterodorsal process of the premaxilla, and the suborbital fenestra is enclosed by only two bones – the palatine and the ectopterygoid. *Clevosaurus* is closely related to the Late Triassic *Brachyrhinodon* from Scotland and *Polysphenodon* from Germany (Carroll, 1985; Fraser and Benton, 1989).

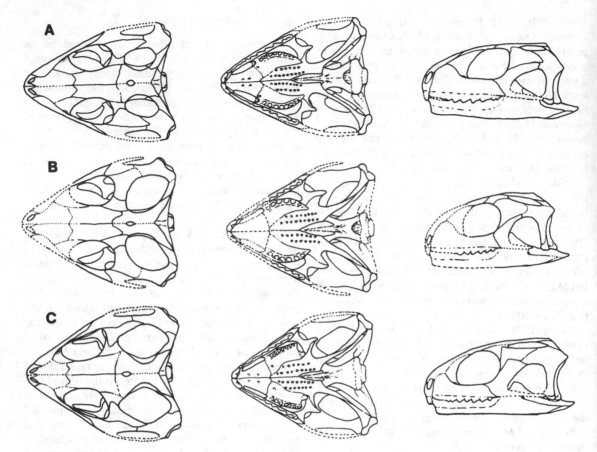

Figure 14.11. Reconstructions of the skulls of the sphenodontians from the Lower Lufeng Formation: (A) *Clevosaurus wangi* (IVPP 8271); (B) *C*. ("*Dianosaurus*") *petilus* (IVPP 4007); (C) *C. mcgilli* (IVPP 8272, 8273). Left column, dorsal views; middle column, ventral views; right column, lateral views. (From Wu, 1991.)

Crocodylomorpha

The most abundant small diapsids in the Lower Lufeng Formation are crocodylomorph archosaurs (sensu Clark, 1986; Benton and Clark, 1988), including both spheno-suchians and protosuchians. Four genera have been described from the Dark Red Beds of the Lower Lufeng Formation. An additional taxon will be described on the basis of recent finds (Wu and Chatterjee, unpublished data).

Sphenosuchia

Dibothrosuchus Simmons, 1965

Dibothrosuchus elaphros Simmons, 1965 is the only sphenosuchian recorded from the Lower Lufeng Formation (Figure 14.12). It is a medium-sized animal with a total length of about 1.3 m (skull length 17 cm). Simmons (1965) suggested that the type specimen of

D. elaphros came from near Dadi ("Tati"), a locality belonging to our stratum 6. However, the rock matrix of the holotype is purplish red in color. This is the dominant color of stratum 5, but is uncharacteristic of stratum 6. Additional material of *D. elaphros* (Wu, 1986) was collected from stratum 5 in the Dark Red Beds near Dawa. Based on this evidence and the stratigraphic position of the new specimen (Wu, 1986), we believe that the holotype of *D. elaphros* (CUP 2081) (Simmons, 1965) was most likely obtained from stratum 5.

Dibothrosuchus was established by Simmons (1965) on the basis of a partial skull and some postcranial bones and was assigned to the family Ornithosuchidae of the order Thecodontia. Crush (1984) reconsidered the systematic position of *Dibothrosuchus* and tentatively placed it in the Sphenosuchidae in the order Croco-dylomorpha. Wu (1986) confirmed the sphenosuchid affinities of *Dibothrosuchus* based on additional material, including an almost complete skull (Figure 14.12) and

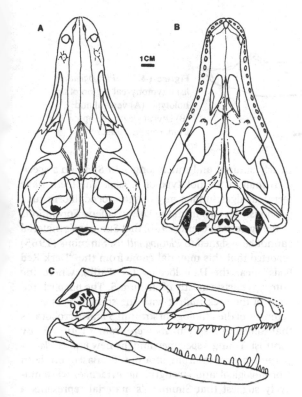

Figure 14.12. Reconstruction of the skull of *Dibothrosuchus elaphros* (IVPP 7907): (A) dorsal, (B) ventral, and (C) lateral views. (Adapted from Wu and Chatterjee, 1993.)

the anterior half of the postcranial skeleton. He erected a new species, *D. xingsuensis*, for that material. Re-examination of the type specimen of *D. elaphros* shows that the original description and identification of a number of characters by Simmons (1965) were not accurate (Wu and Chatterjee, 1993). There seem to be few differences between the type specimens of *D. elaphros* and *D. xingsuensis*, and *D. xingsuensis* is here regarded as a junior synonym of *D. elaphros* (Wu and Chatterjee, 1993).

Dibothrosuchus differs from other sphenosuchians in the following features (Wu and Chatterjee, 1993): Three ridges are present on the frontals, converging at either end. The suture between the frontal and postorbital forms a pronounced ridge. The supratemporal fenestra is transversely broad. The basioccipital condyle is completely sheathed dorsally by the exoccipital. The humerus has an oval ligamental depression on the medial surface below the proximal articular head. On the basis of two apomorphies, Wu and Chatterjee (1993) suggested that *Dibothrosuchus* is most closely related to *Sphenosuchus* from the upper Elliot Formation of South Africa (Walker, 1990); The narrow and elongate posteroventral process of the coracoid is longer than its body; the parietals are fused.

The sphenosuchians usually are grouped together as a suborder Sphenosuchia (Bonaparte, 1982; Crush, 1984). The monophyletic nature of this group has been questioned by Clark (1986; Benton and Clark, 1988) and Parrish (1991), who argued that the Sphenosuchia could not be diagnosed cladistically. However, a recent cladistic analysis by Wu and Chatterjee (1993) indicates that most sphenosuchians can be diagnosed by a set of synapomorphies as a monophyletic group within the Crocodylomorpha.

Crocodyliformes

Platyognathus Young, 1944

Platyognathus hsui Young, 1944 is a small crocodyliform with a skull about 8 cm in length (Figures 14.13 and 14.14). The holotype of *P. hsui* was collected from the Dahuangtian locality ("Huanjiatian" of Young, 1944, and Simmons, 1965), but was lost during World War II. Fortunately, a new specimen referable to *P. hsui* was recovered by Wu from stratum 6 of the Dark Red Beds at the same locality, which indicates that the holotype may also have been collected from stratum 6.

The holotype of *P. hsui* comprised only a short mandibular symphysis with a fused symphyseal suture. These features are also present in the new specimen, which is also represented by the symphyseal portion of the lower jaws that are occluded with partial upper jaws. This new specimen is of the same size as the holotype. In addition to the fused symphysis, *Platyognathus* can be distinguished from other known primitive crocodyliforms by several other features: The antorbital fossa surrounding the antorbital fenestra is divided by a longitudinal ridge; the caniniform tooth of the dentary is polygonal in cross-section; and the premaxillary teeth and six anterior dentary teeth project upward and forward (Young, 1944). These diagnostic features suggest that *P. hsui* is a valid taxon (Wu and Chatterjee, unpublished data).

Young (1946) suggested that *Platyognathus* might have a close affinity to *Sphenosuchus*. Romer (1956) placed *Platyognathus* in the Notochampsidae (Protosuchidae), whereas Piveteau (1955) suggested a relationship to the Stagonolepididae. Simmons (1965) erected a new family Platyognathidae for this taxon, and he followed Piveteau in placing *Platyognathus* in the "pseudosuchians," but went further to suggest that *Platyognathus* might be an intermediate form between the Pseudosuchia and the Protosuchia (sensu Romer, 1956).

We suggest that *Platyognathus* has a close relationship to the Protosuchidae, including *Protosuchus*, *Hemiprotosuchus* and ?*Edentosuchus* from the Kayenta Formation (Clark, 1986). They share a large caniniform tooth in the anterior part of the dentary, fitting

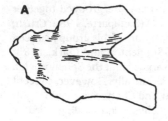

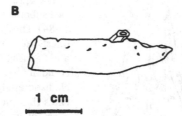

1 cm

Figure 14.13. *Platyognathus hsui*: symphyseal region of the holotype: (A) ventral and (B) lateral views. (Adapted from Young, 1944.)

into a lateral open notch between the premaxilla and the maxilla, and probably a dorsally arched surangular. The jugal of *Platyognathus* is very similar to that of *Protosuchus* in the great depth of the anterior process and the strongly curved ventral margin of the posterior process.

Simmons (1965) described a partial skull, mandible, and some postcranial remains with osteoderms (CUP 2083, 2104, and 2105) and referred these specimens to *Platyognathus hsui*. This, however, was questioned by Clark (1986). According to Clark, most of the diagnostic characters cited by Simmons to justify assignment of his material to *Platyognathus* were not correct. For example, the mandibular symphysis is

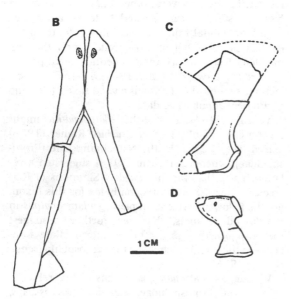

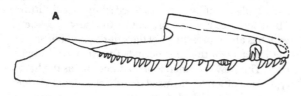

Figure 14.14. ?*Platyognathus hsui* (CUP 2083), anterior portions of the upper and lower jaws. (A) Lateral view. (B) Ventral view. (C) Left scapula. (D) Left coracoid.

clearly sutured and much longer in Simmon's specimen than in the holotype of *P. hsui*. The teeth have serrated edges, unlike any of the crocodyliform taxa previously known from the Lower Lufeng Formation.

We agree with Clark (1986) that Simmons's material cannot be assigned to *Platyognathus*. Simmons (1965) reported that this material came from the "Dark Red Beds" near the Dadi locality (= "Tati"), where the outcrop consists mostly of stratum 5. The purplish red color of the rock matrix with the specimens is characteristic of the lithology of stratum 5 and corroborates that Simmons's specimens were from stratum 5. By contrast, Young's specimen and the new find of *Platyognathus* are both from stratum 6. Based on both morphological and stratigraphic evidence, we tentatively suggest that Simmons's material represents a new crocodyliform taxon, pending publication of a more detailed study (Wu and Chatterjee, unpublished data). In any event, the material assigned by Simmons (1965) to "*Platyognathus*" may also be related to the Protosuchidae. It shares with protosuchids an open notch between the premaxilla and maxilla for reception of the large caniniform tooth of the dentary.

Dianosuchus Young, 1982

Dianosuchus changchiawaensis Young, 1982 is a protosuchid crocodyliform based on an incomplete skull (Figure 14.15) from stratum 6 at the Zhangjiawa ("Changchiawa") locality. This specimen is about half of the size of *Orthosuchus stormbergi* or *Platyognathus hsui*. According to our study, the following characters are diagnostic of *Dianosuchus* among primitive crocodyliforms: The supratemporal fenestra is diagonally oriented; the depression encircling the supratemporal fenestra in other forms is absent; and the supratemporal fenestrae are broadly separated, so that the parietal–squamosal suture is positioned medial to the fenestrae. Two features suggest that *Dianosuchus* is closely related to the Protosuchidae: a dorsally arched surangular and a lateral open notch between the premaxilla and the maxilla for reception of the lower caniniform tooth of the dentary.

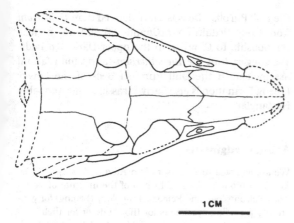

1 CM

Figure 14.15. Reconstruction of the skull of *Dianosuchus changchiawaensis* in dorsal view (IVPP 4730).

Microchampsa Young, 1951

Microchampsa scutata Young, 1951 is a very small animal from stratum 6 of the Dahuangtian locality. Young (1951) established this taxon on the basis of a single specimen comprising several articulated posterior cervical vertebrae, anterior dorsal vertebrae, ribs, and three rows of dorsal osteoderms. According to Young (1951), the most important diagnostic feature is the presence of three rows of dorsal osteoderms. Simmons (1965) assigned three additional specimens (CUP 2085–87) to *Microchampsa* based on what appeared to be three rows of dorsal osteoderms. Clark (1986) pointed out that Simmons's additional specimens all had two rows rather than three rows of dorsal osteoderms. These are similar in all respects to the dorsal osteoderms in the material referred by Simmons (1965) to *Platyognathus*. Therefore, aside from the holotype of *Microchampsa scutata*, no other material from the Lufeng Basin has three rows of osteoderms. Moreover, Clark (1986) speculated that this character in Young's specimen of *Microchampsa* was an artifact. Unfortunately, this holotype was also lost during the war. Whether or not this genus is a valid taxon thus remains uncertain, and we regard *M. scutata* as a nomen dubium.

Archosauria indet.

Strigosuchus Simmons, 1965

Strigosuchus licinus Simmons, 1965 is represented by a single fragment of a left mandibular ramus. Simmons (1965) stated that the specimen is supposedly from the Dadi ("Ta Ti") locality, which exposes stratum 5. The specimen shows some resemblance to the mandible of *Chasmatosaurus* (Young, 1936; Simmons, 1965). Nevertheless, Simmons (1965) concluded that it was

similar to those of *Hesperosuchus* (Colbert, 1952) and *Dibothrosuchus*. Based on the mandibular fenestration, Simmons classified this species in the Ornithosuchidae of the Pseudosuchia, along with *Dibothrosuchus* and *Hesperosuchus*. Recently, the latter two forms have been reinterpreted as sphenosuchian crocodylomorphs (Crush, 1984; Clark, 1986; Wu, 1986; Parrish, 1991; Wu and Chatterjee, 1993). However, Simmons's description shows no diagnostic feature for this species, and the type specimen is very fragmentary. It is difficult to assign *Strigosuchus* to sphenosuchians or to exclude affinities of this form to other archosauromorphs. Under these circumstances, we consider it a nomen dubium.

Pachysuchus Young, 1951

Pachysuchus imperfectus Young, 1951 was based on fragments of the nasals, the middle portion of the maxilla, and a partial basioccipital originally referred to *Lufengosaurus magnus* (Young, 1947b). Young (1947b, 1951) stated that the fossils came from the Dahuangtian locality, which is dominated by stratum 6. Thus, we place *P. imperfectus* in stratum 6. Young classified *P. imperfectus* as a phytosaur. The alleged presence of phytosaurs in the Dark Red Beds was an important part of Young's argument that the Lower Lufeng Formation was Triassic in age. Westphal (1976) argued that *Pachysuchus* was indeterminate because of the poor preservation and subsequent loss of the holotype and only known specimen. It is not possible to clarify the affinities of *Pachysuchus* to the Parasuchia or to any other archosaurs. Thus there is no demonstrable taxonomic evidence for assigning a Triassic age to the Lower Lufeng Formation.

Faunal correlations

Evidence for correlating small tetrapod assemblages of the Lower Lufeng Formation to those from other continents is provided by four groups: morganucodontids, tritylodontids, sphenodontians, and crocodylomorphs. *Sinoconodon* has been reported only from Lufeng and cannot be used for stratigraphic correlation. *Oligokyphus* and *Morganucodon* from Lower Jurassic fissure deposits in Great Britain are considered Hettangian to Sinemurian in age (Kermack et al., 1973, 1981; Clemens et al., 1979; Fraser, Walkden, and Stewart, 1985; Clemens, 1986). The only record of *Morganucodon* that may be older than Hettangian is *Morganucodon peyeri* Clemens, 1980, from the Hallau Bonebed in Switzerland. The Kayenta Formation is considered to be no older than the Sinemurian (Olsen and Sues, 1986; Sues, 1986b). It has yielded *Dinnetherium* (Jenkins, Crompton, and Downs, 1983), which shares many derived dental characters with *Morganucodon* (Crompton and Luo,

1993), and three tritylodontids, all of which can be compared to the taxa in the Lower Lufeng Formation. *Tritylodon* from the upper Elliot Formation of southern Africa is regarded as Early Jurassic (Sues, 1986b). *Clevosaurus* has a more extensive stratigraphic range. It is present in the McCoy Brook Formation of Nova Scotia, which is Hettangian in age (H.-D. Sues, N. H. Shubin, and P. E. Olsen, pers. commun.), and it ranges from the Late Triassic through Early Jurassic in Great Britain (Fraser, 1988; Fraser and Benton, 1989; N. C. Fraser, pers. commun.). The crocodylomorphs in Lufeng are comparable to *Protosuchus* in the Moenave Formation of Arizona and *Sphenosuchus* in the upper Elliot Formation of southern Africa.

We agree with Sun et al. (1985) that the upper faunal assemblage from the Dark Red Beds is Early Jurassic in age. Based on the phylogenetic framework proposed by Clark and Hopson (1985) and Sues (1986b), the diverse tritylodontids of the upper faunal assemblage are fairly advanced. The tritylodontids in the upper faunal assemblage can be correlated to those of the Kayenta Formation of Arizona. The species of *Morganucodon* in the upper faunal assemblage of Lufeng and the fissure deposits of Great Britain strongly indicate that the two faunas are coeval (Kermack et al., 1981). *Morganucodon* can also be related to *Dinnetherium* from the Kayenta Formation (Crompton and Luo, 1993). Several crocodyliforms of the upper faunal assemblage (Wu and Chatterjee, unpublished data) show similarities to *Protosuchus* from the Moenave Formation (Crompton and Smith, 1980; Clark, 1986), which may range from the Hettangian to the Sinemurian (Olsen and Sues, 1986). *Dibothrosuchus* is closely related to *Sphenosuchus* from the upper Elliot Formation of the upper Stormberg Group of southern Africa (Lower Jurassic), and these two genera are the most derived sphenosuchians (Wu, 1986; Wu and Chatterjee, 1993). Therefore, we argue that the upper faunal assemblage is Early Jurassic in age, most probably Sinemurian, and definitely no older than Hettangian.

The lower faunal assemblage from the Dull Purplish Beds shows a much lower diversity of small tetrapods than the upper faunal assemblage. Thus its correlation is more difficult. *Bienotherium* is the only tritylodontid found in the lower Dull Purplish Beds. Its shows more plesiomorphic features than *Lufengia* and other tritylodontids that are present in the upper Dark Red Beds. However, it is more derived than *Oligokyphus* from the fissure deposits of Britain (Hettangian to Sinemurian) and the Kayenta Formation of North America (Sinemurian). It is also more derived than *Tritylodon* from the upper Elliot Formation of southern Africa (Hettangian). Thus the presence of *Bienotherium* indicates that the lower assemblage is Early Jurassic. This correlation is also corroborated by the systematic relationships of *Clevosaurus* (Wu, 1991, and Chapter 3). *Clevosaurus* ("*Dianosaurus*") *petilus* from stratum 4 of

the Dull Purplish Beds is more derived than *C. hudsoni* from Great Britain (Wu, Chapter 3) and has a close relationship to *C. wangi* in the upper Dark Red Beds. We suggest that the *Bienotherium*-saurischian faunal assemblage in the Dull Purplish Beds of the Lower Lufeng Formation is also Early Jurassic in age, probably Hettangian.

Acknowledgments

We are indebted to Professors Sun Ailin and Dong Zhimin, Mr. Cui Guihai, and Mr. Yu Chao of the Institute of Vertebrate Paleontology and Paleoanthropology (Beijing) for generously providing specimens for this study or for their support and assistance in the fieldwork. We also thank Drs. A. W. Crompton, R. L. Carroll, S. Chatterjee, and A. Sun for their support, collaboration, and advice on our research. We benefited from suggestions and comments by Drs. A. W. Crompton, W. A. Clemens, S. Chatterjee, R. L. Carroll, H.-D. Sues, N. H. Shubin, A. Sun, J. A. Hopson, and N. C. Fraser. Drs. Sues and Fraser provided editorial help. Dr. Hopson allowed us to cite his unpublished Ph.D. thesis. Dr. John Bolt (Field Museum of Natural History) granted us the opportunity to study the CUP collection. Work by Luo benefited from support by NSF grants DEB-8818098 and DEB-9020034 to A. W. Crompton and a faculty research and development grant from the College of Charleston. Research by Wu benefited from support of an NSERC grant to R. L. Carroll and a NGS grant to S. Chatterjee and was completed during his tenure as a postdoctoral fellow at the Royal Tyrrell Museum of Palaeontology and the University of Calgary.

References

Benton, M. J., and J. M. Clark. 1988. Archosaur phylogeny and the relationships of the Crocodylia. Pp. 295–338 in M. J. Benton (ed.), *Phylogeny and Classification of the Tetrapods. Vol. 1.* Systematics Association Special Volume 35A. Oxford: Clarendon Press.

Bien, M. N. 1941. "Red Beds" of Yunnan. *Bulletin of the Geological Society of China* 21: 159–198.

Bonaparte, J. F. 1982. Classification of the Thecodontia. *Géobios, Mémoire Spéciale* 6: 99–112.

Carroll, R. L. 1985. A pleurosaur from the Lower Jurassic and the taxonomic position of the Sphenodontida. *Palaeontographica* A189: 1–28.

Carroll, R. L., and P. M. Galton. 1977. Modern lizard from the upper Triassic of China. *Nature* 266: 252–255.

Chow, M. 1962. [A tritylodont specimen from Lufeng, Yunnan.] *Vertebrata Palasiatica* 6: 365–367. [in Chinese]

Chow, M., and C. C. Hu. 1959. [A new tritylodontid from Lufeng, Yunnan.] *Vertebrata Palasiatica* 3: 9–12. [in Chinese]

Clark, J. M. 1986. Phylogenetic relationships of the crocodylomorph archosaurs. Ph.D. dissertation, University of Chicago.

Clark, J. M., and J. A. Hopson. 1985. Distinctive mammal-like reptile from Mexico and its bearings on the phylogeny of Tritylodontidae. *Nature* 315: 398–400.

Clemens, W. A. 1979. A problem in morganucodontid taxonomy (Mammalia). *Zoological Journal of the Linnean Society* 66: 1–14.

1980. Rhaeto-Liassic mammals from Switzerland and West Germany. *Zittel.iana* 5: 51–92.

1986. On the Triassic and Jurassic mammals. Pp. 237–246 in K. Padian (ed.), *The Beginning of the Age of Dinosaurs: Faunal Change across the Triassic–Jurassic Boundary*. Cambridge University Press.

Clemens, W. A., J. A. Lillegraven, E. H. Lindsay, and G. G. Simpson. 1979. Where, when, and what – a survey of known Mesozoic mammal distribution. Pp. 7–58 in J. A. Lillegraven, Z. Kielan-Jaworowska, and W. A. Clemens (eds.), *Mesozoic Mammals: The First Two-thirds of Mammalian History*. Berkeley: University of California Press.

Colbert, E. H. 1952. A pseudosuchian reptile from northern Arizona. *Bulletin of the American Museum of Natural History* 99: 563–592.

Crompton, A. W. 1974. The dentitions and relationships of the southern African Triassic mammals, *Erythrotherium parringtoni* and *Megazostrodon rudnerae*. *Bulletin of the British Museum (Natural History), Geology* 24: 399–437.

Crompton, A. W., and Z. Luo. 1993. Relationships of the Liassic mammals, *Sinoconodon*, *Morganucodon oehleri*, and *Dinnetherium*. Pp. 30–44 in F. S. Szalay, M. J. Novacek, and M. C. McKenna (eds.), *Mammal Phylogeny*. Berlin: Springer-Verlag.

Crompton, A. W., and K. K. Smith. 1980. A new genus and species of crocodilian from the Kayenta Formation (Late Triassic?). Pp. 193–217 in L. L. Jacobs (ed.), *Aspects of Vertebrate History*. Flagstaff: Museum of Northern Arizona Press.

Crompton, A. W., and A. L. Sun. 1985. Cranial structure and relationships of the Liassic mammal *Sinoconodon*. *Zoological Journal of the Linnean Society* 85: 99–119.

Crush, P. 1984. A late Upper Triassic sphenosuchid crocodile from Wales. *Palaeontology* 27: 133–157.

Cui, G. 1976 [*Yunnania*, a new tritylodontid from Lufeng, Yunnan.] *Vertebrata Palasiatica* 25: 1–7. [in Chinese]

1981. [A new genus of Tritylodontidae.] *Vertebrata Palasiatica* 19:5–10. [in Chinese]

1986. [*Yunnanodon*, a replacement name for *Yunnania* Cui, 1976.] *Vertebrata Palasiatica* 24: 9. [in Chinese]

Cui, G., and A. Sun. 1987. [Postcanine root system of tritylodonts.] *Vertebrata Palasiatica* 25: 245–259. [in Chinese]

Evans, S. E., and A. R. Milner. 1989. *Fulengia*, a supposed early lizard reinterpreted as a prosauropod dinosaur. *Palaeontology* 32: 223–230.

Fraser, N. C. 1988. The osteology and relationships of *Clevosaurus* (Reptilia: Sphenodontida). *Philosophical Transactions of the Royal Society of London* B321: 125–178.

Fraser, N. C., and M. J. Benton. 1989. The Triassic reptiles *Brachyrhinodon* and *Polysphenodon* and the relationships of the sphenodontids. *Zoological Journal of the Linnean Society* 96: 413–445.

Fraser, N. C., G. M. Walkden, and V. Stewart. 1985. The first pre-Rhaetic therian mammal. *Nature* 314: 161–163.

Hopson, J. A. 1964. The braincase of the advanced mammal-like reptile *Bienotherium*. *Postilla* 87: 1–30.

1965. Tritylodontid therapsids from Yunnan and the cranial morphology of *Bienotherium*. Ph.D. dissertation, University of Chicago.

Hopson, J. A., and J. W. Kitching. 1972. A revised classification of cynodonts (Reptilia, Therapsida). *Palaeontologia Africana* 14: 71–85.

Jenkins, F. A., Jr. 1984. A survey of mammalian origins. Pp. 32–47 in P. D. Gingerich and C. E. Badgley (eds.), *Mammals: Notes for a Short Course*. University of Tennessee, Department of Geological Sciences, Studies in Geology no. 8.

Jenkins, F. A., Jr., A. W. Crompton, and W. R. Downs. 1983. Mesozoic mammals from Arizona: new evidence on mammalian evolution. *Science* 222: 1233–1235.

Kermack, K. A., F. Mussett, and H. W. Rigney. 1973. The lower jaw of *Morganucodon*. *Zoological Journal of the Linnean Society* 53: 87–175.

1981. The skull of *Morganucodon*. *Zoological Journal of the Linnean Society* 71: 1–158.

Kühne, W. G. 1949. On a triconodont tooth of a new pattern from a fissure-filling in South Glamorgan. *Proceedings of the Zoological Society of London* 119: 345–350.

1956. *The Liassic Therapsid Oligokyphus*. London: British Museum (Natural History).

Mills, J. R. E. 1971. The dentition of *Morganucodon*. Pp. 26–63 in D. M. Kermack and K. A. Kermack (eds.), *Early Mammals. Zoological Journal of the Linnean Society*, Vol. 50, Suppl. no. 1.

Olsen, P. E., and H.-D. Sues. 1986. Correlation of continental Late Triassic and Early Jurassic sediments, and patterns of Triassic–Jurassic transition. Pp. 321–351 in K. Padian (ed.), *The Beginning of the Age of Dinosaurs: Faunal Change across the Triassic–Jurassic Boundary*. Cambridge University Press.

Parrington, F. R. 1971. On the Upper Triassic mammals. *Philosophical Transactions of the Royal Society of London* B261: 231–272.

1973. The dentition of the earliest mammals. *Zoological Journal of the Linnean Society* 52: 85–95.

1978. A further account of the Triassic mammals. *Philosophical Transactions of the Royal Society of London* B282: 177–204.

Parrish, J. M. 1991. A new specimen of an early crocodylomorph (cf. *Sphenosuchus* sp.) from the Upper Triassic Chinle Formation of Petrified Forest National Park, Arizona. *Journal of Vertebrate Paleontology* 11: 198–212.

Patterson, B., and E. C. Olson. 1961. A triconodontid mammal from the Triassic of Yunnan. Pp. 129–191 in *International Colloquium in the Evolution of Lower and Nonspecialized Mammals*. Brussels: Koninklijke Vlaamse Academie voor Wetenschapen, Letteren en Schone Kunsten van Belgie.

Piveteau, J. (ed.). 1955. *Traité de Paléontologie*, Vol. 5. Paris: Masson et Cie.

Rigney, H. W. 1963. A specimen of *Morganucodon* from Yunnan. *Nature* 197: 1122–1123.

Robinson, P. L. 1973. A problematic reptile from the British Upper Trias. *Journal of the Geological Society of London* 129: 457– 479.

Romer, A. S. 1956. *The Osteology of the Reptiles.* University of Chicago Press.

Rowe, T. 1988. Definition, diagnosis, and origin of Mammalia. *Journal of Vertebrate Paleontology* 8: 241–264.

Sigogneau-Russell, D., and A.-L. Sun. 1981. A brief review of Chinese synapsids. *Géobios* 14: 275–279.

Simmons, D. J. 1965. The non-therapsid reptiles of the Lufeng Basin, Yunnan, China. *Fieldiana, Geology* 15: 1–93.

Sues, H.-D. 1985a. First record of the tritylodontid *Oligokyphus* (Synapsida) from the Lower Jurassic of western North America. *Journal of Vertebrate Paleontology* 5: 328–335.

———. 1985b. The relationships of the Tritylodontidae (Synapsida). *Zoological Journal of the Linnean Society* 85: 205–217.

———. 1986a. The skull and dentition of two tritylodontid synapsids from the Lower Jurassic of western North America. *Bulletin of the Museum of Comparative Zoology, Harvard University* 151: 217–268.

———. 1986b. Relationships and biostratigraphic significance of the Tritylodontidae (Synapsida) from the Kayenta Formation of northeastern Arizona. Pp. 279–284 in K. Padian (ed.), *The Beginning of the Age of Dinosaurs: Faunal Change across the Triassic–Jurassic Boundary.* Cambridge University Press.

Sun, A. L. 1984. Skull morphology of the tritylodont genus *Bienotheroides* of Sichuan. *Scientia Sinica* B27: 970–984.

Sun, A. L., and G. Cui. 1986. A brief introduction to the Lower Lufeng saurischian fauna (Lower Jurassic: Lufeng, Yunnan, People's Republic of China). Pp. 275–278 in K. Padian (ed.), *The Beginning of the Age of Dinosaurs: Faunal Change across the Triassic–Jurassic Boundary.* Cambridge University Press.

———. 1987. [Otic region in the tritylodont *Yunnanodon*.] *Vertebrata Palasiatica* 25: 1–7. [in Chinese]

Sun, A. L., G. Cui, Y. Li, and X. Wu. 1985. [A verified list of the Lufeng Saurischian Fauna.] *Vertebrata Palasiatica* 22: 1–12. [in Chinese]

Walker, A. D. 1990. A revision of *Sphenosuchus acutus* Haughton, a crocodylomorph reptile from the Elliot Formation (late Triassic or early Jurassic) of South Africa. *Philosophical Transactions of the Royal Society of London* B330: 1–120.

Westphal, F. 1976. Phytosauria. Pp. 99–120 in O. Kuhn (ed.), *Handbuch der Paläoherpetologie. Vol. 13: Thecodontia.* Stuttgart: Gustav Fischer Verlag.

Wu, X. 1986. [A new species of *Dibothrosuchus* from Lufeng Basin]. *Vertebrata Palasiatica* 24: 42–62. [in Chinese]

———. 1991. The comparative anatomy and systematics of Mesozoic sphenodontidans. Ph.D. dissertation, McGill University.

Wu, X., and S. Chatterjee. 1993. *Dibothrosuchus elaphros,* a crocodylomorph from the Early Jurassic of China and the phylogeny of Sphenosuchia. *Journal of Vertebrate Paleontology* 13: 58–89.

Young, C. C. 1936. On a new *Chasmatosaurus* from Sinkiang. *Bulletin of the Geological Society of China* 15: 291–330.

———. 1940a. Preliminary note on the Mesozoic mammals of Lufeng, Yunnan, China. *Bulletin of the Geological Society of China* 20: 93–111.

———. 1940b. Preliminary note of the Lufeng vertebrate fossils. *Bulletin of the Geological Society of China* 20: 235–240.

———. 1944. On a supposed new pseudosuchian from Upper Triassic saurischian-bearing beds of Lufeng, Yunnan, China. *American Museum Novitates* 1264: 1–4.

———. 1946. The Triassic vertebrate remains of China. *American Museum Novitates* 1324: 1–14.

———. 1947a. Mammal-like reptiles from Lufeng, Yunnan, China. *Proceedings of the Zoological Society of London* 117: 537–597.

———. 1947b. On *Lufengosaurus magnus* Young (sp. nov.) and additional finds of *Lufengosaurus huenei* Young. *Palaeontologia Sinica, n.s.,* C 12: 1–53.

———. 1951. The Lufeng saurischian fauna in China. *Palaeontologia Sinica, n.s.,* C 13: 19–96.

———. 1974. [New material of therapsids from Lufeng, Yunnan.] *Vertebrata Palasiatica* 12: 111–114. [in Chinese]

———. 1978. [New material of *Eozostrodon.*] *Vertebrata Palasiatica* 16: 1–3. [in Chinese]

———. 1982. [*Selected Works of Yang Zhongjian (Young Chung-Chien).*] Beijing: Science Press. [in Chinese]

Zhang, F., and G. Cui. 1983. [New material and new understanding of *Sinoconodon.*] *Vertebrata Palasiatica* 21: 32–41. [in Chinese]

15

Assemblages of small tetrapods from the Early Jurassic of Britain

SUSAN E. EVANS AND KENNETH A. KERMACK

Introduction

The Lower Jurassic of Britain comprises most of the Lias Group, ranging from Hettangian to Toarcian in age, and represents some 25 million years, between about 205 and 180 Ma (Forster and Warrington, 1985; Hallam et al., 1985). The latest Triassic was a time of gentle subsidence and a transgression from the Tethys Ocean into northern Europe. As a result, most Lias Group rocks in that region were deposited under marine conditions. Exceptions are deposits that accumulated in fissure and cave systems on islands that were gradually inundated during the Early Jurassic. These deposits have yielded important assemblages of small terrestrial tetrapods.

Geology and paleoenvironment

In the Vale of Glamorgan, South Wales, and in the Bristol-Mendip region of England, a Mesozoic landscape has been partly disinterred by recent erosion (Strahan and Tiddeman, 1904). The high ground of this old land surface was formed by Carboniferous limestone that (as is the case today) contained systems of caves and fissures formed under the action of meteoric groundwater. These filled with silt and acted as traps for the remains of small animals, mostly tetrapods. The fissure infills were protected from the weight of younger deposits by the Carboniferous limestone within which they were embedded.

Marine limestones of the Dinantian (Early Carboniferous) age form the hills of the present-day landscape. In Late Carboniferous times, Armorican folding produced the Mendip and Cardiff–Cowbridge anticlines. Following active erosion throughout the Permian and most of the Triassic, Mesozoic sediments were laid down on a surface largely composed of Devonian and Carboniferous rocks. In the Late Triassic, the Carboni-

ferous limestone hills formed islands separated by water bodies, in which deposition of red marls occurred (Tucker, 1977; Ager and Edwards, 1985). Dinosaurian footprints discovered at Sully in South Wales (Tucker and Burchette, 1977) were made by animals that came down to the water to drink. The water body was initially seasonal, but subsequently became permanent. The climate was hot and rather arid; infrequent torrential rains carried away much of the surface soil.

With further sinking, the sea flooded the lowland during latest Triassic times, leaving the former hills as an archipelago (Trueman, 1922). The remains of plants and small animals were washed by rain and stream action into cracks in the limestone and thence into the underground water systems, where they were deposited. The subsidence continued, and the sea became deeper, until ultimately the islands were submerged. This did not occur synchronously, however, and the fissure deposits at different localities range from Late Triassic to Early Jurassic in age.

The fissure localities fall into two geographic groups: the quarries of the Bristol-Mendip region to the southeast of the Bristol Channel (notably Emborough, Cromhall, Tytherington, Holwell, and Windsor Hill), and those of the Vale of Glamorgan, to the northwest (Figure 15.1). The principal Welsh fissure localities can be further divided into western and eastern groups. The western localities (notably Pant, Pontalun, Ewenny, Duchy, and Cnap Twt) lie on a small limestone plateau that formed an island, St. Bride's Island (Robinson, 1971), in Early Jurassic times. The eastern fissures (notably Ruthin and Pant-y-ffynon) were formed on one or more islands near Cowbridge (Robinson, 1957) (Figure 15.2).

The fissures of St. Bride's Island, Holwell, and Windsor Hill have yielded abundant synapsids (tritylodontids and mammals). The apparent absence of these faunal elements from the remaining fissures led Robinson

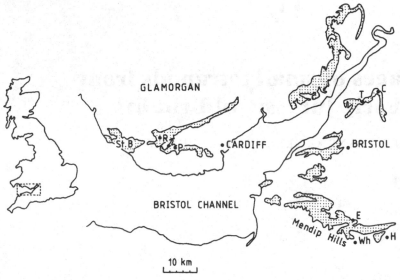

Figure 15.1. Map showing principal outcrops of Carboniferous limestone (stippled) in the Bristol Channel region of Britain. Key to fissure localities: C, Cromhall (Slickstones); E, Emborough; H, Holwell; P, Pant-y-ffynon; R, Ruthin; St. B, St. Bride's; T, Tytherington; Wh, Windsor Hill. (Redrawn from British Geological Survey Sheet, 1979.)

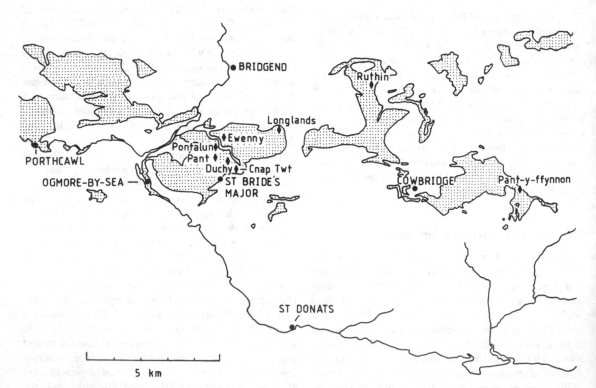

Figure 15.2. Carboniferous limestone outcrops (stippled) in the Vale of Glamorgan and principal fissure localities (diamonds) mentioned in the text.

(1971) to conclude that the fissure assemblages fell into two chronologically and ecologically distinct groups. The first (Emborough, Cromhall, Tytherington, Ruthin, Pant-y-ffynon) represented pretransgressive upland faunas living in arid continental conditions too harsh to support synapsids. The second (the St. Bride's fissures, Holwell, Windsor Hill), by contrast, represented milder posttransgressive conditions afforded by an archipelago fringing a continent.

Robinson's hypothesis has been challenged by the discovery of two teeth of the therian mammal *Kuehneotherium* at Emborough Quarry (Fraser, Walkden, and Stewart, 1985) and by the work of Marshall and Whiteside (1980) and Whiteside (1986), who concluded, on palynological evidence, that the Tytherington fissures (and therefore probably Cromhall, which has a similar tetrapod fauna) are of latest Triassic age (cf. Fraser, 1986, 1988a,b). Nonetheless, there is general agreement that the Bristol-Mendip fissures (except Windsor Hill) are of Late Triassic age, whereas those of St. Bride's Island are of Early Jurassic (Hettangian or Sinemurian) age (Robinson, 1971; Kermack, Mussett, and Rigney, 1973, 1981; Whiteside, 1986; Fraser, 1988b).

The Cowbridge assemblages are more controversial. Geological evidence (Robinson, 1957) suggests that the Cowbridge islands were submerged in the early Hettangian, so their assemblages could be older, but not younger, than that. Robinson (1971) considered them to be of pretransgressive age, a view accepted by Crush (1984) and Fraser et al. (1985). Warrener (1983) and Kermack (née Warrener) (1984) chose Rhaeto-Liassic, with the possibility that the assemblages could be of Hettangian age. On faunal composition (Table 15.1), they relate best to the Bristol-Mendip assemblages.

For the purpose of this review, we have restricted ourselves to those assemblages most generally agreed to be of Early Jurassic age: the fissures of St. Bride's Island and Windsor Hill. Fraser (Chapter 11) has reviewed the other Bristol-Mendip assemblages and those of the Cowbridge localities. Following the recommendations of Cope et al. (1980) and Warrington et al. (1980), the base of the Jurassic is recognized as the level of first appearance of the ammonite *Psiloceras*.

History of discovery of the fissure faunas

Carboniferous limestone is quarried for road metal, for aggregate, for fluxing stone for blast furnaces, and, to a small extent, for burning to lime. In the 1940s and 1950s the limestone was worked by firing small charges in shallow holes drilled in the quarry face. The stone thus obtained was loaded either by hand or by comparatively small mechanical shovels. The slow rate of working ensured that exposed fissures would remain visible for a considerable time. The economically worthless fissure material was separated from the limestone and dumped on a disused area of the quarry floor.

The first discovery of Mesozoic vertebrates in fissures in Carboniferous limestone was made in the middle of the nineteenth century by Charles Moore (1867). Moore's fissure was at Holwell in the Mendip region, and it yielded a vertebrate fauna including haramiyid mammals and small reptiles. Holwell was revisited in 1939 by Walter Kühne, who found additional microvertebrate material, including two triconodontan teeth, later named *Eozostrodon* (Parrington, 1941; Kühne, 1947b). A short time later, Kühne discovered a locality at Windsor Hill Quarry in Somerset. Windsor Hill yielded bones of the tritylodontid *Oligokyphus*. Kühne was interned during part of World War II; he was released in 1943 and spent his time working on the Windsor Hill material at University College London. After the war he and his wife Charlotte resumed prospecting. In 1947, in Duchy Quarry, Glamorgan,

Table 15.1. *Comparison of faunal components in the Bristol Channel assemblages*

Group	Emborough	Cromhall	Tytherington	Ruthin	Pant-y-ffynon	Holwell	St. Bride's	Windsor Hill
Procolophonids	—	present	—	present	—	—	—	—
Tricuspisaurus/ Variodens	present	present	—	present	—	—	—	—
Kuehneosaurus	present	present	—	?present	present	—	—	—
Gephyrosaurus	—	—	—	—	present	—	present	—
Diphydontosaurus	—	present	present	—	—	—	—	—
Planocephalosaurus	—	present	present	—	—	—	—	—
Clevosaurus	present	present	present	—	present	—	—	—
Terrestrisuchus	—	?present	?present	?present	present	—	—	—
Coelurosaur	—	—	?present	—	present	—	present	—
Thecodontosaurus	—	present	present	?present	present	—	—	—
Oligokyphus	—	—	—	—	—	—	present	present
Tritylodontid	—	—	—	—	—	present	—	—
Haramiyids	—	—	—	—	—	present	present	—
Eozostrodon	—	—	—	—	—	present	present	—
Morganucodon	—	—	—	—	—	—	present	—
Kuehneotherium	present	—	—	—	—	—	present	—

Kühne found teeth and mandibular fragments of an early mammal, *Morganucodon* (Kühne, 1949, 1958), and also a symmetrodont tooth, later named *Kuehneon* (Kretzoi, 1960). The material was found on a quarry dump, the original fissure system having been destroyed before Kühne's visit.

Walter Kühne returned to Germany in 1951 and went on to discover the important Upper Jurassic deposits at Guimarota. Pamela Robinson and Kenneth Kermack continued with the fieldwork, Robinson concentrating on the fissure deposits of the Mendips, and Kermack working in South Wales. Over the period 1952–84 the research group at University College found fossiliferous fissures in sixteen different Welsh quarries (Kermack et al., 1973).

Fissure localities

Windsor Hill

This locality (U.K. National Grid Reference ST 615452) lies close to the town of Shepton Mallet in Somerset.

The fissure was a Neptunian dyke (a fissure opening under the sea) and contained a fauna of marine invertebrates and fish (*Acrodus*, *Hybodus*, *Birgeria*, and chimaeroids), as well as a terrestrial component composed entirely of the tritylodontid *Oligokyphus*. The invertebrate assemblage is diverse and includes fragmentary ammonoids, belemnoids, gastropods, bivalves, and brachiopods that indicate a Pliensbachian age (Kühne, 1947a,b, 1956).

Windsor Hill is remarkable in that it contains a single genus of tetrapod – *Oligokyphus* – with over 2,000 specimens representing a minimum of 44 individuals. The adult specimens fall into two size groups, designated *Oligokyphus major* and *Oligokyphus minor* by Kühne (1947a, 1956) (Figure 15.3). The two morphs may represent distinct species of *Oligokyphus*, or perhaps male and female animals of the same species (Kühne, 1947a, 1956).

Kühne made a meticulous examination of the Windsor Hill material, including the fine fractions, and found no trace of other tetrapods. Because the bones of *Oligokyphus* show no evidence of having been

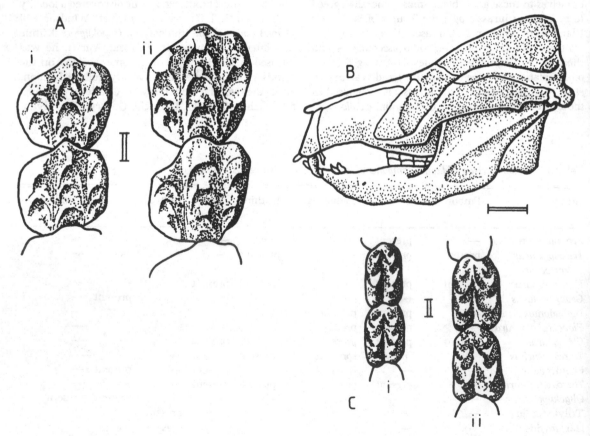

Figure 15.3. Skull and dentition of *Oligokyphus*, Windsor Hill. (A) Upper molariform teeth of (i) *O. minor* (BMNH R.7025) and (ii) *O. major* (BMNH R.7030), in occlusal view. (B) Reconstruction of the skull in lateral view. (C) Lower molariform teeth of (i) *O. minor* (BMNH R.7204) and (ii) *O. major* (BMNH R.7123), in occlusal view. Scale bars equal 1 mm, except in B, where scale bar is 10 mm. (Adapted from Kühne, 1956.)

reworked, the fissure must have been the primary site of deposition. The bones probably were carried out to sea by a small stream. They appear to have been broken and dissociated before transportation, but are only mildly waterworn, suggesting that the fissure lay close to shore (Kühne, 1947a, 1956).

St. Bride's Island

Robinson (1957) gave the name "St. Bride's Island" to a small region of Glamorgan that probably remained above water until early Sinemurian (*bucklandi* zone) times.

During the period 1952–84, a maximum of six quarries worked the Carboniferous limestone in this region: four in the Alun valley, Ewenny (SS 902769), Pontalun (SS 899765), Duchy (SS 906757), and Cnap Twt (SS 910752); one, Pant Quarry (SS 896760), in Pant St. Bride's; and one, Longlands Quarry (SS 928772), on the northern slope of the plateau. With the exception of Longlands, the quarries lie within an area of less than 2 km² and have yielded the same characteristic fauna and flora.

The most widespread fossil from the St. Bride's fissures is the conifer *Hirmeriella muensteri*, which gives its name to the faunal and floral association as a whole. [Note: This conifer has been known under several generic names – *Cheirolepis, Cheirolepidium, Hirmerella.* In the most recent review, Watson (1988) advocated the use of *Hirmeriella.*] *Hirmeriella* has never been found in any of the other fissure systems in the Bristol Channel region, and the *Hirmeriella* association is confined to five of the six quarries mentioned earlier. The remaining quarry, Longlands, lies 2–3 km to the east of the main group. It is close to the site of Brocastle Pit (Moore, 1867) and shows the lower Lias transgressing over the Carboniferous limestone. It has yielded only a few scraps of indeterminate bone and plant debris.

Fissures that contain the *Hirmeriella* association vary in both size and form. Most commonly they are narrow slots formed in open joints in the limestone and can be up to 1 m wide, but usually are narrower, some being less than 0.3 m wide; such pockets (e.g., in Pontalun Quarry) may contain many tons of matrix.

The matrix filling of the fissures ranges from soft clay to hard marl and often is rich in hematite grains. The color of the matrix varies from red to yellow, green, and gray, and that of the bone varies from white to dark gray and brown; most commonly, white bone occurs in a red matrix. The bone shows no evidence of having been reworked, and the fissure infills were the primary sites of deposition. The plant fossils, which usually are preserved as charcoal (fusain), are most common in a gray matrix (which indicates reducing conditions), but they have also been found in both yellow and red deposits.

The conifer *Hirmeriella muensteri* [and its pollen type *Classopollis* (Alvin, 1982)] has been found in most of the St. Bride's quarries but is most abundant at Cnap Twt, where it is associated with the megaspore *Triletes* (now *Bacutriletes*) *tylotus* together with *Lycospora* (now *Kraeuselisporites*) *reissingeri* and other miospores that were referred to the genera *Annulatisporites, Entylissa, Leiotriletes, Pericutosporites,* and *Pityosporites* (Harris, 1957; Lewarne and Pallot, 1957). There are also rare fragments of other plants, including the conifers *Ctenis, Pterophyllum,* and *Cycadolepis.*

In most fissures, the vertebrate component of the *Hirmeriella* association is limited to three principal genera. Nearly 60 percent of the material is referable to the lepidosaur *Gephyrosaurus bridensis* (Evans, 1980, 1981) (Figure 15.4); the remaining material is mammalian, with *Morganucodon watsoni* (Figure 15.5) the dominant form (Kermack, Mussett, and Rigney, 1973, 1981). The bones of both genera are completely dissociated, but virtually every bone in the skeleton is represented, often in exquisite preservation.

The third standard member of the *Hirmeriella* association is the therian mammal *Kuehneotherium praecursoris* Kermack, Kermack, and Mussett, 1968, which is represented only by teeth (Figure 15.6) and rare jaw fragments. This rarity must partly reflect the fact that *Kuehneotherium* was a smaller animal, with more fragile bones, but it also stems from an anatomical peculiarity of *Morganucodon. Morganucodon* has teeth with roots that are expanded at their ends to form a sort of "elephant's foot" (taurodont condition) (Figure 15.5C–E), thereby locking the teeth into the jaw after death and giving it extra strength. The teeth of *Kuehneotherium* have tapered roots, and consequently they dropped out after death, causing the greatly weakened jaw to break up. There is one exception to the general rule that *Morganucodon* is much more common than *Kuehneotherium.* Kermack, Kermack, and Mussett (1968) described a small pocket in Pontalun (fissure I) in which teeth of *Kuehneotherium* were numerous.

In this chapter we have referred all material of *Kuehneotherium* to *K. praecursoris,* but Mills (1984) believed the kuehneotheriids from Pant and Pontalun to be at least specifically distinct. Kühne's symmetrodont tooth from Duchy was named *Kuehneon* (Kretzoi, 1960), but the specimen has been lost, and its relationship to *Kuehneotherium* remains unresolved.

Gephyrosaurus, Morganucodon, and *Kuehneotherium* are the three consistently occurring vertebrate components of the *Hirmeriella* association, but other genera have occasionally been reported. Invertebrates are rare; fragments of a beetle (*Metacupes*) and a gastropod ("*Natica*" *oppeli* Moore) have been recovered from fissures at Cnap Twt and Pant, respectively (Kermack, Mussett, and Rigney, 1973), but the local invertebrate fauna must have been richer than that. Additional

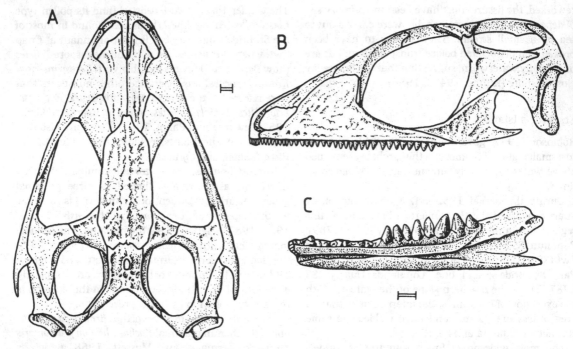

Figure 15.4. *Gephyrosaurus bridensis*, skull and lower jaw. Reconstruction of the skull in (A) dorsal and (B) lateral views. (C) Holotype right dentary (BMNH RU 1503) in medial view. All scale bars equal 1 mm. (Adapted from Evans, 1980.)

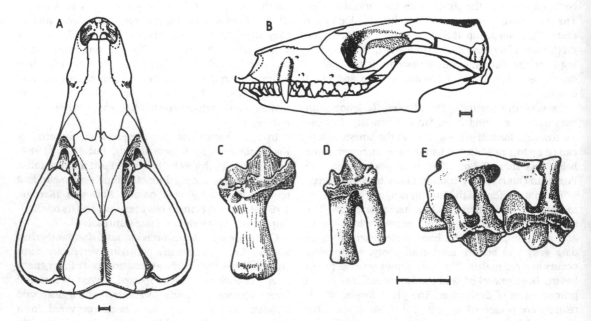

Figure 15.5. *Morganucodon watsoni*, skull and dentition. Reconstruction of the skull in (A) dorsal and (B) lateral views. (C, D) Lower molars in lingual view. (E) Upper molars locked into a fragment of maxilla (buccal view). Scale bars equal 1 mm. (A and B adapted from Kermack et al., 1981; C–E drawn from teeth in the UCL collection.)

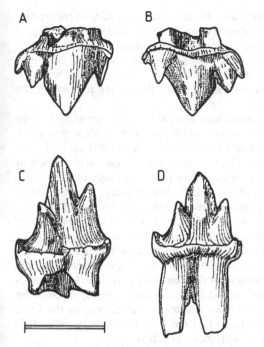

Figure 15.6. *Kuehneotherium praecursoris*, dentition. Holotype (BMNH M. 19165), a left upper molar, in (A) lingual and (B) buccal views. (C) Paratype (BMNH M.19155), a left lower molar, in lingual view. (D) Paratype (BMNH C. 855), a left lower molar, in lingual view. Scale bar equals 1 mm. (Adapted from Kermack et al., 1968.)

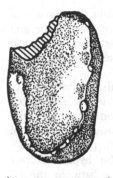

Figure 15.7. *Thomasia anglica* (Haramiyidae), Holwell Quarry. Molariform tooth (Yale Peabody Museum 13622B), conventional anterior end to top of page. Scale bar equals 1 mm. (Adapted from Simpson, 1928.)

vertebrates are known from rare specimens. Kühne (1958) reported finding a fragment of an amphibian dentition and a possible dipnoan tooth fragment at Duchy, but these interesting observations have never been confirmed, and the specimens appear to have been lost.

Until recently, one small archosaur tooth crown from Pontalun was the only evidence of predators on St. Bride's Island (Kermack, Mussett, and Rigney, 1973). More recently, Fraser (1989) has described a small archosaurian fauna from Pontalun (Lithalun), including the premaxilla of a rhamphorhynchid pterosaur and isolated teeth of at least three kinds of archosaurs. Of the latter, those in one group resemble the teeth either of small theropods or of the crocodylomorph *Terrestrisuchus* (= *Saltoposuchus*) (Benton and Clark, 1988) described by Crush (1984) from Pant-y-ffynon. The others may be referable to primitive archosaurs, crocodylomorphs, or pterosaurs.

By comparison with the fissure assemblages of the Bristol-Mendip region, Fraser (1989) described the fauna of St. Bride's Island as depauperate. He suggested that this was due to reduction of the land area by the encroaching sea. This explanation assumes, however, that the typical *Hirmeriella* assemblage is fully representative of the contemporaneous fauna of St. Bride's – it is not.

In 1968, work began to expose a new fissure at Pant Quarry. This fissure, designated Pant 4, has yielded a vertebrate fauna that includes the usual members of the *Hirmeriella* association, in addition to abundant remains of new reptilian and mammalian genera. Pant 4 and its fauna were studied by David Pacey in his Ph.D. dissertation (Pacey, 1978). In addition to *Gephyrosaurus*, *Morganucodon watsoni*, and *Kuehneotherium*, Pant 4 has produced the following: three new species of sphenodontian; several archosaurs, broadly matching the types described by Fraser (1989); several forms of the tritylodontid *Oligokyphus*, including the Windsor Hill morphs *O. minor* and *O. major*; a haramiyid, *Thomasia* (Figure 15.7); and several additional mammals, including *Eozostrodon* and a new, large morganucodontid. This assemblage is clearly different from that of Cromhall or Emborough, but it is not depauperate.

Source of the fissure vertebrates

The Mendip/Glamorgan fissures are remarkable for their high concentrations of small vertebrate bones. The fissure deposits derived from surface material washed into the underground water system during periods of heavy seasonal rains. Because it seems unlikely that hydrologic effects alone could concentrate small vertebrates so effectively, various attempts have been made to explain how the vertebrate material accumulated (e.g., Kermack, Mussett, and Rigney, 1973).

Harris (1958), noting that the majority of plant material in the St. Bride's fissures was in the form fusain (charcoal), suggested that seasonal storms might have sparked forest fires, which could have killed, but not necessarily incinerated, small vertebrates living on the ground. An alternative, or complementary, explanation would be that small animals were caught up and drowned by flash floods. Both suggestions are reasonable, but they do not explain the absence of invertebrates (apart from one carbonized beetle and one snail) nor the differences between the typical *Hirmeriella* association and the contemporaneous fauna from Pant 4.

Kühne (1956) concluded that the concentration of bones referable to *Oligokyphus* at Windsor Hill resulted from predator activity, a suggestion borne out by the pattern of bone damage, the relative proportions of skeletal elements preserved, and the presence of tooth marks on some bones. A similar hypothesis best explains the apparent differences between the St. Bride's fissure assemblages (Kermack, Mussett, and Rigney, 1973; Pacey, 1978).

Recent studies of Holocene fissure assemblages from the Caribbean (Olson and Pregill, 1982; Pregill, 1982; Pregill et al., 1988) show some interesting parallels. These modern fissure assemblages, while containing a broad spectrum of genera usually are dominated by a few common species. The rodent *Geocapromys* is the only native terrestrial mammal on New Providence Island and is the principal fissure component; lizards of the genus *Anolis* are also abundant (Olson and Pregill, 1982). On Antigua (Pregill et al., 1988), *Anolis* is the dominant fissure vertebrate, followed by a bat, *Natalis* (Pregill et al., 1988). On both islands, frogs take third place, followed by a series of rarer forms. The presence or absence of these rare, and probably accidental, components can vary from fissure to fissure on the same island. In both modern assemblages, the predators were owls.

This work provides a possible model for the typical *Hirmeriella* association. If the assemblage resulted from the activity of a selective predator, this would explain why only a small proportion of the contemporaneous fauna is represented. The Pant 4 assemblage is more complex. The Pant 4 fissure lay close to other fossiliferous fissures in the same quarry and probably was roughly contemporaneous (Pacey, 1978). Like the typical *Hirmeriella* association, the dominant faunal components in Pant 4 are a small reptile and a synapsid, but the combination is different. It consists (Pacey, 1978) of a new sphenodontian and the tritylodontid *Oligokyphus*; *Morganucodon watsoni*, *Kuehneotherium praecursoris*, and *Gephyrosaurus bridensis* are comparatively rare. Most of the Pant 4 bone is waterworn (but not reworked) (Pacey, 1978) and clearly has been carried some distance; the only recognizable bones are teeth and jaws. Pant 4 may contain a larger sample of the island vertebrates, partly because the main predator was different, and partly because the fissure had a larger catchment area.

If the fissure assemblages were accumulated by predators, the question of the identity of those predators remains. Tertiary or Quaternary fissure and cave assemblages resulting from owl activity generally show good preservation of individual skeletal components.

Owls typically swallow their prey and then regurgitate a pellet of bone and other indigestible parts, the degree of initial damage and the amount of bone digested varying with the age and species of owl (Mayhew, 1977; Dodson and Wexlar, 1979). Clearly, however, this model has its limits; there were no owls in the Jurassic.

Accumulations of small Mesozoic vertebrates have sometimes been attributed to crocodilian predators, but Fisher (1981a,b) has shown this to be highly unlikely. The very low pH conditions in a crocodilian stomach ensure that little bone survives digestion. Traces of bone in the feces typically are decalcified; teeth, if they survive, are stripped of enamel. This is not unique to crocodiles. Fisher (1981a) presented an excellent review of vertebrate fecal structure. Modern amphibians, lizards, and snakes are much like the modern crocodilians in their digestive processes; the food remains in the stomach a long time. In mammals (Fisher, 1981a), food passes more quickly through the gut, and recognized bones may appear in the feces. Birds also have high metabolic rates, but the indigestible residues are rejected as pellets.

In the St. Bride's fissures, with the exception of Pant 4, all skeletal components, even the most delicate, are conserved, and the tooth enamel is intact. Some jaws of *Gephyrosaurus* retain unankylosed replacement teeth in situ on the alveolar margin. These bones could not, therefore, have passed through a typically reptilian digestive system. This leaves only two possibilities: Either the bones were regurgitated as pellets, or they passed through a mammalian type of digestive system.

Archosaurs, like other reptiles, regularly replace their teeth. Consequently, archosaurian predators often leave shed tooth crowns with the remains of their prey. The presence of shed crowns, some quite large, in the assemblage at Pant 4 and, more rarely, at Pontalun suggests that small theropod dinosaurs may have been the principal predators. This assumes an essentially avian digestive system for these dinosaurs.

A second alternative, which does not necessarily exclude the first, is that the fissure predators were mammalian. Because adult mammals do not replace their teeth, this would explain the apparent rarity of predator remains. The only obvious candidate among the fissure mammals is a large morganucodontid represented by a single tooth from Pant 4.

Age of the fissures of St. Bride's Island

Early in the Jurassic, St. Bride's Island was inundated by the sea. Robinson (1971) concluded that the island remained throughout Hettangian times, being finally submerged early in the Sinemurian (*bucklandi* zone). This is confirmed by the presence, around the island, of littoral deposits of early Sinemurian age (Cope et al.,

1980). The *Hirmeriella* association cannot be younger than that.

The red deposits, which the fissures often contain, resemble in color and texture the local red marls at Merlin Cwcw, which are of latest Triassic ("Rhaetian") age (Strahan and Tiddeman, 1904). However, although no bedded terrestrial deposits (e.g., limestone weathering "soils") are known from the Lias Group of Glamorgan, such deposits could well have been lithologically similar to those of late Norian age. Thus the lithological evidence is useless, except to warn us against the "red is Norian" fallacy.

The Mesozoic is zoned on the basis of marine invertebrates, particularly ammonoids. In the ammonoid faunas, the transition between the Triassic and the Jurassic is clear-cut, because a crisis occurred in ammonoid evolution at that time: progressive extinction during the Norian, with few ammonite genera in the later part of the Norian, being followed by radiation in the Early Jurassic. However, that particular crisis in marine faunas may not have been mirrored for terrestrial vertebrates living in the Bristol Channel area; they may have continued with little change from the Norian to the Early Jurassic. This is the more likely for the fauna and flora of St. Bride's Island, which formed an isolated island community from Norian times.

Early attempts to date the fissure assemblages concentrated on their plant and invertebrate-fossil contents. The botanical evidence has been discussed by Harris (1957), Lewarne and Pallot (1957), and, more recently, by Muir and van Konijnenburg-van Cittert (1970) and Orbell (1973). The conifer *Hirmeriella muensteri* is found in southern Germany, where the sediments can be correlated with a series of lower Liassic plant-bearing beds (*Thaumatopteris* zone) (Harris 1937; Orbell, 1973) in East Greenland, Poland, southern Germany, and Sweden. The megaspore *Bacutriletes* (*Triletes*) *tylotus* is known by rare specimens from Cnap Twt, from one locality in East Greenland dated as "Rhaetian" (*Lepidopteris* zone) (Harris 1937), and from a lacustrine deposit in Airel, France (Muir and van Konijnenburg-van Cittert, 1970), associated with a flora closely similar to that of Cnap Twt.

Orbell (1973), in a study of British Rhaeto-Liassic palynology, proposed two palynofloral zones roughly equivalent to the megafloral zones of Harris (1937): the older *Rhaetopollis* zone of "Rhaetian" age, encompassing the Blue Anchor and Westbury formations, and the younger *Heliosporites* zone, of "late Rhaetian" age, and probably extending into the early Hettangian, encompassing the Lilstock Formation (Cotham Member, "White Lias") and Watchet Beds. The microflora of Cnap Twt, with its relative abundance (6 percent) of *Kraeuselisporites* (*Heliosporites*), relates to the younger zone (Orbell, 1973). Morbey (1975) compared the Rhaeto-Liassic palynology of Austria and Britain. He

recognized a floral subzone (FG) of "latest Rhaetian" age that extended upward into the Hettangian. This zone was characterized by an abundance of *Kraeuselisporites* (*Heliosportites*). Similar findings have been reported by Warrington in palynological studies of the Penarth Group and Lias Group in central and southern Britain, as in the succession at Lavernock Point, Glamorgan (Warrington, in Waters and Lawrence, 1987, p. 61).

These studies suggest a "latest Rhaetian" or Early Jurassic age for the Cnap Twt assemblage, a conclusion further supported by the apparent absence of palynomorphs such as *Ovalipollis*, *Rhaetopollis*, and *Ricciisporites*. These three palynomorphs normally characterize deposits of latest Triassic age up to the top of the Penarth Group or in the basal Lias, within a few meters of the base of the Jurassic (Warrington, in Waters and Lawrence, 1987; G. Warrington, pers. commun.). This is probably the finest degree of stratigraphic resolution attainable with the available palynological evidence. There are no obvious differences between palynomorph assemblages of latest Triassic and Early Jurassic age, from the Pre-*planorbis* Beds and the succeeding (Jurassic) part of the Lias Group (Warrington, in Waters and Lawrence, 1987; Warrington and Ivimey-Cook, 1990; G. Warrington, pers. commun.). There are, however, differences between the palynomorph assemblages from the earlier "Rhaetian" deposits, such as the Westbury Formation and part of the succeeding Lilstock Formation (Penarth Group), and the latest Triassic deposits, including the remainder of the Penarth Group and the Pre-*planorbis* Beds of the Lias Group (Warrington, in Waters and Lawrence, 1987). The available palynological evidence supports an age difference between the faunas of the Tytherington and the St. Bride's fissures (G. Warrington, pers. commun.).

The gastropods from Pant Quarry were identified by H. C. Ivimey-Cook (pers. commun. to KAK) as "*Natica*" *oppeli* Moore. Gastropods of this genus and species are common in the Penarth Group at Lavernock Point, Glamorgan, but have not been recorded from the Lias, although "*Neridomus*"-like taxa, which could be congeneric with "*Natica*," are present (H. C. Ivimey-Cook, pers. commun.). Because gastropods are very rare in the early Hettangian of the Bristol Channel region, the evidence is inconclusive.

With the discovery of the Pant 4 fauna, an additional piece of evidence emerged. The Windsor Hill material of *Oligokyphus* is of Pliensbachian age (Kühne, 1956). According to Pacey (1978), teeth of *Oligokyphus* from Pant 4 closely resemble those from Windsor Hill, even in fine detail. The St. Bride's material cannot be Pliensbachian in age because the island had ceased to exist by that time, but the Pant 4 *Oligokyphus* provides support for the latest possible age: early Sinemurian. This is further supported by the structure of the fissures themselves. Slot fissures, like those of St. Bride's Island, are generally assumed to be immature solution phenomena (Robinson, 1971), suggesting that they formed late in the island's history.

Paleoecology of St. Bride's Island

St. Bride's was a small island, about 20 km² at maximum size, some distance from the closest large landmass, which lay to the north. Shortly after the "Rhaetian" transgression, it formed part of a small archipelago, but may well have been more isolated by late Hettangian times. The island lay at a paleolatitude of about 15° N (Robinson, 1971). In its size, latitude, topography (karst), and climate it must have closely resembled some of the small islands of the present-day Lesser Antilles.

The climate would have been tropical or subtropical, with heavy seasonal rains. The dominant plant was the conifer *Hirmeriella muensteri*, which Jung (1968) has reconstructed as a medium-sized tree, reaching a height of about 6 m. The remaining flora is less certain, because our knowledge of it is based mostly on spores (Harris, 1957) and other fragments. It included club mosses, possible ferns, bennettitaleans (*Pterophyllum*, *Cycadolepis*), and possible cycads. The flora was therefore varied and reasonably abundant, although we have no way of estimating the degree or complexity of the ground cover it provided.

The vertebrate fauna consisted of a broad range of terrestrial genera, including several insectivorous mammals, a morganucodontid at Pant 4 large enough to have been a small carnivore (Pacey, 1978), the herbivorous tritylodontid *Oligokyphus*, an insectivorous lepidosaur (*Gephyrosaurus*), several sphenodontians that may have been insectivores or omnivores, and a range of small archosaurs that may have been either small carnivores or insectivores. The reptiles, on the one hand, and the mammals and *Oligokyphus*, on the other, probably divided up the day as lizards and small mammals do today – the reptiles being diurnal, and the mammals and *Oligokyphus* being nocturnal or crepuscular. Evans (1983) noted skeletal injuries in *Gephyrosaurus* consistent with intraspecific aggression and suggested that *Gephyrosaurus* was a territorial, sit-and-wait feeder, like many modern diurnal lizards (e.g., *Anolis*). That may also have increased its visibility, and therefore vulnerability, to predators.

St. Bride's Island was small, and many comparable modern islands (e.g., the Lesser Antilles) have only small animals among their native fauna. This may be the reason we find only small vertebrates on St. Bride's. However, if the fissure assemblages were a coprocoenosis (sensu Mellett, 1974), then predator selection, as well as the sizes of the fissures, also played a part. The largest animals are found in Pant 4: the large morganucodontid (the size of a small mustelid), a theropod (perhaps 0.5 m in height, based on tooth size), and *Oligokyphus* (rabbit-sized).

Significance of the British Early Jurassic assemblages

The British Early Jurassic fissure material, like that of the Late Triassic, provides a clear picture of small-vertebrate evolution at that important time. The value of this material lies in the contribution it makes not only to our knowledge of individual lineages but also to our knowledge of the evolution of character complexes and our understanding of biological diversity and paleoenvironment at that particular period in the fossil record.

The material from the fissures is dissociated, but the bones are often exquisitely preserved. This factor, in combination with the limited nature of the typical *Hirmeriella* association, is significant. It has permitted detailed reconstruction of skeletal structure.

Morganucodon was not the first early mammal to be found. However, the sheer numbers and the excellent preservation of the Welsh fissure fossils have permitted *Morganucodon* to be described and reconstructed in greater detail than any other early mammal (Figure 15.5A,B). In many ways, *Morganucodon* provides the perfect structural intermediate between advanced theriodonts and true mammals. The material from the Welsh quarries has made possible detailed descriptions of key regions, such as the dentition, the braincase, and the middle-ear/jaw-joint system (Kermack, Mussett, and Rigney, 1973, 1981). The postcranial skeleton has also been described in detail (Jenkins and Parrington, 1976). Additional material of *Morganucodon* is known from the Norian of Switzerland (Hallau) (Clemens, 1986) and the Early Jurassic of Yunnan, China (Rigney, 1963; Yang, 1982; Luo, Chapter 6) and North America (Jenkins, Crompton, and Downs, 1983), and further morganucodontids have been recovered from the Early Jurassic of South Africa (*Megazostrodon*) (Crompton, 1974) and the Late Triassic of Europe (e.g., at Saint-Nicolas-de-Port, France) (Sigogneau-Russell, 1983c).

Kuehneotherium, unfortunately, is known only from its dentition and jaw fragments (Kermack et al., 1968). Before its discovery, the Middle Jurassic Stonesfield Slate mammal *Amphitherium* was generally regarded as the archetypal early therian, the perfect structural ancestor for all later forms (Simpson, 1928). *Kuehneotherium* provides a more ancient, and even more simplified, therian dentition (Figure 15.6) and shows that therian evolution had begun much earlier than was previously imagined. *Kuehneotherium* has subsequently been found at Emborough Quarry in the Mendips (Fraser et al., 1985), and kuehneotheriids are now known to range from the Late Triassic of Britain (Emborough) and France (Saint-Nicolas-de-Port) (Sigogneau-Russell, 1978, 1983a–c) to the Middle

Jurassic of Britain (Kirtlington) (Freeman, 1976, 1979), and they have been recovered from the Lower Jurassic Kota Formation of India (Datta, Yadagiri, and Rao, 1978).

Gephyrosaurus was the first of the fissure reptiles to be described in detail (Evans, 1980, 1981). In the characters of its skull, dentition, and postcranial skeleton (Evans, 1980, 1981), *Gephyrosaurus* is far more "lizard-like" than the supposed early lizard *Kuehneosaurus*; only in the structure and articulation of its quadrate is *Gephyrosaurus* more primitive. It has become clear that most of the lizardlike features of the kuehneosaurid skull were convergent developments.

Placing *Gephyrosaurus* taxonomically proved more difficult. It shares several derived characters with sphenodontians, but lacks the characteristic acrodont dentition of that group. The key to the solution of the problem was provided by a new reptile, *Diphydontosaurus*, from Tytherington and Cromhall (Whiteside, 1986). *Diphydontosaurus* resembles *Gephyrosaurus* in many respects, but with one important difference: Although the anterior teeth are pleurodont, the posterior teeth have an acrodont implantation. *Diphydontosaurus* thus provides the link between *Gephyrosaurus* and typical sphenodontians, and *Gephyrosaurus* is now considered an early member of that group. *Gephyrosaurus* has also been reported from Pant-y-ffynon (Crush, 1981), and a possible gephyrosaurid is present at Holwell (C. J. Duffin, pers. commun.).

The Windsor Hill material of *Oligokyphus* enabled Kühne to make the first detailed description of the skull and postcranial skeleton of a tritylodontid and to demonstrate that the tritylodontids were not mammals (Kühne, 1947a, 1956).

Little can be said at this stage about the significance of the Pant 4 fauna, because, unfortunately, it has yet to be formally described. Its chief contribution, so far, has been to show that the typical *Hirmeriella* association is not representative and that the fauna of St. Bride's Island was far more complex than was originally believed. Once described, the assemblage will provide new information about the diversification of sphenodontians and early mammals.

Conclusions

The British Early Jurassic fissures at Windsor Hill and in the quarries of St. Bride's Island have yielded a rich assemblage of small vertebrates, most notably synapsids. The nature of the fossil material and the limited numbers of species present in most of the fissure assemblages have permitted detailed analyses of the structures and relationships of these genera and of the faunas to which they belong.

The Early Jurassic fissure assemblages differ in their constituent genera from those of the Late Triassic, but the lesson from the quarry at Pant 4 is that fissure assemblages are not necessarily fully representative of the contemporaneous natural fauna. More recent finds in France (Saint-Nicolas-de-Port) (Sigogneau-Russell, 1978, 1983a–c; Sigogneau-Russell, Frank, and Hemmerlé, 1986) and Switzerland (Hallau) (Peyer, 1956) indicate that a varied mammalian fauna existed in Europe by Late Triassic times, and the teeth of *Kuehneotherium* from Emborough are evidence that mammals lived in the Bristol Channel archipelago in the Late Triassic. It may well be that many of the differences between the British fissure assemblages of the Late Triassic and Early Jurassic are artifacts of preservation and sampling (e.g., predator selection).

Acknowledgments

Dr. Doris Kermack, Patricia Lees, and Frances Mussett were part of the active University College London (UCL) team that has surveyed and studied the Welsh fissures and their early Mesozoic vertebrate assemblages for more than twenty years. They must take a large share of the credit for the results obtained. Our thanks are also due to Drs. H. Ivimey-Cook and G. Warrington (British Geological Survey) for advice on the gastropods and palynomorphs, respectively, and for their critical review of this manuscript. We would like to take this opportunity to acknowledge the unique contribution made by Professor W. G. Kühne to the discovery and study of Mesozoic microvertebrates. Regrettably, Walter Kühne died in 1992. A grant from the Royal Society permitted one of us (SEE) to participate in the Virginia meeting, and we would like to thank Drs. Nick Fraser and Hans-Dieter Sues for their invitation.

References

Ager, D. V., and D. Edwards. 1985. The fauna and flora of the Rhaetian of South Wales and adjacent areas. *Nature in Wales* 4: 71–79.

Alvin, K. L. 1982. Cheirolepidiaceae: biology, structure and paleoecology. *Review of Palaeobotany and Palynology* 37: 71–98.

Benton, M. J., and J. M. Clark. 1988. Archosaur phylogeny and the relationships of the Crocodylia. Pp. 295–338 in M. J. Benton (ed.), *The Phylogeny and Systematics of the Tetrapods. Vol. 1: Amphibians, Reptiles and Birds.* Oxford: Clarendon Press.

Clemens, W. A. 1986. On Triassic and Jurassic mammals. Pp. 237–246 in K. Padian (ed.), *The Beginning of the Age of Dinosaurs: Faunal Change across the Triassic–Jurassic Boundary.* Cambridge University Press.

Clemens, W. A., J. A. Lillegraven, E. H. Lindsay, and G. G. Simpson. 1979. Where, when and what – a survey of known Mesozoic mammal distribution. Pp. 7–58 in J. A. Lillegraven, Z. Kielan-Jaworowska, and W. A. Clemens (eds.), *Mesozoic Mammals: The First Two-thirds of Mammalian Evolution.* Berkeley: University of California Press.

Cope, J. C. W., K. L. Duff, C. E. Parsons, H. S. Torrens, W. A. Wimbledon, and J. Wright. 1980. A correlation of Jurassic rocks in the British Isles. Pt. 1: Introduction and Lower Jurassic. *Geological Society Special Report* 14: 1–73.

Crompton, A. W. 1974. The dentitions and relationships of the southern African Triassic mammals, *Erythrotherium parringtoni* and *Megazostrodon rudnerae*. *Bulletin of the British Museum (Natural History)*, *Geology* 24: 397–437.

Crush, P. J. 1981. An early terrestrial crocodile from South Wales. Ph. D. thesis, University of London.

——— 1984. A late upper Triassic sphenosuchid crocodilian from Wales. *Palaeontology* 27: 131–157.

Datta, P. M., P. Yadagiri, and B. R. J. Rao. 1978. Discovery of Early Jurassic micromammals from the Upper Gondwana sequence of Pranhita Godavari Valley, India. *Journal of the Geological Society of India* 19: 64–68.

Dodson, P., and D. Wexlar. 1979. Taphonomic investigations of owl pellets. *Paleobiology* 5: 275–284.

Evans, S. E. 1980. The skull of a new eosuchian reptile from the Lower Jurassic of South Wales. *Zoological Journal of the Linnean Society* 70: 203–264.

——— 1981. The postcranial skeleton of the Lower Jurassic eosuchian *Gephyrosaurus bridensis*. *Zoological Journal of the Linnean Society* 73: 81–116.

——— 1983. Mandibular fracture and inferred behavior in a fossil reptile. *Copeia* 1983: 845–847.

Fisher, D. C. 1981a. Taphonomic interpretation of enamel-less teeth in the Shotgun local fauna (Paleocene, Wyoming). *Contributions from the Museum of Paleontology, University of Michigan* 25: 259–275.

——— 1981b. Crocodilian scatology, microvertebrate concentrations, and enamel-less teeth. *Paleobiology* 7: 262–275.

Forster, S. C., and G. Warrington. 1985. Geochronology of the Carboniferous, Permian and Triassic. Pp. 99–113 in N. J. Snelling (ed.), *The Chronology of the Geological Record*. Memoir 10, The Geological Society. Oxford: Blackwell Scientific.

Fraser, N. C. 1986. Terrestrial vertebrates at the Triassic–Jurassic boundary in South West Britain. *Modern Geology* 10: 147–157.

——— 1988a. The osteology and relationships of *Clevosaurus* (Reptilia: Sphenodontida). *Philosophical Transactions of the Royal Society of London* B321: 125–178.

——— 1988b. Latest Triassic terrestrial vertebrates and their biostratigraphy. *Modern Geology* 13: 125–140.

——— 1989. A new pterosaur record from a Lower Jurassic fissure deposit in South Wales. *Neues Jahrbuch für Geologie und Paläontologie, Monatshefte* 1989: 129–135.

Fraser, N. C., G. M. Walkden, and V. Stewart. 1985. The first pre-Rhaetic therian mammal. *Nature* 314: 161–163.

Freeman, E. F. 1976. Mammal teeth from the Forest Marble (Middle Jurassic) of Oxfordshire, England. *Science* 194: 1053–1055.

——— 1979. A Middle Jurassic mammal bed from Oxfordshire. *Palaeontology* 22: 135–166.

Hallam, A., J. M. Hancock, J. L. LaBrecque, W. Lowrie, and J. E. T. Channell. 1985. Jurassic to Paleogene. Part

I: Jurassic and Cretaceous geochronology and Jurassic to Paleogene magnetostratigraphy. Pp. 118–140 in N. J. Snelling (ed.), *The Chronology of the Geological Record*. Memoir 10, The Geological Society. Oxford: Blackwell Scientific.

Harris, T. M. 1937. The fossil flora of Scoresby Sound, East Greenland. *Meddelelser om Grønland* 112: 1–114.

——— 1957. A Liasso-Rhaetic flora in South Wales. *Proceedings of the Royal Society of London* B147: 289–308.

——— 1958. Forest fire in the Mesozoic. *Journal of Ecology* 46: 447–453.

Jenkins, F. A., Jr., A. W. Crompton, and W. R. Downs. 1983. Mesozoic mammals from Arizona: new evidence on mammalian evolution. *Science* 222: 1233–1235.

Jenkins, F. A., Jr., and F. R. Parrington. 1976. The postcranial skeletons of the Triassic mammals *Eozostrodon*, *Megazostrodon* and *Erythrotherium*. *Philosophical Transactions of the Royal Society of London* B273: 387–431.

Jung, W. 1968. *Hirmeriella münsteri* (Schenk) Jung *nov. comb.*, eine bedeutsame Konifere des Mesozoikums. *Palaeontographica* B122: 55–93.

Kermack, D. 1984. New prosauropod material from South Wales. *Zoological Journal of the Linnean Society* 82: 101–117.

Kermack, D. M., K. A. Kermack, and F. Mussett. 1968. The Welsh pantothere *Kuehneotherium praecursoris*. *Journal of the Linnean Society (Zoology)* 47: 407–423.

Kermack, K. A., F. Mussett, and H. W. Rigney. 1973. The lower jaw of *Morganucodon*. *Zoological Journal of the Linnean Society* 53:87–175.

——— 1981. The skull of *Morganucodon*. *Zoological Journal of the Linnean Society* 71: 1–158.

Kretzoi, M. 1960. Zur Benennung des ältesten Symmetrodonten. *Vertebrata Hungarica* 2: 307–309.

Kühne, W. G. 1947a. The tritylodont reptile *Oligokyphus*. Doctoral dissertation, Rheinische Friedrich-Wilhelms-Universität, Bonn.

——— 1947b. The geology of fissure-filling "Holwell 2"; the age-determination of the mammalian teeth therein; and a report on the technique employed when collecting the teeth of *Eozostrodon* and Microcleptidae. *Proceedings of the Zoological Society of London* 116: 729–733.

——— 1949. On a triconodont tooth of a new pattern from a fissure-filling in South Glamorgan. *Proceedings of the Zoological Society of London* 119: 345–350.

——— 1956. *The Liassic Therapsid Oligokyphus*. London: Trustees of the British Museum.

——— 1958. Rhaetische Triconodonten aus Glamorgan, ihre Stellung zwischen den Klassen Reptilia und Mammalia und ihre Bedeutung für die Reichart'sche Theorie. *Paläontologische Zeitschrift* 32: 197–235.

Lewarne, G., and J. M. Pallot. 1957. Mesozoic plants from fissures in the Carboniferous Limestone of South Wales. *Annals and Magazine of Natural History*, (12) 10: 72–79.

Marshall, J. E. A, and D. I. Whiteside. 1980. Marine influence in the Triassic "uplands." *Nature* 287: 627–628.

Mayhew, D. F. 1977. Avian predators as accumulators of fossil mammal material. *Boreas* 6: 25–31.

Mellett, J. S. 1974. Scatological origin of microvertebrate fossil accumulations. *Science* 185: 349–350.

Mills, J. R. E. 1984. The molar dentition of a Welsh pantothere. *Zoological Journal of the Linnean Society* 82: 189–205.

Moore, C. 1867. On the abnormal condition of Secondary deposits. *Quarterly Journal of the Geological Society* 23: 449–568.

Morbey, S. J. 1975. The palynostratigraphy of the Rhaetian stage, Upper Triassic, in the Kendelbachgraben, Austria. *Palaeontographica* B152: 1–75.

Muir, M., and J. H. A. van Konijnenburg-van Cittert. 1970. A Rhaeto-Liassic flora from Airel, Northern France. *Palaeontology* 13: 433–442.

Olson, S. L., and G. K. Pregill. 1982. Introduction to the paleontology of Bahamanian vertebrates. Pp. 1–7 in S. L. Olson (ed.), *Fossil vertebrates from the Bahamas. Smithsonian Contributions to Paleobiology* 48.

Orbell, G. 1973. Palynology of the British Rhaeto-Liassic. *Bulletin of the Geological Survey of Britain* 44: 1–44.

Pacey, D. E. 1978. On a tetrapod assemblage from a Mesozoic fissure filling in South Wales. Ph.D. thesis, University of London.

Parrington, F. R. 1941. On two mammalian teeth from the Lower Rhaetic of Somerset. *Annals and Magazine of Natural History* (11) 8: 140–144.

Peyer, B. 1956. Ueber Zahne von Haramiyden, von Triconodonten und von wahrscheinlich synapsiden Reptilien aus dem Rhät von Hallau, Kt Schaffhausen, Schweiz. *Schweizerische Paläontologische Abhandlungen* 72: 1–72.

Pregill, G. K. 1982. Fossil amphibians and reptiles from New Providence, Bahamas. Pp. 8–21 in S. L. Olson (ed.), *Fossil vertebrates from the Bahamas. Smithsonian Contributions to Paleobiology* 48.

Pregill, G. K., D. W. Steadman, S. L. Olson, and F. V. Grady. 1988. Late Holocene fossil vertebrates from Burma Quarry, Antigua, Lesser Antilles. *Smithsonian Contributions to Zoology* 463: 1–27.

Rigney, H. W. 1963. A specimen of *Morganucodon* from Yunnan. *Nature* 197: 1122–1123.

Robinson, P. L. 1957. The Mesozoic fissures of the Bristol Channel area and their vertebrate faunas. *Journal of the Linnean Society (Zoology)* 43: 260–282.

——— 1971. A problem of faunal replacement on Permo-Triassic continents. *Palaeontology* 14: 131–153.

Sigogneau-Russell, D. 1978. Découverte de mammifères rhétiens (Trias supérieur) dans l'est de la France. *Comptes Rendus d'Académie des Sciences, Paris* 287: 991–993.

——— 1983a. A new therian mammal from the Rhaetic locality of St-Nicolas-de-Port (France). *Zoological Journal of the Linnean Society* 78: 175–186.

——— 1983b. Characteristiques de la faune mammalienne du Rhétien de Saint-Nicolas-de-Port (Meurthe-et-Moselle). *Bulletin d'Information des Géologues du Bassin de Paris* 20: 51–53.

——— 1983c. Nouveaux taxons de mammiferes rhetiel *Palaeontologica Polonica* 28: 233–249.

Sigogneau Russell, D., R. M. Frank, and J. Hemmerlé. 1986. A new family of mammals from the lower part of the French Rhaetic. Pp. 99–108 in K. Padian (ed.), *The Beginning of the Age of Dinosaurs: Faunal Change across the Triassic–Jurassic Boundary.* Cambridge University Press.

Simpson, G. G. 1928. *A Catalogue of the Mesozoic Mammalia in the Geological Department of the British Museum.* London: Trustees of the British Museum.

Strahan, A., and R. H. Tiddeman. 1904. Rhaetic–Western part. Pp. 99–102 in A. Strahan and T. C. Cantrill (eds.), *The Geology of the South Wales Coal-field. Part 6: The Country around Bridgend. Memoirs of the Geological Survey* 261–262.

Trueman, A. E. 1922. The Liassic rocks of Glamorgan. *Proceedings of the Geologists' Association* 33: 245–284.

Tucker, M. E. 1977. The marginal Triassic deposits of South Wales: continental facies and palaeogeography. *Geological Journal* 12: 169–188.

Tucker, M. E., and T. P. Burchette. 1977. Triassic dinosaur footprints from South Wales: their context and preservation. *Palaeogeography, Palaeoclimatology, Palaeoecology* 22: 195–208.

Warrener, D. 1983. An archosaurian fauna from a Welsh locality. Ph.D. thesis, University of London.

Warrington, G., M. G. Audley-Charles, R. E. Elliott, W. B. Evans, H. C. Ivimey-Cook, P. E. Kent, P. L. Robinson, F. W. Shotton, and F. M. Taylor. 1980. A correlation of Triassic rocks in the British Isles. *Geological Society Special Report* 13: 1–78.

Warrington, G, and H. C. Ivimey-Cook. 1990. Biostratigraphy of the Late Triassic and Early Jurassic: a review of type sections in Southern Britain. *Cahiers d'Université Catholique de Lyon, Sér. Sci.* 3: 207–213.

Waters, R. A., and D. J. D. Lawrence. 1987. *Geology of the South Wales Coalfield. Part III: The Country around Cardiff. Memoir for 1:50 000 Geological Sheet 263 (England and Wales).* British Geological Survey, HMSO, London.

Watson, J. 1988. The Cheirolepidaceae. Pp. 382–447 in C. B. Beck (ed.), *Origin and Evolution of Gymnosperms.* New York: Columbia University Press.

Whiteside, D. I. 1986. The head skeleton of the Rhaetian sphenodontid *Diphydontosaurus avonis* gen. et sp. nov. and the modernizing of a living fossil. *Philosophical Transactions of the Royal Society of London* B312: 379–430.

Yang, Z. (Young, C. C.). 1982. [Two primitive mammals from Lufeng, Yunnan.] Pp. 21–24 in [*Selected Works of Yang Zhungjian.*] Beijing: Science Press. [in Chinese]

16

A review of the Early Jurassic tetrapods from the Glen Canyon Group of the American Southwest

HANS-DIETER SUES, JAMES M. CLARK, AND
FARISH A. JENKINS, JR.

Introduction

Systematic paleontological exploration of the continental sedimentary rocks of the Glen Canyon Group has commenced only quite recently because its exposures are, for the most part, located in rugged, relatively inaccessible terrain. Nevertheless, its assemblages of fossil vertebrates have rapidly become the best-known assemblages (in terms of taxonomic diversity, quantity of identifiable material, and especially number of articulated skeletons) from the Lower Jurassic of North America.

Initial collection of fossil tetrapods from Glen Canyon Group strata resulted not from organized scientific efforts but rather from the serendipitous placement of a replica of a gold mine that was erected in 1929 during the production of a movie, *The Painted Desert* (1931), in which Clark Gable made his talking-film debut. During or soon after the filming, a local Navajo resident of the Goldtooth family discovered a few pieces of fossil bone, and these proved sufficient to attract Barnum Brown of the American Museum of Natural History to the cliff-forming Dinosaur Canyon Member of the Moenave Formation at the remote site east of Cameron, Arizona. Two weeks of collecting in 1931 yielded one of the most nearly complete skeletons of a Mesozoic crocodylomorph ever found, the holotype of *Protosuchus richardsoni* (Brown, 1933).

Fossil finds made by Navajo residents who subsequently led paleontologists to the sites became a pattern repeated on several occasions in the ensuing years. Such discoveries include the unique partial skeleton of the small ceratosaurian theropod *Segisaurus halli* Camp, 1936 from the Navajo Sandstone, a mass burial of skeletons of the tritylodontid *Kayentatherium wellesi* ["*Nearctylodon broomi*"] from the upper portion of the Kayenta Formation (Lewis, 1958, 1986), and several skeletons of the large ceratosaurian theropod *Dilophosaurus wetherilli* (Welles, 1954, 1984).

Fossil collecting in the Glen Canyon Group during the 1940s and 1960s was pursued mainly by field-parties under the direction of C. L. Camp and S. P. Welles from the University of California Museum of Paleontology and was concentrated in the vicinity of Tuba City, Arizona. The vast exposures of the "silty facies" of the Kayenta Formation to the southeast of Tuba City received little attention until 1971, when D. Lawler of the Museum of Northern Arizona discovered the holotype of the small armored ornithischian dinosaur *Scutellosaurus lawleri* Colbert, 1981. The abundance of fossil vertebrate remains in the "silty facies" became apparent only as a result of systematic exploration by F. A. Jenkins, Jr., and his associates from the Museum of Comparative Zoology of Harvard University (MCZ) and the Museum of Northern Arizona from 1977 to 1983. This work culminated in the discovery of a locality very rich in small tetrapods at Gold Spring ("Quarry 1"), which has yielded many at least partially articulated skulls and skeletons (e.g., Jenkins, Crompton, and Downs, 1983; Sues, 1985). Additional collections from the Adeii Eechii Cliffs were made by personnel from the University of California Museum of Paleontology (UCMP) in 1981 and 1983.

In this chapter we present an overview of the diversity and geological context of the tetrapods described to date and compare the assemblages from the Glen Canyon Group to those from other Lower Jurassic continental strata.

Geological background

The Glen Canyon Group is exposed on the Colorado Plateau in the area of the present-day "four corners" states (Arizona, Colorado, New Mexico, and Utah). As is the case for most fluviolacustrine and eolian sequences, the lithological relationships within and among its component formations are varied and complex. The

Glen Canyon Group comprises four formations (in stratigraphic order from oldest to youngest): Wingate Sandstone, Moenave Formation, Kayenta Formation, and Navajo Sandstone. These units are lithologically distinct but reportedly intertongue at their mutual contacts, and they change markedly over their geographic ranges (Harshbarger, Repenning, and Irwin, 1957). Fossils of small tetrapods have been recovered from all these formations, with the exception of the Wingate Sandstone, but most specimens found to date come from strata of the Kayenta Formation, especially the "silty facies" (sensu Harshbarger et al., 1957) exposed along the Adeii Eechii Cliffs on Ward Terrace on lands of the Navajo Nation in northeastern Arizona (Figure 16.1). Clark and Fastovsky (1986) published a more detailed treatment of the geological context of the major localities for vertebrate fossils.

The lower stratigraphic boundary of the Glen Canyon Group has long been uncertain. Most recent studies recognize a widespread unconformity between the cliff-forming Wingate Sandstone and the underlying Chinle Formation (Pipiringos and O'Sullivan, 1978; Peterson and Pipiringos, 1979). The same unconformity separates the Rock Point Member of the Wingate Sandstone from the overlying Lukachukai Member. Unlike earlier authors (e.g., Harshbarger et al., 1957), Peterson and Pipiringos (1979) thus

placed this unit in the Chinle Formation and restricted the Wingate Sandstone to the Lukachukai Member. The Moenave Formation overlies the Wingate Sandstone and comprises three members, two of which, the Whitmore Point and Dinosaur Canyon members, have yielded fossils of small tetrapods. The geographically restricted Whitmore Point Member (Wilson, 1967) consists of less than 40 m of thinly bedded lacustrine deposits, with abundant semionotid fishes; a few tetrapod remains have been collected from this unit (Clark and Fastovsky, 1986), but they have not yet been studied.

The Dinosaur Canyon Member of the Moenave Formation is characterized by bright orange, cliff-forming sandstones and includes both eolian and fluvial deposits (Harshbarger et al., 1957). Edwards (1985) interpreted these strata as having been deposited in an inland desert that underwent occasional flooding. The most common fossil vertebrate from the Dinosaur Canyon Member is *Protosuchus richardsoni*; remains of at least eight individuals have been collected to date (Clark, 1986, in press). Until recently, this was the only taxon represented by skeletal material from this unit although a Newark Type 3 assemblage of tetrapod ichnotaxa (sensu Olsen and Galton, 1977) has been documented from several localities. Ongoing work under the direction of M. Morales from the Museum of

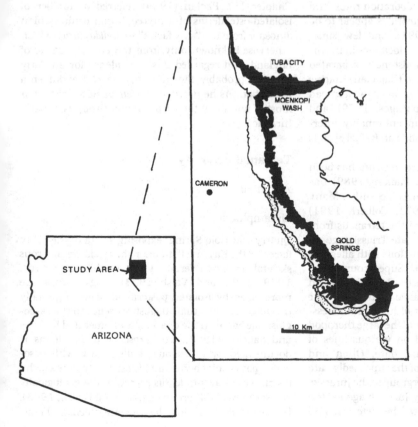

Figure 16.1. Map of the Adeii Eechii Cliffs in northeastern Arizona, showing distribution of Glen Canyon Group strata in this region. Black area denotes Kayenta Formation, and lightly stippled area denotes the Lukachukai Member of the Wingate Sandstone; the unshaded region between these two areas represents the Moenave Formation. (Adapted from Cooley et al., 1969.)

Northern Arizona has recently resulted in the discovery of a partial skeleton of a theropod dinosaur (M. Morales and E. H. Colbert, pers. commun.).

Following Harshbarger et al. (1957), the Kayenta Formation is informally divided into a northern "typical facies," which is dominated by sandstones, and a southern "silty facies," which mainly comprises mudstones and siltstones. These facies were derived from two different source areas, the "typical facies" from the Uncompahgre Uplift in Colorado, and the "silty facies" from the Cordilleran magmatic arc to the southeast (Luttrell, 1987). The "silty facies" differs from the "typical facies" mainly in preserving extensive overbank deposits and occasional eolian sediments and in having smaller channel systems, indicating that it was deposited under more arid conditions (Luttrell, 1987). Most fossils recovered to date have come from the "silty facies," which forms extensive exposures along the Adeii Eechii cliffs on Ward Terrace.

The Navajo Sandstone overlies the Kayenta Formation and locally intertongues with it (Middleton and Blakey, 1983; Luttrell, 1987). It comprises the deposits of an erg that extended over the entire Colorado Plateau (Blakey, Peterson, and Kocurek 1988); correlative eolian sandstones occur in southern Arizona, southern Nevada, and southeastern California. This erg formed under climatic conditions that presumably were comparable to those of the present-day Sahara Desert, with little rainfall, high evaporation rates, and frequent strong winds. Fossil vertebrates appear to be uncommon (Winkler et al., 1991), and few small tetrapods have been found, all from localities in northeastern Arizona. The Navajo Sandstone is separated from the overlying Middle Jurassic (Bajocian) marine deposits of the San Rafael Group by a widespread unconformity (Peterson and Pipiringos, 1979), and the Temple Cap Sandstone is unconformably interposed between the Glen Canyon and San Rafael groups in southern Utah.

The age of the Glen Canyon Group strata has been the subject of much discussion. Padian (1989) has reviewed the pertinent data most recently. Many earlier studies (e.g., Galton, 1971; Colbert, 1981) concluded that most (if not all) of the tetrapods from the Glen Canyon Group were of Late Triassic age on the basis of biostratigraphic correlations with allegedly Late Triassic strata of the Newark Supergroup in the Hartford basin of Connecticut. Welles (1954) suggested an Early or even Middle Jurassic dating for the Kayenta Formation on the basis of his initial assessment of the phylogenetic affinities of the large theropod "Megalosaurus" wetherilli. Based on various lines of evidence, especially palynological data, Olsen and Galton (1977) reassigned much of the supposedly Late Triassic strata of the Newark Supergroup to the Jurassic, and they also suggested an Early Jurassic age for the Glen Canyon Group. Cornet, cited by Peterson and

Pipiringos (1979), recovered well-preserved palynoflorules from the lower part of the Whitmore Point Member of the Moenave Formation at Whitmore Point, Arizona, and dated them as Early Jurassic, based on the extreme abundance of species of *Corollina* (especially *C. torosus*), which make up about 95–99 percent of the samples. More specifically, he proposed correlating the Whitmore Point Member with the "upper-lower to lower-middle part" of the Portland Formation of the Hartford basin, which he considered late Sinemurian to early Pliensbachian in age (Cornet and Traverse, 1975; Litwin, 1986). Consequently, the Kayenta Formation, which overlies the Moenave Formation, could not be older than late Sinemurian. Both Sues (1985, 1986c) and Padian (1989) interpreted fossil tetrapods as providing additional evidence of an Early Jurassic age for this formation. Sues cited the presence of the tritylodontid *Oligokyphus* in the "silty facies" of the Kayenta Formation as an indicator that these strata were time-equivalent to Early Jurassic occurrences of *Oligokyphus* in Europe. Although Padian (1989) questioned this line of evidence because he deemed the stratigraphic position of the records of *Oligokyphus* from southern Germany to be too uncertain, an Early Jurassic age for the abundant English material referable to this genus is firmly established by its association with marine invertebrates in Neptunian dykes (Kühne, 1956; Sues, 1985; Evans and Kermack, Chapter 15). Padian (1989) referred a number of isolated osteoderms of a thyreophoran ornithischian dinosaur from the "silty facies" to *Scelidosaurus*, which otherwise is known only from the Lower Jurassic of England, and regarded this as evidence for an Early Jurassic, probably Sinemurian, age of the Kayenta Formation. As he noted, a consensus now exists that most, if not all, of the Glen Canyon Group is of Early Jurassic age.

Tetrapod diversity

Amphibia

Lissamphibia

Quarry 1 at Gold Spring, exposing strata of the "silty facies" of the Kayenta Formation, has yielded numerous skeletal remains referable to lissamphibians (Curtis, 1989; Jenkins and Walsh, 1993). Frogs (Anura) are represented by isolated postcranial bones, especially the diagnostic ilia, that suggest reference to the Discoglossidae based on the characters enumerated by Estes and Sanchíz (1982). The presence of caecilians is documented by numerous skulls, many with associated postcranial bones, and is especially noteworthy because the Kayenta fossils represent the geologically oldest record of this group (Jenkins and Walsh, 1993). Unlike extant caecilians, the Kayenta caecilian, *Eocae-*

ciliu micropodia, retains reduced girdle and limb elements. The skull of this form, however, already shows many apomorphies of the group, including the presence of a tentacular fossa along the anterior margin of the orbit, an os basale (comprising the fused exoccipitals, otic capsules, and parasphenoid), and two upper rows of pedicellate teeth. The orbit is larger than in extant caecilians. The alleged presence of salamanders (Curtis, 1989) is questionable, because this identification is based mostly on isolated atlantal vertebrae; the supposedly diagnostic interglenoid tubercle also occurs in caecilians (Jenkins and Walsh, 1993).

Amniota

Testudinata

One of the most common fossil tetrapods in the "silty facies" of the Kayenta Formation is the turtle *Kayentachelys aprix* Gaffney, Hutchinson, Jenkins, and Meeker, 1987. It has the diagnostic cryptodiran trochlear mechanism, but retains pterygoid teeth and an interpterygoid vacuity (Gaffney et al., 1987; Gaffney and Meylan, 1988). *Kayentachelys* was interpreted as

the earliest known cryptodiran by Gaffney et al. (1987). Gauthier et al. (1989), however, considered it to have diverged prior to the origin of extant turtle groups, and the cryptodiran type of trochlear system thus would be diagnostic for a more inclusive group containing all turtles aside from the Late Triassic *Proganochelys*. The shell of *Kayentachelys* is low and smooth, with tapered edges, features regarded as indicative of an aquatic mode of life (Gaffney et al., 1987). Turtles are not known from the "typical facies" of the Kayenta Formation, and only a few indeterminate shell fragments are known from the northern part of the "silty facies" near Moenkopi Wash.

Diapsida

Lepidosauria

Sphenodontia

Sphenodontians are documented by an articulated skeleton and numerous isolated jaw fragments from Quarry 1 at Gold Spring (Meszoely, Jenkins, and Schaff 1987) and by jaw fragments from the *Eopneumatosuchus*

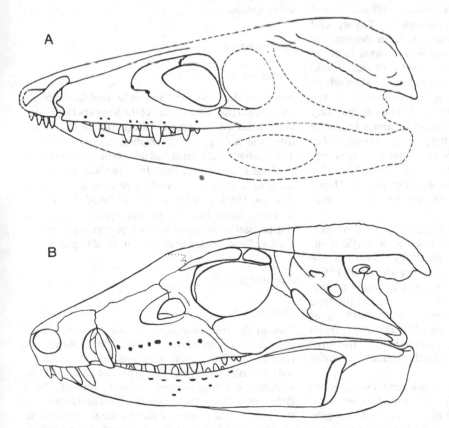

Figure 16.2. (A) Skull of an unnamed sphenosuchian (UCMP 131830) from the Kayenta Formation; reconstruction in left lateral view. (B) Skull of *Protosuchus richardsoni* (Brown, 1933); reconstruction in left lateral view. (Adapted from Clark, 1986.)

locality (Curtis, 1989). No anatomical information is yet available to assess the affinities of this material.

Diapsida

Archosauria

Crocodylomorpha

The Crocodylomorpha of the Glen Canyon Group comprise two named forms, *Protosuchus richardsoni* (Brown, 1933) from the Moenave Formation and *Eopneumatosuchus colberti* Crompton and Smith, 1980 from the Kayenta Formation, and at least four as yet unnamed taxa (Clark, 1986, and Chapter 5). The unnamed forms include a "sphenosuchian"-grade taxon and three "protosuchian"-grade species from the Kayenta Formation (Clark, 1986, in press).

Protosuchus richardsoni is known only from the Dinosaur Canyon Member of the Moenave Formation. Colbert and Mook (1951) studied its skeleton in detail, and Crompton and Smith (1980) briefly described a more recently discovered skull. Clark (1986, in press) presented a detailed anatomical account of the skull of *P. richardsoni* (Figure 16.2B) and synonymized the southern African genera *Lesothosuchus* Whetstone and Whybrow, 1983 and *Baroqueosuchus* Busbey and Gow, 1984 with *Protosuchus* (Clark, in Benton and Clark, 1988). *Protosuchus* was also reported from the Kayenta Formation (Lewis, 1958) and the Navajo Sandstone (Galton, 1971), but the material on which those reports were based is not diagnostic at the generic level. A dentary (MCZ 8813) from the "silty facies" of the Kayenta Formation is similar to that of *P. richardsoni* but differs in the presence of two, rather than one, caniniform teeth; the latter feature is also present in a new species of *Protosuchus* from the McCoy Brook Formation (Lower Jurassic: Hettangian) of Nova Scotia (H.-D. Sues, N. H. Shubin, and P. E. Olsen, unpublished data).

Eopneumatosuchus colberti was described by Crompton and Smith (1980) on the basis of a single well-preserved braincase from the "silty facies" of the Kayenta Formation; see also Busbey and Gow (1984) and Clark (1986). Other material referable to this species includes several isolated bones from the type locality and a partial skeleton (MCZ 8895) from another site in the "silty facies." If these specimens indeed represent *E. colberti*, then the ilium of this form lacks an anterior process, as in the Mesoeucrocodylia (Clark, 1986).

The "sphenosuchian" is known from a single articulated skeleton (UCMP 131830) that has yet to be completely prepared. Its skull (Figure 16.2A) is generally similar to that of *Sphenosuchus acutus* from the upper Elliot Formation of South Africa (Walker, 1990), but unlike the condition in the latter, the squamosal

bears a groove along its lateral edge similar to a groove on the squamosal of crocodyliforms to which the dorsal ear flap attaches. The femur of the new form is unusual in its possession of a well-developed, inturned proximal head and resembles that of *Hallopus victor* from the Middle Jurassic lower Ralston Creek Formation of Colorado (Walker, 1970; Norell and Storrs, 1989).

Two of the three as yet unnamed "protosuchians" are distinguished by the presence of bulbous, crushing postcaniniform teeth that, unlike the teeth of extant Crocodylia, are multicuspid (Figure 16.3). These two species are similar to *Edentosuchus tienshanensis* Young, 1973 from the Lower Cretaceous of Xinjiang, China (Li, 1985), but are clearly closely related to *Protosuchus* (Clark, 1986, and Chapter 5). Specimens of these two taxa were obtained from neighboring localities at different stratigraphic levels. The two species differ in the complexity of the cusp pattern on the postcaniniforms, the species from the higher horizon having more complex teeth. The third unnamed species is less specialized but is poorly known. Its braincase lacks the expanded pneumatic spaces present in the braincases of the other two "protosuchians" and in *Protosuchus*. Its skull is generally more similar to that of *Orthosuchus stormbergi*, but it lacks the reduced dentition of the latter species.

Archosauria

Pterosauria

A crushed skull fragment and an isolated fourth metacarpal from the "silty facies" of the Kayenta Formation were described by Padian (1984). The former constitutes the holotype of *Rhamphinion jenkinsi* Padian, 1984, which is differentiated from other pterosaurs on the basis of its "nearly rounded antorbital and orbital margins next to the ascending process of the jugal" (Padian, 1984, p. 408). In the published illustration, however, these margins do not appear unusual in comparison with those in other pterosaurian genera such as *Dimorphodon* from the Liassic of England.

Archosauria

Dinosauria

Two small ceratosaurian theropods are known from the Glen Canyon Group (Rowe, 1989; Rowe and Gauthier, 1990). *Syntarsus kayentakatae* Rowe, 1989 is documented by the remains of at least 16 individuals, including a well-preserved skeleton, from the "silty facies" of the Kayenta Formation. It is distinguished by the presence of a pair of nasolacrimal crests on its snout and by fusion of the distal end of the fibula to the calcaneum in adult individuals. *Segisaurus halli* Camp, 1936 is known from a single poorly preserved

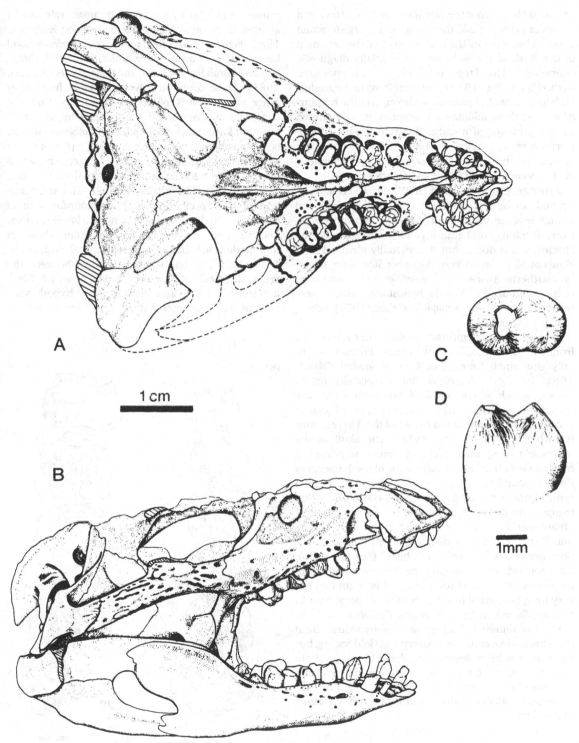

Figure 16.3. Partial skull, with mandible, of an unnamed *Edentosuchus*-like protosuchid (UCMP, unnumbered) in (A) palatal and (B) right lateral views, and enlarged views of "cheek" tooth in (C) occlusal and (D) posterior views. (Drawings by C. Vanderslice, from Clark, 1986.)

partial skeleton from the Navajo Sandstone. Rowe and Gauthier (1990) noted the presence of a large foramen in the ischium below the acetabulum and the flattening of the shaft of the ischium as potentially diagnostic characters. The large ceratosaurian *Dilophosaurus wetherilli* (Welles, 1954) is known from two subadult skeletons, a partial juvenile skeleton, and the remains of at least three additional specimens. It is diagnosed by the presence of a distinctly arched, paired naso-lacrimal crest, cervical neural spines with cruciform apices, and the square distal expansion of the scapular blade (Welles, 1984; Rowe and Gauthier, 1990).

Attridge, Crompton, and Jenkins (1985) briefly described an almost complete but crushed skull and mandible of the basal sauropodomorph *Massospondylus*. Gow, Kitching, and Raath (1990) reinterpreted the specimen and noted that it is virtually identical with skulls of *M. carinatus* from the upper Stormberg Group of southern Africa. It is possible that postcranial remains from the Navajo Sandstone, which were referred to *Ammosaurus major* by Galton (1971), belong to the same taxon.

Several taxa of ornithischian dinosaurs are known from the "silty facies" of the Kayenta Formation, but only one small form, *Scutellosaurus lawleri* Colbert, 1981, has been described. Based especially on its armor, which is composed of osteoderms that are dorsally keeled and ventrally excavated, *Scutellosaurus* is considered a primitive member of the Thyreophora (Gauthier, 1984; Sereno, 1986). The skull of the holotype is very incomplete, but much unpublished material is housed in the collections of the Museum of Comparative Zoology at Harvard University. The currently known material represents a considerable size range. The dentition is similar to that of primitive "fabrosaurid" ornithischians (Sereno, 1991). The maxillary tooth row is distinctly inset, indicating the presence of a well-developed "cheek" (Sereno, 1986). The pointed and diverging, rather than blunt and parallel, distal ends of the pubis and ischium and the very long tail (equal to 2.5 times the presacral length) may be diagnostic features of *Scutellosaurus*.

Skeletal remains of a very small heterodontosaurid have been recovered from Quarry 1 at Gold Spring, but have not yet been described in detail (Attridge et al., 1985). This material has "many of the unusual features of *Heterodontosaurus*" (Sereno, 1986, p. 248) including a tall caniniform tooth with serrated carinae in the dentary.

Synapsida

Tritylodontidae

Skeletal remains of tritylodontid cynodonts are among the most common tetrapod fossils in the "silty facies" of the Kayenta Formation and are represented by numerous excellently preserved specimens referable to at least three taxa: *Kayentatherium wellesi* Kermack, 1982 (Figure 16.4A) (including *Nearctylodon broomi* Lewis, 1986 as a subjective junior synonym), *Dinnebitodon amarali* Sues, 1986a, and *Oligokyphus* sp. (Sues, 1985). These forms are distinguishable from each other on the basis of dental characters that are consistent in juvenile and adult specimens. The most common Kayenta tritylodontid, *Kayentatherium wellesi*, is diagnosed by the presence of a single pair of upper and lower incisors; its upper postcanine teeth have two buccal, three median, and three lingual cusps (Sues, 1986b,c). Sues (1986b) described the skull and dentition of this species in detail. *Dinnebitodon amarali* (Figure 16.4B) has three upper and two lower incisors, and its upper postcanines differ from those of *Kayentatherium* in having only two lingual cusps (Sues, 1986a,b). It also has a sutural contact between the premaxilla and palatine on the palate, which Clark and Hopson (1985) and Sues (1986b,c) hypothesized as a derived character-state among Tritylodontidae.

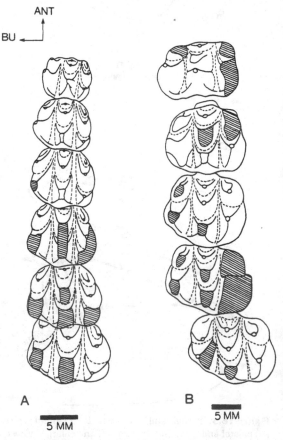

Figure 16.4. Right upper postcanine tooth rows of (A) *Kayentatherium wellesi* Kermack, 1982 (MCZ 8811) and (B) *Dinnebitodon amarali* Sues, 1986 (MNA V3222, holotype) in occlusal view. ANT, anterior direction; BU, buccal direction. (Adapted from Sues, 1986b.)

Oligokyphus, which is known from a well-preserved but dorsoventrally crushed skull and a number of jaws and isolated postcanine teeth from Gold Spring, has upper postcanine teeth with three buccal, four median, and four lingual cusps and lower postcanines with three buccal and three lingual cusps (Sues, 1985). The specific affinities of the Kayenta specimens of *Oligokyphus* cannot be resolved at present because the various European species of this genus are based on inadequate type materials. Two tritylodontid specimens, a partial dentary with teeth (UCMP 130857) and a rostral portion of a snout (UCMP 130858), from the Pumpkin Patch locality (UCMP loc. V6899) in the northernmost part of the "silty facies" of the Kayenta Formation may represent a fourth tritylodontid, but the currently available material is insufficient for the designation of a new taxon. The premaxilla apparently has three incisors, rather than just one as in *Kayentatherium*, but the premaxilla contacts the maxilla on the palate, unlike the condition in *Dinnebitodon*. The two postcanine teeth preserved in the dentary have two cusps each in the buccal and lingual rows, as in *Dinnebitodon* and *Kayentatherium*, but unlike *Oligokyphus*, which has three cusps in each row.

Winkler et al. (1991) briefly described a partial skeleton of a large tritylodontid from an interdune deposit of the Navajo Sandstone in northern Arizona, but, because of the lack of the diagnostically important postcanine dentition, the specimen is not diagnostic at the genus level.

Jenkins et al. (1983) inferred the presence of a haramiyid on the basis of a single, badly worn tooth, but they could not rule out the possibility that it might be a very worn and damaged lower postcanine of a small specimen of *Oligokyphus*. No teeth definitely referable to the Haramiyidae have been recovered to date.

Morganucodontidae

The Morganucodontidae, which at present are known only from Gold Spring Quarry 1, include *Dinnetherium nezorum* Jenkins, Crompton, and Downs, 1983 and a morganucodontid (Jenkins et al., 1983). Both are represented by cranial and postcranial remains.

Crompton and Luo (1993) argue for a close relationship of *Dinnetherium* to *Morganucodon* on the basis of the structure and occlusal pattern of the molariform teeth. The principal cusp of the lower molariform occludes between the anterior accessory and principal cusps of the corresponding upper tooth. Lower molariform teeth of *Dinnetherium* are characterized by anterior and posterior accessory cusps that are positioned symmetrically on either side of the median principal cusp. The lateral ridge of the articular process of the dentary in *Dinnetherium* forms a prominent ventrolateral flange, which Jenkins et al. (1983) interpreted

as an angular process [but see Sues (1986b) and Rowe (1988)]. This feature is also present in *Megazostrodon* from the upper Elliot Formation of South Africa (Gow, 1986).

Tetrapod assemblages from the Glen Canyon Group

The stratigraphic positions of the tetrapod-bearing localities of the Glen Canyon Group are poorly constrained. At present, the fossils themselves provide the only means for inferring the temporal relationships among the sites from which they were recovered. Only two taxa, the Crocodylomorpha and the Tritylodontidae, are widely distributed and have been studied in sufficient detail to indicate possible temporal interrelationships. Due to the inadequacy of the fossil record outside the "silty facies" of the Kayenta Formation, however, only tentative inferences are possible at this point.

Protosuchus richardsoni from the Moenave Formation is the stratigraphically oldest of the well-studied tetrapod taxa from the Glen Canyon Group. No diagnostic material of this species is known from the Kayenta Formation and the Navajo Sandstone.

Most of the fossil vertebrates from the "silty facies" of the Kayenta Formation in Moenkopi Wash were found close to the base of that formation, and these finds are perhaps older than the fossils from the "silty facies" to the south. The crocodylomorphs from the Pumpkin Patch locality are not known from other sites in the formation, and the tritylodontid from Pumpkin Patch described earlier differs from the three species known from more southern outcrops of the "silty facies." The relative temporal position of the "protosuchian" with highly cuspidate postcaniniform teeth is less clear because this form is known only from a single specimen from a stratigraphically higher horizon than that exposed at the Pumpkin Patch site.

The tritylodontid *Kayentatherium wellesi* is known from two localities high in the "typical facies" and throughout the "silty facies" south of Moenkopi Wash. Although many other tetrapod taxa have been recovered from the southern exposures of the "silty facies," they have yet to be found elsewhere in the formation.

The tetrapods of the Navajo Sandstone are represented by only a few partial skeletons (Winkler et al., 1991), nearly all of which lack diagnostic features at low taxonomic levels. The holotype of *Segisaurus halli* Camp, 1936 is the only specimen that has features of potential diagnostic value at the species level, and it differs from the related *Syntarsus kayentakatae* Rowe, 1989 from the Kayenta Formation (Rowe and Gauthier, 1990).

Summarizing, there exists some evidence for the presence of at least three, presumably diachronous, assemblages among the early Mesozoic tetrapods from the Glen Canyon Group. These assemblages occur in (1) the Dinosaur Canyon Member of the Moenave Formation, (2) the Kayenta Formation, and (3) the Navajo Sandstone.

Paleobiogeography

Attridge et al. (1985) emphasized the close faunal similarity between the tetrapod assemblages from the Kayenta Formation and the upper Stormberg Group of southern Africa. Subsequent studies of the "proto-suchian"-grade Crocodylomorpha by Clark (1986, and Chapter 5) and the description of a new species of the theropod dinosaur *Syntarsus* from the Kayenta Formation by Rowe (1989) provide additional support for the conclusions reached by Attridge et al. (1985). Many of the tetrapods from the Kayenta Formation also are closely related to taxa from the Early Jurassic portion of the Newark Supergroup of eastern North America, the Liassic of western Europe, and the Lower Lufeng Formation of Yunnan, China. A review of the currently known continental tetrapod assemblages of Early Jurassic age shows a remarkable faunal homogeneity, often extending to the generic level, across Pangaea during that interval of geological time (Shubin and Sues, 1991). This uniformity of tetrapod assemblages is surprising in view of the fact that the breakup of this supercontinent had already commenced.

Acknowledgments

We thank W. R. Downs (Northern Arizona University), K. Padian (University of California at Berkeley), and T. Rowe (The University of Texas at Austin) for helpful comments on a draft of the manuscript.

References

Attridge, J., A. W. Crompton, and F. A. Jenkins, Jr. 1985. Southern African Liassic prosauropod *Massospondylus* discovered in North America. *Journal of Vertebrate Paleontology* 5: 128–132.

Benton, M. J., and J. M. Clark. 1988. Archosaur phylogeny and the relationships of the Crocodylia. Pp. 295–338 in M. J. Benton (ed.), *The Phylogeny and Classification of the Tetrapods. Vol. 1: Amphibians, Reptiles and Birds.* Oxford: Clarendon Press.

Blakey, R. C., F. Peterson, and G. Kocurek. 1988. Synthesis of late Paleozoic and Mesozoic eolian deposits of the Western Interior of the United States. *Sedimentary Geology* 56: 3–125.

Brown, B. 1933. An ancestral crocodile. *American Museum Novitates* 638: 1–4.

Busbey, A. B., III, and C. E. Gow. 1984. A new protosuchian crocodile from the Upper Triassic Elliot Formation of South Africa. *Palaeontologia Africana* 25: 127–149.

Camp, C. L. 1936. A new type of small bipedal dinosaur from the Navajo Sandstone of Arizona. *University of California Publications in Geological Sciences* 24: 39–56.

Clark, J. M. 1986. Phylogenetic relationships of the crocodylomorph archosaurs. Ph.D. dissertation, University of Chicago.

In press. Cranial anatomy of *Protosuchus richardsoni* and two new protosuchids, and the relationships of "Protosuchia" (Archosauria: Crocodylomorpha). *Bulletin of the American Museum of Natural History*

Clark, J. M., and D. E. Fastovsky. 1986. Vertebrate biostratigraphy of the Glen Canyon Group in northern Arizona. Pp. 285–301 in K. Padian (ed.), *The Beginning of the Age of Dinosaurs: Faunal Change across the Triassic–Jurassic Boundary.* Cambridge University Press.

Clark, J. M., and J. A. Hopson. 1985. Distinctive mammal-like reptile from Mexico and its bearing on the phylogeny of the Tritylodontidae. *Nature* 315: 398–400.

Colbert, E. H. 1981. A primitive ornithischian dinosaur from the Kayenta Formation of Arizona. *Bulletin of the Museum of Northern Arizona* 53: 1–61.

Colbert, E. H., and C. C. Mook. 1951. The ancestral crocodilian *Protosuchus. Bulletin of the American Museum of Natural History* 97: 147–182.

Cooley, M. E., J. W. Harshbarger, J. P. Akers, and W. F. Hardt. 1969. Regional hydrogeology of the Navajo and Hopi Indian Reservations, Arizona, New Mexico and Utah. *United States Geological Survey Professional Paper* 521-A: A12–A61.

Cornet, B., and A. Traverse. 1975. Palynological contributions to the chronology and stratigraphy of the Hartford basin in Connecticut and Massachusetts. *Geoscience and Man* 11: 1–33.

Crompton, A. W., and Z. Luo. 1993. Relationships of the Liassic mammals, *Sinoconodon, Morganucodon oehleri,* and *Dinnetherium.* Pp. 30–44 in F. S. Szalay, M. J. Novacek, and M. C. McKenna (eds.)., *Mammal Phylogeny.* Berlin: Springer-Verlag.

Crompton, A. W., and K. K. Smith. 1980. A new genus and species of crocodilian from the Kayenta Formation (Late Triassic?) of northern Arizona. Pp. 193–217 in L. L. Jacobs (ed.), *Aspects of Vertebrate History: Essays in Honor of Edwin Harris Colbert.* Flagstaff: Museum of Northern Arizona Press.

Curtis, K. M. 1989. A taxonomic analysis of a microvertebrate fauna from the Kayenta Formation (Early Jurassic) of Arizona and its comparison to an Upper Triassic microvertebrate fauna from the Chinle Formation. M. A. thesis, University of California, Berkeley.

Edwards, D. P. 1985. Depositional controls on the Lower Jurassic Dinosaur Canyon Member of the Moenave Formation, southern Utah and northern Arizona. M. S. thesis, Northern Arizona University, Flagstaff.

Estes, R., and B. Sanchiz. 1982. New discoglossid and palaeobatrachid frogs from the Late Cretaceous of Wyoming and Montana, and a review of other frogs from the Lance and Hell Creek formations. *Journal of Vertebrate Paleontology* 2: 9–20.

Gaffney, E. S., J. H. Hutchinson, F. A. Jenkins, Jr., and L. Meeker. 1987. Modern turtle origins: the oldest known cryptodire. *Science* 237: 289–291.

Gaffney, E. S., and P. A. Meylan. 1988. A phylogeny of turtles. Pp. 157–219 in M. J. Benton (ed.), *The*

Phylogeny and Classification of Tetrapods. Vol. 1: Amphibians, Reptiles and Birds. Oxford: Clarendon Press.

Galton, P. M. 1971. The prosauropod dinosaur *Ammosaurus*, the crocodile *Protosuchus*, and their bearing on the age of the Navajo Sandstone of northeastern Arizona. *Journal of Paleontology* 45: 781–795.

Gauthier, J. 1984. A cladistic analysis of the higher systematic categories of the Diapsida. Ph.D. dissertation, University of California, Berkeley.

Gauthier, J., D. Cannatella, K. de Queiroz, A. G. Kluge, and T. Rowe. 1989. Tetrapod phylogeny. Pp. 337–353 in B. Fernholm, K Bremer, and H. Jöurnvall (eds.), *The Hierarchy of Life*. Amsterdam: Elsevier Science Publishers.

Gow, C. E. 1986. A new skull of *Megazostrodon* (Mammalia, Triconodonta) from the Elliot Formation (Lower Jurassic) of southern Africa. *Palaeontologia Africana* 26: 13–23.

Gow, C. E., J. W. Kitching, and M. A. Raath. 1990. Skulls of the prosauropod dinosaur *Massospondylus carinatus* Owen in the collections of the Bernard Price Institute for Palaeontological Research. *Palaeontologia Africana* 27: 45–58.

Harshbarger, J. F., C. A. Repenning, and J. H. Irwin. 1957. Stratigraphy of the uppermost Triassic and Jurassic rocks, Navajo country. *United States Geological Survey Professional Paper* 291: 1–74.

Jenkins, F. A., Jr. A. W. Crompton, and W. R. Downs. 1983. Mesozoic mammals from Arizona: new evidence on mammalian evolution. *Science* 222: 1233–1235.

Jenkins, F. A., Jr., and D. M. Walsh. 1993. An Early Jurassic caecilian with limbs. *Nature* 365: 246–250.

Kermack, D. M. 1982. A new tritylodont from the Kayenta Formation of Arizona. *Zoological Journal of the Linnean Society* 76: 1–17.

Kühne, W. G. 1956. *The Liassic Therapsid Oligokyphus.* London: Trustees of the British Museum.

Lewis, G. E. 1958. American Triassic mammal-like vertebrates. *Bulletin of the Geological Society of America* 69: 1735. [abstract]

———. 1986. *Nearctylodon broomi*, the first Nearctic tritylodont. Pp. 295–304 in N. Hotton III, P. D. MacLean, J. J. Roth, and C. E. Roth (eds.), *The Ecology and Biology of Mammal-like Reptiles.* Washington, D. C.: Smithsonian Institution Press.

Li, J. 1985. [Revision of *Edentosuchus tienshanensis* Young.] *Vertebrata Palasiatica* 23: 196–206. [in Chinese, with English summary]

Litwin, R. 1986. The palynostratigraphy and age of the Chinle and Moenave formations, southwestern U. S. Ph.D. dissertation, Pennsylvania State University.

Luttrell, P. R. 1987. Basin analysis of the Kayenta Formation (Lower Jurassic), central portion Colorado Plateau. M. S. thesis, Northern Arizona University, Flagstaff.

Meszoely, C. A. M., F. A. Jenkins, Jr., and C. R. Schaff. 1987. Early Jurassic sphenodontids from northeastern Arizona. *Journal of Vertebrate Paleontology* 7(Suppl. to 3): 21A [abstract]

Middleton, L. T., and R. C. Blakey. 1983. Processes and controls on the intertonguing of the Kayenta and

Navajo formations, northern Arizona: eolian–fluvial interactions. Pp. 613–634 in M. E. Brookfield and T. S. Ahlbrandt (eds.), *Eolian Sediments and Processes.* Amsterdam: Elsevier.

Norell, M. A., and G. W. Storrs. 1989. Catalogue and review of the type fossil crocodilians in the Yale Peabody Museum. *Postilla* 203: 1–28.

Olsen, P. E., and P. M. Galton. 1977. Triassic–Jurassic tetrapod extinctions: are they real? *Science* 197: 983–986.

Padian, K. 1984. Pterosaur remains from the Kayenta Formation (?Early Jurassic) of Arizona. *Palaeontology* 27: 407–413.

———. 1989. Presence of the dinosaur *Scelidosaurus* indicates Jurassic age for the Kayenta Formation (Glen Canyon Group, northern Arizona). *Geology* 17: 438–441.

Peterson, F., and G. N. Pipiringos. 1979. Stratigraphic relations of the Navajo Sandstone to Middle Jurassic formations, southern Utah and northern Arizona. *United States Geological Survey Professional Paper* 1035-B: B1–B43.

Pipiringos, G. N., and R. B. O'Sullivan. 1978. Principal unconformities in Triassic and Jurassic rocks, Western Interior United States – A preliminary survey. *United States Geological Survey Professional Paper* 1035-A: A1–A29.

Rowe, T. 1988. Definition, diagnosis, and origin of Mammalia. *Journal of Vertebrate Paleontology* 8: 241–264.

———. 1989. A new species of the theropod dinosaur *Syntarsus* from the Early Jurassic Kayenta Formation of Arizona. *Journal of Vertebrate Paleontology* 9: 125–136.

Rowe, T., and J. A. Gauthier. 1990. Ceratosauria. Pp. 151–168 in D. B. Weishampel, P. Dodson, and H. Osmolska (eds.), The *Dinosauria*. Berkeley: University of California Press.

Sereno, P. C. 1986. Phylogeny of the bird-hipped dinosaurs (Order Ornithischia). *National Geographic Research* 2: 234–256.

———. 1991. *Lesothosaurus*, "fabrosaurids," and the early evolution of Ornithischia. *Journal of Vertebrate Paleontology* 11: 168–197.

Shubin, N. H., and H.-D. Sues. 1991. Early Mesozoic continental tetrapod distributions: patterns and implications. *Paleobiology* 17: 214–230.

Sues, H.-D. 1985. First record of the tritylodontid *Oligokyphus* (Synapsida) from the Lower Jurassic of western North America. *Journal of Vertebrate Paleontology* 5: 328–339.

———. 1986a. *Dinnebitodon amarali*, a new tritylodontid (Synapsida) from the Lower Jurassic of western North America. *Journal of Paleontology* 60: 758–762.

———. 1986b. The skull and dentition of two tritylodontid synapsids from the Lower Jurassic of western North America. *Bulletin of the Museum of Comparative Zoology, Harvard University* 151: 217–268.

———. 1986c. Relationships and biostratigraphic significance of the Tritylodontidae (Synapsida) from the Kayenta Formation of northeastern Arizona. Pp. 279–284 in K. Padian (ed.). *The Beginning of the Age of Dinosaurs:*

Faunal Change across the Triassic–Jurassic Boundary.
Cambridge University Press.

Walker, A. D. 1970. A revision of the Jurassic reptile
Hallopus victor (Marsh), with remarks on the
classification of crocodiles. *Philosophical Transactions
of the Royal Society of London* B257: 323–372.

——— 1990. A revision of *Sphenosuchus acutus* Haughton, a
crocodylomorph reptile from the Elliot Formation
(late Triassic or early Jurassic) of South Africa.
*Philosophical Transactions of the Royal Society of
London* B330: 1–120.

Welles, S. P. 1954. New Jurassic dinosaur from the Kayenta
Formation of Arizona. *Bulletin of the Geological
Society of America* 65: 591–598.

——— 1984. *Dilophosaurus wetherilli* (Dinosauria, Theropoda):
osteology and comparisons. *Palaeontographica*
A185: 85–180.

Whetstone, K. N., and P. J. Whybrow. 1983. A "cursorial"
crocodilian from the Triassic of Lesotho
(Basutoland), South Africa. *University of Kansas
Museum of Natural History, Occasional Papers* 106:
1–37.

Wilson, R. F. 1967. Whitmore Point, a new member of the
Moenave Formation in Utah and Arizona. *Plateau*
40: 29–40.

Winkler, D. A., L. L. Jacobs, J. D. Congleton, and W. R.
Downs. 1991. Life in a sand sea: biota from Jurassic
interdunes. *Geology* 19: 889–892.

Young, C. C. 1973. [A new fossil crocodile from Wuerho.]
*Memoirs of the Institute of Vertebrate Paleontology and
Paleoanthropology, Academia Sinica* 11: 37–45. [in
Chinese]

17

An Early or Middle Jurassic tetrapod assemblage from the La Boca Formation, northeastern Mexico

JAMES M. CLARK, MARISOL MONTELLANO,
JAMES A. HOPSON, RENE HERNANDEZ, AND
DAVID E. FASTOVSKY

Introduction

The discovery of fossil vertebrates in Huizachal Canyon, near Ciudad Victoria, Tamaulipas, in 1982 brought to light the oldest known terrestrial tetrapod assemblage in Mexico. The initial find, a nearly complete skull and mandible of the tritylodontid therapsid *Bocatherium mexicanum* Clark and Hopson, 1985, was soon followed by the observation that small bones occur abundantly in these beds. Ongoing fieldwork has gathered a large collection of fossils representing at least ten taxa of terrestrial vertebrates. In this chapter we briefly summarize our current knowledge of these forms and compare the fauna to others from the early part of the Jurassic.

The thick sequence of red siltstones and sandstones in Huizachal Canyon was designated the type section of the Huizachal Formation by Imlay et al. (1948). Subsequent work by Mixon, Murray, and Diaz (1959) divided this sequence into the La Boca Formation at the base, with its type section in La Boca Canyon 30 km to the north, and the La Joya Formation at the top, with its type section in Huizachal Canyon. However, whether or not the beds identified as the La Boca Formation in Huizachal Canyon can indeed be correlated with those at the type section remains unresolved. Fossil vertebrates are not known from the type section.

Fossils occur only in the lowest part of the section in Huizachal Canyon, in beds identified by Mixon et al. (1959) as belonging to the La Boca Formation. The fossil-bearing portion of the formation is about 30 m in thickness. Fossils generally occur in laterally extensive 1–2-m-thick beds of red siltstones and sandy siltstones that are interpreted to have been deposited as water-lain debris flows (Fastovsky, Clark, and Hopson, 1987). The fossils are sparsely distributed throughout the beds, usually as pieces of bone only a few millimeters in length. The extremely hard, silicified matrix precludes the use of screen-washing techniques and requires much painstaking work to gather large samples.

During the formation of the Sierra Madre Oriental, the La Boca Formation became warped and heavily faulted, making it difficult to trace individual beds throughout Huizachal Canyon. Furthermore, isochronous "marker" beds that would allow correlations of noncontiguous sections have not been found. Despite these problems, however, it is clear that the fossil-bearing sites in the canyon, some of which are several kilometers apart, are roughly contemporaneous, because identical taxa are found at most localities. In particular, well-preserved and nearly complete skulls of the distinctive tritylodontid *Bocatherium mexicanum* are now known from four widely separated parts of the canyon.

The state of preservation of the fossils ranges from excellent to poor. The best fossils are nearly undistorted, articulated skulls, but in others diagenesis has proceeded to the point where the bone substance is recrystallized. Articulated skulls are very rare, but isolated partial mandibles and teeth are common. The extreme hardness of the matrix and the softness of the bone require that preparation be accomplished using fine needles under a microscope, greatly slowing the study of this material.

The unfossiliferous type section of the La Joya Formation in Huizachal Canyon underlies the Zuloaga Limestone, which has an Oxfordian (early Late Jurassic) ammonite fauna (Imlay, 1980). The relationships between the beds that Mixon et al. (1959) placed in the La Boca Formation and the underlying beds are much more complex. A series of volcaniclastic and pyroclastic rocks previously considered to be interbedded with the Huizachal Group (Belcher, 1979) actually forms a relict topography over which the red

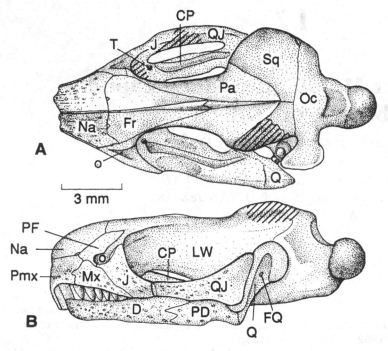

Figure 17.1. Skull of new burrowing diapsid (IGM 6620) in dorsal (A) and left lateral (B) views. The squamosal on the left side is damaged, exposing the underlying ear region and suspensorium. Abbreviations: CP, coronoid process; D, dentary; FQ, foramen within quadrate; Fr, frontal; J, jugal; LW, lateral wall of braincase Mx, maxilla; Na, nasal; o, orbit; oc, fused bones of occiput and otic region; Pa, parietal; PD, postdentary; PF, prefrontal; Pmx, premaxilla; Q. quadrate; QJ, quadratojugal; Sq, squamosal; T, tooth in dentary. Hatching denotes damaged areas.

beds were deposited (Fastovsky et al., 1988). The age of these igneous strata is currently unknown.

In the following section we review the taxa currently known from the La Boca Formation in Huizachal Canyon. In so doing we utilize a cladistic taxonomy without ranks. All specimens described here are catalogued in the collections of the Instituto de Geologia, Universidad Nacional Autonoma de Mexico (IGM).

Systematic paleontology

Diapsida

Comments. Diapsida comprises Lepidosauria (Squamata and Rhynchocephalia) and Archosauria (Crocodylia and Aves) and the descendants of their closest common ancestor (Gauthier, Estes, and de Queiroz, 1988).

Lepidosauria

Comments. Lepidosauria comprises Rhynchocephalia and Squamata and the descendants of their closest common ancestor (Gauthier, Estes, and de Queiroz, 1988).

Lepidosauria incertae sedis

The most surprising discovery among the fossils in Huizachal Canyon concerns the remains of a burrowing diapsid representing a heretofore unknown clade. Two nearly complete skulls of this form are known

(IGM 6620 and 6621), the former in articulation with the anterior six cervical vertebrae.

The skull of this new taxon (Figure 17.1) is remarkably similar to those of amphisbaenians, which use their compact heads in burrowing (Gans, 1974), but several features clearly indicate that it does not belong to this, or any other, group of squamates. As in amphisbaenians, the rostrum is short and compact, the orbit is reduced to a tiny opening, the maxilla and its dentition are reduced, the braincase is walled laterally by a descending process from the parietal, the temporal fossa is extremely large, the premaxillary teeth are large, the coronoid process of the mandible is large, and the inclined quadrate and short mandible combine to give the mouth a very small gape. Unlike all squamates, however, a large quadratojugal completes the lower temporal bar, the squamosal and premaxilla are extremely large, and the metotic fissure is apparently undivided. The vertebrae are very similar to those of squamates, procoelous centra with fused neural arches, but the odontoid process is separate from the axis.

This new form shares many features with squamates: the prominent excavation on the quadrate where the tympanum attached, pleurodont tooth implantation, absence of a ventral process of the squamosal, reduction of the quadrate ramus of the pterygoid, absence of a quadrate foramen between the quadrate and quadratojugal, fusion of the neural arches to the centra, procoelous centra, well-developed hypapophyses, and single-headed cervical ribs. However, it lacks several squamate synapomorphies: loss of the quadratojugal and development of quadrate kinesis (streptostyly), a divided metotic fissure, and fusion of the premaxillae. It thus appears that this new taxon is closely related to squamates, but is not a member of the Squamata as currently defined. The demonstrable convergence of this form with amphisbaenians suggested that some of its squamate features may also be convergent. Because it is so specialized, it undoubtedly represents a group of diapsids with a long, but previously unrecorded, history.

Rhynchocephalia

Comments. The taxon Rhynchocephalia has had several usages in the recent past, but all include the extant *Sphenodon* and a variety of Late Triassic to Early Cretaceous taxa that are more closely related to *Sphenodon* than to squamates. Gauthier, Estes, and de Queiroz (1988) considered the Rhynchocephalia to include the Sphenodontia and *Gephyrosaurus*, but they failed to define this group formally; other authors (e.g., Fraser and Benton, 1989) consider the Sphenodontia to include *Gephyrosaurus*.

Taxon A

A rhynchocephalian with a mandible very similar to that of *Sphenodon* is represented by several dozen partial jaws, but more complete material is lacking. As in rhynchocephalians generally, tooth implantation is acrodont, the mandible is very high and narrow, and the coronoid process is well developed. Several dentaries (e.g., IGM 3495) display a tall, conical caniniform tooth anteriorly, suggesting that this taxon may be more closely related to *Opisthias* and *Sphenodon* than are other rhynchocephalians (Gauthier, Estes, and de Queiroz, 1988). V. H. Reynoso (UNAM) is currently studying this material.

Taxon B

A second taxon of rhynchocephalian is represented by two small skulls, one slightly disarticulated (IGM 3496), and the other somewhat flattened dorsoventrally (IGM 3497). The teeth are similar to those of the other rhynchocephalian, but the mandible is much narrower, and a caniniform tooth is absent. The rostrum of this form appears to be flattened, and the facial process of the maxilla appears to have a more strongly convex posterior edge than that in most rhynchocephalians.

Archosauria

Comments. Archosauria comprises Aves, Crocodylia, and the descendants of their closest common ancestor (Gauthier, 1986).

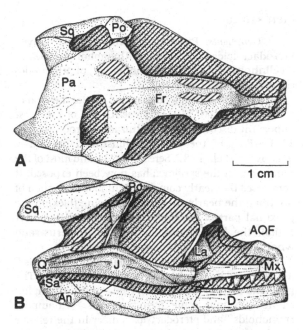

Figure 17.2. Skull of an unidentified crocodyliform (IGM 3498) in dorsal (A) and right lateral (B) views. Abbreviations: An, angular; AOF, antorbital fenestra; D, dentary; Fr, frontal; La, lacrimal; Mx, maxilla; Pa, parietal; Po, postorbital; Q, quadrate; Sa, surangular; Sq, squamosal. Hatching denotes matrix.

Crocodyliformes

Comments. Crocodyliformes comprises the taxa previously included in the Protosuchia, Mesosuchia, and Eusuchia (Benton and Clark, 1988).

?Metasuchia

Comments. Metasuchia comprises all Mesoeucrocodylia except thalattosuchians (Benton and Clark, 1988).

A new crocodyliform is represented by several partial skulls and postcranial skeletons. The most complete skull (IGM 3498) (Figure 17.2) lacks a braincase and palate. It has a large antorbital fenestra and homodont maxillary and dentary tooth rows, as in primitive crocodyliforms, but its postorbital bar is inset and columnar, as in all but the most primitive Metasuchia. Its rostrum does not appear to be as broad as in Metasuchia, but it has been crushed transversely. The lateral edge of the squamosal is peculiarly expanded vertically, and the supratemporal fossa (as best preserved on the right side) appears to be broader than long. A mandibular fenestra is absent. The vertebrae are amphicoelous.

Pterosauria

Comments. Pterosauria comprises the derived Pterodactyloidea and the basal Rhamphorhynchoidea (Wellnhofer, 1991). The monophyly of Rhamphorhynchoidea is uncertain.

An exquisite, virtually uncrushed partial skeleton of a pterosaur (IGM 3494) represents one of the most important discoveries in the Huizachal Group (Figure 17.3). Part of the foot was illustrated previously (Fastovsky et al., 1987; Sereno, 1991), and most of the remainder of the specimen has now been exposed. It comprises the nearly complete right foot and much of the tibia, the nearly complete right wing (lacking the proximal part of the humerus and two of the short digits), the left scapulocoracoid and humerus, and what appears to be a portion of the braincase.

The specimen is typical of "rhamphorhynchoids," as evidenced by the elongate fifth digit on the pes. The proximal carpal bones are fused, as in rhamphorhynchoids and some pterodactyloids. "Rhamphorhynchoids" and pterodactyloids differ in the relative slenderness of the fourth metacarpal, which supports the wing finger. Although the fourth metacarpal of the Huizachal Canyon specimen is not as slender as in pterodactyloids, it is more slender than in most "rhamphorhynchoids."

The specimen is large for a Jurassic pterosaur, although not outside the range of known specimens. When the length of the incomplete humerus is estimated from the intact left humerus, the articulated right wing has a total length of approximately 65 cm. Doubling this figure and adding 10 cm as an estimate of trunk width [based on the proportions in the skeletal reconstruction of *Rhamphorhynchus* by Wellnhofer (1991, p. 46)] indicates a total maximum wingspan of approximately 1.4 m. When compared with Early Jurassic "rhamphorhynchoids," this is similar to the largest specimens of *Dimorphodon macronyx* but smaller than the largest individuals of *Campylognathoides zitteli* (Wellnhofer, 1991), two well-known Early Jurassic rhamphorhynchoids.

Dinosauria

Comments: Dinosauria comprises Ornithischia, Saurischia (including Aves), and the descendants of their closest common ancestor. Note that some authors include certain out-groups of Dinosauria, as here defined, within the group (e.g., Novas, 1992).

Ornithischia

A small ornithischian is known from a few teeth, which are similar to those of *Heterodontosaurus*. These teeth are nearly square in cross-section at the base, with a strong subvertical wear surface on one side and a vertical ridge on the opposite side.

?Saurischia

?Theropoda

Isolated teeth similar to those of theropod dinosaurs occur rarely in the Huizachal Group. These teeth typically are 4–5 cm high, laterally compressed, and recurved, with serrated edges. Their identification as theropod rests in part on their occurrence in Jurassic deposits, but the possibility cannot be ruled out that they derive from another type of large, carnivorous archosaur.

Synapsida

Comments. Synapsida comprises those amniotes more closely related to Mammalia than to Reptilia (Chelonia and Diapsida) (Gauthier, Kluge, and Rowe, 1988).

Tritylodontidae

Bocatherium mexicanum Clark and Hopson, 1985

Subsequent to the type description of this species we have collected two nearly complete skulls with mandibles (IGM 3500 and 3501) and several isolated mandibles and teeth. As noted by Clark and Hopson (1985), *B. mexicanum* has the peculiar derived feature of a maxilla reduced to a cylinder of bone surrounding the teeth, a feature shared with only a few other tritylodontids, *Bienotheroides* (Sun, 1984), *Stereognathus*, and *Dinnebitodon* (Sues, 1986).

Mammaliaformes

Comment. Rowe (1988) restricted Mammalia to the descendants of the closest common ancestor of monotremes and therians, and he erected Mammaliaformes for the taxon previously termed Mammalia (including Morganucodontidae and the descendants of its common ancestor with Mammalia).

Taxon A

A mammaliaform with teeth similar to those of *Dinnetherium nezorum* from the Kayenta Formation of Arizona (Jenkins, Crompton, and Downs, 1983) is known from a single dentary fragment with one partial and two complete molars (IGM 6617). As in *D. nezorum*, classified by Luo (Chapter 6) as a morganucodontid, and many "amphilestids," the molars (Figure 17.4) are dominated by a large central cusp flanked on each side by smaller cusps slightly offset linguallly. A low cingulum is developed on the lingual surface; the labial surface has not yet been prepared. A deep longitudinal internal groove extends across the medial surface

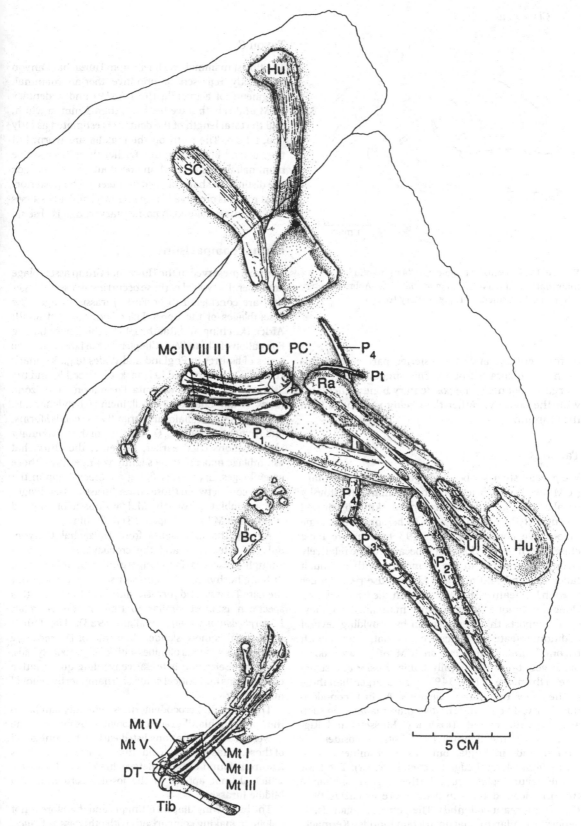

Figure 17.3. Pterosaurian skeleton (IGM 3494) as preserved. Abbreviations: Bc, partial braincase; DC, distal carpal; DT, distal tarsal; Hu, humerus; Mc, metacarpal; Mt, metatarsal; P, wing phalanges of fourth digit; Pt, pteroid; Ra, radius; SC, scapulocoracoid; Tib, tibia; Ul, ulna.

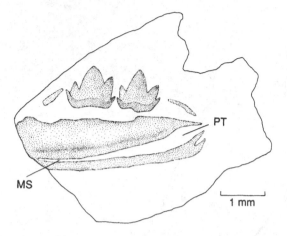

Figure 17.4. Dentary fragment of "amphilestid" mammaliaform (IGM 6617) in medial view. Abbreviations: MS, Meckelian sulcus; PT, postdentary trough.

of the dentary, and the posterior part corresponds to the postdentary trough. This suggests that, as in morganucodontids, the postdentary bones articulated with the dentary, rather than being suspended from the cranium.

Taxon B

We previously described and figured a partial dentary (IGM 3493) of a "triconodontid" mammal (Fastovsky et al., 1987), and an incomplete skull discovered recently (IGM 6618) appears to belong to the same taxon. We considered IGM 3493 to compare most closely with triconodontids (1) because of the relatively steep size gradient in the premolars, with p4 much taller than p3 and m1, and (2) because p4 is longer than m1, a feature shared only with the triconodontid *Priacodon* among "triconodont" mammals. The new skull supports this identification by providing several additional features indicative of relationships with triconodontids. The teeth on IGM 6618 are poorly preserved, but are generally similar in size and structure to the molar of IGM 3493 and are larger than those of the other two mammaliaforms. As in triconodontids, but unlike more primitive mammaliaforms such as morganucodontids (Kermack, Mussett, and Rigney, 1973), the dentary lacks both a postdentary trough and an angle, but has a prominent ridge on its ventrolateral edge (masseteric crest). The floor of the cavum epiptericum, furthermore, is similar to that in triconodontids in being more extensive than that in morganucodontids. The petrosal region is, in general, similar to that in triconodontids (Kermack, 1963).

Taxon C

The tiniest mammal specimens from Huizachal Canyon apparently represent a primitive therian mammal. Fragments of a maxilla (IGM 6619) and a dentary (IGM 6622) both have teeth less than 1 mm in width, and the total length of the dentary is estimated as only about 1 cm. The teeth on the maxilla are worn, but they extend lingually much farther than in primitive mammaliaforms other than docodonts, and they lack the distinctive shape of docodont teeth. The posterior-most molar is preserved on the dentary, and it possesses a rudimentary talonid. A postdentary trough is absent.

Faunal comparisons

The taxa preserved in the Huizachal Group assemblage are generally typical of those occurring in assemblages that are considered to be Early Jurassic in age. The assemblages of the upper Elliot Formation of South Africa (Kitching and Raath, 1984), the Lower Lufeng Formation of China (Wu, Chapter 3), the Lower Jurassic series of fissures in England and Wales (e.g., Kermack et al., 1973; Evans and Kermack, Chapter 15) and the "silty facies" of the Kayenta Formation in Arizona (Sues et al., Chapter 16) all include tritylodontids, primitive mammaliaforms, primitive crocodyliforms, and rhynchocephalians. On the basis of the preliminary identifications cited earlier, however, the Huizachal assemblage appears to be slightly younger than those assemblages, apparently filling a temporal gap in the fossil record between those Lower Jurassic assemblages and assemblages from the Middle Jurassic of England (Evans and Milner, Chapter 18) and China.

Among the archosaurs from Huizachal Canyon, only the pterosaur and the crocodyliform are well enough preserved for comparison with other taxa. "Rhamphorhynchoid" pterosaurs occur throughout the Late Triassic and Jurassic, although the foot of this specimen is most similar to that of the Toarcian *Dimorphodon macronyx* (Padian, 1983). The fourth metacarpal is more slender than that of *D. macronyx* and perhaps suggests affinities with later pterodactyloids, but this observation is tentative pending quantitative comparisons and critical study of "rhamphorhynchoid" monophyly.

The Huizachal crocodyliform is generally similar to the "protosuchian"-grade crocodyliforms found in the Lower Jurassic formations listed earlier, but unlike all of those, it has an inset, columnar postorbital bar. This feature is found only in metasuchian crocodyliforms, which do not appear in the fossil record until the Middle Jurassic.

The burrowing diapsid is unique and therefore is not helpful in making comparisons with other assemblages, but rhynchocephalian A tentatively indicates a somewhat younger age for the assemblage. Unlike the known rhynchocephalians of those formations, rhynchocephalian A possesses a caniniform tooth similar to that present in the Late Jurassic *Opisthias* and extant *Sphenodon*. This, too, is tentative, however, because this feature is variable in those taxa possessing it.

The mammals provide the strongest evidence in support of a younger age for this assemblage. In all known Early Jurassic mammaliaforms, the postdentary bones are still associated with the mandible, as well as with the middle ear, as in mammaliaform A from the Huizachal assemblage. The lack of a groove on the dentary for postdentary bones in the other two mammaliaforms, the similarity of the skull to that in triconodontids s. str. (which do not appear in the fossil record until the Late Jurassic), and the presence of an incipient talonid on the molar of mammaliaform C all strongly suggest that the assemblage is younger than those from the well-known Lower Jurassic formations.

A surprising feature of many of the fossils in Huizachal Canyon is their small size. Although pieces of large bone several centimeters across and isolated teeth of medium-sized theropod dinosaurs indicate the presence of larger animals, the largest intact bones are from animals that probably were no more than 2 m in length. Given the ubiquity of sauropodomorph dinosaurs in Jurassic deposits, it seems likely that their absence from the Huizachal Group reflects a taphonomic bias against larger animals. Taphonomic sorting is not surprising, considering the unusual depositional environment of these occurrences, but it is unclear whether that sorting occurred prior to or during the deposition of the debris flows.

Acknowledgments

This project has been supported by the UNAM Instituto de Geologia, by two grants from the National Geographic Society, and by NSF grant EAR 89-117389. We especially thank F. Ortega, E. Martinez, and O. Carranza for their support and assistance. The senior author's work was facilitated by the Smithsonian Institution. Our fieldwork was greatly expedited by the people of the Huizachal Ejido and in particular by Sr. Fidencio Moreno and his family. Help in the field was provided by V. Reynoso, N. Strater, J. Wible, J. O'Brien, and P. Sereno.

References

Belcher, R. 1979. Depositional environments, paleomagnetism, and tectonic significance of Huizachal red beds (lower Mesozoic), northeastern Mexico. Ph.D. dissertation, University of Texas at Austin.

Benton, M. J., and J. M. Clark. 1988. Archosaur phylogeny and the relationships of the Crocodylia. Pp. 295–338 in M. J. Benton (ed.), *The Phylogeny and Classification of the Tetrapods. Vol. 1: Amphibians, Reptiles and Birds.* Oxford: Clarendon Press.

Clark, J. M., and J. A. Hopson. 1985. Distinctive mammal-like reptile from Mexico and its bearing on the phylogeny of the Tritylodontidae. *Nature* 315: 398–400.

Fastovsky, D. E., J. M. Clark, and J. A. Hopson. 1987. Preliminary report of a vertebrate fauna from an unusual paleoenvironmental setting, Huizachal Group, Early or Mid-Jurassic, Tamaulipas, Mexico. Pp. 82–87 in P. J. Currie and E. H. Koster (eds.), *Fourth Symposium on Mesozoic Terrestrial Ecosystems, Short Papers.* Occasional Papers of the Tyrrell Museum of Palaeontology, 3.

Fastovsky, D. E., O. D. Hermes, J. M. Clark, and J. A. Hopson. 1988. Volcanoes, debris flows, and Mesozoic mammals: Huizachal Group (Early or Middle Jurassic), Tamaulipas, Mexico. *Geological Society of America, Abstracts with Program* 20: 317–318.

Fraser, N. C., and M. J. Benton. 1989. The Triassic reptiles *Brachyrhinodon* and *Polysphenodon* and the relationships of the sphenodontids. *Zoological Journal of the Linnean Society* 96: 413–445.

Gans, C. 1974. *Biomechanics, An Approach to Vertebrate Biology.* Ann Arbor: University of Michigan Press.

Gauthier, J. A. 1986. Saurischian monophyly and the origin of birds. Pp. 1–56 in K. Padian (ed.), *The Origin of Birds and the Evolution of Flight.* California Academy of Sciences, Memoir 8.

Gauthier, J. A., R. Estes, and K. de Queiroz. 1988. A phylogenetic analysis of Lepidosauromorpha. Pp. 15–98 in R. Estes and G. Pregill (eds.), *The Phylogenetic Relationships of the Lizard Families.* Stanford University Press.

Gauthier, J. A., A. G. Kluge, and T. Rowe. 1988. Amniote phylogeny and the importance of fossils. *Cladistics* 4: 105–209.

Imlay, R. 1980. Jurassic paleobiogeography of the coterminous United States in its continental setting. *United States Geological Survey Professional Paper* 1062: 1–134.

Imlay, R., E. Cepeda, M. Alvarez, and T. Diaz. 1948. Stratigraphic relations of certain Jurassic formations of eastern Mexico. *American Association of Petroleum Geologists Bulletin* 32: 1750–1761.

Jenkins, F. A., Jr., A. W. Crompton, and W. R. Downs. 1983. Mesozoic mammals from Arizona: new evidence on mammalian evolution. *Science* 222: 1233–1235.

Kermack, K. A. 1963. The cranial structure of triconodonts. *Philosophical Transactions of the Royal Society of London* B246: 83–103.

Kermack, K. A., F. Mussett, and H. W. Rigney. 1973. The lower jaw of *Morganucodon. Zoological Journal of the Linnean Society* 53: 87–175.

Kitching, J. W., and M. A. Raath. 1984. Fossils from the Elliot and Clarens formations (Karoo sequence) of the northeastern Cape, Orange Free State and Lesotho, and a suggested biozonation based on tetrapods. *Palaeontologia Africana* 25: 111–125.

Mixon, R. B., G. E. Murray, and G. Diaz. 1959. Age and correlation of Huizachal Group (Mesozoic), state of Tamaulipas, Mexico. *American Association of Petroleum Geologists Bulletin* 43: 757–771.

Novas, F. 1992. Phylogenetic relationships of the basal

dinosaurs, the Herrerasauridae. *Palaeontology* 35: 51–62.

Padian, K. 1983. Osteology and functional morphology of *Dimorphodon macronyx* (Buckland) (Pterosauria: Rhamphorhynchoidea) based on new material in the Yale Peabody Museum. *Postilla* 189: 1–44.

Rowe, T. 1988. Definition, diagnosis, and origin of Mammalia. *Journal of Vertebrate Paleontology* 8: 241–264.

Sereno, P. C. 1991. Basal archosaurs: phylogenetic relationships and functional implications. *Society of Vertebrate Paleontology Memoirs* 2: 1–53.

Sues, H.-D. 1986. Relationships and biostratigraphic significance of the Tritylodontidae (Synapsida) from the Kayenta Formation of northeastern Arizona. Pp. 279–284 in K. Padian (ed.), *The Beginning of the Age of Dinosaurs: Faunal Change across the Triassic–Jurassic Boundary.* Cambridge University Press.

Sun, A. 1984. Skull morphology of the tritylodont genus *Bienotheroides* of Sichuan. *Scientia Sinica* B27: 257–268.

Wellnhofer, R. 1991. *The Illustrated Encyclopedia of Pterosaurs.* New York: Crescent Books.

18

Middle Jurassic microvertebrate assemblages from the British Isles

SUSAN E. EVANS AND ANDREW R. MILNER

Introduction

Late Jurassic microvertebrate faunas, such as those of the Purbeck Beds of southern England (Ensom, 1988; Ensom, Evans, and Milner, 1991), differ in many respects from those of the Late Triassic and Early Jurassic, most notably in the appearance of more modern faunal components: frogs, salamanders, true lizards, goniopholidid crocodilians, and a greater diversity of mammals. Until recently, the only continental assemblage bridging the gap between the Early and Late Jurassic was that from the Stonesfield Slates in Oxfordshire. In the past two decades, however, several new Middle Jurassic microvertebrate localities have been identified within the British Isles, and these are deepening our understanding of vertebrate evolution during that period. The new microvertebrate assemblages have a major advantage over those of the Late Triassic and Early Jurassic: They are all found in bedded strata, rather than fissures and cave systems, and many can be referred to known ammonite zones.

Geology

The Middle Jurassic includes the Aalenian, Bajocian, Bathonian, and Callovian stages (Cope et al., 1980). Much of Britain was covered by a shallow epicontinental sea at that time, with areas of high ground forming substantial landmasses. Terrestrial or freshwater deposits yielding vertebrate remains are relatively rare. In Britain, they are largely restricted to the Bathonian, where, during a period of marine regression, lagoonal, deltaic, and coastal swamp deposits were widespread on the European shelf.

Geological evidence (e.g., Ager, 1956; Martin, 1962; Palmer and Jenkyns, 1975; Callomon, 1979) suggests that there were three main land areas in the region of Britain during the Bathonian (Figure 18.1): the

London-Ardennes landmass in the southeast, probably with an island barrier off its western coast (Palmer and Jenkyns, 1975; Ware and Windle, 1981; Ware and Whatley, 1983), the Cornish-Mendip island(s) to the southwest (Martin, 1962), possibly linked to a Welsh land area (Callomon, 1979), and, to the north, the Scottish-Pennine (northern British) land area. These landmasses surrounded a shallow central sea into which sediments were deposited (Figure 18.1).

According to a recent global stratigraphic chart (ICS: IUGS 1989), the Bathonian lasted from about 170 Ma to 159 Ma, but estimates vary (e.g., Cope et al., 1980; Harland et al., 1982). Table 18.1 shows a stratigraphic column taken from Cope et al. (1980) for the Oxfordshire Bathonian. The Oxfordshire horizons can be correlated with horizons in Gloucestershire, Warwickshire, Dorset, and Wiltshire, but the correlation with the Great Estuarine Series of Scotland is less clear-cut.

Localities and assemblages

Some of the most conspicuous terrestrial fossils in the Bathonian of England are the bones of large sauropods such as *Cetiosaurus*. These have been found at several horizons (Woodward, 1894) and usually are preserved in marls or clays, associated with freshwater gastropods and plants. Frequently, but not universally, they indicate quiet, freshwater, or brackish-water environments suitable for the preservation of microvertebrates. However, we have not attempted to catalogue the sites that have yielded only large sauropod bones (many of the old quarries have now disappeared), but rather have restricted ourselves to those localities known to have produced microvertebrates.

All known Bathonian microvertebrate horizons show one of two basic lithologies: (1) "slates" (using this term not in its strict geological sense but to denote a

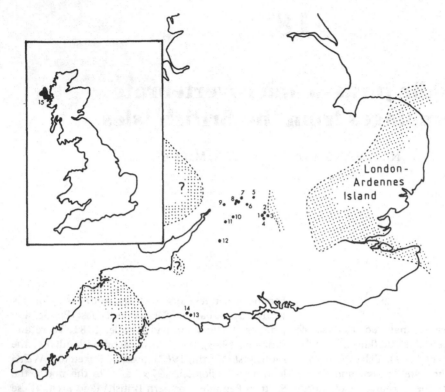

Figure 18.1. Map of Britain showing principal Bathonian microvertebrate localities in relation to Middle Jurassic land areas. Key: 1, Stonesfield; 2, Kirtlington; 3, Woodeaton; 4, Shipton; 5, Sharp's Hill; 6, Sarsden; 7, New Park; 8, Eyford group; 9, Hornsleasow; 10, Ready Token; 11, Tarlton; 12, Leigh Delamere; 13, Swyre; 14, Watton; 15, Skye. (Adapted from Ware and Windle, 1981.)

laminated rock that can be split along bedding planes to produce roofing tiles, or slates) deposited under estuarine or shallow offshore marine conditions and (2) unconsolidated, usually freshwater, marls and clays. Most of the older finds of small vertebrates were in slates, rare terrestrial fossils collected casually by Victorian stoneworkers. Stonesfield provides the classic example of this type of locality. Most of the new finds have been made by bulk-processing unconsolidated clays and marls, a method that yields much richer rewards. To date, the Mammal Bed at Kirtlington is the best-known, and most thoroughly worked, example of this second type of deposit. (In the locality descriptions that follow, the figures in parentheses are U.K. National Grid References).

Stonesfield (SP 392172, 387168, 379172, 387171, a series of small mines and quarries)

    Microvertebrate horizon. Stonesfield Slate (Stonesfield Member of the Sharp's Hill Formation), middle Bathonian (*Procerites progracilis* zone). The Stonesfield Slate consists of layers of fine calcareous sandstone. It is interbedded with thin laminae of oolites, which act as natural parting planes through which the rock splits

during weathering to produce the roofing tiles (slates) from which the formation derives its name. The bulk of the assemblage consists of marine or estuarine invertebrates, fish, plesiosaurs, and the crocodilians *Steneosaurus* and *Teleosaurus* (Phillips, 1871; Woodward, 1894). The deposit is generally thought to have been laid down in quiet estuarine (Arkell, 1933; Savage, 1963) or very shallow offshore (Sellwood and McKerrow, 1974) conditions. Remains of terrestrial plants are quite common, and the rarer terrestrial components of the fauna (including mammals and a tritylodontid, *Stereognathus ooliticus*) may have been either carried in on rafts of vegetation (Savage, 1963) or washed in from strand-line deposits during rough weather (Sellwood and McKerrow, 1974).

    In addition to the crocodiles, the reptiles include the following: scute impressions of a turtle ("*Protochelys*") (Phillips, 1871; Lydekker, 1889); the jaw of a rhamphorhynchid pterosaur (*Rhamphocephalus bucklandi* Meyer, 1832) and pterosaurian phalanges [referred by Owen (1874) to several species of *Pterodactylus*, but almost certainly attributable to *Rhamphocephalus*] (Wellnhofer, 1978); two theropod ilia (*Iliosuchus*) (Galton, 1976); the holotype dentary, referred teeth, and other fragments of *Megalosaurus bucklandi* (Lydekker,

Table 18.1. *Bathonian ammonite zones and the stratigraphy of Oxfordshire*

		ZONES	Log B12: Oxfordshire		
B A T H O N I A N	U P P E R	Clydoniceras discus	Lower Cornbrash		
			Forest Marble Formation		
		Oppelia aspidiodes			
		Procerites hodsoni	W h i t e	L i m e s t o n e	Bladon Member
	M I D D L E				Ardley Member
		Morrisiceras morrisi			Shipton Member
		Tulites subcontractatus			
		Procerites progracilis	Hampen Marly Formation		
			Taynton Limeston		
	L O W E R	Asphinctites tenuiplicatus	S h a r p s H i l l	F o r m a t i o n	Stonesfield Member
					Sharp's Hill Member
		Zigzagiceras zigzag	Chipping Norton Member		
			Swerford Member		
			?		
			Hook Norton Member		

Source: Adapted from Cope et al. (1980).

1888); and a premaxillary tooth of a possible hypsilophodontid (Galton, 1975, 1980). Duffin (1985, p. 144) noted the presence of lepidosaurs, but gave no source.

Kirtlington Cement Quarry (Washford Quarry) (SP 494199)

Microvertebrate horizon. Forest Marble, upper Bathonian (*Oppelia aspidoides* zone). Kirtlington is not far from Stonesfield, but the productive horizons are slightly younger. Beds dated to the middle and late Bathonian are exposed, from the White Limestone to the Cornbrash.

When it was worked commercially in the 1920s and earlier, Kirtlington Cement Quarry occasionally yielded bones of the large sauropod *Cetiosaurus* in a clay layer (the *fimbriata-waltoni* clay, beds 2o, 3l, 4e, 6f of McKerrow, Johnson, and Jakobson, 1969) below the Coral *Epithyris* Limestone of McKerrow et al. (1969). However, in 1974, Eric Freeman, an amateur paleontologist, prospected the quarry and recognized that the unconsolidated marls might produce mammals. One of them, now generally called the Mammal Bed, did, and in quite spectacular numbers (Freeman, 1976a,b, 1979). Subsequently, from 1976 to 1978, the site was worked by a team from University College London (UCL) under the direction of Kenneth Kermack, and with the cooperation of Eric Freeman, Ernest King, and David Ward.

The main microvertebrate deposit, the Mammal Bed, is a thin layer of unconsolidated marly clay (bed 3p of McKerrow et al., 1969) that immediately overlies the Coral *Epithyris* Limestone (beds 3o, 4i, 6j of McKerrow et al., 1969). The geology of Kirtlington has been studied by a number of workers, most notably Odling (1913), Arkell (1931, 1947), McKerrow et al., (1969), Palmer (1973), and Sumbler (1984). There is disagreement about the position of the boundary between the Forest Marble and the underlying White Limestone (Odling, 1913; Arkell, 1931, 1947; McKerrow et al., 1969; Palmer, 1973; Cope et al., 1980; Sumbler, 1984), and most decisions seem arbitrary. In the majority of schemes, the Mammal Bed is placed at or near the base of the Forest Marble (*aspidoides* zone) (Cope et al., 1980) and is of late Bathonian age. Above the Mammal Bed are further marly bands (e.g., bed 3w, about 2 m up) that have yielded a vertebrate fauna similar to that of the Mammal Bed.

At the time of deposition of the Mammal Bed, the Kirtlington locality probably lay on or near the shore of an island barrier some 30 km northwest of the London-Ardennes landmass (Palmer and Jenkyns, 1975). The Mammal Bed vertebrates are associated with lignite, charophytes, and freshwater ostracods and gastropods (*Valvata, Bathonella*) (Ware, 1978; Ware and Whatley, 1980; Jakovides, 1982). These, and the marly deposits themselves, suggest shallow, swampy coastal regions with creeks, lagoons, and freshwater lakes (Palmer, 1979; Ware and Whatley, 1980), not unlike the present-day Everglades of Florida, which lie at a similar subtropical latitude (Briden, in McKerrow et al., 1969).

In the UCL project, about ten tons of matrix were broken down, concentrated, and picked for small-vertebrate remains; regarding methods, see Kermack et al. (1987) and Ward (1984). The resulting collection includes several hundred Middle Jurassic mammal teeth and a wealth of nonmammalian vertebrates representing at least 30 species of tetrapods and many kinds of fish. The assemblage also includes marine invertebrates, such as corals, brachiopods, and echinoderms, but their state of preservation suggests secondary derivation from the underlying White Limestone (E. F. Freeman, pers. commun.). The same may be true for some of the fish. The following is a preliminary list of the vertebrates recovered:

Chondrichthyes
  Selachii
    Hybodontoidea
      Asteracanthus, Hybodus, Lissodus (Duffin, 1985; E. F. Freeman, pers. commun.)
    Batoidei indet.
Osteichthyes
  Semionotoidea: cf. *Lepidotes*
  Pycnodontoidea indet.
  ?Amioidea indet.
Amphibia
  Anura
    Discoglossidae: *Eodiscoglossus* (Evans, Milner, and Mussett, 1990)
  Caudata
    Albanerpetontidae: *Albanerpeton*
    Incertae sedis
      Marmorerpeton Evans et al., 1988
      Salamander A (common form)
      Salamander B (small form)
Sauropsida
  Chelonia
    Cryptodira: cf. Pleurosternidae
  Lepidosauromorpha
    Marmoretta Evans, 1992
    Lepidosauria
      Sphenodontia indet.
      Squamata
        Scincomorpha: *Saurillodon* sp. and others (S. E. Evans, unpublished data)
        Anguinomorpha
        ?Gekkota
  Archosauromorpha
    Choristodera: *Cteniogenys* (Evans, 1989, 1990, 1991a)
    Archosauria
      Crocodylia
        Goniopholididae: cf. *Nannosuchus* (? = juvenile *Goniopholis*)
        Atoposauridae
      Pterosauria

Rhamphorhynchoidea
Pterodactyloidea
Ornithischia
  "Fabrosauridae": cf. *Alocodon* Thulborn, 1973
Saurischia
  Theropoda: Carnosauria and "Coelurosauria," indet.
Synapsida
  Tritylodontidae: *Stereognathus*
  Mammalia
    Triconodonta
      Morganucodontidae: *Wareolestes* Freeman, 1979
    Docodonta
      Docodontidae: *Simpsonodon* Kermack et al., 1987
    Multituberculata (under study by Kermack et al.)
    Symmetrodonta
      Kuehneotheriidae: *Cyrtlatherium* Freeman, 1979
    Eupantotheria
      Peramuridae: *Palaeoxonodon* Freeman, 1979
      ?Dryolestidae: unnamed (Freeman, 1976b, 1979

Plants. Traces of plants are found throughout the layer in the form of indeterminate carbonaceous or limonitic impressions (Freeman, 1979); the only identified structures are charophyte gyrogonites and megaspores of "*Triletes*" (Ware, 1978).

The remaining microvertebrate horizons range in age from early to late Bathonian.

New Park Quarry, Longborough, Stow-on-the-Wold (SP 175282)

Microvertebrate horizon. Hook Norton Member, Chipping Norton Formation, lower Bathonian (*Zigzagiceras zigzag* zone). This is a "slate" horizon, reportedly older than that of Stonesfield (Woodward, 1894; Richardson, 1929). It has yielded *Megalosaurus* and *Cetiosaurus*, as well as stegosaurian plates referred to *Lexovisaurus vetustus* (Galton and Powell, 1983) and specimens of *Steneosaurus* and *Teleosaurus* (Richardson, 1929). Richardson (1929) commented that the quarry was remarkable in the number of crocodilian fragments preserved.

Hornsleasow Quarry, Snowshill (SP 131322)

Microvertebrate horizon. Chipping Norton Member, Chipping Norton Formation, lower Bathonian (*Zigzagiceras zigzag* zone). Hornsleasow has recently been reopened for the quarrying of the Chipping Norton Limestone. In 1987, an amateur paleontologist, Kevin Gardner, located a clay lens within the limestone that contained bones of *Cetiosaurus*. The site was then excavated by the Gloucester Museum in collaboration with the Crickley Hill Archaeological Trust, using archaeological techniques (Darlington, 1988; Vaughan, 1988, 1989). In addition to *Cetiosaurus*, the clay lens has yielded teeth of *Megalosaurus* and numerous microvertebrate remains, including small theropod and ornithischian teeth, crocodilian teeth and osteoderms, turtle plates, *Lepidotes* scales, pterosaurian teeth, and the remains of small reptiles, including lizards, *Cteniogenys* and *Marmoretta* (S. E. Evans and A. R. Milner, personal observations), and tritylodontid teeth (Darlington, 1988; Vaughan, 1988). More recently, mammals and amphibians (frogs and salamanders) have been identified (Vaughan, 1989; S. E. Evans and A. R. Milner, personal observations). In general, the assemblage is like that of Kirtlington, but because the beds are significantly older, detailed differences may emerge when the material has been fully analyzed. In addition, because of the rigorous collecting techniques employed, Hornsleasow may provide valuable taphonomic information. The project has recently been taken over by the University of Bristol, under the direction of Dr. M. J. Benton.

Sharp's Hill (SP 337358)

Microvertebrate horizon. Sharp's Hill Member, bed 18, Sharp's Hill Formation, lower Bathonian (*Procerites progracilis* zone). This horizon is better known for its large vertebrates. Stegosaurian vertebrae, referred to the species *Lexovisaurus vetustus*, have been described by Galton and Powell (1983). More recently, work by an amateur paleontologist, Brian Boneham, has yielded further stegosaurian material and a microvertebrate fauna similar to that of Kirtlington. The vertebrates are associated with freshwater ostracods (Sylvester-Bradley, 1948), freshwater gastropods of the genus *Bathonella* (Richardson, 1911; Arkell, 1947), and charophytes (Sellwood and McKerrow, 1974).

Sarsden (SP 300266) (?Smith's Quarry) (Woodward, 1894, p. 325)

Microvertebrate horizon. Probably Sharp's Hill Beds, Sharp's Hill Formation, layer of marl with *Bathonella*, lower Bathonian. This site yielded the holotype mandible of the rhamphorhynchid pterosaur *Rhamphocephalus depressirostris* (Huxley, 1859; Lydekker, 1888; Wellnhofer, 1978), a tooth referred to *Megalosaurus* (Lydekker, 1888), and a humerus of *Cetiosaurus* (Huxley, 1859; Phillips, 1871; Richardson, 1911).

Eyford group

There is a series of quarries in the Eyford area, north and northeast of Naunton (Woodward, 1894). Like Stonesfield, these quarries were worked for roofing tiles (slates). The Cotswold Slate is generally considered to be of the same age (middle Bathonian, *Procerites progracilis* zone) as the Stonesfield Slate and was deposited under similar conditions (Richardson 1925, 1929; Arkell, 1933; Savage, 1963; McKerrow, Ager, and Donovan, 1964; Sellwood and McKerrow, 1974; Cope et al., 1980). Ripple marks and invertebrate borings suggest very shallow marine conditions (Sellwood and McKerrow, 1974).

Eyford Hill Quarry (SP 132252, 131251)

Microvertebrate horizon. Cotswold Slate, middle Bathonian. Savage (1963) gave a comprehensive account of this site, which has yielded rich and varied flora and fauna comparable to those of Stonesfield, but lacks mammals. A large collection of this material, made by Witts in the 1800s, is held at Gloucester Museum (G.1–G.722). The faunal assemblage combines marine/estuarine invertebrates (bivalves, gastropods, echinoderms, and ammonites), fish (including pycnodonts, *Lepidotes*, and hybodont sharks), and reptiles (plesiosaurs and the crocodiles *Steneosaurus* and *Teleosaurus*) with terrestrial plants (e.g., cycads, ginkgos), insects (beetles, neuropterans), and reptiles (*Megalosaurus, Rhamphocephalus*).

Huntsman's Quarry (SP 125255–126252)

Microvertebrate horizon. Cotswold Slate, middle Bathonian. This quarry is about 0.8 km from Eyford (Richardson, 1929). It has reportedly yielded an assemblage of turtles, pterosaurs, crocodilians, and dinosaurs (Benton, 1988).

Kineton Thorns Quarry (SP 123263)

Microvertebrate horizon. Cotswold Slate, middle Bathonian. The quarry is 0.8 km northwest of Huntsman's Quarry and is said by McKerrow et al. (1964) to be more fossiliferous. Woodward (1894) recorded fish, plants, a tooth of *Megalosaurus*, and bones of the pterosaur *Rhamphocephalus*.

Skye, Loch Scavaig (NG 518168, NG 520157) and Elgol (NG 516147)

Microvertebrate horizon. Ostracod Limestone, Great Estuarine Series, middle Bathonian. Mammals were discovered at two localities in Skye by Michael Waldman and J. B. Dobinson in 1971. The localities were the first of the new (i.e., other than Stonesfield) Middle Jurassic sites (including Kirtlington) to be the subjects of a published account of mammalian material. The Skye sites may be slightly older (middle Bathonian) than Kirtlington. They have yielded two partial jaws of a docodont mammal, *Borealestes* Waldman and Savage, 1972, an unnamed pantothere (Clemens, 1986; Savage, 1984), and a new species of tritylodontid,

Stereognathus hebridicus Waldman and Savage, 1972. These genera are found in association with a richer tetrapod assemblage (Savage, 1984) that includes ?pleurosternid turtles, crocodiles, "lizards," and small dinosaurs, but the fauna has yet to be studied in depth.

Arkell (1933) described a bed toward the bottom of the Great Estuarine Series that yielded ostracods, insects, and freshwater molluscs in association with plesiosaurs, sharks, *Lepidotes*, and fragments of the bones, vertebrae, and teeth of crocodilians, dinosaurs, and pterosaurs. Because Waldman and Savage (1972) described their bed as being toward the top of the series, presumably there is more than one microvertebrate horizon in Skye.

Woodeaton (SP 534122)

Microvertebrate horizon. "Monster Bed," Hampen Marly Formation, middle Bathonian. The Monster Bed takes its name from the bones of *Cetiosaurus* that it has yielded. Palmer (1973) reported finding freshwater ostracods and plants characteristic of an upland biota (T. Harris, pers. commun. to Palmer). On that basis, Palmer suggested that the Monster Bed was deposited in a coastal lagoon that received the drainage of a fairly large river.

Freeman (1979, pers. commun.) recovered a possible mammalian incisor, fragments of a tritylodontid (?*Stereognathus*), crocodilian teeth, and teeth of the freshwater shark *Lissodus* from the Monster Bed. A team from UCL collected from the Hampen Marly Formation in 1983, taking samples from the Monster Bed and layers immediately above it. The samples have yielded plants, fish remains, amphibian vertebrae, crocodilian teeth and osteoderms, a pterosaurian tooth fragment, small reptilian jaws, archosaurian teeth, and scraps of tritylodontid teeth (?*Stereognathus*) (F. Mussett, pers. commun.).

Shipton Cement Works Quarry (SP 477175)

Microvertebrate horizon. White Limestone (*fimbriata-waltoni* clay) and Forest Marble, middle to upper Bathonian. This quarry was examined by the UCL team in 1982 and 1983, and we are indebted to Frances Mussett for the following information. Plant material and fish remains (teeth, scales, vertebrae) were found in clay beds probably equivalent to the Forest Marble. A lower clay bed (?*fimbriata-waltoni* clay) (W. A. Wimbledon, pers. commun. to Frances Mussett) yielded fish remains.

Tarlton Clay Pit, near Cirencester (SO 970001)

Microvertebrate horizon. Forest Marble, upper Bathonian. Timberlake (1982), Ware and Windle (1981), and Ware and Whatley (1983) recorded freshwater ostracods from this horizon, in association with charophytes and megaspores, including "*Triletes.*" The environment is reconstructed as being like that of Kirtlington, an inshore brackish or freshwater lagoon. The closest land surface probably was Palmer and Jenkyns's (1975) island barrier. Tetrapod remains are more fragmentary than those at Kirtlington, but the faunal assemblage is closely similar (Table 18.2). To date, Tarlton has yielded a mammalian incisor and two fragmentary molars, one apparently docodontid, and the other triconodontid (S. E. Evans, personal observation, material collected by Martin Ware and Simon Timberlake).

Ready Token, near Cirencester (SP 100050)

Microvertebrate horizon. Forest Marble (Wytchwood Beds) (shelly limestone facies), upper Bathonian. Microvertebrate remains, including a single indeterminate mammalian molar and assorted fish and tetrapod bones, were found at this locality by an amateur collector, Brian Beveridge (M. J. Simms, pers. commun.).

Watton Cliff (West Cliff), Dorset (SY 451908–453907)

Microvertebrate horizon. Forest Marble, upper Bathonian. The Forest Marble at Watton Cliff contains a thick horizon of shell detritus suggestive of an offshore shell bank (Holloway, 1983); channels cut into these shell deposits during storms became filled with detritus, including wood, other plant debris, and microvertebrates. Freeman (1976b) reported finding a multituberculate tooth in the Forest Marble at Watton Cliff (in association with shark and *Teleosaurus* teeth), and Ensom (1977) described a tritylodontid from the same horizon. A team from UCL also collected from the site (Kermack, 1988). The tetrapod bones from Watton often are rolled and abraded, although there are occasional well-preserved specimens (S. E. Evans, personal observation) (Figures 18.2I–K, 18.7E–G). It was clearly a higher-energy depositional environment than that of Kirtlington (E. F. Freeman, pers. commun.; Holloway, 1983). The fauna is similar, in terms of species, to that of Kirtlington (S. E. Evans, unpublished data), but the assemblages differ in that, for example, shark teeth are generally better preserved and the tetrapod elements less well preserved at Watton (E. F. Freeman, pers. commun.).

Swyre, Dorset (SY 525868)

Microvertebrate horizon. Forest Marble, upper Bathonian. This is a small cliff-top exposure from which the UCL team collected in 1976 and 1977. It has yielded an assemblage similar to that of Kirtlington, including mammals. The paleoenvironment was similar to that at Watton (Holloway, 1983), and the tetrapod material is similarly abraded.

Table 18.2. *Middle Jurassic faunal elements*

Taxon	N[a]	H	Sh	S	E	Sk	K	W	Sw	L	T
Anura		×					×	×	×		
Discoglossidae							×	×	×		
Caudata		×					×	×	×	×	×
Albanerpetontidae		?					×	×			×
Marmorerpeton		?					×	×	×		×
Salamander A							×	×	×	×	×
Salamander B		?					×				
Chelonia		×		×		×	×	×	×	×	×
Cryptodira						×	×	×	×		
?Pleurosternidae						×	×				
Choristodera		×					×	×	×	×	×
Cteniogenys		×					×	×	×	×	×
Lepidosauromorpha		×					×	×	×	×	×
Lepidosauromorph A		×					×	×	×		×
Sphenodontia		×					×	×			
Squamata		×					×	×		×	
Scincomorpha							×	×		×	×
Anguinomorpha							×				
Gekkota							?				
Crocodylia	×	×	×	×	×	×	×	×	×	×	×
Goniopholididae		×	×			×	×	×	×	×	×
Teleosauridae	×			×	×		×				
Atoposauridae							×				
Sauropoda	×	×					×				
Theropoda	×	×		×	×	×	×	×	×		
Megalosauridae	×	×		×	×		×				
Coelurosauria		×					×	×			
Iliosuchus				×							
Ornithischia	×	×	×	×			×	×	×		×
Stegosauridae	×	×	×								
?Fabrosauridae		×	×				×	×	×		×
Hypsilophodontidae		×		×							
Pterosauria		×	×	×	×		×	×	×		
Rhamphorhynchidae		×	×	×	×		×				
Rhamphocephalus				×	×						
Pterodactylidae		×					×				
Therapsida		×		×		×	×	×	×	×	×
Tritylodontidae		×		×		×	×	×	×	×	×
Stereognathus		×		×		×	×	×	×	×	×
Mammalia		×		×		×	×	×	×		×
Triconodonta				×			×	×			×
Morganucodontidae							×				?
Wareolestes							×				?
Amphilestidae				×							
Amphilestes				×							
Phascolotherium				×							
Docodonta						×	×	×			×
Borealestes						×					
Simpsonodon							×	×			
Multituberculata							×	×			
Symmetrodonta							×				
Kuehneotheriidae							×				
Cyrtlatherium							×				

Table 18.2. *(Continued)*

Taxon	N[a]	H	Sh	S	E	Sk	K	W	Sw	L	T
Eupantotheria				×		×	×	×			
Amphitheriidae				×	×	?		×			
Amphitherium				×				×			
Peramuridae							×	×			
Palaeoxonodon							×	×			
Dryolestidae							?				

Note: Resolution varies for different sites, and this is reflected in the table. Thus, for example, frogs (Anura) have been reported from four localities but have been identified as discoglossid at only three.

[a]N, New Park Quarry; H, Hornsleasow; Sh, Sharp's Hill; S, Stonesfield; E, Eyford group; Sk, Skye; K, Kirtlington; W, Watton; Sw, Swyre; L, Leigh Delamere; T, Tarlton.

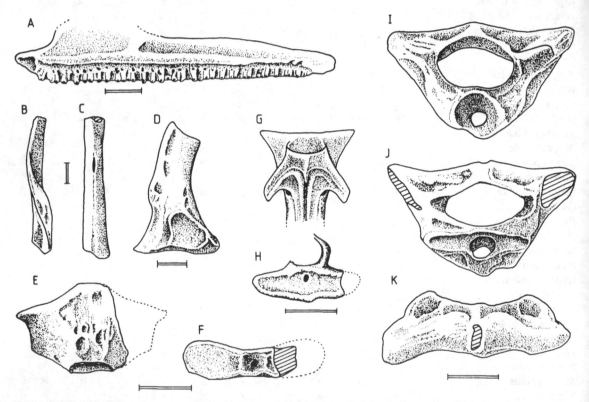

Figure 18.2. Frog *Eodiscoglossus oxoniensis*. (A) Reconstruction of right maxilla in medial view. (B) BMNH R.11706, left angulosplenial in dorsal view. (C) BMNH R.11719, distal shaft of tibiofibula. (D) BMNH R.11701, right ilium in lateral view. (E, F) BMNH R.11707, atlantal centrum in (E) ventral and (F) anterior views. (G, H) Urostyle: (G) partial reconstruction of proximal region in dorsal view; (H) anterior view (UCL collection). (I–K) Sacral vertebra in (I) anterior, (J) posterior, and (K) dorsal views. Scale bars equal 1 mm. All specimens from Kirtlington except the sacral vertebra (I–K) from Watton. (A–F adapted from Evans et al., 1990.)

Leigh Delamere, Wiltshire (ST 890790)

    Microvertebrate horizon. Forest Marble, upper Bathonian. Like Watton Cliff and Swyre, the Forest Marble at Leigh Delamere represents an offshore, high-energy environment, which Holloway (1983)

reconstructed as a tidal delta. Samples collected by the UCL team have yielded rare tetrapod material of poor quality. The bone is broken and very rolled, but the horizon has produced the major components of the Kirtlington fauna, namely, the common salamander A, *Cteniogenys*, *Marmoretta*, lizard, crocodile, turtle, and tritylodontid fragments (S. E. Evans, unpublished data).

Discussion

The new British Bathonian localities are making a significant contribution to our knowledge of Middle Jurassic faunas, both in terms of overall diversity and with respect to the fossil record of individual groups.

The restricted Stonesfield assemblage led to an underestimate of mammalian and nonmammalian tetrapod diversity in the Middle Jurassic. From the evidence to date, the Middle Jurassic appears to have been an important stage linking the essentially archaic faunas of the Triassic and Early Jurassic to the more "modern" faunas of the Late Jurassic and Cretaceous. It provides a record of the survival of some archaic groups and the first appearance of others. However, it must be emphasized that our knowledge of Triassic and Early Jurassic microvertebrates is still incomplete, and many of the groups first recorded in the Bathonian will subsequently be found in earlier deposits. In fact, their appearance in diagnostic form in the Middle Jurassic is, in itself, evidence of a longer fossil history.

Amphibia

Anura

Anuran material is present at Hornsleasow, Kirtlington, Watton, and Swyre. All determinate material either is diagnostically attributable to the Discoglossidae or is consistent with attribution to that family. There is, as yet, no evidence for more than one type of frog at any one locality. The Kirtlington material has been described as *Eodiscoglossus oxoniensis* by Evans et al. (1990); diagnostic elements include ilium, maxilla, atlas, vertebrae, and urostyle (Figure 18.2). The British Bathonian frog material is the earliest certain European frog record, the earliest discoglossid, and the second-oldest crown-group anuran yet described, the oldest being the Early Jurassic *Vieraella* from Argentina (Estes and Reig, 1973).

Caudata

Caudate (salamander) and caudate-like material has been recorded from Hornsleasow, Kirtlington, Watton, Swyre, Leigh Delamere, and Tarlton. At least five taxa can be recognized from atlas vertebrae in the Kirtlington assemblage, and the material from other sites is consistent with attribution to the same taxa. Three of the Kirtlington salamanders are certainly caudates, and the other two are superficially salamander-like lissamphibians with less certain caudate affinities.

The caudates include two large species, *Marmorerpeton kermacki* and *M. freemani*, both Evans, Milner, and Mussett, 1988 (Figure 18.3), with individuals up to 0.4 m in length, and a very small form, here designated

Salamander B. *Marmorerpeton* shows some resemblances to members of the Cretaceous and Early Tertiary Scapherpetontidae, but the absence of intravertebral foramina in the atlas suggests that *Marmorerpeton* is more primitive than any other known salamander. Evans et al. (1988) associated a range of cranial and postcranial material with the vertebrae of *Marmorerpeton*, but it now seems likely that some of the tooth-bearing elements belong to Salamander A, as discussed later. The third crown-group salamander, Salamander B, is a tiny form with quite distinctive atlas and trunk vertebrae. It clearly is not a juvenile of any of the other forms, but its vertebrae do not permit ready assignment to any known family.

Marmorerpeton and Salamander B, together with the contemporaneous *Kokartus* Nessov, 1988 from Kirghizstan, are the oldest indisputable members of the Caudata.

Of the two remaining lissamphibians, one is *Albanerpeton*, a genus now proving to be ubiquitous in the Jurassic and Cretaceous freshwater assemblages of Laurasia. In the British Bathonian, *Albanerpeton* is thus far known only from Kirtlington, Watton, and Tarlton. The British *Albanerpeton* is the second-oldest record of an albanerpetontid, the oldest being a single atlas centrum from the Bajocian of Aveyron (Seiffert, 1969). Finally, Salamander A is the most common lissamphibian at Kirtlington; it is a medium-sized, robustly built form of enigmatic relationships and has yet to be studied in detail. It is known from postcranial and cranial elements. Of the latter, some (jaws and vomers) were previously attributed to *Marmorerpeton*.

Those assemblages, such as Kirtlington, in which diverse salamanders occur in quantity would appear to be largely freshwater in origin.

Amniota

Chelonia

The earliest recorded chelonians are *Proganochelys* and *Proterochersis* from the Upper Triassic of Germany. *Proganochelys* is a stem chelonian, but *Proterochersis* may be an early pleurodire (Gaffney and Meylan, 1988). The first recorded cryptodire is the Early Jurassic *Kayentachelys* from North America. There is then a hiatus in the fossil record between *Kayentachelys* and the turtle assemblages of the Upper Jurassic of North America, Britain, Portugal, Germany, Switzerland, France, and China (Gaffney, 1975; Rieppel, 1980; Gaffney and Meylan, 1988). Until now, the only Middle Jurassic records have been indeterminate turtle scutes, referred to the genus *Protochelys*, in the Stonesfield Slate, and carapace fragments of a ?pleurosternid turtle from the Great Estuarine Series of Skye (Waldman and Savage, 1972; Savage, 1984). However, the Mammal Bed at Kirtlington has yielded skull bones and vertebrae, in addition to shell fragments. Preliminary

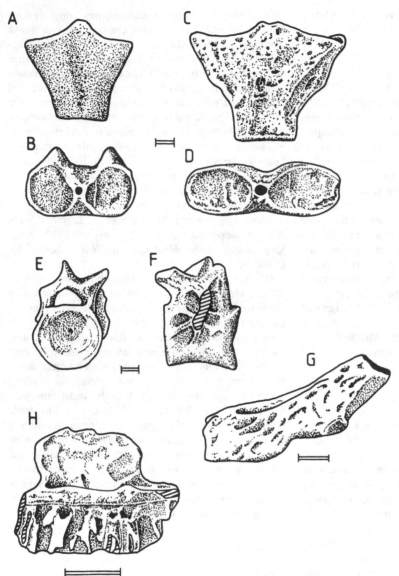

Figure 18.3. Caudate *Marmorerpeton*. (A, B) BMNH R.11361, holotype atlantal centrum of *Marmorerpeton kermacki* in (A) ventral and (B) anterior views. (C, D) BMNH R.11364, holotype atlantal centrum of *Marmorerpeton freemani* in (C) ventral and (D) anterior views. (E) BMNH R.11376, trunk vertebra of *Marmorerpeton* in anterior view. (F) BMNH R.11376, trunk vertebra of *Marmorerpeton* in right lateral view. (G) BMNH R.11375, neural spine of *Marmorerpeton* in left lateral view. (H) BMNH R.11365, left premaxilla of *Marmorerpeton* in lingual view. All scale bars equal 1 mm. (Adapted from Evans et al., 1988.)

work (C. Gillham, pers. commun.) indicates the presence of a cryptodiran turtle related to pleurosternids, but at least specifically different from previously described forms.

Lepidosauromorpha

Lepidosauromorph reptiles (sensu Evans, 1988; Gauthier, 1984; Gauthier, Estes, and de Queiroz, 1988) form a clade encompassing crown-group lepidosaurs (squamates and sphenodontians) and those genera more closely related to them than to any other group (e.g., *Saurosternon* and perhaps *Paliguana* and kuehneosaurs). Sphenodontians are found from the Middle Triassic onward (Fraser and Benton, 1989), but currently there is no record of squamates, although they

must have arisen by that time. (Most workers agree that *Saurosternon*, *Paliguana*, and kuehneosaurs are not lizards and lie outside the lepidosaurian clade.)

Marmoretta. This is one of the most common reptiles at Kirtlington, and has been found at several other sites, including Swyre, Watton, Tarlton, and Hornsleasow (Figure 18.4). Cladistic analysis (Evans, 1991b) places *Marmoretta* as the sister-taxon of known lepidosaurs, suggesting that the genus represents an archaic lineage that split from the lepidosauromorph stem prior to the evolution and radiation of crown-group lepidosaurs. The same genus, represented by a different species (as yet undescribed; S. E. Evans, personal observation), is found at Guimarota, Portugal.

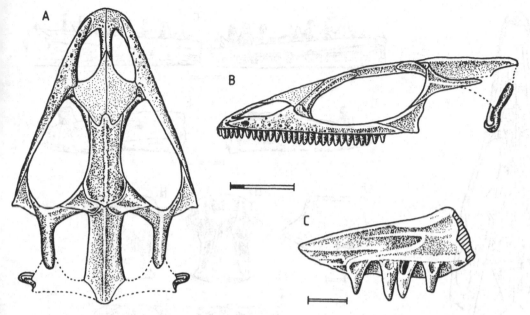

Figure 18.4. Lepidosauromorph *Marmoretta*. (A, B) Reconstruction of skull in (A) dorsal and (B) left lateral views. (C) BMNH R.12020, holotype, anterior region of right maxilla in medial view. Scale bar for A, B equals 5 mm; bar for C equals 1 mm.

Sphenodontia. Sphenodontians are represented by rare jaw fragments from Hornsleasow, Kirtlington, and Watton. These have yet to be analyzed in detail.

Squamata. The earliest recorded crown-group lizards are those of the Upper Jurassic of Europe (Dorset, England; Solnhofen, Germany; Guimarota, Portugal), North America (Como and other Morrison Formation sites), Kazakhstan, and China. Many of these localities yield a diverse assemblage of scincomorph lizards, with occasional anguinomorphs (e.g., Dorsetisauridae from Dorset and Guimarota). Gekkotans are rarer and are represented in the Late Jurassic only by the ardeosaurs (Solnhofen and China) and the bavarisaurs (Solnhofen), although the status of both families remains controversial (Estes, 1983).

The undescribed lizards from the Mammal Bed at Kirtlington, therefore, provide the earliest record of crown-group squamates. They include at least two species of Scincomorpha, an anguinomorph, and a possible primitive gekkotan. The scincomorphs are similar to those previously described from the Late Jurassic, particularly to the genera from Guimarota. At least one Guimarota genus, *Saurillodon*, is represented at Kirtlington.

Archosauromorpha

Choristodera. Choristoderes (most notably *Champsosaurus* and *Simoedosaurus*) are enigmatic gavial-like reptiles known, until recently, only from the Cretaceous and Early Tertiary of North America, Europe, Mongolia, and China. That record has now been extended into the Middle and Upper Jurassic of Britain, Portugal, and North America by new work on the genus *Cteniogenys* (Evans, 1989, 1990, 1991a).

Gilmore (1928) based *Cteniogenys antiquus* on a slender jaw from the Upper Jurassic Morrison Formation of North America. He took it to belong to either a lizard or an amphibian. Seiffert (1973) found more specimens of this taxon at Guimarota and referred the genus to the lizard-like Kuehneosauridae. *Cteniogenys* is abundant at Kirtlington and is represented by most skeletal elements. This new material (Figure 18.5) has shown that *Cteniogenys* is an early, very small (roughly 150 mm snout–vent length) representative of the Choristodera (Evans 1989, 1990, 1991a).

Crocodylia. Crocodilians are common in the British Bathonian localities, either as large skeletal fragments or as abundant teeth (and rarer osteoderms, vertebrae, and skull bones) in the microvertebrate assemblages. The long-snouted crocodilians *Steneosaurus* and *Teleosaurus*, generally associated with some marine influence, are described from Stonesfield and New Park, with some attributed teeth from Watton and from quarries in the Eyford group. The crocodilians from the newer localities have yet to be studied in detail, but goniopholidid teeth (Figure 18.6D,F) and osteoderms are abundant, and at Kirtlington occasional *Theriosuchus*-like atoposaurid teeth have been found (Figure 18.6E). These Bathonian records of the

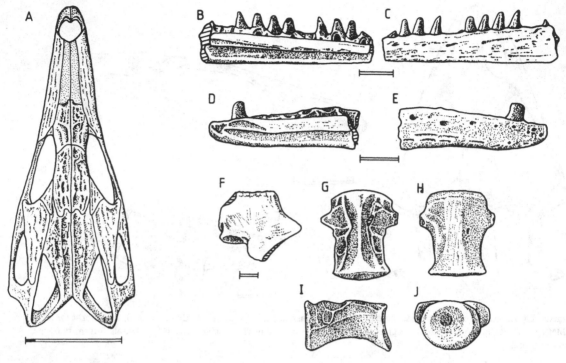

Figure 18.5. Choristodere *Cteniogenys*. (A) Reconstruction of skull in dorsal view. (B, C) BMNH R.11724, juvenile left dentary in (B) medial and (C) lateral views. (D, E) Symphyseal region of a right dentary, BMNH R.11725, in (D) medial and (E) lateral view. (F) BMNH R.11787, dorsal neural spine in right lateral aspect. (G–J) Centrum of a dorsal vertebra, BMNH R.11788, in (G) dorsal, (H) ventral, (I) left lateral, and (J) anterior views. Scale bars for B–J equal 1 mm; bar for A equals 10 mm. (A–E adapted from Evans, 1990; F–J from Evans, 1991a.)

Goniopholididae and Atoposauridae are the first for both families and suggest an earlier diversification of mesosuchian crocodilians than was previously evident.

Pterosauria. "Rhamphorhynchoid" skeletal elements have long been known from Stonesfield and Sarsden and have formed the basis of the ill-defined genus *Rhamphocephalus*, apparently a member of the Rhamphorhynchidae, similar to *Rhamphorhynchus*. At the new localities, pterosaurian teeth are relatively common, and some postcranial elements are recognizable. Slender spikelike teeth corresponding to the anterior teeth of rhamphorhynchids are known from several sites (Figure 18.6C). At Hornsleasow and Kirtlington, short blunt pterosaurian teeth (Figure 18.6A,B) occur that do not resemble those of any known "rhamphorhynchoid" but do resemble those of *Pterodactylus*. If this attribution is correct, even if only at the higher taxonomic level, these teeth represent the earliest records of Pterodactyloidea.

Dinosauria. Much of the Bathonian dinosaurian material comprises isolated large skeletal elements that fall outside the scope of this review. We note the main groups recorded from such elements, as this may

ultimately be useful in identifying small dinosaurian teeth, some of which might belong to juveniles of the larger forms. No ankylosaurians or large ornithopods have yet been reported from the Bathonian of the British Isles.

Ornithopoda (including possible Fabrosauridae). Small ornithopods are represented by teeth in the Bathonian assemblages. Teeth resembling the Guimarota taxon *Alocodon* Thulborn, 1973 are abundant at several localities (Figure 18.6K–M). *Alocodon* is of uncertain systematic position, sometimes attributed to the Fabrosauridae and sometimes treated as an indeterminate primitive ornithischian. Galton (1975, 1980) reported a hypsilophodontid tooth from Stonesfield, and teeth similar to the tooth-genus *Phyllodon* Thulborn, 1973, generally attributed to the Hypsilophodontidae, are also present at Hornsleasow. The Bathonian hypsilophodontid material represents the earliest record of the group.

Stegosauria. The Stegosauridae are represented in the early Bathonian localities, most notably New Park and Sharp's Hill, by isolated skeletal elements. This material was referred to *Lexovisaurus* by Galton

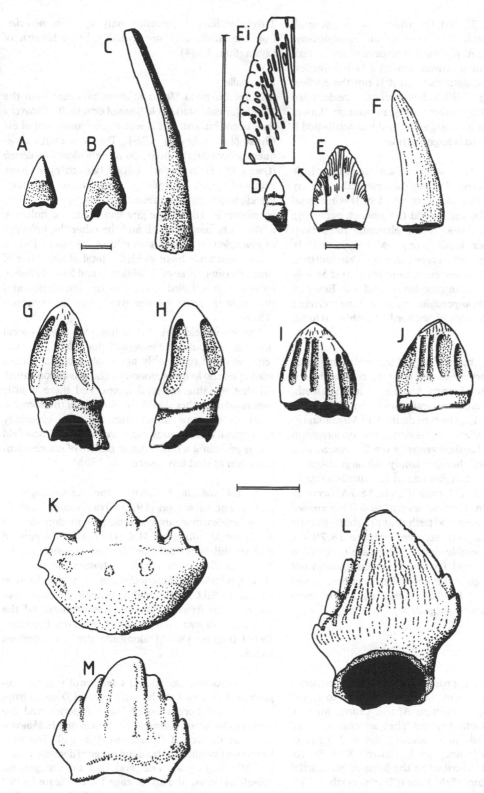

Figure 18.6. Archosaurian teeth from the Forest Marble. (A–C) Pterosauria: (A, B) Pterodactylidae; (C) Rhamphorhynchidae. (D–F) Crocodylia: (D, F) Goniopholididae; (E, Ei) Atoposauridae. (G–J) Unidentified teeth, possibly crocodilian or ornithopod. (K–M) Fabrosauridae. All scale bars equal 1 mm. Uncatalogued teeth in the UCL collection; all from Kirtlington except K (Watton).

and Powell (1983), but the attribution is notional. Stegosaurian teeth are present at Hornsleasow (Vaughan, 1988; A. R. Milner, personal observation). The Bathonian stegosaurian material is the earliest record of the Stegosaurinae, but it is not the earliest stegosaur. Dong (1990) has recently reidentified *Tatisaurus* from the Lower Lufeng Formation (Lower Jurassic) of China as a stegosaur and has attributed it to the subfamily Huayangosaurinae.

Sauropoda. The Bathonian sauropods have little place in this review, but are represented in many localities by isolated elements. Most of this material is attributed to the cetiosaurid *Cetiosaurus*, occurring at numerous localities. A few elements have been assigned to other families (e.g., McIntosh, 1990): the Diplodocidae are represented by *Cetiosauriscus glymptonensis* (a few vertebrae from the Forest Marble of Cogenhoe, Northamptonshire), and the Brachiosauridae by *Bothriospondylus robustus* (one vertebra from the Forest Marble of Bradford, Wiltshire) (Owen, 1875).

Theropoda. The taxonomy of Bathonian theropods is at a rudimentary stage, and the material can be discussed only in broad terms. Apart from isolated teeth, the carnosaurs are represented only by Stonesfield material, comprising the type dentary of *Megalosaurus bucklandi*, together with ilia named *Iliosuchus incognitus* (Galton, 1976). Neither genus is readily placed in a critically diagnosed theropod family. All large theropod teeth from the British Bathonian are attributable to *Megalosaurus*, and such teeth (Figure 18.7A) occur at most localities. In addition, several sites have yielded distinctive small theropod teeth that are short, slightly flattened, and slightly recurved (Figure 18.7B–G). They broadly resemble teeth of the Late Cretaceous Dromaeosauridae and Troodontidae and, though not necessarily belonging to either of these families, almost certainly represent Middle Jurassic relatives of these small maniraptoran theropods.

Theropsida

Tritylodontidae

Tritylodontids are a group of advanced herbivorous cynodonts known from the Upper Triassic to Upper Jurassic of Europe, southern Africa, China, Mexico, Argentina and North America. They are characterized by multiple-rooted, quadrilateral teeth with rows of crescentic cusps (Sues, 1986) (Figure 18.8). *Stereognathus* was first described on the basis of two partial jaws from the Stonesfield Slate (Charlesworth, 1855; Goodrich, 1894; Simpson, 1928). It has now been found at several sites (Table 18.2), but always as isolated teeth. *Stereognathus* is restricted to the Bathonian of Britain and is one of the last representatives of the

Tritylodontidae, superseded only by *Bienotheroides*, from the Middle or, more probably, Upper Jurassic of China (Sun, 1984).

Mammalia

The first recorded Mesozoic mammal came from the Middle Jurassic Stonesfield Slate of Britain. Its discovery date is not known, but it was in the possession of Sir Christopher Sykes by 1764. This specimen subsequently became the holotype of *Amphilestes broderipii* (Owen, 1845). In the early 1800s, two other jaws were acquired by W. J. Broderip, a student of Buckland. Buckland published a preliminary description of these specimens in 1824. One jaw was made the holotype of *Amphitherium prevostii*, and the other the holotype of *Phascolotherium bucklandi* (Owen, 1838a,b, 1842). Subsequent finds have yielded a total of four jaws of *Amphitherium*, three of *Amphilestes*, and four of *Phascolotherium*. A detailed history of the discoveries and the ensuing controversies was given by Simpson (1928).

The Stonesfield Slate facies has only a small local occurrence around the village of Stonesfield. After the removal of all reasonably accessible material, active mining was no longer economical and was discontinued. All that remains are spoil heaps, and there is little likelihood of further significant finds from this locality. Work at the new Middle Jurassic sites, especially Kirtlington, is continuing the work started at Stonesfield and is providing a much better picture of mammalian evolution at that time (Kermack, 1988).

Triconodonta. Following the terminology of Jenkins and Crompton (1979), Triconodonta include three families: morganucodontids, triconodontids, and amphilestids, although Mills (1971) has suggested that amphilestids might be aberrant symmetrodonts. The Stonesfield Slate material showed the presence of amphilestids (*Phascolotherium* and *Amphilestes*) (Figure 18.9B,C) in the Middle Jurassic of Britain, but the material from Kirtlington has documented the survival of morganucodontids (*Wareolestes* Freeman, 1979) (Figure 18.9A) alongside the more derived forms.

Docodonta. Docodonts were, until recently, represented by two Late Jurassic genera: *Docodon* from the Morrison Formation of North America and the Purbeck Limestone Formation of Britain and *Haldanodon* from the lignites of Guimarota, Portugal. The new Bathonian localities extend the range of docodonts into the Middle Jurassic and add two further genera: *Borealestes* (Waldman and Savage,1972) (Figure 18.9D) and *Simpsonodon* (Kermack et al., 1987) (Figure 18.9E). More recently, however, docodonts have been reported from the Upper Triassic of Saint-Nicolas-de-Port, France (Sigogneau-Russell, 1983a,b).

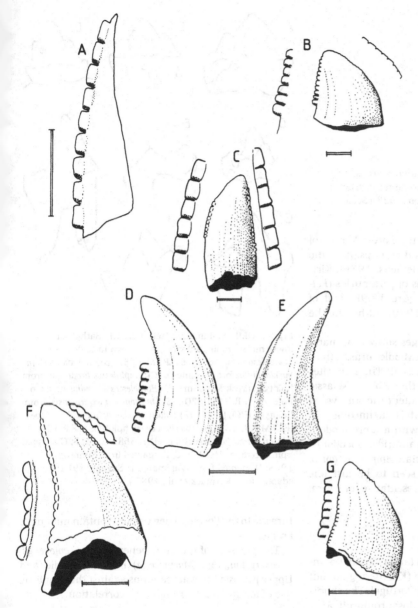

Figure 18.7. Theropod teeth from the Forest Marble. (A) Carnosaurian tooth (cf. *Megalosaurus*), from Kirtlington. (B–G) Dromaeosaur-like teeth: (B) Kirtlington; (C) premaxillary or anterior dentary tooth, Watton; (D, E) two views of the same tooth, Watton; (F) possible troodontid tooth, Watton; (G) posterior tooth possibly allied to (C), Watton. Scale bars equal 1 mm.

Multituberculata. Multituberculates are a major group of herbivorous mammals that survived into the Eocene. They resemble the "Rhaeto"-Liassic haramiyids, and several authors (e.g., Hahn, 1973; Clemens and Kielan-Jaworowska, 1979) consider haramiyids to be ancestral multituberculates. The new Bathonian multituberculates extend the temporal range of the group, but teeth are found only as isolated specimens and, like those of haramiyids, are difficult to orient. The teeth are currently being studied by Kenneth Kermack. Isolated teeth of a similar configuration have been described from the Upper Triassic of Saint-Nicolas-de-Port (Sigogneau-Russell, 1983c).

Symmetrodonta. Symmetrodonts are primitive therians, represented in the Upper Jurassic of Britain (Purbeck) by *Spalacotherium* and *Peralestes*. *Kuehneotherium* from the "Rhaeto"-Liassic may be an early symmetrodont (Cassiliano and Clemens, 1979; cf. Kermack and Kermack, 1984). *Cyrtlatherium* from Kirtlington (Figure 18.9G) was referred to the Kuehneotheriidae by Freeman (1979) and therefore extends the range of this family to the Bathonian.

Eupantotheria. This group includes amphitheriids, peramurids, paurodontids, and dryolestids. Amphitheriids (*Amphitherium*) are known from the Stones-

Figure 18.8. Tritylodontid *Stereognathus ooliticus*, Stonesfield. Right upper molariform tooth reconstructed from holotype. (Adapted from Simpson, 1928.) Scale equals 1 mm.

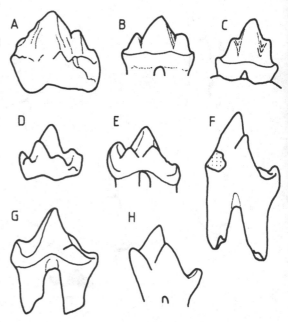

Figure 18.9. Mammalian teeth from the Bathonian (all lower molars in lingual view, not drawn to scale). (A) *Wareolestes rex*, holotype (EF FM/K25). (B) *Phascolotherium bucklandi* (Oxford specimen). (C) *Amphilestes broderipii*, from holotype (York Museum). (D) *Borealestes serendipitus*, from holotype (UBGM 20570). (E) *Simpsonodon oxfordiensis*, from holotype (UCL J.100). (F) *Palaeoxonodon ooliticus*, holotype (EF FM/K8). (G) *Cyrtlatherium canei*, holotype (EF FM/K11). (H) *Amphitherium prevostii* (BMNH 36822). (A, F, G adapted from Freeman, 1979; B, C, H adapted from Simpson, 1928; D adapted from Waldman and Savage, 1972; E adapted from Kermack et al., 1987.)

field Slate (Figure 18.9H), from the Forest Marble of Watton Cliff (Freeman, 1979), and perhaps, from the Great Estuarine Series of Skye (Clemens, 1986). Kirtlington has provided new records of peramurids (*Palaeoxonodon* Freeman, 1979) (Figure 18.9F) and of possible dryolestids (Freeman, 1979), although the latter remain unconfirmed.

The new Bathonian assemblages show that mammals were far more diverse in the Middle Jurassic than the Stonesfield assemblage suggested. The new sites provide an intermediate between the "Rhaeto"-Liassic and the Late Jurassic. They document the survival of archaic groups like morganucodontids, kuehneotheriids, and tritylodontids, in company with a more modern fauna including dryolestids, peramurids, docodonts, and multituberculates. The therian *Amphitherium* of Stonesfield (Simpson, 1928) is seen to be just one element in a fauna containing representatives of several therian lineages.

Conclusions

In many respects, the Kirtlington fauna closely resembles that from the Upper Jurassic (Kimmeridgian, but possibly Oxfordian) of Guimarota, Portugal. The choristodere *Cteniogenys* and the lepidosauromorph *Marmoretta* (Evans, 1991b) are found at Guimarota, along with *Albanerpeton*, a large salamander cf. *Marmorerpeton*, similar lizards, small dinosaurs, a discoglossid frog, and a small goniopholidid crocodile. Only the theropsids are significantly different, with the apparent absence of morganucodontids, kuehneotheriids, and tritylodontids and a greater diversity of multituberculates. Whether the differences are due to age or to a difference in preservation or ecology, however, is difficult to judge. *Cteniogenys* is also found in the slightly younger, but generally similar, Como Bluff fauna. In fact, this freshwater/lagoonal assemblage of choristoderes, turtles, crocodiles, albanerpetontid and scapherpetontid salamanders, and discoglossid frogs appears to be highly stable. Although individual genera change, faunas of this type have been recorded from the Middle Jurassic to the Eocene. They usually contain mammals and lizards.

The presence of the same genera (e.g., *Cteniogenys*, *Albanerpeton*, and *Marmoretta*) in both Middle and Upper Jurassic horizons cautions against the simplistic use of tetrapods for stratigraphic correlation.

Acknowledgments

Our thanks are due to Dr. D. Dartnell, Mr. C. Gillham, Professor K. A. Kermack, Ms. P. Lees, and Mrs. F. Mussett, who have provided additional information about specimens and localities, and to Professor R. Whatley for access to material from Tarlton collected by Dr. Martin Ware and Mr. Simon Timberlake. We would also like to record our thanks to Mr. Eric Freeman, who discovered and first worked the Kirtlington Mammal Bed; he has made his own collections available for study and has been a valuable source of information on Bathonian localities. Last, but not least, our thanks go to Drs. Nick Fraser and Hans-Dieter Sues for the invitation to participate in the Virginia workshop.

A grant from the Royal Society of London enabled one of us (SEE) to attend the Virginia meeting; ARM was supported by a Birkbeck College Research Conference Grant.

References

Ager, D. V. 1956. Field meeting in the Central Cotswolds. *Proceedings of the Geologists' Association* 66: 242–260.

Arkell, W. J. 1931. The Upper Great Oolite, Bradford Beds and Forest Marble of South Oxfordshire, and the succession of gastropod faunas in the Great Oolite. *Quarterly Journal of the Geological Society* 87: 563–629.

1933. *The Jurassic System in Great Britain.* Oxford: Clarendon Press.

1947. *The Geology of Oxford.* Oxford: Clarendon Press.

Benton, M. J. 1988. British fossil reptile sites. *Special Papers in Palaeontology* 40: 73–84.

Buckland, W. 1824. Notice on the *Megalosaurus* or great fossil lizard of Stonesfield. *Transactions of the Geological Society of London* 2: 390–396.

Callomon, J. H. 1979. Marine boreal Bathonian fossils from the North Sea and their palaeogeographical significance. *Proceedings of the Geologists' Association* 90: 163–169.

Cassiliano, M. L., and W. A. Clemens. 1979. Symmetrodonta. Pp. 150–161 in J. A. Lillegraven, Z. Kielan-Jaworowska, and W. A. Clemens (eds.), *Mesozoic Mammals.* Berkeley: University of California Press.

Charlesworth, E. 1855. *Report to the British Association 1854 (Liverpool), Abstracts:* 80.

Clemens, W. A. 1986. On Triassic and Jurassic mammals. Pp. 237–246 in K. Padian (ed.), *The Beginning of the Age of Dinosaurs: Faunal Change across the Triassic–Jurassic Boundary.* Cambridge University Press.

Clemens, W. A., and Z. Kielan-Jaworowska. 1979. Multituberculata. Pp. 99–149 in J. A. Lillegraven, Z. Kielan-Jaworowska, and W. A. Clemens (eds.), *Mesozoic Mammals.* Berkeley: University of California Press.

Cope, J. C. W., K. L. Duff, C. F. Parsons, H. S. Torrens, W. A., Wimbledon, and J. Wright. 1980. *A Correlation of Jurassic Rocks in the British Isles. Pt. 2: Middle and Upper Jurassic. Geological Society Special Report* 15.

Darlington, J. 1988. The Hornsleasow dinosaur excavation. *British Archaeology* 8: 8–10.

Dong, Z.-M. 1990. Stegosaurs of Asia. Pp. 255–268 in K. Carpenter and P. J. Currie (eds.), *Dinosaur Systematics: Approaches and Perspectives.* Cambridge University Press.

Duffin, C. J. 1985. Revision of the hybodont selachian genus *Lissodus* Brough (1935). *Palaeontographica* A188: 105–152.

Ensom, P. C. 1977. A therapsid tooth from the Forest Marble (Middle Jurassic) of Dorset. *Proceedings of the Geologists' Association* 88: 201–205.

1988. Excavations at Sunnydown Farm, Langton Matravers, Dorset. Amphibians discovered in the Purbeck Limestone Formation. *Proceedings of the Dorset Natural History and Archaeological Society* 109: 148–150.

Ensom, P. C., S. E. Evans, and A. R. Milner. 1991. Amphibians and reptiles from the Purbeck Limestone Formation (Upper Jurassic) of Dorset. Pp. 19–20 in Z. Kielan-Jaworowska, N. Heintz, and H.-A. Nakrem (eds.), *Fifth Symposium on Mesozoic Continental Ecosystems and Biota.* Contributions from the Palaeontological Museum, University of Oslo.

Estes, R. 1983. *Sauria terrestria, Amphisbaenia.* In P. Wellnhofer (ed.), *Handbuch der Paläoherpetologie,* Vol. 10A. Stuttgart: Gustav Fischer Verlag.

Estes, R., and O. A. Reig. 1973. The early fossil record of frogs: a review of the evidence. Pp. 11–63 in J. L. Vial (ed.), *Evolutionary Biology of Anurans.* Columbia: University of Missouri Press.

Evans, S. E. 1988. The early history and relationships of the Diapsida. Pp. 221–260 in M. J. Benton (ed.), *The Phylogeny and Classification of the Tetrapods, Vol. 1: Amphibians, Reptiles, Birds.* Oxford: Clarendon Press.

1989. New material of *Cteniogenys* (Reptilia; Diapsida) and a reassessment of the systematic position of the genus. *Neues Jahrbuch für Geologie und Paläontologie, Monatshefte* 1989: 577–589.

1990. The skull of *Cteniogenys,* a choristodere (Reptilia: Archosauromorpha) from the Middle Jurassic of Oxfordshire. *Zoological Journal of the Linnean Society* 99: 205–237.

1991a. The postcranial skeleton of the choristodere *Cteniogenys* (Reptilia: Diapsida) from the Middle Jurassic of England. *Géobios* 24: 187–199.

1991b. A new lizard-like reptile (Diapsida: Lepidosauromorpha) from the Middle Jurassic of England. *Zoological Journal of the Linnean Society* 103: 391–412.

Evans, S. E., A. R. Milner, and F. Mussett. 1988. The earliest known true salamanders (Amphibia: Caudata): a record from the Middle Jurassic of England. *Géobios* 21: 539–552.

1990. A discoglossid frog from the Middle Jurassic of England. *Palaeontology* 33: 299–311.

Fraser, N. C., and M. J. Benton. 1989. A redescription of the Triassic reptiles *Brachyrhinodon* and *Polysphenodon. Zoological Journal of the Linnean Society* 96: 413–445.

Freeman, E. F. 1976a. A mammalian fossil from the Forest Marble of Dorset. *Proceedings of the Geologists' Association* 87:231– 235.

1976b. Mammal teeth from the Forest Marble (Middle Jurassic) of Oxfordshire, England. *Science* 194: 1053–1055.

1979. A Middle Jurassic mammal bed from Oxfordshire. *Palaeontology* 22: 135–166.

Gaffney, E. S. 1975. A taxonomic revision of the Jurassic turtles *Portlandemys* and *Plesiochelys. American Museum Novitates* 2574: 1–33.

Gaffney, E. S., and P. A. Meylan. 1988. A phylogeny of turtles. Pp. 157–219 in M. J. Benton (ed.), *The Phylogeny and Classification of the Tetrapods. Vol. 1: Amphibians, Reptiles, Birds.* Oxford: Clarendon Press.

Galton, P. M. 1975. English hypsilophodontid dinosaurs (Reptilia: Ornithischia). *Palaeontology* 18: 741–752.

1976. *Iliosuchus,* a Jurassic dinosaur from Oxfordshire and Utah. *Palaeontology* 19: 587–589.

1980. European Jurassic ornithopod dinosaurs of the families Hypsilophodontidae and Iguanodontidae. *Neues Jahrbuch für Geologie und Paläontologie, Abhandlungen* 160: 73–95.

Galton, P. M., and H. P. Powell. 1983. Stegosaurian dinosaurs from the Bathonian (Middle Jurassic) of England, the earliest record of the family Stegosauridae. *Géobios* 16: 219–229.

Gauthier, J. 1984. A cladistic analysis of the higher systematic categories of the Diapsida. Ph.D. thesis, University of California, Berkeley.

Gauthier, J., R. Estes, and K. de Queiroz. 1988. A phylogenetic analysis of Lepidosauromorpha. Pp. 15–98 in R. Estes and G. Pregill (eds.), *Phylogenetic Relationships of the Lizard Families.* Stanford University Press.

Gilmore, C. W., 1928. Fossil lizards of North America. *Memoirs of the National Academy of Sciences* 22: 1–169.

Goodrich, E. S. 1894. On the fossil Mammalia of the Stonesfield Slate. *Quarterly Journal of Microscopical Science, n.s.* 35: 407–432.

Hahn, G. 1973. Neue Zähne von Haramiyiden aus der deutschen Ober–Trias und ihre Beziehungen zu den Multituberculaten. *Palaeontographica* A142: 1–15.

Harland, W. B., A. V. Cox, P. G. Llewellyn, C. A. G. Pickton, A. G. Smith, and R. Walters. 1982. *Geological Time Scale.* Cambridge University Press.

Holloway, S. 1983. The shell-detrital calcirudites of the Forest Marble Formation (Bathonian) of South West England. *Proceedings of the Geologists' Assocation* 94: 259–266.

Huxley, T. H. 1859. On *Rhamphorhynchus bucklandi*, a pterosaurian from the Stonesfield Slate. *Quarterly Journal of the Geological Society* 15: 658–670.

Jakovides, J. 1982. A palaeoecological study of Bathonian Ostracoda from Kirtlington, Oxfordshire. M.Sc. thesis, University College of Wales, Aberystwyth.

Jenkins, F. A., Jr., and A. W. Crompton. 1979. Triconodonta. Pp. 74– 90 in J. A. Lillegraven, Z. Kielan-Jaworowska, and W. A. Clemens (eds.), *Mesozoic Mammals.* Berkeley: University of California Press.

Kermack, D. M., and K. A. Kermack. 1984. *The Evolution of Mammalian Characters.* London: Croom Helm.

Kermack, K. A. 1988. British Mesozoic mammal sites. *Special Papers in Palaeontology* 40: 85–93.

Kermack, K. A., A. J. Lee, P. M. Lees, and F. Mussett. 1987. A new docodont from the Forest Marble. *Zoological Journal of the Linnean Society* 89: 1–39.

Lydekker, R. 1888. *Catalogue of Fossil Reptiles and Amphibians in the British Museum, London. Part 1: Containing the Orders Ornithosauria, Crocodilia, Dinosauria, Squamata, Rhynchocephalia and Proterosauria.* London: Trustees of the British Museum.

    1889. *Catalogue of Fossil Reptiles and Amphibians in the British Museum, London. Part 3: Containing the Order Chelonia.* London: Trustees of the British Museum.

McIntosh, J. S. 1990. Sauropoda. Pp. 345–401 in D. B. Weishampel, P. Dodson, and H. Osmolska (eds.), *The Dinosauria.* Berkeley: University of California Press.

McKerrow, W. S., D. V. Ager, and D. T. Donovan. 1964. *Geology of the Cotswold Hills.* Geologists' Association Guide 36. Colchester: Benham and Co.

McKerrow, W. S., R. T. Johnson, and M. E. Jakobson. 1969. Palaeoecological studies in the Great Oolite at Kirtlington, Oxfordshire. *Palaeontology* 12: 56–83.

Martin, A. J. 1962. Bathonian sedimentation in southern

England. *Proceedings of the Geologists' Association* 78: 473–488.

Mills, J. R. E. 1971. The dentition of *Morganucodon*. Pp. 29–63 in D. M. Kermack and K. A. Kermack (eds.), *Early Mammals.* London: Academic Press.

Nessov, L. A. 1988. Late Mesozoic amphibians and lizards of Soviet Middle Asia. *Acta Zoologica Cracoviensis* 31: 475–486.

Odling, M. 1913. The Bathonian rocks of the Oxford District. *Quarterly Journal of the Geological Society of London* 69: 484–513.

Owen, R. 1838a. On the jaws of the *Thylacotherium prevostii* (Valenciennes) from Stonesfield. *Proceedings of the Geological Society of London* 3: 5–9.

    1838b. A paper on the "*Phascolotherium*," being the second part of the "Description of the remains of marsupial Mammalia from the Stonesfield Slate." *Proceedings of the Geological Society of London* 3: 17–21.

    1842. Observations on the fossils representing the *Thylacotherium prevostii* (Valenciennes), with reference to the doubts of its mammalian and marsupial nature recently promulgated; and on the *Phascolotherium bucklandi. Transactions of the Geological Society of London* (2) 6: 47–65.

    1845. *Odontography*, 2 vols. London: Hippolyte Bailliere.

    1874. Monographs of the fossil Reptilia of the Mesozoic formations. Pt. 1: Pterosauria. *Palaeontographical Society Monographs* 27: 1–14.

    1875. Monographs of the fossil Reptilia of the Mesozoic formations. Pt. 2: Genera *Bothriospondylus, Cetiosaurus, Omosaurus. Palaeontographical Society Monographs* 29: 15–94.

Palmer, T. J. 1973. Field meeting in the Great Oolite of Oxfordshire. *Proceedings of the Geologists' Association* 84: 53–64.

    1979. The Hampen Marley and White Limestone Formations: Florida-type carbonate lagoons in the Jurassic of Central England. *Palaeontology* 22: 189–228.

Palmer, T. J., and H. C. Jenkyns. 1975. A carbonate island barrier from the Great Oolite (Middle Jurassic) of Central England. *Sedimentology* 22: 125–135.

Phillips, J. 1871. *Geology of Oxford and the Valley of the Thames.* Oxford: Clarendon Press.

Richardson, L. 1911. On the sections of Forest Marble and Great Oolite on the railway between Cirencester and Chedworth, Gloucestershire. *Proceedings of the Geologists' Association* 22: 95–115.

    1925. Excursion to the North Cotteswolds (Stonesfield Slate, Naeran Beds and Chipping Norton Limestone). *Proceedings of the Cotteswold Naturalists' Field Club* 22: 67–73.

    1929. The country around Moreton-in Marsh. *Memoirs of the Geological Survey of England and Wales, Sheet* 217.

Rieppel, O. 1980. The skull of the Upper Jurassic cryptodire turtle *Thalassemys*, with a reconsideration of the chelonian braincase. *Palaeontographica* A171: 105–140.

Savage, R. J. G. 1963. The Witts Collection of Stonesfield Slate fossils. *Proceedings of the Cotteswold Naturalists' Field Club* 33: 177–182.

1984. Mid-Jurassic mammals from Scotland.
Pp. 211–213 in W. E. Reif and F. Westphal (eds.),
Third Symposium on Mesozoic Terrestrial Ecosystems
Tübingen: Attempto.

Seiffert, J. 1969. Urodelen-Atlas aus der obersten Bajocien
von SE-Aveyron (Südfrankreich). *Paläontologische
Zeitschrift* 43: 32–36.

1973. Upper Jurassic lizards from Central Portugal.
Memorias Serviços Geologicos de Portugal 22: 7–85.

Sellwood, B. W., and W. S. McKerrow. 1974. Depositional
environments in the lower part of the Great Oolite
Group of Oxfordshire and North Gloucestershire.
Proceedings of the Geologists' Association 85: 189–210.

Sigogneau-Russell, D. 1983a. A new therian mammal
from the Rhaetic locality of Saint-Nicolas-de-Port
(France). *Zoological Journal of the Linnean Society*
78: 175–186.

1983b. Caractéristiques de la faune mammalienne
du Rhétien de Saint-Nicolas-de-Port
(Meurthe-et-Moselle). *Bulletin d'Information des
Géologues du Bassin de Paris* 20: 51–53.

1983c. Nouveaux taxons de mammiféres rhétiens. *Acta
Palaeontologica Polonica* 28: 233–249.

Sigogneau-Russell, D., R. M. Frank, and J. Hemmerlé.
1986. A new family of mammals from the lower
part of the French Rhaetic. Pp. 99–108 in
K. Padian (ed.), *The Beginning of the Age of Dinosaurs:
Faunal Change across the Triassic–Jurassic Boundary.*
Cambridge University Press.

Simpson, G. G. 1928. *A Catalogue of Mesozoic Mammalia in
the Geological Department of the British Museum.*
London: Trustees of the British Museum.

Smith, A. G., and J. C. Briden. 1977. *Mesozoic and Cenozoic
Continental Maps.* Cambridge University Press.

Sues, H.-D. 1986. Relationships and biostratigraphic
significance of the Tritylodontidae (Synapsida) from
the Kayenta Formation of northeastern Arizona.
Pp. 279–284 in K. Padian (ed.), *The Beginning of
the Age of Dinosaurs: Faunal Change across the
Triassic–Jurassic Boundary.* Cambridge University
Press.

Sumbler, M. G. 1984. The stratigraphy of the Bathonian
White Limestone and Forest Marble formations of
Oxfordshire. *Proceedings of the Geologists' Association*
95: 51–64.

Sun, A.-L. 1984. Skull morphology of the tritylodont genus
Bienotheroides of Sichuan. *Scientia Sinica* B27:
970–984.

Sylvester-Bradley, P. C. 1948. *Bathonella* and *Viviparus.*
Geological Magazine 85: 367.

Thulborn, R. A. 1973. Teeth of ornithischian dinosaurs
from the Upper Jurassic of Portugal. *Memorias
Serviços Geologicos de Portugal* 22: 89–134.

Timberlake, S. 1982. Taxonomic and population study of
Limnocytheridae (Ostracoda) from the Forest Marble
of Tarlton, Gloucestershire. M.Sc. thesis, University
College of Wales, Aberysthwyth.

Vaughan, R. F. 1988. Cotswold dinosaur excavation.
Geology Today 4: 150–151.

1989. *The Excavation at Hornsleasow Quarry: An Interim
Report Prepared for the Nature Conservancy Council.*
Gloucester: City Museum and Art Gallery.

Waldman, M., and R. J. G. Savage. 1972. The first Jurassic
mammal from Scotland. *Journal of the Geological
Society of London* 128: 119–125.

Ward, D. J. 1984. Collecting isolated microvertebrate
fossils. *Zoological Journal of the Linnean Society*
82: 245–259.

Ware, M. 1978. Palaeoecology and Ostracoda of a
Bathonian mammal bed in Oxfordshire. M.Sc.
thesis, University College of Wales, Aberystwyth.

Ware, M., and R. Whatley. 1980. New genera and species
of Ostracoda from the Bathonian of Oxfordshire,
England. *Revista Española de Micropaleontologia*
12: 199–230.

1983. Use of serial ostracod counts to elucidate
the depositional history of a Bathonian clay.
Pp. 131–164 in R. F. Maddocks (ed.), *Applications of
Ostracoda.* University of Houston Geosciences.

Ware, M., and T. M. F. Windle. 1981. Micropalaeontological
evidence for land near Cirencester, England, in
Forest Marble (Bathonian) times: a preliminary
account. *Geological Magazine* 118: 415–420.

Wellnhofer, P. 1978. *Pterosauria.* In P. Wellnhofer (ed.),
Handbuch der Paläoherpetologie, Vol. 19. Stuttgart:
Gustav Fischer Verlag.

Woodward, H. B. 1894. The Jurassic rocks of Britain.
Vol. 4. The Lower Oolitic rocks of England
(Yorkshire excepted). *Memoirs of the Geological
Survey of the United Kingdom* 1894: 1–628.

19

A new Bathonian microvertebrate locality in the English Midlands

SARA J. METCALF AND RACHEL J. WALKER

Introduction

The Middle Jurassic of the English Midlands is yielding information on fossil terrestrial vertebrates during an important period of their evolution that is not well represented elsewhere in the world. Until some recent discoveries in China, Britain had virtually the only prolific sites of that age worldwide. Indeed, the first dinosaurian fossil recorded, the distal extremity of a femur of *Megalosaurus* sp., was discovered near Chipping Norton, Oxfordshire, in the Bajocian Inferior Oolite sequence (Plot, 1677).

The present locality (*zigzag* zone) at Hornsleasow Quarry, North Gloucestershire, was discovered in 1987 when an amateur geologist, Kevin Gardner, recovered fragments of long bone and vertebrae (since identified as belonging to the sauropod dinosaur *Cetiosaurus* sp.) from the debris of a recently blasted clay lens within the limestone succession. Subsequent sampling of the site has revealed a wealth of vertebrate fossils, both large and small.

The terrestrial tetrapod record of the Middle Jurassic is very poorly known. The Middle Jurassic was the time of radiation of many of the classic dinosaurian groups, replacing Triassic and Early Jurassic taxa. Other groups were becoming more diversified, such as frogs, salamanders, turtles, and mammals. The English Midlands has already produced the oldest known salamander remains (Evans, Milner, and Mussett, 1988), the earliest frog of modern aspect (Evans, Milner, and Mussett, 1990), and the oldest choristoderan, *Cteniogenys* sp. (Evans, 1989, 1990), from the upper Bathonian locality at Kirtlington, Oxfordshire. The Cotswolds also offer much potential for new finds, as many of the 30 or so vertebrate sites documented in the eighteenth, nineteenth, and early twentieth centuries have been lost or were not prospected for microvertebrate remains. In fact, the only sites currently producing micro-

vertebrate material are Hornsleasow, Kirtlington, and Stonesfield (middle Bathonian of Oxfordshire).

Hornsleasow is important as being the oldest of these Bathonian sites, and preliminary sampling of the vertebrate material has revealed taxa similar to those discovered at Kirtlington, increasing the range of these groups downward by several million years. The aims of this chapter are to outline current research at the Hornsleasow locality and to state the aims for future work, in the hope that this study may add to the knowledge of Middle Jurassic vertebrate evolution being gleaned from the other Cotswold sites and from the material recovered from localities in Portugal (Guimarota lignite mines, Leira, of Oxfordian–Kimmeridgian age) (Milner and Evans, 1991), China, Kazakhstan (Nessov, 1988), and Mexico (Clark and Hopson, 1985).

Location

Hornsleasow Quarry (or Snowshill Quarry, as it was known in the earlier literature) is situated in a down-faulted block of the Chipping Norton Limestone Formation, against the deposits of the Inferior Oolite (Figure 19.1). The site is located in the northern Cotswold hills, some 5 km northwest of the market town Stow-in-the-Wold, and to the east of Gloucester and Cheltenham (U.K. National Grid Reference SP 131322). It is worked occasionally for the Chipping Norton Limestone, an oolitic limestone used for building and facing work. The blasting of this limestone dislodged much of a large clay lens, from which the scattered remains were collected; it also exposed the lens lying in situ within the limestone outcrop (Figure 19.2). The original finds were reported to Gloucester City Museum, which undertook an excavation to remove the whole lens for sieving and microvertebrate extraction.

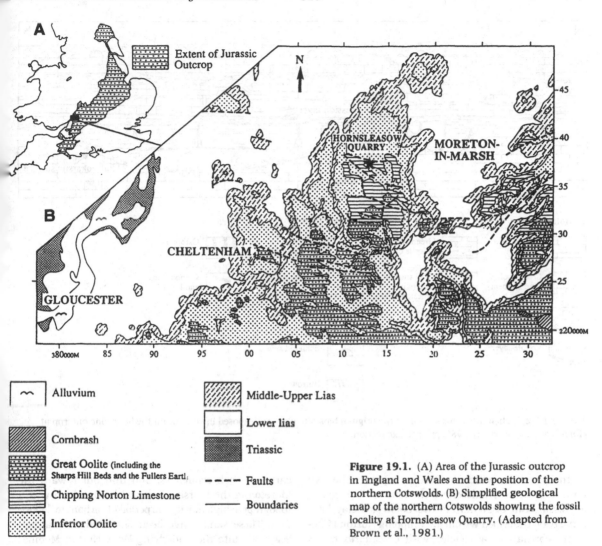

Figure 19.1. (A) Area of the Jurassic outcrop in England and Wales and the position of the northern Cotswolds. (B) Simplified geological map of the northern Cotswolds showing the fossil locality at Hornsleasow Quarry. (Adapted from Brown et al., 1981.)

Geology

Stratigraphy of Hornsleasow Quarry

The succession at Hornsleasow Quarry is well known to workers on the British Jurassic and has been designated an SSSI (site of special scientific interest) because of the complete succession of the Sharp's Hill Beds (which overlie the Chipping Norton Limestone and the clay lens). The position of the lens within the section is shown in Figure 19.3, and within the context of the Bathonian succession of the Cotswolds in Figure 19.4. The Hornsleasow succession has been described by several authors (Richardson, 1929; Channon, 1950; Torrens, 1968, 1969; Vaughan, 1989). The section was most fully described by Torrens, who produced a most detailed synthesis of the geology (Torrens, 1969, pp. 16–18).

The Chipping Norton Limestone Formation was first described by Hudleston (1878), from a succession at Chipping Norton (to the southeast of Hornsleasow). He intended that the name should apply to the intervening limestone between the Sharp's Hill Beds and the underlying *Clypeus* Grit. The limestone is rather variable in lithology, ranging from a pure white oolitic limestone in the southwest to a rather sandy deposit eastward. It has been split into two distinct members on the basis of the succession at Hook Norton (to the east of the present locality), where the soft, sandy oolitic Chipping Norton Member lies above a more massive, evenly bedded limestone, the Hook Norton Member (Richardson, 1911). Whereas the division between the two members is distinct at Hook Norton, elsewhere it can be proved only by using paleontological evidence.

It is therefore interesting to note that Channon (1950, p. 249) described a "large lenticle of tough black clay"

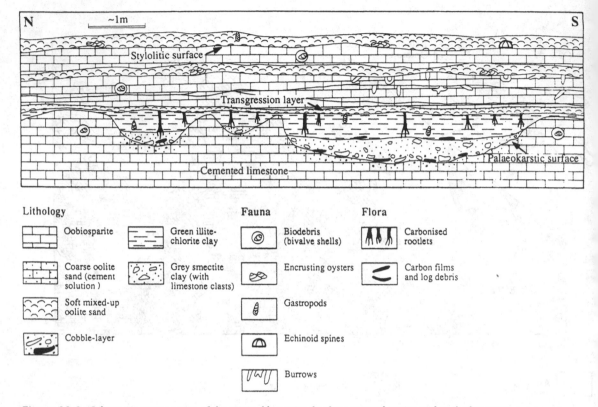

Figure 19.2. Schematic cross-section of the original bipartite clay lens exposed in situ within the limestone outcrop at Hornsleasow Quarry, in 1987, prior to excavation.

at the junction, which he used to separate the two limestone units. He described the lens as "barren" of all fossils, but elsewhere in the Cotswolds clay beds have been described within the Chipping Norton Limestone Formation, associated with *Cetiosaurus* bones (Beesley, 1877; Phillips, 1878; Richardson, 1911). A lignitic clay lens within the limestone at Sarsden (SP 300266), to the south of Chipping Norton, has been described as containing the ?freshwater gastropod *Viviparus* (Arkell, 1947, p. 65) and has been interpreted as a "temporary fresh-water pool within the carbonate dominated area" (Sellwood and McKerrow, 1974, p. 198). Except for the lens at Hornsleasow described by Channon (1950), it would be imprudent to suggest a correlation for the other Cotswold clay deposits with the present lens, but fieldwork being undertaken within the Cotsworlds should help to elucidate this.

In the regional context, the Chipping Norton Limestone has been described by Richardson (1911, 1929), Arkell (1933), Sellwood and McKerrow (1974), and Horton (1977). The Swerford Beds are included in the succession as the lowest part of the Chipping Norton Member (Figure 19.4). These cross-bedded silica sands and clay drapes are land-derived, from the London-Brabant landmass, some 20 km to the east of Hornsleasow. They grade westward into more carbonate-rich sequences, and thin out toward the Moreton-in-the Marsh structural high, which was a topographic high during the period of carbonate deposition. These sands have been described as channeling down into the underlying Hook Norton Member (Torrens, 1968). At Hornsleasow, Vaughan (1989) has reported a cross-bedded transgressive sand unit overlying the clay lens (and the Hook Norton Member in parts of the quarry section where the clay has pinched out). This appears to be the lateral equivalent to the Swerford Beds, but is in fact a marine carbonate sand packed full of biodebris, rather than the land-derived silica sands in the east.

The Hornsleasow Clay Unit is placed stratigraphically between the Chipping Norton and Hook Norton members (Figures 19.3 and 19.4). The lens (Figure 19.2) lies upon the weathered, undulating upper surface of the Hook Norton Member, within a hollow developed within the limestone. The surface shows much iron staining and dissolution. It has been interpreted as a paleokarst, and the undulating topography of the surface is an original feature of the karstic landscape (M. J. Simms, unpublished data). This surface can be traced throughout the quarry. The sharpness of the karstic features has led Simms (pers. commun.) to suggest that development of the karst was not con-

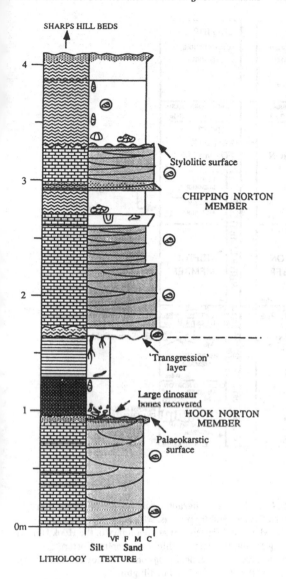

temporaneous with deposition of the clay, but was initiated several thousand years earlier.

The clay unit is not homogeneous, but is divided into two distinct layers (Figure 19.2), the upper "Green Clay" and the lower "Grey Clay." The Grey Clay overlies the karstic surface and is 0.3 m thick. It is matrix-supported, but contains large, irregularly sub-rounded limestone cobbles (concentrated in the basal portion, which is interpreted as a regolith), with associated lime and fine-fraction silica sand. The clay is smectitic, with a high organic content, and it contains much partially lignitized wood (particularly in the basal layer) that has retained its original textures. This suggests that the Grey Clay was deposited under somewhat anoxic conditions, halting the decay of the logs.

The Grey Clay layer is finely laminated. This lamination has been somewhat disrupted, possibly by bioturbation, which could also account for the fragmentary nature of the fossil vertebrate material. The fauna of the deposit is predominantly terrestrial. Of 8,000 microfossils thus far retrieved through sieving, the majority are from the Grey Clay. The large dinosaurian bones and the log material were recovered from the base of the regolith layer, and their nondirectional nature suggests low paleocurrent activity within the deposit (Vaughan, 1989).

The upper Green Clay (0.4 m thick) is more homogeneous than the Grey Clay. Limestone clasts and sand make up only 3 percent of the composition by weight. The deposit has more silica sand and is an illite/chlorite clay. Vertical rootlets pervade the layer, extending from the mud-cracked, upper surface, which apparently was covered in a thick, lycopodaceous flora (J. Cole, pers. commun.), into the underlying Grey Clay (Figure 19.2). The plant material is somewhat oxidized, suggesting more oxidizing conditions than those evident during Grey Clay deposition.

Depositional environment for the Hornsleasow Clay Unit

The Hornsleasow Clay Unit is interpreted as a paleosol. The depositional environment proposed for the Grey Clay is one of marshy, low-pH type. The partially decayed organic material within this deposit suggests relatively reducing conditions, and the presence of pyrolusite (MnO_2), which forms a coating on the upper surfaces of the pebbles and bones within the basal regolith, is suggestive of bog or lake deposition. The invertebrate assemblage therein is composed of derived marine material and some autochthonous gastropods and ostracods. The former have been identified as *Viviparus* sp. and *Valvata* sp. (R. J. Clements, pers. commun.), which suggests that salinity was low to moderate.

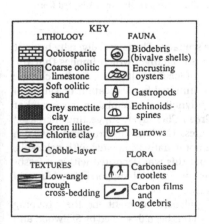

Figure 19.3. Graphic log of the section exposed in Hornsleasow Quarry, showing the position of the clay lens unit within the Chipping Norton Formation.

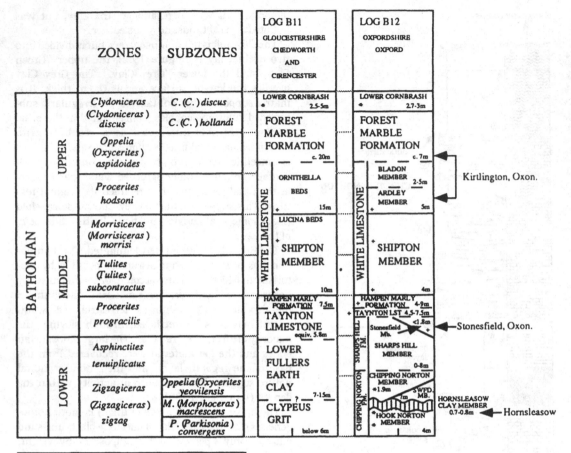

	ZONES	SUBZONES	LOG B11 GLOUCESTERSHIRE CHEDWORTH AND CIRENCESTER	LOG B12 OXFORDSHIRE OXFORD	

Figure 19.4. Bathonian statigraphic columns for the Oxfordshire and northern Gloucestershire regions, with Hornsleasow Clay Member placed at the "nonsequential" boundary of the Hook Norton and Chipping Norton members within the Chipping Norton Formation (*mascrescens* subzone, *zigzag* zone of the lower Bathonian). The positions of the Stonesfield and Kirtlington microvertebrate-bearing localities are also shown. (Adapted from Cope et al., 1980.)

The Grey Clay is a waterlogged, boggy paleosol developing within a freshwater to brackish-water pond. The absence of dispersed kerogen within the clay indicates a low-energy depositional environment, with little fluvial influence, and the palynofloral assemblage suggests very low sedimentation rates within a low-nutrient (oligotrophic) lacustrine environment (J. Cole, pers. commun.).

The Green Clay is also interpreted as a paleosol, deposited under more oxidizing, high-pH conditions. The boundary between the two clay units is relatively sharp, and the layers are easily separated along this contact. There is a considerable change in mineralogy of the clays across this boundary: a smectitic clay could be altered to an illite/chlorite clay during diagenesis, but the sedimentological differences suggest that a rapid, localized change in depositional environment is a more plausible scenario. The faunal diversity has not been affected by this change, which implies that the effects were only of local significance. Such a change could have been initiated by desiccation or silting of the pond. The palynofloral assemblage is somewhat barren in the Green Clay layer, suggesting increased sedimentation rates. The lens has produced a wealth of diverse fauna not usually associated with paleosols, which implies that this was deposited within a freshwater to brackish-water pool supporting a fauna of terrestrial animals.

The regional paleoenvironment for the Chipping Norton Formation has been described by Sellwood and McKerrow (1974), who postulated a reduced-salinity estuarine environment bordering the London-Brabant landmass to the east, grading westward into more open, high-energy, shallow-marine carbonate facies.

Hardgrounds were developed within the oolitic lime-stones, which formed a stable substrate for encrusting bivalves. The shallow nature of this platform is attested to by the terrestrial environment (and faunal/floral assemblages) encountered at Hornsleasow. Sea-level curves suggest a stillstand or slight fall in sea level within the lower Bathonian (Haq, Hardenbol, and Vail, 1988), and Hallam (1978) has implied a relatively regressive episode during the earlier part of the stage.

Age of the deposit

The Hornsleasow Clay Unit is placed within the *mascrescens* subzone of the *zigzag* zone of the lower Bathonian (Figure 19.4). Hornsleasow is thus the earliest Bathonian mammal-producing site. Prior to the discovery, the oldest Bathonian mammalian remains were recovered from the Stonesfield Slate Member of the Sharp's Hill Formation (*progracilis* zone of the middle Bathonian) at Stonesfield, Oxfordshire. The site also predates the Mammal Bed at the Kirtlington locality currently being studied by S. E. Evans and A. R. Milner, which is of upper Bathonian age (*aspidoides* and *hodsoni* zones).

Details on excavation of the site have been given by Darlington (1988) and Vaughan (1988, 1989), and the static sieving techniques described by Ward (1981) have been employed to process the material.

Fauna

The skeletal remains of large reptiles are ubiquitous within the Chipping Norton Limestone Formation of the Cotswolds, and many of the old collections of European Middle Jurassic dinosaurian and crocodilian remains came from this region (Beesley, 1877; Phillips, 1878; Richardson, 1911). Yet the diversity of forms known from this sequence was, until recently, quite low. That was due in part to a preservational bias

toward the larger vertebrate remains and in part to the fact that many of the classic Chipping Norton Limestone localities (such as the quarries surrounding Stow-in-the-Wold and Chipping Norton) and the later Bathonian sites were prospected only for the large vertebrate remains.

Hornsleasow, to date, has produced a spectacular variety of terrestrial and freshwater aquatic vertebrate material. Figure 19.5 shows the percentages of the various types of finds recorded, whether an isolated tooth or semiarticulated jaw fragment, and thus it does not indicate the actual faunal diversity of the deposit. The following provides a summary of the material recovered to date.

Fish

Mainly, isolated dermal scales, particularly of the semionotid *Lepidotes* sp. have been found. There are also many small teeth of hybodont sharks (*Asteracanthus* sp.) and pycnodonts.

Amphibia

So far, jaw fragments, vertebrae, and limb elements similar to the salamander remains from the Kirtlington site (Evans et al., 1988; S. E. Evans and A. R. Milner, pers. commun.) have been recovered. There are also possible frog remains.

Nonmammalian amniotes

Testudines. Mostly, undetermined turtle cara-pace material and some postcranial bones (dissociated) have been found.

Lepidosauromorpha. Many small jaw fragments and a few limb elements attributed to the Sphenodontia,

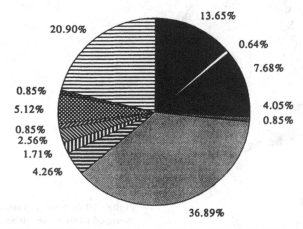

FI ☐ AM ■ TE ■ LE ▨ CP ▨ CR ⊟ PT ⊞ DI ⧅ TH
▨ ON ▨ ML ▨ MA ⊟ BO

Figure 19.5. Pie chart showing a representative sample (500 specimens) of the microvertebrate material recovered from the Hornsleasow deposit. Abbreviations: AM, amphibian material (see text for description); BO, unspecified bone elements and fragmented bone; CP, choristoderan bone material; CR, crocodilian material; DI, unspecified dinosaurian material; FI, fish tooth plates and scales; LE, lepidosauromorph material (see text); MA, mammal teeth; ML, tritylodont teeth; ON, ornithischian teeth; PT, pterosaurian material; TH, theropod teeth.

and some possibly with squamate affinities, have been found.

Choristodera. A tiny proportion (Figure 19.5) of the fragmentary "lizardlike" jaw elements found in this deposit are similar to those of the small diapsid reptile *Cteniogenys* sp. described from Kirtlington (Evans, 1989) and the Upper Jurassic (Tithonian) of Como Bluff (Wyoming).

Crocodylia. Many thousands of small crocodilian teeth and osteoderms attributable to the "goniopholids" make up nearly 40 percent of the small finds from the present site (Figure 19.5). There is also some undetermined cranial and postcranial material.

Pterosauria. The pterosaurian material is largely made up of isolated broken tooth crowns ("rhamphorhynchoid" and ?pterodactyloid) and fragmentary limb elements and vertebrae.

Sauropod dinosaurs. The original finds made at Hornsleasow were referred to *Cetiosaurus*, and additional bones of this genus were recovered from the excavation and are retained at Gloucester City Museum. The bone is in a poor state of preservation, often being fragmented and quite porous. The upper surfaces (i.e., the surfaces "face-up" in the Grey Clay layer) are in the poorest condition and are covered in a fine layer by pyrolusite (MnO_2) and iron hydroxides. This possibly reflects the rather acidic conditions that prevailed in the depositional environment of the Grey Clay.

Theropod dinosaurs. Theropod dinosaurs are represented by at least five large (> 10 mm) tooth crowns. These show finely serrated carinae and are slightly recurved. They are identified as *Megalosaurus* sp. with considerable confidence, as *Megalosaurus bucklandi* was recovered from the Stonesfield Slate Member of the middle Bathonian of Oxfordshire (Buckland, 1824), and many similar teeth, some with jaws, have been assigned to this genus from the Cotswold Bathonian. The remaining teeth are much smaller (2–10 mm). Some of these may in fact prove to be juvenile megalosaurid teeth, because they show the same characters as the larger teeth. The majority of the small theropod teeth have been bracketed as "Theropoda indet." There are a few small teeth (Figure 19.6) that are more triangular in shape and bear much coarser serrations, and these resemble the juvenile and replacement teeth of the dromaeosaurs (C. L. Chandler, pers. commun.; S. J. Metcalf and R. J. Walker, personal observations). They may be maniraptoran in nature, but only character analysis will elucidate this. The deposit also contains some large and small disarticulated bones attributable to theropods.

Ornithischian dinosaurs. The site has produced a few small (2–10 mm) ornithischian teeth (Figure 19.5), some of which are quite worn, and none are associated

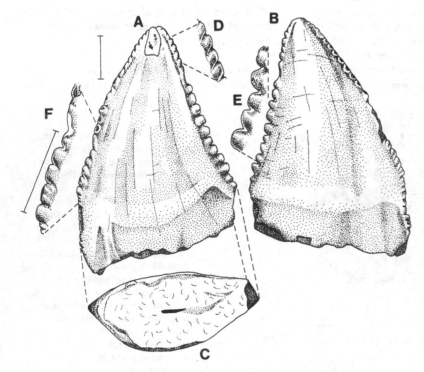

Figure 19.6. Small triangular theropod tooth in side views (A, B) and basal view (C). (D–F) Coarse serrations. Scale bars equal 1 mm.

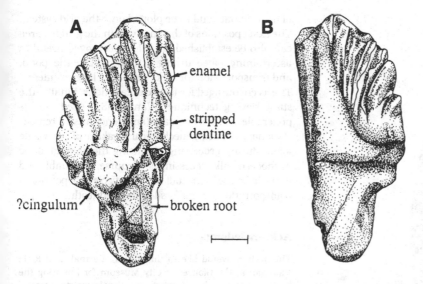

enamel

stripped dentine

?cingulum — broken root

Figure 19.7. Rather broken and eroded isolated ornithischian tooth of probable stegosaurian affinities: (A) ?lateral view with bulbous crown base, above broken root; (B) ?medial view with most of the crown broken off. Scale bar equals 1 mm.

with any jaw material. It is almost impossible to go beyond classifying them as Ornithopoda indet. They could be described as "fabrosaurid" or "hypsilophodontid" (possibly *Phyllodon* sp.), but such distinctions are not very meaningful. Stegosaurian teeth have also been recovered, but these, too, are extremely eroded (Figure 19.7).

Tritylodontid synapsids. The sheer numbers (around 20 complete teeth and a further 50 fragments) (Figure 19.5) of recovered teeth attributed to *Stereognathus* sp. from the deposit must make this one of the richest localities for Bathonian tritylodontids.

Mammalia

The locality has yielded several tiny (2–5 mm) incisors and premolars of eupantothere mammals. There is also an undescribed small jaw, currently housed at Gloucester Museum.

Significance of the site

The Middle Jurassic record of terrestrial tetrapods is poor and constitutes a significant gap in the fossil record of vertebrates. British sites are thus of extreme importance, particularly as equivalent-age Chinese sites are exploited only for the macrovertebrate material. As small forms have not been extensively sampled elsewhere, work at Hornsleasow should produce new and possibly unique taxa that will supplement those discovered at Kirtlington.

It was during the Middle Jurassic that a major "watershed" in terrestrial vertebrate evolution occurred. Many groups of amphibians, turtles, archosaurs, and therapsids with Late Triassic affinities were declining. New taxa were becoming established, such as advanced amphibians, lizards, dinosaurs, pterodactyloid

pterosaurs, and birds. The Upper Jurassic record of these groups is well represented, and the British Bathonian finds should help to establish phylogenetic links between the groups. This will be particularly significant for the smaller finds, which on preliminary examination have yielded remains of salamanders, frogs, and possibly true ?anguimorph lizards.

Preservation

The bone material is disarticulated, and many of the larger elements are in a poor state of preservation. A plot of these elements and the log materials as they lay in situ on the paleokarstic base of the deposit (the bottom of the pool) has not produced any indication of paleocurrent direction(s) within the pond (Vaughan, 1989, fig. 8). The palynological interpretation and textures within the clays suggest that there was little transport of the sediments, but the dissociation and fragmentation of the bone material may indicate transport of these elements and/or vertebrate bioturbation within the pond.

The microvertebrate assemblage is dominated by teeth and unidentified fragments of bone and dermal armor. The condition of these smaller elements is somewhat better than that of the large bone material. Although breakage of some of the elements has occurred, they have not been rounded or suffered a great amount of erosion. The skeletal material is disarticulated, but there is occasionally an association of elements (particularly small dinosaurian teeth) of one taxon (possibly one individual), which may indicate that there has been a concentration of these elements, possibly within fecal pellets. This is supported by the eroded nature of these teeth, which display stripping of enamel (Figures 19.7 and 19.8), possibly related to ingestion. The conditions prevalent during deposition of the Grey Clay were low-pH, and that could also have

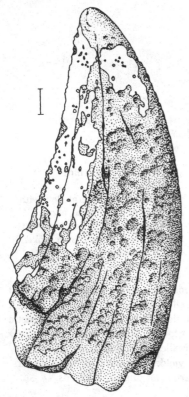

Figure 19.8. Lateral view of a possibly ingested theropod tooth. Enamel is shown in white. Scale bar equals 1 mm.

of the original pond to be plotted onto the grid system. The exact positions of the finds within the gridded area can also be established, and this has proved useful in ascertaining paleocurrent activity within the pond, and transportation (or lack of it) of the large elements. This is encouraged for future excavations. Finally, the static sieving technique (Ward, 1981) used here is preferable, as the sieve trays (and hence fossils) remain stationary, reducing mechanical breakage of the fragile fossils during processing. Reducing the damage done by modern bulk processing is, of course, invaluable and also will be useful in studies of the effects of depositional transportation on the fossil bones and teeth.

Acknowledgments

The authors would like to thank D. L. Dartnall and R. F. Vaughan at the Gloucester City Museum for initiating the project and for the loan of specimens; the Geologists' Association, through the Curry Fund, for providing two substantial grants for the excavation described here; the Crickley Hill Archaeological Trust for facilitating the expert techniques employed at Hornsleasow; Huntsman's Quarry Ltd. for allowing the excavation to disrupt work and for retaining the site for future research; the Natural Environment Council for contributing financial support for M. J. Benton to transfer the project to Bristol University and to extend the work (grant GR3/7691) for RJW. SJM is supported by the Tratman Scholarship (Bristol University). We thank M. J. Benton, J. Cole, M. J. Simms, and R. J. G. Savage for comments and access to unpublished information.

contributed to erosion of the bone, particularly the larger elements, but the presence of lime sand and cobbles within this deposit suggests that the acidic conditions were transient.

Conclusions

This is a preliminary report on the recent discoveries made at Hornsleasow. Much more work on the clays and the underlying limestone is needed before a true picture of the paleoenvironment can be presented. Correlation with other potential sites in the Cotswolds will help to establish whether or not this indeed represents a period of regression off the Bathonian carbonate platform.

The microfauna is currently being studied by the authors, with the intention of providing an examination of the faunal assemblage and the nature of preservation. This will provide much information on the paleoecology of the deposit, in addition to improving our knowledge of Middle Jurassic terrestrial vertebrates. The Hornsleasow assemblage is much older than those of the other Cotswold sites and promises to yield taxa similar to those at Kirtlington.

The careful stratigraphic nature of the excavation of the clay units has enabled a topographic representation

References

Arkell, W. J. 1933. *The Jurassic System in Great Britain.* Oxford: Clarendon Press.

———. 1947. *The Geology of Oxford.* Oxford: Clarendon Press.

Beesley, T. 1877. On the geology of the eastern portion of the Banbury and Cheltenham Direct Railway. *Proceedings of the Geological Association* 5(4): 185.

Brown, G. M., H. G. Dines, H. H. Howell, E. Hull, L. Richardson, and A. Whittaker. 1981. *Geological Survey of Great Britain (England and Wales), Moreton-in-Marsh Sheet 217, 1:50000 Series, solid and drift.*

Buckland, W. 1824. Notice on the *Megalosaurus* or great fossil lizard of Stonesfield. *Transactions of the Geological Society of London* 1: 390–396.

Channon, P. J. 1950. New and enlarged Jurassic sections in the Cotswolds. *Proceedings of the Geological Association* 61: 242–260.

Clark, J. M., and J. A. Hopson. 1985. Distinctive mammal-like reptile from Mexico and its bearing on the phylogeny of the Tritylodontidae. *Nature* 315: 398–400.

Cole, J. 1989. Palynofacies report of a sample from a vertebrate bearing clay lens from Hornsleasow Quarry. Unpublished report, Aberdeen.

Cope, J. C. W., K. L. Duff, C. F. Parsons, H. S. Torrens, W. A. Wimbledon, and J. K. Wright. 1980. *A Correlation of*

Jurassic Rocks in the British Isles. Geological Society of London, Special Report no. 15. Oxford: Blackwell Scientific.

Darlington, J. 1988. *Hornsleasow Quarry Archive Report*. Gloucestershire: Crickley Hill Archaeological Trust.

Evans, S. E. 1989. New material of *Cteniogenys* (Reptilia: Diapsida, Jurassic) and a reassessment of the phylogenetic position of the genus. *Neues Jahrbuch für Geologie und Paläontologie, Monatshefte* 1989: 577–587.

——— 1990. The skull of *Cteniogenys*, a choristodere (Reptilia: Archosauromorpha) from the Middle Jurassic of Oxfordshire. *Zoological Journal of the Linnean Society* 99: 205–237.

Evans, S. E., A. R. Milner, and F. Mussett. 1988. The earliest known salamanders (Amphibia, Caudata): a record from the Middle Jurassic of England. *Géobios* 21: 539–552.

——— 1990. A discoglossid frog from the Middle Jurassic of England. *Palaeontology* 33: 299–311.

Hallam, A. 1978. Eustatic cycles in the Jurassic. *Palaeogeography, Palaeoclimatology, Palaeoecology* 23: 1–32.

Haq, B. U., J. Hardenbol, and P. R. Vail. 1988. Mesozoic and Cenozoic chronostratigraphy, and cycles of sea-level change. Pp. 71–108 in C. Wilgus (ed.), *Sea-Level Changes – An Integrated Approach*. Society of Economic Paleontologists and Mineralogists, Special Publication no. 42.

Horton, A. 1977. The age of the Middle Jurassic "white sands" of North Oxfordshire. *Proceedings of the Geological Association* 88: 147–162.

Hudleston, W. H. 1878. Excursions to Chipping Norton. *Proceedings of the Geological Association* 10: 166–172.

Milner, A. R., and S. E. Evans. 1991. The upper Jurassic diapsid *Lisboasaurus estesi* – a maniraptoran theropod. *Palaeontology* 34: 503–514.

Nessov, L. A. 1988. Late Mesozoic amphibians and lizards of Soviet Middle Asia. *Acta Zoologica Cracoviensis* 31: 475–486.

Phillips, J. 1878. *Geology of Oxford and the Valley of the Thames*. Oxford: Clarendon Press.

Plot, R. 1677. *The Natural History of Oxfordshire, being an Essay toward the Natural History of England*. Oxford.

Richardson, L. 1911. The Inferior Oolite and contiguous deposits of the Chipping Norton district. *Proceedings of the Cotteswolds Naturalists' Field Club* 17: 203–206.

——— 1929. *The Country around Moreton-in-the-Marsh*. Memoirs of the Geological Survey of Great Britain, no. 217.

Sellwood, B. W., and W. S. McKerrow. 1974. Depositional environments in the lower part of the Great Oolite Group of Oxfordshire and North Gloucestershire. *Proceedings of the Geological Association* 85: 190–210.

Torrens, H. S. 1968. The Great Oolite Series. Pp. 227–263 in P. C. Sylvester-Bradley and T. D. Ford (eds.), *Geology of the East Midlands*. Leicester University Press.

——— (ed.). 1969. *International Field Symposium on the British Jurassic, Part V (North Somerset, Gloucestershire and South Wales)*. Keele University Press.

Vaughan, R. F. 1988. Cotswold Dinosaur Excavation. *Geology Today* 4(5): 150–151.

——— 1989. *The Excavation at Hornsleasow Quarry*. Nature Conservancy Council Interim Report no. 1, Gloucester City Museum.

Ward, D. 1981. A simple machine for bulk processing of clays and silts. *Tertiary Research* 3: 121–124.

PART III

Faunal change

Colbert (1958) first hypothesized that a major mass extinction marked the end of the Triassic period. Olsen and Galton's (1977) major reassessment of the ages of many early Mesozoic assemblages as Early Jurassic, rather than Late Triassic, challenged that notion, and a more gradual faunal change across the boundary became the more widely accepted view. However, recent research, although supporting the revised chronostratigraphic framework proposed by Olsen and Galton, has shifted opinion back toward the hypothesis of a mass-extinction event (Benton, 1986, 1991; Olsen and Sues, 1986; Olsen, Shubin, and Anders, 1987). This has been largely the result of the exploration of new localities and of more rigorous analyses of the faunal material from classic localities. Although the majority of workers now recognize the occurrence of a major extinction among tetrapods in the Late Triassic, the exact timing of that event remains very much in dispute. Some authors have even claimed the existence of two such events: one at the end of the Carnian, and one at the close of the Norian (e.g., Benton, 1991, and Chapter 22). Clearly, individual tetrapod assemblages need to be dated with greater precision, preferably at the substage level, in order to resolve these divergent views. Only then can any meaningful comparisons of tetrapod faunas be attempted on a global scale. One of the most important concerns for research on early Mesozoic tetrapods is the establishment of a reliable biostratigraphic framework. Attempts have been made to develop palynological zonation schemes with tie-ins into the ammonoid zones of the marine Triassic sequences, and such schemes may ultimately prove valuable in correlating the European and North American sections (e.g., Litwin, Traverse, and Ash, 1991). It is becoming increasingly apparent that Late Triassic and Early Jurassic terrestrial vertebrate faunas were remarkably homogeneous in their taxonomic composition worldwide, particularly during the latter time

interval. Consequently there is potential for utilizing tetrapods as biostratigraphic markers for the major early Mesozoic continental strata worldwide. Based primarily on their work in the American Southwest, Hunt and Lucas (e.g., 1991) have recently proposed novel correlation schemes based on the distribution of phytosaurs and stagonolepidids. As with other index fossils, the taxa must be widespread and sufficiently restricted in temporal duration. Hunt and Lucas have argued that these criteria are met by the phytosaurs *Paleorhinus* (late Carnian), *Rutiodon* (latest Carnian), and *Pseudopalatus* (early Norian) and by the stagonolepidids *Longosuchus* (middle to late Carnian), *Stagonolepis* (latest Carnian), *Typothorax* (early to middle Norian), and *Redondasuchus* (middle to late Norian). One of the major obstacles to using large tetrapods for biostratigraphic correlation is the inability to recover large numbers of relatively complete fossils. Thus, reported occurrences of such forms frequently are based on isolated bones and/or incompletely prepared material. Only when details of all ontogenetic stages of a particular taxon are available will it be possible to evaluate whether or not diagnostic characters exist for each of the individual bones of the skeleton. Many more discoveries of the phytosaurian and stagonolepidid archosaurs listed earlier will be required before their reliability as biostratigraphic indicators can be fully established.

The considerable abundance of bones of small tetrapods at many localities gives them a great potential advantage as biostratigraphic indicators over larger forms. It is therefore of interest that the abundance and widespread distribution of sphenodontian lepidosaurs in strata of Late Triassic and Jurassic age have been recognized in the past few years. More importantly, it would appear that certain sphenodontian taxa are both temporally restricted and geographically widely distributed (e.g., *Clevosaurus*) (Fraser, Chapter 11; Wu,

Chapter 3, and Shubin et al., Chapter 13) and thus might prove to be useful biostratigraphic indicators. Preliminary observations indicate that additional groups of early Mesozoic tetrapods are spatially widespread and temporally restricted. These include synapsids, procolophonids, and certain crocodylomorph archosaurs. It is hoped that future work on the taxonomy and distribution of early Mesozoic tetrapods will help to test and refine recently proposed biostratigraphic schemes discussed in this section.

References

Benton, M. J. 1986. The Late Triassic tetrapod extinction events. Pp. 303–320 in K. Padian (ed.), *The Beginning of the Age of Dinosaurs: Faunal Change across the Triassic–Jurassic Boundary*. Cambridge University Press.
    1991. What really happened in the Late Triassic? *Historical Biology* 5: 263–278.
Colbert, E. H. 1958. Triassic tetrapod extinctions at the end of the Triassic period. *Proceedings of the National Academy of Sciences USA* 44: 973–977.
Hunt, A. G., and S. G. Lucas. 1991. The *Paleorhinus* biochron and the correlation of the nonmarine Upper Triassic of Pangaea. *Palaeontology* 34: 487–501.
Litwin, R. J., A. Traverse, and S. R. Ash. 1991. Preliminary palynological zonation of the Chinle Formation, southwestern U.S.A., and its correlation to the Newark Supergroup (eastern U.S.A.). *Review of Palaeobotany and Palynology* 68: 269–287.
Olsen, P. E., and P. M. Galton. 1977. Triassic–Jurassic extinctions: are they real? *Science* 197: 983–986.
Olsen, P. E., N. H. Shubin, and M. H. Anders. 1987. New Early Jurassic tetrapod assemblages constrain Triassic–Jurassic tetrapod extinction event. *Science* 237: 1025–1029.
Olsen, P. E., and H.-D. Sues. 1986. Correlation of continental Late Triassic and Early Jurassic sediments, and patterns of the Triassic–Jurassic tetrapod transition. Pp. 321–351 in K. Padian (ed.), *The Beginning of the Age of Dinosaurs: Faunal Change across the Triassic–Jurassic Boundary*. Cambridge University Press.

20

The chronology and paleobiogeography of mammalian origins

SPENCER G. LUCAS AND ADRIAN P. HUNT

Introduction

During the past decade there has been renewed interest in the origin of mammals, prompted by new discoveries and by new analyses, especially cladistic, of mammalian phylogeny. Recent hypotheses concerning the phylogeny of the earliest mammals (e.g., Rowe 1988; Hopson, 1991; Wible, 1991; Crompton and Luo, 1993) have been based solely on morphological data and have not been discussed in light of the chronological and paleogeographic distribution of non-mammalian cynodonts and mammals. To facilitate such discussion, we shall review the chronology and geographic distribution of the earliest mammals (those of Late Triassic and Early Jurassic age) and the most advanced nonmammalian cynodonts, the Traversodontidae, Tritylodontidae, and Tritheledontidae. We first review the geographic and stratigraphic distributional data for these groups, present a group-by-group summary, and conclude with a discussion of the implications of these data for paleobiogeography and current phylogenies of mammalian origins.

Triassic–Jurassic time scale

Before reviewing the chronology of mammalian origins, we must briefly clarify the Late Triassic–Early Jurassic time scale employed here. This time scale (Figure 20.1) is the SGCS (standard global chronostratigraphic scale), and our numerical calibration for the Late Triassic follows Menning (1990), and that for the Early Jurassic follows Harland et al. (1990).

We emphasize here that geochronometry of the Late Triassic–Early Jurassic is poorly constrained because of a general lack of reliable radiometric dates that can be related unambiguously to biochronology. The numerical ages of stage boundaries in this interval usually are obtained by interpolation based on assump-

tions about the durations of ammonoid zones (e.g., Westermann, 1984; Harland et al., 1989) and thus are provisional estimates pending the availability of age-constraining radiometric dates.

We follow Menning (1990) in placing the Ladinian–Carnian (Middle–Late Triassic) boundary at 229 Ma, a placement consistent with most published time scales. Most of these time scales estimate the Carnian–Norian boundary at 220–223 Ma by interpolation, so Menning's (1990) placement of this boundary at 222 Ma is reasonable (Olsen et al., 1989).

Tozer (e.g., 1979, 1988) has argued against the use of the Rhaetian as a stage and believes it should be subsumed into the Norian as a "substage" at best. However, the Subcommission on Triassic Stratigraphy (STS) of the International Union of Geological Sciences (IUGS) and various workers (e.g., Wiedmann et al., 1979; Ager, 1987; Dagys and Dagys, 1990; Golebiowski, 1990; Krystyn, 1990) favor retention of the Rhaetian as a stage distinct from the Norian. We (unlike other authors in this volume) recognize the Rhaetian as a stage, and we define its base (as recommended by the STS) as the base of the *Choristoceras marshi* ammonite zone, estimated at about 210 Ma by Harland et al. (1990). [Menning (1990) offered no numerical estimate for the beginning of Rhaetian time.] We note, however, that the term "Rhaetian" is applied in western Europe (discussed later) to the generally transgressive deposits above the German Knollenmergel and equivalents (e.g., Mercia Mudstone Group) and may encompass more time than the *Choristoceras marshi* biochron.

The Triassic–Jurassic boundary is the Rhaetian–Hettangian boundary, usually defined by the first-appearance datum (FOD) of the ammonoid *Psiloceras planorbis*, though this definition is not without its problems (Hallam, 1990; Cope and Hallam, 1991). Numerical estimates of the Triassic–Jurassic boundary during the past decade have ranged from 184 Ma

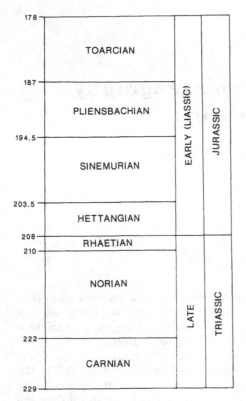

178			
	TOARCIAN		
187			
	PLIENSBACHIAN	EARLY (LIASSIC)	JURASSIC
194.5			
	SINEMURIAN		
203.5			
	HETTANGIAN		
208	RHAETIAN		
210			
	NORIAN	LATE	TRIASSIC
222			
	CARNIAN		
229			

Figure 20.1. Time scale of the Late Triassic–Early Jurassic employed in this chapter.

(Seidemann, 1988) to 213 Ma (Harland et al., 1982). A review of the available relevant radiometric ages in British Columbia by Armstrong (1982) suggested an age of about 208 Ma, which is also the age arrived at by ammonite interpolation from calibrated tie points in both the Jurassic (Kent and Gradstein, 1985) and the Triassic (Harland et al., 1990).

Olsen et al. (1989; Olsen, Fowell, and Cornet, 1990) placed the Triassic–Jurassic boundary at 201 Ma, based largely on the radiometric ages of intrusives that apparently are related to flow basalts of the Newark Supergroup. However, these intrusives have generally been regarded as of Jurassic age (Froelich and Gottfried, 1988; Sutter, 1988) and are related to flows that are encased within Lower Jurassic strata. For example, in the Newark basin, the Palisades Sill (202 Ma) presumably is coeval with the Preakness Basalt, which is within Hettangian strata (Ratcliffe, 1988; Olsen et al., 1990). Also, the U-Pb zircon age of 202 ± 1 Ma for the North Mountain Basalt in Nova Scotia similarly provides a minimum age for the Triassic–Jurassic boundary (Hodych and Dunning, 1992). At most, these data demonstrate that the Triassic–Jurassic boundary is older than 202 Ma (Sutter, 1988).

In other words, counting presumed Milankovitch cycles between the base of the Preakness Basalt and the palynologically located base of the Jurassic must

yield a Triassic–Jurassic boundary date older than 202 Ma. Nevertheless, this is problematic, because (1) the nonconformity between the basalt and underlying sediments represents an undetermined amount of time, (2) the continuity of Milankovitch cycles between the basalt and the base of the Jurassic cannot be demonstrated on outcrops, and (3) the coincidence of the palynologically based Triassic–Jurassic boundary with the base of the marine Hettangian is not well demonstrated. In light of these factors, and the afore-mentioned considerations, we do not accept any of the relatively young (184–202 Ma) numerical estimates for the Triassic–Jurassic boundary based on rocks from the Newark Supergroup.

The numerical time scale for the Early Jurassic of Harland et al. (1990) is also poorly constrained geo-chronometrically. It is used here without modification, although clearly there are some problems with it. For example, the Carmel Formation in Utah contains middle Bajocian marine invertebrates (Peterson and Pipiringos, 1979), but yields radiometric ages of about 165 Ma (Marvin, Wright, and Walthall, 1965; Everett, Kowallis, and Christiansen, 1989; Marzolf, 1990). Clearly, the Early Jurassic numerical time scale needs some adjustment.

Localities and horizons

Late Triassic and Early Jurassic fossil localities are shown in Figures 20.2 and 20.3, and Figure 20.4 shows a correlation of the Upper Triassic and Lower Jurassic strata involved.

Europe

Switzerland. Traversodontids, tritylodontids, and tritheledontids are not known from Switzerland. Rhaetic bonebeds at Hallau, Kanton Schaffhausen, have yielded a number of mammalian specimens that have been reviewed by Clemens (1980). These specimens represent the ?mammal *Tricuspes* cf. *T. tuebingensis*, the hara-miyids cf. *Thomasia antiqua*, *Thomasia anglica*, *?Thomasia* sp., cf. *Thomasia* sp., *Haramiya moorei*, and *?Hara-miyidae*, the morganucodontids *Morganucodon peyeri* and *?Morganucodon* sp., the ?morganucodontid *Helve-tiodon schuetzi* and the "triconodontan" *Hallautherium schalchi* (Clemens, 1980). Clemens (1980) reviewed the evidence for the age of the Hallau bonebed, which indicates that it postdates the late Norian (younger than the Knollenmergel) and predates the *Psiloceras johnstoni* zone (early, but not earliest, Hettangian).

Germany. Traversodontids and tritheledontids have not been described from Germany, but unde-scribed traversodontid teeth of Ladinian age are under study by J. A. Hopson and H.-D. Sues (pers. commun.). Several taxa of tritylodontids have been described from

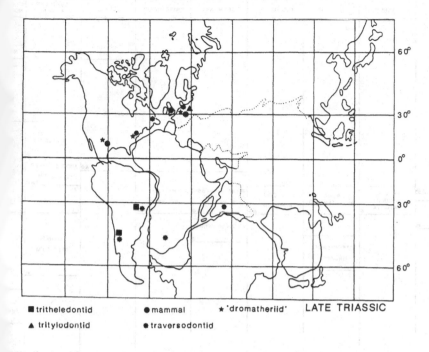

Figure 20.2. Late Triassic localities yielding traversodontids, tritylodontids, tritheledontids, "dromatheriids," and mammals.

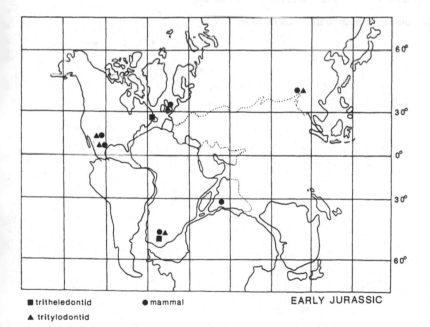

Figure 20.3. Early Jurassic localities yielding tritheledontids, tritylodontids, and mammals.

the "Rhaeto-Liassic" bonebeds of Baden-Württemberg, including *Oligokyphus triserialis* (= *O. biserialis*, = *Mucrotherium cingulatum*, = *Uniserium enigmatum*), *Tritylodon fraasi*, and *Chalepotherium plieningeri* (Hennig, 1922; Butler, 1939; Kühne, 1950, 1956; Clemens et al., 1979; Sues, 1985a).

Late Triassic mammals have been recovered from two areas in Germany. Hahn (1973) described a single tooth of *Thomasia* sp. 1 from the uppermost middle Keuper (Knollenmergel) of Halberstadt. Dietrich (1937)

described an indeterminate, fragmentary ulna from the same locality as the putative mammal *Eoraetia siegerti*.

Mammalian teeth have also been found in the "Rhaeto-Liassic" bonebeds of Baden-Württemberg. These teeth represent the ?mammal *Tricuspes tuebingensis* and the haramiyids *Thomasia antiqua* and ?*Thomasia* sp. (Clemens, 1980).

Advanced nonmammalian cynodonts and mammals thus occur at two stratigraphic horizons in Germany.

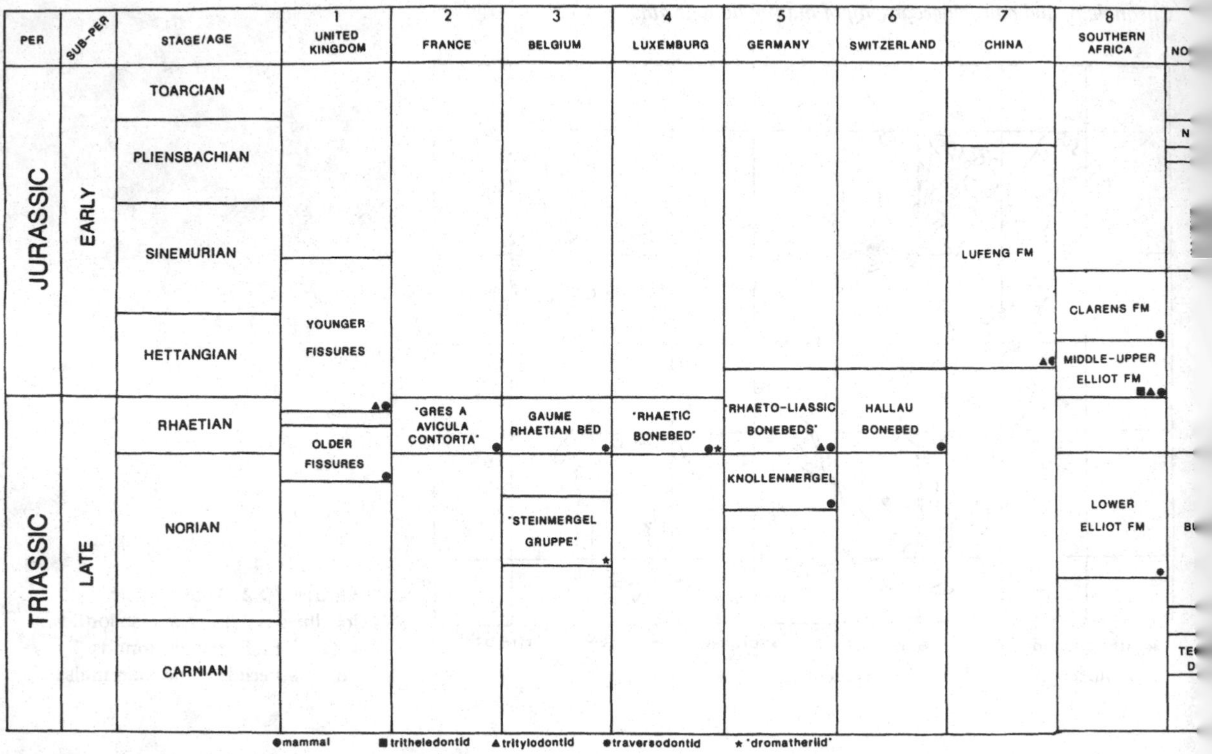

PER	SUB-PER	STAGE/AGE	1 UNITED KINGDOM	2 FRANCE	3 BELGIUM	4 LUXEMBURG	5 GERMANY	6 SWITZERLAND	7 CHINA	8 SOUTHERN AFRICA	NO
JURASSIC	EARLY	TOARCIAN									
		PLIENSBACHIAN									N
		SINEMURIAN							LUFENG FM		
		HETTANGIAN	YOUNGER FISSURES							CLARENS FM	
										MIDDLE-UPPER ELLIOT FM ■▲●	
TRIASSIC	LATE	RHAETIAN	OLDER FISSURES ▲●	'GRES A AVICULA CONTORTA'	GAUME RHAETIAN BED ●	'RHAETIC BONEBED' ●★	'RHAETO-LIASSIC BONEBEDS' ▲●	HALLAU BONEBED ●			
		NORIAN			'STEINMERGEL GRUPPE' ★		KNOLLENMERGEL ●			LOWER ELLIOT FM	BU
		CARNIAN									TE D

● mammal   ■ tritheledontid   ▲ tritylodontid   ● traversodontid   ★ 'dromatheriid'

Figure 20.4. Correlation of Upper Triassic and Lower Jurassic strata discussed in the text.

The Halberstadt locality is in the Knollenmergel of late Norian age. Clemens (1980) reviewed the age of the "Rhaetic-Liassic" bonebeds. The bonebeds overlie the upper Norian Knollenmergel, and *Psiloceras planorbis* (an early Hettangian marker) has been found in nearby sections. In addition, Huene (1933) reported juvenile ammonites in one locality in the bonebed. Thus, the bonebeds can be no older than late Norian and no younger than early Hettangian. Aepler (1974) argued that the juvenile ammonites might indicate an age older than the *planorbis* zone.

France. No specimens of traversodontids, tritylodontids, or tritheledontids have been recovered from France. Sigogneau-Russell (1983a,b; Sigogneau-Russell, Frank, and Hemmerlé, 1986) described a number of taxa of mammals from the "Grés à *Avicula contorta*" at Saint-Nicolas-de-Port, including the morganucodontid *Brachyzostrodon coupatezi*, the kuehneotheriid *Woutersia mirabilis*, and the theroteinid *Theroteinus nikolai*.

Sigogneau-Russell and associates (e.g., Sigogneau-Russell et al., 1986) consider the Saint-Nicolas-de-Port locality early Rhaetian in age, whereas Buffetaut and Wouters (1986) consider it a Knollenmergel equivalent (late Norian). The main basis for a Knollenmergel age is the supposed presence of the phytosaur *Rutiodon ruetimeyeri* (Buffetaut and Wouters, 1986). However, this identification is based solely on isolated teeth (Buffetaut and Wouters, 1986), which only indicate the presence of a heterodont phytosaur. Several species of phytosaurs could produce a similar assemblage of teeth, including *Brachysuchus megalodon* and *Rutiodon gregorii*. The remainder of the fauna from Saint-Nicolas-de-Port includes, with the identifications by Buffetaut and Wouters (1986) in parentheses, Temnospondyli indet. (?Capitosauridae), Plagiosauridae indet. (cf. *Plagiosaurus*), *Plateosaurus* sp., and Theropoda indet. (Procompsognathidae indet.).

The main lines of evidence for a Rhaetian age are the presence of early to middle Rhaetian pollen in the "Grés à *Avicula contorta*" (Schurmann, 1977) and the fact that elsewhere this stratigraphic unit is overlain by the Rhaetian "Argiles de Levallois." The presence of *Plateosaurus* indicates a late Norian or Rhaetian age (Hunt, 1991), and the presence of plagiosaurs indicates a pre-Liassic age. We conclude that the Saint-Nicolas-de-Port mammals are of Rhaetian age.

Belgium. Tritylodontids and Late Triassic–Early Jurassic mammals are not known from Belgium. Hahn, Lepage, and Wouters (1984) described isolated teeth from a bonebed in the "Steinmergel-Gruppe" of Medernach as *Pseudotriconodon wildi*. Although this taxon is nominally a dromatheriid (Hahn et al., 1984), we include it here because dromatheriids are essentially tooth taxa that may prove to be chiniquodontids

or mammals with additional anatomical information (Lucas and Oakes, 1988). The bonebed at Medernach is part of the "Steinmergel-Gruppe," which is considered middle Norian in age (Hahn et al., 1984).

The only Belgian traversodontid is *Microscalenodon nanus* from the Rhaetian at Gaume in southern Belgium (Hahn et al., 1988). This is the youngest known traversodontid.

Luxembourg. Traversodontids and tritylodontids are not known from Luxembourg. Hahn, Wild, and Wouters (1987) described the "dromatheriid" *Pseudotriconodon* sp. from a Rhaetic bonebed at Habay-la-Vieille. Wouters, Sigogneau-Russell, and Lepage (1984) described a tooth of the mammal *Haramiya* sp., also from Habay-la-Vieille. The bonebed at Habay-la-Vieille is considered early Rhaetian in age (Hahn et al., 1987).

United Kingdom. Traversodontids and tritheledontids are not known from the United Kingdom. Kühne (1956) named two species of the tritylodontid *Oligokyphus*, *O. major* and *O. minor*, from a fissure-fill deposit at Windsor Hill Quarry in Somerset. These taxa probably are sexual dimorphs of *O. major* (Sues, 1985b). Another taxon of tritylodontid is present at the nearby Holwell Quarry (Savage and Waldman, 1966; Savage, 1971). Two tritylodontids are known from Middle Jurassic strata in the United Kingdom: *Stereognathus ooliticus* from the Stonesfield Slate of Oxfordshire, England (Charlesworth, 1855), and *S. hebridicus* from the Kilmalaug Formation (upper Great Estuarine Group) of Skye, Scotland (Waldman and Savage, 1972; Cope et al., 1980).

Mammals are known from several Late Triassic or Early Jurassic fissure-filling deposits in southwestern England and southern Wales. The Welsh quarries (Duchy, Ewenny, Pant, Pontalun) have yielded specimens of the morganucodontid *Morganucodon watsoni* and the kuehneotheriids *Kuehneotherium praecursoris* and *Kuehneon duchyense* (Clemens et al., 1979; Fraser, 1985). Holwell Quarry in England has produced the haramiyids *Haramiya moorei*, *H. fissurae*, ?*Haramiya* sp., and *Thomasia anglica* and the morganucodontid *Holwellconodon problematicus* (Clemens et al., 1979; Lucas and Hunt, 1990). Fraser, Walkden, and Stewart (1985) reported *Kuehneotherium* sp. from the Emborough Quarry in England. An isolated tooth (now lost) was named *Hypsiprimnopsis rhaeticus* by Dawkins (1864). Clemens et al. (1979) suggest that this tooth probably was from a tritylodontid, rather than a mammal, apparently based on its large size. We consider this taxon a nomen dubium and are unable to assess its affinities based on the published information.

Except for an occurrence of dubious age in China (Chow and Rich, 1984a), all Middle Jurassic mammal localities occur in the United Kingdom. The fauna of the Stonesfield Slate in Oxfordshire includes the amphilestids *Amphilestes broderipii* and *Phascolotherium buck-landi* and the amphitheriid *Amphitherium prevosti* (Clemens et al., 1979). During the past 15 years, Freeman has reported three new localities in the Forest Marble of England (Freeman, 1976a,b, 1979), two of which are near Oxford, and the third is on the Dorset coast. The fauna from the Oxford-area localities includes a multituberculate, the morganucodontid *Wareolestes rex*, the docodont *Simpsonodon*, the kueh-neotheriid *Cyrtlatherium canei*, dryolestids, and the peramurid *Palaeoxonodon ooliticus* (Freeman, 1979). Evans and Milner (1991; Evans, Milner, and Mussett, 1990) reported a tritylodontid (*Stereognathus*) and several mammals (*Wareolestes, Cyrtlatherium, Simpsonodon, Palaeoxonodon*, a docodontid, a dryolestid, and a multituberculate) from an upper Bathonian horizon of the Forest Marble at Kirtlington. Waldman and Savage (1972) named *Borealestes serendipitus* for jaw fragments of a docodontid from the Kilmalaug Formation of Skye, Scotland, and Savage (1984) noted the presence of a new pantothere from the same locality.

The ages of the fissure fills in southwestern Britain have always been unclear (e.g., Fraser, 1985, 1986). Robinson (1957, 1971) distinguished two age groups of fissures based on fauna, flora, and lithology. She postulated that Rhaeto-Liassic fissures could be distinguished from pre-Rhaetian fissures on the basis of (1) the presence of mammalian fossils, (2) a tetrapod assemblage dominated by *Gephyrosaurus* and mammals instead of *Clevosaurus, Kuehneosaurus, Planocephalosaurus*, and primitive crocodyomorphs, (3) the absence of Norian-aspect (Mercia Mudstone Group) sediments, and (4) the presence of *Classopollis* (Fraser, 1985; Fraser et al., 1985). This dichotomy was accepted by most workers until the 1980s, when Rhaetian pollen (Marshall and Whiteside, 1980) and mammalian remains (Fraser et al., 1985) were found in fissures that formerly had been considered pre-Rhaetian on the basis of other evidence.

The fissures represent both solution caves and tectonic slots within Carboniferous limestone. Like all cave deposits, the fissure fills have complicated stratigraphic sequences, with evidence of mixing of strata. It is clear that parts of some fissures must be Late Triassic in age, as they contain aetosaurs and rauisuchians (Fraser, 1985, and Chapter 11). However, if these fissures started to form during the Carnian pluvial interval of Simms and Ruffell (1989, 1990), some of these faunas could be as old as middle Carnian, which is significantly older than the conventional dates for these deposits. It also worth noting that the fissure deposits never occur in superposition above the Mercia Mudstone Group of pre-Rhaetian age.

Some fissures apparently are Rhaetian–Sinemurian in age, on the basis of the occurrence of pollen referable to *Hirmeriella* (*Cheirolepis*) (Kermack, Mussett, and Rigney, 1973). The presence of *Morganucodon* and

Oligokyphus in some fissures is also strong evidence for an Early Jurassic age.

There are several major factors that need to be considered in assessing the age of the fissure faunas: (1) The internal stratigraphy of fissure fillings is complicated, particularly in the "collapsed-cave" type of fissure. (2) All the fossils in one fissure need not be the same age. (3) There is evidence of disturbance and mixing of strata within fissures (Fraser, 1985), which undoubtedly affected pollen (Fraser et al., 1985), but also small vertebrate fossils. (4) The supposed Late Triassic strata of the Penarth Group, whose relationships to fissures have been used to justify the ages of fissures, may be Early Jurassic in age (Donovan, Curtis, and Curtis, 1989). We conclude that a single age should not necessarily be assigned to the fossils from one fissure and that individual fossils from the fissures may range in age from middle Carnian to Sinemurian. Furthermore, we are not convinced by arguments for a pre-Rhaetian age of Emborough Quarry (where the "oldest" *Kuehneotherium* supposedly occurs), which invoke extrapolation of the supposed trend of the Rhaetian marine strata above the fissure (Fraser et al., 1985).

In contrast to the fissure-filling faunas, the Middle Jurassic mammalian assemblages from England and Scotland have well-constrained ages because of their placement within marine sequences. The Forest Marble is from the middle upper part of the upper Bathonian; the Kilmalaug Formation is from the lower middle part of the upper Bathonian; the Stonesfield Slate is from the lower middle Bathonian (Cope et al., 1980).

China

No traversodontids have been reported from China. *Traversodontoides wangwuensis*, described by Young (1974) as a traversodontid, is actually a bauriamorph (Sun, 1981).

In contrast, tritylodontids are abundant and diverse in the Lower Lufeng Formation of western Yunnan. Tritylodontid genera from Lufeng are *Bienotherium, Lufengia, ?Oligokyphus* (based on a dentary fragment, possibly of an immature *Bienotherium*) (Sues, 1986b), *Yunnania*, and *Dianzhongia* (Young, 1947, 1974; Chow and Hu, 1959; Chow, 1962; Cui, 1976, 1981). Fossils of these tritylodontids are derived from the Dark Red Beds of the Lower Lufeng Formation, except for some specimens of *Bienotherium*, which occur in the underlying Dull Purplish Beds (Sun and Cui, 1986).

Other Chinese tritylodontids include those of Middle Jurassic (Bajocian) age from the lower Shaximiao Formation of Sichuan: *Bienotheroides zigongensis* and *Polistodon chuannanensis* (Sun, 1984, 1986; Sun and Li, 1985; Cui and Sun, 1987). *Bienotheroides zigongensis* is also known from the Bajocian-age Wucaiwan Formation in the Junggur basin of Xinjiang (Sun and Cui,

1989). The youngest known tritylodontid is *Bieno-theroides wanshienensis* (Young, 1982), from the upper Shaximiao Formation in Sichuan, of Tithonian age (Chen et al., 1982a,b; Lucas, 1991d).

The oldest known Chinese fossil mammals are of Early Jurassic age and also come from the Dark Red Beds of the Lufeng Formation. These are specimens of *Morganucodon* and *Sinoconodon* (Patterson and Olson, 1961; Rigney, 1963; Young, 1978, 1982; Zhang and Cui, 1983). *Kunminia* (the only specimen of which has been lost), from Lufeng, described by Young (1947), may also be a mammal (e.g., Hopson and Kitching, 1972). Chow and Rich (1984b) mentioned a mammalian tooth fragment from Luzhang in Sichuan that may be of Early Jurassic age.

Long debate as to whether the Lower Lufeng Formation is Late Triassic or Early Jurassic (cf. Colbert, 1986) was resolved by Chen et al. (1982a,b) based on microfossils. Thus, the Lower Lufeng Formation (Fengjiahe Formation of their usage; these strata have produced the "*Lufengosaurus* fauna," which encompasses the tritylodontids and mammals) contains the *Gomphocythere-Darwinula* ostracod fauna of the Lower Jurassic, though no more precise age within the Hettangian–Toarcian interval is indicated. Furthermore, *Lufengosaurus* is also known from the Zhengzhuchong Formation of the Ziliujing Group in Sichuan (Dong, Zhou, and Zhang, 1983), a unit that has produced conchostracans (*Palaeolimnadia*) and bivalves (*Qiyangia*, *Apseudocardina*) indicative of a Hettangian to Pliensbachian age.

Southern Africa

Several Middle Triassic traversodontids occur in southern Africa. These include *Scalenodon angustifrons*, "*Theropsodon njalilus*," "*Scalenodon*" *attridgei*, "*Scalenodon*" *hirschoni*, and "*Scalenodon*" *charigi* from the Manda Formation of Zambia, as well as *Luangwa drysdalli* from the Ntawere Formation of Zambia (Crompton, 1972; Hopson and Kitching, 1972; Kemp, 1980). The only Late Triassic traversodontid from Africa is *Scalenodontoides macrodontes* Crompton and Ellenberger, 1957 from the lower Elliot Formation of Lesotho (Hopson, 1984).

Three species of tritylodontids have been named from the middle and upper portions of the Elliot Formation: *Tritylodon longaevus* Owen, 1884, *Likhoelia ellenbergi* Ginsburg, 1961, and *Tritylodontoideus maximus* Fourie, 1962 (= *Tritylodontoides maximus* Fourie, 1963). The latter two taxa probably are synonymous with *Tritylodon longaevus* (Kitching and Raath, 1984; Sues, 1986b).

Three taxa of tritheledontids are present in the middle and upper Elliot Formation: *Tritheledon riconoi* Broom, 1912, *Pachygenelus monus* Watson, 1913, and *Diarthrognathus broomi* Crompton, 1958 (Gow,

1980; Kitching and Raath, 1984). *Pattsia likhoelensis* Lees and Mills, 1983 from the upper Stormberg Group of Lesotho may be a tritheledontid, possibly a synonym of *Pachygenelus monus*.

Three localities in Lesotho (upper Elliot Formation) and South Africa (Clarens Formation) yield early mammalian fossils (Clemens et al., 1979). Two species of Morganucodontidae, *Erythrotherium parringtoni* Crompton, 1964 and *Megazostrodon rudnerae* Crompton and Jenkins, 1968, are represented by complete upper and lower dentitions and the greater part of postcranial skeletons. Hopson and Reif (1981) demonstrated that the holotype of the putative mammal *Archaeodon* ("*Archaeotherium*") *reuningi*, from the Upper Triassic of Namibia, represents the chalcedony infilling of a vesicle in lava.

Except for the Middle Triassic traversodontids, the advanced nonmammalian cynodonts and mammals from the Late Triassic–Early Jurassic of southern Africa are from either the lower (lower Elliot Formation) or upper (middle and upper Elliot and Clarens formations) Stormberg Group. Traditionally, these formations were considered to be of Late Triassic age, but Olsen and Galton (1977, 1984) argued convincingly that the upper Stormberg is actually Early Jurassic in age. Thus, the lower Elliot Formation probably represents most of Norian time, and the middle to upper Elliot and Clarens formations are Hettangian–Sinemurian in age (Olsen and Galton, 1984, fig. 2).

North America

Upper Triassic–Lower Jurassic strata in North America most relevant to mammalian origins are the Newark Supergroup of the eastern United States and Canada and the Chinle Group and Glen Canyon Group in the western United States (Lucas, 1992). In addition, a tritylodontid and mammals are known from the Lower Jurassic of northeastern Mexico.

Traversodontids are known only from the Newark Supergroup. *Arctotraversodon plemmyridon* is a traversodontid, based on dentaries and teeth from the Wolfville Formation of the Fundy Group in the Minas basin, Nova Scotia, Canada (Hopson, 1984; Sues, Hopson, and Shubin, 1992). The Wolfville Formation produces a small vertebrate fauna that includes a large metoposaurid, "*Scaphonyx*" (H.-D. Sues, pers. commun., 1992, states that this is a different form), a *Stagonolepis*-like aetosaur, and a dicynodont, all indicative of a late Carnian age (Hopson, 1984; Hunt and Lucas, 1990, 1991b). *Boreogomphodon jeffersoni* Sues and Olsen, 1990 is a traversodontid from the Turkey Branch Formation of the Richmond basin of Virginia. Sues and Olsen (1990) noted that pollen and spores from this unit indicate that it is of early to middle Carnian age.

No tritylodontids are known from the Newark Supergroup. Three genera of Tritylodontidae are present in the Kayenta Formation of the Glen Canyon Group in northeastern Arizona: *Kayentatherium, Oligokyphus,* and *Dinnebitodon* (Kermack, 1982; Sues, 1985a, 1986a,b). Palynomorphs and vertebrate fossil evidence indicate that the Kayenta Formation is Early Jurassic, but no older than late Sinemurian and possibly as young as late Pliensbachian (e.g., Peterson and Pipiringos, 1979; Clark and Fastovsky, 1986; Sues, 1986b; Padian, 1989a,b). Indeed, a high abundance of *Corollina* in palynomorph samples from the Whitmore Point Member of the Moenave Formation, beneath the Kayenta, has been interpreted to indicate a Sinemurian age (Cornet, in Peterson and Pipiringos, 1979). The dinosaurs *Scelidosaurus* and *Syntarsus* from the Kayenta Formation, as well as the tritylodontids themselves, suggest an Early Jurassic age.

The only other named North American tritylodontid is *Bocatherium mexicanum* from the La Boca Formation of Tamaulipas, northeastern Mexico (Clark and Hopson, 1985). Previously, independent age constraints on *Bocatherium* were poor. The La Boca Formation rests unconformably between Permian and Upper Jurassic (Oxfordian) marine strata and has produced putative Late Triassic plants at a locality different from the tritylodontid locality (Mixon, Murray, and Diaz, 1959). Recent collecting by Clark and others has yielded a diverse vertebrate fauna that includes potentially biochronologically useful taxa such as a neosuchian, a ?thyreophoran, and a ?heterodontosaurid (J. M. Clark, pers. commun.). These taxa indicate an Early Jurassic age (Clark et al., 1991).

Winkler et al. (1991) reported a postcranial skeleton of a tritylodontid from the Navajo Sandstone in Arizona. The Navajo Sandstone is of Early Jurassic, probably Pliensbachian, age (Marzolf, 1990).

The only North American tritheledontid is *Pachygenelus* from the Lower Jurassic (Hettangian) McCoy Brook Formation of Nova Scotia (Shubin et al., 1991). An earlier report of *Pachygenelus* from West Texas (Chatterjee, 1983) apparently was a misidentification (Shubin et al., 1991).

The "dromatheriids" *Dromatherium* and *Microconodon,* both originally described by Emmons (1857), may be chiniquodontids. Both are based on dentary fragments from the upper Carnian Cumnock Formation in the Sanford basin of North Carolina. A *Microconodon*-like dromatheriid is present at the type locality of the traversodontid *Boreogomphodon jeffersoni* (Sues and Olsen, 1990). Lucas and Oakes (1988) described the "dromatheriid" *Pseudotriconodon chatterjeei* from the lower Norian Bull Canyon Formation in eastern New Mexico, but identification of this taxon as a cynodont has been questioned (Sues and Olsen, 1990).

Adelobasileus cromptoni is the only North American Triassic mammal (Lucas and Hunt, 1990). The Chinle Group sequence in West Texas, where the only known specimen of *A. cromptoni* was collected, pertains to the Dockum Formation, consisting of four members (in ascending order): Camp Springs, Tecovas, Trujillo, and Cooper (Chatterjee, 1986; Lucas, 1993).

The holotype of *A. cromptoni* was collected at the head of the south fork of Home Creek north of Kalgary, Crosby County, Texas (UTM 3701500N, 298850E, zone 14). This locality is in the lower part of the Tecovas Member, 11.25 m above the base of the Dockum Formation (unconformable contact with the underlying Permian Quartermaster Formation). It has produced an assemblage of darwinulid ostracods that support correlation with the lower part of the Chinle Group (Bluewater Creek and Monitor Butte formations) on the Colorado Plateau (Kietzke and Lucas, 1991a), units of late Carnian age (Lucas and Hayden, 1989; Lucas and Hunt, 1989; Hunt and Lucas, 1990, 1991a; Lucas, 1990a, 1991b).

The associated vertebrate fauna includes *Xenacanthus moorei,* an indeterminate redfieldiid, two unidentified perleidids, a semionotid, a large metoposaurid, two taxa of sphenodontians, *Trilophosaurus* sp., indeterminate phytosaurs, a rauisuchid, a *Eudimorphodon*-like pterosaur, and teeth of saurischian and ornithischian dinosaurs. Also present are numerous vertebrate coprolites, unionid bivalves, "spirorbid worms" (probably vermiform gastropods) (Kietzke and Lucas, 1991b), and fragments of petrified wood. This fauna is very similar to that of a nearby (and stratigraphically equivalent) locality in the Tecovas Member described by Murry (1989). The tetrapod assemblage of the Tecovas Member is of late Carnian age, as discussed later.

Dunay and Fisher (1979, p. 65) collected palynomorphs and megafossil plants from the Tecovas Formation "in the headwaters of Home Creek, on the Lewis Ranch, approximately 4.8 km (3 miles) north of the town of Kalgary, Crosby County in a dark gray to brown clay intercalated between red and white sandstone and shale horizons 0.9 m (3 ft) above the Permian–Triassic unconformity." Ash (1980) identified the megafossil plants as *Otozamites powelli* and *Dinophyton spinosum.* These plants fit Ash's (1980) megafossil-plant zone of *Dinophyton,* of late Carnian age, as do megafossil plants from other localities in the Tecovas Member of the Dockum Formation in Crosby County (Ash, 1980). The palynoflora reported by Dunay and Fisher (1979) compares well with other palynomorph assemblages from the Tecovas and Trujillo members of the Dockum Formation in West Texas and is of late Carnian age (Litwin, Traverse, and Ash, 1991).

Three temporally successive tetrapod faunachrons (=land-vertebrate "ages") (Lucas, 1990b) can be recognized from fossils collected in the Dockum Formation of West Texas. The oldest faunachron ("A") is based on fossils from the Camp Springs Member in Randall, Crosby, and Scurry counties, as well as the lower, mudrock-dominated Dockum Formation in

Howard County. These strata produce fossils of the primitive phytosaurs *Paleorhinus* and *Angistorhinus* (Case, 1922; Langston, 1949; Gregory, 1962; Hunt and Lucas, 1991a), rhynchosaurs (Hunt and Lucas, 1991b), the aetosaurs *Longosuchus* and *Desmatosuchus* (Hunt and Lucas, 1990), and an abundance of large metoposaurids. Particularly significant is the occurrence of *Paleorhinus*, as this taxon is known from late Carnian (Tuvalian) marine strata in Austria (Hunt and Lucas, 1991a). The late Carnian (Tuvalian) age of the faunachron-A tetrapods in the Dockum Formation is also consistent with correlations based on ostracods and palynomorphs (Kietzke, 1989; Litwin et al., 1991).

Tetrapods from the Tecovas Member, including those from the type locality of *Adelobasileus cromptoni*, represent faunachron B. Biochronologically significant tetrapods are the phytosaur *Rutiodon* (sensu Ballew, 1989), the aetosaurs *Desmatosuchus* and *Stagonolepis*, and large metoposaurids. Tetrapod biochronology suggests a late Carnian age for faunachron B, by default; it apparently lacks taxa (e.g., *Pseudopalatus*, *Typothorax*) known from the next-youngest faunachron (C) that indicate an early Norian age. Palynomorphs also support a late Carnian date for faunachron B (Dunay and Fisher, 1979; Litwin et al., 1991). Most of the known tetrapod fauna of the Dockum Formation, especially specimens collected and described by Case (e.g., 1922), are from the Tecovas Member of the Dockum Formation.

Faunachron C is represented by fossils from the Cooper Member of the Dockum Formation, principally from the Post Quarry in Garza County (Chatterjee, 1986, 1991; Small, 1989). Biochronologically significant tetrapods are the phytosaur *Pseudopalatus* (sensu Ballew, 1989), the aetosaur *Typothorax*, and a small metoposaurid (*Anaschisma* of some authors). The phytosaur *Nicrosaurus* from the Norian portion of the German Keuper is closely related to *Pseudopalatus* (Ballew, 1989). The Norian age of faunachron C is also supported by palynostratigraphy (Dunay and Fisher, 1979; Litwin et al., 1991).

Tetrapod biochronology and correlation of the Dockum Formation in West Texas with Chinle Group strata on the Colorado Plateau and in east-central New Mexico are consistent with correlations based on fossil fish (Huber, Lucas, and Hunt, 1992), plant megafossils (Ash, 1980), palynomorphs (Litwin et al., 1991), ostracods, and charophytes (Kietzke, 1989). Furthermore, Lucas (1991a,b) has presented sequence-stratigraphic correlations of the Chinle Group nonmarine Upper Triassic with marine Upper Triassic strata of the shelfal terrane (upper Star Peak Group and Auld Lang Syne Group) in western Nevada. These correlations are remarkably consistent with correlation with the Carnian and Norian ages based on tetrapod biochronology and palynostratigraphy, and they provide a basis for correlating tetrapod faunachrons of the Chinle Group with Late Triassic ammonite zones (Figure 20.5).

Thus, tetrapod biochronology, supported by correlations based on other Chinle Group fossils and sequence-stratigraphic correlations, has identified the age of *A. cromptoni* as late Carnian (late Tuvalian). Sequence stratigraphy and the marine occurrence of *Paleorhinus*

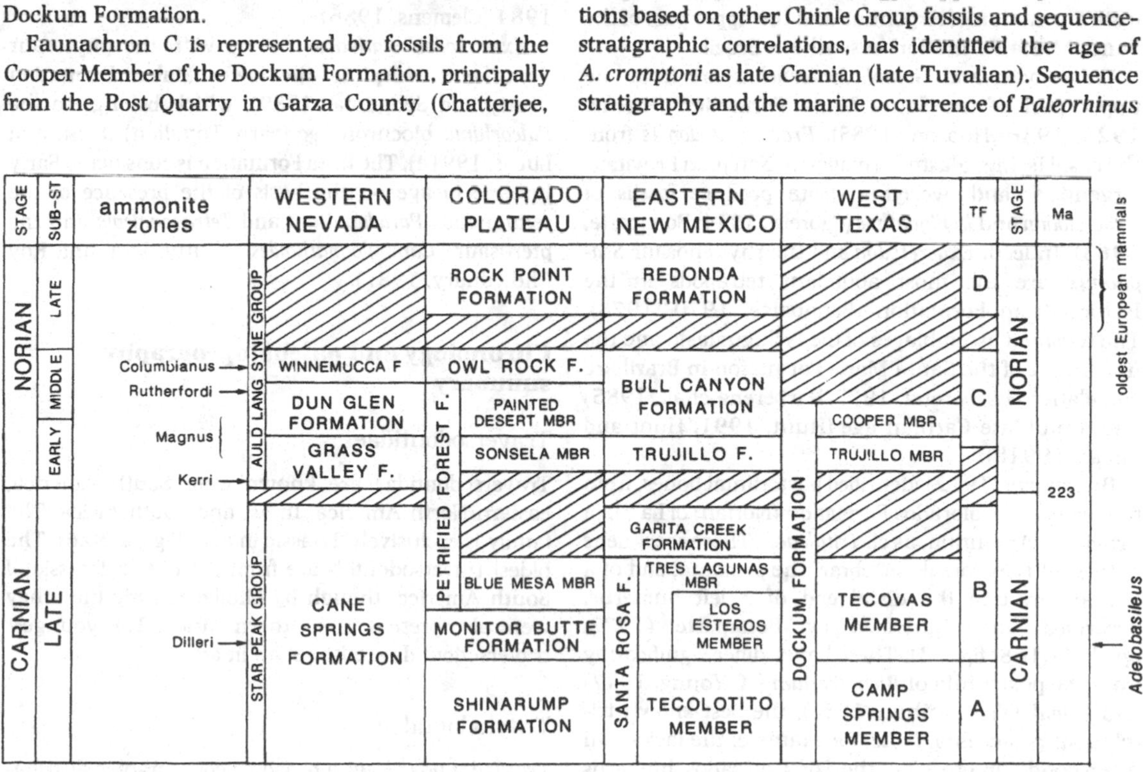

Figure 20.5. Correlation of Upper Triassic Chinle Group strata with some Upper Triassic strata in Nevada and with ammonoid biochronology. (Adapted from Lucas, 1991c.)

in Austria suggest correlation with the *welleri* and *Macrolobatus* zones of ammonite biochronology.

Early Jurassic mammals from North America were first described from the Kayenta Formation by Jenkins, Crompton, and Downs (1983). They are *Morganucodon* sp., *Dinnetherium nezorum*, and a "haramiyid" (probably a worn lower postcanine of a small *Oligokyphus*; H.-D. Sues, pers. commun.) (Clemens, 1986). Clark et al. (1991) note that at least three species of mammals are present in the Lower Jurassic La Boca Formation of northeastern Mexico.

South America

A variety of traversodonts are known from Triassic strata in Argentina. Several of these are older than Late Triassic: (1) Early Triassic *Pascualagnathus* from the upper part of the Puesto Viejo Formation of Mendoza Province, Argentina (Bonaparte, 1967); (2) *Andescynodon* from the coeval Rio Mendoza Formation of Argentina (Bonaparte, 1967); (3) *Rusconiodon*, also from the Mendoza Formation (Bonaparte, 1970); and (4) Middle Triassic *Massetognathus* from the lower part of the Ischichuca Formation in La Rioja Province, Argentina, and the Santa Maria Formation of southern Brazil (Romer, 1967; Barberena, Araújo, and Lavina, 1985). The diversity of Late Triassic traversodontids is similar; material comes from two units, the Ischigualasto Formation of Argentina and the upper part of the Santa Maria Formation in southern Brazil.

Traversodon and *Gomphodontosuchus* are from the upper part of the Santa Maria Formation (Huene, 1928, 1936; Hopson, 1985). *Proexaeretodon* is from the basal Ischigualasto Formation in San Juan Province, Argentina, and overlying strata produce fossils of *Exaeretodon* and *Ischignathus* (Cabrera, 1944; Bonaparte, 1963). Indeed, *Exaeretodon* and the rhynchosaur *Scaphonyx* are the most abundant tetrapods in the Ischigualasto Formation (Bonaparte, 1970, 1978). The tetrapod assemblages of the Ischigualasto and the upper part of the Santa Maria Formation in Brazil are correlative (Barberena, 1977; Barberena et al., 1985) and are of late Carnian age (Hunt, 1991; Hunt and Lucas, 1991b).

Bonaparte (1972) identified postcranial bones, from the upper Los Colorados Formation (Norian) of La Rioja Province, Argentina, as cf. *Tritylodon*. These specimens consist of two dorsal vertebrae, the proximal end of a left femur, and the distal end of a left humerus, described and illustrated by Bonaparte (1972, pp. 167–168, fig. 73). These bones differ significantly from the postcrania of *Bienotherium* (cf. Young, 1947) and *Oligokyphus* (Kühne, 1956), the best-known tritylodont postcrania. Note, for example, the lack of an ectepicondylar crest on the Los Colorados humerus and the weakly developed greater trochanter of the femur. These differences are so marked that we hesitate

to identify the Los Colorados specimens as tritylodontid, and we suspect that they pertain to the tritheledontid *Chaliminia*, which is known from skulls from the Los Colorados Formation.

No Late Triassic–Early Jurassic mammals are known from South America. The only South American tritheledontid is *Chaliminia* Bonaparte, 1980 from the Norian Los Colorados Formation. *Therioherpeton* from the upper Santa Maria Formation (late Carnian) of Rio Grande do Sul, Brazil (Bonaparte and Barberena, 1975) is probably a chiniquodontid.

Chiniquodontids (including *Probainognathus*) may be closely related to tritheledontids (e.g., Kemp, 1982) and are known from the Middle Triassic (Candelaria, Chiniquá) of Brazil and the Middle and Late Triassic (late Carnian) of Argentina (Bonaparte, 1974; Barberena et al., 1985).

India

Chatterjee (1982) described *Exaeretodon statisticae* from the Maleri Formation, which is the only Late Triassic traversodontid known from India. Tritylodontids and tritheledontids are absent from India. Early Jurassic mammals are known from the Kota Formation and include the kuehneotheriid *Kotatherium haldanei* Datta, 1981 and the ?kuehneotheriids *Trishulotherium kotaensis* Yadagiri, 1984 and *Indotherium pranhitai* Yadagiri, 1984 (Clemens, 1986).

Exaeretodon statisticae occurs with the phytosaur *Paleorhinus hislopi* in the lower Maleri Formation (Kutty and Sengupta, 1989), which belongs to the *Paleorhinus* biochron age (early Tuvalian) (Hunt and Lucas, 1991a). The Kota Formation is considered Early Jurassic in age on the basis of the presence of the semionotids *Paradapedium* and *Tetragonolepis* and the pterosaur "*Campylognathoides*" (Kutty, Jain, and Roy Chowdhury, 1987).

Chronology and paleobiogeography: summary

Traversodontidae

Traversodontidae are known from South America, eastern North America, India, and South Africa. The family is exclusively Triassic in age (Figure 20.6). The oldest traversodontids are from the Lower Triassic of South America, though by Middle Triassic time they were also present in southern Africa. The youngest traversodontids are Rhaetian in age.

Tritylodontidae

Tritylodontidae have a broad paleogeographic distribution and are principally Jurassic in age. Indeed, for the Sinemurian–Pliensbachian, tritylodontids are known

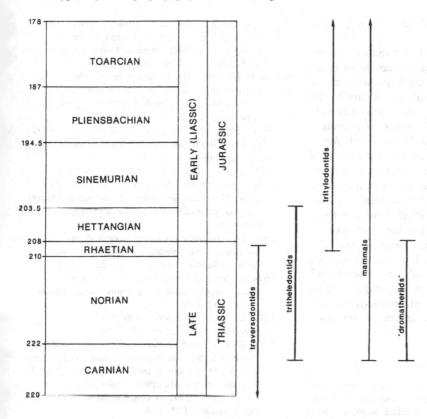

Figure 20.6. Temporal ranges of the advanced nonmammalian cynodonts and the oldest mammals.

from Europe, South Africa, China, and North America. The oldest tritylodontids are of possible Rhaetian age in Germany, and the youngest are from the Tithonian of Sichuan, China (Figure 20.6).

Tritheledontidae

Tritheledontidae range in age from late Norian to Sinemurian–Pliensbachian (Figure 20.6). The oldest is *Chaliminia*, from the late Norian of Argentina. "Dromatheriids" and chiniquodontids are stratigraphically older than the tritheledontids. The youngest tritheledontid is *Diarthrognathus* from the Sinemurian–Pliensbachian-age Clarens Formation in southern Africa.

Mammalia

The oldest mammal is *Adelobasileus* from the upper Carnian of West Texas (Figure 20.6). Other Late Triassic mammals are late Norian–Rhaetian in age and are known only from western Europe. Early Jurassic mammals have a broader distribution encompassing the western United States, Mexico, southern China, western Europe, India, and southern Africa.

Discussion

This review documents the extent to which new discoveries during the 1980s have rewritten the paleogeo-graphic and stratigraphic distribution of the advanced nonmammalian cynodonts and Late Triassic–Early Jurassic mammals. These discoveries reflect an intensive search on several continents for small Late Triassic and Early Jurassic vertebrates and the consistent application of screen-washing techniques. Because new discoveries have so rapidly altered our understanding of advanced synapsid distribution, caution must be exercised when interpreting these aspects of the record of mammalian origins. Yet a few clear patterns seem to have emerged.

The broad geographic distribution of tritylodontids and of some mammals and tritheledontids during the Early Jurassic reflects the apparent homogeneity/cosmopolitanism of tetrapod faunas at that time, as already noted by other workers (e.g., Olsen and Galton, 1984; Shubin et al., 1991). The Late Triassic distribution of traversodontids suggests a comparable broad geographic range, though Late Triassic traversodontids are rare except in inland faunas like that from the Ischigualasto Formation in Argentina.

Late Triassic tritheledontids are restricted to South America. Late Triassic mammal distribution is patchy – a late Carnian occurrence in western Texas, and much younger, late Norian–Rhaetian occurrences in western Europe. Late Triassic traversodontid distribution certainly does not mirror the tetrapod paleoprovinciality of the Late Triassic inferred by some workers (e.g., Olsen and Galton, 1984; Battail, 1991). It fits better published models of tetrapod cosmopolitanism during the Late Triassic (e.g., Kalandadze and Rautian, 1991; Shubin and Sues, 1991).

The stratigraphic (temporal) distribution of the advanced cynodonts and early mammals casts an interesting light on two current hypotheses about which group of nonmammalian cynodonts is the sister-taxon of the Mammalia. Those who advocate that tritylodontids are the sister-taxon of mammals (e.g., Kemp, 1983; Rowe, 1988; Wible, 1991) face a 15-million-year gap in the Late Triassic fossil record of Tritylodontidae (Figure 20.6). In effect, their hypothesis argues that this gap is due to the incompleteness of the fossil record or sampling and that tritylodontids appeared long before their oldest known fossils. Similarly, those who hypothesize tritheledontids as the sister-taxon of mammals (e.g., Hopson, 1991; Crompton and Luo, 1993) face a gap (late Carnian–late Norian) between the oldest mammal and the oldest tritheledontid, though it is a somewhat shorter gap (Figure 20.6). Clearly, the fossil record of the advanced synapsids is still incomplete and in need of much more sampling.

Acknowledgments

We thank Zhexi Luo and Ailin Sun for access to unpublished information, and the National Geographic Society and Petrified Forest Museum Association for financial support. Zhexi Luo and Hans-Dieter Sues provided helpful reviews of the manuscript.

References

Aepler, R. 1974. Der Rhätsandstein von Tübingen – ein kondensiertes Delta. *Neues Jahrbuch für Geologie und Paläontologie, Abhandlungen* 147: 113–162.

Ager, D. 1987. A defence of the Rhaetian stage. *Albertiana* 6: 4–13.

Armstong, R. L. 1982. Late Triassic–Early Jurassic time-scale calibration in British Columbia, Canada. Pp. 509–513 in G. S. Odin (ed.), *Numerical Dating in Stratigraphy*. New York: Wiley.

Ash, S. R. 1980. Upper Triassic floral zones of North America. Pp. 153–270 in D. L. Dilcher and T. M. Taylor (eds.), *Biostratigraphy of Fossil Plants*. Stroudsburg: Dowden, Hutchinson, and Ross.

Ballew, K. L. 1989. A phylogenetic analysis of Phytosauria from the Late Triassic of the western United States. Pp. 309–339 in S. G. Lucas and A. P. Hunt (eds.), *The Dawn of the Age of Dinosaurs in the American Southwest*. Albuquerque: New Mexico Museum of Natural History.

Barberena, M. C. 1977. Bioestratigrafia preliminar da Formacao Santa Maria. *Pesquisas (Porto Alegre)* 7: 111–129.

Barberena, M. C., D. C. Araújo, and E. L. Lavina. 1985. Late Permian and Triassic tetrapods of southern Brazil. *National Geographic Research* 1985: 5–20.

Battail, B. 1991. Triassic terrestrial ecosystems and the biogeography of the Triassic. Pp. 3–4 in Z. Kielan-Jaworowska, N. Heintz, and H. A. Nakrem (eds.), *Fifth Symposium on Mesozoic Terrestrial Ecosystems and Biota, Extended Abstracts*. University of Oslo Palaeontological Museum.

Bonaparte, J. F. 1963. Descripcion de *Ischignathus sudamericanus*, nuevo cinodonte gonfodonte del Triásico Medio superior de San Juan, Argentina. *Acta Geologica Lilloana* 4: 111–128.

——— 1967. Sobre nuevos terapsidos triásicos hallados en el centro de la Provincia de Mendoza, Argentina. *Acta Geologica Lilloana* 8: 91–100.

——— 1970. Annotated list of the South American Triassic tetrapods. Pp. 665–682 in S. H. Haughton (ed.), *Second Gondwana Symposium: Proceedings and Papers*. Pretoria: Council for Scientific and Industrial Research.

——— 1972. Los tetrapodos del sector superior de la formacion Los Colorados, La Rioja, Argentina (Triásico Superior). *Opera Lilloana* 22: 1–183.

——— 1974. Edades/reptil para el Triásico de Argentina y Brasil. *Congreso Geologico Argentino, Actas (Buenos Aires)* 5: 93–129.

——— 1978. El Mesozoico de América del Sur y sus tetrapodos. *Opera Lilloana* 26: 1–596.

——— 1980. El primer ictidosaurio (Reptilia – Therapsida) de América del Sur, *Chaliminia musteloides*, del Triásico Superior de La Rioja, Republica Argentina. *Actas del Segundo Congreso Argentino de Paleontologia y Bioestratigrafia y Primer Congreso Latinoamericano de Paleontologia* 1: 123–133.

Bonaparte, J. F., and M. C. Barberena. 1975. A possible mammalian ancestor from the Middle Triassic of Brazil (Therapsida – Cynodontia). *Journal of Paleontology* 49: 931–936.

Broom, R. 1912. On a new type of cynodont from the Stormberg. *Annals of the South African Museum* 7: 334–336.

Buffetaut, E., and G. Wouters. 1986. Amphibian and reptile remains from the Upper Triassic of Saint-Nicolas-de-Port (eastern France) and their biostratigraphic significance. *Modern Geology* 10: 133–145.

Butler, P. M. 1939. The post-canine teeth of *Tritylodon longaevus* Owen. *Annals and Magazine of Natural History* (11) 4: 514–520.

Cabrera, A. 1994. El primer hallazago de terapsidos en la Argentina. *Notas del Museo de La Plata, Paleontologia* 55: 317–331.

Case, E. C. 1922. New reptiles and stegocephalians from the Upper Triassic of western Texas. *Carnegie Institution of Washington Publication* 321: 1–84.

Charlesworth, E. 1855. Notice of new vertebrate fossils. *Report of the British Association for the Advancement of Science* 25: 80.

Chatterjee, S. 1982. A new cynodont reptile from the Triassic of India. *Journal of Paleontology* 56: 203–214.

——— 1983. An ictidosaur fossil from North America. *Science* 220: 1151–1153.

——— 1986. The Late Triassic Dockum vertebrates: their stratigraphic and paleobiogeographic significance. Pp. 139–150 in K. Padian (ed.), *The Beginning of* the Age of Dinosaurs: Faunal Change across the Triassic–Jurassic Boundary. Cambridge University Press.

1991. Cranial anatomy and relationships of a new Triassic bird from Texas. *Philosophical Transactions of the Royal Society of London* B332: 277–342.

Chen, P., W. Li, J. Chen, C. Ye, Z. Wang, Y. Shen, and D. Sen. 1982a. Sequence of fossil biotic groups of Jurassic and Cretaceous in China. *Scientia Sinica* B25: 1011–1020.

1982b. Stratigraphical classification of Jurassic and Cretaceous in China. *Scientia Sinica* B25: 1227–1248.

Chow, M. 1962. [A tritylodontid specimen from Lufeng, Yunnan.] *Vertebrata Palasiatica* 6: 365–367. [in Chinese]

Chow, M., and C. C. Hu. 1959. [A new tritylodont from Lufeng, Yunnan.] *Vertebrata Palasiatica* 3: 9–12. [in Chinese]

Chow, M., and T. H. V. Rich. 1984a. A new triconodontan (Mammalia) from the Jurassic of China. *Journal of Vertebrate Paleontology* 3: 226–231.

1984b. New Mesozoic mammal sites from China. *Alcheringa* 8: 304.

Clark, J. M., and D. E. Fastovsky. 1986. Vertebrate biostratigraphy of the Glen Canyon Group of northern Arizona. Pp. 285–301 in K. Padian (ed.), *The Beginning of the Age of Dinosaurs: Faunal Change across the Triassic–Jurassic Boundary.* Cambridge University Press.

Clark, J. M., and J. A. Hopson. 1985. Distinctive mammal-like reptile from Mexico and its bearing on the phylogeny of the Tritylodontidae. *Nature* 315: 398–400.

Clark, J. M., M. Montellano, J. A. Hopson, and R. Hernandez. 1991. Mammals and other tetrapods from the Early Jurassic La Boca Formation, northeastern Mexico. *Journal of Vertebrate Paleontology* 11 (Suppl. to no. 3): 23A.

Clemens, W. A. 1980. Rhaeto-Liassic mammals from Switzerland and West Germany. *Zitteliana* 5: 51–92.

1986. On Triassic and Jurassic mammals. Pp. 237–246 in K. Padian (ed.), *The Beginning of the Age of Dinosaurs: Faunal Change across the Triassic–Jurassic Boundary.* Cambridge University Press.

Clemens, W. A., J. A. Lillegraven, E. H. Lindsay, and G. G. Simpson. 1979. Where, when, and what – a survey of known Mesozoic mammal distribution. Pp. 7–58 in J. A. Lillegraven, Z. Kielan-Jaworowska, and W. A. Clemens (eds.), *Mesozoic Mammals: The First Two-thirds of Mammalian History.* Berkeley: University of California Press.

Colbert, E. H. 1986. Historical aspects of the Triassic–Jurassic boundary problem. Pp. 9–19 in K. Padian (ed.), *The Beginning of the Age of Dinosaurs: Faunal Change across the Triassic–Jurassic Boundary.* Cambridge University Press.

Cope, J. C. W., T. A. Getty, M. K. Kingsley, N. Morton, and H. S. Torrens. 1980. *A Correlation of Jurassic Rocks in the British Isles. Part 1: Introduction and Lower Jurassic.* Geological Society of London Special Report 14.

Cope, J., and A. Hallam. 1991. Discussion on correlation of the Triassic–Jurassic boundary in England and

Austria. *Journal of the Geological Society, London* 148: 420–422.

Crompton, A. W. 1958. The cranial morphology of a new genus and species of ictidosaurian. *Proceedings of the Zoological Society of London* 130: 183–216.

1964. A preliminary description of a new mammal from the Upper Triassic of South Africa. *Proceedings of the Zoological Society of London* 142: 441–452.

1972. Postcanine occlusion in cynodonts and tritylodonts. *Bulletin of the British Museum (Natural History), Geology* 21: 29–71.

Crompton, A. W., and F. Ellenberger. 1957. On a new cynodont from the Molteno Beds and the origin of the tritylodontids. *Annals of the South African Museum* 44: 1–14.

Crompton, A. W., and F. A. Jenkins, Jr. 1968. Molar occlusion in Late Triassic mammals. *Biological Reviews* 1968: 427–458.

Crompton, A. W., and Z. Luo. 1993. Relationships of the Liassic mammals *Sinoconodon, Morganucodon oehleri* and *Dinnetherium.* Pp. 30–44 in F. S. Szalay, M. Novacek, and M. C. McKenna (eds.), *Mammalian Phylogeny.* Berlin: Springer-Verlag.

Cui, G. 1976. [*Yunnania*, a new tritylodont genus from Lufeng, Yunnan.] *Vertebrata Palasiatica* 14: 85–90. [in Chinese]

1981. [A new genus of Tritylodontoidea.] *Vertebrata Palasiatica* 19: 5–10. [in Chinese]

Cui, G., and A. Sun. 1987. [Postcanine root system in tritylodonts.] *Vertebrata Palasiatica* 25: 245–259. [in Chinese]

Dagys, A. S., and A. A. Dagys. 1990. In favour of the Rhaetian. *Geologiya i Geofisika* 1990: 35–44.

Datta, P. M. 1981. The first Jurassic mammal from India. *Zoological Journal of the Linnean Society* 73: 307–312.

Dawkins, W. B. 1864. On the Rhaetic Beds and White Lias of western and central Somerset, and on the discovery of a new fossil mammal in the grey marlstones below the bone-bed. *Quarterly Journal of the Geological Society of London* 20: 396–412.

Dietrich, W. O. 1937. Ueber eine Säugetierelle aus dem Rät von Halberstadt. *Neues Jahrbuch für Mineralogie, Geologie und Paläontologie, Beilage-Band* 77: 310–319.

Dong, Z., S. Zhou, and Y. Zhang. 1983. The dinosaurian remains from Sichuan basin, China. *Palaeontologia Sinica, new series C,* 23: 1–145.

Donovan, D. T., M. T. Curtis, and S. A. Curtis. 1989. A psiloceratid ammonite from the supposed Triassic Penarth Group of Avon, England. *Palaeontology* 32: 231–235.

Dunay, R. E., and M. J. Fisher. 1979. Palynology of the Dockum Group (Upper Triassic), Texas, U.S.A. *Review of Palaeobotany and Palynology* 28: 61–92.

Emmons, E. 1857. *American Geology. Part VI.* Albany, NY: Sprague.

Evans, S. E., and A. R. Milner. 1991. Middle Jurassic microvertebrate faunas from the British Isles. Pp. 21–22 in Z. Kielan-Jaworowska, N. Heintz, and H. A. Nakrem (eds.), *Fifth Symposium on Mesozoic Terrestrial Ecosystems and Biota, Extended Abstracts.* University of Oslo Palaeontological Museum.

Evans, S. E., A. R. Milner, and F. Mussett. 1990. A discoglossid frog from the Middle Jurassic of England. *Palaeontology* 33: 299–311.

Everett, B. H., B. J. Kowallis, and E. H. Christiansen. 1989. Correlation of Jurassic sediments of southern Utah using bentonite characteristics. *Geological Society of America, Abstracts with Programs* 21: 77.

Fourie, S. 1962. Notes on a new tritylodontid from the Cave Sandstone of South Africa. *Navorsinge van die Nasionale Museum Bloemfontein* 2: 7–19.

——— 1963. A new tritylodontid from the Cave Sandstone of South Africa. *Nature* 198: 201.

Fraser, N. C. 1985. Vertebrate faunas from Mesozoic fissure deposits of south west Britain. *Modern Geology* 9: 273–300.

——— 1986. Terrestrial vertebrates at the Triassic–Jurassic boundary in south west Britain. *Modern Geology* 10: 147–157.

Fraser, N. C., G. M. Walkden, and V. Stewart. 1985. The first pre-Rhaetic therian mammals. *Nature* 314: 161–162.

Freeman, E. F. 1976a. A mammalian fossil from the Forest Marble (Middle Jurassic) of Dorset. *Proceedings of the Geologists' Association* 87: 231–235.

——— 1976b. Mammal teeth from the Forest Marble (Middle Jurassic) of Oxfordshire, England. *Science* 194: 1053–1055.

——— 1979. A Middle Jurassic mammal bed from Oxfordshire. *Palaeontology* 22: 135–166.

Froelich, A. J., and D. Gottfried. 1988. An overview of early Mesozoic intrusive rocks in the Culpeper basin, Virginia and Maryland. *U.S. Geological Survey Bulletin* 1776: 151–165.

Ginsburg, L. 1961. Un nouveau tritylodonte du Trias supérieur du Basutoland (Afrique du sud). *Comptes Rendus des Séances de l'Académie des Sciences, Paris* 252: 3853–3854.

Golebiowski, R. 1990. The Alpine Kossen Formation, a key for European topmost Triassic correlations. *Albertiana* 8: 25–35.

Gow, C. E. 1980. The dentitions of Tritheledontidae (Therapsida: Cynodontia). *Proceedings of the Royal Society of London* 208: 461–481.

Gregory, J. T. 1962. The genera of phytosaurs. *American Journal of Science* 260: 652–690.

Hahn, G. 1973. Neue Zähne von Haramiyiden aus der deutschen Ober-Trias und ihre Beziehungen zu den Multituberculaten. *Palaeontographica* A142: 1–15.

Hahn, G., J. C. Lepage, and G. Wouters. 1984. Cynodontier-Zähne aus der Ober-Trias von Medernach, Grossherzogtum Luxembourg. *Bulletin de la Société belge de Géologie* 93: 357–373.

——— 1988. Traversodontiden-Zähne (Cynodontia) aus der Ober-Trias von Gaume (Süd-Belgien). *Bulletin du Institut Royal des Sciences Naturelles de Belgique, Sciences de la Terre* 58: 177–186.

Hahn, G., R. Wild, and G. Wouters. 1987. Cynodontier-Zähne aus der Ober-Trias von Gaume (S-Belgien) *Mémoires pour servir à l'Explication des Cartes Géologiques et Minières de la Belgique* 24: 1–32.

Hallam, A. 1990. Correlation of the Triassic–Jurassic boundary in England and Austria. *Journal of the Geological Society, London* 147: 421–424.

Harland, W. B., R. L. Armstrong, A. V. Cox, L. E. Craig, A. G. Smith, and D. V. Smith. 1990. *A Geologic Time Scale 1989.* Cambridge University Press.

Harland, W. B., A. V. Cox, P. G. Llewellyn, C. A. G. Pickton, A. G. Smith, and R. Walters. 1982. *A Geologic Time Scale.* Cambridge University Press.

Hennig, E. 1922. Die Säugerzähne des württembergischen Rhät-Lias-Bonebeds. *Neues Jahrbuch für Mineralogie, Geologie und Paläontologie, Beilage-Band* 46: 181–267.

Hodych, J. P., and G. R. Dunning. 1992. Did the Manicouagan impact trigger end-of-Triassic mass extinction? *Geology* 20: 51–54.

Hopson, J. A. 1984. Late Triassic traversodont cynodonts from Nova Scotia and southern Africa. *Palaeontologia Africana* 25: 181–201.

——— 1985. Morphology and relationships of *Gomphodontosuchus brasiliensis* von Huene (Synapsida, Cynodontia, Tritylodontoidea) from the Triassic of Brazil. *Neues Jahrbuch für Geologie und Paläontologie, Monatshefte* 1985: 285–299.

——— 1991. Convergence in mammals, tritheledonts, and tritylodonts. *Journal of Vertebrate Paleontology (Suppl.)* 11(3): 36A.

Hopson, J. A., and J. W. Kitching. 1972. A revised classification of cynodonts (Reptilia; Therapsida). *Palaeontologia Africana* 14: 71–85.

Hopson, J. A., and W.-E. Reif. 1981. The status of *Archaeodon reuningi* von Huene, a supposed Late Triassic mammal from southern Africa. *Neues Jahrbuch für Geologie und Paläontologie, Monatshefte* 1981: 307–310.

Huber, P., S. G. Lucas, and A. P. Hunt. 1993. Late Triassic fish assemblages of the North American Western Interior and their biochronological significance. *Museum of Northern Arizona Bulletin* 95: 51–66.

Huene, E. von. 1933. Zur Kenntnis des württembergischen Rhätbonebeds mit Zahnfunden neuer Säuger und säugerähnlicher Reptilien. *Jahreshefte des Verein für vaterländische Naturkunde Württemberg* 84: 65–128.

Huene, F. von. 1928. Ein Cynodontier aus der Trias Brasiliens. *Centralblatt für Mineralogie, Geologie und Paläontologie* B1928: 251–270.

——— 1936. *Die fossilen Reptilien des südamerikanischen Gondwanalandes. Lieferung 2.* Tübingen: F. Heine.

Hunt, A. P. 1991. The early diversification pattern of dinosaurs in the Late Triassic. *Modern Geology* 16: 43–60.

Hunt, A. P., and S. G. Lucas. 1990. Re-evaluation of "*Typothorax*" *meadei*, a Late Triassic aetosaur from the United States. *Paläontologische Zeitschrift* 64: 317–328.

——— 1991a. The *Paleorhinus* biochron and the correlation of the non-marine Upper Triassic of Pangaea. *Palaeontology* 34: 487–501.

——— 1991b. A new rhynchosaur from the Upper Triassic of West Texas, and the biochronology of Late Triassic rhynchosaurs. *Palaeontology* 34: 927–938.

Jenkins, F. A., Jr., A. W. Crompton, and W. R. Downs. 1983. Mesozoic mammals from Arizona: new evidence on mammalian evolution. *Science* 222: 1233–1235.

Kalandadze, N. N. and A. S. Rautian. 1991. Late Triassic

zoogeography and a reconstruction of the terrestrial tetrapod fauna of North Africa. *Paleontological Journal* 1991(1): 1–12.

Kemp, T. S. 1980. Aspects of the structure and functional anatomy of the Middle Triassic cynodont *Luangwa. Journal of Zoology, London* 191: 193–239.

——— 1982. *Mammal-like Reptiles and the Origin of Mammals.* London: Academic Press.

——— 1983. The relationships of mammals. *Zoological Journal of the Linnean Society* 77: 353–384.

Kent, D. V., and F. M. Gradstein. 1985. A Cretaceous and Jurassic geochronology. *Geological Society of America Bulletin* 96: 1419–1427.

Kermack, D. M. 1982. A new tritylodont from the Kayenta Formation of Arizona. *Zoological Journal of the Linnean Society* 76: 1–17.

Kermack, K. A., F. Mussett, and H. W. Rigney. 1973. The lower jaw of *Morganucodon. Zoological Journal of the Linnean Society* 53: 87–175.

Kietzke, K. K. 1989. Calcareous microfossils from the Triassic of the southwestern United States. Pp. 223–232 in S. G. Lucas and A. P. Hunt (eds.), *Dawn of the Age of Dinosaurs in the American Southwest.* Albuquerque: New Mexico Museum of Natural History.

Kietzke, K. K., and S. G. Lucas. 1991a. Ostracoda from the Upper Triassic (Carnian) Tecovas Formation near Kalgary, Crosby County, Texas. *Texas Journal of Science* 43: 191–197.

——— 1991b. Triassic nonmarine "*Spirorbis*": gastropods not worms. *New Mexico Geology* 13: 93.

Kitching, J. W., and M. A. Raath. 1984. Fossils from the Elliot and Clarens formations (Karoo sequence) of the northeastern Cape, Orange Free State and Lesotho, and a suggested biozonation based on tetrapods. *Palaeontologia Africana* 25: 111–125.

Krystyn, L. 1990. A Rhaetian stage-chronostratigraphy: subdivisions and their intercontinental correlation. *Albertiana* 8: 15–24.

Kühne, W. G. 1950. *Mucrotherium* und *Uniserium* E. von Huene sind Fragmente unterer Backenzähne eines Tritylodontiers. *Neues Jahrbuch für Geologie und Paläontologie, Monatshefte* 1950: 187–191.

——— 1956. *The Liassic Therapsid* Oligokyphus. London: British Museum (Natural History).

Kutty, T. S., S. L. Jain, and T. Roy Chowdhury. 1987. Gondwana sequence of the northern Pranhita-Godavari valley: its stratigraphy and vertebrate fauna. *The Palaeobotanist* 36: 214–229.

Kutty, T. S., and D. P. Sengupta. 1989. The Late Triassic formations of the Pranhita-Godavari valley and their faunal succession – a reappraisal. *Indian Journal of Earth Sciences* 16: 189–206.

Langston, W. 1949. A new species of *Paleorhinus* from Triassic of Texas. *American Journal of Science* 247: 324–341.

Lees, P. M., and R. Mills. 1983. A quasi-mammal from Lesotho. *Acta Palaeontologica Polonica* 28: 171–180.

Litwin, R., A. Traverse, and S. R. Ash. 1991. Preliminary palynological zonation of the Chinle Formation, southwestern U.S.A., and its correlation to the Newark Supergroup (eastern U.S.A.). *Review of Palaeobotany and Palynology* 68: 269–287.

Lucas, S. G. 1990a. The rise of the dinosaur dynasty. *New Scientist* 1737: 44–46.

——— 1990b. Toward a vertebrate biochronology of the Triassic. *Albertiana* 8: 36–41.

——— 1991a. Sequence stratigraphy of nonmarine Upper Triassic sediments, Western Interior United States. *Geological Society of America, Abstracts with Programs* 23: 44.

——— 1991b. Correlation of Triassic strata of the Colorado Plateau and southern High Plains, New Mexico. *New Mexico Bureau of Mines and Mineral Resources Bulletin* 137: 47–56.

——— 1991c. Sequence stratigraphic correlation of nonmarine and marine Late Triassic biochronologies, western United States. *Albertiana* 9: 11–18.

——— 1991d. Dinosaurs and Mesozoic biochronology. *Modern Geology* 16: 127–138.

——— 1993. The Chinle Group: revised stratigraphy and chronology of Upper Triassic nonmarine strata in the western United States. *Museum of Northern Arizona Bulletin* 59: 27–50.

Lucas, S. G., and S. N. Hayden. 1989. Triassic stratigraphy of west-central New Mexico. *New Mexico Geological Society Guidebook* 40: 191–211.

Lucas, S. G., and A. P. Hunt. 1989. Vertebrate biochronology of the Late Triassic. *28th International Geological Congress, Abstracts* 2: 335–336.

——— 1990. The oldest mammal. *New Mexico Journal of Science* 30: 41–49.

Lucas, S. G., and W. Oakes. 1988. A Late Triassic cynodont from the American South-West. *Palaeontology* 31: 445–449.

Marshall, J. E. A., and D. I. Whiteside. 1980. Marine influences in the Triassic "uplands." *Nature* 287: 627–628.

Marvin, R. F., J. C. Wright, and F. G. Walthall. 1965. K-Ar and Rb-Sr ages of biotite from the Middle Jurassic part of the Carmel Formation, Utah. *United States Geological Survey Professional Paper* 525-B: 104–107.

Marzolf, J. E. 1990. Reconstruction of extensionally dismembered early Mesozoic sedimentary basins; southwestern Colorado Plateau to the eastern Mojave Desert. *Geological Society of America Memoir* 176: 477–500.

Menning, M. 1990. A new scheme for the Permian and Triassic succession of central Europe. *Permophiles* 16: 11.

Mixon, R. B., G. E. Murray, and T. G. Diaz. 1959. Age and correlation of the Huizachal Group (Mesozoic), state of Tamaulipas, Mexico. *American Association of Petroleum Geologists Bulletin* 43: 757–771.

Murry, P. A. 1989. Geology and paleontology of the Dockum (Upper Triassic), West Texas and eastern New Mexico. Pp. 102–145 in S. G. Lucas and A. P. Hunt (eds.), *Dawn of the Age of Dinosaurs in the American Southwest.* Albuquerque: New Mexico Museum of Natural History.

Olsen, P. E., S. J. Fowell, and B. Cornet. 1990. The Triassic/Jurassic boundary in continental rocks of eastern North America; a progress report. *Geological Society of America Special Paper* 247: 585–593.

Olsen, P. E., S. Fowell, B. Cornet, and W. K. Witte. 1989. Calibration of Late Triassic–Early Jurassic time scale based on orbitally induced lake cycles. *28th International Geological Congress, Abstracts* 2: 547–548.

Olsen, P. E., and P. M. Galton. 1977. Triassic–Jurassic extinctions: are they real? *Science* 197: 983–986.

——— 1984. A review of the reptile and amphibian assemblages from the Stormberg of southern Africa, with special emphasis on the footprints and the age of the Stormberg. *Palaeontologia Africana* 25: 87–110.

Owen, R. 1884. On the skull and dentition of a Triassic mammal (*Tritylodon longaevus*) from South Africa. *Quarterly Journal of the Geological Society of London* 40: 146–152.

Padian, K. 1989a. Did "thecodontians" survive the Triassic? Pp. 401–414 in S. G. Lucas and A. P. Hunt (eds.), *Dawn of the Age of Dinosaurs in the American Southwest*. Albuquerque: New Mexico Museum of Natural History.

——— 1989b. Presence of the dinosaur *Scelidosaurus* indicates Jurassic age for the Kayenta Formation (Glen Canyon Group, northern Arizona). *Geology* 17: 438–441.

Patterson, B., and E. C. Olsen. 1961. A triconodontid mammal from the Triassic of Yunnan. Pp. 129–191 in *International Colloquium on Evolution of Lower and Nonspecialized Mammals*. Brussels.

Peterson, F., and G. N. Pipiringos. 1979. Stratigraphic relations of the Navajo Sandstone to Middle Jurassic formations, southern Utah and northern Arizona. *United States Geological Survey Professional Paper* 1035-B: B1–B43.

Ratcliffe, N. M. 1988. Reinterpretation of the relationship of the western extension of the Palisades sill to the lava flows at Ladentown, New York, based on new core data. *United States Geological Survey Bulletin* 1737: 113–135.

Rigney, H. W. 1963. A specimen of *Morganucodon* from Yunnan. *Nature* 197: 1122–1123.

Robinson, P. L. 1957. The Mesozoic fissures of the Bristol Channel area and their vertebrate faunas. *Zoological Journal of the Linnean Society* 43: 260–282.

——— 1971. A problem of faunal replacement on Permo-Triassic continents. *Palaeontology* 14: 131–153.

Romer, A. S. 1967. The Chañares (Argentina) Triassic reptile fauna. III. Two new gomphodonts, *Massetognathus pascuali* and *M. teruggi. Breviora* 264: 1–25.

Rowe, T. 1988. Definition, diagnosis, and origin of Mammalia. *Journal of Vertebrate Paleontology* 8: 241–264.

Savage, R. J. G. 1971. Tritylodontid incertae sedis. *Proceedings of the Bristol Naturalists' Society* 32: 80–83.

——— 1984. Mid-Jurassic mammals from Scotland. Pp. 211–213 in W.-E. Reif and F. Westphal (eds.), *Third Symposium on Mesozoic Terrestrial Ecosystems, Short Papers*. Tübingen: ATTEMPTO.

Savage, R. J. G., and M. Waldman. 1966. *Oligokyphus* from Holwell Quarry, Somerset. *Proceedings of the Bristol Naturalists' Society* 31: 185–192.

Schurmann, W. M. L. 1977. Aspects of Late Triassic palynology. 2. Palynology of the "Grés et Schiste à Avicula contorta" and "Argiles de Levallois" (Rhaetian) of northeastern France and southern Luxembourg. *Review of Palaeobotany and Palynology* 23: 159–253.

Seidemann, D. E. 1988. Age of the Triassic–Jurassic boundary; a view from the Hartford basin. *American Journal of Science* 289: 553–562.

Shubin, N. H., A. W. Crompton, H.-D. Sues. and P. E. Olsen. 1991. New fossil evidence on the sister-group of mammals and early Mesozoic faunal distributions. *Science* 251: 1063–1065.

Shubin, N. H., and H.-D. Sues. 1991. Biogeography of early Mesozoic continental tetrapods: patterns and implications. *Paleobiology* 17: 214–230.

Sigogneau-Russell, D. 1983a. A new therian mammal from the Rhaetic locality of Saint-Nicolas-de-Port (France). *Zoological Journal of the Linnean Society* 78: 175–186.

——— 1983b. Nouveaux taxons de mammifères rhétiens. *Acta Palaeontologia Polonica* 28: 233–249.

Sigogneau-Russell, D., R. M. Frank, and J. Hemmerlé. 1986. A new family of mammals from the lower part of the French Rhaetic. Pp. 99–108 in K. Padian (ed.), *The Beginning of the Age of Dinosaurs: Faunal Change across the Triassic–Jurassic Boundary*. Cambridge University Press.

Simms, M. J., and A. H. Ruffell. 1989. Synchroneity of climatic change and extinctions in the Late Triassic. *Geology* 17: 265–268.

——— 1990. Climatic and biotic change in the Late Triassic. *Journal of the Geological Society, London* 147: 321–327.

Small, B. J. 1989. Post quarry. Pp. 145–148 in S. G. Lucas and A. P. Hunt (eds.), *Dawn of the Age of Dinosaurs in the American Southwest*. Albuquerque: New Mexico Museum of Natural History.

Sues, H.-D. 1985a. The relationships of the Tritylodontidae (Synapsida). *Zoological Journal of the Linnean Society* 85: 205–217.

——— 1985b. First record of the tritylodont *Oligokyphus* (Synapsida) from the Lower Jurassic of western North America. *Journal of Vertebrate Paleontology* 5: 328–335.

——— 1986a. *Dinnebitodon amarali*, a new tritylodontid (Synapsida) from the Lower Jurassic of western North America. *Journal of Paleontology* 60: 758–762.

——— 1986b. Relationships and biostratigraphic significance of the Tritylodontidae (Synapsida) from the Kayenta Formation of northeastern Arizona. Pp. 279–284 in K. Padian (ed.), *The Beginning of the Age of Dinosaurs: Faunal Change across the Triassic–Jurassic Boundary*. Cambridge University Press.

Sues, H.-D., J. A. Hopson, and N. H. Shubin. 1992. Affinities of ?*Scalenodontoides plemmyridon* Hopson, 1984 (Synapsida: Cynodontia) from the Upper Triassic of Nova Scotia. *Journal of Vertebrate Paleontology* 12: 168–171.

Sues, H.-D., and P. E. Olsen. 1990. Triassic vertebrates of Gondwanan aspect from the Richmond basin of Virginia. *Science* 249: 1020–1023.

Sun, A. 1981. [Reidentification of *Traversodontoides wangwuensis* Young.] *Vertebrata Palasiatica* 19: 1–4. [in Chinese]

1984. Skull morphology of the tritylodont genus *Bienotheroides* of Sichuan. *Scientia Sinica* B27: 970–984.

1986. [New material of *Bienotheroides* (tritylodont reptile) from Shaximiao Formation of Sichuan.] *Vertebrata Palasiatica* 24: 165–170. [in Chinese]

Sun, A. L., and G. Cui. 1986. A brief introduction to the Lower Lufeng saurischian fauna (Lower Jurassic: Lufeng, Yunnan, People's Republic of China). Pp. 275–278 in K. Padian (ed.), *The Beginning of the Age of Dinosaurs: Faunal Change across the Triassic–Jurassic Boundary*. Cambridge University Press.

1989. [Tritylodont reptile from Xinjiang.] *Vertebrata Palasiatica* 27: 1–8. [in Chinese]

Sun, A. L., and Y. Li. 1985. [The postcranial skeleton of the last tritylodont *Bienotheroides*.] *Vertebrata Palasiatica* 23: 135–151. [in Chinese]

Sutter, J. F. 1988. Innovative approaches to dating of igneous events in the early Mesozoic basins of the eastern United States. *United States Geological Survey Bulletin* 1776: 194–200.

Tozer, E. T. 1979. Latest Triassic ammonoid faunas and biochronology, western Canada. *Geological Survey of Canada Paper* 79-1B: 127–135.

1988. Rhaetian: a substage, not a stage. *Albertiana* 7: 9–15.

Waldman, M., and R. J. G. Savage. 1972. The first Jurassic mammal from Scotland. *Journal of the Geological Society of London* 128: 119–125.

Watson, D. M. S. 1913. On a new cynodont from the Stormberg. *Geological Megazine* (5) 10: 145–148.

Westermann, G. 1984. Gauging the duration of stages: a new approach for the Jurassic. *Episodes* 7: 26–28.

Wible, J. R. 1991. Origin of Mammalia: the craniodental evidence reexamined. *Journal of Vertebrate Paleontology* 11: 1–28.

Wiedmann, J., F. Fabricius, L. Krystyn, J. Reitner, and M. Ulrichs. 1979. Ueber Umfang und Stellung des Rhaet. *Newsletters in Stratigraphy* 8: 133–152.

Winkler, D. A., L. L. Jacobs, J. D. Congleton, and W. R. Downs. 1991. Life in a sand sea: biota from Jurassic interdunes. *Geology* 19: 889–892.

Wouters, G., D. Sigogneau-Russell, and J. C. Lepage. 1984. Decouverté d'une dent d'Haramiyide (Mammalia) dans des niveaux rhétiens de la Gaume (en Lorraine belge). *Bulletin Société belge de Géologie* 93: 351–355.

Yadagiri, P. 1984. New symmetrodonts from Kota Formation (Early Jurassic) India. *Journal of the Geological Society of India* 25: 514–521.

Young, C. C. 1947. Mammal-like reptiles from Lufeng, Yunnan, China. *Proceedings of the Zoological Society of London* 117: 537–597.

1974. [A new genus of Traversodontidae in Jiyuan, Honan.] *Vertebrata Palasiatica* 12: 203–211. [in Chinese]

1978. [New materials of *Eozostrodon*.] *Vertebrata Palasiatica* 16: 1–3. [in Chinese]

1982. [*Selected Works of Yang Zhungjian*.] Beijing: Science Press. [in Chinese]

Zhang, F., and G. Cui. 1983. [New material and new understanding of *Sinoconodon*.] *Vertebrata Palasiatica* 21 :32–41. [in Chinese]

21

Biotic and climatic changes in the Carnian (Triassic) of Europe and adjacent areas

MICHAEL J. SIMMS, ALASTAIR H. RUFFELL, AND
ANDREW L. A. JOHNSON

Introduction

The Triassic period represents one of the more important episodes of biotic change in earth history. Following the disappearance of many characteristic Paleozoic groups around the end of the Permian, surviving taxa in many cases experienced major diversification in the ensuing Triassic. However, by the beginning of the Jurassic, many distinctive taxa that had appeared during the Triassic, as well as some of the survivors from the Paleozoic, had become extinct. The final disappearance of groups such as the ceratitid ammonites, conodonts, and some bivalve taxa took place very close to the Triassic–Jurassic boundary (Hallam, 1981; Benton, 1986a; Clark, 1987), but it is also clear that many other higher taxa that did not survive into the Jurassic had already become extinct long before the end of the Triassic. Raup and Sepkoski (1986) noted a late Carnian extinction peak in their analysis of marine animal genera, but regarded it as an artifact of sampling. In contrast, Benton's (1986a) analyses of various taxa suggested that the Carnian "extinction event" was real and that, moreover, there was a clear nonsynchroneity between the peak of marine extinctions at the end of the early Carnian and the maximum turnover among nonmarine groups at the Carnian–Norian boundary.

However, it is difficult to constrain the timing of these changes with any precision. This is certainly true for the terrestrial fauna and flora because correlation between the marine Triassic, on which the standard sequence is based, and terrestrial successions has proved problematic. Although some success has been achieved with palynological correlation in the European Triassic (Visscher and Van der Zwan, 1981; Van der Eem, 1983), it has proved difficult to extend this further afield (Cornet and Olsen, 1990).

In North America, interbedded basaltic lava flows in the Newark Supergroup have provided absolute dates for parts of the succession. On the assumption that the Van Houten and other cycles reflect Milankovitch-type climatic cycles, Olsen (1986) has been able to date much of the remainder of the Newark Supergroup with considerable precision. However, because absolute dates for much of the European marine Triassic are unavailable, precise correlation between the Newark Supergroup and the standard marine Triassic succession is still not possible.

Evidence for biotic change

Independent documentation of various taxonomic groups provides strong evidence for an episode of major biotic change during the Carnian. The evidence from well-constrained groups and successions does lend support to Benton's (1986a) suggestion of non-synchroneity between the marine and nonmarine biotas. However, even allowing for the correlation problems discussed earlier, precise dating of diversity changes is lacking for several groups, such as gastropods, conodonts, and fishes, owing to the recording of data only to the stage level, rather than to the substage level.

Among Triassic crinoids, the Encrinidae, which superficially resemble certain Paleozoic crinoid taxa, are the most conspicuous elements of Anisian to early Carnian faunas. They reached a peak of diversity in the early Carnian, but had disappeared entirely by the middle Carnian. Other crinoid groups also showed significant declines in diversity between the early and middle Carnian, although the minute, supposedly planktonic Somphocrinidae apparently were unaffected and continued to flourish throughout the Carnian, before experiencing a dramatic decline early in the Norian (Johnson and Simms, 1989; Simms, 1990a). Similar patterns of an abrupt decline in diversity following an early Carnian peak have been documented among echinoids (Smith, 1990) and scallops (Johnson and Simms, 1989). In both instances, distinctive clades, in some cases morphologically "archaic" in character, failed to survive into the middle Carnian. However, in contrast to that, Newton et al. (1987) have found that a considerable number of bivalve taxa from an early Norian coral reef in Oregon are also present in the early Carnian Cassian Formation, suggesting that the Carnian extinction may have been fairly selective in the taxa that were affected within some groups, as has already been indicated by the crinoid data. Ammonites experienced an elevated extinction rate at the end of the early Carnian, although that was only one of several comparable events that affected the group during the Triassic (Benton, 1986a). Other groups, such as bryozoans, brachiopods, and gastropods, are more poorly documented. Bryozoans and brachiopods appear to show similar patterns (Schäfer and Fois, 1987; Stanley, 1988), but gastropods experienced a steady increase in extinction rate from Ladinian to Norian times, without any significant peak in the Carnian (Erwin, 1990). In each case it is quite possible that the pattern would be rather different if plotted at the level of substage rather than stage, but this must remain conjectural in the absence of more precise data.

The data from conodonts and fishes are more equivocal, with the former experiencing merely a decrease in diversity during the Carnian due to a drop in species origination rate (Clark, 1987; Aldridge, 1988). Among fishes, Patterson and Smith (1987) found only a single extinction peak, in the Carnian, between the Induan (lowermost Triassic) and the Tithonian (uppermost Jurassic), although subsequently (Smith and Patterson, 1988) they dismissed all but the Campanian, Maastrichtian, and Eocene peaks as statistically insignificant.

Scleractinian corals increased enormously in their ecological importance within the marine biotas during the Carnian (Stanley, 1988). First known from the Anisian, they remained a minor element of reef communities throughout the Middle Triassic and into the early Carnian. Such reefs remained dominated by holdovers from Permian reef communities, particularly calcisponges and *Tubiphytes*. Late Carnian reefs are rarely developed, but by Norian times scleractinians had assumed a dominant role as framework builders within the reef, while many of the Permian holdovers were lost or greatly reduced in importance. As Stanley (1988, fig. 3) demonstrated, there is some non-synchroneity of response among marine taxa. One of the dominant reef-builders, *Tubiphytes*, experienced a severe decline in the middle Carnian, whereas calcisponges, bryozoans, rhodophytes, and chlorophytes declined in the late Carnian, this being roughly synchronous with the diversification of foraminifera, tabulozoans, spongiomorphs, and corals. Interestingly, the early Norian reef described from Oregon by Stanley

and Whalen (1989) is of this advanced type, but it supports some relict Cassian bivalves (Newton et al., 1987).

The Carnian extinction episode and subsequent diversification represent a critical episode in the early Mesozoic evolution of reef communities, with the arid climate of Tethys during much of the Triassic providing a relatively clastic-free seaway in which such organisms could flourish. Whereas environmental changes in the Late Permian and Early Jurassic had severe, long-lasting effects on the reef biotas, the Carnian was the time of an episode of extinction and rapid rediversification that may require a different explanation.

Changes in the terrestrial biota were no less spectacular than those in the marine realm, although they appear to have been concentrated around the Carnian–Norian boundary, with no significant turnover at the end of the early Carnian. Nonmarine tetrapod groups such as various synapsids, rhynchosaurs, "thecodontians," and labyrinthodonts experienced a dramatic decline in diversity at the end of the Carnian, whereas dinosaurs, which are first known from the latter part of the Carnian, increased greatly in diversity and abundance during the ensuing Norian (Tucker and Benton, 1982; Benton, 1986a,b, 1989; Olsen and Sues, 1986).

A similar pattern has been documented for Late Triassic floras by Boulter, Spicer, and Thomas (1988). They recognized a middle Carnian episode of increased originations compared with extinctions, followed by a considerable increase in the extinction rate around the Carnian–Norian boundary. In the eastern Urals, Kiritchkova (1990) noted a change from a pteridosperm-dominated flora in the Ladinian and Carnian to a conifer-fern-ginkgo assemblage in the Norian and Early Jurassic. An almost identical change has been found to mark the onset of aridity in Late Jurassic and Barremian (Early Cretaceous) times (Ruffell and Batten, 1990). Visscher and Van der Zwan (1981) reported an increased incidence of hygrophytic palynomorphs in some middle to late Carnian successions.

In the most detailed paleobotanical work on the Newark Supergroup, Cornet and Olsen (1990) identified an episode of elevated palynofloral turnover near or at the end of the middle Carnian. In terms of percentage turnover, it was only slightly higher than the change from the late Carnian to the Norian indicated by their tabulated data (Cornet and Olsen, 1990, p. 57), yet they consider the palynofloral change across the latter boundary not to have been significant.

Causes of change in Carnian biotas

Various suggestions have been made as to the cause of the extinctions at the end of the Triassic, but there have been few previous attempts to account for the Carnian extinctions. However, several independent lines of evidence, discussed later, indicate that a major change in climate occurred almost synchronously with the biotic changes, with the widespread aridity of much of the Triassic being interrupted by a significantly more humid episode during the middle to late Carnian. Direct and indirect effects of this climatic change provide the most parsimonious explanation for the great breadth of the biotic changes observed. The terrestrial fauna and flora may have been affected directly by the return to aridity at the beginning of the Norian, which presumably would have had a more traumatic effect than the onset of a more humid climate, but that is less likely for the marine macrofauna.

Habitat loss, as a result of late Carnian regression and clastic influx, has been suggested as a cause of the marine invertebrate turnover (Stanley, 1988; Johnson and Simms, 1989). A regional abiotic change such as that might perhaps account for the apparent persistence of Cassian faunal elements into the early Carnian of Oregon. This site is recognized as one of a number of accreted terranes on the western coast of North America. As such, its Triassic position relative to the North American craton is difficult to assess, but it may have been significantly isolated from the effects of whatever was causing the biotic changes in the European area and hence may have acted as a refugium for Cassian taxa.

The sudden rise to dominance of scleractinian corals, as well as many of the changes seen in other elements of the marine biotas across that interval, suggests another possibility. Such changes may perhaps be attributable to a fundamental change in the composition of the plankton, the lowest level in the marine food chain. One of the most important post-Paleozoic plankton groups, the coccolithophores, are first known from the latter part of the Carnian (Perch-Nielsen, 1985). Similarly, there was an important Carnian diversification among the dinoflagellates, with the appearance of both S-cysts and G-cysts to supplement the R-cysts, the latter being first recorded from the Anisian. Only a small proportion of the modern plankton has any significant preservation potential in the fossil record. Hence the diversification of both of these groups almost simultaneously with other major changes in the marine realm suggests that there may have been a major, though largely unpreserved, change in plankton composition. The changing role of scleractinians through the Carnian may perhaps be a manifestation of this change in the plankton. Dinoflagellates include the symbiotic zooxanthellate algae found in hermatypic corals. These zooxanthellae considerably enhance the growth rate of such corals by comparison with ahermatypic types, and it is the initial acquisition of that symbiosis during the Carnian that Stanley (1981, 1988) has suggested may account for their changing role between the early Carnian and the Norian. Al-

though it is almost impossible to prove conclusively that any changes in the marine macrofauna were direct consequences of changes further down the food chain, the approximate synchroneity of the changes in the plankton and macrofauna would appear to indicate a causal link, as is supported by the work of Aebischer, Coulson, and Colebrook (1990). However, such a change might be expected to have been global in extent, and hence it is somewhat difficult to account for the apparent geographic selectivity shown by the responses of some taxa.

Evidence for climatic change

Terrestrial sedimentary sequences

Throughout much of Europe and North America, Middle and Upper Triassic terrestrial successions are dominated by facies typical of arid and semiarid environments. In Britain, this comprises the Mercia Mudstone Group (Warrington et al., 1980), a monotonous succession of red and green mudstones and dolomitic siltstones, with subordinate sandstone or evaporite horizons. Exact depositional environments for much of the Mercia Mudstone Group are still poorly defined, having variously been described as "evaporitic" or "playa lakes" (Jeans, 1978; Petrie et al., 1990; Simms and Ruffell, 1990) or "epeiric seaways" (Warrington, 1981). Klein (1962) suggested, on the basis of various sedimentary structures, that the Keuper Marl (= Mercia Mudstone Group) was deposited in a predominantly lacustrine environment of varying depth with occasional episodes of emergence, a view also supported by Tucker (1977), Arthurton (1980), and Whittaker and Green (1983). Arthurton (1980) favored eustatic control for the alternation of the laminated and blocky mudstones that he observed, with the thick (typically more than 100 m) halite beds resulting from evaporation in the more central parts of the basins following repeated episodes of marine flooding. Although the halite beds are best accounted for by this mechanism, the regularity and scale (typically about 1–5 m thick) of the laminated/blocky mudstone cycles are more reminiscent of climatically induced cyclicity. The general paleoenvironment would, therefore, appear to be one of an arid plain subject to flooding during fairly frequent episodes of higher runoff and more occasional marine inundation.

Within this succession is found an extremely widespread, though locally discontinuous, sandstone unit of late Carnian age known under a variety of names in different areas of Britain, including the Arden Sandstone, Butcombe Sandstone, and North Curry Sandstone members. Eastward its correlative can be traced across France and into Germany, southern Poland, and northern Switzerland. In parts of Germany, this unit, termed the Schilfsandstein, attains its maximum

of the North Atlantic continental margins, western Europe, and European Paleotethys. Note how the Carnian
ions. The Eastern North Atlantic includes the Celtic Seas, Western Approaches, and Wessex basins.

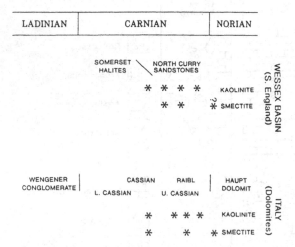

Figure 21.2. Comparison of kaolinite/smectite abundances within the "Germanic" Trias succession of southern England and the Tethyan Trias of northern Italy.

development of more than 35 m in thickness. These sandstones occupy channels up to several kilometers wide, eroded into the underlying sediments, and they pass laterally into thin mudstones or siltstones. Wurster (1964) considered the Schilfsandstein to represent deltaic deposits in a very shallow sea. Evidence for diachroneity of the unit (Hahn, 1984), which becomes younger in the direction of flow, from north to south, also suggests a slowly prograding delta, although Bechstädt and Schweizer (1991) consider the Schilfsandstein to represent incised valley fills that prograded southward. Warrington and Williams (1984) interpreted the thicker, more arenaceous sequences of the North Curry Sandstone Member as having formed in delta distributaries or fluvial channels, whereas the thinner, more argillaceous lateral correlatives represent interdistributary or overbank deposits.

The existence of such a widespread development of upper Carnian sandstones (Figure 21.1) containing substantial channeled sand bodies indicates a significant change of environment, with increased levels of runoff. We interpret this as the result of a temporary change in the prevailing climate, from the predominantly arid climate during deposition of the typical Mercia Mudstone facies to a significantly more humid episode. Such a climatic change has been alluded to for this unit on a number of occasions previously (Green and Welch, 1965; Wills, 1970; Lott et al., 1982). Further support for the theory of climatic change comes from clay mineral analyses across this interval. Throughout southern England, where these sandstones have been sampled, they have been found to have elevated detrital kaolinite contents, in comparison with the contiguous Mercia Mudstone above and below, indicating a warm, humid climate during

their deposition (Simms and Ruffell, 1989, 1990). However, our most recent analyses indicate that at certain sites, kaolinite is present in the basal conglomerate of the sandstone or in the underlying mudstones for a short distance below (Figure 21.2). This suggests that the climate in the region was already becoming more humid prior to development of these fluvial systems. The difficulties associated with dating the Mercia Mudstone mean that it is effectively impossible to determine accurately the time interval represented by the basal lag of these sandstones, although Warrington (1971) found palynomorphs indicating a middle to late Carnian age at a level 42 m below the Weston Mouth Sandstone, of late Carnian age, on the south Devon coast.

On the western side of the North Atlantic, along the eastern margin of North America, are thick basinal successions of lacustrine sediments spanning the Carnian–Norian interval. Smoot and Olsen (1988) found clear evidence in the Newark Supergroup for a change from a humid climate in the Carnian to a more arid one in the Norian. Similarly, Gore (1989) noted that flooding and drying cycles, indicative of a wet monsoonal climate, are particularly well developed in the Carnian part of the succession in some of these basins. These humid Carnian sediments are exposed principally in the southern basins, whereas the more northern ones expose only the more arid Norian and Lower Jurassic sequences. Olsen (1986) contends that this indicates the existence of a narrow humid belt around the equator, with more arid zones symmetrically disposed to the north and south. However, the successions in most of the more northerly basins do not extend below the Norian, whereas even in the southern basins there is a clear change from humid sediments in the middle Carnian to arid successions later in the Carnian and in the Norian.

Well-documented paleoclimatic data from elsewhere are few, and considerable uncertainty may surround their dating. Evidence for a wet monsoonal climate has been claimed for late Carnian to late Norian sediments of the Chinle Formation in the southwestern United States (Dubiel, 1984, 1987). In contrast, Blakey and Gubitosa (1984) contend that the climate was hot and semiarid across much of the Chinle basin, but that the climate in the upland source terranes must have been humid to generate continuous stream flow over large distances. However, like Dubiel (1984, 1987), they recognize an increase in aridity in late Norian times. In the northern North Sea, the Lomvi Formation is a thick fluviatile sandstone that Fossen (1989) considers to be Carnian in age, whereas Nystuen et al. (1989) assign it to the Ladinian. Our previous suggestion (Simms and Ruffell, 1989, 1990) that the Fownes Head Member in southern New Brunswick might also indicate a climatic change in the middle to late Carnian has since been disproved by the discovery of an

Anisian palynoflora in that unit (Cornet and Olsen, 1990).

Marine sedimentary successions

For much of the Middle Triassic, marine successions in southern Europe were dominated by thick carbonate successions (Figure 21.1). These persisted into the early Carnian and suggest low runoff and an arid climate (Cecil, 1990). However, Bosellini (1984) has proposed that these limestones were deposited as semi-isolated carbonate platforms prograding into fairly deep basins containing mud and clastic/carbonate turbidites (Cassian Formation). This model suggests that there was no hinterland to act as a source of clastics on the platforms and that precipitation was sufficient to supply and deposit the basinal clastics.

In the middle and late Carnian, carbonate deposition in many areas was greatly reduced in significance, at the expense of influxes of coarser terrigenous clastic detritus, before carbonate deposition on a large scale resumed in the Norian. These clastics originated from sources probably equivalent to those of the Schilfsandstein and other fluviodeltaic systems developed at the southern margin of the Germanic facies belt. The sand was transported to the site of Raibl deposition by longshore drift at times of highstand, rather than by direct input, with development of carbonate platforms at times of lowstand, when the clastics were deposited at the shelf margin or in basinal areas (Bechstädt and Schweizer, 1991). Although there are clear eustatic and tectonic controls on these facies changes (Bosellini, 1984; Brandner, 1984; Bechstädt and Schweizer, 1991), an important component of the facies changes could perhaps also relate to the development of a significantly wetter climate (Cecil, 1990). In the Northern Calcareous Alps the presence of coals testifies to a humid climatic regime, but in the Southern Calcareous Alps the lower sandy units are overlain by fine-grained red beds with calcrete horizons and evaporites. Bechstädt and Schweizer (1991) concluded that the latter succession indicated a change from a warm humid climate in the early part of the middle Carnian to an increasingly more arid climate in later Carnian times. This was also suggested by Jerz (1966) and is indicated by clay mineral analysis (Figure 21.2). This shows kaolinite first appearing near the base of the upper Cassian Formation, continuing through into the Raibl Beds, but disappearing before the close of Raibl deposition.

Very little relevant geochemical work has been undertaken on Triassic marine successions although stable-isotope data indicating a change to a more humid climate were reported by Magaritz and Druckman (1984). They found an extreme $\delta^{13}C$ depletion in the middle and upper Carnian part of a marine carbonate and evaporite succession in Israel and interpreted this as evidence for influxes of fresh water during middle to late Carnian times.

Caves

Sediment-filled "fissures" of Late Triassic and Early Jurassic age have attracted the attention of paleontologists for more than a century because of the rich terrestrial vertebrate faunas that they sometimes contain. However, it is clear that these sites include at least two distinct types of fissures: those that are essentially of tectonic origin (Jenkyns and Senior, 1991), and those that are karst conduits (Simms, 1990b). The presence of conduit caves is of paleoclimatic significance, because cave development within a limestone mass is largely under the control of three variables: the amount of water available, usually equated with precipitation; the amount of carbon dioxide available, which usually is related to plant and soil cover; and the temperature (White, 1988). Although caves are known from many of the world's desert regions, they are much less common than in more humid regions and rarely are more than intermittently active. Waltham, Brown, and Middleton (1985) suggested that most of the caves in Oman had in fact developed during an earlier, wetter episode in Plio-Pleistocene times, for which there is independent evidence (Maizels, 1990). Ancient conduit cave systems are developed on a large scale in Britain wherever Paleozoic limestones are overlain unconformably by Late Triassic terrestrial sediments, and this indicates a period of high rainfall at some time when the limestone was subaerially exposed. However, determining exactly when the conduits formed and were hydrologically active is highly problematic; only the contained sediments can be dated, and these must, necessarily, be younger than the cavity itself, which is undateable. The total age range for fissure infills of Triassic sediments in Carboniferous (and occasionally earlier) limestones extends from perhaps the late Carnian (Crush, 1984) to the Middle Jurassic (Duff, McKirdy, and Harley, 1985), but the age of the true conduit infills, based on palynomorphs and the rich vertebrate faunas that they sometimes contain, as well as less direct evidence from stratimorphic relationships and mineralized caves, would appear to be virtually confined to the Norian (Ivimey-Cook, 1974; Marshall and Whiteside, 1980; Fraser, 1985; Simms, 1990b). Consideration of conduit morphology indicates that these cave systems often have experienced long and complex histories prior to their abandonment. In the case of the keyhole-shaped passage described by Marshall and Whiteside (1980) and Whiteside and Robinson (1983), it is clear that an initial period of phreatic development and subsequent vadose drawdown preceded marine inundation in late Norian times. The infilling of an abandoned conduit is also a complex process that may involve repeated episodes of deposition and erosion over an extended period of time. Indeed, examples are known of Triassic cave passages that are still at least partly open (Simms, 1990b). However, the absence of any Carnian or pre-Carnian sediments in these caves, and the abundance of Norian sediments, would seem to indicate that the caves developed and were hydrologically active during the latter part of the Carnian, a supposition consistent with the other evidence for a more humid climate during that interval.

Excellent examples of Triassic caves choked with sediment are exposed in quarries to the north and south of Bristol, in South Wales, and in Leicestershire, but there has been little documentation of such phenomena outside of Britain. However, Mutti (1991 and pers. commun.) has described large-scale karstification of the Middle Triassic Esino Limestone in the southern Alps. Her geochemical analyses of meteoric cements within this karst indicate that precipitation rates were high enough to maintain a stable phreatic lens within the carbonate platform. However, the humid episode that she recognized can be dated as latest Ladinian to early Carnian, although she has also observed extensive karstification in some of the middle Carnian units. Similarly, Klau and Mostler (1986) and Bechstädt and Dohler-Hirner (1983) described significant karstification of carbonates in the latest Ladinian and early Carnian of the eastern Alps.

For some of these ancient cave systems in Britain it is still possible to estimate their approximate catchment areas, because the current topography is, in part, an exhumed Triassic landscape (Simms and Ruffell, 1990; Simms, 1990b). By comparing catchment area with phreatic conduit diameter at these sites, it may be possible to estimate runoff during their development. Hence, although the dating of this karstic evidence for a more humid climate remains somewhat imprecise, nonetheless these caves may provide a means of determining actual levels of precipitation during that interval.

Discussion

Evidence from various abiotic sources indicates that the climate during part of the Carnian was significantly more humid than in the ensuing Norian and probably more so than in much of the preceding Ladinian stage. In this account, the fossil fauna and flora have not been used as directly recording paleoclimatic changes, but they do show significant levels of turnover approximately synchronous with the (independently determined) climatic changes between aridity and humidity. However, there appear to be significant differences in the timing of the climatic changes in different regions and, to a lesser extent, differences in the timing of the biotic changes.

Obviously the most precise dating is that of the

marine sequences of the Alpine Triassic. Here the change from predominantly carbonate deposition to terrigenous clastics occurs in the middle and upper Carnian. However, as already mentioned, there is some evidence, from clay minerals and karstification, for a humid climate in the latest Ladinian and early Carnian, with a return to arid conditions already by late Carnian times. In contrast, Magaritz and Druckman (1984) found evidence for high runoff throughout the middle and late Carnian, while the terrestrial Triassic sequence in Europe also shows evidence for a humid climate in late Carnian times.

In North America the evidence is more equivocal. The Newark Supergroup strata show evidence for a humid climate in the middle Carnian, followed by a change to more arid sequences in the late Carnian and Norian, but the Chinle Formation shows evidence that may indicate a much longer period of humidity, from late Carnian to late Norian.

The range of organisms affected during the Carnian–Norian interval indicates that this apparent episode of major turnover cannot be attributed merely to sampling bias, as has been suggested by Raup and Sepkoski (1986). Not only do the affected organisms cover a wide range of trophic groups, both marine and terrestrial, but a variety of distinct paleontological sampling methods have been employed in their documentation. For example, the techniques used to recover conodonts and palynomorphs are quite different from those employed in the recovery of marine macroinvertebrates or terrestrial vertebrates, and yet they show essentially the same pattern of turnover, albeit with a degree of nonsynchroneity between the marine and terrestrial changes. Benton (Chapter 22) has argued that changes documented for the terrestrial vertebrate fauna across the Carnian–Norian boundary are real. The number of lower Norian vertebrate-bearing sites is only 18 percent less than the figure for the upper Carnian, whereas the drop in diversity is very much greater than this, implying that the early Norian fauna is genuinely depauperate rather than poorly sampled.

Changes in the marine realm appear to be concentrated around the boundary between the early and middle Carnian, although for the scleractinian corals and some other groups the precise timing of the most significant changes remains unclear, but appears to have occurred sometime in the middle to late Carnian. Although many accounts place the terrestrial biotic changes at the Carnian–Norian boundary, this is disputed by others (Cornet and Olsen, 1990), who instead place the major floral change close to the boundary between middle and late Carnian. The diachroneity of the climatic changes would appear, at first, to undermine any suggestions of a causal link with the biotic changes. However, for both the marine and terrestrial biotas these changes correlate with the latest timing of the onset of the humid regime or the

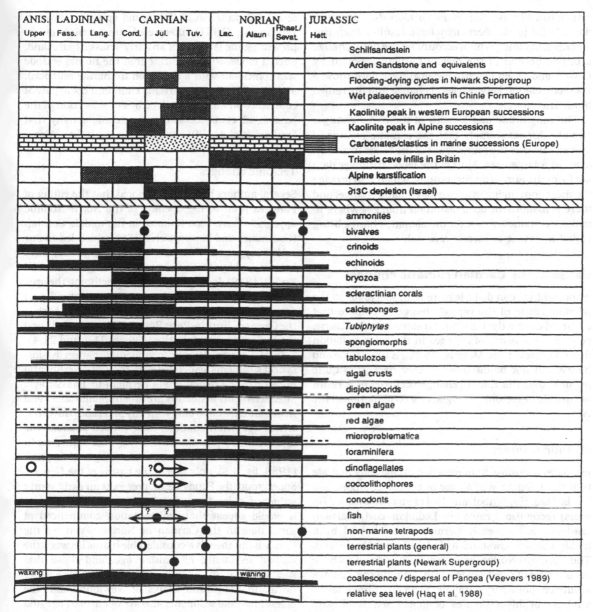

Figure 21.3. Correlation chart for sources of evidence of climatic change and episodes of biotic turnover. Filled circles indicate extinction events; open circles indicate origination and/or high turnover. Questionable dating is indicated by arrows and question marks. For other taxa, relative diversity levels are indicated by thicknesses of bars. Sources of biotic data as follows: ammonites (Benton, 1986a), bivalves (Johnson and Simms, 1989; Hallam, 1981, 1990), crinoids (Simms, 1990a), echinoids (Smith, 1990), bryozoans (Schäfer and Fois, 1987), scleractinian corals to foraminifera (Stanley, 1988), dinoflagellates, coccolithophores (Perch-Nielson, 1985), conodonts (Clark, 1987), fishes (Patterson and Smith, 1987), nonmarine tetrapods (Benton, 1986a), terrestrial plants (Boulter et al., 1988; Cornet and Olsen, 1990). Substages of the Middle and Late Triassic as follows: Fassanian, Langobardian, Cordevolian, Julian, Tuvalian, Lacian, Alaunian, Sevatian/Rhaetian; Anis. = Anisian stage; Hett. = Hettangian stage.

return to an arid one (Figure 21.3). For example, although the evidence from some areas indicates that the climate was already becoming more humid in the late Ladinian, by middle Carnian times this climatic regime appears to have affected all of the areas studied. Similarly, with the possible exception of the Chinle

Formation, aridity seems to have become prevalent by the beginning of the Norian. The changes in the marine biotas correlate fairly closely with the latest onset of the humid climate, around the beginning of the middle Carnian, whereas the turnover among the terrestrial biotas appears to correlate with the latest return to arid conditions at the Carnian–Norian boundary. Local biotic effects may have resulted from this diachroneity and perhaps account for the floral change around the end of the middle Carnian reported by Cornet and Olsen (1990), because their data show a good correlation between floral change and the return to an arid climate.

The magnitude and breadth of the biotic changes already discussed imply that the climatic change, if indeed this was the prime cause of the biotic turnover, affected a large area. Most of the biotic turnover data are from the European successions, which also provide the most convincing evidence for climatic change, and there is insufficient documentation from elsewhere to make a convincing case for the climatic change being global in extent or merely regional.

Cause of the Carnian climatic changes

At present we can do little more than speculate on the possible cause of the radical change in climate that occurred during the Carnian. An extraterrestrial cause, perhaps as a result of changes in the solar radiation flux (Wetherald and Manabe, 1975), is untestable and hence will not be discussed further here. It is quite possible that the climate change occurred as a result of eustatic or tectonic changes, or a combination of the two.

Eustatic changes

Estimates of eustasy from sequence stratigraphy indicate that the Carnian was a time of rapidly changing sea levels. Haq, Hardenbol, and Vail (1988) and Ruffell (in press) recognize a rapid late Ladinian–early Carnian sea-level fall, an early Carnian rise, followed by a middle Carnian lowstand, and a late Carnian highstand. The Carnian–Norian boundary is considered to have been a time of rapid sea-level fall, forming the 224-Ma sequence boundary. Hallam (1984) utilized facies analysis in his estimates of global sea level, and accordingly also recognized a middle Carnian lowstand.

Doglioni, Bosellini, and Vail (1990) found a good correlation between the sedimentary sequences of the Dolomites and the Haq et al. (1988) chart, thereby substantiating suggestions of early and late Carnian sea-level rises and a middle Carnian regression. Interestingly, Doglioni et al. (1990) also indicate that tectonic "enhancement" of the Ladinian–Carnian and late Carnian sequence boundaries (the product of falls in sea level) also occurred, suggesting a more complex scenario for Carnian events than models of eustasy would indicate.

It is quite possible that the short-term changes in sea level did have a causal relationship with the changes in climate, because the middle Carnian sea-level fall and subsequent rise occurred at similar times as the onset and cessation of the humid episode. However, Warrington (1981) interpreted the Carnian as representing the last part of an Early Triassic highstand. Such a model would indicate that the humid episode was coincident with a reduction in marine seaboard, and therefore also in maritime climate. This apparent contradiction may be resolved by consideration of the possible effects of tectonic and volcanic activity that Doglioni et al. (1990) recognized in the Italian Alps.

Tectonic changes

Several authors have already noted that the rifting of Pangaea prior to the opening of the North Atlantic appears to have been initiated in the middle Carnian (Cousminer and Manspeizer, 1976), which led to our earlier suggestion that the climatic change may have been caused by some aspect of that rifting (Simms and Ruffell, 1989). In a discussion of the tectonics, the area under consideration can be divided into three broad regions: the North Atlantic continental-margin basins, the German–British–Paris Basin continental shelf, and the Italian–Carpathian Alps (Figures 21.1 and 21.4).

In the North Atlantic continental-margin basins, synrift sediments associated with the early continental breakup of the region are of Permian to Triassic age. Such sediments frequently are of continental facies and hence are difficult to date with any precision. However, where age determinations have proved possible, the first sediments preserved as early synrift sequences in many areas are of Carnian age. According to Manspeizer (1988, fig. 3.5), only one of the basins of the Newark Supergroup, the Fundy basin, preserves an early synrift sequence older than Carnian in age, and in most it would appear that sedimentation commenced at some time in the Carnian or Norian. Even for the Fundy basin there is a considerable hiatus between the earliest sediments, of Anisian age, and those that succeed them, of Carnian age (Figure 21.1). Farther north, in the Grand Banks of Newfoundland, the oldest Mesozoic sediments also are of Carnian age and represent the earliest phase of rifting (Tankard and Welsink, 1988). Welsink, Srivastava, and Tankard (1989) examined the extensional tectonic fabric of the Scotian basin (offshore Newfoundland) and regarded intense rift-related faulting as having commenced in the Carnian (Figure 21.4). Similar rifting is thought to have begun in the latest Ladinian of the adjacent Whale basin (Balkwill and Legall, 1989). On the eastern margin of the North Atlantic, a similar situation exists. Carnian clastic sediments rest on basement rocks in the Lusitanian basin of Portugal (Wilson et al., 1989), and the adjacent Jeanne d'Arc basin also contains synrift sediments of late Carnian age overlying basement rocks (Figure 21.4). Those of Morocco are predominantly Ladinian or Carnian in age (Manspeizer, 1988).

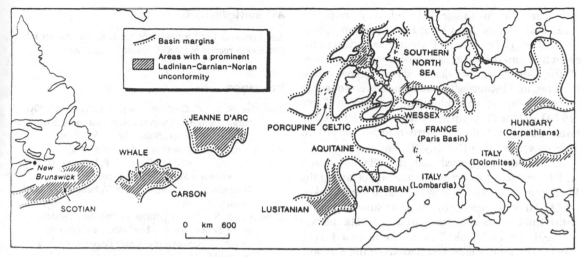

Figure 21.4. Present-day distribution of the Triassic basins (stratigraphy in Figure 21.1). The Ladinian–Carnian unconformity is a product of uplift of basins within the North Atlantic and basin margins in Europe.

On the United Kingdom continental shelf and in onshore basins in England, France, and Germany, synrift sediments were preserved as early as the Permian, with Carnian sediments being found as part of a conformable Triassic succession in basin depocenters (Chadwick, Livermore, and Penn, 1989; Petrie et al., 1990). The basin margins to such areas as the Wessex basin in southern England do show a Middle Triassic unconformity on Paleozoic basement (the "Dolomitic Conglomerate" unconformity of the Mendip area and elsewhere) (Green and Welch, 1965), but this has proved impossible to date precisely and probably spans a considerable time interval.

Beyond the North Atlantic–Alps–Carpathians area, Veevers (1989) gives a summary of many tectonic changes in the Carnian. The change from Hercynian to Alpine tectonics in the western USSR is dated as Carnian, whereas major Carnian unconformities occur in the middle Absaroka Formation of North America and in the Triassic deltaic successions of Brazil. A Ladinian–Carnian unconformity is also recognized in the middle of the Karoo Series of South Africa, the Gondwana Series of India, and the Innamincka of Australia.

The Carnian of the Italian–Carpathian Alps occurs in a generally conformable Triassic succession of carbonates and minor clastics (Figure 21.1). A prominent unconformity occurs only on platforms and local uplifts (Zapfe, 1974). Doglioni et al. (1990) suggested that early Anisian, late Ladinian, and late Carnian tectonic pulses occurred and that they were coincident with global falls in sea level.

In an earlier account (Simms and Ruffell, 1989), we suggested that the climatic change during the Carnian may have been precipitated by some aspect of the initial stages of rifting of the North Atlantic, and in particular by volcanism associated with that rifting. Further support for this hypothesis was provided by Veevers (1989), who identified an episode of radical change in the Phanerozoic history of continental displacement. According to Veevers, the final coalescence of Pangaea and the almost immediate onset of continental rifting would have caused, among other things, a rise in mean surface air temperature as a result of the greenhouse effect of greater input of volcanic CO_2. Veevers (1989) estimated the timing of that event to lie around the Middle–Late Triassic (Ladinian–Carnian) boundary, at about 230 ± 5 Ma. The Carnian humid episode identified here and in earlier papers (Simms and Ruffell, 1989, 1990) occurred at some time between 230 and 225 Ma on the DNAG time scale (Palmer, 1983) and hence lies well within Veevers's 5-Ma margin of error. Such a close correlation between Veevers's predicted development of a greenhouse climate and the climatic change documented is strongly in support of a causal link between large-scale climatic change and the onset of continental rifting in Pangaea.

However, despite this suggestion, there are rather few documented large-scale igneous sites of appropriate age. In the Newark Supergroup, on the western Atlantic margin, the volcanics are confined to several tholeiitic basalt flows in the Hettangian and Sinemurian. Lavas of similar age are also encountered on the African continental margin of the North Atlantic, and in the Oranian Messeta of Morocco, andesites of Ladinian age represent the earliest part of the synrift succession (Manspeizer, 1988). Volcanics present in the Lusitanian and Cantabrian basins (Wilson et al., 1989; Garcia-Mondejar, 1990) have been considered to be early Norian in age, although they are poorly dated and may perhaps be of latest Carnian age. A "major" late Ladinian–early Carnian magmatic/tectonic event had a marked effect on the sedimentary sequences of the Dolomites, and late Ladinian pillow lavas, breccias, and hyaloclastic basalts have been shown in many regions of the sub-Alpine Triassic by Zapfe (1974) and the Carpathians by Baldi et al. (1983). The latter also identified early Carnian to Norian bedded volcanics in western Hungary (Figure 21.1).

It has been suggested that the Siberian Traps, which represent the largest outpourings of lava in the Phanerozoic, are at least partly Carnian in age (Nalivkin, 1973), but more recent work tends to suggest that the entire thickness was extruded in a 5–10-Ma interval in the Early to Middle Triassic (Baksi and Farrar, 1991). The other major flood basalts in the Amazon, Karoo, and Transantarctic-Tasmanian provinces were all extruded somewhat later, between rifting and sea-floor spreading (Veevers, 1989).

Thus, although considerable volcanism occurred during the initial stages of the rifting of the North Atlantic, the climatic changes documented do not appear to show a significant correspondence to any one particular volcanic episode during that interval. Consequently, the ultimate cause of the climatic changes in the Late Triassic remains unclear, although a link with rifting seems probable.

Conclusions

Documentation of a wide variety of fossil groups, both terrestrial and marine, provides clear evidence of an important episode of biotic change in middle to late Carnian times. Independent evidence from terrestrial and marine sedimentary successions, karstic features, and stable isotopes indicates a major climatic change, from arid to humid conditions, approximately synchronous with the biotic changes. The timing of the onset and cessation of that climatic event shows some correlation with episodes of high turnover among both the marine and terrestrial biotas. The case for a causal link between climatic and biotic changes is strengthened by this nonsynchroneity between marine and terrestrial turnover. The terrestrial faunal and floral changes may have been in response to the return of arid conditions at the start of the Norian, and the marine fauna may have been affected by a fundamental, climatically induced change in plankton composition, with resultant effects on the entire marine food chain. The ultimate cause of the climatic change during the Carnian may relate to a rise in mean surface air temperature and development of a greenhouse climatic state. That could have occurred as a result of an increase in the output of volcanic CO_2 associated with the initiation of continental rifting in Pangaea prior to the opening of the North Atlantic.

Acknowledgments

Attendance at the Virginia meeting was undertaken while MJS was in receipt of a NERC research fellowship.

References

Aebischer, N. J., J. C. Coulson, and J. M. Colebrook. 1990. Parallel long-term trends across four marine trophic levels and weather. *Nature* 347: 753–755.

Aldridge, R. 1988. Extinction and survival in the Conodonta. Pp. 231–256 in G. P. Larwood (ed.), *Extinction and Survival in the Fossil Record.* Systematics Association Special Volume 34. London: Academic Press.

Arthurton, R. S. 1980. Rhythmic sedimentary sequences in the Triassic Keuper Marl (Mercia Mudstone Group) of Cheshire, northwest England. *Geological Journal* 15: 43–58.

Bakker, R. T. 1977. Tetrapod mass extinctions – a model of the regulation of speciation rates and immigration by cycles of topographic diversity. Pp. 439–468 in A. Hallam (ed.), *Patterns of Evolution as Illustrated in the Fossil Record.* Amsterdam: Elsevier.

Baksi, A. K., and E. Farrar. 1991. $^{40}Ar/^{39}Ar$ dating of the Siberian Traps, USSR: evaluation of the ages of the two major extinction events relative to episodes of flood-basalt volcanism in the USSR and the Deccan Traps, India. *Geology* 19: 461–464.

Baldi, T., K. Balogh, A. Barabas, E. Dudich, J. Fulop, B. Geczy, G. Hamor, A. Jambor, B. Jantsky, A. Ronai, and T. Szederkenyi (compilers). 1983. *Magyarorszag litosztratigrafiai formacioi.* Budapest: Kartografiai Vallalat.

Balkwill, H. R., and F. D. Legall. 1989. Whale Basin, offshore Newfoundland: extension and salt diapirism. Pp. 233–246 in A. J. Tankard and H. R. Balkwill (eds.), *Extensional Tectonics and Stratigraphy of the North Atlantic Margins.* American Association of Petroleum Geologists Memoir 46.

Bechstädt, T., and B. Dohler-Hirner. 1983. Lead-zinc deposits of Bleiberg-Kreuth. Pp. 55–63 in P. A. Scholle, D. G. Bebout, and C. H. Moore (eds.), *Carbonate Depositional Environments.* American Association of Petroleum Geologists Memoir 33.

Bechstädt, T., and T. Schweizer. 1991. The carbonate-clastic cycles of the east Alpine Raibl Group: result of third-order sea-level fluctuations in the Carnian. *Sedimentary Geology* 70: 241–270.

Benton, M. J. 1986a. More than one event in the Late Triassic extinction. *Nature* 321: 857–859.

1986b. The Late Triassic tetrapod extinction events. Pp. 303–320 in K. Padian (ed.), *The Beginning of the Age of Dinosaurs: Faunal Change across the Triassic–Jurassic Boundary.* Cambridge University Press.

1989. Mass extinctions among tetrapods and the quality of the fossil record. *Philosophical Transactions of the Royal Society of London* B325: 369–386.

Blakey, R. C., and R. Gubitosa. 1984. Controls of sandstone body and architecture in the Chinle Formation (Upper Triassic), Colorado Plateau. *Sedimentary Geology* 38: 51–86.

Bosellini, A. 1984. Progradation geometries of carbonate platforms: examples from the Triassic of the Dolomites, northern Italy. *Sedimentology* 31: 1–24.

Boulter, M. C., R. A. Spicer, and B. A. Thomas. 1988. Patterns of plant extinction from some palaeobotanical evidence. Pp. 1–36 in G. P. Larwood (ed.), *Extinction and Survival in the Fossil Record*. Systematics Association Special Volume 34. London: Academic Press.

Brandner, R. 1984. Meeresspiegelschwankungen und Tektonik in der Trias der NW-Tethys. *Jahrbuch der Geologischen Bundesanstalt* 126: 435–475.

Cecil, C. B. 1990. Paleoclimate controls on stratigraphic repetition of chemical and siliciclastic rock. *Geology* 18: 533–536.

Chadwick, R. A., R. A. Livermore, and I. A. Penn. 1989. Continental extension in southern Britain and surrounding areas and its relationship to the opening of the North Atlantic. Pp. 411–424 in A. J. Tankards and H. R. Balkwill (eds.), *Extensional Tectonics and Stratigraphy of the North Atlantic Margins*. American Association of Petroleum Geologists Memoir 46.

Clark, D. L. 1987. Conodonts: the final fifty million years. Pp. 165–174 in R. J. Aldridge (ed.), *Palaeobiology of Conodonts*. Chichester: Ellis Horwood.

Cornet, B., and P. E. Olsen. 1990. Early to Middle Carnian (Triassic) flora and fauna of the Richmond and Taylorsville basins, Virginia and Maryland, USA. *Virginia Museum of Natural History Guidebook* 1: 1–83.

Cousminer, H. L., and W. Manspeizer. 1976. Triassic pollen date Morrocan High Atlas and the incipient rifting of Pangaea in middle Carnian. *Science* 191:943.

Crush, P. J. 1984. A late Triassic sphenosuchid crocodilian from Wales. *Palaeontology* 27: 131–157.

Doglioni, C., A. Bosellini, and P. R. Vail. 1990. Stratal patterns: a proposal of classification and examples from the Dolomites. *Basin Research* 2: 83–95.

Dubiel, R. F. 1984. Evidence for wet paleoenvironments, Upper Triassic, Chinle Formation, Utah. *Geological Society of America, Abstracts with Programs* 16: 220.

1987. Sedimentology of the Upper Triassic Chinle Formation: paleoclimatic implications. *Journal of the Arizona and Nevada Academy of Science* 22: 35–45.

Duff, K. L., A. P. McKirdy, M. J. Harley. (eds.). 1985. *New Sites for Old. A Students' Guide to the Geology of the East Mendips*. Peterborough: Nature Conservancy Council.

Erwin, D. H. 1990. Carboniferous–Triassic gastropod diversity patterns and the Permo-Triassic mass extinction. *Paleobiology* 16: 187–203.

Fossen, H. 1989. Indication of transgressional tectonics in the Gullfaks oil-field, northern North Sea. *Marine and Petroleum Geology* 6: 22–30.

Fraser, N. 1985. Vertebrate faunas from Mesozoic fissure deposits of southwest Britain. *Modern Geology* 9: 273–300.

Garcia-Mondejar, J. 1990. The northern Spain Basque–Cantabrian basin during the Aptian–Albian interval: a strike-slip setting related to the Bay of Biscay opening. Pp. 133–142 in A. J. Tankard and

H. Balkwill (eds), *Extensional Tectonics and Stratigraphy of the North Atlantic Margins*. American Association of Petroleum Geologists Memoir 46.

Gore, P. J. W. 1989. Toward a model for open- and closed-basin deposition in ancient lacustrine sequences: the Newark Supergroup (Triassic–Jurassic), eastern North America. *Palaeogeography, Palaeoclimatology Palaeoecology* 70: 29–51.

Green, G. W., and F. B. A. Welch. 1965. Geology of the country around Wells and Cheddar. *Memoir of the Geological Survey of Great Britain, Sheet 280.*

Hahn, G. 1984. Palaeomagnetische Untersuchungen im Schilfsandstein (Trias, Km2). *Geologische Rundschau* 73: 499–516.

Hallam, A. 1981. The end-Triassic bivalve extinction event. *Palaeogeography, Palaeoclimatology, Palaeoecology* 39: 1–44.

1984. Continental humid and arid zones during the Jurassic and Cretaceous. *Palaeogeography, Palaeoclimatology, Palaeoecology* 47: 195–223.

1990. The end-Triassic mass extinction event. *Geological Society of America Special Paper* 247: 577–583.

Haq, B. U., J. Hardenbol, and P. R. Vail. 1988. Mesozoic and Cenozoic chronostratigraphy and cycles of sea-level change. Pp. 71–108 in C. K. Wilgus (ed.), *Sea-Level Changes: An Integrated Approach*. Society of Economic Paleontologists and Mineralogists Special Publication 42.

Ivimey-Cook, H. C. 1974. The Permian and Triassic rocks of Wales. Pp. 295–321 in T. R. Owen (ed.), *The Upper Palaeozoic and post-Palaeozoic Rocks of Wales*. Cardiff: University of Wales Press.

Jeans, C. V. 1978. The origin of the Triassic clay assemblages of Europe with special reference to the Keuper Marl and Rhaetic of parts of England. *Philosophical Transactions of the Royal Society of London* A289: 549–639.

Jenkyns, H. C., and J. R. Senior. 1991. Geological evidence for intra-Jurassic faulting in the Wessex Basin and its margins. *Journal of the Geological Society, London* 148: 245–260.

Jerz, H. 1966. Untersuchungen über Stoffbestand, Bildungsbedingungen und Paläogeographie der Raibler Schichten zwischen Lech und Inn (Nördliche Kalkalpen). *Geologica Bavarica* 56: 3–102.

Johnson, A. L. A., and M. J. Simms. 1989. The timing and cause of late Triassic marine invertebrate extinctions: evidence from scallops and crinoids. Pp. 174–194 in S. K. Donovan (ed.), *Mass Extinctions: Processes and Evidence*. New York: Columbia University Press.

Kiritchkova, A. I. 1990. [Triassic–Early Jurassic flora of the eastern Urals.] *Paleontologicheskiy Zhurnal* 1990: 110–119. [in Russian]

Klau, W., and H. Mostler. 1986. On the formation of Alpine Middle and Upper Triassic Pb-Zn deposits, with some remarks on Irish carbonate-hosted base metal deposits. Pp. 663–675 in C. J. Andrew, R. W. Crowe, S. Finlay, W. M. Pennell, and J. F. Pyne (eds.), *Geology and Genesis of Mineral Deposits in*

Ireland. Dublin: Irish Association for Economic Geology.

Klein, G. de V. 1962. Sedimentary structures in the Keuper Marl (Upper Triassic). *Geological Magazine* 99: 137–144.

Lott, G. K., R. A. Sobey, G. Warrington, and A. Whittaker. 1982. The Mercia Mudstone Group (Triassic) in the western Wessex Basin. *Proceedings of the Ussher Society* 5: 340–346.

Magaritz, M., and Y. Druckman. 1984. Carbon isotope composition of an Upper Triassic evaporite section in Israel; evidence for meteoric water influx. *American Association of Petroleum Geologists Bulletin* 68: 502.

Maizels, J. 1990. Raised channel systems as indicators of palaeohydrologic change: a case study from Oman. *Palaeogeography, Palaeoclimatology, Palaeoecology* 76: 241–277.

Manspeizer, W. 1988. Triassic–Jurassic rifting and opening of the Atlantic: an overview. Pp. 41–79 in W. Manspeizer (ed.), *Triassic–Jurassic Rifting: Continental Breakup and the Origin of the Atlantic Ocean and Passive Margins*. Amsterdam: Elsevier.

Marshall, J. E. A., and D. I. Whiteside. 1980. Marine influences in the Triassic "uplands." *Nature* 287: 627–628.

Mutti, M. 1991. A Middle Triassic paleokarst, Esino Limestone Formation, southern Alps, Italy. *Terra Abstracts* 3: 240–241.

Nalivkin, D. V. 1973. *Geology of the U.S.S.R.* Edinburgh: Oliver and Boyd.

Newton, C. R., M. T. Whalen, J. B. Thompson, N. Prins, and D. Dellalla. 1987. *Systematics and Paleoecology of Norian (Late Triassic) Bivalves from a Tropical Island Arc: Wallowa Terrane, Oregon*. Paleontological Society Memoir 22.

North, C. P. 1988. Structure and sedimentology of the Mercia Mudstone Group (Upper Triassic) of the Severn Estuary region, SW Britain. Ph.D. thesis, University of Bristol.

Nystuen, J. P., R. Knarud, K. Jorde, and K. O. Stanley. 1989. Correlation of Triassic to Lower Jurassic sequences, the Snorre Field, northern Sea and adjacent areas. Pp. 273–290 in J. D. Collinson (ed.), *Correlation in Hydrocarbon Exploration*. London: Graham and Trotman.

Olsen, P. E. 1986. A 40-million-year lake record of early Mesozoic climatic forcing. *Science* 234: 842–848.

Olsen, P. E., and H.-D. Sues. 1986. Correlation of continental Late Triassic and Early Jurassic sediments and patterns of the Triassic–Jurassic tetrapod transition. Pp. 321–351 in K. Padian (ed.), *The Beginning of the Age of Dinosaurs: Faunal Change across the Triassic–Jurassic Boundary*. Cambridge University Press.

Palmer, A. R. 1983. Decade of North American Geology (DNAG) geologic time scale. *Geology* 11: 503–504.

Patterson, C., and A. B. Smith. 1987. Is the periodicity of extinctions a taxonomic artefact? *Nature* 330: 248–252.

Perch-Nielsen, K. 1985. Mesozoic calcareous nannofossils. Pp. 329–426 in H. M. Bolli, J. B. Saunders, and K.

Perch-Nielsen (eds.), *Plankton Stratigraphy*. Cambridge University Press.

Petrie, S. H., J. R. Brown, P. J. Granger, and J. P. B. Lovell. 1990. Mesozoic history of the Celtic Sea Basins. In A. J. Tankard and H. R. Balkwill (eds.), *Extensional Tectonics and Stratigraphy of the North Atlantic Margins*. American Association of Petroleum Geologists Memoir 46.

Raup, D. M., and J. J. Sepkoski, Jr. 1986. Periodic extinction of families and genera. *Science* 231: 833–836.

Ruffell, A. H. In press. Palaeoenvironmental analysis of the Late Triassic succession in the Wessex Basin and correlation with surrounding areas. *Proceedings of the Ussher Society* 8.

Ruffell, A. H., and D. J. Batten. 1990. The Barremian–Aptian arid phase in western Europe. *Palaeogeography, Palaeoclimatology, Palaeoecology* 80: 197–212.

Ruffell, A. H., and G. Warrington. 1989. An arenaceous member in the Mercia Mudstone Group (Triassic) west Taunton, Somerset. *Proceedings of the Ussher Society* 7: 102–103.

Schäfer, P., and E. Fois. 1987. Systematics and evolution of Triassic Bryozoa. *Geologica et Palaeontologica* 21: 173–225.

Simms, M. J. 1990a. Crinoid diversity and the Triassic–Jurassic boundary. *Cahiers de l'Université Catholique de Lyon, Séries Scientifiques* 3: 67–77.

——— 1990b. Triassic palaeokarst in Britain. *Cave Science* 17: 93–101.

Simms, M. J., and A. H. Ruffell. 1989. Synchroneity of climatic change and extinctions in the late Triassic. *Geology* 17: 265–268.

——— 1990. Climatic and biotic change in the late Triassic. *Journal of the Geological Society, London* 147: 321–327.

Smith, A. B. 1990. Echinoid evolution from the Triassic to Lower Liassic. *Cahiers de l'Université Catholique de Lyon, Séries Scientifiques* 3: 79–117.

Smith, A. B., and C. Patterson. 1988. The influence of taxonomic method on the perception of patterns of evolution. *Evolutionary Biology* 23: 127–216.

Smoot, J. P., and P. E. Olsen. 1988. Massive mudstones in basin analysis and paleoclimatic interpretation of the Newark Supergroup. Pp. 249–274 in W. Manspeizer (ed.), *Triassic–Jurassic Rifting: Continental Breakup and the Origin of the Atlantic Ocean and Passive Margins*. New York: Elsevier.

Stanley, G. D. 1981. Early history of scleractinian corals and its geological consequences. *Geology* 9: 507–511.

——— 1988. The history of early Mesozoic reef communities: a three-step process. *Palaios* 3: 170–183.

Stanley, G. D., and M. T. Whalen. 1989. Triassic corals and spongiomorphs from Hells Canyon, Wallowa Terrane, Oregon, *Journal of Paleontology* 63: 800–819.

Tankard, A. J., and H. J. Welsink. 1988. Extensional tectonics, structural styles and stratigraphy of the Mesozoic Grand Banks of Newfoundland. Pp. 129–165 in W. Manspeizer (ed.), *Triassic–Jurassic Rifting: Continental Breakup and* the Origin of the Atlantic Ocean and Passive Margins. Amsterdam: Elsevier.

Tucker, M. E. 1977. The marginal Triassic deposits of South Wales: continental facies and palaeogeography. *Geological Journal* 12: 169–188.

Tucker, M. E., and M. J. Benton. 1982. Triassic environments, climates and reptile evolution. *Palaeogeography, Palaeoclimatology, Palaeoecology* 40: 361–379.

Van der Eem, J. G. L. A. 1983. Aspects of Middle and Late Triassic palynology. 6. Palynological investigations in the Ladinian and lower Carnian of the western Dolomites, Italy. *Review of Palaeobotany and Palynology* 39: 189–300.

Veevers, J. J. 1989. Middle/Late Triassic (230 ± 5 Ma) singularity in the stratigraphic and magmatic history of the Pangean heat anomaly. *Geology* 17: 784–787.

Visscher, H., and C. J. Van der Zwan. 1981. Palynology of the circum-Mediterranean Triassic; phytogeographical and palaeoclimatological implications. *Geologische Rundschau* 70: 625–634.

Waltham, A. C., R. D. Brown, and T. C. Middleton. 1985. Karst and caves in the Jabal Akhder, Oman. *Cave Science* 12: 69–79.

Warrington, G. 1971. Palynology of the New Red Sandstone sequence of the south Devon coast. *Proceedings of the Ussher Society* 2: 307–314.

1981. The indigenous micropalaeontology of the British Triassic shelf sea deposits. Pp. 61–70 in J. W. Neale and M. D. Brasier (eds.). *Microfossils from Recent and Fossil Shelf Seas*. Chichester: Ellis Horwood Ltd.

Warrington, G., et al. 1980. A correlation of Triassic rocks in the British Isles. *Special Report of the Geological Society of London* 13: 1–78.

Warrington, G., and B. J. Williams. 1984. The North Curry Sandstone Member (Late Triassic) near Taunton, Somerset. *Proceedings of the Ussher Society* 6: 82–87.

Welsink, H. J., S. P. Srivastava, and A. J. Tankard. 1989. Basin architecture of the Newfoundland continental margin and its relationship to ocean crust fabric

during extension. Pp. 197–214 in A. J. Tankard and H. R. Balkwill (eds.), *Extensional Tectonics and Stratigraphy of the North Atlantic Margins*. American Association of Petroleum Geologists Memoir 46.

Wetherald, R. T., and S. Manabe. 1975. The effects of changing the Solar Constant on the climate of a General Circulation Model. *Journal of the Atmospheric Sciences* 32: 2044–2059.

White, W. B. 1988. *Geomorphology and Hydrology of Karst Terrains*. Oxford: Clarendon Press.

Whiteside, D. I., and D. Robinson. 1983. A glauconitic clay mineral from a speleological deposit of late Triassic age. *Palaeogeography, Palaeoclimatology, Palaeoecology* 41: 81–85.

Whittaker, A., and G. Green. 1983. Geology of the country around Weston-Super-Mare. *Geological Survey of Great Britain Memoir for Sheet 279 and Parts of Sheets 263 and 295.*

Wills, L. J. 1970. The Triassic succession in the central Midlands in its regional setting. *Quarterly Journal of the Geological Society, London* 126: 225–285.

Wilson, R. C. L., R. N. Hiscott, M. G. Wills, and F. M. Gradstein. 1989. The Lusitanian Basin of west-central Portugal: Mesozoic and Tertiary tectonic, stratigraphic, and subsidence history. Pp. 341–361 in A. J. Tankard and H. R. Balkwill (eds.), *Extensional Tectonics and Stratigraphy of the North Atlantic Margins*. American Association of Petroleum Geologists Memoir 46.

Witte, W. K., and D. V. Kent. 1989. A middle Carnian to early Norian (~225 Ma) paleopole from sediments of the Newark Basin, Pennsylvania. *Geological Society of America Bulletin* 101: 1118–1126.

Wurster, P. 1964. Geologie des Schilfsandsteins. *Mitteilungen aus dem Geologischen Staatsinstitut in Hamburg* 33: 1–140.

Zapfe, H. 1974. Trias in Oesterreich. Pp. 245–251 in H. Zapfe (ed.), *The Stratigraphy of the Alpine–Mediterranean Triassic*, Vol. 2. Schriftenreihe der Erdwissenschaftlichen Kommission, Oesterreichische Akademie der Wissenschaften.

22

Late Triassic to Middle Jurassic extinctions among continental tetrapods: testing the pattern

MICHAEL J. BENTON

Introduction

Several attempts have been made to document patterns of diversity change and postulated extinctions of amphibians and reptiles through the Triassic and Early Jurassic interval, both on a global scale (e.g., Benton, 1983, 1986a,b, 1991, 1993a; Olsen and Sues, 1986; Zawiskie, 1986; Lucas, 1990; Hunt, 1991) and based on localized sequences of faunas in Germany (e.g., Benton, 1986b, 1993a; Olsen and Sues, 1986) and North America (e.g., Olsen and Sues, 1986; Olsen, Shubin, and Anders, 1987, 1988; Olsen, Fowell, and Cornet, 1990; Hunt, 1991). The results have been equivocal, with strong arguments being presented both for the existence of two extinction events (one in the late Carnian, and one at the Triassic–Jurassic boundary) and in favor of a single event (at the end of the Triassic). In the record of tetrapods, little evidence has been found for mass extinctions during the Early and Middle Jurassic, although such extinctions have been predicted based on analyses of the marine record by Sepkoski (1989, 1990) at the Pliensbachian–Toarcian boundary and in the Bajocian and/or Callovian.

The debate about tetrapod extinctions has ramifications for the wider question of whether there was a Carnian extinction event among marine life (Stanley, 1988; Sepkoski, 1989; Simms and Ruffell, 1989, 1990; Simms et al., Chapter 21) or merely a single mass extinction at the end of the Triassic period, as well as the question whether or not there is any significance to the (admittedly low) peaks of extinction reported by Sepkoski (1989,1990) in the Early and Middle Jurassic. In addition, there is considerable relevance to the debate over the suggestion of impact-produced mass extinctions and the postulated periodicity of extinction crises.

The evidence in these debates has been reviewed by

Hallam (1990), who favors a single terminal-Triassic extinction event, and by Benton (1991) and Simms et al. (Chapter 21), who favor two, of which the late Carnian one, they argue, was critical in wiping out terrestrial and marine life. Olsen et al. (1990), on the other hand, focused on the role of the end-Triassic event as it relates to the diversity of terrestrial tetrapods. As for the postulated Early and Middle Jurassic events, Hallam (1986) argued that the Pliensbachian extinction was merely a European affair, and Benton (1987b) noted the rather incomplete nature of the tetrapod fossil record during much of the Jurassic, and hence the difficulty of identifying extinction events during that time interval. The debates no doubt have a great deal of running in them yet, and the purpose of this chapter is not to reiterate previous arguments.

The data on extinctions

Stratigraphy

The rationale behind the stratigraphic scheme used here is based on several independent approaches that give relatively confident age assignments for some tetrapod-bearing units. The biostratigraphy of the Late Triassic is founded on the temporal distribution of ammonoids from the Alpine region (Figure 22.1). Detailed correlations are possible with marine sequences in other parts of the world, such as western Canada (Tozer, 1974, 1979). Attempts are being made to correlate the palynological zonation of the Late Triassic with this marine standard, by studies of the marginal and terrestrial sequences around the Alpine marine area, but the temporal acuity of the palynological biostratigraphic zones is poorer than that of the marine ammonoid zones: 6 palynological zones, compared with 13 ammonoid zones in the Carnian and Norian (including Rhaetian), giving mean durations for the

		Ammonoid zones	Palynological zones	Tetrapod zones	Chinle pollen zones	Newark pollen zones	Phytosaur biochrons	Aetosaur biochrons
NORIAN	(Rhaetian)	Crickmayi	(Rhaetian) Sevatian	NOR L2		'Upper Balls Bluff-Upper Passaic' Palynofloral zone		?
NORIAN	U.	Amoenum	(Rhaetian) Sevatian	NOR L2		'Upper Balls Bluff-Upper Passaic' Palynofloral zone		?
NORIAN	U.	Cordilleranus	(Rhaetian) Sevatian	NOR L1		'Upper Balls Bluff-Upper Passaic' Palynofloral zone		?
NORIAN	M.	Columbianus	Upper Norian (Alaunian)	NOR M2	?			Redondasuchus Biochron
NORIAN	M.	Rutherfordi	Upper Norian (Alaunian)	NOR M1	III			Redondasuchus Biochron
NORIAN	M.	Magnus	Lower Norian (Lacian)	NOR M1	III	'Lower Passaic-Heidlersburg' Palynofloral zone		Redondasuchus Biochron
NORIAN	L.	Dawsoni	Lower Norian (Lacian)	NOR E	III	'Lower Passaic-Heidlersburg' Palynofloral zone	Pseudopalatus Biochron	Typothorax Biochron
NORIAN	L.	Kerri	Lower Norian (Lacian)	NOR E	III	'Lower Passaic-Heidlersburg' Palynofloral zone	Pseudopalatus Biochron	Typothorax Biochron
CARNIAN	U.	Macrolobatus	Tuvalian	CRN L2	II	'New Oxford-Lockatong' Palynofloral zone	Rutiodon Biochron	Stagonolepis Biochron
CARNIAN	U.	Welleri	Tuvalian	CRN L2	II	'New Oxford-Lockatong' Palynofloral zone	Rutiodon Biochron	Stagonolepis Biochron
CARNIAN	U.	Dilleri	Tuvalian	CRN L1	I	'Chatham-Richmond-Taylorsville' Palynofloral zone ?	Paleorhinus Biochron	Longosuchus Biochron
CARNIAN	L.	Nanseni	Julian	CRN M	?	'Chatham-Richmond-Taylorsville' Palynofloral zone ?		
CARNIAN	L.	Obesum	Cordevolian	CRN E		'Chatham-Richmond-Taylorsville' Palynofloral zone ?		
LADINIAN	U.	Sutherlandi	Langobardian	LAD				
LADINIAN	U.	Maclearni	Langobardian	LAD				
LADINIAN	U.	Meginae	Langobardian	LAD				

Figure 22.1. Zonation of the Late Triassic, by means of ammonoids, palynomorphs, and tetrapods. Based on data from Tozer (1974, 1979), Visscher and Brugman (1981), Litwin et al. (1991), and Hunt and Lucas (1990, 1991a, b, c, 1992).

zones of 4.2–5.0 and 1.9–2.3 million years, respectively, depending upon the accepted total duration of the Late Triassic – 25 million years (Forster and Warrington, 1985; Cowie and Bassett, 1989), 27 million years (Harland et al., 1990), or 30 million years (Olsen, Schlische, and Gore, 1989).

Lithostratigraphic techniques have allowed a relative correlation of the Middle and Late Triassic sediments of the Germanic Basin (Figure 22.2), extending from Bavaria and Thuringia in eastern Germany to Baden-Württemberg in southwestern Germany, as well as northwestern Switzerland, Luxembourg, and Lorraine, France (e.g., Brenner, 1973, 1979; Gwinner, 1980; Brenner and Villinger, 1981). These terrestrial sediments are geographically close to the marine rocks of the Alps, and attempts have been made to establish detailed unit-by-unit correlations between the two using ostracods, bivalves, gastropods, fish, amphibians, palynomorphs, and charophytes (Kozur, 1975; Dockter et al., 1980; Blendinger, 1988). In addition, attempts are being made to establish standard palynological zones for the Alpine succession that will correlate with the ammonoid zones (e.g., Klaus, 1960; Mädler, 1964; Schulz, 1967; Scheuring, 1970; Morbey, 1975; Dunay and Fisher, 1978; Schuurman, 1979; Visscher,

Schuurman, and Van Erve, 1980; Visscher and Brugman, 1981; Van der Eem, 1983; Blendinger, 1988; Weiss, 1989). The results have been reasonably good for the Late Triassic (Figures 22.1 and 22.2), in which a number of direct tie points between palynomorphs and ammonoids have been possible, but Weiss (1989) was unable to extend this kind of scheme into the Early Jurassic.

There currently are two ways of interpreting the position of the Carnian–Norian boundary in the German Keuper (Figure 22.2). According to what we shall call interpretation A, the Rote Wand and Kieselsandstein are early Norian in age (Geiger and Hopping, 1968; Fisher, 1972; Fisher and Bujak, 1975; Dunay and Fisher, 1979; Dockter et al., 1980; Anderson, 1981; Schröder, 1982), whereas according to interpretation B, those two horizons are late Carnian (Kozur, 1975; Gall, Durand, and Muller, 1977; Olsen, McCune, and Thomson, 1982). Paleoclimatic evidence tends to favor interpretation B, according to Dockter et al. (1980, p. 960). The oberer Gipskeuper (= Rote Wand + Kieselsandstein, or Rote Wand + Blasensandstein) contains numerous evaporitic horizons, some of which carry gypsum, which is true also of the southern Alpine Torrer Schichten, Opponitzer Schichten, and

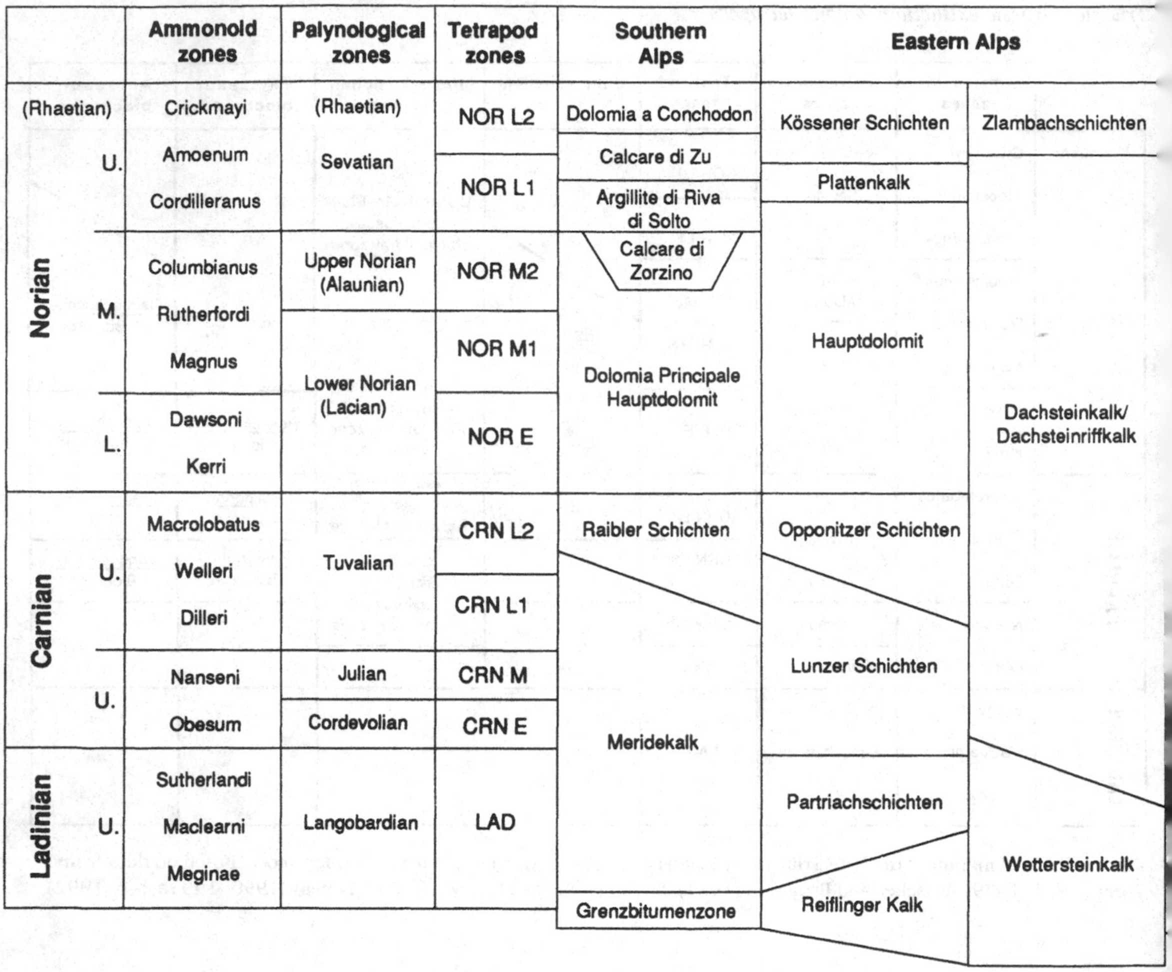

	Ammonoid zones	Palynological zones	Tetrapod zones	Southern Alps	Eastern Alps	
(Rhaetian)	Crickmayi	(Rhaetian)	NOR L2	Dolomia a Conchodon	Kössener Schichten	Zlambachschichten
Norian U.	Amoenum	Sevatian		Calcare di Zu	Plattenkalk	
	Cordilleranus		NOR L1	Argillite di Riva di Solto		
Norian M.	Columbianus	Upper Norian (Alaunian)	NOR M2	Calcare di Zorzino	Hauptdolomit	Dachsteinkalk/ Dachsteinriffkalk
	Rutherfordi					
	Magnus	Lower Norian (Lacian)	NOR M1	Dolomia Principale Hauptdolomit		
Norian L.	Dawsoni		NOR E			
	Kerri					
Carnian U.	Macrolobatus	Tuvalian	CRN L2	Raibler Schichten	Opponitzer Schichten	
	Welleri		CRN L1			
	Dilleri				Lunzer Schichten	
Carnian U.	Nanseni	Julian	CRN M			
	Obesum	Cordevolian	CRN E	Meridekalk		
Ladinian U.	Sutherlandi	Langobardian	LAD		Partriachschichten	Wettersteinkalk
	Maclearni					
	Meginae			Grenzbitumenzone	Reiflinger Kalk	

Figure 22.2. Stratigraphic chart of the major Late Triassic formations of the terrestrial Germanic Basin and marine Alpine are 1979) and palynological zones (after Visscher and Brugman, 1981) are indicated as standards. Two schemes for the relative pl Germany are noted, and scheme B is preferred here. The sequences in Bavaria and Thuringia are indicated in line with scheme Kozur (1975), Gwinner (1978, 1980), Dockter et al. (1980), Brenner and Villinger (1981), and Benton (1993a).

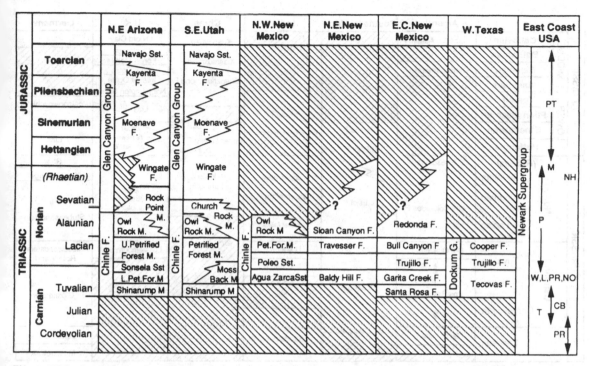

Figure 22.3. Stratigraphy of some Late Triassic and Early Jurassic vertebrate-bearing sequences in North America. The expansion of the Chinle Formation and the Dockum Group through the late Carnian and much of the Norian is based on comparisons of tetrapods with the German sequence by Olsen and Sues (1986), Hunt and Lucas (1990, 1991a, b, c, 1992), and others, and on palynological work by Litwin et al. (1991). Abbreviations: CB, Cow Branch Formation; L, Lockatong Formation; M, McCoy Brook Formation; NH, New Haven Arkose; NO, New Oxford Formation; P, Passaic Formation; PK, Pekin Formation; PT, Portland Formation; T, Turkey Branch Formation; W, Wolfville Formation.

Raibler Gips of Austria. The Opponitzer Schichten are dated as uppermost Carnian (Tuvalian) by their brackish-water fauna, via ammonoids and palynomorphs, and the oberer Gipskeuper is given the same date by its rich ostracod fauna, including *Costatoria vestita* (Dockter et al., 1980). The remaining gypsiferous horizons of the Alpine and Germanic basins are then correlated on paleoclimatic grounds. Magnetostratigraphic evidence, on the other hand, indicates that the Schilfsandstein is latest Carnian in age (Hahn, 1982), and this favors interpretation A. In addition, Wild (1989) implied support for scheme A because he dated both the Untere and Mittlere Stubensandstein as middle Norian on the basis of the shared presence of *Aetosaurus* in these units and in the Calcare de Zorzino of northern Italy, dated by ammonoids as Alaunian (Figure 22.2). In this chapter interpretation B is followed; if A had been selected, the results would have been little changed. Other aspects of the dating in detail of the Keuper are still unclear; for example, the lower boundary of the Gipskeuper, in north Württemberg at least, falls in the uppermost Ladinian (Bachmann and Gwinner, 1971).

Litwin, Traverse, and Ash (1991) have extended the palynological scheme to the Chinle Formation and Dockum Group of the southwestern United States

(Figures 22.1 and 22.3). These units, in New Mexico, Arizona, and Utah, include palynomorphs of three zones, termed I, II, and III, which correspond to the early part of the Tuvalian, the later Tuvalian (both late Carnian), and the Lacian (?early Norian, pre-Rhaetian). Litwin et al. (1991) correlate these palynological zones with those of the eastern United States: They regard the "Chatham-Richmond-Taylorsville Palynofloral Zone" of Cornet (1977,1989; Cornet and Olsen, 1985) as partially equivalent to their zone I, but the Newark zone extends lower, into the Julian, and possibly into the Cordevolian (early Carnian). Litwin et al. (1991) equate their zone II with Cornet's (1977) "New Oxford–Lockatong Palynofloral Zone," and their zone III with Cornet's "Lower Passaic–Heidlersburg Palynofloral Zone." Cornet's (1977) youngest zone, renamed by Litwin et al. (1991) the "Upper Balls Bluff–Upper Passaic Palynofloral Zone," is not represented by an equivalent in the southwestern United States (Figure 22.1). These palynological zones are tied to European, and other, schemes. It is interesting to note that the palynological work of Litwin et al. (1991) confirms the expansion of the Triassic formations of the American Southwest from a limited late Carnian duration (e.g., Dunay and Fisher, 1979; Olsen and Sues, 1986) to a

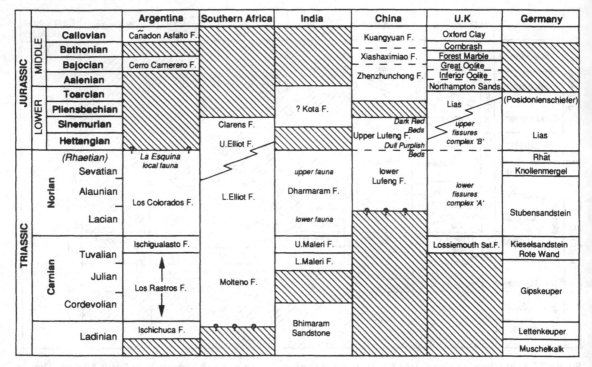

			Argentina	Southern Africa	India	China	U.K	Germany
JURASSIC	MIDDLE	Callovian	Cañadon Asfalto F.			Kuangyuan F.	Oxford Clay	
		Bathonian				Xiashaximiao F.	Cornbrash	
		Bajocian	Cerro Carnerero F.				Forest Marble	
		Aalenian				Zhenzhunchong F.	Great Oolite / Inferior Oolite	
							Northampton Sands	
	LOWER	Toarcian			? Kota F.		Lias	(Posidonienschiefer)
		Pliensbachian						
		Sinemurian		Clarens F.		Dark Red Beds	upper fissures complex 'B'	Lias
		Hettangian		U.Elliot F.		Upper Lufeng F. / Dull Purplish Beds		
TRIASSIC	Norian	(Rhaetian)	La Esquina local fauna			lower Lufeng F.		Rhät
		Sevatian			upper fauna			Knollenmergel
		Alaunian	Los Colorados F.	L.Elliot F.	Dharmaram F.		lower fissures complex 'A'	Stubensandstein
		Lacian			lower fauna			
	Carnian	Tuvalian	Ischigualasto F.		U.Maleri F.		Lossiemouth Sst.F.	Kieselsandstein Rote Wand
		Julian	Los Rastros F.	Molteno F.	L.Maleri F.			Gipskeuper
		Cordevolian						
		Ladinian	Ischichuca F.		Bhimaram Sandstone			Lettenkeuper
								Muschelkalk

Figure 22.4. Stratigraphy of vertebrate-bearing sequences from the Late Triassic to Middle Jurassic for various parts of Gondwana and Europe. The dates are based largely on comparisons of tetrapods with the German sequence by Olsen and Sues (1986) and others, with some sporadic information from palynology, invertebrates, and absolute age dates. Based on data from Anderson and Cruickshank (1978), Benton (1983, 1993a), Olsen and Galton (1984), Kutty and Sengupta (1989), Olsen et al. (1989), Weishampel (1990), and Hunt and Lucas (1990, 1991a,b,c, 1992).

potential span from late Carnian to the latest Triassic; the latter view was indicated independently by studies of the tetrapods.

Hunt and Lucas (1990, 1991a,b,c, 1992) have been developing global correlation schemes for Late Triassic terrestrial units based on phytosaurs and aetosaurs, arguing that certain genera of both groups are sufficiently restricted in temporal duration, sufficiently identifiable, and sufficiently widespread to permit their use as index fossils. The phytosaur *Paleorhinus* (synonyms: "*Mesorhinus*," *Promystriosuchus*, *Francosuchus*, *Ebrachosuchus*, *Parasuchus*) is represented in several parts of the world (Germany, Austria, Morocco, India, North America), and Hunt and Lucas (1991a) use it to define the *Paleorhinus* Biochron (Figure 22.1). This is tied to the marine ammonoid sequences by a specimen from the lower part of the Opponitzer Schichten of southern Austria, which are dated as Tuvalian (Zapfe, 1974), and lower Tuvalian for the phytosaur horizon. This is then used by Hunt and Lucas (1991a) to assign an early Tuvalian age to all other beds containing *Paleorhinus*, namely the Popo Agie Formation of Wyoming, the lower part of the Petrified Forest Member of Arizona, the Camp Springs

Member of the Tecovas Formation of Texas, the lower Dockum Group of Texas, the Blasensandstein of Bavaria, the Argana Formation of Morocco, and the Maleri Formation of India (Figures 22.3 and 22.4). The *Paleorhinus* Biochron is followed by the *Rutiodon* Biochron (latest Carnian) and the *Pseudopalatus* Biochron (early Norian) (Hunt, 1991; Hunt and Lucas, 1991b).

The aetosaurs give less datable zones, but Hunt and Lucas (1990, 1992) established a sequence, the *Longosuchus* Biochron (middle to late Carnian), the *Stagonolepis* Biochron (latest Carnian), the *Typothorax* Biochron (early to middle Norian), and the *Redondasuchus* Biochron (middle to late Norian). Hence, the *Longosuchus* Biochron is partly equivalent to the *Paleorhinus* Biochron, the *Stagonolepis* Biochron is broadly equivalent to the *Rutiodon*, and the *Typothorax* broadly to the *Pseudopalatus* (Figure 22.1). This scheme has not yet been fully developed, nor has it been tested, but it offers some promise.

The zones used here for terrestrial tetrapod-bearing units during the Late Triassic (Figure 22.1) take advantage of the new palynological work, especially that tied to ammonoid zonations, and the new tetrapod-based schemes. The zones are based on the broad palynological divisions of the Carnian and Norian, where early, middle, and late time slices are recognized; further, the late Carnian, middle Norian, and late Norian are each divided into two subunits, reflecting the suggestions of various authors, based on ammonoids, palynomorphs, and tetrapods. The tetrapod-based time units are not formally named, but are coded CRN E, CRN M, CRN L1, and CRN L2 for early Carnian, middle Carnian, early part of the late Carnian, and later part of the late Carnian, respectively, and similarly for the Norian. Note, in particular, that the Rhaetian stage is not used here, following recommendations by Tozer (1974) and others, because it is applicable only to the marine Rhaetic facies of Britain and central Europe. The later part of the Sevatian substage (NOR L2) is essentially equivalent to the "Rhaetian" of older usage.

The stratigraphic assignments of tetrapod-bearing formations from all parts of the world in the interval from the end of the Middle Triassic to the end of the Middle Jurassic are listed in Appendix 22.1 and summarized in Figures 22.2–22.4. The age assignments are based on numerous references, including Anderson and Cruickshank (1978), Benton (1983, 1993a), Olsen and Galton (1984), Kutty and Sengupta (1989), Olsen et al. (1989), Weishampel (1990), Hunt and Lucas (1990, 1991a,b,c, 1992), and Hunt (1991).

Taxonomy

For the present compilation, the familial assignments of Late Triassic to Middle Jurassic tetrapods are based, so far as possible, on current cladistic studies, such as Milner (1988, 1993b) on amphibians, Gauthier, Kluge, and Rowe (1988) on basal reptiles, Gaffney and Meylan (1988) on turtles, Benton (1985) and Evans (1988) on basal diapsids, Benton and Clark (1988) on Triassic archosaurs and crocodylomorphs, Gauthier (1986) and Weishampel, Dodson, and Osmólska (1990) on dinosaurs, Wellnhofer (1978) and Unwin (1991) on pterosaurs, Estes (1983) and Estes, de Queiroz, and Gauthier (1988) on lizards, Kemp (1982), Hopson and Barghusen (1986), and King (1988) on therapsids, and Hahn, Sigogneau-Russell, and Wouters (1989) and Stucky and McKenna (1993) on mammals. Exclusively marine groups, such as the Sauropterygia (Pachypleurosauria, Placodontia, "Nothosauria," Plesiosauria), Ichthyosauria, Askeptosauridae, Claraziidae, Thalattosauridae, and Pleurosauridae are omitted. The stratigraphic distribution data for each family are based on information from Milner (1993a), Benton (1993b), and Stucky and McKenna (1993) for amphibians, reptiles, and mammals, respectively, as well as numerous comments by contributors to this volume. The ranges are plotted in Figure 22.5, and the firsts and lasts for each family are summarized in Appendix 22.2.

Patterns

The diversity of terrestrial tetrapod families through the Middle Triassic–Middle Jurassic interval is shown in Figure 22.6a, with calculated metrics of origination and extinction for families shown in Figures 22.6B and 22.6C, respectively. In all cases, two curves are shown, one for all families documented in Figure 22.5, and one for the nonsingleton families only (i.e., those families based on more than a single genus – often a single species, or even a single specimen).

The graphs of diversity change show significant drops at the end of the Carnian and in the Early Jurassic. The magnitudes of these decreases are greater than it may seem from the graphs, because they are masked to some extent by the origin of new families in the succeeding stages; for example, the diversity drop at the Triassic–Jurassic boundary does not appear clearly in Figure 22.6A because an equivalent number of new families apparently originated during the Hettangian time interval. High rates of origination and extinction occur in the two late Carnian substages and in the "Rhaetian." High origination rates also occur in the Hettangian, Sinemurian, and Bathonian, but these may be partly *Lagerstätten* effects, in that these stages follow gaps in the record during the Sinemurian and the Toarcian–Bajocian interval. High extinction rates in the Sinemurian, Pliensbachian, and Toarcian may be connected with these same gaps.

Quality of the data

Stratigraphy

Accuracy of dating. The assignment of precise ages to Triassic and Jurassic terrestrial faunas is very difficult. Indeed, recent reviews of the stratigraphy of these faunas (e.g., Olsen and Galton, 1977, 1984; Olsen and Sues, 1986) introduced dramatic reappraisals of ages, with many units previously dated as "Late Triassic" being reassigned to the Early and even Middle Jurassic (a jump of four to seven stages, or 10–35 million years) on the basis of exact age dates from associated volcanic horizons, fossil fish, palynomorphs, footprints, and comparisons of tetrapod faunas.

Other stratigraphic approaches that may be of assistance in the future include chronostratigraphy and magnetostratigraphy. Exact ages have been reported for volcanic horizons in earliest Jurassic rocks in several basins in the Newark Supergroup and later in the Early Jurassic in southern Africa. Other data points are needed within the Late Triassic, associated with tetrapod-bearing sediments, to supplement the poorly documented ?Carnian date from Argentina (Forster and Warrington, 1985, p. 107). Outline magnetostratigraphic schemes are available for the Germanic Basin Late Triassic (Hahn, 1982) and the Newark Supergroup

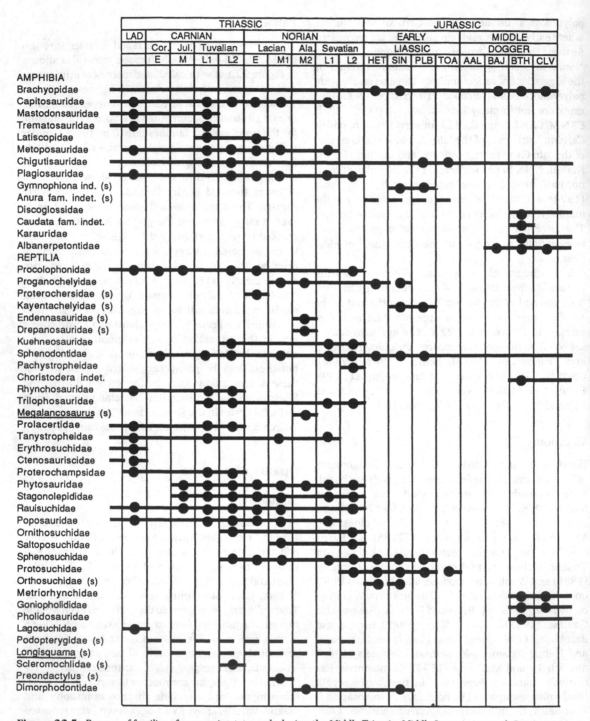

Figure 22.5. Ranges of families of nonmarine tetrapods during the Middle Triassic–Middle Jurassic interval. Ranges are shown based on data in Appendix 22.2. Black dots indicate that fossil material of the family in question is known from the stratigraphic unit indicated. Singleton taxa are denoted by (s).

(Olsen et al., 1989, p. 7; Witte, Kent, and Olsen, 1991). Current work in the Newark Supergroup should greatly enhance the usefulness of the latter.

Dramatic as many of the recent revisions of Late Triassic and Early Jurassic terrestrial biostratigraphy have been, the different approaches are tending to

confirm the new schemes (Figures 22.1–22.4). Hence, it seems unlikely that these will be heavily revised in the future, at least not to the extent of the changes set in train by Olsen and Galton (1977). Recent revisions have concerned fine-scale stratigraphic reassignments, generally from one substage to another, involving

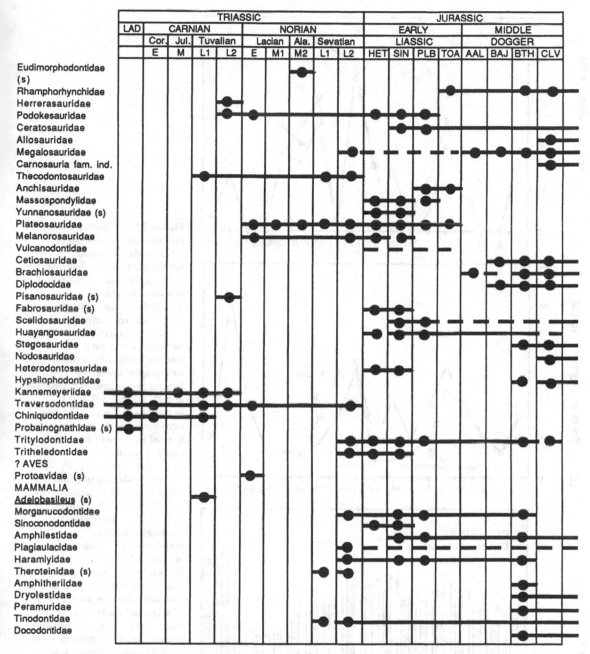

Figure 22.5 (*Continued*)

timespans estimated at 2–5 million years. Nevertheless, such revisions are crucial, and we await further confirmations of the marine–terrestrial biostratigraphic link in Europe and in North America and the proper integration of southern-continent sequences into such schemes.

There are several outstanding stratigraphic problems. For example, there is debate over the dating of the "fissure Complex-A" of southwest Britain (see Fraser, Chapter 11, and Evans and Kermack, Chapter 15). In addition, as mentioned earlier, the faunas in the Late Triassic and Early Jurassic of South America, India, and China, require clearer definitions and firmer correlations with the European and North American formations. New data are still revealing how poorly defined many such units are, as illustrated, for example, by the splitting of the fauna of the Santa Maria Formation of Brazil into two (Barberena, Araújo, and Lavina, 1985), and of the faunas of the Maleri Formation and the Dharmaram Formation of the Pranhita-

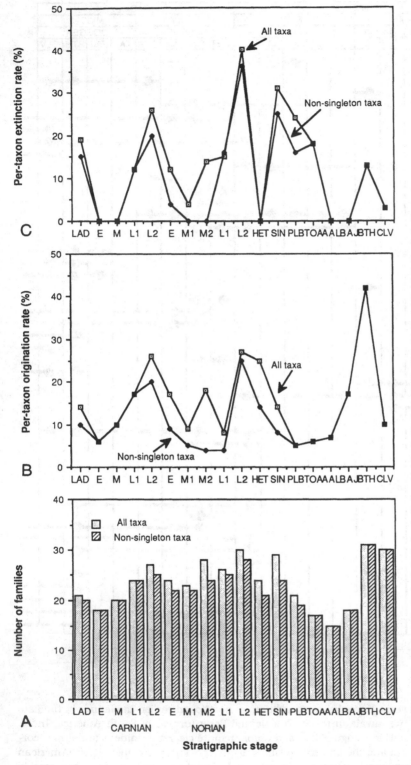

Figure 22.6. Diversity and evolutionary rates for Middle Triassic–Middle Jurassic nonmarine tetrapods, indicated by stratigraphic stage or substage. Measures are given separately for nonsingleton taxa and for all taxa (including singleton families). Where a particular formation is assigned tentatively to two stratigraphic stages, the families are counted as if present in both. For less well-dated formations that span more than two time units, the families are ignored altogether (this applies only to four families here: the Anura fam. indet., Podopterygidae, *Longisquama*, and Vulcanodontidae). Questionable extensions to familial ranges, indicated by dashed lines (for the Megalosauridae, Scelidosauridae, Plagiaulacidae) are counted as confirmed. Rates are percentages, scaled to the total numbers of taxa present in a time unit. They are not scaled to time because the stratigraphic stages and substages are of rather variable durations, depending on the time scale employed; in any case, the durations are roughly comparable, ranging from 3–11 million years (mean: 4.6 million years), according to dates by Cowie and Bassett (1989). (A) Total diversity. (B) Per-taxon origination rate. (C) Per-taxon extinction rate.

Godavari Valley in India into two each (Kutty and Sengupta, 1989). It is astonishing how little we know of such seemingly fundamental matters.

Continuity of sections. Adequate tests of patterns of diversity change through time, including mass extinctions, will require fossiliferous sections that span the interval in question as completely as possible. There are many such sequences of Late Triassic and Early Jurassic terrestrial rocks in various parts of the world that would seem to be suitable on superficial inspection. However, many of them are not so good when examined in detail. This issue is explored by Benton (1993a) and is summarized here.

The sequences in the Germanic Basin are thick, up to 1,750 m (Schröder, 1982), and relatively well dated by palynology and comparisons with the marine Alpine sequences. Fossil tetrapods have been found throughout the sequences (Brenner 1973; Benton 1986a, 1993a), but are rare in the Carnian. Hence, those sequences are barely adequate to test any postulated late Carnian extinction event. They are of no use for testing the end-Triassic event because the relatively rich "Rhaetian" faunas (mixed marine and derived terrestrial material) are followed by a major facies change to fully marine conditions, and hence the terrestrial faunas of the time are very poorly sampled.

The Late Triassic and Early Jurassic sequences in the southwestern United States (Figure 22.3) seem to hold more promise, for analysis of the postulated late Carnian event at least. The Chinle Formation, Dockum Group, and equivalents in Texas, Arizona, New Mexico, and Utah are now know to span from late Carnian times well into the Norian, possibly to the top (Hunt and Lucas, 1990, 1991a,b,c; Litwin et al., 1991; Lucas, 1991), and many diverse tetrapod faunas are known from all levels of the succession. However, the total thicknesses of these successions are not great, about 300–550 m, and more information is needed regarding possible unconformities. The transition from the Chinle Formation to the Glen Canyon Group seems to continue the succession fairly conformably across the Triassic–Jurassic boundary, but better biostratigraphic control will be required for these upper units in order to determine the value of such sequences for testing the nature of the postulated end-Triassic extinction event.

The Newark Supergroup of the eastern United States and Canada covers a time span from the middle Carnian, or earlier, to the Pliensbachian (Figure 22.3), within a thickness of over 6,000 m of lacustrine and fluviatile sediments (Olsen et al., 1989, 1990). It has been argued that this represents the best succession for testing the nature of tetrapod extinctions during the Late Triassic–Early Jurassic interval (Olsen and Sues, 1986; Olsen et al., 1987, 1990; Hallam, 1990). However, skeletal fossils from the various basins within the Newark Supergroup offer little hope of testing such events. Faunas in the middle and late Carnian are relatively diverse, as are those in the earliest Jurassic. Indeed, Olsen et al. (1987, 1988) and Shubin et al. (1991) have argued that the basal Jurassic fauna of the McCoy Brook Formation of Nova Scotia provides a crucial test of the effects of the postulated end-Triassic extinction event. However, the Norian interval in the Newark Supergroup is nearly devoid of tetrapod skeletal fossils, having yielded only about ten specimens in all from the extensive Passaic Formation and the New Haven Arkose. The near absence of Norian fossils makes it impossible to test either the after effects of the postulated late Carnian event or the nature of pre-extinction faunas for the postulated end-Triassic event. Footprint faunas are richer during the Norian interval in the Newark Supergroup, and they may offer some possibility of assessing extinction events (Olsen and Sues, 1986; Olsen et al., 1990), but there is always the serious problem of assigning tracks to the correct trackmakers, as noted by Olsen et al. (1990). For example, tracks of the ichnogenus *Rhynchosauroides* extend into the Norian, well beyond the disappearance of the rhynchosaurs at the end of the Carnian as documented by skeletal fossils. However, it is likely that *Rhynchosauroides*-type tracks were made by a wide range of terrestrial diapsids; indeed, probably rather few, such as those from the Middle Triassic of England (Benton et al., Chapter 7), were actually made by rhynchosaurs. Hence the documented stratigraphic range of *Rhynchosauroides* prints is difficult to interpret in terms of the appearance and disappearance of particular animal groups.

In Britain, the fissures of the Bristol region and South Wales (Figure 22.4) may offer hope for testing such events. They are classified into those of Complex A, dated broadly as "Late Triassic," and those of Complex B, dated more securely by associated palynomorphs as Sinemurian (Fraser 1986, and Chapter 11; Evans and Kermack, Chapter 15). One Complex-A fissure in Tytherington Quarry has been assigned a Rhaetian (late Sevatian) age on the basis of palynomorphs (Marshall and Whiteside, 1980), and Whiteside (1986) suggested that all fissure Complex-A sites were late Sevatian in age. However, as Fraser (Chapter 11) notes, this is not a necessary conclusion. It is not even demonstrated that all the fissures at Tytherington are late Sevatian in age; indeed, fieldwork suggests that many quarries in Carboniferous limestone in South Wales and around Bristol contain fissure fills of varying ages. Simms (1990a) has argued that many of the fissures could extend back in age to the middle or late Carnian, based on the assumption of their formation during an early to middle Carnian pluvial episode. Benton (1993a) concurred on the basis of the tetrapod faunas, some elements of which are closely comparable with animals from the geographically isolated Lossiemouth Sandstone Formation (late Carnian), and others with animals from the German Stubensandstein (early to middle Norian). If these dates are confirmed, and if the fissures can be arranged in a stratigraphic sequence, they may offer detailed samples of the smaller elements of Carnian to Sinemurian tetrapod faunas and hence be of tremendous potential for testing the nature of Late Triassic events.

The Late Triassic sequence in the Ischigualasto basin in the province of La Rioja, Argentina (Figure 22.4), may offer potential for testing at least the postulated late Carnian event. The 1,500-m-thick, seemingly conformable succession through the Ischichuca, Ischigualasto, and Los Colorados formations has yielded rich tetrapod faunas at several levels. In particular, the late Carnian Ischigualasto Formation passes continuously into the base of the Los Colorados Formation, but tetrapod fossils are rare in the latter. The La Esquina fauna of dinosaurs and other reptiles comes from the top 100 m of the Los Colorados Formation and is dated as late to latest Norian. It is not followed by Early Jurassic faunas, so the postulated end-Triassic event cannot be studied. Independent palynological dating of this succession is urgently needed.

Independent dating is also needed for the Late Triassic sequence of the Pranhita-Godavari Valley in India (Figure 22.4). Here, the Bhimaram, Maleri, Dharmaram, and Kota formations make up a 1,230-m-thick sequence spanning in age from the Ladinian to the Sinemurian/Toarcian. The Maleri Formation has yielded two faunas, of late and latest Carnian age (Kutty and Sengupta, 1989), and the Dharmaram Formation has a lower fauna of early Norian age. These might be used to constrain aspects of the postulated late Carnian event. The upper Dharmaram fauna appears to be late, or latest, Norian in age, but it is followed by a considerable unconformity before the Kota Formation, which is often dated as Sinemurian, but might be Toarcian, or younger, in age. Palynological evidence is limited at present, and more detailed studies will be required to firm up the age assignments of the tetrapod-bearing formations.

In southern Africa, the Stormberg succession (i.e., the Elliot and Clarens formations) (Figure 22.4) forms a sequence only 250 m thick that spans in age from the early Norian to the Sinemurian. Both formations have yielded a sequence of faunas (Kitching and Raath, 1984) that may provide evidence of the nature of the postulated end-Triassic event. However, vertebrate fossils are absent from the underlying Carnian-age Molteno Formation, and hence nothing can be said of the postulated late Carnian event. Again, more palynological control is needed on the ages of the Elliot and Clarens formations.

The Lower Lufeng Formation of Yunnan, China (Figure 22.4), also appears to straddle the Triassic–Jurassic boundary and could therefore be used to test the nature of the postulated end-Triassic event. The sequence is 750 m thick and has yielded separate faunas in the Dull Purplish Beds and the Dark Red Beds. The former may be latest Norian or earliest Jurassic in age (Zhen et al., 1985), and the latter appear to be Hettangian or Sinemurian in age (Sun and Cui, 1986). The ages are confirmed to some extent by ostracods and molluscs, but more refined biostratigraphic work is required.

In conclusion of this section, the terrestrial sequences of the Germanic basin, possibly the British fissures, the sequences of the American Southwest, and possibly those from the Ischigualasto basin of Argentina and the Pranhita-Godavari Valley of India may allow testing of the nature of the postulated late Carnian event. Further, the terrestrial sequences of the British fissures, and possibly the Stormberg sequence of southern Africa and the Lower Lufeng Formation of China, may allow testing of the nature of the postulated end-Triassic event. Other sequences cannot allow study of the latter event because terminal-Triassic faunas are followed by marine Jurassic sequences (the Germanic Basin, the British, French, and Swedish "Rhaetic") or by an apparent gap (India, Argentina), or else the terminal-Triassic record is lacking or inadequate (?American Southwest, Newark Supergroup, and possibly also southern Africa and China).

Gaps and collection failure

Qualitative arguments. The key question to be tackled in any study that purports to identify extinction events is, are they real, or merely the result of gaps in the record? This criticism has been leveled by Olsen and Sues (1986, p. 343), Sepkoski (1986, p. 286), Sepkoski and Raup (1986, p. 11), and others at the postulated late Carnian peak of extinction for both marine and terrestrial organisms. A qualitative counterargument is simply to assert that it is more parsimonious to read the fossil record literally, to accept the appearance of a sharp drop in diversity as real, than to argue for special cases of variable preservation conditions. Olsen et al. (1987, 1988) took such a literal view in arguing that the McCoy Brook Formation of Nova Scotia, dated as earliest Jurassic, could be used to constrain the nature of the postulated end-Triassic event. Because certain Triassic tetrapods, such as tanystropheids, procolophonids, rhynchosaurids, and traversodontids, were not found in the McCoy Brook Formation, Olsen et al. (1988) argued plausibly that they had died out previously.

Benton (1991) has argued similarly that the absence from the early Norian of groups of readily fossilizable tetrapods that had been abundant in the late Carnian, such as mastodonsaurids, trematosaurids, rhynchosaurids, proterochampsids, kannemeyeriids, and chiniquodontids, actually proves that they had already

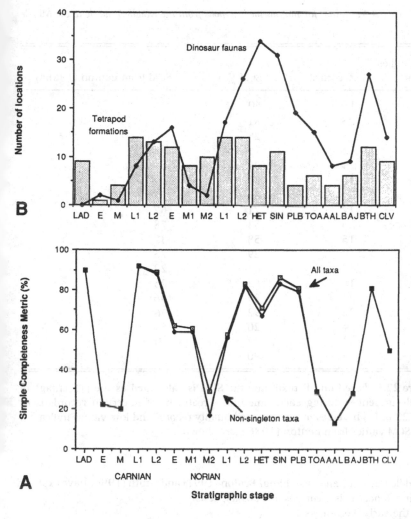

Figure 22.7. Measures of the possible effects of completeness of the fossil record on the patterns of diversification, origination, and extinction recorded. (A) The SCM (a measure of relative completeness; see Table 22.1) measured by time units for nonmarine tetrapod families for the Middle Triassic–Middle Jurassic time interval. High values indicate a good record, low values a poorer one. (B) The availability of rocks containing nonmarine tetrapods during the same time interval. Two measures are shown, a count of the geological formations enumerated in Appendix 22.1, and counts of dinosaur skeleton and footprint localities listed in Table 22.2. The peaks and troughs in dinosaur formations, in particular, roughly mimic the SCM values in part A.

died out. Olsen and Sues (1986, p. 343), on the other hand, state that "early Norian vertebrate assemblages are very poorly known, and, therefore, it is difficult to place much faith in the peak of Carnian extinctions." Such diverging assertions require a more precise test of whether the late Carnian extinction peak truly represents a mass extinction or is merely an artifact of a subsequent gap.

Measuring relative completeness: finding gaps. A first approach is to attempt to quantify completeness. The relative completeness of a fossil record may be estimated by calculating the proportions of "actual" and "assumed" fossil groups represented within each time unit. Actual groups represented are those for which fossils have been found in rocks of the age in question, and the assumed groups are the actual groups plus those that are known to span the time interval under study. Hence the difference between assumed and actual numbers represents the number

of Lazarus taxa. The ratio, as a percentage, has been termed the simple completeness metric (SCM) by Benton (1987b), and it ranges from zero (i.e., no fossils found) to 100 percent (all groups assumed to be present are represented by actual fossil finds). The SCM does not take account of taxa that arise, die out, or have their duration within a gap, and the effectiveness of the SCM as an estimator of completeness diminishes as the gap size increases (Benton, 1987b). Such problems can be partially overcome by probabilistic modifications to the SCM, based on measures of the general spottiness of the record of a particular clade and on aspects of the relevant rock record (Strauss and Sadler, 1989).

SCM values for the stages and substages employed in this tabulation of taxa (Figure 22.5) are listed in Table 22.1 and shown in Figure 22.7A. These show high SCM values in the Ladinian, late Carnian (both substages), early Norian, middle Norian (1), late Norian to Pliensbachian, Bathonian, and Callovian,

Table 22.1. *Simple completeness metric (SCM) values for nonmarine tetrapods from the Middle Triassic to the Middle Jurassic[a]*

Time unit	No. of families Apparent	Recorded	SCM (%)	SCM from Benton (1986a)
Callovian	30	15	50	
Bathonian	31	25	81	
Bajocian	18	5	28	
Aalenian	15	2	13	
Toarcian	17	5	29	
Pliensbachian	21	17	81	
Sinemurian	29	25	86	
Hettangian	24	17	71	67
Late Norian 2	30	25	83	78
Late Norian 1	26	15	58	37
Middle Norian 2	28	8	29	75
Middle Norian 1	23	14	61	22
Early Norian	24	15	62	25
Late Carnian 2	27	24	89	(96)
Late Carnian 1	24	22	92	(96)
Middle Carnian	20	4	20	54
Early Carnian	18	4	22	0
Ladinian	21	19	90	

[a]The data are derived from Figure 22.5, based on all taxa, and the SCM is calculated as the percentage of recorded fossils to apparent family presences during each time unit. Treatment of uncertain records is as explained in the legend to Figure 22.6. High values indicate a good-quality record, and low values indicate a poor record. For comparison, the SCM values from Benton (1986a) are shown.

and low values in the early to middle Carnian, middle Norian, and Toarcian to Bajocian. Hence, it is clear that there was no "gap" during the early Norian, as had been stated (e.g., Olsen and Sues, 1986), despite earlier SCM figures that seemed to indicate such a gap (Benton, 1986a, 1991). Redating of the Chinle and Dockum sequences (e.g., Litwin et al., 1991) and restudy of the Germanic Basin early Norian have filled the early Norian "gap." There is, however, an apparent gap in the middle Norian record of nonmarine tetrapods, when groups that are known to have spanned that interval are rather poorly represented by fossils. This does not prove that the late Carnian extinction peak is an artifact of a poor fossil record.

Gaps do not necessarily indicate poor preservation. Gaps in the fossil record can suggest either poor sampling of a diverse fauna or excellent sampling of a depauperate fauna. The gap could indicate genuinely depauperate faunas in the middle Norian (and in the Toarcian–Bajocian interval), following the trauma of a preceding mass extinction. Is it possible to distinguish between the literal reading of a gap as a time of low biotic diversity and the less parsimonious reading of such as the result of poor sampling of a fully diverse

biota? Benton (1991) and Smith (1990) have explored some possible tests.

Interpreting gaps: preservation failure or post-extinction biotas? The first approach, applied by both Benton (1991) and Smith (1990), was to test the assertion that the diversity decline from late Carnian to early Norian times simply mirrored the decline in fossiliferous deposits. This is seemingly partially true for the marine case, but not so for the terrestrial situation. A crude impression can be obtained by examination of a histogram of numbers of tetrapod-yielding formations recorded in Appendix 22.1 (Figure 22.7B). This shows similar numbers of formations dated as late Carnian 2 and early Norian (13 and 12, respectively), although it takes no account of the available areas of outcrop, the available rock volume, nor the proportion of potentially fossiliferous sedimentary facies in rocks of different ages. It is not, however, evident that any of these factors vary significantly between the late Carnian 2 and the early Norian occurrences.

Benton (1991) provided a similar test based on the independent data compilation by Weishampel (1990) in which he listed major basins that have yielded

Table 22.2. *Numbers of sedimentary rocks of different ages, from Middle Triassic to Middle Jurassic, that have produced dinosaur skeletal fossils or dinosaur footprints*[a]

Time unit	N. America	Europe	Asia	S. America	Africa	Australasia	Total
Callovian	0	8	2	3	1	0	14
Bathonian	0	21	2	0	4	0	27
Bajocian	0	6	1	1	0	1	9
Aalenian	0	7	1	0	0	0	8
Toarcian	7	2	2	3	1	0	15
Pliensbachian	9	1	3	3	3	0	19
Sinemurian	6	5	4	3	13	0	31
Hettangian	9	8	5	3	9	0	34
Late Norian 2	2	19	2	3	0	0	26
Late Norian 1	0	16	1	0	0	0	17
Middle Norian 2	1	1	0	0	0	0	2
Middle Norian 1	2	2	0	0	0	0	4
Early Norian	5	2	1	0	8	0	16
Late Carnian 2	7	2	1	3	0	0	13
Late Carnian 1	3	2	2	0	1	0	8
Middle Carnian	?1	0	0	0	0	0	?1
Early Carnian	1	?2	0	0	0	0	?3
Ladinian	0	0	0	0	0	0	0

[a]These figures are taken from Weishampel's (1990) state-by-state listing and include all formations dated certainly, or tentatively, by him. Tentatively dated units are assigned to precise time units by counting them twice (if dated across two of the time divisions used here). Some very poorly dated formations are excluded. Note that dinosaur fossils are not known clearly before the late Carnian, and hence the low numbers of faunas during that early time do not correspond to the overall number of nonmarine tetrapod-bearing formations.
Source: Data from Weishampel (1990).

remains of dinosaurs, both skeletal and ichnological. A histogram of the numbers of basins per time unit through the Late Triassic and Early to Middle Jurassic (Table 22.2; Figure 22.7B) confirms that there was no major drop in numbers of fossiliferous basins between the late Carnian and the early Norian: The totals fall from 21 to 16 (or rise from 13 to 16 if one counts only the late Carnian 2 substage formations), hardly a significant change. Across the postulated end-Triassic event, the totals rise from 26 latest Triassic ("Rhaetian") fossiliferous basins to 34 Hettangian, which is of significance for those attempting to chart the apparent diversity decline through that time interval.

Smith's (1990) other tests provided convincing evidence that the late Carnian diversity decline among marine echinoids was real. Some of these tests might be applicable to the tetrapod record, and should be so applied in the future. These tests are as follows:

1. Does the pattern of decline continue through several time units after the supposedly most highly fossiliferous horizons (*Lagerstätten*)? If there is no such longer-term decline, the highly fossiliferous interval may truly be followed by an episode of preservation failure. This is hard to determine for the tetrapod record because of the coarseness of stratigraphic acuity. Smith (1990) found such a pattern of decline in echinoid species diversity after the time of the highly fossiliferous Cassian Beds.

2. How many Lazarus taxa are there in the interval where a gap is postulated? If there are many taxa that apparently disappear and then reappear after the postulated gap, then preservation failure is implicated. If Lazarus taxa are not abundant, then the interval may truly represent a depauperate postextinction fauna. Smith (1990) found that many major echinoid lineages disappeared in the late Carnian, never to reappear, and he concluded from the low number of Lazarus taxa that the early Norian was not a time of serious preservation failure. For the terrestrial tetrapod record, the early Norian interval includes 12 Lazarus family-level taxa, compared with only 4 actually represented by fossils, and 7 extinctions at the end of the Carnian, all at the family level, according to Benton (1986a, p. 315). However, the new data set (Figure 22.4) indicates 7 Lazarus taxa in the early Norian time

interval, compared with 16 families actually represented by fossils, the change being the result of more precise dating of the Chinle and Dockum groups. Note, however, as stressed before, that Lazarus taxa may represent either preservation or collection failure, as is usually assumed, or they may indicate times of depauperate faunas when individuals were so rare that we hardly ever find them.

3. Are there any indications of poorer-quality preservation during the gap interval? Smith (1990) argued that if diversity levels were actually high during the gap interval, but fossils were rare, one would expect to find a higher proportion of incomplete or poorly preserved fossils. He found that that was not the case for the early Norian interval and that the proportion of disarticulated echinoid remains compared with complete remains did not increase. This test could be applied to the tetrapod record by comparing the proportions of complete and incomplete skeletons recovered from different time intervals.

4. Do different indicators of diversity reveal the same patterns? Smith (1990) found that graphs of diversity based on whole echinoid tests and those based on isolated spines showed the same patterns, and therefore he assumed, on the reasoning of Sepkoski et al. (1981), that these were close to the true patterns. It may be possible to apply such a test to the terrestrial tetrapod record by comparing the data from skeletal fossils and trace fossils. However, much more work on the patterns of the global appearances and disappearances of tetrapod footprint types is needed, more importantly, clear cases for equating particular footprint types with particular animal groups are required. It is likely that certain footprint types, having been produced by an array of taxonomic groups, would have to be omitted from such compilations.

Taxonomy

The assignment of fossil tetrapod specimens to species, genera, and families is a process fraught with problems at various levels. Certain groups in the Triassic and Jurassic appear to be well-defined cohesive clades, but many others are somewhat less tangible. For these latter taxa there may be only scattered fossil material available; in some cases such material may be difficult to study or poorly described, the material may indicate an assortment of taxonomic characters that defy clear-cut identification of the taxon, or the attributes of the beast may defy phylogenetic analysis. Problems of these sorts at all levels of systematic study are typical and they make the job of assessing biotic diversity in the past (as well as the present) quite difficult. One approach is to ignore such problems, because they introduce nonsystematic errors to any macroevolutionary analysis; that is, they will introduce a certain amount of background noise to the data set, but will

not necessarily distort it in any particular direction, a point made by Sepkoski (1989). The errors will be stochastic, nondirectional; that is, in popular terms, they may cancel each other out.

In the compilations of tetrapod data presented here, three measures have been taken in an attempt to improve the taxonomic quality of the information:

1. The families are all based on recent cladistic phylogenetic analyses and, so far as possible, are all monophyletic clades. Interestingly, this produces little change at the family level when one compares these family lists with older, pre-cladistic lists (Maxwell and Benton, 1990). For a long time, most tetrapod families have been defined by sharply indicated unique characters that are now called synapomorphies. The revolution wrought by the application of cladistic techniques to fossil and living tetrapods has generally affected our understanding of relationships at suprafamilial levels: the refinement of our views of early amphibian relationships (including the abolition of the "labyrinthodonts" and "lepospondyls"), the recognition that prolacertiforms are archosauromorph diapsids, the shift of rhynchosaurs from the lepidosaurs to the archosauromorphs (including the abolition of the "Rhynchocephalia" and the "Eosuchia"), the refinement of classifications of Triassic archosaurs (including the recognition of the monophyly of Dinosauria and the postulated sister-group relationship of Dinosauria and Pterosauria), the clarification of relationships within Cynodontia (including recognition of a monophyletic taxon Mammalia), and many more. These changes have not generally affected our interpretation of the boundaries of the families, but other taxonomic revisions have.

2. The contents of most families of Triassic and Jurassic tetrapods have been reassessed at the alpha-taxonomic level. In other words, paleontologists have examined much of the original material on which species, genera, and families were founded, and they have been obliged to synonymize many such taxa, or to declare them nomina dubia. This has been particularly true for the Permo-Triassic "mammal-like reptiles," for which many new taxa were erected with great enthusiasm earlier this century, but also for Triassic dinosaurs and for other groups to a lesser extent. The results of alpha-taxonomic revisions of these groups have had dramatic effects on the shape of current data bases, even when compared with those of the 1960s (Maxwell and Benton, 1990).

3. Singleton taxa are identified and treated in two ways, in order to highlight their existence. Many families of Triassic and Jurassic tetrapods have been erected on the basis of single genera, even single species or single specimens from single localities. Many such families have been synonymized with others, but others remain, and these are probably largely valid and distinct. Some day they may acquire new bedfellows, and hence cease

to be singletons. For the present, however, singleton families can be argued to distort macroevolutionary data bases, for four main reasons: (a) They are not equivalent, in ecological terms, to nonsingleton families, many of which were diverse, were abundant in individual faunas, and had global distributions. (b) They are commonly associated with fossil *Lagerstätten*, times of exceptional preservation, and hence distort the background signal coming from more typically fossilized groups. (c) They have point distributions in time, so that they do not have a true duration, and hence cannot strictly contribute to calculations of rates of change. (d) Many of them may be dubious and will disappear on further taxonomic revision.

None of these arguments against the inclusion of singleton taxa is devastating. In the first case (a), of course, singleton taxa are just one end of a continuum of family diversities, ranging from families containing a single species, families with two, three, or four, up to families with, say, 50 species. There is therefore no reason to draw a qualitative distinction between all singleton families and all nonsingleton families. In the second case (b), fossil *Lagerstätten* actually give us a more nearly true picture of the life of the past than do normal kinds of fossil deposits (Briggs, 1991), and they should really be used exclusively in attempts to plot past biotic diversities, while the data from non-*Lagerstätten* intervals should be corrected upward to take account of preservation loss. In the third case (c), this is a semantic quibble: Singleton families probably are singletons only because of our limited knowledge of the fossil record. It is unlikely that the Archaeopterygidae were really an isolated family of six or seven individual birds existing for an instant of time in the Tithonian of southern Germany, and having no forebears, contemporaries, or descendants. Finally, in the fourth case (d), it is true that many singleton families of tetrapods have already been discarded, because they were based on hopelessly inadequate or ill-defined material, but most of the singleton families in current lists are based on good, complete specimens of bizarre creatures (e.g., Kayentachelyidae, Drepanosauridae, Podopterygidae, Eudimorphodontidae, Protoavidae, Theroteinidae), and their families doubtless will continue to be deemed distinct from all others.

Therefore, for the purposes of my analyses, I have identified the singleton families and carried out the analyses both with and without them. Their inclusion tends to raise the rates of both origination and extinction during times of peaks, as would be expected, but in no case do the singletons alone generate such a peak.

Data bases and the future

Changing data bases. Comparison of the changing shapes of mass-extinction peaks through the years indicates no clear trend (Maxwell and Benton, 1990).

The main conclusion seems to be that the data bases may change fundamentally, by 50 percent or more over the past two decades for tetrapods, and yet the extinction events and their magnitudes remain fairly static. Sepkoski (1990) reports similar findings from analyses of his evolving family-level and generic-level data bases on marine animals.

Two opposite kinds of predictions can be made about the effects of future collecting on the nature of extinction peaks. At a detailed stratigraphic level, it might be expected that more collecting will sharpen up the shape of extinction peaks. Certainly, the studies by Ward (1990) on ammonite species distributions before the Cretaceous–Tertiary boundary show that as more collecting is done, the sharper the extinction becomes, because the Signor-Lipps effect (backward smearing of an instantaneous mass-extinction event) is being diminished. An opposite prediction might be that, in some cases, increasing knowledge will tend to broaden out the peaks of extinction intensity, on the assumption that "firsts" will be extended back in time, and "lasts" will be extended forward in time. So far, it is not yet clear which way we are heading with the Late Triassic extinction peaks.

Nevertheless, our views on the systematics and biostratigraphy of Triassic and Jurassic tetrapods have changed fundamentally in the past two decades, and many new fossil finds have been made. Surely there have been changes in the identification and the nature of the mass extinctions revealed? Surely, comparisons may also indicate the main reasons for such changes in our understanding and may hint at avenues for future research. An attempt is made here (Figure 22.8) to compare the results from data bases over the past two decades, starting with Olsen's (1982) paper, which is essentially based on Romer (1966).

The results show that the most dramatic cause of change in the patterns discovered over the past two decades has been rather prosaic: It is simply that the stratigraphic divisions in use have been refined from the rather crude "Middle Triassic" and "Upper Triassic" used by Olsen (1982) to the smaller substage revisions used in more recent analyses. Other changes have resulted from the discovery of new taxa during the past decade (e.g., Gymnophiona indet., Kayentachelyidae, Endennasauridae, Drepanosauridae, *Megalancosaurus*, Protoavidae, *Adelobasileus*, Theroteinidae), nearly all in "known" tetrapod-bearing sedimentary basins, however. Ranges have also been extended by new finds (e.g., Jurassic temnospondyls, Discoglossidae, Caudata fam. indet., Albanerpetontidae, Choristodera indet., Traversodontidae, Plagiaulacidae), again largely from "known" localities. New localities have been discovered recently, but they have yet to yield taxa that are entirely new or that will dramatically alter known ranges. Other recent changes in the ranges are results of alpha-level taxonomy, the familial reassignment of

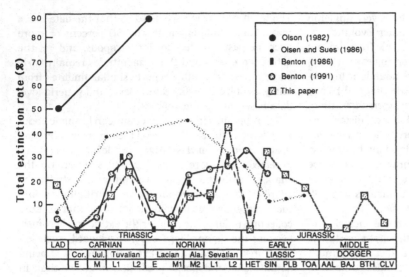

Figure 22.8. Changing perceptions of the nonmarine tetrapod extinction events during the Late Triassic and Early to Middle Jurassic. Earlier works did not clearly discriminate time intervals. Post-1985 results show improving time precision and reveal two peaks of high extinction rates in the Late Triassic, and possibly one in the Early Jurassic.

individual specimens, and small-scale stratigraphic reshuffling. The sample of data used here is too small to quantify usefully the roles of such factors in effecting changes in our ideas of macroevolutionary patterns among Late Triassic–Middle Jurassic tetrapods.

The major changes resulting from cladistic revisions, confirmed also by the findings of Patterson and Smith (1987) on data bases of echinoderms and fish, may be restricted to the current phase in the history of our science. Large-scale cladistic reviews of most tetrapod groups have now been carried out and will no doubt continue. It may be that the major revolution has passed, for the vertebrates at least, and future cladistic work will be restricted to minor adjustments that will not feed through to the data bases in such a radical way. In 50 years' time, it will be interesting to see whether we have experienced a short intense phase of major phylogenetic revision of higher-level tetrapod relationships (1975–90) or whether the rate of discovery of new phylogenetically deep nodes will continue. For eutherian mammals, at least, Novacek (1992) notes a growing concordance of phylogenetic trees produced from molecular data, morphological traits, and the fossil record. Of course, as he also notes, the congruence of several independently produced phylogenetic trees does not indicate that systematists are close to the truth: All of the analyses may be incorrect.

Future research directions. I predict that information on the Triassic and Jurassic extinctions of terrestrial tetrapods will improve along a number of lines:

1. A great deal remains to be learned by further discoveries of fossils, in the Jurassic interval in particular. New collection techniques, and the new focus on smaller tetrapod fossils, often sampled by screen-washing techniques, will doubtless continue to reveal many taxa.

2. A major desideratum is refined knowledge of the stratigraphy of sequences in Europe and North America, and the closer association of the southern-continent sequences with the new biostratigraphic schemes. This will require a closer link between the refined marine stratigraphic standards and the terrestrial palynological (and ostracod, conchostracan, fish, footprint, and other) schemes. New work in magneto-stratigraphy and chronometric data may also contribute, but the ammonoid zonal scheme is considerably more refined in terms of stratigraphic acuity (units of about 1 million years in the Late Triassic, and less than 1 million years in the Jurassic), and such precision hardly seems likely in the foreseeable future from either magnetostratigraphy or chronometry.

3. Continuing alpha-taxonomy and higher-level cladistic revisions of existing collections will greatly assist in resolving difficulties over family definitions.

These avenues of research may permit vertebrate paleontologists to attempt serious studies of the Triassic and Jurassic events based on genera, or even species, in the future. At present, this is not possible, because too many genera and species have only point distributions in time. As Padian and Clemens (1985) noted (and it is true for most Mesozoic vertebrates), the dinosaurs actually went extinct globally at the generic level dozens of times: so few genera span from one stratigraphic stage to the next. This problem of the mismatch of our taxonomic acuity and our stratigraphic acuity can be overcome only by a great improvement in the precision of dating faunas. At present, the time intervals in use are too crude, and their durations are greater than the mean generic durations for most vertebrates (1–5 million years) (Stanley, 1979).

Causation of Late Triassic extinctions. There are, broadly, three views on the causes of the major faunal changes that took place among tetrapods during the Late Triassic. One, a firming-up of the "classical" view of Romer (1970), is that the replacement of mammal-like reptiles and rhynchosaurs by thecodontians, and then by dinosaurs, was a long-term competitive process (e.g., Bonaparte, 1982; Charig, 1984).

That viewpoint has been refuted by Benton (1983, 1986a,b, 1987a,b, 1991), who has argued for a second model: that the dominant mammal-like reptiles and rhynchosaurs died out during a late Carnian extinction event, possibly related to a major floral change, and possibly to climatic stresses (Simms et al., Chapter 21, this volume), and that the dinosaurs (and turtles, sphenodontians, crocodylomorphs, pterosaurs, and mammals) radiated opportunistically during the Norian to fill the ecological void.

The third view, argued by Olsen et al. (1987, 1988, 1990) and Hallam (1990), is that the Carnian event was nonexistent, or at least is difficult to detect, and that the end-Triassic event was the key one. This has been linked explicitly to extraterrestrial causation by Olsen et al. (1987, 1988, 1990) and Sepkoski (1989), among others, specifically to the impact of the asteroid that produced the giant Manicouagan crater in Canada. However, this crater dates, if anything, closer to the Carnian–Norian than to the Triassic–Jurassic boundary (Olsen et al., 1987; Hodych and Dunning, 1992), and the Triassic–Jurassic boundary sections have not yielded up the expected impact indicators (e.g., shocked quartz, glass spherules) found in such abundance at the Cretaceous–Tertiary boundary. The recent report (Bice et al.,1992) of shocked quartz from a Triassic–Jurassic boundary section in the Il Fiume gorge in northern Italy is not such strong evidence for impact as might at first be thought. The "shocked" quartz grains occur in the putative Triassic–Jurassic boundary layer, but also in two layers 1–3 m below. In addition, the grains do not have more than four sets of planar deformation features (indeed, most have only single sets), and the angular distribution of the planar deformation features is rather diffuse. As Bice et al. (1992, p. 445) note, "these differences [from classic K-T shocked quartz] make it impossible to demonstrate unambiguously that the grains at the T-J [sic] boundary have a shock-metamorphic origin.... An alternative hypothesis would be that these grains contain highly unusual Böhm lamellae" presumably produced by normal earthbound tectonism.

Hence, I find little in favor of the impact-induced extinction model, and I adhere firmly to the reality of the late Carnian extinction as being ecologically the key event for terrestrial tetrapods, and as having real significance in the sea as well (e.g., Stanley, 1988; Simms and Ruffell, 1989, 1990; Simms, 1990b; Smith, 1990; Simms et al., Chapter 21). In recent reviews, both Sepkoski (1990) and Hallam (1992) accept the significance of the late Carnian extinction event. This is not to deny the reality of the end-Triassic event: It was a great catastrophe for marine life, but ecologically seemingly less significant for tetrapods.

Acknowledgments

I thank Nick Fraser and Hans-Dieter Sues for the invitation to attend the Front Royal Workshop in May 1991, and the Royal Society for travel funds to attend. Numerous people have looked at the data base and have supplied helpful comments: Susan Evans, Nick Fraser, Gerhard Hahn, Adrian Hunt, Kenneth Kermack, Spencer Lucas, Andrew Milner, Denise Signogneau-Russell, and Xiao-chun Wu. I thank others for commenting on various drafts of the manuscript: Nick Fraser, Mike Simms, and Glenn Storrs. This project was funded by grants from the Leverhulme Trust, SERC (grant GRF 2362.0), and NERC (grants GR3/7691 and GR9/372).

References

Anderson, J. M. 1981. World Permo-Triassic correlations: their biostratigraphic basis. Pp. 3–10 in M. M. Cresswell and P. Vella (eds.), *Gondwana Five.* Rotterdam: A. A. Balkema.

Anderson, J. M., and A. R. I. Cruickshank. 1978. The biostratigraphy of the Permian and Triassic. Part 5. A review of the classification and distribution of Permo-Triassic tetrapods. *Palaeontologia Africana* 21: 15–44.

Bachmann, G. J., and M. P. Gwinner. 1971. *Nordwürttemberg. Sammlung Geologischer Führer.* Berlin: Borntraeger.

Barberena, M. C., D. C. Araújo, and E. L. Lavina. 1985. Late Permian and Triassic tetrapods of southern Brazil. *National Geographic Research* 1: 5–20.

Benton, M. J. 1983. Dinosaur success in the Triassic: a noncompetitive ecological model. *Quarterly Review of Biology* 58: 29–55.

1985. Classification and phylogeny of the diapsid reptiles. *Zoological Journal of the Linnean Society* 84: 97–164.

1986a. More than one event in the Late Triassic mass extinction. *Nature* 321: 857–861.

1986b. The Late Triassic tetrapod extinction events. Pp. 303–320 in K. Padian (ed.), *The Beginning of the Age of Dinosaurs: Faunal Change across the Triassic–Jurassic Boundary.* Cambridge University Press.

1987a. Progress and competition in macroevolution. *Biological Reviews* 62: 305–338.

1987b. Mass extinctions among families of non-marine tetrapods: the data. *Mémoires de la Société Géologique de France* 150: 21–32.

1991. What really happened in the Late Triassic? *Historical Biology* 5: 263–278.

1993a. Late Triassic terrestrial vertebrate extinctions: stratigraphic aspects and the record of the Germanic Basin. *Memorie della Società Italiana di Scienze Naturali e del Museo Civico di Storia Naturale di Milano.*

1993b. Reptilia. In M. J. Benton (ed.), *The Fossil Record,* Vol. 2. London: Chapman & Hall.

Benton, M. J., and J. M. Clark. 1988. Archosaur phylogeny and the relationships of the Crocodylia. Pp. 289–332 in M. J. Benton (ed.), *The Phylogeny and Classification of the Tetrapods. Vol. 1: Amphibians, Reptiles, Birds.* Oxford: Clarendon Press.

Bice, D. M., C. R. Newton, S. McCauley, P. W. Reiners, and C. A. McRoberts. 1992. Shocked quartz at the Triassic–Jurassic boundary in Italy. *Science* 255: 443–446.

Blendinger, E. 1988. Palynostratigraphy of the late Ladinian and Carnian in the southeastern Dolomites. *Review of Palaeobotany and Palynology* 53: 329–348.

Bonaparte, J. F. 1982. Faunal replacement in the Triassic of South America. *Journal of Vertebrate Paleontology* 21: 362–371.

Brenner, K. 1973. Stratigraphie und Paläogeographie des Oberen Mittelkeupers in Südwest-Deutschland. *Arbeiten aus dem Institut für Geologie und Paläontologie an der Universität Stuttgart* 68: 101–222.

1979. Paläogeographische Raumbilder Südwestdeutschlands für die Ablagerungszeit von Kiesel- und Stubensandstein. *Jahresbericht des oberrheinischen geologischen Vereins, Neue Folge* 61: 331–335.

Brenner, K., and E. Villinger. 1981. Stratigraphie und Nomenklatur des südwestdeutschen Sandsteinkeupers. *Jahreshefte des geologischen Landesamts von Baden-Württemberg* 23: 45–86.

Briggs, D. E. 1991. Extraordinary fossils. *American Scientist* 79: 130–141.

Charig, A. J. 1984. Competition between therapsids and archosaurs during the Triassic period: a review and synthesis of current theories. *Symposia of the Zoological Society of London* 52: 597–628.

Clemens, W. A. 1980. Rhaeto-Liassic mammals from Switzerland and West Germany. *Zitteliana* 5: 51–92.

Cornet, B. 1977. The palynostratigraphy and age of the Newark Supergroup. Ph.D. dissertation, Pennsylvania State University.

1989. Richmond basin lithostratigraphy and paleoenvironments. Pp. 47–53 in P. E. Olsen, R. W. Schlische, and P. J. W. Gore (eds.), *Tectonic, Depositional, and Paleoecological History of Early Mesozoic Rift Basins.* 28th International Geological Congress, Field Trip Guidebook T351. Washington, D. Ç.: American Geophysical Union.

Cornet, B., and P. E. Olsen. 1985. A summary of the biostratigraphy of the Newark Supergroup of eastern North America with comments on early Mesozoic provinciality. Pp. 67–81 in R. Weber (ed.), *Simposio sobre Floras del Triasico Tardio, su Fitogeografia y Paleoecologia.* III Congresso Latinoamericano de Paleontologia, Mexico.

Cowie, J. W., and M. G. Bassett. 1989. *International Union of Geological Sciences, 1989 Global Stratigraphic Chart, Episodes 22, Supplement.*

Cuny, G. 1991. Nouvelles données sur la faune et l'age de Saint Nicolas de Port. *Revue de Paléobiologie* 10: 69–78.

Dockter, J., P. Puff, G. Seidel, and H. Kozur. 1980. Zur Triasgliederung und Symbolgebung in der DDR. *Zeitschrift für Geologische Wissenschaften* 8: 951–963.

Dunay, R. E., and M. J. Fisher. 1978. The Karnian palynofloral succession in the northern Calcareous Alps, Lunz-am-See, Austria. *Pollen et Spores* 20: 177–187.

1979. Palynology of the Dockum Group (Upper Triassic), Texas, U. S. A. *Review of Palaeobotany and Palynology* 28: 61–92.

Estes, R. 1983. *Sauria terrestria, Amphisbaenia. Handbuch der Paläoherpetologie* 10A: 1–249.

Estes, R., K. de Queiroz, and J. Gauthier. 1988. Phylogenetic relationships within Squamata. Pp. 119–281 in R. Estes and G. Pregill (eds.), *Phylogenetic Relationships of the Lizard Families.* Stanford University Press.

Evans, S. E. 1988. The early history and relationships of the Diapsida. Pp. 221–260 in M. J. Benton (ed.), *The Phylogeny and Classification of the Tetrapods. Vol. 1: Amphibians, Reptiles, Birds.* Oxford: Clarendon Press.

Fisher, M. J. 1972. The Triassic palynofloral succession in England. *Geoscience and Man* 4: 101–109.

Fisher, M. J., and J. Bujak. 1975. Upper Triassic palynofloras from Arctic Canada. *Geoscience and Man* 11: 87–94.

Forster, S. C., and G. Warrington. 1985. Geochronology of the Carboniferous, Permian and Triassic. Pp. 99–113 in N. J. Snelling (ed.), *The Chronology of the Geological Record.* London: The Geological Society.

Fraser, N. C. 1986. Terrestrial vertebrates at the Triassic–Jurassic boundary in south west Britain. *Modern Geology* 10: 147–157.

Gaffney, E. S., and P. A. Meylan. 1988. A phylogeny of turtles. Pp. 157–219 in M. J. Benton (ed.), *The Phylogeny and Classification of the Tetrapods. Vol. 1: Amphibians, Reptiles, Birds.* Oxford: Clarendon Press.

Gall, J.-C., M. Durand, and E. Muller. 1977. Le Trias de part et d'autre du Rhin. Corrélations entre les marges et le centre du bassin germanique. *Bulletin du Bureau de Recherches Géologiques et Minières, sér. 2,* 3: 193–204.

Gauthier, J. A. 1986. Saurischian monophyly and the origin of birds. *Memoirs of the California Academy of Sciences* 8: 1–55.

Gauthier, J. A., A. G. Kluge, and T. Rowe. 1988. The early evolution of the Amniota. Pp. 103–155 in M. J. Benton (ed.), *The Phylogeny and Classification of the Tetrapods. Vol. 1: Amphibians, Reptiles, Birds.* Oxford: Clarendon Press.

Geiger, M. E., and C. A. Hopping. 1968. Triassic stratigraphy of the southern North Sea Basin. *Philosophical Transactions of the Royal Society of London* B254: 1–36.

Gwinner, M. P. 1978. *Geologie der Alpen.* Stuttgart: E. Schweizerbart'sche Verlagsbuchhandlung.

1980. Eine einheitliche Gliederung des Keupers (Germanische Trias) in Süddeutschland. *Neues Jahrbuch für Geologie und Paläontologie, Monatshefte* 1980: 229–234.

Hahn, G. G. 1982. Paläomagnetische Untersuchungen im Schilfsandstein (Trias, Km2) Westeuropas. *Geologische Rundschau* 73: 499–516.

Hahn, G., D. Sigogneau-Russell, and G. Wouters. 1989. New data on Theroteinidae. Their relations with Paulchoffatiidae and Haramiyidae. *Geologica et Palaeontologica* 23: 205–215.

Hahn, G., R. Wild, and G. Wouters. 1987. Cynodontier-Zähne aus der Ober-Trias von Gaume (S-Belgien). *Mémoires pour servir à l'Explication des Cartes Géologiques et Minières de la Belgique* 24: 1–33.

Hallam, A. 1986. The Pliensbachian and Tithonian extinction events. *Nature* 319: 765–768.

1990. The end-Triassic mass extinction event. *Geological Society of America Special Paper* 247: 577–583.

1992. Major bio-events in the Triassic and Jurassic. In O. H. Walliser (ed.), *Patterns and Causes of Global Bio-Events in the Phanerozoic.* Berlin: Springer-Verlag.

Harland, W. B., R. L. Armstrong, A. V. Cox, L. E. Craig, A. G. Smith, and D. G. Smith. 1990. *A Geologic Time Scale 1989.* Cambridge University Press.

Hodych, J. P., and G. R. Dunning. 1992. Did the Manicouagan impact trigger end-of-Triassic mass extinction? *Geology* 20: 51–54.

Hopson, J. A., and H. R. Barghusen. 1986. An analysis of therapsid relationships. Pp. 83–106 in N. Hotton III, P. D. MacLean, J. J. Roth, and E. C. Roth (eds.), *The Ecology and Biology of Mammal-like Reptiles.* Washington, D. C.: Smithsonian Institution Press.

Hunt, A. P. 1991. The early diversification pattern of dinosaurs in the Late Triassic. *Modern Geology* 16: 43–60.

Hunt, A. P., and S. G. Lucas. 1989. Late Triassic vertebrate localities in New Mexico. Pp. 72–101 in S. G. Lucas and A. P. Hunt (eds.), *The Dawn of the Age of Dinosaurs in the American Southwest.* Albuquerque: New Mexico Museum of Natural History.

1990. Re-evaluation of "*Typothorax*" *meadei*, a Late Triassic aetosaur from the United States. *Paläontologische Zeitschrift* 64: 317–328.

1991a. The *Paleorhinus* biochron and the correlation of the nonmarine Upper Triassic of Pangaea. *Palaeontology* 34: 487–501.

1991b. A new rhynchosaur from the Upper Triassic of West Texas, and the biochronology of Late Triassic rhynchosaurs. *Palaeontology* 34: 927–938.

1991c. A new aetosaur from the Redonda Formation (Late Triassic: middle Norian) of east-central New Mexico, USA. *Neues Jahrbuch für Geologie und Paläontologie, Monatshefte* 1991: 728–736.

1992. The first occurrence of the aetosaur *Paratypothorax andressi* (Reptilia, Aetosauria) in the western United States and its biochronological significance. *Paläontologische Zeitschrift* 66: 147–157.

Kemp, T. S. 1982. *Mammal-like Reptiles and the Origin of Mammals.* London: Academic Press.

King, G. M. 1988. *Anomodontia. Handbuch der Paläoherpetologie* 17C: 1–174.

Kitching, J. W., and M. A. Raath. 1984. Fossils from the Elliot and Clarens formations (Karoo sequence) of the northeastern Cape, Orange Free State and Lesotho, and a suggested biozonation based on tetrapods. *Palaeontologia Africana* 25: 111–125.

Klaus, W. 1960. Sporen der Karnischen Stufe der ostalpinen Trias. *Jahrbuch der Geologischen Bundesanstalt, Sonderbericht* 5: 107–182.

Kozur, H. 1975. Probleme der Triasgliederung und Parallelisierung der germanischen und tethyalen Trias. Teil II. Anschluss der germanischen Trias an die internationale Triasgliederung. *Freiberger Forschungshefte* C304: 51–77.

Kutty, T. S., and D. P. Sengupta. 1989. The Late Triassic formations of the Pranhita-Godavari Valley and their vertebrate faunal succession – a reappraisal. *Indian Journal of Earth Sciences* 16: 189–206.

Litwin, R. J., A. Traverse, and S. R. Ash. 1991. Preliminary palynological zonation of the Chinle Formation, southwestern U. S. A., and its correlation to the Newark Supergroup (eastern U. S. A.). *Review of Palaeobotany and Palynology* 68: 269–287.

Lucas, S. G. 1990. The rise of the dinosaur dynasty. *New Scientist* 127: 44–46.

1991. Sequence stratigraphic correlation of nonmarine and marine Late Triassic biochronologies, western United States. *Albertiana* 9: 11–18.

Lucas, S. G., and A. P. Hunt (eds.). 1989. *The Dawn of the Age of Dinosaurs in the American Southwest.* Albuquerque: New Mexico Museum of Natural History.

Mädler, K. 1964. Bemerkenswerte Sporenformen aus dem Keuper und Lias. *Fortschritte der Geologie von Rheinland und Westfalen* 12: 169–200.

Marshall, J. E. A., and D. I. Whiteside. 1980. Marine influences in the Triassic "uplands." *Nature* 287: 627–628.

Maxwell, W. D., and M. J. Benton. 1990. Historical tests of the absolute completeness of the fossil record of tetrapods. *Paleobiology* 16: 322–335.

Milner, A. R. 1988. The relationships and origin of living amphibians. Pp. 59–102 in M. J. Benton (ed.), *The Phylogeny and Classification of the Tetrapods. Vol. 1: Amphibians, Reptiles, Birds.* Oxford: Clarendon Press.

1993a. Amphibia. In M. J. Benton (ed.), *The Fossil Record,* Vol. 2. London: Chapman & Hall.

1993b. *Temnospondyli. Handbuch der Paläoherpetologie.*

Molnar, R. E., S. M. Kurzanov, and Z. Dong. 1990. Carnosauria. Pp. 169–209 in D. B. Weishampel, P. Dodson, and H. Osmólska (eds.), *The Dinosauria.* Berkeley: University of California Press.

Morbey, S. J. 1975. The palynostratigraphy of the Rhaetian stage, Upper Triassic in the Kendelbachgraben. *Palaeontographica* B152: 1–75.

Novacek, M. J. 1992. Mammalian phylogeny: shaking the tree. *Nature* 356:121–125.

Olsen, P. E., S. J. Fowell, and B. Cornet. 1990. The Triassic/Jurassic boundary in continental rocks of eastern North America: a progress report. *Geological Society of America Special Paper* 247: 585–593.

Olsen, P. E., and P. M. Galton. 1977. Triassic–Jurassic extinctions: are they real? *Science* 197: 983–986.

1984. A review of the reptile and amphibian assemblages from the Stormberg of southern Africa, with special emphasis on the footprints and the age of the Stormberg. *Palaeontologia Africana* 25: 87–110.

Olsen, P. E., A. R. McCune, and K. S. Thomson. 1982. Correlation of the early Mesozoic Newark Supergroup by vertebrates, principally fishes. *American Journal of Science* 282: 1–44.

Olsen, P. E., R. W. Schlische, and P. J. W. Gore. 1989. *Tectonic, Depositional, and Paleoecological History of Early Mesozoic Rift Basins, Eastern North America.* Washington, D. C.: American Geophysical Union.

Olsen, P. E., N. H. Shubin, and M. H. Anders. 1987. New Early Jurassic tetrapod assemblages constrain Triassic–Jurassic tetrapod extinction event. *Science* 237: 1025–1029.

1988. Triassic–Jurassic extinctions [reply to comment by Padian]. *Science* 241: 1359–1360.

Olsen, P. E., and H.-D. Sues. 1986. Correlation of continental Late Triassic and Early Jurassic sediments, and patterns of the Triassic–Jurassic tetrapod transition. Pp. 321–351 in K. Padian (ed.), *The Beginning of the Age of Dinosaurs: Faunal Change across the Triassic–Jurassic Boundary.* Cambridge University Press.

Olson, E. C. 1982. Extinctions of Permian and Triassic non-marine vertebrates. *Geological Society of America Special Paper* 190: 501–511.

Padian, K., and W. A. Clemens. 1985. Terrestrial vertebrate diversity: episodes and insights. Pp. 41–96 in J. W. Valentine (ed.), *Phanerozoic Diversity Patterns: Profiles in Macroevolution.* Princeton University Press.

Patterson, C., and A. B. Smith. 1987. Is periodicity of mass extinctions a taxonomic artefact? *Nature* 330: 248–252.

Romer, A. S. 1966. *Vertebrate Paleontology*, 3rd ed. University of Chicago Press.

1970. The Triassic faunal succession and the Gondwanaland problem. Pp. 375–400 in *Gondwana Stratigraphy.* IUGS Symposium Buenos Aires 1967. Paris: UNESCO.

Scheuring, B. W. 1970. Palynologische und palynostratigraphische Untersuchungen des Keupers im Bolchentunnel (Solothurner Jura). *Schweizerische Paläontologische Abhandlungen* 88: 1–119.

Schröder, B. 1982. Entwicklung des Sedimentbeckens und Stratigraphie der klassischen Germanischen Trias. *Geologische Rundschau* 71: 783–794.

Schulz, E. 1967. Sporenpaläontologische Untersuchungen rätoliassischer Schichten im Zentralteil des Germanischen Beckens. *Paläontologische Abhandlungen* B2: 544–633.

Schuurman, W. M. L. 1979. Aspects of Late Triassic palynology. 3. Palynology of latest Triassic and earliest Jurassic deposits of the Northern Limestone Alps in Austria and southern Germany, with special reference to a palynological characterization of the Rhaetian Stage in Europe. *Review of Palaeobotany and Palynology* 27: 53–75.

Sepkoski, J. J., Jr. 1986. Phanerozoic overview of mass extinctions. Pp. 277–296 in D. M. Raup and D. Jablonski (eds.), *Patterns and Process in the History of Life.* Berlin: Springer-Verlag.

1989. Periodicity in extinction and the problem of catastrophism in the history of life. *Journal of the Geological Society, London* 146: 7–19.

1990. The taxonomic structure of periodic extinction. *Geological Society of America Special Paper* 247: 33–44.

Sepkoski, J. J., Jr., R. K. Bambach, D. M. Raup, and J. W. Valentine. 1981. Phanerozoic marine diversity and the fossil record. *Nature* 293: 435–437.

Sepkoski, J. J., Jr., and D. M. Raup. 1986. Periodicity in marine extinction events. Pp. 3–36 in D. K. Elliott (ed.), *Dynamics of Extinction.* New York: Wiley.

Shishkin, M. A. 1991. [A labyrinthodont from the Jurassic of Mongolia.] *Paleontologischeskiy Zhurnal* 1991(1): 81–95. [in Russian]

Shubin, N. H., A. W. Crompton, H.-D. Sues, and P. E. Olsen. 1991. New fossil evidence on the sister-group of mammals and early Mesozoic faunal distributions. *Science* 251: 1063–1065.

Simms, M. J. 1990a. Triassic palaeokarst in Britain. *Cave Science* 17: 93–101.

1990b. Crinoid diversity and the Triassic/Jurassic boundary. *Cahiers de l'Université Catholique de Lyon, Séries Scientifique* 3: 67–77.

Simms, M. J., and A. H. Ruffell. 1989. Synchroneity of climatic change and extinctions in the late Triassic. *Geology* 17: 265–268.

1990. Climatic and biotic change in the late Triassic. *Journal of the Geological Society, London* 147: 321–327.

Smith, A. B. 1990. Echinoid evolution from the Triassic to Lower Liassic. *Cahiers de l'Université Catholique de Lyon, Séries Scientifique* 3: 79–117.

Stanley, G. D., Jr. 1988. The history of early Mesozoic reef communities: a three-step process. *Palaios* 3: 170–183.

Stanley, S. M. 1979. *Macroevolution.* San Francisco: Freeman.

Strauss, D., and P. M. Sadler. 1989. Classical confidence intervals and bayesian probability estimates for ends of local taxon ranges. *Mathematical Geology* 21: 411–427.

Stucky, R. K., and M. C. McKenna. 1993. Mammalia. In M. J. Benton (ed.), *The Fossil Record*, Vol. 2. London: Chapman & Hall.

Sun, A. L., and K. H. Cui. 1986. A brief introduction to the Lower Lufeng saurischian fauna (Lower Jurassic: Lufeng, Yunnan, People's Republic of China). Pp. 275–278 in K. Padian (ed.), *The Beginning of the Age of Dinosaurs: Faunal Change across the Triassic–Jurassic Boundary.* Cambridge University Press.

Tozer, E. T. 1974. Definitions and limits of Triassic stages and substages: suggestions prompted by comparisons between North America and the Alpine-Mediterranean region. Pp. 195–206 in H. Zapfe (ed.), *Die Stratigraphie der alpin-mediterranen Trias.* Wien: Springer-Verlag.

1979. Latest Triassic ammonoid faunas and biochronology, western Canada. *Geological Survey of Canada Paper* 79–1B: 127–135.

Unwin, D. M. 1991. The morphology, systematics, and evolutionary history of pterosaurs from the Cretaceous Cambridge Greensand of England. Ph.D. thesis, University of Reading.

Van der Eem, J. G. L. A. 1983. Aspects of Middle and Late Triassic palynology. 6. Palynological investigations in the Ladinian and Lower Karnian of the western Dolomites. *Review of Palaeobotany and Palynology* 39: 189–300.

Visscher, H., and W. A. Brugman. 1981. Ranges of selected palynomorphs in the Alpine Triassic of Europe. *Review of Palaeobotany and Palynology* 34: 115–128.

Visscher, H., W. M. L. Schuurman, and A. W. Van Erve. 1980. Aspects of a palynological characterisation of Late Triassic and Early Jurassic "standard" units of chronostratigraphical classification in Europe. *Proceedings of the Fourth International Palynological Conference, Lucknow (1976–77)* 2: 281–287.

Ward, P. D. 1990. The Cretaceous/Tertiary extinctions in the marine realm; a 1990 perspective. *Geological Society of America Special Paper* 247: 425–432.

Weishampel, D. B. 1990. Dinosaurian distribution. Pp. 63–139 in D. B. Weishampel, P. Dodson, and H. Osmólska (eds.), *The Dinosauria.* Berkeley: University of California Press.

Weishampel, D. B., P. Dodson, and H. Osmólska (eds.) 1990. *The Dinosauria.* Berkeley: University of California Press.

Weiss, M. 1989. Die Sporenfloren aus Rät und Jura Südwest-Deutschlands und ihre Beziehung zur Ammoniten-Stratigraphie. *Palaeontographica* B215: 1–168.

Wellnhofer, P. 1978. *Pterosauria. Handbuch der Paläoherpetologie* 19: 1–82.

Whiteside, D. I. 1986. The head skeleton of the Rhaetian sphenodontid *Diphydontosaurus avonis* gen. et sp. nov. and the modernizing of a living fossil. *Philosophical Transactions of the Royal Society of London* B312: 379–430.

Wild, R. 1989. *Aetosaurus* (Reptilia: Thecodontia) from the Upper Triassic (Norian) of Cene near Bergamo, Italy, with a revision of the genus. *Rivista del Museo Civico di Scienze Naturali "Enrico Caffi," Bergamo* 14: 1–24.

Witte, W. K., D. V. Kent, and P. E. Olsen. 1991. Magnetostratigraphy and paleomagnetic poles from Late Triassic–earliest Jurassic strata of the Newark basin. *Geological Society of America Bulletin* 103: 1648–1662.

Zapfe, H. 1974. Trias in Oesterreich. Pp. 245–251 in H. Zapfe (ed.), *Die Stratigraphie der alpin-mediterranen Trias.* Wien: Springer-Verlag.

Zawiskie, J. M. 1986. Terrestrial vertebrate faunal succession during the Triassic. Pp. 353–362 in K. Padian (ed.), *The Beginning of the Age of Dinosaurs: Faunal Change across the Triassic–Jurassic Boundary.* Cambridge University Press.

Zhen, S., B. Zhen, N. J. Mateer, and S. G. Lucas. 1985. The Mesozoic reptiles of China. *Bulletin of the Geological Institute of the University of Uppsala* 11: 133–150.

Appendix 22.1

Assignments of terrestrial tetrapod-bearing formations to stages and "substages" of the Middle and Late Triassic and the earliest Jurassic, based on data from Anderson and Cruickshank (1978), Benton (1983,1993a), Olsen and Galton (1984), Kutty and Sengupta (1989), Olsen et al. (1989), Weishampel (1990), and Hunt and Lucas (1990, 1991a,b,c, 1992). The dating of the Complex-A fissure fills from southwest Britain is problematic; they are assigned here a conservative late Norian (Sevatian) age.

Ladinian

Lettenkeuper, SW Germany
Grenzdolomit, SE Germany
Oberer Muschelkalk, Germany
Grenzbitumenzone, Switzerland (ANS/LAD)
Tschermakfjellet Member, Spitsbergen
Sol'lletsk Series (Zone VII), Russian Platform
Ischichuca (Chañares) Formation, Argentina
Bhimaram Sandstone, India
?Karamay Formation, Junggar, Xinjiang, China
?Batung Formation, Hunan, China

Early Carnian (Cordevolian)

Turkey Branch Formation, Virginia, USA

Middle Carnian (Julian)

Unterer Schilfsandstein, Germany
Hosselkus Limestone, California, USA
Pekin Formation, North Carolina, USA
Cumnock Formation, North Carolina, USA

Late Carnian 1 (early Tuvalian)

Oberer Schilfsandstein, Germany
Blasensandstein (lower part), SW Germany
Opponitzer Schichten (lower part), Austria
Cow Branch Formation, Virginia, USA
Wolfville Formation, Nova Scotia, Canada
Popo Agie Formation, Wyoming, USA
Chinle Formation (Shinarump Member), Arizona, USA
Santa Rosa Formation, New Mexico, USA
Tecovas Formation (Camp Springs Member), Texas, USA
Lower Dockum Group, Texas, USA
Santa Maria Formation (*Dinodontosaurus* Assemblage Zone), Brazil
Argana Formation, Morocco
Maleri Formation (lower fauna), India
Tiki Formation, India

Late Carnian 2 (late Tuvalian)

Blasensandstein (upper part), Germany
Kieselsandstein, SW Germany
Rote Wand, Lehrbergschichten, untere Bunte Mergel, Germany
Lossiemouth Sandstone Formation, Scotland, UK
New Oxford Formation, Pennsylvania, USA
Lockatong Formation, Pennsylvania, USA
Chinle Formation (Petrified Forest Member, lower part, and Moss Back Member), Arizona, USA
Garita Creek Formation, New Mexico, USA
Tecovas Formation (post-Camp Springs Member), Texas, USA
Santa Maria Formation (*Scaphonyx* Assemblage Zone), Brazil
Caturrita Formation, Brazil
Ischigualasto Formation, Argentina
Maleri Formation (upper fauna), India

Early Norian (early Lacian)

Unterer Stubensandstein, SW Germany
Unterer Burgsandstein, Germany
Unterer Dolomitmergelkeuper, E Germany
Passaic Formation (lower part), Pennsylvania and New Jersey, USA
Chinle Formation (Petrified Forest Member, upper part, above the Sonsela Sandstone), Arizona, USA
Trujillo Formation, Texas and New Mexico, USA
Bull Canyon Formation, New Mexico, USA
Cooper Formation, West Texas, USA
Lower Elliott Formation, Lesotho, South Africa
Mpandi Formation, Zimbabwe
Bushveld Sandstone Formation (Springbok Flats Member), South Africa
Dharmaram Formation (lower fauna), India

Middle Norian 1 (late Lacian)

Mittlerer Stubensandstein, SW Germany
Mittlerer Burgsandstein, SE Germany
Mittlerer Dolomitmergelkeuper, E Germany
Passaic Formation (lower and middle parts), Pennsylvania and New Jersey, USA
New Haven Arkose (?lower and middle parts), Connecticut, USA
Chinle Formation (Owl Rock Member), Arizona, Utah, and New Mexico, USA
Redonda Formation (lower part), New Mexico, USA
Sloan Canyon Formation (lower part), New Mexico, USA

Middle Norian 2 (Alaunian)

Oberer Stubensandstein, SW Germany
Oberer Burgsandstein, SE Germany
Oberer Dolomitmergelkeuper, E Germany
Calcare di Zorzino, N. Italy
Dolomia di Forni, N. Italy
Passaic Formation (middle part), Pennsylvania and New Jersey, USA
New Haven Arkose (middle part), Connecticut, USA
Chinle Formation (Rock Point Member), New Mexico, USA
Chinle Formation (Church Rock Member), Utah, USA

Redonda Formation (middle part), New Mexico, USA
Sloan Canyon Formation (middle part), New Mexico, USA

Late Norian 1 (early Sevatian)

Fissure fills, SW England, South Wales, UK (Complex A) (late NOR–RHT?)
Knollenmergel, Germany, Switzerland
Lehrbergstufe, SW Germany
Feuerletten, SE Germany
Obere Bunte Mergel, Switzerland
Argillite di Riva di Solto, Italy
Magnesian Conglomerate, England, UK
Grès à *Avicula contorta*, France
Marnes irisées supérieures, France (= Steinmergelkeuper, Germany)
Passaic Formation (upper part), Pennsylvania and New Jersey, USA
New Haven Arkose (middle and upper parts), Connecticut, USA
Redonda Formation (upper part), New Mexico, USA
Sloan Canyon Formation (upper part), New Mexico, USA
Dharmaram Formation (upper fauna), India

Late Norian 2 (late Sevatian; Rhaetian)

Rhaetic, England, Wales, UK
Fissure fills, SW England, South Wales, UK (Complex A) (late NOR–RHT?)
Westbury Formation, England, UK
Rhätsandstein, Germany, Switzerland
Rhétien, France
Saint-Nicolas-de-Port, France
Rhaetic, Scania, Sweden
Passaic Formation (upper part), Pennsylvania and New Jersey, USA
New Haven Arkose (upper part), Connecticut, USA
Wingate Formation, Arizona and Utah, USA
Los Colorados Formation (La Esquina local fauna), Argentina
Quebrada del Barro Formation, Argentina
El Tranquilo Formation, Argentina
Dull Purplish Beds, Lower Lufeng Formation, Yunnan, China

Hettangian

McCoy Brook Formation, Nova Scotia, Canada
Wingate Formation, Arizona, USA
Vulcanodon Beds, Zimbabwe

Hettangian/Sinemurian

"Fissure complex B," South Wales, UK
Lower Portland Formation, Connecticut, USA
Upper Elliott Formation, Lesotho, South Africa
Forest Sandstone, Zimbabwe
Dark Red Beds, Lower Lufeng Series, Yunnan, China

Sinemurian

Lower Lias, Dorset, Warwickshire, Leicestershire, England, UK

Unnamed unit, Beiro Litoral, Portugal
Bushveld Sandstone Formation (Zoutpansberg Member),
    South Africa
Clarens Formation, South Africa

Sinemurian/Pliensbachian

Moenave Formation, Arizona, USA
Kayenta Formation, Arizona, USA

Pliensbachian/Toarcian

Navajo Sandstone, Arizona, USA
Portland Formation (upper part), Connecticut, USA

Toarcian

Lias epsilon, Germany
Posidonienschiefer, SW Germany
Kota Formation, India

Toarcian/Bajocian

Zhenzhunchong Formation, Sichuan, China

Aalenian

Northampton Sands Formation, Northamptonshire, England

Aalenian/Bajocian

Inferior Oolite, Northamptonshire, Gloucestershire, Dorset,
    and Wiltshire, England

Aalenian–Callovian

Dapuka Group, Xizang Zizhqu, China

Bajocian

Inferior Oolite, Yorkshire and Oxfordshire, England
Cerro Carnerero Formation, Chubut, Argentina
Injune Creek Beds, Queensland, Australia

Bathonian

Sharp's Hill Formation, Oxfordshire, England (early Bathonian)
Chipping Norton Formation, Gloucestershire and Oxfordshire,
    England (early Bathonian)
Stonesfield Slate, Oxfordshire, England (middle Bathonian)
Great Oolite, Nottinghamshire, Northamptonshire,
    Buckinghamshire, and Wiltshire, England (late
    Bathonian)
Forest Marble, Northamptonshire, Gloucestershire,
    Oxfordshire, Wiltshire, and Dorset, England (late
    Bathonian)
Cornbrash Formation, Oxfordshire, England (late Bathonian)
Calcaire de Caen, Normandy, France (early Bathonian)
Guettioua Sandstone, Morocco
Isalo Formation, Madagascar

Bathonian/Callovian

Xiashaximiao Formation, Sichuan, China
Kuangyuan Series, Sichuan, China

Callovian

Lower Oxford Clay, Northamptonshire and Dorset, England
    (middle Callovian)
Oxford Clay, Cambridgeshire and Oxfordshire, England
    (middle–late Callovian)
Middle Oxford Clay, Buckinghamshire, England (late
    Callovian)
Marnes d'Argences, Calvados, France (middle Callovian)
Marnes de Dives, Calvados, France (late Callovian)
Cañadon Asfalto Formation, Chubut, Argentina

Appendix 22.2

Documentation of firsts and lasts for all families of terrestrial tetrapods recorded during the Middle Triassic–Middle Jurassic interval. The amphibian data are authored by Andrew R. Milner; for greater detail, see Milner (1993a). The reptile data are based on Benton (1993b), and the mammalian data on Stucky and McKenna (1993). For further details, the reader should consult these compilations. Paraphyletic taxa are indicated by (p).

Amphibia

Temnospondyli Zittel, 1888

Family Brachyopidae Lydekker, 1885
P.(KAZ/TAT)–J. (CLV)

    First: *Bothriceps major* Woodward, 1909, Lithgow Coal Measures, Airly, New South Wales, Australia.
    Last: *Ferganobatrachus riabinini* Nessov, 1990, Balabansay Formation, Kirghizia.
    Comment: *Ferganobatrachus* was described as a "capitosauroid," but the holotype clavicle appears to be brachyopid (Shishkin, 1991). A brachyopid, *Gobiops desertus*, has been described from the Upper Jurassic (stage uncertain) of Shara Teg, Mongolia (Shishkin, 1991).

Family Capitosauridae Watson, 1919
Tr.(GRI/DIE–NOR)

    First: *Parotosuchus rewanensis* Warren, 1980, *P. gunganj* Warren, 1980, and *P. aliciae* Warren and Hutchinson, 1988, Arcadia Formation, Queensland, Australia; and *P. madagascariensis* (Lehman, 1961), Sakamena Formation, Madagascar.
    Last: *Cyclotosaurus carinidens* (Jaekel, 1914), Knollenmergel, Halberstadt, Germany.

Family Mastodonsauridae Lydekker, 1885
Tr.(SPA/ANS–CRN)

    First: *Mastodonsaurus cappelensis* Wepfer, 1923, oberer Buntsandstein, Kappel, Baden-Württemberg, Germany.

Last: *Mastodonsaurus keuperinus* Fraas, 1889, Schilfsandstein, Stuttgart, Baden-Württemberg, Germany.

Family Trematosauridae Watson, 1919
Tr.(GRI–CRN)

First: *Gonioglyptus longirostris* Huxley, 1865, *Glyptognathus fragilis* Lydekker, 1882, and *Panchetosaurus panchetensis* Tripathi, 1969, Panchet Formation, Bengal, India.

Last: *Hyperokynodon keuperinus* Plieninger, 1852, Schilfsandstein, Baden-Württemberg, Germany.

Family Latiscopidae Wilson, 1948
Tr.(CRN–NOR)

First: *Almasaurus habbazi* Dutuit, 1972, Argana Formation, Argana Valley, Morocco.

Last: *Latiscopus disjunctus* Wilson, 1948, Cooper Formation, upper Dockum Group, Texas, USA.

Family Metoposauridae Watson, 1919
Tr.(LAD–NOR)

First: *Trigonosternum latum* Schmidt, 1931, Lettenkeuper, Germany, and an undescribed skull, Baden-Württemberg, Germany.

Last: "new, small metoposaurid," upper Redonda Formation, New Mexico, USA (Hunt and Lucas, 1989).

Comment: Slightly older material includes unnamed metoposaurids from lower in the Redonda Formation and from the Sloan Canyon Formation of New Mexico, USA (Hunt and Lucas, 1989), as well as species of *Metoposaurus*, *Anaschisma*, and *Kalamoiketer*, upper Petrified Forest Member, Arizona, Bull Canyon Formation, New Mexico, and Cooper Formation, Texas, USA (all early Norian). The youngest European material is *Metoposaurus stuttgartensis* Fraas, 1913, Lehrbergstufe, late Carnian, Baden-Württemberg, Germany.

Family Chigutisauridae Rusconi, 1951
Tr.(GRI/DIE)–K.(BER/ALB)

First: *Keratobrachyops australis* Warren, 1981, Arcadia Formation, Queensland, Australia.

Last: Unnamed material, Strzelecki Formation, Victoria, Australia.

Family Plagiosauridae Jaekel, 1914
Tr.(GRI/DIE–RHT)

First: *Plagiobatrachus australis* Warren, 1985, Arcadia Formation, Queensland, Australia.

Last: *Gerrothorax rhaeticus* Nilsson, 1934, Rhaetic, Scania, Sweden.

Lissamphibia Haeckel, 1866

Family unnamed J.(SIN/PLB)

First and last: Undescribed gymnophionan, Kayenta Formation, Arizona, USA.

Family Unnamed J.(HET/TOA)

First and last: *Vieraella herbstii* Reig, 1961, Roca Blanca Formation, Santa Cruz Province, Argentina.

Comment: *Vieraella* has been placed in the Leiopelmatidae (= Ascaphidae), but it is more likely a stem-anuran with no immediate relationship to any living family (Milner, Chapter 1).

Family Discoglossidae Guenther, 1859
J.(BTH)–Rec.

First: *Eodiscoglossus oxoniensis* Evans, Milner, and Mussett, 1990, Forest Marble Formation, Oxfordshire, England.

Family Unnamed J.(BTH)

First and last: *Marmorerpeton kermacki* and *M. freemani* Evans, Milner, and Mussett, 1988, Forest Marble Formation, Oxfordshire, England.

Family Karauridae Ivakhnenko, 1978 J.(BTH–KIM)

First: *Kokartus honorarius* Nessov, 1988, black and red shales, Kizylsu River, Kirghizia.

Last: *Karaurus sharovi* Ivakhnenko, 1978, Karabastau Formation, Kazakhstan.

Family Albanerpetontidae Fox and Naylor, 1982
J.(BAJ)–T.(BUR/LAN)

First: Atlas centrum referred to *Albanerpeton megacephalus*, Aveyron, France.

Last: *Albanerpeton inexpectatum* Estes and Hoffstetter, 1976, Miocene fissures, La Grive St. Alban, France.

Reptilia Laurenti, 1768

Family Procolophonidae Cope, 1889
P.(KAZ)–Tr.(NOR)

First: *Owenetta rubidgei* Broom, 1939, *Aulacephalodon-Cistecephalus* Assemblage Zone, South Africa.

Last: *Hypsognathus fenneri* Gilmore, 1928, upper Passaic Formation, New Jersey and Pennsylvania, USA.

Comment: *Sphodrosaurus pennsylvanicus* Colbert, 1960, Hammer Mill Formation, Pennsylvania, USA, seems to be a diapsid (H.-D. Sues and D. Baird, pers. commun.), while the Rhaetian or latest Norian "procolophonoid" described by Cuny (1991) from the Saint-Nicolas-de-Port locality in France is incorrectly identified (P. S. Spencer, pers. commun., 1992).

Testudines Batsch, 1788

Family Proganochelyidae Baur, 1888
Tr.(NOR)–J.(HET)

First: *Proganochelys quenstedtii* Baur, 1887, mittlerer and oberer Stubensandstein, Germany.

Last: Unnamed proganochelyid, upper Elliot Formation (Red Beds), Orange Free State, South Africa.

Comment: The age of *P. ruchae* is assumed to be equivalent to the German formations, but that is not certain.

Family Proterochersidae Nopcsa, 1928 Tr.(NOR)

First and last: *Proterochersis robusta* E. Fraas, 1913, unterer Stubensandstein, Baden-Württemberg, Germany.

Family Kayentachelyidae Gaffney, Hutchison, Jenkins, and Meeker, 1987 J.(SIN/PLB)

First and last: *Kayentachelys aprix* Gaffney, Hutchison, Jenkins, and Meeker, 1987, Kayenta Formation, Arizona, USA.

Diapsida Osborn, 1903

Diapsida incertae sedis

Family Endennasauridae Carroll, 1987 Tr.(NOR)

First and last: *Endennasaurus acutirostris* Renesto, 1984, Calcare di Zorzino, Bergamo, Italy.

Family Drepanosauridae Carroll, 1987 Tr.(NOR)

First and last: *Drepanosaurus unguicaudatus* Pinna, 1980, Calcare di Zorzino, Bergamo, Italy.

Lepidosauromorpha Benton, 1983

Family Kuehneosauridae Romer, 1966 Tr.(CRN–RHT)

First: *Icarosaurus siefkeri* Colbert, 1966, Lockatong Formation, New Jersey, USA; and "?kuehneosaur jaw fragments," lower unit of Petrified Forest Member, Chinle Formation, Arizona, USA.

Last: *Kuehneosaurus latus* Robinson, 1962, Pant-y-ffynon Quarry, Glamorgan, Wales.

Comment: Pant-y-ffynon Quarry is dated as Rhaetian. The type material of *K. latus* comes from Emborough Quarry, Somerset, England, whose age is probably Norian, but this is not certain. Later supposed kuehneosaurs, or close relatives, such as *Cteniogenys antiquus* Gilmore, 1928 from the Upper Jurassic and *Litakis gilmorei* Estes, 1964 from the Upper Cretaceous are very doubtful. *Cteniogenys* has been reclassified as a choristodere.

Family Sphenodontidae Cope 1870 (p) Tr.(CRN)–Rec.

First: "sphenodontian," Turkey Branch Formation, Virginia, USA; *Brachyrhinodon taylori* Huene, 1912, Lossiemouth Sandstone Formation, Elgin, Scotland. Extant.

Comment: Other "late" late Carnian sphenodontids have been reported from Arizona, New Mexico, and Texas. Older supposed sphenodontids, such as *Palacrodon* from the Early Triassic of South Africa, and *Anisodontosaurus* from the Middle Triassic of Arizona, may be procolophonids.

Elachistosuchus is an archosauromorph. The family Sphenodontidae, as presented here, is paraphyletic because of the exclusion of the Pleurosauridae. *Sapheosaurus* and *Gephyrosaurus* are included here within the Sphenodontidae and are not given separate families.

Archosauromorpha Huene, 1946

Choristodera Cope, 1876

Family Pachystropheidae Kuhn, 1961 Tr.(RHT)

First and last: *Pachystropheus rhaeticus* E. von Huene, 1935, Rhaetic bonebed, southern England, Germany.

Family Unnamed J.(BTH–KIM)

First: *Cteniogenys antiquus* Gilmore, 1928, Chipping Norton Formation, lower Bathonian, Gloucestershire, England; Forest Marble Formation, upper Bathonian, Oxfordshire, England.

Last: *Cteniogenys antiquus* Gilmore, 1928, Morrison Formation, Wyoming, USA.

Comment: The familial assignment of these early choristoderes has not been confirmed, and relationships to the pachystropheids and to later champsosaurs are unclear at present. Earlier choristoderes, perhaps belonging to this group, have been noted from the Kayenta Formation (J. Clark, pers. commun., 1991).

Rhynchosauria Osborn, 1903

Family Rhynchosauridae Huxley, 1887 (Cope, 1870) Tr.(SCY–CRN)

First: *Howesia browni* Broom, 1905, and *Mesosuchus browni* Watson, 1912, Cynognathus-Diademodon Assemblage Zone, Karoo Basin, South Africa.

Last: *Hyperodapedon gordoni* Huxley, 1859, Lossiemouth Sandstone Formation, Scotland; *Scaphonyx sanjuanensis* Sill, 1970, Ischigualasto Formation, San Juan, Argentina; *Scaphonyx sulcognathus* Azevedo and Schultz, 1988, Caturrita Formation, Rio Grande do Sul, Brazil; *Otischalkia elderae* Hunt and Lucas, 1991, lower Dockum Group, Texas, USA; undescribed rhynchosaur, Wolfville Formation, Nova Scotia, Canada.

Comment: *Noteosuchus colletti* (Watson, 1912) from the *Lystrosaurus-Procolophon* Assemblage Zone of South Africa has been called the oldest rhynchosaur, but it lacks diagnostic characters of the group. Other late Carnian rhynchosaurs are known, but these are dated as "early" late Carnian by Hunt and Lucas (1991b), while the "Lasts" listed earlier are given as "late" late Carnian.

Family Trilophosauridae Gregory, 1945 Tr.(CRN–RHT)

First: *Trilophosaurus buettneri* Case, 1928, lower Dockum Group, Crosby County, Texas, USA.

Last: *Tricuspisaurus thomasi* Robinson, 1957, Late Triassic (Norian?), Ruthin Quarry fissure, Glamorgan, Wales.

Comment: Earlier supposed trilophosaurids, such as the Triassic taxa *Doniceps* and *Anisodontosaurus*, as well as

Toxolophosaurus from the Lower Cretaceous, probably are not trilophosaurids. It is unclear whether or not *Tricuspisaurus* and *Variodens*, both from the English-Welsh fissures, are trilophosaurids.

Family Unnamed Tr.(NOR)

First and last: *Megalancosaurus preonensis* Calzavara, Muscio, and Wild, 1980, Dolomia di Forni, Udine, Italy.

Prolacertiformes Camp, 1945

Family Prolacertidae Parrington, 1935 Tr.(SCY–CRN)

First: *Prolacerta broomi* Parrington, 1935, *Lystrosaurus-Procolophon* Assemblage Zone, Karoo Basin, South Africa, and Fremouw Formation, Antarctica.

Last: *Malerisaurus robinsonae* Chatterjee, 1980, Maleri Formation, Andhra Pradesh, India; and *M. langstoni* Chatterjee, 1986, Tecovas Formation, lower Dockum Group, Howard County, Texas, USA.

Family Tanystropheidae Romer, 1945 Tr.(ANS–NOR)

First: *"Tanystropheus" conspicuus* Huene, 1931, oberer Buntsandstein, southern Germany.

Last: *Tanystropheus fossai* Wild, 1980, Argillite di Riva di Solto, Val Brembana, Itlay.

Archosauria Cope, 1869

Family Erythrosuchidae Watson 1917 Tr.(SCY–LAD)

First: *Fugusuchus hejiapensis* Cheng, 1980, He Shanggou Formation, Shanxi Province, North China, and *Garjainia prima* Ochev, 1958, Yarenskian Horizon, upper part of Zone V, Orenburg region, Russia, both middle to late Scythian.

Last: *Cuyosuchus huenei* Reig, 1961, Cacheuta Formation, Mendoza Province, Argentina.

Family Ctenosauriscidae Kuhn, 1964 Tr.(SCY–ANS/LAD)

First: *Ctenosauriscus koeneni* (Huene, 1902), mittlerer Buntsandstein, Germany.

Last: *Lotosaurus adentus* Zhang, 1975, Batung Formation, Hunan, China.

Comment: These two taxa of archosaurs share long dorsal neural spines, but their systematic position is uncertain. It is not clear whether they are related to each other or not.

Family Proterochampsidae Sill, 1967 Tr.(LAD–CRN)

First: *Chanaresuchus bonapartei* Romer, 1971, and *Gualosuchus reigi* Romer, 1971, Chañares Formation, La Rioja Province, Argentina.

Last: *Proterochampsa barrionuevoi* Reig, 1959, Ischigualasto Formation, San Juan Province, Argentina.

Family Phytosauridae Lydekker, 1888 Tr.(CRN–RHT)

First: *"Rutiodon sp.,"* Pekin Formation, middle Carnian, North Carolina, USA (Olsen et al., 1989).

Last: *Rutiodon* sp., Rhät, Switzerland, North Germany; "phytosaurs," upper Passaic Formation, New Jersey, upper New Haven Arkose, Connecticut, USA.

Comment: Apparently older phytosaurs, *Mesorhinosuchus fraasi* (Jaekel, 1910) from the mittlerer Buntsandstein (Scythian) of Bernburg, Germany, and others from the Muschelkalk of Germany (Anisian–Ladinian) are all doubtful records. There are numerous late Carnian phytosaurs, *Paleorhinus bransoni* Williston, 1904, Popo Agie Formation, Fremont County, Wyoming, USA, and other species of *Paleorhinus* from Arizona and Texas, USA, Morocco, West Germany, Austria, and India (Hunt and Lucas, 1991a).

Family Stagonolepididae Lydekker, 1887 Tr.(CRN–RHT)

First: *Longosuchus*, Pekin Formation, North Carolina, USA (Olsen et al. 1989); *Longosuchus meadei* (Sawin, 1947), lower Dockum Group, Howard County, Texas; Salitral Member, Chinle Formation, Rio Arriba County, New Mexico, USA (Hunt and Lucas, 1990).

Last: *Neoaetosauroides engaeus* Bonaparte, 1969, upper Los Colorados Formation, La Rioja, Argentina; aetosaur elements, Penarth Group ("Rhaetian"), SW England.

Comment: There are numerous late Carnian stagonolepidids: *Stagonolepis robertsoni* Agassiz, 1844, Lossiemouth Sandstone Formation, Scotland; *Aetosauroides scagliai* Casamiquela, 1960, and *Argentinosuchus bonapartei* Casamiquela, 1960, Ischigualasto Formation, San Juan, Argentina; *Desmatosuchus haplocerus* (Cope, 1892), lower units of the Chinle Formation and Dockum Group, New Mexico and Texas, USA; unnamed stagonolepidid, Wolfville Formation, Nova Scotia, Canada.

Family Rauisuchidae Huene, 1942 Tr.(ANS–RHT)

First: *Wangisuchus tzeyii* Young, 1964, and *Fenhosuchus cristatus* Young, 1964, Er-Ma-Ying Series, Shansi, China; *Vjushkovisaurus berdjanensis* Ochev, 1982, Donguz Series, Orenburg region, Russia; *Stagonosuchus major* (Haughton, 1932) and *"Mandasuchus,"* upper bonebed of the Manda Formation, Ruhuhu region, Tanzania; "rauisuchid," Yerrapalli Formation, India.

Last: *Fasolasuchus tenax* Bonaparte, 1978, upper Los Colorados Formation, La Rioja, Argentina.

Family Poposauridae Nopcsa, 1928 Tr.(ANS–NOR)

First: *Bromsgroveia walkeri* Galton, 1985, Bromsgrove Sandstone Formation, Warwick, England.

Last: Poposaurid, upper Redonda Formation, New Mexico, USA.

Comment: If the "last" record is not confirmed, there are several early and middle Norian poposaurids: *Teratosaurus suevicus* Meyer, 1861, mittlerer Stubensandstein, Baden-Württemberg, Germany; *Postosuchus kirkpatricki* Chatterjee, 1985, upper Dockum Group, Texas, USA.

Family Ornithosuchidae Huene, 1908 Tr.(CRN–RHT)

First: *Ornithosuchus longidens* Newton, 1894, Lossiemouth Sandstone Formation, Scotland, and *Venaticosuchus rusconii* Bonaparte, 1971, Ischigualasto Formation, La Rioja, Argentina.

Last: *Riojasuchus tenuiceps* Bonaparte, 1969, upper Los Colorados Formation, La Rioja, Argentina.

Crocodylomorpha Walker, 1968

Family Saltoposuchidae Crush, 1984 Tr.(NOR–RHT)

First: *Saltoposuchus connectens* Huene, 1921, mittlerer Stubensandstein, Wüttemberg, Germany.

Last: *Terrestrisuchus gracilis* Crush, 1984, Ruthin Quarry, Glamorgan, Wales.

Family Sphenosuchidae Huene, 1922 Tr.(CRN)–J.(SIN/PLB) (p)

First: *Hesperosuchus agilis* Colbert, 1952, lower Petrified Forest Member, Chinle Formation, Arizona, USA.

Last: Unnamed form, Kayenta Formation, Arizona, USA (Sues et al., Chapter 16).

Comment: *Hallopus victor* (Marsh, 1877) is a crocodylomorph that may belong to this clade (Clark, in Benton and Clark, 1988). It is probably from the lower Ralston Creek Formation (Callovian) of Freemont County, Colorado.

Family Protosuchidae Brown, 1934 Tr.(RHT)–J.(PLB/TOA)

First: *Hemiprotosuchus leali* Bonaparte, 1969, upper Los Colorados Formation, La Rioja, Argentina.

Last: Unnamed forms, Kayenta Formation, Arizona, USA (Clark, in Benton and Clark, 1988); *Stegomosuchus longipes* Lull, 1953, upper Portland Group, Connecticut, USA.

Comment: The range of Protosuchidae could be much greater if one includes *Dyoplax arenaceus* Fraas, 1867, Schilfsandstein, Germany, as Walker (1961) suggests, and *Edentosuchus tienshanensis* Young, 1973, Wuerho, China, (Early Cretaceous), as Clark (in Benton and Clark, 1988) suggests.

Family Orthosuchidae Whetstone and Whybrow, 1983 J.(HET/SIN)

First and last: *Orthosuchus stormbergi* Nash, 1968, upper Elliot Formation, Lesotho, South Africa.

Family Metriorhynchidae Fitzinger, 1843 J.(BTH)–K.(HAU)

First: *Teleidosaurus calvadosi* (J. A. Eudes-Deslongchamps, 1866), *T. gaudryi* Collot, 1905, and *T. bathonicus* (Mercier, 1933), Bathonian, Normandy and Burgundy, France.

Last: *Dakosaurus maximus* (Plieninger, 1846), Hauterivian, Provence, France.

Family Goniopholididae Cope, 1875 J.(BTH)–K.(MAA)

First: "Goniopholids," Ostracod Limestone, Skye, Scotland (Savage, 1984), Chipping Norton, White Limestone, and Forest Marble formations, Gloucestershire and Oxfordshire, England.

Last: "*Goniopholis*" *kirtlandicus* Wiman, 1931, Maastrichtian, New Mexico, USA.

Family Pholidosauridae Eastman, 1902 J.(BTH)–K.(CEN)

First: *Anglosuchus geoffroyi* (Owen, 1884), *A. laticeps* (Owen, 1884), White Limestone Formation, Oxfordshire, England.

Last: *Teleorhinus mesabiensis* Erickson, 1969, Cenomanian, Iron Range, Minnesota, USA.

Ornithodira Gauthier, 1986

Family Lagosuchidae Arcucci, 1987 Tr.(LAD)

First and last: *Lagosuchus talampayensis* Romer, 1971, *Lagerpeton chanarensis* Romer, 1971, and *Pseudolagosuchus major* Arcucci, 1987, all Chañares Formation, La Rioja, Argentina.

Family Podopterygidae Sharov, 1971 Tr.(CRN/NOR)

First and last: *Sharovipteryx mirabilis* (Sharov, 1971), Madyigenskaya Svita, Fergana, Kirghizia.

Family unnamed Tr.(CRN/NOR)

First and last: *Longisquama insignis* Sharov, 1970, Madyigenskaya Svita, Fergana, Kirghizia.

Family Scleromochlidae Huene, 1914 Tr.(CRN)

First and last: *Scleromochlus taylori* Woodward, 1907, Lossiemouth Sandstone Formation, Morayshire, Scotland.

Pterosauria Owen, 1840 (Kaup, 1834)

Family Unnamed Tr.(NOR)

First and last: *Preondactylus buffarini* Wild, 1983, lower middle part of the "Dolomia Principale," Udine, Italy.

Family Dimorphodontidae Seeley, 1870 Tr.(NOR)–J.(SIN)

First: *Peteinosaurus zambellii* Wild, 1978, upper half of the Calcare di Zorzino, Bergamo, Italy.

Last: *Dimorphodon macronyx* (Buckland, 1829), upper Blue Lias, Dorset, England.

Family Eudimorphodontidae Wellnhofer, 1978 Tr.(NOR)

First and last: *Eudimorphodon ranzii* Zambelli, 1973, upper half of the Calcare di Zorzino, Bergamo, Italy.

Family Rhamphorhynchidae Seeley, 1870 J.(TOA–TTH)

First: *Parapsicephalus purdoni* (Newton, 1888), upper Lias, Yorkshire, England; *Dorygnathus banthensis* (Theodori, 1930), upper Lias, Germany.

Last: *Rhamphorhynchus longicaudus* (Münster, 1839), *R. intermedius* Koh, 1937, *R. muensteri* (Goldfuss, 1831), *R. gemmingi* Meyer, 1846, *R. longiceps* Woodward, 1902, *Scaphognathus crassirostris* (Goldfuss, 1831), and *Odontorhynchus aculeatus* Stolley, 1936 (?nom. nud.), Solnhofener Schichten, Bavaria, Germany.

Dinosauria Owen, 1842 (p)

Family Herrerasauridae Benedetto, 1973 Tr.(CRN)

First and last: *Staurikosaurus pricei* Colbert, 1970, *Scaphonyx* Assemblage Zone, Santa Maria Formation, Rio Grande do Sul, Argentina; *Herrerasaurus ischigualastensis* Reig, 1983, Ischigualasto Formation, San Juan, Aregentina.

Family Podokesauridae Huene, 1914 Tr(CRN)–J.(PLB)

First: *Coelophysis bauri* (Cope, 1889), lower part of Petrified Forest Member, Chinle Formation, Arizona, USA.

Last: *Syntarsus kayentakatae* Rown, 1989, Kayenta Formation, Arizona, USA.

Comment: The famous *Coelophysis* quarry at Ghost Ranch, New Mexico, USA, is in the upper part of the Petrified Forest Member, dated lower Norian.

Family Ceratosauridae Marsh, 1884 (p) J.(SIN–KIM/TTH)

First: *Sarcosaurus woodi* Andrews, 1921, Lias, Leicestershire, England.

Last: *Ceratosaurus nasicornis* Marsh, 1884, Morrison Formation, Colorado, USA.

Family Allosauridae Marsh, 1879 J. (CLV)–K.(ALB)

First: *Piatnitzkysaurus floresi* Bonaparte, 1979, Cañadon Asfalto Formation, Chubut, Argentina.

Last: *Chilantaisaurus marotuensis* Hu, 1964, unnamed unit, Nei Mongol Zizhiqu, China.

Family Megalosauridae Huxley, 1869 Tr.(RHT)?–K.(VLG/ALB)Terr.

First: *Megalosaurus cambrensis* (Newton, 1899), Rhaetic, Glamorgan, Wales.

Last: *Kelmayisaurus petrolicus* Dong, 1973, Lianmugin Formation, Xinjiang Uygur Zizhiqu, China.

Comment: The family Megalosauridae is not accepted by Molnar et al. (1990), although they suggest that *Megalosaurus*, *Magnosaurus*, and *Kelmayisaurus* may be related. There is little evidence that *M. cambrensis* is a true megalosaur. If not, the earliest records of *Megalosaurus* are Aalenian and Bajocian.

Family Unnamed J.(CLV–KIM/TTH)

First: *Eustreptospondylus oxoniensis* Walker, 1964, Oxford Clay, Oxfordshire and Buckinghamshire, England.

Last: *Torvosaurus tanneri* Galton and Jensen, 1979, Morrison Formation, Colorado, USA.

Comment: This family is hinted at by Molnar et al. (1990, p. 209), in suggesting a phyletic link among *Eustreptospondylus*, *Torvosaurus*, and *Yangchuanosaurus*.

Family Thecodontosauridae Huene, 1908 Tr.(CRN–RHT)

First: *Azendohsaurus laaroussi* Dutuit, 1972, Argana Formation, Argana Valley, Morocco.

Last: *Thecodontosaurus antiquus* Riley and Stutchbury, 1836, Magnesian Conglomerate, Avon, England; fissure fillings, Glamorgan, Wales.

Family Anchisauridae Marsh, 1885 J.(PLB/YOA)

First and last: *Anchisaurus polyzelus* (Hitchcock, 1865), upper Portland Formation, Connecticut and Massachusetts, USA.

Family Massospondylidae Huene, 1914 J.(HET–SIN/PLB)

First: *Massospondylus carinatus* Owen, 1854, upper Elliot Formation, Clarens Formation, and Bushveld Sandstone, South Africa; Forest Sandstone, Zimbabwe; upper Elliot Formation, Lesotho.

Last: *Massospondylus* sp., Kayenta Formation, Arizona, USA.

Family Yunnanosauridae Young, 1942 J.(HET/SIN)

First and last: *Yunnanosaurus huangi* Young, 1942, upper Lower Lufeng Series, Yunnan, China.

Family Plateosauridae Marsh, 1895 Tr.(NOR)–J.(PLB/TOA)Terr.

First: *Sellosaurus gracilis* Huene, 1907–8, unterer and mittlerer Stubensandstein, Baden-Württemberg, Germany.

Last: *Ammosaurus major* (Marsh, 1891), upper Portland Formation, Connecticut; Navajo Sandstone, Arizona, USA.

Family Melanorosauridae Huene, 1929 Tr.(NOR–HET/SIN)

First: *Euskelosaurus browni* Huxley, 1866, lower Elliot Formation and Bushveld Sandstone, South Africa;

lower Elliot Formation, Lesotho; Mpandi Formation, Zimbabwe; *Melanorosaurus readi* Haughton, 1924, lower Elliot Formation, South Africa.

Last: *Lufengosaurus huenei* Young, 1941, upper Lower Lufeng Series, Yunnan, China.

Comment: cf. *Lufengosaurus* is noted from the Zhenzhunchong Formation, Sichuan, China, dated as Toarcian/Bajocian (Weishampel, 1990).

Family Vulcanodontidae Cooper, 1984
(p?) J.(HET–TOA)

First: *Vulcanodon karibaensis* Raath, 1972, *Vulcanodon* Beds, Mashonaland North, Zimbabwe.

Last: *Ohmdenosaurus liasicus* Wild, 1978, Posidonienschiefer, Baden-Württemberg, Germany.

Family Cetiosauridae Lydekker, 1888
(p) J.(BAJ–KIM/TTH)

First: *Cetiosaurus medius* Owen, 1842, Inferior Oolite, West Yorkshire, England; *Amygdalodon patagonicus* Cabrera, 1947, Cerro Carnerero Formation, Chubut, Argentina; ?*Rhoetosaurus brownei* Longman, 1925, ?Injune Creek Beds, Queensland, Australia.

Last: *Haplocanthosaurus priscus* (Hatcher, 1903) and *H. delfsi* McIntosh and Williams, 1988, Morrison Formation, Colorado and Wyoming, USA.

Family Brachiosauridae Riggs, 1904
J.(?AAL/BTH)–K.(ALB)

First: "brachiosaurid," Northamptonshire Sand Formation, Northamptonshire, England.

Last: *Brachiosaurus nougaredi* Lapparent, 1960, "Continental Intercalaire," Wargla, Algeria; *Chubutisaurus insignis* Corro, 1974, Gorro Frigio Formation, Chubut, Argentina.

Comment: If the Northamptonshire brachiosaurid is not confirmed, definite Bathonian examples include the following: *Bothriospondylus robustus* Owen, 1875, Forest, Marble, Wiltshire, England; *B. madagascariensis* Lydekker, 1895 and *Lapparentosaurus madagascariensis* Bonaparte, 1986, Isalo Formation, Madagascar.

Family Diplodocidae Marsh, 1884
J.(BAJ)–K.(CMP/MAA)

First: *Cetiosauriscus longus* (Owen, 1842), Inferior Oolite, West Yorkshire, England.

Last: *Nemegtosaurus mongoliensis* Nowinski, 1971, Nemegt Formation, Omnogov, Mongolia.

Family Pisanosauridae Casamiquela, 1967 Tr.(CRN)

First and last: *Pisanosaurus mertii* Casamiquela, 1967, Ischigualasto Formation, La Rioja Province, Argentina.

Family Fabrosauridae Galton, 1972 J.(HET/SIN)

First and last: *Lesothosaurus diagnosticus* Galton, 1978, upper Elliot Formation, Mafeting District, Lesotho.

Comment: Other supposed fabrosaurids such as *Technosaurus* and *Revueltosaurus* (CRN), *Scutellosaurus* (HET), *Fabrosaurus*, *Tawasaurus*, and *Fulengia* (HET/HIN), *Xiaosaurus* (BTH), *Alocodon* and *Trimucrodon* (OXF), *Nanosaurus* (KIM), and *Echinodon* (BER) are not regarded as fabrosaurids, but merely Ornithischia incertae sedis, or thyreophorans (e.g., *Scutellosaurus*), or prosauropods (e.g., *Fulengia*, *Tawasaurus*, *Technosaurus* in part).

Family Scelidosauridae Huxley, 1869
(p?) J.(SIN–TTH?)

First: *Scelidosaurus harrissoni* Owen, 1861, lower Lias, Dorset, England.

Last: *Echinodon becklesi* Owen, 1861, middle Purbeck Beds, Dorset, England.

Comment: The family Scelidosauridae is equated here with the "basal Thyreophora." If *Echinodon* is not a "basal thyreophoran," the family range becomes SIN–PLB?, with *Scutellosaurus lawleri* Colbert, 1981, as the youngest member.

Family Huayangosauridae Dong, Tang, and Zhou, 1982 J.(HET/PLB–BTH/CLV)

First: *Tatisaurus oehleri* Simmons, 1965, Dark Red Beds of the Lower Lufeng Group, Yunnan, China.

Last: *Huayangosaurus taibaii* Dong, Tang, and Zhou, 1982, Xiashaximiao Formation, Sichuan, China.

Family Stegosauridae Marsh, 1880
J.(BTH)–K.(CON)

First: Unnamed stegosaur, Chipping Norton Formation, lower Bathonian, Gloucestershire, England, and from other Bathonian localities in Gloucestershire and Oxfordshire.

Last: *Dravidosaurus blanfordi* Yadagiri and Ayyasami, 1979, Trichinopoly Group, Tamil Nadu, India.

Family Nodosauridae Marsh 1890
J.(CLV)–K.(MAA)

First: *Sarcolestes leedsi* Lydekker, 1893, lower Oxford Clay, Cambridgeshire, England.

Last: "*Struthiosaurus transilvanicus*" Nopcsa, 1915, Sinpetru Beds, Hunedoara, Romania; Gosau Formation, Niederösterreich, Austria; "*Denversaurus schlessmani*" Bakker, 1988, Lance Formation, South Dakota, USA.

Family Heterodontosauridae Romer, 1966
J.(HET/SIN–SIN)

First: *Lycorhinus angustidens* Haughton, 1924, *Lanasaurus scalpridens* Gow, 1975, and *Abrictosaurus consors* (Thulborn, 1975), upper Elliot Formation, South Africa and/or Lesotho.

Last: *Heterodontosaurus tucki* Crompton and Charig, 1962, Clarens Formation, Cape Province, South Africa.

Family Hypsilophodontidae Dollo, 1882
J.(BTH/CLV)–K.(MAA)

First: *Yandusaurus honheensis* He, 1979, Xiashaximiao Formation, Sichuan, China.
Last: *Thescelosaurus neglectus* Gilmore, 1913, Lance Formation, Wyoming, USA; Hell Creek Formation, Montana and South Dakota, USA; Scollard Formation, Alberta, Canada; ?*T. garbanii* Morris, 1976, Hell Creek Formation, Montana, USA.

Synapsida Osborn, 1903

Therapsida Broom, 1905 (p)

Anomodontia Owen, 1859

Family Kannemeyeriidae Huene, 1948
Tr (SCY = CRN)

First: *Kannemeyeria simocephalus* (Weithofer, 1888), lower Etjo Beds, southwest Africa; *K. wilsoni* Broom, 1937, *Cynognathus-Diademodon* Assemblage Zone, South Africa; *K. argentinensis* Bonaparte, 1966, Puesto Viejo Formation, Mendoza Province, Argentina; *Vinceria andina* Bonaparte, 1967, Cerro de Las Cabras Formation, Mendoza Province, Argentina.
Last: *Jachaleria colorata* Bonaparte, 1971, boundary between Ischigualasto Formation and lower Los Colorados Formation, La Rioja Province, Argentina.

Cynodontia Owen, 1860

Family Traversodontidae Huene, 1936
Tr.(SCY–RHT)

First: *Pascualgnathus polanskii* Bonaparte, 1966, Puesto Viejo Formation, and *Andescynodon mendozensis* Bonaparte, 1967, and *Rusconiodon mignonei* Bonaparte, 1972, Rio Mendoza Formation, Mendoza Province, Argentina (Bonaparte, 1982).
Last: *Microscalenodon nanus* Hahn et al. 1988, lower "Rhaetian" bonebed, Gaume, Belgium.

Family Chiniquodontidae Huene, 1948
Tr.(?ANS–CRN)

First: *Aleodon brachyramphus* Crompton, 1955, Manda Formation, Ruhuhu Valley, Tanzania. If this is not a chiniquodontid, the oldest representatives are *Probelesodon lewisi* Romer, 1969, and *Chiniquodon* sp. from the Chañares Formation, La Rioja Province, Argentina (Ladinian).
Last: *Chiniquodon theotonicus* Huene, 1936, *Dinodontosaurus* Assemblage Zone, Santa Maria Formation, Rio Grande do Sul, Brazil.
Comment: A chiniquodontid tooth, *Lepagia gaumensis* Hahn, Wild, and Wouters, 1987, has been reported from the lower Rhaetian bonebed of Gaume, southern Belgium (Hahn, Wild, and Wouters, 1987).

Family Probainognathidae Romer, 1973 Tr.(LAD)

First and last: *Probainognathus jenseni* Romer, 1970, lower beds of Chañares Formation, La Rioja Province, Argentina.

Family Tritylodontidae Cope, 1884
Tr.(RHT)–J.(BTH/CLV)

First: "cf. *Tritylodon*," upper beds of Los Colorados Formation, La Rioja Province, Argentina.
Last: *Bienotheroides wanhsienensis* Young, 1982, upper Xiashaximiao Formation, Sichuan, China.

Family Tritheledontidae Broom, 1912
Tr.(RHT)–J.(SIN)

First: *Chaliminia musteloides* Bonaparte, 1980, Los Colorados Formation, La Rioja Province, Argentina.
Last: *Pachygenelus monus* Watson, 1913, Clarens Formation, South Africa, Lesotho.
Comment: *Therioherpeton cargnini* Bonaparte and Barberena, 1975, Santa Maria Formation, Parana basin, Brazil (Carnian), is sometimes classified as the oldest tritheledontid, but Shubin et al. (1991) show that this assignment is incorrect.

?Aves

Family Protoavidae Chatterjee, 1991 Tr.(NOR)

First and last: *Protoavis texensis* Chatterjee, 1991, Cooper Formation, Texas, USA.
Comment: Whether or not *Protoavis* is a bird, it may well prove to represent a unique family-level taxon.

Mammalia Linnaeus, 1758

Triconodonta Osborn, 1888

Family Morganucodontidae Kühne, 1958
Tr.(RHT)–J.(BTH)

First: *Eozostrodon parvus* Parrington, 1941, and other species, fissure fillings, Somerset, England; *Morganucodon watsoni* Kühne, 1949, fissure fillings, Glamorgan, Wales; *M. peyeri* Clemens, 1980, and *Helvetiodon schuetzi* Clemens, 1980, Rhaetic bonebed, Hallau, Switzerland; *Brachyzostrodon coupatezi* Sigogneau-Russell, 1983, Saint-Nicolas-de-Port, France.
Last: *Wareolestes rex* Freeman, 1979, Forest Marble Formation, Oxfordshire, England.
Comment: Other species of *Eozostrodon* and *Morganucodon*, as well as *Erythrotherium*, from southern Africa and China, are all probably Early Jurassic in age.

Family Sinoconodontidae Mills, 1971
J.(HET/SIN–SIN)

First: *Sinoconodon rigneyi* Patterson and Olson, 1961, *Lufengoconodon changchiawaensis* Young, 1982, and other species, Dark Red Beds, Lower Lufeng Formation, Yunnan, China.

Last: *Megazostrodon rudnerae* Crompton and Jenkins, 1968, Clarens Formation, Lesotho.

Family Amphilestidae Osborn, 1888
J.(SIN/PLB)–K.(?CMP)

First: *Dinnetherium nezorum* Jenkins, Crompton, and Downs, 1983, Kayenta Formation, Arizona, USA.

Last: *Guchinodon hoburensis* Trofimov, 1978 and *Gobiconodon borissiaki* Trofimov, 1978, both Khovboor locality, Mongolia.

Multituberculata Cope, 1884

Family Plagiaulacidae Gill, 1872
?Tr.(RHT)/J.(OXF)–K.(BER/APT/ALB)

First: *Pseudobolodon oreas* Hahn, 1977, *Paulchoffatia delgadoi* Kühne, 1961, *Guimarotodon leiiensis* Hahn, 1977, all Guimarota, Portugal.

Last: *Paulchoffatia* sp. and *Bolodon* sp., Galve local fauna, Spain; plagiaulacid, Trinity Sands, Texas, USA.

Comment: A possible paulchoffatiid (= plagiaulacid), *Mojo usuratus* Hahn, Lepage, and Wouters, 1987, has been described from the lower Rhaetian bonebed of Gaume, Belgium (Hahn et al., 1987).

Haramiyoidea Hahn, 1973

Family Haramiyidae Simpson, 1947
Tr.(RHT)–J.(BTH)

First: *Haramiya moorei* (Owen, 1871), *H. fissurae* (Simpson, 1928), and *Thomasia anglica* Simpson, 1928, Holwell Quarry, Somerset, England; *Haramiya* and *Thomasia* species, Rhaetic bonebeds, Stuttgart area, Germany, and Hallau Bonebed, Switzerland (Clemens, 1980).

Last: Haramiyid, Forest Marble Formation, Oxfordshire, England.

Allotheria incertae sedis

Family Theroteinidae Sigogneau-Russell, Frank, and Hemmerlé, 1986 Tr.(NOR/RHT)

First and last: *Theroteinus nikolai* Sigogneau-Russell, Frank, and Hemmerlé, 1986, Saint-Nicolas-de-Port, Lorraine, France.

Comment: The age of this locality has been disputed, being assigned to the lower Rhaetian, or to the late Norian, as an equivalent of the Knollenmergel.

Dryolestoidea Butler, 1939

Family Amphitheriidae Owen, 1846 J.(BTH)

First and last: *Amphitherium prevostii* (von Meyer, 1832), Stonesfield Slate, Oxfordshire, England.

Family Dryolestidae Marsh, 1879
J.(BTH)–K.(CMP/MAA)

First: Dryolestid, Forest Marble Formation, Oxfordshire, England.

Last: *Leonardus cuspidatus* Bonaparte, 1990, *Groeberitherium stipanicici* Bonaparte, 1986, and *G. novasi* Bonaparte, 1986, all Los Alamitos Formation, Neuquén Province, Argentina; dryolestid, Mesa Verde Formation, Wyoming, USA.

Incertae sedis

Family Peramuridae Kretzoi, 1946
J.(BTH)–K.(?ALB)

First: *Palaeoxonodon ooliticus* Freeman, 1976, Forest Marble Formation, Oxfordshire, England.

Last: *Arguimus khosbajari* Dashzeveg, 1979, Mongolia.

Family Tinodontidae Marsh, 1887
Tr.(NOR)–K.(CMP)

First: *Kuehneotherium praecursoris* Kermack, and Mussett, 1968, Bridgend, Glamorgan, Wales; *Kuehneotherium* sp., Emborough Quarry, Somerset, England.

Last: *Bondesius ferox* Bonaparte, 1990, El Molino Formation, Neuquén Province, Argentina; *Mictodon simpsoni* Fox, 1984, Milk River Formation, Alberta, Canada.

Comment: Emborough Quarry is dated as pre-Rhaetian on topographic evidence by Fraser (1986, and Chapter 11).

Family Docodontidae Simpson, 1947 J.(BTH–KIM)

First: *Borealestes serendipitus* Waldman and Savage, 1972, Ostracod Limestone, Isle of Skye, Scotland; *Simpsonodon oxfordensis* Kermack, Lee, Lees, and Mussett, 1987, Forest Marble Formation, Oxfordshire, England.

Last: *Docodon victor* (Marsh, 1890), and other species, Morrison Formation, Colorado and Wyoming, USA.

23

Comments on Benton's "Late Triassic to Middle Jurassic extinctions among continental tetrapods"

NICHOLAS C. FRASER AND HANS-DIETER SUES

Studies of mass extinctions in the fossil record and possible periodicities of such events typically are based on data at the taxonomic level of family and the stratigraphic level of stage (e.g., Olsen and Sues, 1986; Benton, 1991, and Chapter 22). Despite the many problems inherent in any assessment of faunas at such coarse levels of resolution, this kind of analysis can be very useful in defining the future research agenda. In the first instance, such studies draw a great deal of attention to the possibility that there were major extinction events. As studies on any putative mass extinction progress, however, the data must be analyzed at increasingly finer geographic, stratigraphic, and taxonomic levels in order to provide rigorous tests of the hypotheses in question. Consequently, recent efforts in this direction have increasingly focused on periods of apparently increased extinction rates (e.g., Archibald and Bryant, 1990).

Regarding the end-Triassic biotic changes, Olsen and Sues (1986) found substantial evidence to support the hypothesis of a worldwide extinction of continental tetrapods at the end of the period (i.e., at the end of the Norian stage), as first proposed by Colbert (1958). Recent discoveries in the field have spawned new analyses of faunal turnover that have provided some rather different perspectives, but have also generated new problems. The most recent appraisal of the record of early Mesozoic continental vertebrates by Benton (1991, and Chapter 22) has adopted the point of view that the apparent paucity of fossils in the Norian is a real phenomenon that does not merely reflect a "gap" in the fossil record.

The patchiness of the fossil record is well known and makes it particularly difficult to test hypotheses of global mass extinctions. For instance, in any situation where one postulates worldwide extinction of a particular taxon, the possibility arises that populations persisted in small refugia for some time. In terms of the fossil record, documentation of such an extinction may well be limited to a few localities at which the taxon in question does indeed suddenly disappear (e.g., extinction of nonavian dinosaurs in the Western Interior at the end of the Cretaceous). In this case, inferences concerning a global extinction of this taxon would appear well corroborated. By contrast, if the handful of available sites included, by coincidence, one of the very rare localities documenting a refugium, we would, of course, not accept the idea that the sudden extinction of that taxon had occurred. Conversely, we could obtain an incorrect impression of a sudden extinction through the sampling of areas where localized extinctions of a taxon had taken place, when the taxon actually survived elsewhere (e.g., brachyopid and chigutisaurid temnospondyls; Milner, Chapter 1). It is thus all too easy to invoke the limitations of the fossil record when the data do not support a particular hypothesis. Usually a paucity of fossils in a given sequence of sedimentary rocks is regarded as indicative of a poor fossil record rather than as an accurate reflection of a depauperate fauna.

Benton (1991) marshaled various lines of evidence to support the view that the fossil record of Norian-age continental vertebrates is indeed good and reflects taxonomically relatively impoverished faunas. Although he concurred with Olsen and Sues (1986) that a number of tetrapod families disappear from the fossil record at the end of the Norian ("Rhaetian"), he argued that a much greater extinction event occurred at the end of the Carnian. Furthermore, Benton suggested that there was a marked decrease in diversity among many of those families that ranged across the Carnian–Norian boundary to the end of the Norian ("Rhaetian"). He forcefully argues this case again in the preceding chapter. His contribution offers two competing hypotheses, the relative merits of which can be tested in the future. Considerable differences of opinion concerning both the pattern and timing of early Mesozoic faunal changes were expressed by participants at the 1991 workshop. Therefore, in order to provide the reader with a different view, we should like to address some of the issues raised by Benton.

As Benton (1991) noted, sampling problems form an obstacle to all studies of major biotic changes, but he made the valid point that one should not always assume that a gap in the fossil record is just an artifact. Nevertheless, it is our belief that his assertion that "it is clear that there was not a gap during the early Norian" ignores a number of problems.

First, the available amounts of exposed vertebrate-bearing strata must be considered. In the case of the Upper Triassic, it is frequently assumed that there are extensive outcrops worldwide. That is incorrect. The stratigraphic controls on many of the occurrences in question are, in fact, quite poor, notably in the American Southwest and for the fissure fillings of southwest Britain. Although the Newark Supergroup strata are excellently constrained stratigraphically, there is an apparent paucity of Norian vertebrate fossils. Outcrops of strata from this stage in eastern North America are quite substantial, but they have not been extensively prospected for fossils, and many of the formations represent depositional environments that were not ideal for the preservation of fossil vertebrate remains. Nevertheless, ongoing work is revealing increasing diversity among Norian tetrapods (e.g., Sues and Baird, 1993), and much of that information is still unpublished and thus could not be considered in Benton's analysis.

Benton (1991) examined the significance of Lazarus taxa during the Late Triassic and concluded that the low numbers of such taxa in the Norian reflect depauperate faunas during that time interval. This may be a plausible argument for gaps of relatively short duration. But poorly fossiliferous strata may represent significant amounts of time. The longer the time interval in question, the more equivocal become the statements concerning extinctions [according to Harland et al. (1990), the Norian stage spans some 13.9 million years, and the "Rhaetian" an additional 1.5 million years]. Placing an extinction event at the beginning or the end of a given stage with a poorly fossiliferous rock record is difficult, and there is a tendency to bring back extinctions to the beginning of the time interval in question (Signor-Lipps effect).

One of the most controversial topics in the discussion of mass-extinction events is the significance of "singleton" families. Singletons sensu Benton contain just one taxon and typically are based on specimens (or sometimes a single fossil) from a single locality. Benton (1991) argued that it is unacceptable to treat such taxa in the same way as more diversified groups. We consider his distinction between singletons and more diversified taxa artificial. The taxa in question do

not represent instances of spontaneous origination and extinction, but instead form part of a genealogical nexus (Norell, 1992). We should like to consider one example: the Early Jurassic rhynchocephalian *Gephyrosaurus* (and the monotypic and thus redundant "family" Gephyrosauridae). Well-corroborated cladistic analyses (e.g., Wu, Chapter 3) place this taxon as the sister-group of the Sphenodontia. The oldest definite representatives of the latter group described to date are late Carnian in age, and thus the lineage represented by *Gephyrosaurus* must be extended to the point of divergence of the two sister-taxa earlier in the Carnian or possibly in pre-Carnian times. Consequently, *Gephyrosaurus* does not represent an Early Jurassic origination, but rather an otherwise paleontologically as yet undocumented lineage that dates back to the early Late Triassic. Thus a distinction between this "singleton" and the more diversified sister-taxon Sphenodontia is unjustified. We should like to note, however, that the recognition of a number of apparent singletons in a given time interval is suggestive of a poor fossil record.

In order to achieve better resolution of the faunal turnover(s) in the Late Triassic, Benton (1991) attempted to plot diversity levels within families during the Carnian and Norian. His data indicate that the loss in the number of families at the end of the Carnian was somewhat greater than that at the end of the Norian, but that diversity levels within families dropped dramatically at the end of the Carnian when compared with the end of the Norian. We note, however, that many of the original diversity estimates probably are artificially raised, because many taxa at the level of the genus and species are based on fragmentary and often virtually indeterminate material. This is especially true for large tetrapods, for which, as in most other geological formations, reasonably complete skeletons are uncommon. Consequently, we would revise downward the species-level diversity estimates for certain Carnian tetrapod families provided by Benton (1991, and Chapter 22). For instance, our survey of the literature yielded only 11 stagonolepidid taxa of late Carnian age, whereas Benton counted 14.

Benton claims that the apparent drop in the number of tetrapod species recorded from the Norian does not reflect a paucity of tetrapod-producing localities for that stage. It is worth noting, however, that although the number of Norian tetrapod-bearing localities known to date is considerable, many tend to sample the same geological time interval and general geographic provenance (e.g., Hallau and Saint-Nicolas-de-Port), leading to autoreplication in statistical analyses.

Differences in preservation and in collecting methods from locality to locality can have pronounced effects on the taxonomic composition of the assemblages reported from any given site. Sigogneau-Russell and Hahn (Chapter 10) note an apparently abrupt change in the composition of faunal assemblages from the middle to the upper Norian in the Germanic and Paris basins. Much of the collecting from strata of middle Norian age was undertaken at a time when only the larger vertebrate fossils attracted the attention of paleontologists. Most of the late Norian fossils were collected more recently and comprise small-vertebrate remains obtained by using modern methods of screen-washing. Thus Sigogneau-Russell and Hahn believe that prospecting for small tetrapods in middle Norian strata of the Germanic and Paris basins would decrease the apparent differences between the middle and late Norian assemblages known at present from these areas.

Sigogneau-Russell and Hahn's contribution also serves to emphasize that different classifications can produce very different data sets for the determination of stratigraphic ranges of higher taxa. On the basis of isolated postcanine teeth these authors recognize the presence of Chiniquodontidae (which, according to Benton, disappear at the end of the Carnian) in the upper Norian, and they also maintain families such as the Therioherpetidae, which Benton omits from consideration.

When comparing extinction events, it is crucial to assess the respective data by comparable means. In recording the numbers of families disappearing at the end of the Carnian, Benton combines the numbers of those that disappeared during or at the end of two palynological zones (L1 and L2), whereas for the Norian tetrapod families he records only those that disappeared at the end of the last palynological zone (latest Sevatian). If he had combined the last two palynological zones of the Norian, Benton could have added up to six additional families to his list of Norian extinctions.

The implication that a gap in the Norian fossil record really reflects depauperate assemblages should be viewed with caution at present. The recent upsurge in interest in Late Triassic continental tetrapods, coupled with fieldwork specifically aimed at the recovery of small vertebrate remains, has resulted in many revisions to faunal lists and will continue to do so. Furthermore, the interrelationships of the various tetrapod groups are now being elucidated by the increasing use of explicitly cladistic analyses. Although Benton's analysis reflects the nature of the currently available data concerning both the phylogenetic affinities and stratigraphic distributions of early Mesozoic tetrapods, there exists considerable evidence from ongoing research that the picture will prove to be much more complex.

References

Archibald, J. D., and L. J. Bryant. 1990. Differential Cretaceous/Tertiary extinctions of nonmarine vertebrates: Evidence from northeastern Montana. Pp. 549–562 in V. L. Sharpton and P. D. Ward (eds.), *Global Catastrophes in Earth History: An Interdisciplinary Conference on Impacts, Volcanism, and Mass Mortality*. Geological Society of America Special Paper 247.

Benton, M. J. 1991. What really happened in the Late Triassic? *Historical Biology* 5: 263–278.

Colbert, E. H. 1958. Triassic tetrapod extinctions at the end of the Triassic period. *Proceedings of the National Academy of Sciences USA* 44: 973–977.

Harland, W. B., R. L. Armstrong, A. V. Cox, L. E. Craig, A. G. Smith, and D. G. Smith. 1990. *A Geologic Time Scale 1989*. Cambridge University Press.

Norell, M. A. 1992. Taxic origin and temporal diversity: the effect of phylogeny. Pp. 89–118 in M. J. Novacek and Q. D. Wheeler (eds.), *Extinction and Phylogeny*. New York: Columbia University Press.

Olsen, P. E., and H.-D. Sues. 1986. Correlation of continental Late Triassic and Early Jurassic sediments, and the Triassic–Jurassic tetrapod transition. Pp. 321–351 in K. Padian (ed.), *The Beginning of the Age of Dinosaurs: Faunal Change across the Triassic–Jurassic Boundary*. Cambridge University Press.

Sues, H.-D., and D. Baird. 1993. A skull of a sphenodontian lepidosaur from the New Haven Arkose (Upper Triassic) of Connecticut. *Journal of Vertebrate Paleontology* 13: 370–372.

24

What were the tempo and mode of evolutionary change in the Late Triassic to Middle Jurassic?

KEVIN PADIAN

In 1986, more than 30 vertebrate paleontologists collaborated to produce a volume summarizing diverse aspects of the problem of what happened at the beginning of the "Age of Dinosaurs" over 200 million years ago (Padian, 1986). Three years later, another volume appeared (Lucas and Hunt, 1989) that concentrated on new research on Triassic vertebrate taxa and faunas. This volume is a third installment that returns to the theme of the first: analyzing the vertebrate groups, faunas, and stratigraphic relationships of terrestrial vertebrates and environments around the world across the Triassic–Jurassic boundary.

Summarizing the first of these volumes, Padian (1986) noted several topics that had been advanced by new research but would still need further analysis. One was the stratigraphic reassignment of some deposits from the Late Triassic to the Early Jurassic. In virtually all cases these revisions have withstood the test of time. A second question was whether or not dinosaurs would prove to have been diverse and abundant in the Late Triassic. Phylogenetic analysis has demonstrated that dinosaurs must have diversified rapidly during the Late Triassic, because even in the Ischigualasto sediments that contain the most primitive known dinosaurs, ornithischians are present (*Pisanosaurus*), which means that both saurischians and ornithischians had evolved by that time (Hunt, 1991). However, both groups, particularly ornithischians, are thus far elusive and rare in Triassic deposits. The theropod *Coelophysis* and the sauropodomorph *Plateosaurus* are the only Triassic eudinosaurs that can be said to be well known (whether one regards "fabrosaurids," however they are defined, as monophyletic or not, their diversity in the Late Triassic is questionable and is based nearly entirely on teeth; Hunt and Lucas, Chapter 12). On the basis of present evidence, it appears that the dinosaurian radiation did not get into full swing until the Early Jurassic, with the first thyreophoran

(*Scutellosaurus, Scelidosaurus*) and ornithopod (*Heterodontosaurus, Abrictosaurus*) ornithischians, and the size diversity seen in theropod (*Syntarsus, Dilophosaurus*) and sauropodomorph (*Massospondylus, Vulcanodon*) saurischians.

A third question is whether or not the Triassic–Jurassic faunal transition can be said to have occurred in lockstep globally, and here we must allow that the jury is still out, as discussed later. As a corollary, one might ask whether there was one major Late Triassic extinction event or two (end-Carnian and end-Norian), and here also the debate has strong proponents on both sides, but hardly a final resolution. Finally, there is the question of whether or not phylogenetic analysis can improve our concept of the stratigraphic record. I have discussed this idea in detail (Padian, 1989b), and the chapters in the first section of this volume have considerable potential to improve this picture.

Biostratigraphy of the Early Jurassic

The events at the Triassic–Jurassic boundary have become a central focus of vertebrate paleontology since the 1970s. This can be traced perhaps most readily to the seminal paper by Olsen and Galton (1977), which proposed a drastic realignment of strata traditionally considered Late Triassic in age. Olsen and Galton showed, on the basis of vertebrate fossils and footprints, palynology, and radiometric dates, that about half of the rocks assigned to the Late Triassic properly belonged to the Jurassic. The latter formations lacked phytosaurs, rauisuchians, poposaurs, metoposaurs, aetosaurs, and other typical Triassic "marker taxa." Instead, they were dominated by dinosaurs, crocodylomorphs, turtles, pterosaurs, and mammals.

In the ensuing years, a lively discussion centered on the validity of these correlations worldwide (Colbert,

1986). The ages of individual rift basins in the Newark Supergroup of eastern North America were revised by Olsen, McCune, and Thomson (1982), and revision continues (e.g., Olsen and Sues, 1986; Olsen, Shubin, and Anders, 1987, 1988; Olsen, Schlische, and Gore, 1989; Shubin, Olsen, and Sues, Chapter 13). Olsen and Galton (1984) analyzed the implications of temporal revisions for the ages of traditionally Triassic formations in the Stormberg Series of South Africa, which met a fate similar to those of the Newark Supergroup. Elements of the Chinese Lufeng faunas were reviewed and revised by Sun and Cui (1986) and Luo and Wu (Chapter 14).

The Glen Canyon Group of the North American Southwest has historically posed problems for stratigraphers, who have inferred on the basis of various lines of evidence that the group is wholly Triassic, wholly Jurassic, or partly Triassic and partly Jurassic in age (Colbert, 1981; Clark and Fastovsky, 1986). Following the stratigraphic revision of Olsen and Galton (1977), who argued that the entire Glen Canyon Group should be assigned to the Jurassic, discussion centered on the criteria for this conclusion and the question of whether all or only part of the Glen Canyon was Jurassic in age. Some reported evidence maintained that "thecodontians," archosaurs common in the Triassic, were present in the Glen Canyon Group and therefore that the Glen Canyon strata spanned the Triassic–Jurassic boundary. Pipiringos and O'Sullivan (1978) examined unconformities in the Glen Canyon Group and showed that the Rock Point Member of the Moenave Formation, the basal member of the Glen Canyon Group, lay below their "J-O" unconformity and therefore was not clearly part of the Moenave. This was important because the Rock Point Member contains phytosaurs, which are typical Triassic marker taxa found nowhere else in the Glen Canyon Group. There now seems to be consensus that the Rock Point Member should be reassigned to the underlying Chinle Formation.

Other discoveries helped to nail down a Jurassic age for the Glen Canyon Group. Sues (1985) described remains of the tritylodontid Oligokyphus from the Kayenta Formation and suggested an Early Jurassic correlation. Olsen and Padian (1986) described the ichnotaxon Batrachopus from the Moenave Formation and correlated it with occurrences in the Jurassic portion of the Newark Supergroup. For some years it had been alleged that aetosaurs, typical "thecodontian" Triassic marker taxa, were found in the Kayenta Formation, but Padian (1989a) showed that these occurrences did not belong to aetosaurs, but to the ornithischian dinosaur Scelidosaurus, which previously was known only from British deposits assigned a Jurassic age on the basis of ammonite zones. That had the effect of changing a Triassic indicator into a Jurassic one, and it removed most of the last evidence

based on fossil vertebrates that the Glen Canyon Group might be at least partly Triassic. Padian (1989b) further pointed out that the use of the term "thecodontian," which described only a paraphyletic grouping of archosaurs, was inappropriate both for systematics and for biostratigraphy, as it avoided the precision necessary to tease apart stratigraphic ranges of individual archosaurian groups. In most cases, "thecodontians" had been allowed to include noncrocodyliform crocodylomorphs (such as sphenosuchians), whose range uniquely spans the Triassic–Jurassic boundary, thus dragging with it by implication of larger hierarchical level all the other groups (phytosaurs, aetosaurs, rauisuchians, and other traditional Triassic marker taxa) that did not have specific records in the Jurassic (Sun and Cui, 1986). Sues, Clark, and Jenkins (Chapter 16) give an updated summary of Kayenta Formation vertebrates; they report no traditional Triassic marker taxa.

As a result, we may conclude that in general the picture of reassigned Triassic deposits and faunas to the Early Jurassic has withstood review and analysis. But a great deal of finer-scaled work is needed, because we still do not have a very good idea of how the various terrestrial faunas and deposits of the Early Jurassic can be assigned to geologic stages and absolute ages. And, as usual, the paucity of cosmopolitan taxa worldwide continues to make intercontinental correlations difficult. These are clearly areas for future research.

Biostratigraphy of the Late Triassic

Following removal of a substantial number of formations and faunas from the Late Triassic to the Jurassic, the patterns in the Late Triassic would have been expected to be simplified. But in the intervening years, nothing has seemed further from reality. We can rejoice in the increased attention to those questions while despairing at the lack of decisive evidence at this point. In other words, the good news is that we have a Late Triassic record, but the bad news is that we have so many different ones.

Problems with Late Triassic biostratigraphy include the correlation of intercontinental records, the pace and timing of faunal change, and the diversification of the dinosaurs. These problems are exacerbated by the historical fact that the Triassic standard was established on marine strata in Europe (Colbert, 1986; Olsen and Sues, 1986). Terrestrial sedimentary rock thicknesses in the Germanic Basin are not as great as elsewhere, and representation of fossil vertebrates is, like everywhere else, selective. However, the largest remaining problem outside temporal resolution is clearly systematic resolution.

Olsen and Sues (1986), in a seminal paper, reviewed the lithostratigraphic and biostratigraphic grounds on which correlations have been made from the European type sequence to North America and Africa, as well as South America, India, China, and Australia. They explored both stratigraphic and macroevolutionary patterns that followed from the reevaluation of global taxa and faunas and concluded that probably there was not one, but rather two terrestrial tetrapod extinction "events" in the Late Triassic (end-Carnian and end-Norian), a point later taken up by Benton (1986b). However, they cautioned that any interpretation of the pattern depended on the use of metrics: Stratigraphic resolution at the stage or formation level and taxonomic resolution at the family level likely would be too coarse to give reliable answers, in which respects they disagreed with Benton (1986a), who did not find two extinction events in the Late Triassic. As later papers by Benton (1991, and Chapter 22), Olsen et al. (1987, 1988), Padian (1988), Weems (1992), and Fraser and Sues (Chapter 23) show, the question is still highly debatable.

There are really two major issues here that often are difficult to separate because of their entwining implications. The first is the question of how intercontinental deposits are being correlated: the quality of the evidence, the degree of confidence in identifications of taxa and comparability of biotas, and the differences in depositional and ecological regimes. These factors conspire to confuse the faunal picture in virtually all geographic regions. The second is the question of whether or not extinction events in the Late Triassic were synchronous, and how severe they really were. The second cannot be answered without recourse to the first, and it depends on fine-scale stratigraphy and taxonomy, including the strict use of low-level monophyletic groups (Padian, 1989b; Fraser and Sues, Chapter 23). Furthermore, mass extinctions cannot be declared without some metric of normalized rates of origination and extinction for the taxa under study, in order to see if the proposed "event" is really unusual (Padian et al., 1984; Padian and Clemens, 1985; Weems, 1992). These differences characterize in some respects the contrasting viewpoints of Benton (1991, and Chapter 22) and Fraser and Sues (Chapter 23).

Before dealing with this second issue, however, we need to return to the first, which is the reliability of intercontinental correlation and dating for the Late Triassic vertebrate-bearing terrestrial sediments. This question is enormously difficult. It is as if one were being asked to understand how an internal-combustion engine works by being presented with six different blueprints of engines, with the added problem that the plans have been made into jigsaw puzzles, and about half of the pieces are missing for each plan. Furthermore, the plans have salient differences: One is a V-8, another a slant-6, a third a Wankel rotary engine, and so on. Any combination of players could get several neat solutions to the puzzle, but could they make the engine run? At this point we are comparing spark plugs and gaskets. There are a great many holes and a great many parts that we cannot yet identify or use sensibly.

Since 1986 a great deal of work has been done on Late Triassic vertebrate faunas and correlations, notably in the North American East (by P. E. Olsen, H.-D. Sues, D. Baird, N. C. Fraser, and colleagues) and Southwest (by S. G. Lucas, A. P. Hunt, J. M. Parrish, and P. Murry) and in South America (A. Arcucci, F. Novas, P. C. Sereno, and colleagues). Lucas and Hunt in particular have proposed sweeping correlational changes in Late Triassic formations (e.g., Lucas and Hunt, 1989; Hunt and Lucas 1990, 1991a–d, 1992; Hunt, 1991; Lucas, 1991, in press; Lucas, Hunt, and Long, 1992) that will serve as the focus of biostratigraphic debate for some time to come. At present it is premature to evaluate this scheme in full, because parts of it are not yet published, and the authors should be given every opportunity to document the details supporting their correlational revision. At present, only some general review can be given.

One theme of recent revisions has been the compression of the tempo of the early evolution of dinosaurs. Traditional views had dinosaurs first appearing in the Early Triassic or even the Permian, based on the diversification already evidenced in the Late Triassic. More recent views (Padian, 1986) have used lines of evidence including paleomagnetism, tectonic events, radiometric dating, palynology, and vertebrate fossil correlations to conclude that the first dinosaurs probably evolved in the middle to late Carnian stage of the Late Triassic. It has been generally concluded that the South American dinosaur-bearing formations, such as the Santa Maria and Ischigualasto, which contain *Herrerasaurus, Staurikosaurus, Spondylosoma, Ischisaurus,* and *Frenguellisaurus,* are older than any others in the world that bear dinosaurs, with the exception of some formations (e.g., the Argana sequence of Morocco and lower formations in some basins of the Newark Supergroup) that have very fragmentary dinosaurian remains. However, Hunt (1991) has suggested that the picture was not so gradual, and instead that most of the first radiation of dinosaurs was explosive and was concentrated around the Carnian-Norian boundary. This has had the effect of compressing some of the age correlations of pre-Jurassic sediments in the Newark Supergroup (Olsen and Sues, 1986) based on palynology and correlation with the European type section, which is largely marine. Recently, Litwin, Traverse, and Ash (1991) have supported this compression, proposing that the earliest dates posited for the Newark Supergroup basins (e.g., Cornet, 1977; Cornet and Olsen, 1985) may not be quite so old as Ladinian, but are more likely middle Carnian and later. Radiometric dates recently obtained from the Ischigualasto Formation of South America (Rogers et al., 1993) may further support this compression. So far, strata from the North American Southwest have escaped radiometric dating, partly because of the difficulty of obtaining reliable dates from the bentonite-laden mudstones and siltstones that have expanded and contracted with moisture and desiccation over millions of years.

Hunt and Lucas (1990, 1991b–d, 1992; Lucas and Hunt, 1989; Hunt, 1991; Lucas et al., 1992) have developed the concept of the "biochron" to establish far-reaching correlations of Triassic terrestrial vertebrates. They begin by recognizing a deep division between typical "terrestrial" and "aquatic" vertebrate faunas spanned by few taxa. Their proposal for revised correlations centers on using the most conspicuous and common taxa from each environment to characterize biochrons, and using other unique taxa (such as rhynchosaurs and dicynodonts) to tie in geographically disjunct strata and faunas. Their phytosaur biochrons proceed through a sequence of *Paleorhinus*, *Rutiodon*, *Pseudopalatus*, and unnamed phytosaur taxa spanning the Carnian–Norian; aetosaurs proceed through a sequence of *Longosuchus*, *Stagonolepis*, *Typothorax*, and *Redondasuchus*; and among rhynchosaurs, *Otischalkia* and an undescribed/undiagnostic form in the early late Carnian predate *Scaphonyx*, from the late Carnian, while *Hyperodapedon* seems to span both intervals.

There is no question that this series of complex correlations and revisions has a great deal to recommend it, and at present it is the scheme that will require the most intensive examination and testing. The question is how this testing should be done. It is all too easy to accept the broad-brush picture wholesale, figuring that we are not likely to gain better resolution at our present distance, or alternatively to dismiss the scheme as too general and too poorly resolved both stratigraphically and temporally (not to mention taxonomically) to be viable. Either extreme would be a mistake. It is certainly too early to propose macroevolutionary hypotheses of extinction and turnover, though Benton (Chapter 22) is certainly justified in regarding these correlations as fair game for erecting such patterns. Fraser and Sues (Chapter 23) object to the broad scale of family- and stage-level data that Benton uses; if we have learned anything from the past decade (especially from the controversies over dinosaurian extinction at the Cretaceous–Tertiary boundary!), it is that coarse-grain analyses are bound to give us artificially concentrated and truncated ranges. Fraser and Sues understandably object to some of Benton's compilations, because, for example, the uppermost Norian in the southwestern United States is very poorly known and very nearly barren of vertebrate fossils, so we have a very poor idea of what was going on there. Conversely, there is supposed to be a rich fauna in the uppermost Norian of the Newark Supergroup that argues for a catastrophic end-Triassic extinction (Olsen et al., 1987), but inasmuch as this

has not been published and documented, we do not know on what evidence we are supposed to base our estimates of macroevolutionary change at this boundary (Padian, 1988; Benton, Chapter 22).

These questions are those of a field that is healthy and in the throes of attacking its major problems. We all want the answers to larger questions. But Darwin had little patience for sweeping generalizations in advance of all the facts. In his view, a theory, if it is wrong, can do little harm, because it will soon be forgotten; but a "fact" that is wrong does far more harm, because it is picked up and repeated and so is more difficult to dislodge from consensual understanding.

The facts on which biostratigraphy is based essentially comprise the taxonomic identifications of specimens found in faunas. "Facts," which basically are confirmed observations (which frequently can be found to be incorrect), in this instance are exemplified by species identifications of specimens in faunas. These frequently are based on poorly preserved and incomplete specimens, assigned to taxa whose validity is argued among specialists. Overlain on this shifting base are the differing approaches to biostratigraphic correlation: Should we use overall similarity of faunas, or "index fossils" ("marker taxa")? Are first appearances better than last appearances? Are some taxa more reliable than others? Are other lines of evidence (palynology, radiometric dates with their inevitable margin of error, magnetostratigraphy with its problems of correlation and contamination) more reliable, and (or) how should they be invoked or tested against vertebrate fossil biostratigraphy?

Lucas and Hunt have proposed a broad-ranging and internally consistent view of Late Triassic terrestrial biostratigraphy, and for this they deserve credit for advancing the status of the problem. Like any such extensive scheme, the finer points probably will be debated for a long time to come, and these debates can have greater or lesser effects on the overall picture. For example, they separate the *Paleorhinus* and *Rutiodon* biochrons in most areas, while acknowledging the overlap of these taxa in some areas. Overlap would be expected, certainly. But, for example, in the lower Petrified Forest Member of the Chinle Formation (Blue Mesa Formation of the Chinle Group in Lucas's terminology) of northern Arizona, Hunt and Lucas (1991c; Lucas et al., 1992) report the co-occurrence of these two taxa, based on a single specimen of *Paleorhinus* reported from the Downs Quarry (contiguous with the Placerias Quarry) near St. Johns. This specimen, which is partial and very small, is the only example reported from anywhere in the area, even though more than 3,000 macrofossil bones and 4,000 microvertebrates have been recovered from this quarry (Kaye and Padian, Chapter 9), and there are no records of *Paleorhinus* from the lower unit of the Petrified Forest

Member at the Petrified Forest National Park. This is not to say that the identification of the Downs Quarry specimen is incorrect, but it has to be weighed against other lines of evidence. For example, the size of the specimen suggests that it may well be juvenile. *Paleorhinus* is the only phytosaur in which the nares are anterior to the antorbital fenestrae; but of course this is the condition in all other vertebrates. It would be expected, though we do not have a good ontogenetic sequence in phytosaurs, that through ontogeny the nares would migrate posteriorly and dorsally, as they do in fact phylogenetically (Ballew, 1989; Hunt and Lucas, 1991c). Young phytosaurs may well have retained the out-group condition, perhaps leading us incorrectly to identify a juvenile of *Rutiodon* or another phytosaur as *Paleorhinus*. The scarcity of other specimens of the latter taxon, in my view, calls this into question.

Lucas et al. (1992) go a step further in temporally equating nearly all Carnian dinosaur-bearing deposits [following Hunt (1991) and other references in this series]. They infer that deposits bearing *Paleorhinus*, or putatively correlative with other horizons that bear taxa that are found in *Paleorhinus*-bearing horizons elsewhere, are all equally old. This is a fair enough first step, and they have tried to support this "one-taxon" biostratigraphy by showing the ranges of other taxa that usually co-occur. However, the presence of rhynchosaurs in the Placerias–Downs Quarry has been disputed (Parrish, 1989, p. 237; Hunt and Lucas, 1991d, p. 930), *Placerias* is thus far not known from other southwestern horizons, several aetosaur genera co-occur in the lower unit of the Petrified Forest Member (Long and Ballew, 1985; Kaye and Padian, Chapter 9), and some dinosaurian identifications from this unit are questionable or unsubstantiated at present. [Benton (Chapter 22, Figure 22.1) correlates the successive phytosaurid and aetosaurian biochrons of Hunt and Lucas, but the foregoing evidence supports the contention of the original authors that these may not be readily equated.]

Kaye and Padian (Chapter 9) documented remains of theropod, "prosauropod," and ornithischian dinosaurs from the Placerias–Downs Quarry, but found no evidence for "basal" dinosaurian taxa in the microvertebrate assemblage. Murry and Long (1989, p. 41) report the presence of a "staurikosaurid," without documentation; this identification, whether or not it is valid, has been repeated uncritically by other authors. It would seem to be important to the advance of this discussion to substantiate especially the most potentially contentious or unusual occurrences with voucher numbers and illustrations; here, as frequently happens, publication of some results understandably outstrips that of others, but the result is that not all the playing pieces are yet on the table (Olsen et al., 1987, 1988; Padian, 1988). On balance, for example, even if

"staurikosaurids" or other basal dinosaurs turn out to be present in the Placerias Quarry, it will make little difference to biostratigraphy, because a specimen tentatively referred to as "staurikosaurid" (Murry and Long, 1989, p. 35) was collected from the upper unit of the Petrified Forest Member in 1985, thus extending the range of this taxon into the Norian [according to the palynostratigraphic scheme of Litwin et al. (1991)].

Furthermore, a fauna such as that of the Placerias Quarry, which contains theropod, "prosauropod," ornithischian, and possibly basal dinosaurs, would not be expected to be as old as a fauna that contained only basal dinosaurs, based on phylogenetic grounds. The basal dinosaurs had to evolve before the appearance of eudinosaurs (including saurischians and ornithischians), even though certainly basal dinosaurs could (and did) persist into times populated by eudinosaurs. The South American Santa Maria and Ischigualasto faunas contain basal dinosaurs, with only suggestive remains of an ornithischian (*Pisanosaurus*), though the presence of ornithischians indicates that their sister-group, the saurischians, must have evolved by then. This example, following the work of Norell (1992), underscores the importance not only of well-documented, fine-scale taxonomy but also of phylogenetic analysis to the understanding of biostratigraphic trends and evolutionary patterns.

Testing the pattern: progress and prognosis

Padian (1986) suggested five major avenues of research, based on progress in Triassic–Jurassic faunal questions as of 1985. First, Jurassic faunas would require more study, with finer taxonomic analyses. Some progress on this question has already been made in the publications cited earlier, and this volume contains chapters discussing further advances in Early Jurassic faunas in Britain, Canada, Mexico, and southern Africa. Several papers have documented Middle Jurassic faunas, another important avenue of interest noted by Padian (1986). Two other avenues concern the tempo and mode of events across the Triassic–Jurassic boundary, which have been addressed in many papers (e.g., Olsen et al., 1987, 1988; Padian, 1988; Benton, 1991, and Chapter 22; Fraser and Sues, Chapter 23), and the utility of testing patterns of evolutionary change against potential biases of environmental change and preservation (e.g., Padian, 1988; Benton, 1991, and Chapter 22; Hunt, 1991; Weems, 1992).

Perhaps the question on which the least progress has been made is that of discerning the biological factors that may have unified the groups that became extinct at the end – or perhaps it would be better to say "by the end" – of the Triassic. Hunt (1991) has recently reviewed the hypotheses of competition and opportunism that could account for such change at the boundary; an extraterrestrial correlation with end-Triassic extinctions has been suggested by Olsen et al. (1987, 1988) and discussed most recently by Weems (1992). Weems has taken particular issue with the patterns elucidated by Olsen and Sues (1986), compared with the catastrophic extinction scenario discussed by Olsen and his colleagues in subsequent articles. The central difficulty, it seems, is that some of the baseline data on which the scenario is built have not yet been published or are at odds with previously published data. We have already seen that at the time of this writing, similar difficulties are faced with the sweeping revision of Late Triassic correlations of continental strata proposed by Lucas and Hunt. Again, the most productive course would seem to be the documentation of specimens and faunas that support these pictures, as well as frank discussion of apparent anomalies and alternative scenarios. These issues require close attention before unifying biological factors can be realistically assessed.

Documenting specimens and applying consistent taxonomic revision to faunal components can pave only part of the road to progress on these questions. Also needed is a thorough consideration of the phylogenetic relationships of taxa in each group used for biostratigraphic correlation. The sequence of diversification of taxa in a group necessarily implies the sequence of first appearances in the fossil record. The fact that our faunas may not record these sequences exactly is one of the puzzles left to all biostratigraphers; "primitive" taxa may persist, but more "derived" taxa cannot appear before more "primitive" ones first appear. These factors must be considered along with the assessment of *all* taxa in a fauna, not just isolated ones, if questions about faunal change across the Triassic–Jurassic boundary are to advance beyond their current ambiguous status.

References

Ballew, K. 1989. A phylogenetic analysis of Phytosauria from the Late Triassic of the western United States. Pp. 309–339 in S. G. Lucas and A. P. Hunt (eds.), *The Dawn of the Age of Dinosaurs in the American Southwest*. Albuquerque: New Mexico Museum of Natural History.

Benton, M. J. 1986a. The Late Triassic tetrapod extinction events. Pp. 303–320 in K. Padian (ed.), *The Beginning of the Age of Dinosaurs: Faunal Change across the Triassic–Jurassic Boundary*. Cambridge University Press.

1986b. More than one event in the Late Triassic mass extinction. *Nature* 321: 857–861.

1991. What really happened in the Late Triassic? *Historical Biology* 5: 263–278.

Clark, J. M., and D. E. Fastovsky. 1986. Vertebrate biostratigraphy of the Glen Canyon Group in northern Arizona. Pp. 285–302 in K. Padian (ed.), *The Beginning of the Age of Dinosaurs: Faunal Change*

across the Triassic–Jurassic Boundary. Cambridge University Press.

Colbert, E. H. 1981. A primitive ornithischian dinosaur from the Kayenta Formation of Arizona. *Bulletin of the Museum of Northern Arizona* 53: 1–61.

1986. Historical aspects of the Triassic–Jurassic boundary problem. Pp. 9–20 in K. Padian (ed.), *The Beginning of the Age of Dinosaurs: Faunal Change across the Triassic–Jurassic Boundary*. Cambridge University Press.

Cornet, B. 1977. The palynostratigraphy and age of the Newark Supergroup. Ph.D. dissertation, Pennsylvania State University.

Cornet, B., and P. E. Olsen. 1985. A summary of the biostratigraphy of the Newark Supergroup of eastern North America, with comments on early Mesozoic provinciality. Pp. 67–81 in *III Congreso Latinoamericano de Paleontologia, Mexico, Memoria*.

Hunt, A. P. 1991. The early diversification pattern of dinosaurs in the Late Triassic. *Modern Geology* 16: 43–60.

Hunt, A. P., and S. G. Lucas. 1990. Re-evaluation of "*Typothorax*" *meadei*, a Late Triassic aetosaur from the United States. *Paläontologische Zeitschrift* 64: 317–328.

1991a. *Rioarribasaurus*, a new name for a Late Triassic dinosaur from New Mexico (USA). *Paläontologische Zeitschrift* 65: 191–198.

1991b. A new aetosaur from the Redonda Formation (Late Triassic: middle Norian) of east-central New Mexico, USA. *Neues Jahrbuch für Geologie und Paläontologie, Monatshefte* 1991: 728–736.

1991c. The *Paleorhinus* biochron and the correlation of the non-marine Upper Triassic of Pangaea. *Palaeontology* 34: 487–501.

1991d. A new rhynchosaur from the Upper Triassic of West Texas, and the biochronology of Late Triassic rhynchosaurs. *Palaeontology* 34: 927–938.

1992. The first occurrence of the aetosaur *Paratypothorax andressi* (Reptilia, Aetosauria) in the western United States and its biochronological significance. *Paläontologische Zeitschrift* 66: 147–157.

Litwin, R. J., A. Traverse, and S. R. Ash. 1991. Preliminary palynological zonation of the Chinle Formation, southwestern U.S.A., and its correlation to the Newark Supergroup (eastern U.S.A). *Review of Palaeobotany and Palynology* 68: 269–287.

Long, B. A., and K. L. Ballew. 1985. Aetosaur dermal armor from the Late Triassic of southwestern North America, with special reference to material from the Chinle Formation of Petrified Forest National Park. *Bulletin of the Museum of Northern Arizona* 54: 45–68.

Lucas, S. G. 1991. Dinosaurs and Mesozoic biochronology. *Modern Geology* 16: 127–138.

In press. The Chinle Group: Revised stratigraphy and chronology of Upper Triassic nonmarine strata in the western United States. *Bulletin of the Museum of Northern Arizona*.

Lucas, S. G., and A. P. Hunt (eds.). 1989. *The Dawn of the Age of Dinosaurs in the American Southwest*. Albuquerque: New Mexico Museum of Natural History.

Lucas, S. G., A. P. Hunt, and R. A. Long. 1992. The oldest dinosaurs. *Naturwissenschaften* 79: 171–172.

Murry, P. A., and R. A. Long. 1989. Geology and paleontology of the Chinle Formation, Petrified Forest National Park and vicinity, Arizona, and a discussion of vertebrate fossils of the southwestern Upper Triassic. Pp. 29–64 in S. G. Lucas and A. P. Hunt (eds.), *The Dawn of the Age of Dinosaurs in the American Southwest.* Albuquerque: New Mexico Museum of Natural History.

Norell, M. A. 1992. Taxic origin and temporal diversity: the effect of phylogeny. Pp. 89–118 in M. J. Novaceck and Q. D. Wheeler (eds.), *Extinction and Phylogeny.* New York: Columbia University Press.

Olsen, P. E., and P. M. Galton. 1977. Triassic–Jurassic extinctions: Are they real? *Science* 197: 983–986.

1984. A review of the reptile and amphibian assemblages from the Stormberg of southern Africa, with special emphasis on the footprints and the age of the Stormberg. *Palaeontologia Africana* 25: 87–110.

Olsen, P. E., A. R. McCune, and K. S. Thomson. 1982. Correlation of the early Mesozoic Newark Supergroup by vertebrates, principally fishes. *American Journal of Science* 282: 1–44.

Olsen, P. E., and K. Padian. 1986. Earliest records of *Batrachopus* from the southwestern United States, and a revision of some early Mesozoic crocodylomorph ichnogenera. Pp. 259–274 in K. Padian (ed.), *The Beginning of the Age of Dinosaurs: Faunal Change across the Triassic–Jurassic Boundary.* Cambridge University Press.

Olsen, P. E., R. W. Schlische, and P. J. W. Gore (eds.). 1989. *Tectonic, Depositional, and Paleoecological History of Early Mesozoic Rift Basins, Eastern North America.* Field Trip Guidebook T351, 28th International Geological Congress. Washington, D. C.: American Geophysical Union.

Olsen, P. E., N. H. Shubin, and M. H. Anders. 1987. New Early Jurassic tetrapod assemblages constrain Triassic–Jurassic tetrapod extinction event. *Science* 237: 1025–1029.

1988. Triassic–Jurassic extinctions [reply]. *Science* 241: 1359–1360.

Olsen, P. E., and H.-D. Sues. 1986. Correlation of continental Late Triassic and Early Jurassic sediments, and patterns of the Triassic–Jurassic tetrapod transition. Pp. 321–351 in K. Padian (ed.), *The Beginning of the Age of Dinosaurs: Faunal Change across the Triassic–Jurassic Boundary.* Cambridge University Press.

Padian, K. (ed.). 1986. *The Beginning of the Age of Dinosaurs: Faunal Change across the Triassic–Jurassic Boundary.* Cambridge University Press.

1988. Triassic–Jurassic extinctions. *Science* 241: 1358–1359.

1989a. Presence of the dinosaur *Scelidosaurus* indicates Early Jurassic age for the Kayenta Formation (Glen Canyon Group, northern Arizona). *Geology* 17: 438–441.

1989b. Did "thecodontians" survive the Triassic? Pp. 401–414 in S. G. Lucas and A. P. Hunt (eds.), *The Dawn of the Age of Dinosaurs in the American Southwest.* Albuquerque: New Mexico Museum of Natural History.

Padian, K., and W. A. Clemens. 1985. Terrestrial vertebrate diversity: episodes and insights. Pp. 41–96 in J. W. Valentine (ed.), *Phanerozoic Diversity Patterns.* Princeton University Press.

Padian, K., et al. 1984. The possible influence of sudden events on biological radiations and extinctions (group report). Pp. 77–102 in H. D. Holland and A. F. Trendall (eds.), *Patterns of Change in Earth Evolution.* Berlin: Springer-Verlag.

Parrish, J. M. 1989. Vertebrate paleoecology of the Chinle Formation (Late Triassic) of the southwestern United States. *Palaeogeography, Palaeoclimatology, Palaeoecology* 72: 227–247.

Pipiringos, G. N., and R. B. O'Sullivan. 1978. Principal unconformities in Triassic and Jurassic rocks, western interior United States – a preliminary survey. *United States Geological Survey Professional Paper* 1035B: B1–B43.

Rogers, R. R., C. C. Swisher, P. C. Sereno, A. M. Monetta, C. A. Forster, and R. N. Martinez. 1993. The Ischigualasto tetrapod assemblage (Late Triassic, Argentina) and ^{40}Ar/^{39}Ar dating of dinosaur origins. *Science* 260: 794–797.

Sues, H.-D. 1985. First record of the tritylodontid *Oligokyphus* (Synapsida) from the Lower Jurassic of western North America. *Journal of Vertebrate Paleontology* 5: 328–335.

Sun, A. L., and K. H. Cui. 1986. A brief introduction to the Lower Lufeng saurischian fauna (Lower Jurassic: Lufeng, Yunnan, People's Republic of China). Pp. 275–278 in K. Padian (ed.), *The Beginning of the Age of Dinosaurs: Faunal Change across the Triassic–Jurassic Boundary.* Cambridge University Press.

Weems, R. E. 1992. The "terminal Triassic catastrophic extinction event" in perspective: a review of Carboniferous through Early Jurassic terrestrial vertebrate extinction patterns. *Palaeogeography, Palaeoclimatology, Palaeoecology* 94: 1–29.

25

Field guide to Late Triassic tetrapod sites in Virginia and North Carolina

PAUL E. OLSEN and ANNIKA K. JOHANSSON

Geological, climatological, and environmental context of Triassic vertebrate localities in Virginia and North Carolina

The tetrapod-bearing localities visited during the Workshop on Small Early Mesozoic Tetrapods are in the early Mesozoic Newark Supergroup of eastern North America. These units are the exposed fill of rift basins, mostly half-graben, that resulted from the incipient breakup of Pangaea (Figures 25.1–25.3). Recent paleomagnetic studies (Witte and Kent, 1989; Witte, Kent, and Olsen, 1991) indicate that Virginia lay on or near the equator during the early Mesozoic. The coals and well-developed lacustrine strata preserved in Newark basins in Virginia and North Carolina permit the interpretation of a narrow humid zone around the equator. Symmetrically disposed more arid zones existed to the north and south, rather than the previously hypothesized much more complex pattern (Manspeizer, 1982; Olsen, Schlische, and Gore, 1989) that invoked orographic and altitudinal effects.

About 12 degrees of latitude are represented by the Newark Supergroup (Figure 25.1 and 25.2), and a full range of humid to fully arid basins is preserved. This is reflected most clearly in the lacustrine strata of the basins, which can be divided into three basic types of sequences (in order of increasing aridity): the Richmond, Newark, and Fundy types of lacustrine sequences (Olsen, 1991) (Figure 25.4). These different types of lacustrine sedimentary complexes are characterized by the magnitude and frequency of their lake-level changes. Vertebrate localities visited during the conference are in lacustrine sequences of the Richmond type transitional to the Newark type (Stop 2, Tomahawk) and Newark type (Stop 1, Culpeper Crushed Stone Quarry, and Stop 3, Solite Quarry).

Richmond-type sequences have been recognized in the Richmond, Taylorsville, Farmville (with adjacent small basins), and Deep River basins (Figures 25.2 and 25.3). These sequences are characterized by considerable thicknesses (> 25 m) of deep-water shales showing no signs of subaerial exposure. They typically also contain relatively extensive coals and highly bioturbated shallow-water and fluvial sequences, suggesting persistently humid conditions. Published interpretations of seismic data from the Richmond basin (Cornet and Olsen, 1990) and proprietary data from the Taylorsville basin indicate that high-relief sedimentary features such as large-scale deltas and fan deltas developed during lake highstands, and large-scale unconformities developed during lowstands.

Newark-type sequences occur in the Dan River/ Danville, Culpeper, Gettysburg, Newark, Pomperaug, Hartford, and Deerfield basins (Figure 25.2). Changes in precipitation resulted in dramatic changes in lake levels from perhaps 200 m or more in depth to complete exposure, producing repetitive sequences of sedimentary cycles called Van Houten cycles (Olsen, 1986, 1991). Lacustrine sequences transitional in kind between the Newark and Richmond types occur in the Richmond and Taylorsville basins.

Van Houten cycles were produced by the rise and fall of lake levels, apparently controlled by tropical monsoonal precipitation under the effects of Milankovitch-type orbitally driven changes in solar insolation (Olsen, 1986). The 21,000-year cycle of the precession of the equinoxes controlled the Van Houten cycles, and the larger-scale cycles were produced by modulation of the precession cycle by eccentricity cycles of 98,000, 123,000, and 413,000 years (Figure 25.4). This hierarchy of cycles is well displayed in power spectra of Newark-type lacustrine sequences in general, and specifically the sections at Stop 1 and Stop 3. The extreme fluctuations and high frequency of lake-level cycles apparently inhibited the buildup of high-relief sedimentary features or the development of large-scale unconformities, in dramatic contrast to Richmond-

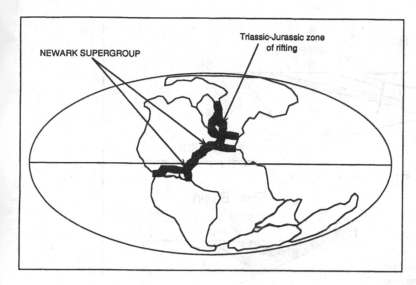

Figure 25.1. Pangaea, showing the position of the Triassic–Jurassic rifting zone. Continental positions from Witte et al. (1991).

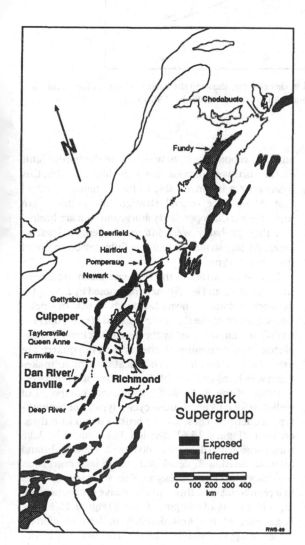

Figure 25.2. The distribution of the Newark Supergroup. (From Olsen et al., 1989.)

type sequences. Consequently, lacustrine strata are characterized by extreme lateral continuity and very slow lateral facies transitions, as was unequivocally demonstrated in the recent coring of the Newark basin (Olsen, 1991).

Fundy-type lacustrine sequences occur in eastern North America only in the Fundy basin of Nova Scotia and New Brunswick, Canada, and are characterized by the presence of sand-patch cycles (Smoot and Olsen, 1988). Sand-patch cycles are lacustrine sequences produced by alternating shallow perennial lakes and playas with well-developed efflorescent salt crusts. Eolian and minor fluvial clastics trapped in the efflorescent crusts produced a characteristic sand-patch fabric, which gives the cycles its name. Associated deposits include abundant gypsum nodules, bedded salts and salt-collapse structures, and eolian sand dunes. As in the Newark basin, depositional relief probably was very low. The Fundy basin was not visited during the workshop field trips. However, while vertebrates are very rare in the sand-patch cycles, associated fluvial and alluvial sequences have proved to be relatively fossiliferous (Baird and Take, 1959; Olsen, Shubin, and Anders, 1987). Fundy-type sequences definitely occur in the Argana and adjacent basins of Morocco (Smoot and Olsen, 1988) and are very similar to the Triassic-age sequences of the Germanic basin, the North Sea, Greenland, and Spitzbergen, where they have also been fossiliferous.

Most Newark Supergroup vertebrate localities are in lacustrine facies or lake-margin deposits, most likely because these deposits make up the majority of the preserved basin fill. The foregoing classification of facies types serves as an overall context for these vertebrate localities and provides a framework for discussing the similarities and differences among Triassic–Jurassic vertebrate assemblages.

The field-stop descriptions in this guide are in part

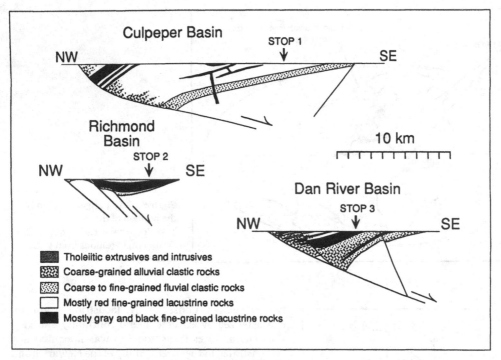

Figure 25.3. Cross-sections of the basins with sites described in this chapter, showing the basic half-graben form and the relative position of each site. Vertical and horizontal scales equal. (Adapted from Schlische, 1990.)

modified from Olsen et al. (1989) and Cornet and Olsen (1990); other authors who contributed to the text in those two field guides are credited as follows: J. P. Smoot, P. J. W. Gore, and R. W. Schlische for Stop 1; H.-D. Sues, B. Cornet, and P. J. W. Gore for Stop 2; and P. J. W. Gore for Stop 3.

Stop 1: Culpeper Crushed Stone Quarry

Exposures in the quarry of the Culpeper Crushed Stone Company comprise about 80 m of lacustrine, thermally altered mudstone and siltstone of the Balls Bluff Siltstone of apparent Norian age (B. Cornet and S. Fowell, pers. commun., 1991) (Figure 25.5, Table 25.1). The quarry lies in the southern part of the Culpeper basin (Figure 25.5), making it the southernmost Newark Supergroup basin known to contain sedimentary rocks and basalt flows of Early Jurassic age. The basin contains the longest depositional record in the southern part of the rift system, with strata of Carnian to late Hettangian age, an interval of about 25 million years. The maximum thickness of the sedimentary basin fill is calculated at 7,900–9,000 m (Table 25.1), although such a thickness might not be preserved at any one place.

The Culpeper basin forms the southern part of the Newark-Gettysburg-Culpeper basin system, which probably represents a single, very large rift basin with adjacent subbasins, all of which have the border fault on the northwest or west side of the basin. In size, this basin system is comparable to the Tanganyika rift of East Africa. The overall stratigraphic patterns are similar in the Culpeper, Gettysburg, and Newark basins, and they probably were interconnected, at least at times. All lacustrine strata of the Culpeper basin are of the Newark type.

Triassic-age strata of the Culpeper basin are extensively intruded and locally metamorphosed by a complex network of diabase sheets (as much as 600 m thick). The sequence in the Culpeper Crushed Stone Quarry is slightly metamorphosed by the nearby Mt. Pony diabase pluton. As a consequence, the rocks in the quarry are extensively cemented with silicates, which makes them very hard and suitable for construction purposes.

Strata of the Balls Bluff Siltstone are composed of well-developed Van Houten cycles typical of Newark-type sequences. These can be broken down into three divisions (Figure 25.6): division 1, produced by lake transgression and deepening; division 2, a lake highstand interval; division 3, developed during lake regression and lowstand. The expression of these divisions varies in a regular fashion through successive cycles, producing a hierarchy of compound cycles (Figure 25.6), the fingerprint of the Milankovitch pattern of climate cycles of approximately 21,000, 109,000, 413,000, and 2,000,000 years. The quarry section can be

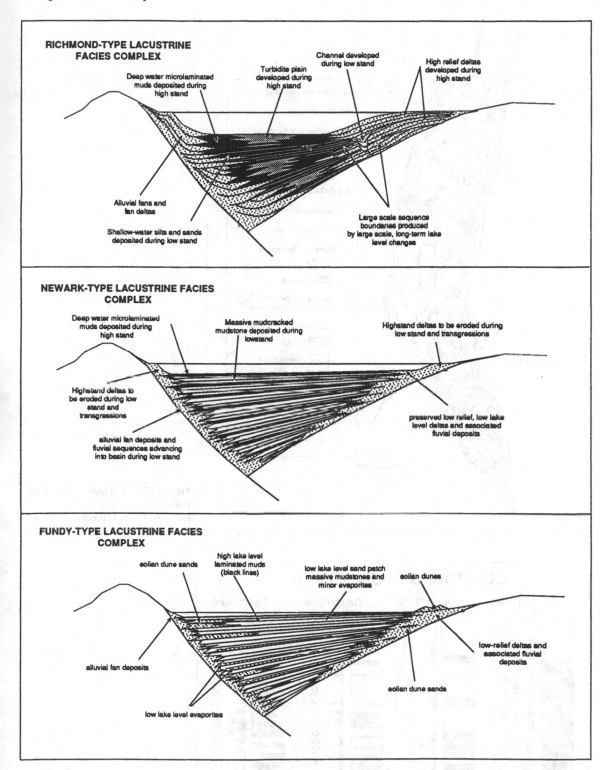

Figure 25.4. Major types of lacustrine complexes in the Newark Supergroup. These cross-sections are highly diagrammatic and show only the relationships of different facies at times when lake level was, on average, high. Other than cyclicity, no historical change through the basin history is implied.

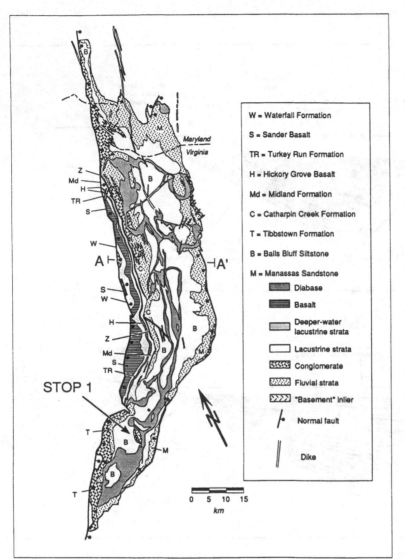

Figure 25.5. Geological map of the Culpeper basin showing the location of Stop 1. (Courtesy of Roy W. Schlische.)

W = Waterfall Formation

S = Sander Basalt

TR = Turkey Run Formation

H = Hickory Grove Basalt

Md = Midland Formation

C = Catharpin Creek Formation

T = Tibbstown Formation

B = Balls Bluff Siltstone

M = Manassas Sandstone

Diabase

Basalt

Deeper-water lacustrine strata

Lacustrine strata

Conglomerate

Fluvial strata

"Basement" inlier

Normal fault

Dike

0 5 10 15
km

STOP 1

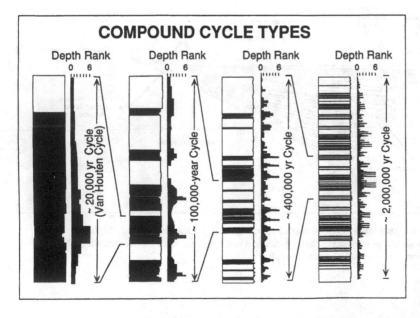

COMPOUND CYCLE TYPES

Depth Rank
0   6

Depth Rank
0   6

Depth Rank
0   6

Depth Rank
0   6

~ 20,000 yr Cycle (Van Houten Cycle)

~ 100,000-year Cycle

~ 400,000 yr Cycle

~ 2,000,000 yr Cycle

Figure 25.6. Hierarchy of cycle types in Newark-type lacustrine sequences.

Table 25.1. *Stratigraphy of the Culpeper basin, Virginia and Maryland*

Units		Thick-	Age	Description
Lindholm (1979)	Lee and Froelich (1989)	ness (m)		
Waterfall Fm.	Waterfall Fm.	1500–1719	Hettangian	Mostly red, fine to coarse, cyclical shallow-water lacustrine clastics; some gray to black deep-water lacustrine clastics and minor limestone
	Millbrook Quarry Mb.	450	Hettangian	Alluvial-fan conglomerate; contains Jurassic basalt clasts
Buckland Fm.	Sander Basalt	140–600	Hettangian	Tholeiitic basalt flow (high Fe); interbedded sediments
	Turkey Run Fm.	150–330	Hettangian	Mostly red and gray, fine to coarse fluviolacustrine clastics; one gray to black lacustrine claystone and limestone
	Hickory Grove Basalt	80–212	Hettangian	Tholeiitic flows; interbedded clastics and basalt-clast conglomerate
	Midland Formation	150–300	Hettangian	Mostly red and gray, fine to coarse fluviolacustrine clastics; some gray to black cyclical lacustrine claystone and limestone
	Mt. Zion Church Bst.	3–180	Hettangian	Tholeiitic basalt flows; interbedded red clastics
Bull Run Fm.	Catharpin Creek Fm.	500	Early Norian	Red to gray, fine to coarse fluvial(?) clastics
	Goose Creek Mb.	500	Early Norian	Fluvial conglomerate; greenstone clasts derived from Blue Ridge
	Tibbstown Fm.	300–640	Early Norian	Fluvial(?) arkosic sandstone
	Mountain Run Mb.	0–640	Early Norian	Fluvial and alluvial conglomerate containing greenstone clasts
	Handricks Mtn. Mb.	500	Early Norian	Fluvial and alluvial conglomerate
	Balls Bluff Siltstone	80–1690	Norian	Mostly red, fine to coarse, cyclic shallow-water lacustrine clastics; some gray to black, cyclic, deep-water lacustrine clastics and carbonates
	Leesburg Mb.	40–1070	Norian	Fluvial and alluvial conglomerate; Predominantly carbonate clasts
Manassas Ss.	Manassas Ss. Poolesville Mb.	200–1000	Early Carnian(?)	Red, fine to coarse, arkosic fluvial clastics
	Tuscarora Creek Mb.	21–67	Early Carnian(?)	Fluvial/alluvial conglomerate; predominantly carbonate clasts
	Rapidan Mb.	70–140	Early Carnian(?)	Fluvial/alluvial conglomerate; clasts derived from Blue Ridge
Reston Fm.	Reston Mb.	3–100	Early Carnian(?)	Fluvial/alluvial conglomerate; clasts derived from Piedmont

Source: From Olsen et al. (1989).

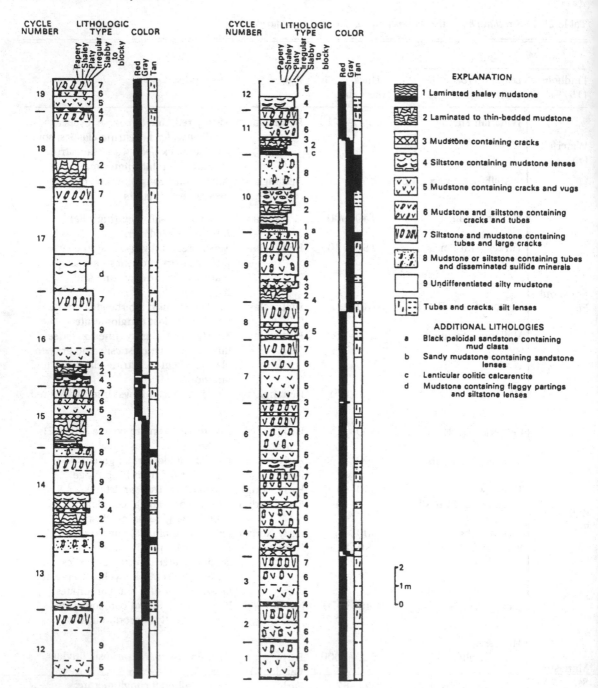

Figure 25.7. Section in Culpeper Crushed Stone Quarry showing different lithologic types. (Adapted from Smoot, in Olsen et al., 1989.)

divided into nine lithologic types (Figure 25.7) that reflect this cyclicity: (1) Dark gray to black laminated shaley mudstones without mudcracks are the deposits of a deep, perennial lake. (2) Dark gray or purplish red platy mudstones with internal pinch-and-swell layering, reflecting oscillatory ripples, and large sinuous mudcracks are shallow-lake deposits that were inter-

mittently subaerially exposed. (3) Gray or purplish red mudstones with layering similar to lithology 2 but broken into breccialike blocks by numerous polygonal mudcracks – these are deposits of a subaerially exposed lake sequence that was wetted and dried repeatedly. (4) Thin beds of tan-weathering siltstone with mudstone partings broken into polygonal, concave-upward

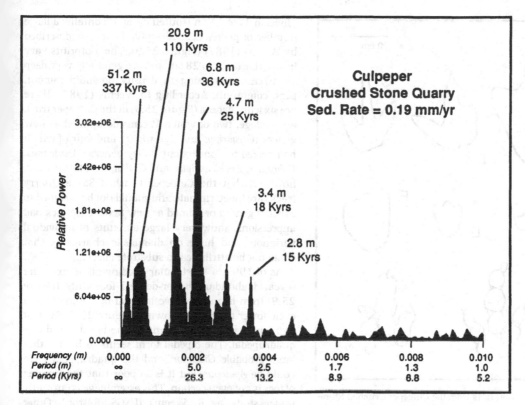

Figure 25.8. Power spectrum for Culpeper Crushed Stone Quarry based on the ranking of lithologies seen in Figure 25.7.

curls reflect a shallow lake or sheetflood deposit. (5) Massive red or gray mudstone with abundant narrow, jagged mudcracks and spheroidal to flattened cement-filled vugs was produced in a playalike mudflat. The vugs probably are vesicles formed from air trapped during flooding events that brought in the sediments. (6) Massive red or gray mudstone similar to lithology 5, but also containing abundant cement-filled tubes that are root structures is a playalike deposit disrupted by roots. (7) Massive red or gray mudstones with abundant millimeter- to centimeter-scale cement-filled tubes (root structures) and deep, narrow, sinuous mudcracks are the products of a vegetated mudflat. The deeper cracks suggest the mudflat was wetter than that of lithology 5. (8) Massive gray to black sandy mudstone or siltstone with tubular root structures and deep, narrow, sinuous mudcracks, sulfide mineralization, and nodules is a vegetated, shallow-water, shoreline deposit that was intermittently subaerially exposed. (9) Undifferentiated, masssive red or gray mudstones, inaccessible and examined by binoculars only, appear to be combinations of lithologies 5 and 6. Another important lithological type ("c" of Figure 25.7) is a lenticular oolitic limestone with stromatolites and abundant sulfide minerals. This is the unit described by Young and Edmundson (1954) and Carozzi (1964),

and it appears to be a shoreline tufa deposit, possibly related to springs.

The foregoing lithologies are organized into Van Houten cycles, with layering that grades upward into massive rocks. Nineteen Van Houten cycles are shown on the stratigraphic column, with two end-member types: 2-1-2-3-4-6-7-8 (thick gray cycles) and 4-5-6-7 (thin red cycles). The thinner red cycles represent drier conditions and shallower lakes than the thicker gray cycles. Fourier analysis of the sediment fabrics outlined earlier reveals a strong hierarchy of cycle thicknesses, comparable in pattern to those seen in the Cow Branch Formation (Stop 3) (Figure 25.8) (Olsen, 1986; Olsen et al., 1989). Most important are the peaks at 4.7 m and 20.9 m, which correspond, respectively, to one mode of the precession cycle (~ 21,000 years) and the average of the high-frequency elements of the eccentricity cycle (~ 109,000 years), if a sedimentation rate of 0.19 mm/year is assumed. A peak at 51.2 m is present that probably corresponds to the 413,000-year cycle of eccentricity, but the section is too short (80 m) to resolve it. Interestingly, a peak at 6.8 m (~ 36,000 years) is present in the spectrum, but no corresponding cycle is obvious in the outcrop. This peak could correspond to the obliquity cycle of 41,000 years, although the match is not especially good. A

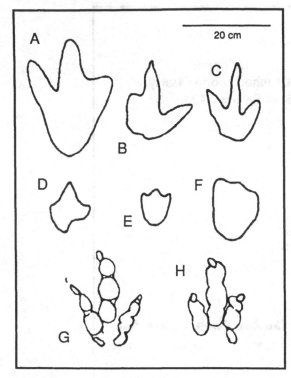

Figure 25.9. Footprints from the Culpeper Crushed Stone Quarry. (A–D) Probably *Grallator* spp.; (E, F) Indeterminate ?nondinosaurian forms. (G, H) *Grallator* (From Olsen et al., 1989, based on Weems, 1987.)

longer section is needed to resolve reliably the periodic elements present; however, correspondence with the predictions of orbital forcing is nonetheless rather good.

The Balls Bluff Siltstone is lithologically similar to the Gettysburg Shale of the Gettysburg basin and to the Passaic Formation of the Newark basin. This resemblance extends to the details of Van Houten cycles and to overall facies trends. The ages of the units are broadly comparable as well, reflecting the former connection of the basins and probably the lakes they contained. The type of sequence at the Culpeper Crushed Stone Quarry exposures of the Balls Bluff Siltstone appears to be restricted to the southern part of the Culpeper basin. The cycles here are generally thinner and have much more evidence of prolonged desiccation than the deposits in the northern part of the basin. The lacustrine deposits at this stop may have formed in a sediment-starved portion of the same lakes that produced the cycles to the north, or they may indicate a different climatic setting in a subbasin of the Culpeper.

Vertebrate ichnology

Lithology 8, just below the horizon marked "c" (Figure 25.7), is the lower part of a transgressive portion

(division 1) of a Van Houten cycle. It contains a large number of poorly defined reptile footprints described by Weems (1987) (Figure 25.9). The footprints vary in length from 8 to 28 cm and are generally very deep (3–10 cm), decreasing in density downdip (Smoot, pers. commun.). According to Weems (1987), there are six ichnogenera (Figure 25.9) in the Culpeper track assemblage, two of which Weems designated as new genera (*Gregaripus* and *Agrestipus*), and four of which he referred to Connecticut Valley genera (*Apatichnus*, *Eubrontes*, *Anchisauripus*, and *Grallator*). It is our view, however, that the Culpeper Crushed Stone Quarry tracks are indeterminate and should not be assigned to existing genera or named as new taxa. They lack pad impressions, show very large amounts of individual variation, and have no diagnostic characters that could not be attributed to substrate effects.

As of 1992, a spectacular bedding-plane exposure revealed abundant better-defined footprints (Figure 25.9) from a transgressive interval of a Van Houten cycle lower than that shown in Figure 25.7. At least two ichnotaxa are represented: one bipedal and one quadrupedal. The bipedal form seems to be a rather large probable *Grallator*, and the quadrupedal form could be *Apatopus*, but it is so poor that the number of toes is not even certain. This assemblage is currently under study by R. Weems (USGS, Reston). Other horizons produce footprints in the quarry, and one of us (Olsen) has found definitive *Rhynchosauroides* sp. and *Grallator* sp. on transported blocks. *Grallator*-type footprints, the size of those seen at this stop, are not known from pre–middle Norian strata of the Newark basin and suggest that the strata in this quarry could be as young as late middle Norian. This age assignment is supported by preliminary graphic correlation of the Culpeper section with the Newark basin (S. Fowell, pers. commun., 1991) and analysis of sedimentary-cycle fabrics and structures (J. P. Smoot, pers. commun., 1992).

Stop 2: Tomahawk locality, Richmond basin

The Tomahawk site (Figure 25.10) is located just east of the east branch of Tomahawk Creek on what used to be the Tomahawk Plantation (Sues and Olsen, 1990). The excavation site is in the Tomahawk Creek Member of the Turkey Branch Formation, which makes up the upper major lacustrine interval in the Richmond basin (Table 25.2). Excavations for the Powhite Parkway exposed most of the Turkey Branch Formation during 1991. The Smithsonian-Columbia excavations are only about 300 m north of this portion of the Powhite Parkway, and strata at the site are easily correlated with the section along the highway. The entire sequence resembles Newark-style lacustrine sequences in being strongly

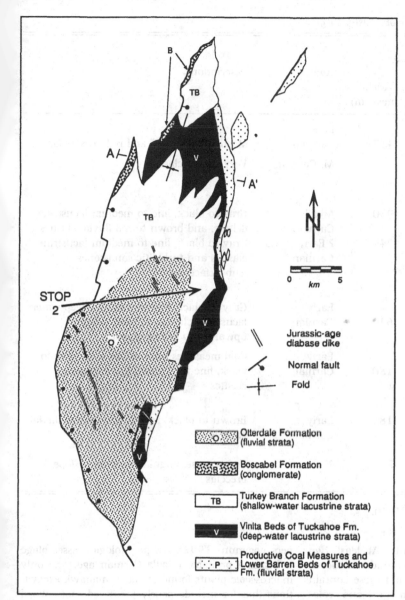

STOP 2

N

0 km 5

Jurassic-age diabase dike

Normal fault

Fold

Otterdale Formation (fluvial strata)

Boscabel Formation (conglomerate)

Turkey Branch Formation (shallow-water lacustrine strata)

Vinita Beds of Tuckahoe Fm. (deep-water lacustrine strata)

Productive Coal Measures and Lower Barren Beds of Tuckahoe Fm. (fluvial strata)

Figure 25.10. Geological map of the Richmond basin showing Location of Stop 2. (Courtesy of R. W. Schlische.)

cyclical, consisting of alternating laminated black shales and sandstones with cross-bedding and root horizons. As is typical of Newark-style sequences, a clear hierarchy of cycles is present, with thinner black shales or thick zones lacking any black shales occurring at rough multiples of 5 and 20 times the average thickness of most black shale–sandstone couplets, which themselves average 9 m. In analogy with better-known Newark-type sequences, the black shale–sandstone couplets could reflect the 20,000-year precession cycles, and the larger black-shale-rich and black-shale-poor compound cycles would reflect the 100,000 and 400,000-year eccentricity cycles. However, unlike typical Newark-style sequences, evidence of exposure is much less common, and

bioturbation is much more common, in these strata. Strata at the Smithsonian-Columbia excavations are the regressive portion of the uppermost black-shale-bearing cycle in the wetter portion of a 100,000-year cycle, within the drier portion of a 400,000-year cycle within the middle of the Turkey Branch Formation.

Strata at the Smithsonian-Columbia excavations themselves consist of a sequence of laminated dark gray claystone containing clam shrimp that grade upward into massive mudstone and nodular limestone containing root structures. These sequences make up the lower part of a single Van Houten cycle. Tetrapod bones are most abundant in the massive mudstones and occur as disarticulated but partially associated jaws, skulls, and skeletons, as well as numerous

Table 25.2. *Stratigraphy of the Richmond basin, Virginia*

Units			Age	Description
Shaler and Woodworth (1899)	Olsen et al. (1989)	Thick-ness (m)		
Otterdale Sandstone Fm.	Otterdale Fm.	415	Late Carnian	Mostly coarse, buff and red fluvial clastics
	Boscabel Fm.		M. Carnian	Very coarse clastics
	Turkey Branch Fm.			
	Tomahawk Creek Mb.	950	Middle Carnian	Gray to black, fine to medium lacustrine, deltaic, and brown to red fluvial clastics
	Hidden Mb.	244	? Early–middle Carnian	Gray to black, fine to medium lacustrine clastics and turbiditic sandstones (subsurface unit)
Vinita Beds Fm.	Tuckahoe Fm. Vinita Beds Mb.	640	Early Carnian	Gray to black, fine to medium deep-water lacustrine clastics at base, passing upward into deltaic clastics
Productive Coal Measures Fm.	Productive Coal Measures Mb.	180	Early Carnian	Coal measures interbedded with gray to black, fine to medium fluviolacustrine clastics
Lower Barren Beds Fm.	Lower Barren Beds Mb.	185	Early Carnian	Brown to black, fine to very coarse fluvial clastics
Boscabel Boulder Beds Fm.		?	?	Very coarse, angular-fault talus-slope breccias

Source: Adapted from Olsen et al. (1989).

isolated bones and bone fragments. At least two irregular bands of more silty, fissile mudstone occur within these massive mudstones, and these contain extremely abundant dissociated fish bones and scales (Figure 25.11), as well as tetrapod bones and teeth. Unlike most other Newark bone localities, the bone is hard, almost uncrushed, and separates easily from the rock. Very small vertebrate bones have been recovered by bulk processing in kerosene and hot water (or simply hot water), followed by sieving and examination of the residue under a microscope. It is worth noting that the bone-producing units (or facies) are lithologically similar to those containing *Doswellia* in the Taylorsville basin; both localities contain gastropod fossils that have not yet been identified. The vertebrate material is reviewed by Sues et al. (Chapter 8).

The claystones underlying the massive mudstones have produced a palynoflora (Cornet and Olsen, 1990) typical of the middle Tomahawk Creek Member of the Turkey Branch Formation, as seen in outcrops along the Powhite Parkway and in well cuttings (B. Cornet,

pers. commun., 1992). The palynological assemblage indicates a probable middle Carnian age. The only macroscopic plants found at the Tomahawk excavations thus far include poorly preserved roots, wood scraps, and a single, poorly preserved fern pinnule. However, a rich plant assemblage was recovered from the uppermost cycle exposed in the Powhite section. This assemblage is dominated by a *Zamites*-like cycadophyte and is the youngest leaf-bearing assemblage known from the basin (B. Cornet, pers. commun., 1992).

Vertebrates from the Tomahawk assemblage shed light on a striking and long-standing paleobiological conundrum revolving around the apparent presence of strong faunal and floral provinciality during the Late Triassic – a time when continental unity would imply that cosmopolitan assemblages should be the rule. Two basic faunal provinces have long been recognized: a Laurasian assemblage, dominated by phytosaurs and metoposaur amphibians, and a Gondwanan assemblage, dominated by nonmammalian synapsids; for a review,

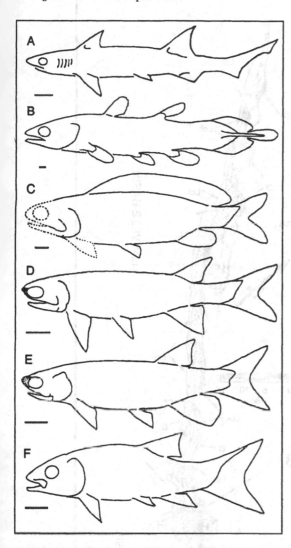

Figure 25.11. Fish from the Richmond basin. (A) Minute hybodont shark *Lissodus* n. sp. (B) Indeterminate coelacanth. (C) Palaeonisciform *Tanaocrossus* sp. (D) Redfieldiid palaeonisciform *Cionichthys*, (E) Redfieldiid palaeonisciform *Dictyopyge*. (F) Holostean *Semionotus*. A based on *Lissodus africanus*; B–F based on outlines in Olsen et al. (1982). Scale is 2 cm.

see Shubin and Sues (1991). The synapsids indicate very close faunal ties to the Gondwanan realm, specifically South America and southern Africa. With the possible exception of the phytosaurs, the tetrapod fauna from the Tomahawk locality shows no similarity to the classic assemblages from the Late Triassic Chinle and Dockum formations of the western United States or to the rest of the Newark Supergroup. Because this new assemblage is slightly older than other well-known Laurasian Late Triassic assemblages, and has close affinities to Gondwanan faunules, it is clear that the differences between the Carnian assemblages of

Laurasia and Gondwana were not caused exclusively by geographic isolation. The apparent faunal provinciality might be more a function of time than of geography: an artifact caused by poor stratigraphic sampling of the Middle Triassic–Late Triassic transition in Laurasia.

Despite the possible elimination of apparent faunal provinciality, the floral provinciality in the early part of the Late Triassic is clearly real, with a *Dicroidium*-dominated megaflora and Ipswich-Onslow-type palynoflora in Gondwana, and a cycadophyte/conifer-dominated flora in Laurasia (Cornet and Olsen, 1985). This floral provinciality does not correspond geographically or temporally in a simple way with the faunas. The best examples of provinciality include Indian occurrences of typical Laurasian-type assemblages with abundant phytosaurs and metaposaurs associated with *Dicroidium*-dominated megafossil florules and Ipswich-Onslow-type palynoflorules (Kumaran and Maheswari, 1980; Cornet and Olsen, 1985), and, of course, the Richmond-Taylorsville tetrapod assemblage associated with Laurasian florules.

Stop 3: Solite Quarry

Some of the best examples of fossil *Lagerstätten* in eastern North America are the outcrops in the quarry of the Virginia Solite Corporation near Eden, North Carolina (Figure 25.12). Most unusual is the extreme abundance of the tanystropheid reptile *Tanytrachelos* and the preservation of soft tissue on vertebrates as well as invertebrates. This quarry actively produces lightweight aggregate from shales in the Cow Branch Formation of the Dan River basin.

The Solite Quarry exposes several hundred meters of the late Carnian Cow Branch Formation of the Dan River basin of Virginia and North Carolina (Robbins and Traverse, 1980; Olsen, McCune, and Thomson, 1982) (Figures 25.1 and 25.12). The Cow Branch Formation comprises possibly two gray and black lacustrine clastic complexes interbedded with the fluvial and deltaic clastics of the Pine Hall and Stoneville formations (Figure 25.12, Table 25.3). The exact relationship of these formations to each other is poorly known, as are details of the depositional environments of formations other than the Cow Branch. Exposures of the Cow Branch Formation in the Solite Quarry are excellent and form the basis of most of what is known about the formation.

The Dan River basin contains lithological sequences typical of the Newark-type lacustrine facies complex. Specifically, the Cow Branch Formation is composed of very well developed Van Houten cycles (Figures 25.13–25.15). The transitions between the three divisions of Van Houten cycles tend to be unidirectional in more northern Newark basins. However, in the Dan River basin, division 2 often contains a regressive interval near its middle (Figure 25.14). As

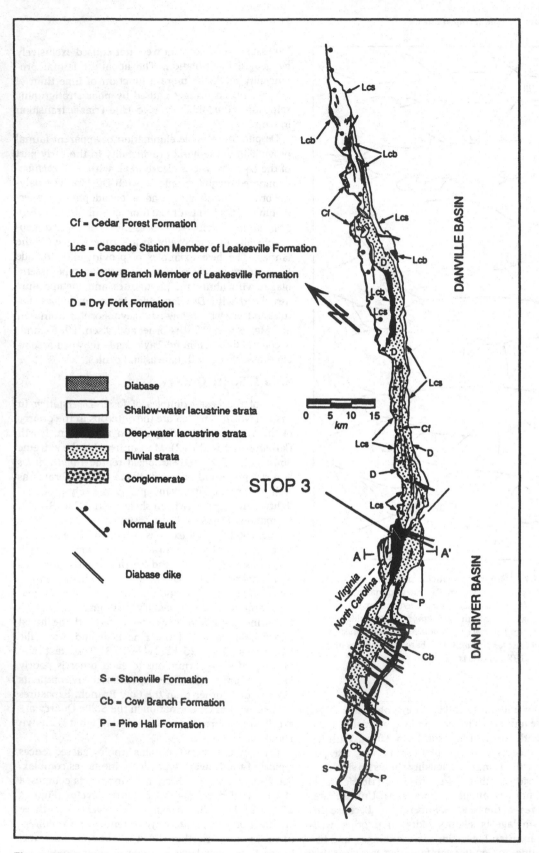

Figure 25.12. Geological map of the Dan River basin showing the position of Stop 3. (Courtesy of R. W. Schlische.)

Table 25.3. *Stratigraphy of the Dan River basin, North Carolina*

Units	Thick-ness (m)	Age	Description
Stoneville Fm.	~1,000	Norian	Red and gray lacustrine, fluvial, and alluvial clastics
Cow Branch Fm. (upper member)	~1,000	Late Carnian	Gray to black, fine to coarse cyclical lacustrine clastics
Pine Hall Fm. (upper tongue)	~800	Carnian	Mostly tan, brown, and red, coarse to fine fluvial and alluvial clastics
Cow Branch Fm. (lower member)	~700	Middle Carnian	Gray to black, fine to coarse cyclical lacustrine clastics; thin coal seams at base
Pine Hall Fm. (lower tongue)	~500	Carnian	Mostly tan, brown, and red, coarse to fine fluvial and alluvial clastics

Source: Adapted from Olsen et al. (1978).

a consequence, power spectra of the Solite Quarry sequence show a prominent peak at roughly one-half the period of the Van Houten cycle (Figure 25.14). This subcycle may reflect the combined effects of the northern and southern hemisphere monsoons near the equator, which should be opposite in phase, while the eccentricity cycles of the two hemispheres would be in phase. Previously published interpretations of the frequency properties of the Solite section failed to recognize the possibility of a ~10,000-year cycle and were based on the assumption that the highest-frequency elements of the power spectrum were in the 21,000-year range (Olsen et al., 1989).

Vertebrate–lithology associations

Biostratonomy (Schäfer, 1972) of the Cow Branch Formation shows strong patterns important not only for interpretation of organism–environment interactions but also for fossil prospecting. A classification of sediment fabrics from the Solite section (Table 25.4) arranged along an inferred lake-depth gradient shows a strong association with fossil preservation and fossil type. Specifically, there is a very strong tendency for the completeness of vertebrate remains to increase in better-laminated units. Abundant complete fish and reptiles (*Tanytrachelos*) are present only in micro-laminated calcareous clay-rich siltstones. This pattern is a common one (lethal-pantostrat biofacies of Schäfer, 1972) and is seen in many famous vertebrate-bearing *Lagerstätten* such as the Devonian Caithness Flagstones of Scotland, the Jurassic Posidonia shales of Germany, the Lower Cretaceous Araripe beds of Brazil, and the Eocene Green River Formation of western United States.

Articulated fish, reptiles, and insects occur only in the most finely laminated (depth rank 6), organic-carbon-rich (T.O.C. ≥ 1 percent) portions of division 2. Plant foliage compressions occur in these units and in

surrounding units that are slightly less well laminated but still organic-carbon-rich. Root structures, burrows, and casts of in situ plant stems occur in poorly laminated units (depth rank 3–0), as do carbonized scraps of mostly conifer wood and foliage. Reptile footprints and plant fragments are present on surfaces of low-depth-rank (2–1) mudstones and sandstones that have polygonal cracks on thin bedding planes. Isolated bones and teeth of phytosaurs can occur in all lithologies. Besides producing extremely large numbers of articulated skeletons of *Tanytrachelos* (Figure 25.16), the Solite Quarry has yielded some of the very oldest true flies and the oldest true water bugs (Figure 25.17). A faunal and floral list is given in Table 25.5.

The taphonomic pattern seen in division 2 of Van Houten cycles fits a chemically stratified lake model (Bradley, 1929, 1963; Ludlam, 1969; Boyer, 1981) in which bioturbation is perennially absent from the deeper parts of the lake bottom because the bottom waters lack the oxygen necessary for almost all macroscopic benthic organisms. Chemical stratification, often called meromixis, can arise by a number of mechanisms, but the main physical principle involved is the exclusion of turbulence from the lower reaches of a water column. This tremendously decreases the rate at which oxygen diffuses from the surface waters and retards the upward movement of other substances. The main source of water turbulence is wind-driven wave mixing. This turbulence usually extends below one-half the wavelength of surface wind waves. If the lake is deeper than the extent of the turbulent zone, the lake becomes stratified, with a lower nonturbulent zone and an upper, turbulently mixed zone. The thickness of the upper mixed zone is also dependent on density differences between the upper waters (epilimnion) and lower waters (hypolimnion) that can be set up by salinity differences (saline meromixis) or by temperature differences, as in many temperate lakes. In the absence of saline or temperature stratification, chemical

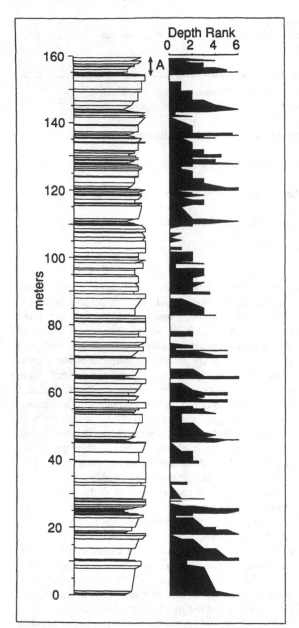

Figure 25.13. Measured section of the Solite Quarry sequence and compound cycle pattern. Key to depth ranks in Table 25.4. "A" shows position of cycle seen in more detail in Figure 25.15.

difference between the epilimnion and hypolimnion. In spite of that, chemical stratification occurs, with the exclusion of oxygen below 200 m (Olsen, 1991), a pattern common in deep tropical lakes. The preservation of microlaminations and fossils in Cow Branch cycles may have been a function of great water depth relative to a small surface area of the lake.

Van Houten cycles with a microlaminated division 2 thus reflect the alternation of (1) shallow, ephemeral lakes or subaerial flats and (2) deep perennial lakes, with an anoxic hypolimnion set up by turbulent stratification under conditions of relatively high primary productivity and low organic consumption (e.g., low ecosystem efficiency). The low organic content of divisions 1 and 3 of the cycles probably reflects higher ecosystem efficiency caused by shallow water depths, rather than lower total organic productivity.

Study of a single cycle

A large (ca. 10 m^2) excavation in a single cycle ("A" in Figure 25.13; see also Figure 25.15) developed during 1975–8 by a team from Yale University produced over 150 skeletons of *Tanytrachelos*, dozens of fish, and over 300 insects, as well as abundant conchostracans and plant fossils (Olsen et al., 1978). The microstratigraphy of this excavation reveals details of the lake transgression and regression. As in other Solite section cycles, fossil preservation tracks lithology, with abundant whole vertebrates being present only in the microlaminated parts of division 2. In most cycles in the Solite section and similar cycles in the rest of the Newark Supergroup, articulated reptiles are limited to the lower few centimeters of division 2. The reasons for this common asymmetry are not clear, but might involve increasing salinity as the lake evaporated. In the Yale quarry, however, fossil preservation is symmetrical around the center of the microlaminated part of division 2 (Figure 25.15). Despite this overall symmetry, fossil preservation is truly exceptional only in the basal 3 cm of the microlaminated portion of this cycle (Figure 25.15). Complete insects are abundant, and several vertebrates, including a specimen of *Tanytrachelos* (Figure 25.16), have carbonaceous films outlining soft tissue.

Tectonic deformation

Many bedding planes in division 2 show bedding-plane-parallel ductile deformation that affected almost all fossils, especially the skeletons of reptiles (*Tanytrachelos*, Figure 25.16) and the insects. Bones of *Tanytrachelos* show numerous parallel cracks. On a microscopic scale, the organic films comprising insect remains show a preferred orientation of light reflectance. This deformation is also visible as a lineation resulting from numerous small deformed fossils, such as conchostracans and

stratification can still arise in deep lakes with relatively high levels of organic productivity. Because oxygen is supplied slowly by diffusion, bacterial consumption of abundant organic matter sinking into the hypolimnion, plus oxidation of bacterial by-products, will eliminate oxygen from the hypolimnion. Lakes Tanganyika and Malawi in East Africa are excellent examples of very deep lakes in which there is little temperature or density

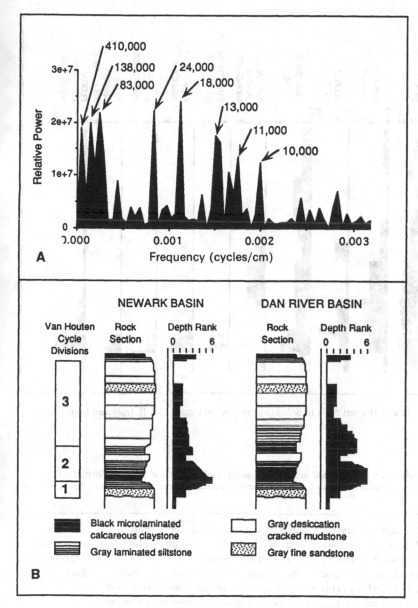

Figure 25.14. Cycles from the Cow Branch Formation of the Solite Quarry. (A) Power spectrum based on depth-rank curve shown in Figure 25.13. (B) Comparison between Van Houten cycles seen in more northerly Newark-type lacustrine sequences (on left) and the type seen in the Solite Quarry (on right); the strong double shape of the depth-rank curve is probably due to the proximity of the equator.

ostracods, and the crinkling of laminae. Small-scale folds are also present, commonly associated both with bedding-parallel shortening and the development of duplexes (Olsen et al., 1989). Polygonal cracks within divisions 1 and 3 of Van Houten cycles do not show as obvious evidence of this kind of ductile deformation, however. The direction of elongation is strikingly parallel to the dip direction, which is itself close to the regional-extension direction. This style of ductile deformation probably occurred at shallow depths while there was still enough water in the strata to allow some plastic flow. Similar deformation of fossils has been described at several other Newark Supergroup vertebrate localities (Olsen et al., 1989; Silvestri, 1991). Special

care must therefore be taken when describing the morphology of Newark fossils, especially those preserved as bedding-plane relief (e.g., fish and footprints), because other signs of deformation often are very subtle.

Comparisons with other parts of the Newark

The biostratonomic patterns seen in the Van Houten cycles of the Solite Quarry are, in general, characteristic of what is seen in other sequences of Van Houten cycles in various Newark-type lacustrine sequences. In fact, skeletons of *Tanytrachelos* were first discovered in the Lockatong Formation of the Newark basin by concentrating collection efforts in the basal 2 cm of

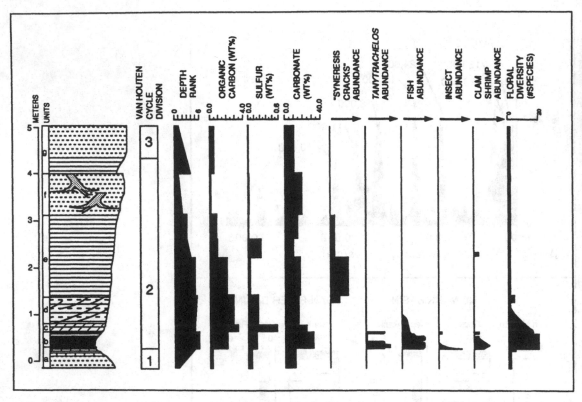

Figure 25.15. Lower three-quarters of Van Houten cycle in Solite Quarry ("A" in Figure 25.13). (Adapted from Olsen et al., 1989.)

Table 25.4. *Description of depth ranks for fine-grained lacustrine sediments for the Cow Branch Formation*

Depth rank	Description	
0	Massively bedded calcareous claystone and siltstone with root, tube, crumb, and vesicular fabric; faint remnant parent fabric present; abundant clay cutans; intensely desiccated fabric; very rare trace fossils	Shallow, oxygenated high-energy
1	Intensely brecciated and cracked calcareous claystone and siltstone; burrows can be common, but obvious remnant parent fabric present; mud curls and cracks with vesicular fabric sometimes present; footprints sometimes present	lake ↑
2	Thin-bedded calcareous claystone and siltstone with desiccation cracks; large patches of uncracked matrix preserved; infrequent desiccation cracks; reptile footprints and burrows often present	
3	Thin-bedded calcareous claystone and rare siltstone with very small scale burrows; rare to absent desiccation cracks; pinch-and-swell lamination often present; rare fish bones and scales, locally common plant debris	Time-averaged conditions
4	Evenly laminated calcareous claystone with some small-scale burrows; abundant discontinuous laminations; no desiccation cracks; fish bones and scales locally abundant, as are ostracods and clam shrimp	↓
5	Evenly and finely laminated calcareous claystone or limestone; abundant discontinuous laminations; no desiccation cracks; partial to complete fish skeletons, single valves of ostracods, and whole clam shrimp locally common	Deep, anoxic
6	Microlaminated calcarous siltstone with only rare disruptions; no desiccation cracks; complete fish and reptile skeletons, single valves of ostracods, and whole clam shrimp locally common	low-energy lake

Source: Adapted from Olsen et al. (1989).

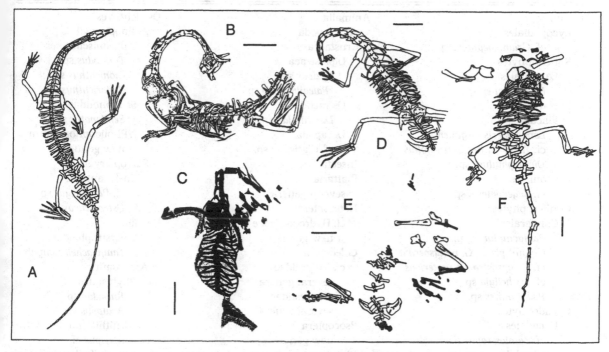

Figure 25.16. *Tanytrachelos ahynis.* (A) Reconstruction of male. (B–E) Line drawings of various specimens. Scale is 2 cm. (From Olsen, 1979.)

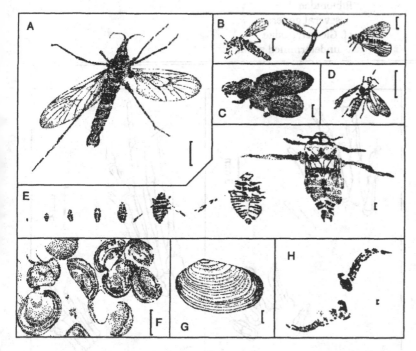

Figure 25.17. Camera lucida drawings of insects and other invertebrates from the Solite Quarry. (A, B) True flies (Diptera, ?Tipulidae). (C) Beetle. (D) Psocopteran. (E) Partial growth series of hydrocoricid or bdelostomatid water bugs. (F) Clam shrimp of the *Palaeolimnadia* type. (G) Clam shrimp of the *Cyzicus* type. (H) Possible phyllocarids. Scale is 0.5 mm. (From Olsen, 1988.)

Van Houten cycles, based on the patterns seen first in the Solite section. However, the most conspicuous difference between the cycles of the Solite Quarry and similar cycles from other parts of the Newark Supergroup, especially those of the Lockatong Formation of the Newark basin, is the great abundance and diversity of plant foliage.

Currently, no post-Carnian cycles are known that yield articulated small reptiles. In Triassic-age sequences this probably is a function of a lack of units that bear articulated fish. There are disarticulated remains of *Tanytrachelos* at a number of Norian-age localities in the Newark basin (Olsen and Flynn, 1989). In contrast, there are many well-collected Early Jurassic lacustrine

Table 25.5. *Plants and animals from the Solite Quarry*

Plants	Animalia	Osteichthyes
Lycopodiales	Arthropoda	Actinopterygii
cf. *Grammaephoios* sp.	Crustacea	Palaeonisciformes
Sphenophytes	Diplostraca	*Turseodus* spp.
Equisetales	*Cyzicus* sp.	*Cionichthys* sp.
Neocalamites sp.	?*Paleolimnadia* sp.	*Synorichthys* sp.
Pteridophytes	Ostracoda	Semionotidae
Filicales	*Darwinula* spp.	*Semionotus* sp.
Lonchopteris virginiensis	Decapoda	??Pholidophoridiformes
cf. *Acrostichites linnaeafolius*	cf. *Clytiopsis* sp.	new genus
Dictyophyllum sp.	Insecta	Sarcopterygii
Caytoniales	Blattaria	Coelacanthini
cf. *Sagenopteris* sp.	several genera	cf. *Pariostegus* sp.
Coniferophytes	Heteroptera	*Osteopleurus* sp.
Coniferales	cf. Hydrocoricidae	Reptilia
Pagiophyllum spp.	new genus	Tanystropheidae
Glyptolepis cf. *G. platysperma*	Coleoptera	*Tanytrachelos ahynis*
cf. *Compsostrobus neotericus*	cf. Nitidulidae	Archosauria
cf. *Dechellyia* sp.	several genera	Phytosauria
Podozamites sp.	cf. Buprestidae	*Rutiodon* sp.
Cycadophytes	several genera	*Apatopus* sp.
Cycadales	Psocoptera	?Ornithischia
cf. *Zamiostrobus lissocardus*	new genus	*Atreipus* cf.
Glandulozamites sp.	Diptera	*A. milfordensis*
large androsporophylls	Tipulidae	Saurischia
Bennettitales (cycadeoids)	several genera	?*Grallator* sp.
Zamites powelli	Bibionidae	
Pterophyllum	several genera	
cf. *Ctenophyllum giganteum*	cf. Glosselytrodae	
	undetermined genus	

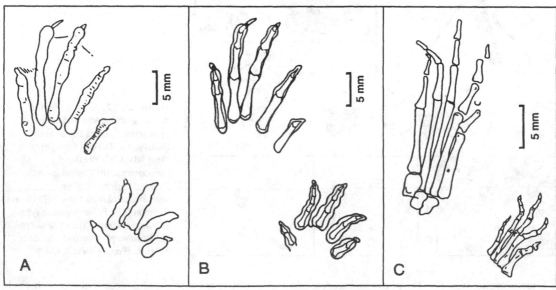

Figure 25.18. Comparison of *Gwyneddichnium majore* and manus and pes of *Tanytrachelos ahynis*. (A) Composite of left manus-pes set of *G. majore* based on the type material. (B) Diagrammatic composite outline of manus and pes of *G. majore* with reconstructed manus and pes skeleton. (C) Left manus and pes of *T. ahynis*, based on YPM 8600. (From Olsen and Flynn, 1989.)

sequences with abundant articulated fish. These cycles are also lithologically similar to those of the Cow Branch and Lockatong formations, but they lack any traces of small reptiles such as *Tanytrachelos*. We believe this absence is real and reflects the Triassic–Jurassic extinction event.

Vertebrates of the Solite Quarry

Tanytrachelos ahynis

The most abundant vertebrate in the Solite Quarry is *Tanytrachelos ahynis* (Figure 25.16), and articulated skeletons of this little reptile appear much more commonly than articulated fish. Olsen (1979) named and described *T. ahynis* on the basis of the abundant remains from the Solite Quarry. However, more recently collected material from the Lockatong Formation of the Newark basin of New Jersey and Pennsylvania clears up some critical parts of the anatomy. The description that follows is based on the Solite and Newark basin material.

Description. Tanytrachelos is a small (less than 50 cm long) tanystropheid with a skull rather similar to that of *Tanystropheus*, but with nostrils and a profile more like those of *Macrocnemus*. The teeth are small and sharply pointed, with thecodont or subthecodont implantation. There are 12 cervical, 13 dorsal, 2 sacral, and at least 25 caudal vertebrae, all of which are strongly procoelous. The neck is nearly as long as the trunk. Recently prepared material shows that cervical vertebrae 2–4 bear variably interdigitated processes that lock the successive vertebrae together. These same vertebrae bear closely spaced, parallel, splintlike ribs similar to the cervical ribs of *Tanystropheus*. The other cervicals are shorter and bear plow-shaped ribs that become larger on more posterior vertebrae. Anterior cervical ribs are double-headed, and more posterior ribs are single-headed. The last three dorsal vertebrae lack ribs and bear slightly anteriorly directed transverse processes. Anterior caudal vertebrae bear long transverse processes that shorten and disappear by the sixth or seventh caudal. The total length of the tail is about twice that of the trunk.

Details of the structure of the pectoral girdle of *Tanytrachelos* were obscure in Solite material and originally appeared consistent with descriptions of *Tanystropheus* (Olsen, 1979). Newark basin material shows that a very peculiar pectoral girdle is present instead. Most distinctive is the relatively huge quadrangular sternum, which is three times the area of the coracoid or the scapula. This element was misidentified as the interclavicle by Olsen and Flynn (1989), but its position as a sternum can be seen, albeit not very clearly, in the Solite specimen YPM 7622. Disarticulated and uncrushed material from the upper Lockatong Formation of Chalfont, Pennsylvania, shows that this sternum is strongly bowed ventrally and bears a thin

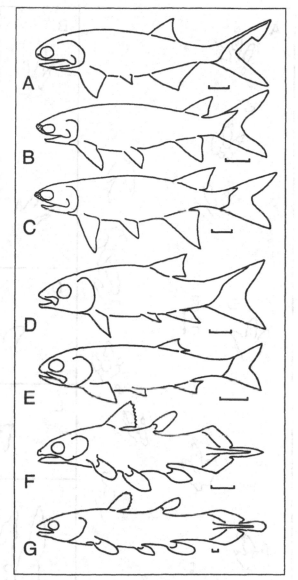

Figure 25.19. Fish from the Solite Quarry. (A) Palaeoniscoid *Turseodus*. (B) Redfieldiid palaeonisciform *Synorichthys*. (C) Redfieldiid palaeonisciform *Cionichthys*. (D) Holostean *Semionotus*. (E) An unnamed neopterygian allied to pholidophorids. (F) Small coelacanth *Diplurus* cf. *newarki*. (G) A large coelacanth, probably *Pariostegus*. Scale is 2 cm. (From Olsen et al., 1989.)

distinct keel. Such a structure is as yet unknown in other tanystropheids. The semilunate scapula and coracoid, however, appear very similar to those of *Tanystropheus*, as does the pelvic girdle.

The forelimbs are relatively short; based on measurements published by Olsen (1979), the humerus is 0.84 times the femur length, with the ulna 0.46 times the length of the humerus. The hindlimbs are relatively

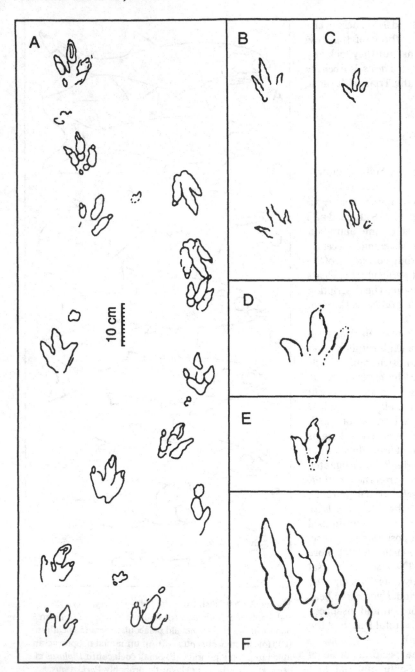

Figure 25.20. Reptilian footprints from the Solite Quarry. (A) Probable ornithischian dinosaur ichnite cf. *Atreipus* sp. (B, C) Theropod dinosaur footprints belonging to *Grallator* sp. (D, E) probably dinosaurian footprints. (F) *Apatopus* cf. *lineatus*, pes impression.

long; the femur is 5.2 times the length of a single dorsal vertebra, and the tibia is 0.68 times the length of the femur. The manus (Figure 25.18) is more primitive than in *Tanystropheus*; the phalangeal formula in *Tanytrachelos* is 2, 3, 4, 5, 3, while that in *Tanystropheus* is 2, 3, 4, 4, 3 (Wild, 1973). The carpus is poorly preserved in all available specimens. Digit I is most robust, and the others become progressively more delicate laterally. The pes (Figure 25.18) has a phalangeal formula of 2, 3, 4, 4?, 4, and digit III is the longest; in *Tanystropheus*, the phalangeal formula is 2, 3, 4, 5, 3, and digit IV is longest. Critically, the proximal phalanx of digit V is elongate, while metatarsal V is short, exactly as in *Tanystropheus*.

About half of the specimens of *Tanytrachelos* have paired hemipenial ossifications (Figure 25.16), closely comparable to those in *Tanystropheus* (Olsen, 1979), and thus almost all specimens can be sexed. It would be interesting to see what other sex-associated skeletal differences are present. One specimen from the Solite

Quarry has a body outline (Figure 25.16). The organic film residue of the soft tissue lacks a distinct texture, and it seems likely that the skin of *Tanytrachelos* was smooth.

Rather common footprints from the Newark basin called *Gwyneddichnium* Bock, 1952, have a reconstructed manus and pes identical with those of *Tanytrachelos* (Olsen and Flynn, 1989) (Figure 25.18). The footfall pattern of this ichnite is identical with what would be expected of *Tanytrachelos*, with a strong tendency for overstepping of manus by the pes impressions. There is little doubt that *Tanytrachelos* was the trackmaker of *Gwyneddichnium*, because of the shared highly derived characters present in the pedal skeletal remains and the inferred structure of the pedal impressions of the ichnite.

Other vertebrates
The Solite fish assemblage is very similar to that of the Lockatong Formation of the Newark basin (Figure 25.19). The most common fish in the Solite Quarry is the palaeoniscoid *Turseodus*, which is easily recognized in even very poorly preserved material because of its characteristic ringlike hemicentra and strongly ridged scales. The semionotid neopterygian *Semionotus* sp. is second in abundance, and it is most readily distinguished by its robust nonsculptured skull bones and its dorsal ridge scales. At least one bed ("A" in Figure 25.13) produces numerous articulated specimens of a small (< 15 cm) neopterygian with apparent pholidophoroid affinities (Olsen, 1988). The large coelacanth *Pariostegus* is present in low numbers, as are the redfieldiid palaeoniscoids *Cionichthys* and *Synorichthys*.

Phytosaurs are represented by rare teeth, bones, and footprints in the form of *Apatopus* (Figure 25.20). Other footprint taxa include the apparently ornithischian ichnite *Atreipus* cf. *milfordensis* (Olsen and Baird, 1986), cf. *Grallator* sp., and undetermined, larger, probably dinosaurian footprints (Figure 25.20). A *Tanytrachelos*-bearing division 2 of a cycle low in the Solite Quarry section has produced an additional reptile taxon represented by a fragmentary but originally articulated specimen. Although too incomplete for certain identification, it appears to be an archosauromorph (N. C. Fraser, pers. commun., 1992).

Acknowledgments

We and all of the people involved with the Workshop on Early Mesozoic Small Tetrapods are very grateful to the owners and managers of the Culpeper Crushed Stone Quarry, the Tomahawk site, and the Solite Quarry for permission to work on their property and for permission for the May 1991 field trip to visit the sites. Work on this chapter was funded by grants from the National Science Foundation (EAR 89-16726 to P. Olsen and D. Kent, and EAR-9016677 to H.-D. Sues and P. Olsen) and the National Geographic Society (to H.-D. Sues and P. Olsen). N. Fraser is thanked for his invaluable aid in organizing the field trip. We also thank H.-D. Sues for many stimulating and enlightening discussions on Triassic–Jurassic tetrapods, and R. Schlische for use of his geological base maps of the Culpeper, Richmond, and Dan River basins.

References

Baird, D., and W. Take. 1959. Triassic reptiles from Nova Scotia. *Geological Society of America Bulletin* 20: 1565–1566.

Bock, W. 1952. Triassic reptilian tracks and trends of locomotive evolution. *Journal of Paleontology* 26: 395–433.

—— 1969. The American Triassic flora and global distribution. *Geological Center Research Series 2–3*: 1–406.

Boyer, B. W. 1981. Tertiary lacustrine sediments from Sentinel Butte, North Dakota, and the sedimentary record of ectogenic meromixis. *Journal of Sedimentary Petrology* 51: 429–440.

Bradley, W. H. 1929. Varves and climate of the Green River epoch. *United States Geological Survey Professional Paper* 158E: 86–110.

—— 1963. Paleolimnology. Pp. 621–648 in D. Frey (ed.), *Limnology in North America*. Madison: University of Wisconsin Press.

Carozzi, A. V. 1964. Complex oöids from Triassic lake deposits, Virginia. *American Journal of Science* 262: 231–241.

Cornet, B. 1977. The palynostratigraphy and age of the Newark Supergroup. Ph.D. dissertation, Pennsylvania State University.

Cornet, B., and P. E. Olsen. 1985. A summary of the biostratigraphy of the Newark Supergroup of eastern North America, with comments on early Mesozoic provinciality. Pp. 67–81 in R. Weber (ed.), *Simposio Sobre Floras del Triasico Tardio, su Fitogeografía y Paleoecologia, Memoria*. III Congreso Latinoamericano de Paleontologia, Mexico.

—— 1990. *Early to Middle Carnian (Triassic) Flora and Fauna of the Richmond and Taylorsville Basins, Virginia and Maryland, U.S.A.* Guidebook 1, Virginia Museum of Natural History, Martinsville.

Kumaran, K. P. N., and H. K. Maheswari. 1980. Upper Triassic sporae dispersae from the Tiki Formation. 2: Miospores from the Janar Nala Section, South Gondwana Basin, India. *Palaeontographica* B173: 26–84.

Lee, K. Y., and Froelich, A. J. 1989. Triassic–Jurassic stratigraphy of the Culpeper and Barboursville basins, Virginia and Maryland. *United States Geological Survey Professional Paper* 1472: 1–52.

Lindholm, R. C. 1979. Geologic history and stratigraphy of the Triassic–Jurassic Culpeper basin, Virginia: summary. *Geological Society of America, Bulletin* 90: 995–997.

Ludlam, S. D. 1969. Fayetteville Green Lake, New York. III. The laminated sediments. *Limnology and Oceanography* 14: 848–857.

Manspeizer, W. 1982. Triassic–Liassic basins and climate

of the Atlantic passive margins. *Geologische Rundschau* 61: 895–917.

Olsen, P. E. 1979. A new aquatic eosuchian from the Newark Supergroup (Late Triassic–Early Jurassic) of North Carolina and Virginia. *Postilla* 176: 1–14.

——— 1986. A 40-million-year lake record of early Mesozoic orbital climatic forcing. *Science* 234: 842–848.

——— 1988. Paleoecology and paleoenvironments of the continental early Mesozoic Newark Supergroup of eastern North America. Pp. 185–230 in W. Manspeizer (ed.), *Triassic–Jurassic Rifting: Continental Breakup and the Origin of the Atlantic Ocean and Passive Margins.* Amsterdam: Elsevier.

——— 1991. Tectonic, climatic, and biotic modulation of lacustrine ecosystems: examples from the Newark Supergroup of eastern North America. Pp. 209–224 in B. J. Katz (ed.), *Lacustrine Basin Exploration—Case Studies and Modern Analogs.* Memoirs of the American Association of Petroleum Geologists, Vol. 50.

Olsen, P. E., and D. Baird. 1986. The ichnogenus *Atreipus* and its significance for Triassic biostratigraphy. Pp. 61–87 in K. Padian (ed.), *The Beginning of the Age of Dinosaurs: Faunal Change across the Triassic–Jurassic Boundary.* Cambridge University Press.

Olsen, P. E., and J. J. Flynn. 1989. Field guide to the vertebrate paleontology of Late Triassic rocks in the southwestern Newark basin (Newark Supergroup, New Jersey and Pennsylvania). *The Mosasaur* 4: 1–43.

Olsen, P. E., A. R. McCune, and K. S. Thomson. 1982. Correlation of the early Mesozoic Newark Supergroup by vertebrates, principally fishes. *American Journal of Science* 282: 1–44.

Olsen, P. E., C. L. Remington, B. Cornet, and K. S. Thomson. 1978. Cyclic change in Late Triassic lacustrine communities. *Science* 201: 729–733.

Olsen, P. E., R. W. Schlische, and P. J. W. Gore. 1989. *Tectonic, Depositional, and Paleoecological History of Early Mesozoic Rift Basins, Eastern North America.* Guidebook T351, 28th International Geological Congress. Washington, D. C.: American Geophysical Union.

Olsen, P. E., N. H. Shubin, and M. H. Anders. 1987. New Early Jurassic tetrapod assemblages constrain Triassic–Jurassic extinction event. *Science* 237: 1025–1029.

Robbins, E. I., and A. Traverse. 1980. Degraded palynomorphs from the Dan River (North Carolina)–Danville (Virginia) basin. Pp. 1–11 in V. Price, Jr., et al. (eds.), *Geological Investigations of Piedmont and Triassic Rocks, Central North Carolina*

and Virginia. Carolina Geological Society Field Trip Guidebook B10.

Schäfer, W. 1972. *Ecology and Paleoecology of Marine Environments.* (English translation of 1962 German edition.) University of Chicago Press.

Schlische, R. W. 1990. Aspects of the structural and stratigraphic development of early Mesozoic rift basins of eastern North America. Ph.D. dissertation, Columbia University.

Shaler, N. S., and J. B. Woodworth. 1899. Geology of the Richmond basin, Virginia. *United States Geological Survey, Annual Report* 19: 385–515.

Shubin, N. H., and H.-D. Sues. 1991. Biogeography of early Mesozoic continental tetrapods: patterns and implications. *Paleobiology* 17: 214–230.

Silvestri, S. 1991. Ichnofauna of the last seven million years of the Triassic from the Jacksonwald syncline, Newark basin, Pennsylvania. M. S. thesis, Rutgers University.

Smoot, J. P., and P. E. Olsen. 1988. Massive mudstones in basin analysis and paleoclimatic interpretation of the Newark Supergroup. Pp. 249–274 in W. Manspeizer (ed.), *Triassic–Jurassic Rifting: Continental Breakup and the Origin of the Atlantic Ocean and Passive Margins.* Amsterdam: Elsevier.

Sues, H.-D., and P. E. Olsen. 1990. Triassic vertebrates of Gondwanan aspect from the Richmond basin of Virginia. *Science* 249: 1020–1023.

Weems, R. E. 1980. An unusual newly discovered archosaur from the Upper Triassic of Virginia, U.S.A. *Transactions of the American Philosophical Society* 70(7): 1–53.

——— 1987. A Late Triassic footprint fauna from the Culpeper basin, northern Virginia (USA). *Transactions of the American Philosophical Society* 77(1): 1–79.

Wild, R. 1973. Die Triasfauna der Tessiner Kalkalpen. XXIII. *Tanystropheus longobardicus* (Bassani) (Neue Ergebnisse). *Schweizerische Paläontologische Abhandlungen* 95: 1–162.

Witte, W. K., and D. V. Kent. 1989. A middle Carnian to early Norian (~ 225 Ma) paleopole from sediments of the Newark basin, Pennsylvania. *Geological Society of America, Bulletin* 101: 1118–1126.

Witte, W. K., D. V. Kent, and P. E. Olsen. 1991. Magnetostratigraphy and paleomagnetic poles from the middle Carnian to Hettangian rocks of the Newark basin, Pennsylvania and New Jersey. *Geological Society of America Bulletin* 103: 1648–1662.

Young, R. S., and R. S. Edmundson. 1954. Oolitic limestone in the Triassic of Virginia. *Journal of Sedimentary Petrology* 24: 275–279.

Taxonomic index

This index lists only the vertebrate taxa based on skeletal material. Species names are not included. Taxa are indexed according to author's usages, which may differ from chapter to chapter. The index also includes traditional names for non-monophyletic groupings, which are placed in quotation marks by some authors in this volume.

Printed in the United States
By Bookmasters